ENCYCLOPEDIA OF PHYSICS

CHIEF EDITOR

S. FLÜGGE

VOLUME XLIX/3

GEOPHYSICS III

PART III

EDITOR

K. RAWER

WITH 261 FIGURES

SPRINGER-VERLAG BERLIN HEIDELBERG GMBH
1984

HANDBUCH DER PHYSIK

HERAUSGEGEBEN VON

S. FLÜGGE

BAND XLIX/3

GEOPHYSIK III

TEIL III

BANDHERAUSGEBER

K. RAWER

MIT 261 FIGUREN

SPRINGER-VERLAG BERLIN HEIDELBERG GMBH
1984

ISBN 978-3-642-46279-5   ISBN 978-3-642-46277-1 (eBook)
DOI 10.1007/978-3-642-46277-1

Das Werk ist urheberrechtlich geschützt. Die dadurch begründeten Rechte, insbesondere die der Übersetzung, des Nachdruckes, der Entnahme von Abbildungen, der Funksendung, der Wiedergabe auf photomechanischem oder ähnlichem Wege und der Speicherung in Datenverarbeitungsanlagen bleiben, auch bei nur auszugsweiser Verwertung, vorbehalten. Bei Vervielfältigungen für gewerbliche Zwecke ist gemäß § 54 UrhG eine Vergütung an den Verlag zu zahlen, deren Höhe mit dem Verlag zu vereinbaren ist.

© by Springer-Verlag Berlin Heidelberg 1984
Ursprünglich erschienen bei Springer-Verlag Berlin Heidelberg New York 1984
Softcover reprint of the hardcover 1st edition 1984

Library of Congress Catalog Card Number A 56-2942.
Die Wiedergabe von Gebrauchsnamen, Handelsnamen, Warenbezeichnungen usw. in diesem Werk berechtigt auch ohne besondere Kennzeichnung nicht zu der Annahme, daß solche Namen im Sinne der Warenzeichen- und Markenschutz-Gesetzgebung als frei zu betrachten wären und daher von jedermann benutzt werden dürften.

Contents.

Introductory Remarks 1

Morphology of Magnetic Disturbance. By Professor Dr. TAKESI NAGATA, Geophysical Institute, University of Tokyo, Tokyo (Japan) and Professor Dr. NAOSHI FUKUSHIMA, Geophysics Research Laboratory, University of Tokyo, Tokyo (Japan). (With 100 Figures) . 5

   A. General introduction 5

   B. Normal geomagnetic variation free from disturbance 16

   C. Characterization of magnetic disturbance 26

   D. Gross characteristics of geomagnetic disturbance over the world 33

   E. Sudden commencement of magnetic storms 42

   F. Average storm-time variation and disturbance-daily variation of geomagnetic field 55

   G. Development and decay of magnetic storms 83

   H. Instantaneous and average disturbance field 90

   I. Geomagnetic bays and other worldwide disturbances of short duration 104

   J. Geomagnetic pulsations 115

   K. Inter-relation among geomagnetic disturbance and associated phenomena, and outline of theoretical interpretation 124

   General references 129

Theoretical Aspects of the Worldwide Magnetic Storm Phenomenon. By Professor Dr. EUGENE N. PARKER, Laboratory for Astrophysics and Space Research, University of Chicago, Chicago (USA), and Professor Dr. VINCENZO C. A. FERRARO, Queen Mary College, University of London, London (Great Britain). (With 16 Figures) 131

   A. Introduction 131

   B. The space environment of Earth 135
      I. General description 135
      II. Processes and problems 140

   C. The geomagnetic storm 159

   D. Confinement and compression of the geomagnetic field 169

   E. Inflation of the geomagnetic field 182

   F. Historical development 199

   General references 205

Maßzahlen der erdmagnetischen Aktivität. Von Professor Dr. MANFRED SIEBERT, Institut für Geophysik der Universität, Göttingen (BRD). (Mit 20 Figuren) 206

   A. Einleitung 206

   B. Die BARTELSsche Kennziffer K und daraus abgeleitete Maßzahlen 208
      I. Überblick und Grundlagen 208

II. Definition und praktische Bedeutung der aus K abgeleiteten lokalen und planetarischen Maßzahlen 217

C. Andere Aktivitätsmaße 235

D. Statistische Aussagen der Maßzahlen über Gesetzmäßigkeiten im Auftreten erdmagnetischer Aktivität und deren Ursache 250

Literatur 275

Classical Methods of Geomagnetic Observations. By Dr. VIGGO LAURSEN and Dr. JOHANNES OLSEN, Det Danske Meteorologiske Institut, Charlottenlund (Denmark). (With 15 Figures) 276

A. Geomagnetic observatories and the recording of variations in the geomagnetic elements 276

B. Direct determination of the geomagnetic elements 297

General references 322

Neuere Meßmethoden der Geomagnetik. Von Dr. habil. HERBERT SCHMIDT und Dipl.-Phys. VOLKER AUSTER, Zentralinstitut Physik der Erde, Adolf-Schmidt-Observatorium, Niemegk (DDR). (Mit 31 Figuren) 323

A. Komponentenmessung mit Hilfe von Stabmagneten als Meßelemente in Verbindung mit elektronischen Anzeige- und Regelorganen 326
  I. Kompensierende Magnetometer 327
  II. Andere Verfahren 335

B. Komponentenmessung mit Hilfe von Induktivitäten 339
  I. Induktionsspulen als Meßelement 339
  II. Mit magnetischen Wechselfeldern ausgesteuerte hochpermeable Kerne als Meßelemente 346

C. Atomphysikalische Feld-Meßmethoden 358
  I. Kernpräzessionsmagnetometer zur Bestimmung der Totalintensität 359
  II. Magnetometer mit optisch gepumpten Gasen und Dämpfen zur Bestimmung der Totalintensität 368
  III. Komponentenmessung mit Totalintensitäts-Magnetometern 376
  IV. Andere Methoden 379

D. Informationsverarbeitung im Geomagnetismus 380

Three-Component Airborne Magnetometers. By Dr. PAUL H. SERSON and Dr. KENNETH WHITHAM, Earth Physics Branch, Department of Energy, Mines & Resources, Ottawa (Canada). (With 5 Figures) 384

Aeromagnetic Surveying with the Fluxgate Magnetometer. By Professor JAMES R. BALSLEY, Wesleyan University, Middletown, Connecticut (USA). (With 8 Figures) 395
  I. Fluxgate magnetometers 395
    a) Basic fluxgate magnetometer 396
    b) Special modifications 404
    c) Associated equipment 406
  II. Field survey technique 408
    a) Operation of airborne magnetometers 408
    b) Compilation of field data 410
    c) Interpretation of results 412
  III. Discussion of advantages and limitations 419

Geophysical Applications of High Resolution Magnetometers. By Dr. Peter J. Hood, Geological Survey of Canada, Department of Energy, Mines & Resources, Ottawa (Canada). (With 30 Figures) 422

    Abstract 422

    A. Nuclear magnetometers 423

    B. Ground applications of optical absorption magnetometers 432

    C. Airborne and satellite applications of optical absorption magnetometers 440

Phénomènes T.B.F. d'origine magnétosphérique. Par Dr. Roger Gendrin, Groupe de Recherches Ionosphériques, CNET, Issy-les-Moulineaux (France). (Avec 36 Figures) 461

    Bibliographie 525

Sachverzeichnis (Deutsch-Englisch) 527

Subject Index (English-German) 532

Index (Français) 537

Introductory Remarks.

According to the original plan, this Encyclopedia should contain three volumes (47, 48 and 49) covering the whole field of geophysics. Vol. 49 was to deal with all aspects of the upper atmosphere including magnetic variations and aeronomy. When Julius Bartels was considering this volume geophysicists all over the world were preparing for the International Geophysical Year. Bartels thought this first world-wide operation in geophysics would yield such important results that he decided to wait until these were available. He subdivided Vol. 49 and himself edited Vols. 49/1 and 49/2. After his unexpected and regretted death a few manuscripts were left which had been written years ago. Considering the rapid growth of space science, new plans had to be made. With respect to the existing manuscripts, it was found that most of these needed some changes and one had become obsolete due to new scientific developments. Succeeding Bartels as scientific editor, I had to persuade the authors to revise their manuscripts. They agreed to do so and, in one case a contribution was entirely rewritten. I wish to congratulate and thank these authors for this kind of work which is much less satisfying than writing the first draft.

Therefore some of the contributions to this volume have been conceived more recently than others (as can be seen from a footnote on the first page of each). All contributions refer to the magnetic fields of Earth. Vol. 48 dealt with the main field and this volume describes the various techniques for measuring magnetism, surveys included. Magnetic disturbances are discussed, current theories are described and the conventional data figures characterizing magnetic activity are explained. Finally, the natural very low frequency phenomena, which can only be understood by the presence of a magnetic field in the natural plasma in the environment of Earth, are described and explained. The magnetosphere as such will be dealt with in Vol. 49/4, and aeronomy and related problems in Vol. 49/5.

Unavoidably, we were involved with the old problem of units to be used when writing equations. There is now a clear trend towards the "Système International" or SI. (The same system is referred to in the literature as m.k.s.a. or Giorgi units.) General use of SI units has been recommended by the International Union of Pure and Applied Physics (IUPAP). In an increasing number of countries this system is being introduced as the only approved one and is to be used exclusively in schools. Thus, looking to the future, we would have preferred to have all equations written in SI units. On the other hand, most of the existing literature in geomagnetism is in one or the other of the c.g.s. systems. In order to deal with both future and tradition, we finally decided to write all equations in a form general enough to be valid in all systems of units which are in use. The necessary transcription has been made by the editor and so is primarily his responsibility.

This generalization implied that we could not admit the particular simplifications used in the different c.g.s. systems, namely that the electric or magnetic permittivities of free space, ε_0 and μ_0, are made dimensionless (and equal to 1). In other words, we had to consider ε_0 and μ_0 as physical quantities, which they are by nature. The three most-used c.g.s. systems are then obtained by specifying the numerical values (c_0 being the free space velocity of light):

	Electrostatic	Electromagnetic	Gauss
$\varepsilon_0 =$	1	$1/c_0^2$	1
$\mu_0 =$	$1/c_0^2$	1	1

So that the product:

$c_0^2\,\varepsilon_0\,\mu_0 =$	1	1	c_0^2

In SI units, which are basically electromagnetic, we have also

$$c_0^2\,\varepsilon_0\,\mu_0 = 1.$$

Unfortunately, this is not the only difference between the systems. The c.g.s. systems which are most often used in the literature are non-rationalized while SI is rationalized (as was the special c.g.s. system introduced by H. A. Lorentz). In a rationalized system of units the famous factor 4π appears only in spherical problems, for example, in Coulomb's law, while in non-rationalized systems the natural factor 4π is artificially eliminated from Coulomb's law, although it appears in planar problems. Therefore, even when admitting ε_0, μ_0 and c_0 without specialization, we had still to take care of the rationalization problem. This can be done by introducing a dimensionless numerical constant, u, which takes the values:

$$u \equiv 1 \quad \text{in rationalized systems,}$$
$$u \equiv 4\pi \quad \text{in non-rationalized systems.}$$

In Vol. 49/2 of this Encyclopedia, in a contribution entitled "Radio Observations of the Ionosphere", K. Rawer and K. Suchy have used such generalized rules for writing equations. In their Appendix (p. 535 of Vol. 49/2) they give detailed explanations and discuss 'transformations' between any two systems of units, as well as the relevant 'invariants' (physical quantities which must be equivalent in all systems, energy expressions, for example). While we refer to this text for details, it may be useful to repeat here the most important equations of electromagnetic theory.

With the definitions

electric field intensity (field strength)	electric flux density (displacement)	magnetic field intensity (field strength)	magnetic flux density (induction)	current density
$\boldsymbol{E}$	$\boldsymbol{D}$	$\boldsymbol{H}$	$\boldsymbol{B}$	$\boldsymbol{J}$

we have

$$\boldsymbol{D} = \varepsilon \boldsymbol{E}; \quad \boldsymbol{B} = \mu \boldsymbol{H}; \quad \boldsymbol{J} = \sigma \boldsymbol{E},$$

and in vacuum

$$\varepsilon = \varepsilon_0; \quad \mu = \mu_0, \quad \sigma = 0.$$

Maxwell's equations connecting the different field quantities are then

$$\frac{\partial}{\partial \boldsymbol{r}} \times \boldsymbol{H} \equiv \nabla \times \boldsymbol{H} = \frac{1}{c_0 \sqrt{\varepsilon_0 \mu_0}} \frac{\partial}{\partial t} \boldsymbol{D} + \frac{u}{c_0 \sqrt{\varepsilon_0 \mu_0}} \boldsymbol{J}$$

$$\frac{\partial}{\partial \boldsymbol{r}} \cdot \boldsymbol{D} \equiv \nabla \cdot \boldsymbol{D} = u \varrho$$

$$\frac{\partial}{\partial \boldsymbol{r}} \times \boldsymbol{E} \equiv \nabla \times \boldsymbol{E} = -\frac{1}{c_0 \sqrt{\varepsilon_0 \mu_0}} \frac{\partial}{\partial t} \boldsymbol{B}$$

$$\frac{\partial}{\partial \boldsymbol{r}} \cdot \boldsymbol{B} \equiv \nabla \cdot \boldsymbol{B} = 0,$$

ϱ being the charge density.

The two systems most used in geomagnetism are that of Gauss and SI. In simplified language we may say that in SI units we have $c^2 \varepsilon_0 \mu_0 = 1$ and $u = 1$, so that in Maxwell's equations the factors can be disregarded. In the Gauss system, only ε_0 and μ_0 can be disregarded, but c_0 remains and $u = 4\pi$.

In a volume dealing with magnetic phenomena, the definition of the magnetic moment **M** must be discussed in particular. Unfortunately, there are two different definitions in use. The potential energy of a dipole magnet in a magnetic field is given by the scalar product of a vector **M** characterizing orientation and strength of the dipole, and a field vector. The question now is whether **H** or **B** is to be taken as field vector. In the older literature Gauss units have mostly been used and the distinction between **H** and **B** has not been made clear because in this particular system where $\mu_0 = 1$, they both have the same dimension. If, however, the distinction is made, **M** has different dimensions according to the choice. IUPAP has decided in favour of **B**, so that the energy must be written

$$\mathscr{E} = -\mathbf{M} \cdot \mathbf{B}$$

and **M** has the dimension Joule/Tesla $= Am^2$ for SI units (and not Vms as would be found with the other choice). The IUPAP definition is used throughout this volume[1].

Note that by definition the product must be invariant so that the general 'transformation' (see Rawer and Suchy, Vol. 49/2, p. 535) now reads for **M**

$$\sqrt{u \mu_0} \, \mathbf{M} = \sqrt{u' \mu_0'} \, \mathbf{M}'.$$

Consequently, the free space field of a dipole is

$$B_{\text{dipole}} = \frac{u}{4\pi} \mu_0 \frac{M}{r^3} (1 + 3 \sin^2 \phi)^{\frac{1}{2}}$$

where M is the magnetic moment of the dipole, r the distance and ϕ a "latitude" angle.

All equations are, of course, usable in any system of units because they are written in physical quantities. The accepted definition of a physical quantity is

(numerical value) · (dimension)

such that by dividing each term through the dimension a purely numerical equation can be obtained. We tend to write such equations, if at all, so that each physical quantity is divided by its own dimension. Where other units are to be used, the numerical change is obtained by *algebraic substitution*, e.g.

$$c_0 = 3 \cdot 10^8 \text{ m s}^{-1}; \quad 1 \text{ (nt. mile)} = 1.853 \text{ km} \ldots 1 \text{ m} = \frac{10^{-3}}{1 \cdot 853} \text{ (nt. mile)} \ldots$$

$$c_0 = 3 \cdot 10^8 \frac{10^{-3}}{1.853} \text{ (nt. mile) s}^{-1} = 1.619 \cdot 10^5 \text{ (nt. mile) s}^{-1}.$$

In equations, division may be expressed by bar or stroke. According to IUPAP rules, the stroke / has priority over all other operations such that $a \cdot b/c \cdot d \equiv \frac{ab}{cd}$. It is worth noting that this is not so in computer languages like ALGOL.

It is a special convention in this Encyclopedia that the natural logarithm is designated by log (not by ln).

---

[1] The same definition is used in the well-known book "Electromagnetic Theory" by Stratton, who puts forward arguments against the second choice. Unfortunately, H. Ebert's "Physikalisches Taschenbuch" uses the second choice in disagreement with the IUPAP definition.

Differential operations in vector fields are normally written with the symbolic vector

$$\frac{\partial}{\partial \boldsymbol{r}} \equiv \boldsymbol{\nabla}$$

which in cartesian coordinates x_1, x_2, x_3 reads

$$\left(\frac{\partial}{\partial x_1}, \frac{\partial}{\partial x_2}, \frac{\partial}{\partial x_3}\right).$$

IUPAP proposes two different ways of denoting tensors: use of sanserif letters as symbols (e.g. T), or the analytical one with reference to cartesian coordinates (e.g. T_{ik}). In the latter case the summation rule is to be applied. Though symbolic writing is preferred, we give both presentations in most cases. The unit tensor is written as U or δ_{ik}. The tensorial product of two vectors $\boldsymbol{a}\,\boldsymbol{b}$ must be distinguished from the vector product $\boldsymbol{a} \times \boldsymbol{b}$ and the scalar product $\boldsymbol{a} \cdot \boldsymbol{b}$.

Measured data and other numerical values are normally given in SI units, but for magnetic fields (induction) Gauss (Gs or Γ) and gamma (γ) are quite generally applied. The relations to the SI unit Tesla (Vsm^{-2}) are

$$1 \text{ Gs} \equiv 1 \text{ } \Gamma = 10^{-4} \text{ T}; \quad 1 \text{ } \gamma \equiv 10^{-5} \text{ } \Gamma = 10^{-9} \text{ T}.$$

In this volume measurable elements of the Earth's magnetic field are printed in special letters in order to be easily identified, e.g. $\mathcal{D}, \mathcal{H}, \mathcal{Z}$.

The editors also considered that a few widely used designations should be replaced by more appropriate terms. One of these is "magneto-hydro-dynamic" (or "hydro-magnetic"). It is misleading because it appears to refer to a fluid instead of a gas. Incompressibility was, infact, assumed in ALFVEN's original theory, and this justified the expression at that time. In the wide field of waves in plasma, however, the plasma has the character of a compressible, gaseous medium, though strongly affected by the presence of a magnetic field. Therefore we prefer the terms "magneto-gas-dynamic", or "gas-magnetic" or simply "magneto-dynamic".

Freiburg, 22 July 1971   KARL RAWER

Morphology of Magnetic Disturbance.

By

TAKESI NAGATA and NAOSHI FUKUSHIMA*.

With 100 Figures.

A. General introduction.

1. Geomagnetic variations and planetary physics. The earth's magnetic field is almost always changing its intensity and direction, regularly or irregularly. The varying component of the geomagnetic field amounts at most to a few percent only of the intensity of the permanent field. Geomagnetic variations recorded at the earth's surface originate primarily in the space above the solid earth, and partly in the earth's interior. On the other hand, the origin of the earth's permanent magnetic field, and its secular variation too, is found deep in the earth's interior. This conclusion was reached after spherical harmonic analysis of the observed geomagnetic field and its variations over the world.

Studies of geomagnetic variations have played an important rôle in the research of the upper atmosphere, because most regular and irregular geomagnetic changes can be attributed to electric currents in the ionosphere or to other electromagnetic phenomena in the outer space around the earth. The presence of electric currents in the upper atmosphere was first proposed by STEWART[1], as early as 1882, well before the ionosphere was experimentally found by radio techniques in 1925. When combined with direct sounding of the ionosphere, analysis of geomagnetic variations has been very useful in the study of the electric properties of the upper atmosphere. In recent years it has become possible to make direct measurements of magnetic fields in the earth's outer atmosphere by rockets (since 1946) and by artificial satellites (since 1957), so that our knowledge of the earth's atmosphere has greatly increased.

About $1/3$ in magnitude of the geomagnetic variation observed on the earth's surface originates in the earth's interior. It is now generally believed that this part of the geomagnetic variation is caused by electric currents under the earth's surface forced to flow by the electromagnetic induction provoked by changes in the external magnetic field[2]. The underground electric conductivity near the earth's surface is not uniform over the world, because of the sea and land distribution and different subterranean structure in the earth's crust and upper mantle. As the penetration depth of electromagnetic field-changes of shorter periods is shallow, the effect of local non-uniformity of the underground electric conductivity becomes more remarkable for short-period geomagnetic variations. For this reason, geomagnetic variations observed on the earth's surface are influenced by the underground electric conditions, and the spatial dependence of the geomagnetic variation is now also studied in order to estimate the global distribution of underground electric properties.

---

* Original manuscript received Aug. 1957, revised manuscript Jan. 1967.

References of greater importance are summarized at the end of this paper. They are given by italic numbers in brackets. Other references are given at the bottom of each page. The book "Geomagnetism" [1] has two chapters (Chap. IX and X) concerning the morphology of magnetic disturbances; these are the essential basis of the present article. See also [40].

[1] B. STEWART: Terrestrial Magnetism, Ency. Brit., 9th ed. 16, 181 (1882).

[2] The problem of electromagnetiic nduction within the earth is explained in Chap. XXII of [1], Chap. II-3 of [40], and in a recent book by T. RIKITAKE [11].

In the present article, the recent knowledge about the morphological characteristics of geomagnetic disturbances is reviewed, and some related phenomena are briefly described. As geomagnetic disturbances originate primarily in the space around the earth, the problems discussed in this article are mainly related with the upper atmosphere. The authors attempt not only to describe various observed facts in detail, but also to discuss briefly the theoretical interpretations of the results[3]. It will be rewarding for the writers, if this article helps to increase the understanding of the nature of geomagnetic disturbance, which is one of the important phenomena to be studied in the field of so-called space science.

In the following sections, a brief description is given of the earth's environment, where electric currents flow, or gasmagnetic waves[4] propagate, all of which are directly connected with the geomagnetic disturbances recorded on the earth's surface.

2. Ionosphere and electric current.

The earth's atmosphere, from the ground up to about 90 km height, is an electrical insulator, because of the very low density and mobility of ions and electrons contained in the gas. On the other hand, the tenuous upper atmosphere above the 90 km level is an electrical conductor and a reflector for h.f. (high-frequency) radio waves, because of the presence of a sufficient number of free electrons and ions. Under normal conditions the maximum electron density in the E-region (90 to 120 km height) is of the order of 10^{11} m^{-3}, and that in the F-region (200 to 350 km) 10^{12} m^{-3}. Although the number of ions and electrons in the ionosphere is still very small compared with that of neutral particles (the ratio is 10^{-7} to 10^{-8} in the E-layer, and 10^{-3} to 10^{-4} in the F2-layer), the main body of the ionosphere behaves as a plasma, and the lowest part of the ionosphere is a transition region from neutral gas to plasma conditions.

The electric conductivity of an ionized gas in a magnetic field depends on the following physical quantities:

N: number density of charged particles,

q: electric charge,

m: mass,

ν: mean collision frequency,

$\omega_B/2\pi$: gyro-frequency,

$$\left(|\omega_B| = \frac{1}{c_0\sqrt{\varepsilon_0\mu_0}} \frac{|q|}{m} B \qquad \text{in a magnetic field of induction } B\right)^*.$$

When the magnetic field is in z-direction, the electric conductivity is written in tensor form as

$$((\sigma)) = \begin{pmatrix} \sigma_1 & \sigma_2 & 0 \\ -\sigma_2 & \sigma_1 & 0 \\ 0 & 0 & \sigma_0 \end{pmatrix} \tag{2.1}$$

where

$$\sigma_1 = \sum \frac{q^2 N}{m\nu} \frac{\nu^2}{\nu^2+\omega_B^2}, \qquad \sigma_2 = \sum \frac{q^2 N}{m\nu} \frac{\nu\omega_B}{\nu^2+\omega_B^2}, \qquad \sigma_0 = \sum \frac{q^2 N}{m\nu}. \tag{2.2}$$

---

* As explained in the "Introductory remarks" we write equations in a generalized form so that any of the usual systems of units may be applied.

[3] See contribution by E. N. PARKER and V. C. A. FERRARO in this volume p. 131.

[4] For reasons explained in the "Introductory remarks" we use the terms "gas magnetic" and "magneto-gas-dynamic" instead of the usual ones "hydromagnetic" and "magneto-hydro-dynamic".

Here $\sum$ denotes the summation over all kinds of charged particles, i.e., positive and negative ions and electrons. σ_1 is called the *Pedersen conductivity*, while σ_2 is the *Hall conductivity*.

When we discuss electric currents in the upper atmosphere which are responsible for ground-level geomagnetic variations, we usually deal with horizontal electric currents in the ionosphere. The vertical component of the current is assumed to be impeded by surface charges on both the upper and lower boundaries of the ionosphere[1,2]. Under this condition, the three-dimensional tensor, $((\sigma))$, can be reduced to a two-dimensional one, viz.

$$((\sigma)) \to \begin{pmatrix} \sigma_{xx} & \sigma_{xy} \\ \sigma_{yx} & \sigma_{yy} \end{pmatrix}. \quad (2.3)$$

Fig. 1. Schematic picture for the horizontal conducting layer permeated by the geomagnetic field.

We take x- and y-axes towards south and east, and we assume that the geomagnetic field is in the xz-plane, as shown in Fig. 1. Then, the tensor elements at a point with geomagnetic dip angle I are given by

$$\left. \begin{aligned} \sigma_{xx} &= \frac{\sigma_0 \sigma_1}{\sigma_0 \sin^2 I + \sigma_1 \cos^2 I}, \\ \sigma_{xy} = -\sigma_{yx} &= \frac{\sigma_0 \sigma_2 \sin I}{\sigma_0 \sin^2 I + \sigma_1 \cos^2 I}, \\ \sigma_{yy} &= \frac{\sigma_0 \sigma_1 \sin^2 I + (\sigma_1^2 + \sigma_2^2) \cos^2 I}{\sigma_0 \sin^2 I + \sigma_1 \cos^2 I}. \end{aligned} \right\} \quad (2.4)$$

The horizontal electric current $\boldsymbol{i} = (i_x, i_y)$ is given by *

$$\left. \begin{aligned} i_x &= \sigma_{xx} E_x + \sigma_{xy} E_y, \\ i_y &= \sigma_{yx} E_x + \sigma_{yy} E_y. \end{aligned} \right\} \quad (2.5)$$

Assuming that the electric field $\boldsymbol{E}$ is independent of the height within the ionosphere, the "total electric current" $\boldsymbol{J} \equiv (J_x, J_y)$ in the ionosphere is obtained by integration over the height as

$$\left. \begin{aligned} J_x &= K_{xx} E_x + K_{xy} E_y, \\ J_y &= K_{yx} E_x + K_{yy} E_y, \end{aligned} \right\} \quad \boldsymbol{J} = \boldsymbol{K} \cdot \boldsymbol{E} \quad (2.6)$$

$\boldsymbol{K} \equiv ((K))$, a two-dimensional tensor, is called the integrated electric conductivity of the ionosphere. Fig. 2 shows an example of numerical evaluation of the tensor elements of $((K))$. In the above equation, $K_{xx} E_x$ in J_x and $K_{yy} E_y$ in J_y are customarily called the *Pedersen current*, and $K_{xy} E_y$ in J_x and $K_{yx} E_x (= -K_{xy} E_x)$ in J_y the *Hall current*.

We see from Fig. 2 that the Hall conductivity is greater than the Pedersen conductivity, except near the equator. With this exception the electric current

---

* i is a current density (measured in A/m²), while $\boldsymbol{J}$ is a "surface current density" (A/m); the corresponding units are S/m for σ, but S for K (unit S $\equiv$ Siemens $=$ A/V).

[1] M. HIRONO: J. Geomag. Geoelectr. 2, 1, 113 (1950); 4, 7 (1952).
[2] W. G. BAKER, and D. F. MARTYN: Nature 170, 1090 (1952); — Phil. Trans. Roy. Soc. London A 246, 281 (1953).

flows nearly perpendicularly to the electric field. In other words, the electric current flows nearly along the equipotential lines. Seen from above the ionosphere, the current flow is clockwise in the northern hemisphere, encircling the region of highest electric potential.

At the geomagnetic equator, the electric conductivity has a very high value, and $K_{xx} > K_{yy} > K_{xy}$. In fact K_{xx} is the electric conductivity along the magnetic field, and the magnetic field does not impede any motion of charged particles in the x-direction. K_{yy} is given by $\int \left(\sigma_1 + \frac{\sigma_2^2}{\sigma_1} \right) dh$, and $K_{yy} \gg K_{xy}$. It is worth noting that the value of K_{yy} would become identical with K_{xx}, if the ionized gas merely

Fig. 2. An estimated value of the height-integrated electric conductivity in the ionosphere, as function of magnetic latitude. (Right-hand scale: electro-magnetic cgs-units; left-hand: international system of units.)

consisted of either positive or negative charged particles alone. In the ionosphere, heavy ions coexist with light electrons, and $K_{yy} < K_{xx}$. The high conductivity in the equatorial ionosphere explains the great amplitude of geomagnetic variations near the geomagnetic equator, as will be seen in later chapters.

In this section, we may state that the electric conductivity in the upper atmosphere is greatest at the ionospheric level. By rocket experiments it has been shown that the electric current responsible for geomagnetic variations does flow in the ionosphere. It is of fundamental importance that the electric conductivity in the ionosphere is anisotropic, because of the presence of the geomagnetic field. In earlier discussions on the relation between electric current and electromotive force in the ionosphere, the anisotropic nature of the ionospheric conductivity was not taken into consideration, and the conductivity value was substituted by the Pedersen conductivity. In Sect. 3, the electric current in the ionosphere is again discussed for the case when an electric field extends from the magnetosphere down into the ionosphere.

3. Magnetosphere. Above the ionosphere the number density of both ionized and neutral particles decreases with height, the latter more quickly. In deep space, more than 10,000 km beyond the earth's surface, the constituents are mostly protons and electrons, so that the charged particles overwhelm the neutral particles in number density. Beyond a geocentric distance of 3...4 a (a earth radius) the number density of particles falls even down to $10^8 ... 10^7$ m^{-3}. The energy density of the earth's permanent magnetic field*, $B^2/2\mathrm{u}\,\mu_0$, in the space above the ionosphere is at least one order of magnitude larger than the kinetic energy of particles contained (nkT). Because of the presence of the geomagnetic field and the high electric conductivity of the ionized medium, the macroscopic motion in the earth's outer atmosphere is controlled by magneto-gasdynamics. Therefore, the region above the ionosphere is now called the magnetosphere[1], because of the fundamental dependence of its behaviour on the earth's magnetic field. Of the magnetosphere, it can for convenience be said that the material and the magnetic field-line move together, or in other words, the material is "frozen in" the mag-

Fig. 3. Motion of ions and electrons, and the resultant electric current in the upper atmosphere, under the influence of a horizontal electric field E and a vertical magnetic field B.

netic field. Accompanying the magnetospheric plasma motion v an electric field E is produced so as to satisfy the condition**

$$c_0 \sqrt{\varepsilon_0 \mu_0}\, E + v \times B = 0 \qquad (3.1)$$

and an electric charge distribution arises so that the electric "polarization" field E is just produced. Since the electric conductivity is extremely great along a magnetic field line, the electric potential can be considered to be constant along any geomagnetic field line, from the magnetosphere down to the ionosphere. In the following paragraph, the relationship between the electric current in the upper atmosphere and the electric field, or magnetospheric motion is schematically illustrated.

For simplicity, the magnetic field is assumed to be vertical, and a horizontal electric field is applied throughout the upper atmosphere, as shown in Fig. 3. In the magnetosphere where charged particles can move without collisions, both ions and electrons drift in the $E \times B$ direction with the same velocity, so that no electric current appears. Below about 150 km, the motion of ions has components in both the $E \times B$- and E-directions, while electrons still move almost parallel to the $E \times B$-direction; this difference in the motions of ions and electrons results in an electric current. Decisive is the increase of the relative collision frequency for ions, v_i/ω_{Bi}, while electrons show almost collisionless motion ($v_e/\omega_{Be} \ll 1$) even at this level of the ionosphere. Below the ionosphere, even for electrons the collision frequency becomes greater than their gyro-frequency, so that only a very weak Pedersen current flows. As a result of this dependence of ion and electron

---

* u≡1 in rationalized, ≡4π in non-rationalized systems of units, see "Introductory Remarks."
  ** $c_0 \sqrt{\varepsilon_0 \mu_0} \equiv 1$ in SI-units (and these are rationalized).
  [1] T. GOLD: J. Geophys. Res. 64, 1219 (1959). See also contributions by W. H. HESS and by H. POEVERLEIN in Vol. 49/4 of this Encyclopedia.

collisions on altitude, the integrated electric current in the ionosphere is nearly antiparallel to the $\boldsymbol{E} \times \boldsymbol{B}$ direction.

Let us now consider the current flow in the ionosphere from a different viewpoint, supposing that the line of magnetic force is moving horizontally in the magnetosphere, as is shown in Fig. 4. In the magnetosphere, both ions and electrons follow the motion of magnetic field-lines, because of the "frozen-in condition". Below the ionosphere, the magnetic field-line is fixed to the earth, so that there may be a slipping between magnetic field-lines and material at ionospheric heights.

In the region between about 90 km and 150 km height, ions cannot follow the motion of magnetic field-lines, whereas electrons still do. Consequently, an electric

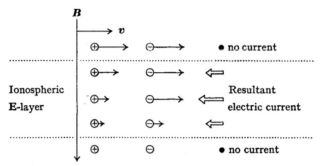

Fig. 4. Motion of ions and electrons accompanying a horizontal movement of lines of magnetic force, and the resulting horizontal electric current in the upper atmosphere.

current flows in the direction opposite to the field-line movement. At the same time, a small bulk displacement of ions and electrons perpendicular to the field-line motion is revealed by a weak Pedersen current.

The above explanations show that the production of a current flow in the ionosphere can be explained with two models, i.e. by an electric field, or by a motion of magnetic field lines, both originating in the magnetosphere. These two viewpoints are not independent of each other; really, they are the same interpretation, the connection being given by the "frozen-in condition"

$$c_0 \sqrt{\varepsilon_0 \mu_0}\, \boldsymbol{E} + \boldsymbol{v} \times \boldsymbol{B} = 0 \tag{3.1}$$

in the magnetosphere.

The magnetosphere is a compressible medium of high electric conductivity, so that gas-magnetic waves[2] of two modes, a torsional one (Alfvén wave) and a compressional mode (magneto-acoustic wave) can propagate in it. Such waves generated in the magnetosphere are propagated down to the ionosphere, and after conversion to u.l.f. (ultra low frequency) electromagnetic waves, they cause geomagnetic pulsations near the earth surface. Not only magneto-gas-dynamic waves are propagated in the magnetosphere, but also whistler waves in the e.l.f. (extremely low frequency) range. Whistler waves can also reach the earth.

We consider now the motion of individual charged particles in the magnetosphere under collisionless conditions. The charged particles gyrate around the magnetic field-lines, producing a diamagnetic effect. Within several earth radii (geocentric distance) the magnetic field configuration is approximately of dipole type. In such a field, charged particles spiralling around field lines make yet two other large-scale motions, namely an oscillating motion along the field line, crossing the equator, and simultaneously a slow perpendicular drift, westwards for positive ions and eastward for negative ions and electrons. As a consequence of this opposed drifting of particles of both signs, an electric current surrounding the earth is produced, provided there is a space-gradient of the density of charged particles. This current around the earth produces a uniform magnetic field in the vicinity of the earth, which is directed parallel to the geomagnetic axis and southward. We know now that in the so-called radiation belts (Van Allen belts encircling the earth) high-energy particles, mostly protons and electrons, are

---

[2] See contribution by V. L. GINZBURG and A. A. RUHADSE in Vol. 49/4 of this Encyclopedia.

trapped in the earth's permanent magnetic field[3]. The number density of high-energy particles detected in the radiation belt is very small in comparison with that of low-energy particles, so that the magnetic effect of charged particles trapped in the magnetosphere is caused mainly by the *low-energy particles*. The combined magnetic effect[4] due to the drift and gyration of charged particles on the earth surface is estimated to be about 40 γ under normal conditions, i.e. on magnetically quiet days[5].

4. Geomagnetic cavity in the solar wind. The sun continuously emits a plasma stream, though its intensity varies considerably from time to time. The plasma stream from the sun is called the solar wind, and its existence is now also known from direct measurements by space probes, such as Mariner II[1]. The solar wind

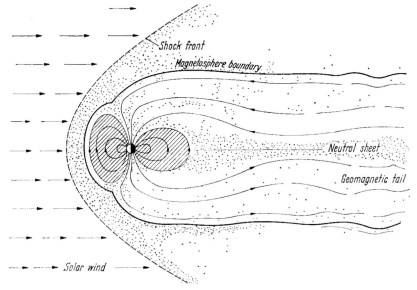

Fig. 5. Schematic illustration of the magnetic-field topology in the earth's magnetosphere in the noon-midnight meridian plane. Charged particles trapped by the geomagnetic field are mainly found in region ▨ including radiation belts. ▦ designates magnetospheric plasma, ⇒ solar (wind) plasma.

consists mainly of protons and electrons with a velocity of $10^5 \ldots 10^6$ m/sec, a number density of $3 \cdot 10^5 \ldots 10^7$ m^{-3}, and a temperature of about 10^5 °K. The earth's magnetic lines of force are not extended to infinity but are confined to a limited region around the earth, which is called the geomagnetic cavity. Under normal conditions (on geomagnetically quiet days), the geocentric distance of the cavity boundary on the side facing the sun is about 15 earth radii, as shown by a number of recent direct measurements by space probes. The size is dependent on the external pressure of the solar wind. Fig. 5 illustrates by a sketch the meridional cross-section of the geomagnetic cavity[2]. Due to the solar wind pressure, the magnetic field distribution in the geomagnetic cavity is not axisymmetric, and the region of charged particle trapping is also asymmetric on the day and night sides.

According to recent measurement by satellites, a westward electric current flows near the geomagnetic equator on the night side, where a neutral sheet

---

[3] See contribution by W. N. Hess in Vol. 49/4 of this Encyclopedia.
[4] In the field of geomagnetism, the magnetic field intensity is customarily given in units Γ (Gauss) or γ (gamma). $1\gamma = 10^{-5} \Gamma = 10^{-9}$ Vs m$^{-2} \equiv 10^{-9}$ T (Tesla).
[5] S.-I. Akasofu, J. C. Cain, and S. Chapman: J. Geophys. Res. **66**, 4013 (1961).
[1] C. W. Snyder, M. Neugebauer, and U. R. Rao: J. Geophys. Res. **68**, 6361 (1963).
[2] N. F. Ness: J. Geophys. Res. **70**, 2989 (1965).

appears, and this makes the night-side cavity elongated like a comet tail. The direction of the geomagnetic tail varies with season because of the solar declination change. The leeside of the magnetic cavity is extremely extended, at least to several tens of earth radii, perhaps even to the orbit of the moon[3].

The presence of a shock-wave front on the sunlit side of the geomagnetic cavity has been inferred from direct measurements of the magnetic field and its time-changes, in the space just outside the magnetosphere. The possible presence of a shock-wave in a tenous, collisionless plasma in the space around the geomagnetic cavity has caught the interest of theoretical and experimental physicists.

The interaction of a plasma stream with a magnetic field of dipole type has also been studied experimentally, and a model of the geomagnetic cavity has been presented by several groups of workers[4-8]. The experimental results are applicable, to some extent, to the large-scale natural phenomenon of the interaction of the solar wind with the geomagnetic field, with the aid of a similarity law. However, due to some technical limitations in the model experiments, the natural large-scale phenomenon cannot yet be fully simulated in the laboratory, especially with regard to the thinness of the cavity boundary in comparison with the earth's dimension, and to the extreme lack of collisions in the magnetosphere and in interplanetary space.

5. Magnetic disturbance and earth storm.

When the earth's magnetic field is disturbed, there are other simultaneous geophysical phenomena, such as auroral display at high latitudes, ionospheric disturbance, increase in activity of natural electromagnetic radiation in the e.l.f. and v.l.f. range, and, occasionally, a change of intensity of primary cosmic rays, all of which are observable on the earth surface (Chap. K). In recent years, some knowledge has been gathered also about simultaneous changes in the earth's environmental conditions accompanying geomagnetic disturbances; such changes are: a redistribution of high-energy charged particles trapped in the magnetosphere, deformation of the geomagnetic cavity, intensity change in solar plasma flux, and so forth. They have not directly been perceived by ground observation, but were detected by space probes. A magnetic disturbance recorded on the earth's surface is, therefore, one consequence of the disturbance phenomena in the earth's upper atmosphere in and above the ionosphere, all of which are caused by an intensified solar corpuscular stream attacking the earth. A complete physical picture for a magnetic disturbance could only be obtained when the simultaneous disturbance phenomena in the earth's environment were understood. The expression "earth storm" is used nowadays to designate the disturbance in the earth's environment as whole, including geomagnetic and other associated phenomena which take place in a vast space in and above the ionosphere.

Below the ionospheric region, a definite correlation between geomagnetic disturbance and physical changes in the atmosphere (such as ozone content or temperature change) has not yet been established. Magnetic disturbances apparently have no correlation with the general meteorological conditions all over the world. The only effect hitherto found is a propagation of infrasonic pressure variations (with periods of 10 ... 100 sec) of very small amplitude, 0.1 ... 1 Nm^{-2} (1 ... 10 dyn cm^{-2}), at the time of magnetic disturbances, near the auroral zone[1,2]. We may say that the agent responsible for magnetic disturbances does not have energy great enough to produce a significant disturbance in the nonconducting, dense lower atmosphere.

---

[3] See contribution by H. POEVERLEIN in Vol. 49/4 of this Encyclopedia.

[4] H. ALFVÉN: Symp. Plasma Space Science, Catholic Univ. of America, Washington, D. C., June 1963.

[5] W. H. BOSTIK, H. BYFIELD, and M. BRETTSCHNEIDER: J. Geophys. Res. **68**, 5315 (1963).

[6] J. B. CLADIS, T. D. MILLER, and J. R. BASKETT: J. Geophys. Res. **69**, 2257 (1964).

[7] N. KAWASHIWA and N. FUKUSHIMA: Planet. Space. Sci. **12**, 1187 (1964).

[8] F. J. F. OSBORNE, M. P. BACHYNSKI, and J. V. GORE: Appl. Phys. Letters **5**, 77 (1964); — J. Geophys. Res. **69**, 4441 (1964).

[1] P. CHRZANOWSKI, G. GREENE, K. T. LEMON, and J. M. YOUNG: J. Geophys. Res. **66**, 3727 (1961).

[2] K. MAEDA, and T. WATANABE: J. Atmosph. Sci. **21**, 15 (1964).

Fig. 6. Example of ordinary (upper) and storm magnetograms (lower) on the same day taken at College, Alaska. [Nov. 25, 08 h ... Nov. 26, 08 h, 1950.]

6. Observation of geomagnetic variations at the ground. In ordinary records of geomagnetic variations, three components of the geomagnetic field: horizontal intensity H, declination D, and vertical component Z (or north-component X, east-component Y, and Z) are recorded. The sensitivity on the record is usually several γ/mm[1], and the paper speed is 20 or 15 mm/hour. Ordinary magnetograms are now taken at about 150 stations all over the world. About half of them have a long history of observation beginning with the time of the Second International Polar Year 1932/33, or even earlier. The International Geophysical Year 1957/58

---

[1] 1 γ (gamma) $= 10^{-5}$ Γ (Gauss).

gave an opportunity to intensify the world network of geomagnetic observations. Although we have quite a large number of magnetic observatories over the world, the present distribution of observatories is still unsatisfactory for the detailed analysis of geomagnetic disturbances, as they show a very complicated dependence on the location, especially at high latitudes, as will be seen in later chapters.

At high latitudes, where geomagnetic disturbances are particularly pronounced, low-sensitivity magnetograms are taken at many stations in addition to the ordinary magnetograms. On days of magnetic storms, it is often very difficult to trace the geomagnetic field change at high latitudes on ordinary magnetograms, because of large and rapid fluctuation, and of occasional off-scaling. The geomagnetic change on stormy days is better seen on low-sensitivity magnetograms, which are therefore called "storm magnetograms". Fig. 6 shows an example of the ordinary and storm magnetograms on the same day at College, Alaska (64° 51′ N, 147° 50′ W).

In recent years recording techniques for rapid geomagnetic variations of small amplitude have been developed. High-sensitivity magnetographs such as the LaCour type and/or the induction magnetograph are used for this purpose with a high paper speed, from several mm/min. to several mm/sec. The induction magnetograph, which has coils with or without cores of highly permeable material, picks up the time-derivative of the geomagnetic field; it is very effective in recording rapid geomagnetic variations[2]. Earth current observations are also used for the study of rapid geomagnetic changes. Most of the earth current changes are caused by electromagnetic induction within the earth, produced by changes in the external magnetic field, so that they are particularly sensitive to short-period variations. Therefore, the measurement of earth currents, or potential differences between two points on the earth's surface, is now widely used for studying geomagnetic variations of shorter periods.

If the resolving power of the magnetometer is increased more and more, the record shows the presence of variations with extremely short periods, of very small amplitude. The frequency of these oscillatory geomagnetic changes even approaches the range of e.l.f. (extremely low frequency) electromagnetic waves. They have now caught the attention of many research workers[3], and the existing gap in the observations of electromagnetic phenomena between nearly static (from the electromagnetic standpoint) geomagnetic variations and low-frequency radio waves of natural origin will very likely be filled in the near future.

7. Equivalent overhead electric current for geomagnetic variations. When studying magnetic disturbances we often discuss the electric current in the upper atmosphere, which is responsible for the geomagnetic variation observed on the earth surface. In this section, the principles for depicting this electric current are briefly described. The three components of the geomagnetic field $(\mathcal{X}, \mathcal{Y}, \mathcal{Z})$ at the earth surface (radius a) are assumed to be derivable from a scalar potential V, by means of

$$\mathcal{X} = \frac{\partial V}{a\, \partial \theta}, \qquad \mathcal{Y} = -\frac{\partial V}{a \sin\theta\, \partial \varphi}, \qquad \mathcal{Z} = \frac{\partial V}{\partial r}, \qquad (7.1)$$

in spherical coordinates r, θ, φ. The potential V may be written as

$$V = V^{(e)} + V^{(i)}, \qquad (7.2)$$

with

$$V^{(e)} = a \sum_{n=1}^{\infty} \left(\frac{r}{a}\right)^n T_n^{(e)}, \qquad V^{(i)} = a \sum_{n=1}^{\infty} \left(\frac{a}{r}\right)^{n+1} T_n^{(i)} \qquad (7.3)$$

where $V^{(e)}$ and $V^{(i)}$ are the parts of the potential V, of external and internal origins with respect to the earth's surface, respectively. Developing:

$$T_n^{(e,i)} = \sum_{m=0}^{n} \{g_n^{m\,(e,i)} \cos m\varphi + h_n^{m\,(e,i)} \sin m\varphi\}\, P_n^m(\theta) \qquad (7.4)$$

---

[2] See contributions by V. LAURSEN and J. OLSEN, p. 276, and by H. SCHMIDT and V. AUSTER, p. 323 in this volume.

[3] See contribution by E. SELZER in Vol. 49/4 of this Encyclopedia.

with the associated Legendre function, P_n^m, at the earth surface, $r=a$, the following relations must hold.

$$\begin{aligned}
X &= \sum_{n=1}^{\infty} \sum_{m=0}^{n} [\{g_n^{m(e)} + g_n^{m(i)}\} \cos m\varphi + \{h_n^{m(e)} + h_n^{m(i)}\} \sin m\varphi] \frac{dP_n^m(\theta)}{d\theta}, \\
Y &= \sum_{n=1}^{\infty} \sum_{m=0}^{n} [m\{g_n^{m(e)} + g_n^{m(i)}\} \sin m\varphi - m\{h_n^{m(e)} + h_n^{m(i)}\} \cos m\varphi] P_n^m(\theta)/\sin\theta, \\
Z &= \sum_{n=1}^{\infty} \sum_{m=0}^{n} [\{n g_n^{m(e)} - (n+1) g_n^{m(i)}\} \cos m\varphi + \{n h_n^{m(e)} - (n+1) h_n^{m(i)}\} \sin m\varphi] P_n^m(\theta).
\end{aligned} \quad (7.5)$$

When the observed geomagnetic field is expanded into spherical harmonics of the form

$$\begin{aligned}
X &= \sum_n \sum_m (a_n^m \cos m\varphi + b_n^m \sin m\varphi) X_n^m(\theta), \\
Y &= \sum_n \sum_m (-b_n^m \cos m\varphi + a_n^m \sin m\varphi) Y_n^m(\theta), \\
Z &= \sum_n \sum_m (a_n^m \cos m\varphi + b_n^m \sin m\varphi) P_n^m(\theta),
\end{aligned} \quad (7.6)$$

with

$$X_n^m(\theta) = \frac{1}{n} \frac{dP_n^m(\theta)}{d\theta}, \quad Y_n^m(\theta) = \frac{m}{n} \frac{P_n^m(\theta)}{\sin\theta}, \quad (7.7)$$

the following mutual relations can be found

$$g_n^{m(e)} = \frac{(n+1) a_n^m + n a_n^m}{n(2n+1)}, \quad g_n^{m(i)} = \frac{a_n^m - a_n^m}{2n+1}, \\
h_n^{m(e)} = \frac{(n+1) b_n^m + n b_n^m}{n(2n+1)}, \quad h_n^{m(i)} = \frac{b_n^m - b_n^m}{2n+1}. \quad (7.8)$$

Thus, the geomagnetic field and its variation can be represented by spherical harmonic functions, and the coefficients of the terms can be separated into parts of internal and external origin with respect to the earth surface[1].

The part of external origin may be represented by an electric current flowing in the earth's upper atmosphere. If the current is assumed to flow in a thin spherical sheet of radius r, a current function R can be introduced by

$$J_\theta = \frac{\partial R}{r \sin\theta \, \partial\varphi}, \quad J_\varphi = -\frac{\partial R}{r \, \partial\theta}. \quad (7.9)$$

In practice, R can be expressed as

$$R = \sum_n R_n = -\sum_n \frac{1}{u} \frac{2n+1}{n+1} \left(\frac{r}{a}\right)^n V_n^{(e)}. \quad (7.10)$$

In this way, the geomagnetic variation originating outside of the earth can be expressed by a thin current shell at an appropriate height. The distribution of current flow depends on the assumed height of the shell. Therefore, the height of the current sheet cannot uniquely be determined from ground observations. In most cases we assume that the current flows at the level of the ionospheric E-layer, because of the distribution of electric conductivity described in Sect. 2 and Fig. 2.

---

[1] From such spherical analysis, we know that the earth's permanent magnetic field originates within the earth. However, the origins of varying magnetic fields dealt with in this article, are both external and internal with respect to the Earth's surface, the amplitude of the former being about twice that of the latter (see Sect. 1 and [11]).

If the distribution of magnetic observatories were dense enough, and an accurate spherical harmonic expansion of the geomagnetic variations on the earth surface could be made, then the detailed structure of the equivalent overhead current could be obtained. If one deals with average characteristics of typical geomagnetic variations, this can be achieved. However, when a world-wide description for a specific instant is intended, it is impossible to derive an accurate world-wide overhead current-system, because the distribution of magnetic observatories over the world is unsatisfactory. In such a case, the following conventional method is used to obtain a drawing of the overhead current.

A plane sheet of uniform current intensity J (in A/m) produces everywhere outside the sheet a uniform magnetic field H, which is perpendicular to the direction of current flow, and the relation $H = \frac{1}{2} u\, J/c_0 \sqrt{\varepsilon_0 \mu_0}$ holds. Since the height of the current-system above the earth's surface will be negligibly small in comparison with the earth's radius, the magnetic disturbance field observed on the earth's surface may approximately be assumed to be produced by the electric current flowing just above the observing point, i.e. the influence of the electric currents distant from this point may be ignored. The equivalent overhead current is, therefore, approximately perpendicular to the direction of the horizontal disturbing force vector H (according to Biot-Savart's law). The intensity of the overhead current vector is indicated by the length of an arrow, proportional to the external part of the disturbing force. Mostly, the external part is assumed to be a fixed fraction of the observed force H, say $^2/_3$ or so. Then stream lines are drawn, which must be parallel to the current arrow near each observing point; the distance between two lines should be reciprocal to the length of the current arrow. Thus the current lines can partially be drawn viz. near each observatory. These partial current lines are then connected over the world, so that each current line is closed, and div $J = 0$ everywhere.

Here it is desirable to take account (at least qualitatively) of the vertical disturbing force when the stream lines are drawn over each station; the interval between the stream lines is broader on the left or right side of the station, according to whether the vertical disturbing force at that station is negative or positive. This trial-and-error method is fairly successful in obtaining an approximate view of the current-system, especially for cases where the disturbance field has sharp spatial gradients. This is, for example, applicable for instantaneous field distributions at high latitudes.

B. Normal geomagnetic variation free from disturbance.

Before discussing geomagnetic disturbances in more detail, it may be useful to review the normal, undisturbed geomagnetic variations. Since disturbance is a deviation from the normal time-variation, we shall have to compare quite often the change due to a disturbance with the regular variations. Therefore, Chap. B is devoted to a brief description of regular geomagnetic variations.

The magnitude of regular geomagnetic variations is only about $1/_{1000}$ of the intensity of the permanent geomagnetic field, depending on season and solar activity. During years of minimum solar activity, the amplitude of regular geomagnetic variations is a few tens of percents smaller than that for years of high solar activity. The regular geomagnetic variations originate primarily in the ionosphere and the magnetosphere.

8. The solar-daily variations on quiet days, S_q. When looking at a series of daily magnetograms, it is evident that the regular geomagnetic variation depends primarily on local time. The regular geomagnetic variation on quiet days as controlled by solar time is abbreviated as S_q (S stands for "solar-daily" and q for "quiet").

The geomagnetic variations, the origin of which is located above the earth's surface, are conveniently explained by an equivalent electric current flowing somewhere in the upper atmosphere, as already explained in Sect. 7. From an analysis of geomagnetic variations recorded all over the world, it is possible to derive a world-wide current-system responsible for the ground-level geomagnetic variations. As to the level of this current, it was simply assumed to be about 100 km above ground in the early investigations. This estimation has been proved to be correct when the ionosphere was experimentally discovered. Furthermore,

Sect. 8. The solar-daily variations on quiet days, S_q. 17

some rocket observations[1,2] in the equatorial ionosphere proved the presence of a horizontal electric current flowing at the level of the E-layer, roughly at a height of 100 km. Fig. 2, Sect. 2 shows the heigth-integrated value of the electric conductivity of the ionospheric regions determined from the actual distribution of electrons and ions with height.

The equivalent overhead current-system for S_q during the IGY[3], a period of high solar activity, is shown by Fig. 7. The current distribution in the northern

Fig. 7. External S_q current-systems averaged worldwide for northern winter, equinox, and northern summer months, and their yearly average. The current intensity between two consecutive lines is $25 \cdot 10^3$ A. The numbers near the dot marks are the total current intensity of these vortices in 10^3 A. (After MATSUSHITA and MAEDA.)

and southern hemispheres is not exactly symmetric with respect to the equator, even in the equinoctial season. In the solstitial seasons, the current intensity in the summer hemisphere is much stronger than that in the winter hemisphere. Throughout the year, a strong eastward electric current flows along the geomagnetic equator in the daytime, corresponding to a large-amplitude diurnal variation of the H-component at the stations situated very near to the geomagnetic equator. This current is called the *equatorial electrojet*. The current-system of Fig. 7 is fixed to the sun, and the earth rotates once a day inside the current-shell. Because of the 11.5° difference of the earth's geomagnetic and rotation axes, the intensity and form of the average current-system varies systematically within a day.

---

[1] S. F. SINGER, E. MAPLE, and W. A. BOWEN jr.: J. Geophys. Res. **56**, 265 (1951).
[2] L. J. CAHILL jr.: J. Geophys. Res. **64**, 489 (1959).
[3] S. MATSUSHITA, and H. MAEDA: J. Geophys. Res. **70**, 2535 (1965).

9. Variability of S_q.

Under quiet geomagnetic conditions, the curves on the magnetograms are not exactly the same on different days. For example, in middle latitudes, the local times of maximum eastward deviation in the morning hours and of the westward deflection in the early afternoon vary slightly from one day to another. The amplitude of the daily geomagnetic variation also differs from day to day. Instantaneous current-systems sometimes show considerable deviations from the average pattern of quiet-day current flow. The day-to-day variability of S_q is an interesting and important problem. A good example of the S_q variability was first depicted by HASEGAWA[1], who showed that the focus of the current-vortex in the sunlit hemisphere may be shifted by more than 10° in latitude on consecutive days. Of course, the S_q variability is partially due to the superposition of the lunar-daily variation (see Sect. 11), but this has much smaller amplitude; it is therefore reasonable to conclude that the current distribution of S_q itself changes considerably in intensity and form with time.

When we discuss geomagnetic disturbances, which are the deviations from the quiet-day variations, the variability of S_q must be taken into consideration. The disturbance field can quite accurately be defined when the disturbance takes place only for a short time, with long intervals of quiet conditions before and after it. However, in other cases, there is some uncertainty in defining the disturbance field, because we must assume a normal-day variation from which the deviation is scaled. When the disturbance becomes very great (for example several times the amplitude of the regular diurnal variation), the ambiguity is no longer a serious problem.

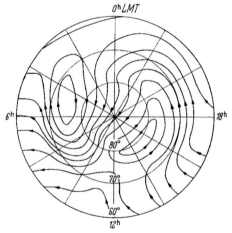

Fig. 8. Equivalent current-systems of mean daily variations in the north polar region on quiet days in June solstice. Electric current between adjacent stream lines is $2 \cdot 10^4$ A. (After NAGATA and KOKUBUN.)

10. Additional solar-daily variations on quiet days, S_q^p, in polar regions.

Approaching the pole, the regular quiet-day diurnal variation of the geomagnetic field changes its phase gradually, and the amplitude slightly increases. This systematic latitudinal dependence of the quiet-day geomagnetic variation was pointed out in a clear form by NAGATA and KOKUBUN [2] after a detailed analysis of IGY data. The result is shown by Fig. 8 describing it by an equivalent overhead current-system. The analysis of the Second Polar Year data by VESTINE et al. [3] and by HASEGAWA [4] also shows this tendency. One may consider that a particular

---
[1] M. HASEGAWA: Proc. Imp. Acad. Tokyo **12**, 88 (1936).

Sect. 10. Additional solar-daily variations on quiet days, S_q^p, in polar regions. 19

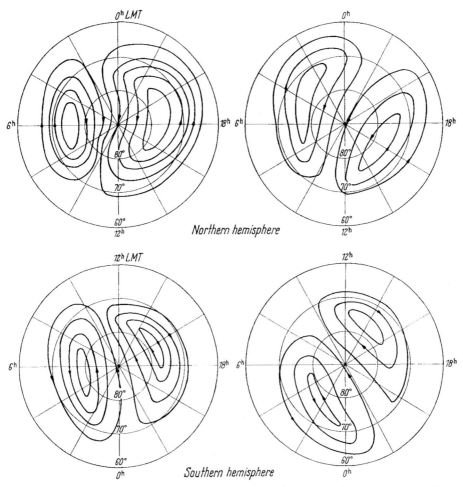

Fig. 9. Equivalent ionospheric current pattern of the additional daily variation field, S_q^p, on quiet days for the sunlit polar regions (left) and for the dark polar regions (right). Electric current between adjacent current lines is 10^4 A. (After NAGATA and KOKUBUN.)

current-system confined to the polar region is superposed on the S_q current shown already in Fig. 7.

This additional solar-daily variation on quiet days in the polar regions is named S_q^p after NAGATA and KOKUBUN; the equivalent current-system is shown in Fig. 9. The pattern of S_q^p consists of two current vortices, a clockwise one on the early morning side and a counter-clockwise one on the afternoon side (seen from above the pole, in the northern hemisphere), (insofar as the diurnally varying part is concerned). The geomagnetic variation on the earth produced by this current-flow is about 100 γ, slightly greater than the amplitude of S_q in temperate latitudes. The current intensity is increased to about 150 γ in the summer season when the entire polar cap is exposed to the solar radiation, while it is reduced to about 50 γ in the dark winter season. These conclusions have been derived from the analysis of hourly mean values of the geomagnetic elements on international quiet days. However, on extremely quiet days, especially in years of low solar

2*

activity, S_q^p seems to be almost negligible, and the solar-daily variation in high latitudes becomes just similar to that in middle latitudes.

11. The lunar-daily variation, L. There is a component of regular geomagnetic variation, which is dependent on lunar time. The amplitude of the lunar-daily variation **L** is about $1/10$ of that of S_q, so that **L** can be derived only by a statistical study of data over a fairly long interval of time. Because S_q is variable from day to day, it is pratically impossible to isolate the lunar-daily variation for individual days. The lunar-daily variation consists mainly of a semi-diurnal component. A statistical treatment of a sufficient quantity of data in various phases of the moon shows that the amplitude of the semi-diurnal variation of the geomagnetic **L**-field is enhanced during sunlit hours. The equivalent electric current-system for the

Fig. 10. External L current-system averaged yearly and worldwide. The current intensity between two consecutive lines is 10^3 A. The numbers near the white circles are the total intensity of the vortices in 10^3 A. (After MATSUSHITA and MAEDA.)

lunar-daily variation (average over data obtained from all phases) is given in Fig. 10[1]. Like the solar-daily variation, the lunar variation is also enhanced in the summer season, and is smaller in winter. At the discussion of geomagnetic disturbances, the presence of the lunar variation is always neglected, because of its small amplitude. In other words, it is practically impossible to determine the influence of magnetic disturbances on the lunar-daily geomagnetic variation.

12. Solar-flare effect on geomagnetic variations. At the time of a solar flare the wave radiation from the sun is suddenly increased, especially in the X-ray or extreme ultraviolet (EUV) region. These solar radiations of short wavelength contribute to the formation of the ionosphere, where the electric current responsible for the ground-level geomagnetic variation flows. Therefore, during a solar flare the geomagnetic variation in the sunlit hemisphere is intensified for a short time — with development during several minutes and decay during about an hour[1-3]. It is understandable that the effect is greatest near the subsolar point; it becomes smaller with increasing solar zenith distance of the observing point on the earth. This kind of geomagnetic variation is called a solar-flare effect (abbreviated: s.f.e. or sfe). This effect, which we designate also by S_{qa} (augmentation of S_q), is observed simultaneously with the optical flare on the solar disk. The geomagnetic s.f.e. is usually excluded from what is called magnetic disturbance,

---

[1] S. MATSUSHITA, and H. MAEDA: J. Geophys. Res. **70**, 2559 (1965).
[1] A. G. McNISH: Terr. Mag. **42**, 109 (1937).
[2] S. IMAMITI: Memoirs Kakioka Mag. Obs. **1**, 13 (1938).
[3] D. VAN SABBEN: J. Atmosph. Terr. Phys. **22**, 32 (1961).

because it is produced by a temporary increase in the solar *wave* radiation, whereas magnetic disturbances are generally considered to be caused by *corpuscular* radiation from the sun. Since the s.f.e. phenomenon is very useful for studying the relation between geomagnetic variations and the ionosphere, the morphology of the s.f.e. is briefly described in the following, using a detailed analysis of many s.f.e.'s during IGY [5].

An example of the progressive changes in the current-system of geomagnetic s.f.e. is shown by Fig. 11. The geomagnetic s.f.e. is detectable not only in the sunlit hemisphere but often even in the dark hemisphere, although its magnitude at nighttime is only about one-tenth of that near the subsolar point. The time of maximum development of a geomagnetic s.f.e. is not always simultaneous all over the world; sometimes there is a systematic time lag in the sunlit hemisphere, i.e. the maximum occurs later with increasing distance from the subsolar point. This tendency is particularly notable for weak solar-flare effects; the time lag between the subsolar point and the day-night boundary may be as much as 10 ... 15 minutes. Similar systematic time lags in the sunlit hemisphere are found for ionospheric solar-flare effects such as the sudden increase in cosmic radio-noise absorption produced by an increase in the electron density of the lower ionosphere[4]. This systematic time lag may be interpreted as being due to different increase and decay characteristics valid for the increased ionization in the ionosphere which is provoked by the sudden enhancement of the short-wave radiation from the sun, at places of different solar zenith angle.

The level where (near the subsolar point) the electric current flows during an s.f.e. is now known as to be the lower E-region or the upper D-region; the height is a little lower than that where the S_q current flows. The s.f.e. current height seems to increase slightly with increasing distance from the subsolar point. The geomagnetic s.f.e. found in the dark hemisphere seems to be caused by a world-wide electric current, which is provoked by the sudden increase of electric conductivity in the sunlit ionosphere.

A solar-flare effect on the geomagnetic variation is sometimes but not necessarily followed by a magnetic storm after a day or so. Usually, no storm follows when solar flares are observed near the limb of the sun, because the corpuscular stream emitted from a source situated at the solar limb may not hit the earth. On the other hand, most solar flares occurring near the centre of the sun's disk are followed by magnetic storms on the earth, with a time delay of not more than a few days.

13. Outline of theoretical interpretations for regular daily variations. The regular geomagnetic variations **S** and **L** are considered to be caused by electric currents in the ionosphere at about E-layer heights. It has been supposed that the electromotive force producing the current flow is caused by an air motion in the ionospheric region.

α) Since the ionosphere is an electrically conducting medium of specific conductivity, $((\sigma))$, a mass motion $\boldsymbol{v}$ across the earth's permanent magnetic field $\boldsymbol{B}$ produces an *induced electromotive force* $\boldsymbol{E}_i$, which results in an electric current of density $\boldsymbol{i}$ in the ionosphere, by the relation [see Eqs. (2.1) (3.1)]*

$$\boldsymbol{i} = ((\sigma)) \cdot \boldsymbol{E} = ((\sigma)) \cdot \{\boldsymbol{E}_i + \boldsymbol{E}_s\} = ((\sigma)) \cdot \left\{ \frac{\boldsymbol{v} \times \boldsymbol{B}}{c_0 \sqrt{\varepsilon_0 \mu_0}} + \boldsymbol{E}_s \right\}. \tag{13.1}$$

---
* $c_0 \sqrt{\varepsilon_0 \mu_0} = 1$ in SI-units.
[4] K. RAWER, and K. SUCKY: this Encyclopedia, vol. 49/2, Sect. 46.

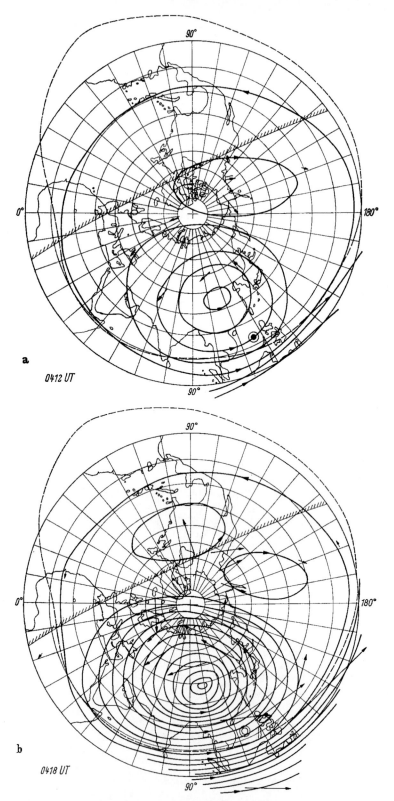

Fig. 11a—d. An example of the progressive change in the equivalent overhead current-system for geomagnetic s.f.e. on May 5, 1958. (After Ohshio et al.)

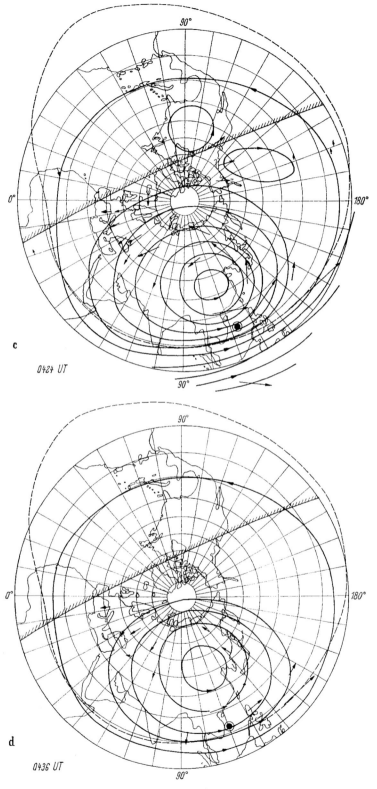

Fig. 11 c and d.

E_s denotes the static electric polarization field due to a non-uniform charge distribution in the conducting ionosphere. This field is needed in order to get a stable current distribution satisfying div $\boldsymbol{i} = 0$.

β) The mechanism described here is just like that of an electric dynamo, so that the theory based on this line of thought is called the *"dynamo theory"*. The dynamo theory was developed by STEWART[1], SCHUSTER[2], CHAPMAN[3] and later investigators. It explains the variations of the magnetic field by the additional field produced by the current after Eq. (3.1), i.e. by motion $\boldsymbol{v}$, of plasma with electric conductivity, $((\sigma))$, in the upper atmosphere in the presence of the permanent geomagnetic field, $\boldsymbol{B}_{\ddot{o}}$. The term

$$((\sigma))\, \boldsymbol{v} \times \boldsymbol{B}_{\ddot{o}}/c_0 \sqrt{\varepsilon_0 \mu_0} \qquad (13.2)$$

is the dynamo current. [It may be increased or decreased by local polarization fields $\boldsymbol{E}_s$ after Eq. (13.1).] By the combined action of the world-wide system of winds (with velocity $\boldsymbol{v}$) on one side, and/or of the distribution of conductivity $((\sigma))$ over the world, this theory is able to explain the regular solar and lunar diurnal variations **S** and **L**, and to some extent also the diurnal part $\mathbf{S_D}$ (or $\mathbf{D_S}$) (explained later in Sect. 38) under disturbed conditions. Finally, the variations occurring during solar flares are also explained, mainly by changes of $((\sigma))$.

After Fig. 2 (in Sect. 2), the height-integrated electric conductivity of the ionosphere under normal conditions is now estimated to be about 10 ... 100 S (or 10^{-8} ... 10^{-7} e.m.u.)*. To obtain an electric current of sufficient intensity to cause the geomagnetic $\mathbf{S_q}$ variation, the wind speed in the ionosphere should be several tens of m/sec. Motions of this magnitude have actually been observed by rocket soundings.

γ) If the $\mathbf{S_q}$ and **L** variations (see Sects. 8 and 11) are really produced by dynamo action in the ionosphere, there should be a *systematic wind motion* in the ionosphere over the entire earth. Analyses of daily geomagnetic variations have been widely used for estimating the air motion in the ionospheric region. The presence of such systematic global air motion in the ionosphere has not yet been experimentally proved by rocket measurement, but radio measurement of the drift of ionospheric irregularities may give some useful information[4]. One must, however, bear in mind that the motion of electrons in the ionospheric region is not the same as that of neutral particles, so that appropriate correction is necessary for deriving the air motion from ionospheric drift data.

It is not possible to infer the air motion in the ionosphere from barometric changes, because the pressure variation corresponding to the regular daily (both solar and lunar) changes at ionospheric heights are masked by the pressure change due to daily meteorological changes in the tropospheric region.

δ) The ion and electron distribution in the ionosphere and the air motion at ionospheric levels depend on the intensity of solar radiation. This means that both the electric conductivity and the electromotive force contributing to the electric current in the ionospheric region may change from day to day. The *variability* of the $\mathbf{S_q}$-field (see Sect. 9) may be a consequence of the slow time-variation of solar wave radiation, at least insofar as daily changes are concerned. The seasonal

---

* S ≡ Siemens = A/V (in SI-units) = $m^{-2}\,kg^{-1}\,s^3\,A^2$ (in the MKSA-system); the corresponding unit in the electromagnetic cgs-system is $1\,cm^{-1}\,s = 10^9$ S, but it is $1\,cm\,s^{-1} = 1.11\cdot 10^{-12}$ S in the electrostatic cgs-system as well as in the Gauss-system.

[1] B. STEWART: Encyclopedia Britannica (9th ed.), London 1882, p. 36.
[2] A. SCHUSTER: Phil. Trans. Roy. Soc. London A **180**, 467 (1889); A **208**, 163 (1908).
[3] S. CHAPMAN: Phil. Trans. Roy. Soc. London A **218**, 1 (1919).
[4] K. RAWER, and K. SUCHY: this Encyclopedia, vol. 49/2, Sects. 50 and 52.

change in the amplitude of the S_q, S_q^p and L variations is related to the seasonal variation of both the electric conductivity and the air motion in the ionosphere. It is also worth noting that the electrostatic potential due to the charge accumulation in the ionosphere accompanying the electric current flow in the ionosphere in the northern hemisphere might be linked with that in the southern hemisphere, because the electric conductivity along magnetic field-lines in the magnetosphere is quite high. Therefore, a three-dimensional consideration is necessary for a precise discussion of the dynamo theory; some preliminary calculations have been made along this line of thought[5–9]. However, it has been estimated that an electric current flowing along the lines of magnetic force in the magnetosphere will contribute only about a few percent to the geomagnetic field variation on the Earth[9]. Therefore, we are allowed to deal with ionospheric currents only, i.e. with a two-dimensional current flow in a thin layer (viz. the ionosphere).

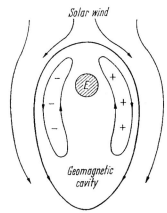

Fig. 12. Schematic illustration of the circulation of the low-energy magnetospheric plasma in the magnetic equatorial plane after the 'closed magnetosphere' model of AXFORD and HINES. (E = earth.)

This does not mean that the electric current flows to the same horizontal direction at every height within the ionosphere. In fact, the direction of electric current flow varies with height, because of the height-dependence of the ratio σ_2/σ_1. Furthermore, solenoidal electric currents are expected in the meridional cross-section, which give rise to a toroidal magnetic field observable only within the ionosphere. However, insofar as the magnetic effect observed on the earth's surface is concerned, we need not consider a complicated three-dimensional current flow in the ionosphere, but may approximate it by a two-dimensional electric current sheet.

ε) Off-hand we cannot reject the idea that the S_q^p variation (see Sect. 10) may wholly be due to an (assumed) particular air motion in the ionosphere in the *polar regions*. However, there is a more promising explanation for the S_q^p-field by a quasi-stationary interaction of the solar wind with the geomagnetic field. An outline of this theory for the production of the S_q^p-field is given in the following.

Fig. 12 shows a possible convective motion of the tenuous plasma with the "frozen-in" magnetic field within the geomagnetic cavity due to the shearing stress near the cavity boundary, caused by the flow of the solar wind around the geomagnetic cavity. Accompanying this magnetospheric convection, electric space charges must be produced so that the resulting electric field finally satisfies the *equilibrium condition*

$$c_0 \sqrt{\varepsilon_0 \mu_0}\, \boldsymbol{E} + \boldsymbol{v} \times \boldsymbol{B} = 0 \qquad [3.1]$$

in the magnetospheric plasma. If the electric potential is propagated along the lines of magnetic force down to the ionosphere, an electric current just similar in form to the S_q^p current-pattern flows, according to the reason explained in Sect. 4. In the E-layer (except near the equator) the Hall conductivity is much greater than the Pedersen conductivity, so that the flow of electric current is almost per-

---

[5] J. P. DOUGHERTY: J. Geophys. Res. **68**, 2383 (1963).
[6] E. M. WESCOTT, R. N. DE WITT, and S.-I. AKASOFU: J. Geophys. Res. **68**, 6377 (1963).
[7] H. MAEDA: J. Atmosph. Sci. **23**, 363 (1966).
[8] D. VAN SABBEN: J. Atmosph. Terr. Phys. **28**, 965 (1966).
[9] K. MAEDA, and H. MURATA: Rep. Ionos. Space Res. Japan **19**, 272 (1965).

pendicular to the electric field; therefore it turns around the points of maximum and minimum electric potential. The direction of electric current flow is clockwise or counter-clockwise (seen from above the ionospheric region, in the northern hemisphere) according to whether the centre of the current vortex has positive or negative electric potential. The seasonal dependence of the amplitude of S_q^p may be attributable mainly to the seasonal change of the electric conductivity in the polar ionosphere.

However, in order to explain S_q^p by this mechanism alone, the general circulation in the magnetosphere should vanish when S_q^p disappears on an extremely quiet day. Another idea for explaining S_q^p is to attribute it to the direct impinging of the solar plasma stream upon the polar ionosphere. When the solar plasma carries a southward magnetic field, the solar particles may easily penetrate into the magnetosphere. On the other hand, solar plasma with a northward magnetic field will be ineffective.

ζ) The *solar flare effect* (see Sect. 12) is mainly due to an increasing of conductivity in the lower ionosphere; flare (wave) radiation comes from the disturbed sun and is absorbed there. As the air motions are probably not seriously influenced, the flare effect appears as a sudden enhancement of the normal diurnal behaviour, followed by a slower decrease.

C. Characterization of magnetic disturbance.

14. Definition of the disturbance field. A continuous record of the earth's magnetic field usually shows more or less irregular variations, which are superposed on the regular daily variation of the geomagnetic field. It is quite rare that an extremely smooth magnetogram trace is obtained throughout a day, especially in high latitudes. While the amplitude of regular daily geomagnetic variation is only about 1/1000 of the permanent field, that of the irregular variations occasionally amounts to as much as several percent, on days of severe magnetic storms.

When the earth's magnetic field is disturbed, one may say that a disturbance field is superposed upon the regular variation of the geomagnetic field. A simple and satisfactory definition of the disturbance field $\mathbf{D}$ is

$$\mathbf{D} = \mathbf{F} - \overline{\mathbf{F}} - \mathbf{S}_q - \mathbf{L}, \tag{14.1}$$

where $\mathbf{F}$ is the observed instantaneous value of the geomagnetic field; $\overline{\mathbf{F}}$, the value of the undisturbed geomagnetic field free from the variable influences; $\mathbf{S}_q$ and $\mathbf{L}$ denote respectively the solar daily variation (on magnetically quiet days, see Sect. 8), and the lunar-daily variation (see Sect. 11). Here the symbol $\mathbf{S}_q$ is considered to include the S_q^p-term at high latitudes as described in Sect. 10. Since after Sect. 11 the amplitude of $\mathbf{L}$ is so small in comparison with that of $\mathbf{S}_q$, $\mathbf{L}$ may be ignored in the above equation. Then the Eq. (14.1) is reduced to

$$\mathbf{D} = \mathbf{F} - \overline{\mathbf{F}} - \mathbf{S}_q. \tag{14.2}$$

During a period of magnetic disturbance, this additional disturbance field varies with time in a complicated manner according to the location on the earth; particularly large variations occur at high latitudes, around the auroral zones and in the polar caps.

Although the disturbance field is defined by Eq. (14.1) or (14.2), it is a great practical problem to determine the undisturbed level of the geomagnetic field, $\overline{F} + S_q$. When we deal with a disturbance field of large magnitude or with a short-period oscillatory disturbance, it is not difficult to identify the disturbance component. On the other hand, for a small disturbance, there is some ambiguity in choosing the 'quiet' level of the geomagnetic field, because of the day-to-day variability of S_q (Sect. 9), and for other reasons. This is in fact a crucial problem in the analysis of geomagnetic variations. Therefore, a conventional assumption on the undisturbed field is sometimes made. For example, $\overline{F}$ is taken as an average value of the observed field over a fairly long interval of time, say a month, excluding disturbed days. The mean daily variation of the quiet days before and after a day of disturbance is often taken as S_q on that day.

15. Character-figures for geomagnetic activity. A great amount of information on time variations of the geomagnetic field is contained in the original magnetograms. A suitable reduction of this abundant information is therefore needed. Many kinds of character-figures for geomagnetic activity have been introduced and conventionally used, especially for studying relationships between geomagnetic activity and other associated phenomena, such as solar activity, ionospheric disturbances, auroral displays, intensity of primary cosmic rays, etc. Even the general development of a geomagnetic disturbance is sometimes well described with appropriate magnetic character-figures; however, any character-figure for geomagnetic activity hitherto proposed is a pale abstract describing a certain aspect of the real variations given by the magnetograms. Hourly values, which are reported regularly by most permanent magnetic observatories, may be considered as to be a first "abstract" of the records. A brief explanation only will be given here of character-figures for geomagnetic activity, which are now widely used[1].

The three-hour-range index K. This index expresses the geomagnetic disturbance in a quasi-logarithmic scale for each three-hour interval (in UT) by the deviation of the magnetogram-trace from the quiet-day curve. The sum of eight indices of a day is often used as a daily character-figure. The K-index has been introduced to give a measure of particle emission from the sun, which has an effect quite different from that of solar wave radiation. Therefore, the regular variation $(S_q + L)$ caused by solar wave radiation must be eliminated in scaling K; solar-flare effects should also be excluded. Also post-perturbation effects must be excluded. These are most clearly seen as a depression of the $\mathcal{H}$-level after a magnetic storm, and a slow recovery within the next several days.

The equivalent three-hour-range a_k, and the equivalent daily amplitude A_k. These indices have a linear scale, different from that of K; they can however, be derived from K-indices. a_k is the total range of the deviation (in units 2γ) around geomagnetic latitude $50°$ during a three-hour interval; A_k is the daily average of 8 values of a_k.

The planetary three-hour-index K_p. K_p has been introduced by international agreement to characterize world-wide geomagnetic activity[2]; it is obtained by averaging standardized K values (called K_s) of 12 observatories situated at geomagnetic latitudes $48°$ to $63°$. This index describes the average world-wide geomagnetic activity very well, except for the regions very near to the northern and southern poles. The index is subdivided into 28 grades, 0_0, 0_+, 1_-, 1_0, 1_+, 2_-, ..., 8_+, 9_-, 9_0. At present the planetary three-hour-index K_p is most widely and generally used for examining the relation between geomagnetic disturbance and associated pheonmena. The daily sum of eight K_p-values is often taken as a daily character-figure for the world-average magnetic disturbance grade [3,4].

The three-hourly equivalent planetary amplitude, a_p, and the daily equivalent planetary amplitude, A_p. Similarly to the relation between the K-index and a_k, a_p can be obtained from K_p with the aid of a conversion table. The value of a_p is used when a linear scale is preferred to the quasi-logarithmic scale of K_p. The average of the eight values of a_p for a day is the daily equivalent planetary amplitude, A_p.

The daily international character-figure C_i, and the daily planetary character-figure, C_p. Since 1884 each individual observatory reported the disturbance grade simply in three classes: quiet (0), moderately disturbed (1) and unusually disturbed (2). C_i is the world-wide average of these character figures, given in 21 grades from 0.0 to 2.0. A substitute consistent with C_i is C_p, which is based on the daily sum of a_p. The maximum grade of C_p is 2.5 (to avoid the saturation at 2.0 in C_i).

---

[1] For details see M. SIEBERT: p. 206 in this volume; also J. BARTELS: Annals of the IGY **4**, Pt. 2 (1957).

[2] J. BARTELS: IATME Bull. No. 12b, 97 (1949); — J. Geophys. Res. **54**, 296 (1949).

[3] K_p-indices for more than 30 years (together with the a_p and A_p values) have been published by the 'International Association of Geomagnetism and Aeronomy' in the IAGA Bulletin No. 12 series.

[4] For typographical convenience, a_k, A_k, K_p, etc. are mostly written ak, Ak, Kp, etc.

The quarter-hourly index, Q. The quarter-hourly index Q is based upon the maximum absolute deviation (in H and D, or X and Y) of the magnetogram trace from the normal-day curve[5], in 12 grades from 0 to 11. The Q-index is reported from high-latitude stations only. This index may be said to indicate the overhead ionospheric current intensity at high latitudes in a quasi-logarithmic scale.

Index for the equatorial ring current. During magnetic storms, the horizontal intensity H of geomagnetic field on the earth's surface is depressed in general, especially in low latitudes. The observed facts can be described by a circular current flowing high above the equator (see Chap. F, Sects. 39 and 40). This depression may last for several days. The K-index and some of the other character-figures do not account for this slow change, although the depression and the slow recovery of the horizontal geomagnetic field is one of the essential characteristics of great geomagnetic disturbances. At high latitudes this effect is largely masked by local disturbance of much greater magnitude. For the data during the IGY 1957/58, two kinds of indices for the equatorial ring current have been proposed, one on hourly basis by SUGIURA[6], and the other one for three-hour intervals by KERTZ[7]. The indices called "u-measure" and "u_1-measure" used in old publications describe the day-to-day change of the equatorial ring current [1].

Hourly range. To examine the geomagnetic disturbance on an hourly basis, the range within one hour of the principal component of geomagnetic variation is very often used. This index has the advantage that it is easily read, without referring to the absolute level of the geomagnetic field.

In addition to these character-figures, there are of course some other geomagnetic indices conveniently used. For example, *storminess*[8] has been used mainly by Scandinavian workers for studying magnetic disturbance in high latitudes on hourly basis. The wax and wane of the westward and eastward auroral electrojets (see Chapters F and G) during magnetic storms is revealed by the AE-*index*[9] given in 2.5 minute intervals. An index $\Sigma Kc3$ is proposed[10] to show the geomagnetic pulsation activity on a daily basis.

16. International disturbed days and quiet days.
On the basis of daily character-figures as described above, the five days of lowest geomagnetic activity and the five most disturbed days are selected in each calendar month by an international comissioner; they are customarily called "the international quiet days" and "the international disturbed days" in that month. After this definition, it may even happen that the international quiet days in a certain month are more disturbed than the international disturbed days in another month, if the former month were unusually disturbed and the magnetic activity in the latter month were extremely low. However, the difference of the mean daily variation on disturbed days and that on quiet days in the same period, at any rate, characterizes disturbance conditions, and is often practical for a statistical study of geomagnetic disturbances.

17. Statistical analyses of long time-series of magnetic character-figures.
A good correlation between geomagnetic disturbance and solar activity has been demonstrated with the aid of daily magnetic character-figures and their monthly or yearly mean values. In general, average geomagnetic activity varies almost parallel with solar activity during the 11-year cycle; a typical example[1] is given in Fig. 13.

There is a statistical semi-annual variation in geomagnetic activity, showing two maxima in the equinoctial seasons, as shown in Fig. 14.

Figs. 15 and 16 show, respectively, K_p-diagrams in years of maximum and minimum solar activity. Each line consists of 27 days' data, in order to demon-

---

[5] J. BARTELS, and N. FUKUSHIMA: Abh. Akad. Wiss. Göttingen, Math.-phys. Kl., Sonderheft Nr. 3 (1956).
[6] M. SUGIURA: Annals of the IGY 35, 9 (1964).
[7] W. KERTZ: Annals of the IGY 35, 49 (1964).
[8] G. GJELLESTAD, and H. DALSEIDE: Arbok Univ. Bergen, Mat.-Naturv. Ser. 1963, No. 7; see also Ref. [28].
[9] T. N. DAVIS, and M. SUGIURA: J. Geophys. Res. 71, 785 (1966).
[10] T. SAITO: Rep. Ionos. Space Res. Japan 18, 260 (1964).
[1] E. J. CHERNOSKY: J. Geophys. Res. 71, 965 (1966).

Sect. 17. Statistical analyses of long time-series of magnetic character-figures.

Fig. 13. Comparison of annual means of geomagnetic activity and sunspot number for the years 1884 ... 1963 (after CHERNOSKY). [Top: yearly mean international character figure C_i; center: number of days per year with $C_i \geq 1$; bottom: yearly mean Zürich sunspot number R_Z.]

Fig. 14. Comparison of semi-annual variation in geomagnetic activity and disturbed-day occurrence for alternate eleven-year cycles, 1884 ... 1963 (after CHERNOSKY). [Upper: monthly mean international character figure C_i; lower: number of days with $C_i \geq 1$, in per cent.]

Fig. 15. K_p-diagram for 1957/58, a period of high solar activity, arranged after solar rotational intervals.

Sect. 17. Statistical analyses of long time-series of magnetic character-figures. 31

strate frequent recurrence of high and low magnetic activity after 27 days. The 27-day recurrence is usually best seen in the years of low solar activity.

The time series of the geomagnetic disturbance indices (such as C_i, K_p, A_p etc.) have been subjected to power spectrum analysis. An example of the applica-

Fig. 16. K_p-diagram for 1964, a period of low solar activity, arranged after solar rotational intervals.

tion of this technique for a long series of geomagnetic indices demonstrates the meaningful existence of the 6-month period (semi-annual variation), the period of 27 days and its harmonics (periods $27/N$ with $N = 2, 3, 4, \ldots, 6$, i.e. approximately 14, 9, 7, 5 and 4 days)[2], as is seen in Fig. 17.

[2] F. W. WARD jr.: J. Geophys. Res. **65**, 2359 (1960).

As for the lunar control of geomagnetic activity, a unanimous conclusion has not yet been attained. Attention has been recently paid to this problem, since satellite observations indicate that the geomagnetic tail may be extended even to the orbit of the moon.

Fig. 17a—c. Autocorrelation function, R_k (Top), and power spectra of the planetary three-hour-index K_p, April 1951 ... March 1956. L_h (in the center) means directly computed power values, U_h (bottom) has been obtained by smoothing. (after WARD)

18. Solar corpuscular streams and geomagnetic disturbance. A typical magnetic storm with sudden commencement (see Sect. 20) occurs in most cases about one day after a solar flare appears near the centre of the sun's disk. Hence, it has been supposed that the sun emits not only visible wave radiation but also a corpuscular stream, which travels with a speed of about 10^6 m/sec from the sun and reaches the earth, where it causes a magnetic storm. However, some magnetic storms occur without corresponding observable optical event on the sun, especially during years of minimum solar activity. These storms have a tendency to return after 27 days, the period of solar rotation. For such magnetic storms one assumes that a solar corpuscular stream of long life is flowing out from a limited region on the sun, which is conventionally called an "M-region". The physical nature of the M-regions is still under discussion. The most active areas on the sun are usually located about 15 ... 30° north and south of the solar equator. Near the equinoxes the rotation axis of the sun is so inclined that the earth is most easily immersed in the corpuscular stream from the active areas in the northern or southern hemisphere of the sun. (One supposes that the stream is ejected with a narrow solid angle.) This is considered to be the reason for the semi-annual variation in geomagnetic activity.

The sun is rotating and emitting the corpuscular stream, like a water sprinkler in the garden, and world-wide geomagnetic disturbance is observed as long as the earth is immersed in the strong 'solar wind', the ejection of which may last for several solar rotations. This will result in the 27-day recurrence of geomagnetic activity. While the corpuscular stream was earlier considered to be ejected only from some particular regions of the sun into a vacuum interplanetary space, it is now believed that a material flow from the solar surface is always existing. It is called the solar wind[1], and its kinetic pressure confines the earth's magnetosphere in a limited space, the so-called 'geomagnetic cavity', as already illustrated by Fig. 5 (Sect. 4). The intensity of the solar wind varies considerably from time to time. Sometimes a new intense stream is ejected, which pushes interplanetary medium aside. Recent direct measurement of the solar plasma flow by space probes (Mariner II and others) show good correlation between solar wind velocity and geomagnetic activity on the earth[2].

D. Gross characteristics of geomagnetic disturbance over the world.

19. Examples of magnetograms at various places. Fig. 18 shows examples of the time-variation in the horizontal intensity, H, on magnetograms taken at various stations over the world during a typical magnetic storm. The simultaneous variations in the declination, D, and the vertical component, Z, are in general smaller than that of H. One notices here that the record at high-latitude stations is different from that at low latitudes, indicating higher geomagnetic activity in high latitudes. The records at two stations not far from each other often show remarkable differences in details. The magnetogram trace at a given observatory differs greatly from one storm to another. In the following, some typical characteristics of magnetic storms are discussed, which have been derived from the analysis of the complicated time-variation of the geomagnetic field.

20. Typical variation of the geomagnetic field during magnetic storms in middle and low latitudes. A typical magnetic storm begins with a sudden change in the geomagnetic elements all over the world. In middle and low latitudes this "sudden commencement" (abbreviated to **SC** or **ssc**, sometimes **s.s.c.**) is usually very distinct, characterized in particular by an increase in the horizontal intensity, H. The morphology of **SC**'s is described in detail in Chap. E.

Fig. 19 gives schematically the time-variation in H after an **SC** at middle or low latitude. The H-value is for some hours above the pre-storm level, beginning with the sudden commencement, and this stage is called the *initial phase* of a magnetic storm. Then the H-value begins to decrease, sometimes only slightly and sometimes very remarkably. This second characteristic stage is called the *main phase*, it lasts about half a day or so. The depression of H is usually 100 ... 500 γ (which is about one per cent of the geomagnetic field intensity on the earth's surface). The magnetic storms are usually classified as moderate, moderately severe, and severe ones, according to whether the maximum K-index during the storm is as great as 5, or 6 to 7, or 8 to 9, respectively. The period following the main phase is the stage of recovery of H to its normal value. It is called the *last* or *recovery phase*, and several days are needed for a complete recovery of H to the pre-storm level.

---

[1] See the contribution by E. N. Parker and V. C. A. Ferraro, p. 131 in this volume; see also that of H. Poeverlein in Vol. 49/4 of this Encyclopedia.

[2] C. W. Snyder, M. Neugebauer, and U. R. Rao: J. Geophys. Res. **68**, 6361 (1963).

Some magnetic storms do not follow the typical time-variation of the geomagnetic field described here. There are cases where a large and long initial-phase variation is followed by

Fig. 18a. Example of three-component magnetograms at the time of a moderate magnetic storm 27/28 Nov. 1959, at different places: GODhavn (Greenland, magnetic latitude 79.8° N), COLlege (Alaska, 64.7° N), HONolulu (Hawaii, 21.0° N), HUAncayo (Peru, 0.6° S). (After Sugiura and Heppner.)

Sect. 20. Typical variation of the geomagnetic field during magnetic storms. 35

Fig. 18b. Horizontal-component magnetograms during the storm of 29 Sept. 1957 with SC at 0016 UT recorded at different stations in the northern auroral zone: JULianehaab (Greenland, 71.0° N), REYkjavik (Iceland, 70.3° N), ABIsko (Sweden, 65.9° N), DIKson (63.0° N) and WELlen (61.8° N) (Siberia), COLlege (Alaska, 64.7° N), MEAnook (61.8° N) and CHUrchill (68.6° N) (Canada). (After AKASOFU and CHAPMAN.)

no significant main phase at all, or where a very small and short initial phase is recorded before a large main-phase variation. Some storms have no noticeable sudden commencement,

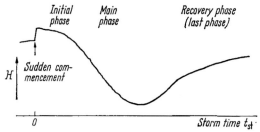

Fig. 19. Schematic time-variation in the horizontal intensity H after an **SC**, for middle and low latitudes.

Fig. 20. Latitudinal dependence of the average magnitude of geomagnetic disturbance along the noon-midnight meridian for different graces of magnetic activity, and for two different solar declinations, during 1957/58, the International Geophysical Year.

but typical main and recovery phases. They are designated as gradually commencing storms by Sg or sg. These storms occur most frequently in years of low solar activity, sometimes with marked 27-day recurrence tendency[1].

21. Dependence of geomagnetic activity on latitude. The magnitude of geomagnetic disturbance depends on the latitude of the observing station, as immediately seen from Fig. 18. Fig. 20 shows the average latitude-dependence of the magnitude of geomagnetic disturbance along the noon-midnight meridian during

Fig. 21 a and b. Magnitude of persistent polar magnetic disturbance near noon and midnight meridians when K_p is 0o for four different inclination angles of the solar wind with respect to the geomagnetic equator. (a) for the International Geophysical Year (high solar activity), (b) for the Second Polar Year (low solar activity) (see next page).

the IGY (1957-58), a period of high solar activity [6]. The magnitude of disturbance is very great at places of geomagnetic latitude higher than 60°, as compared with that at middle and low latitudes. The disturbance has its maximum magnitude around geomagnetic latitudes of 65° ... 70°. This region of maximum geomagnetic activity is identical with the *auroral zone*, where aurorae appear at the time of magnetic disturbance (see Chap. G). The latitudinal distribution is asymmetric with respect to the equator, unless the sun is near to the geomagnetic equatorial plane. In general, on the day side at high latitudes, the geomagnetic disturbance in the summer hemisphere is considerably stronger than that in the winter hemisphere. On the other hand, the disturbance on the night side is of nearly the same magnitude in the summer and winter hemispheres, especially for large K_p values.

---

[1] H. W. NEWTON, and A. S. MILSOM: J. Geophys. Res. **59**, 203 (1954).

22. Residual geomagnetic agitation in the polar-cap regions. There is a very particular polar-cap disturbance occurring even on days of extremely low K_p values. Looking at the original magnetograms at polar-cap stations, one notices that irregular variations are almost always superposed on the diurnal variation, even on days of extremely low K_p values[1-4] [6, 7]. Fig. 21 gives the average magnitude of this irregular agitation of the geomagnetic field in the polar regions, along the noon-midnight meridians. Days with $K_p=0_o$, for the IGY-period (high

Fig. 21 b.

solar activity) and for IIPY (Second Polar Year 1932—33, small solar activity) have been selected. The residual polar-cap disturbance is of the order of several tens of γ; it varies with the season, i.e. with the angle of incidence of the solar wind on the geomagnetic dipole. In the summer season, the residual polar-cap disturbance on quiet days is particularly large on the daylight side around the geomagnetic latitude 80°; it amounts to as much as 150 ... 200 γ during the IGY, and about 50 ... 70 γ during the IIPY.

A dependence of polar-cap disturbance on the "solar inclination", i.e. the angle of arrival of the solar wind, will be seen in the following example. In the monthly mean diurnal curves of the geomagnetic Q-indices reported during IGY from the USSR antarctic station Mirnyj

---

[1] P. N. MAYAUD: IAGA Bull. No. 12, 269 (1961).
[2] M. S. BOBROV: Results of Res. IGY Program, VI. Sect., No. 1, 36, Moscow (1961).
[3] L. A. JUDOVIČ: Geomagnetizm i Aeronomija [Translation: Geomagnetism and Aeronomy], **2**, 1113 (1962).
[4] M. S. BOBROV, N. F. KOROLEVA, and R. M. NOVIKOVA: Geomagnetizm i Aeronomija [Translation: Geomagnetism and Aeronomy], **4**, 333 (1964).

(-77.0^c geomagnetic latitude), the highest value always occurs around local noon (5...6 h UT), while the minimum appears near local midnight (17...18 h UT) in every month. Fig. 22 shows this tendency together with the monthly average sunspot number and the geomagnetic A_p-indices [6]. A distinct seasonal variation of the geomagnetic activity exists at this polar station, and it is only slightly dependent on solar activity or planetary geomagnetic activity, A_p. It is also of great interest to point out that the Q-index at 5...6 h UT in June has nearly the same value as that at 17...18 h UT in December, the solar elevation angle at both these times being just 0° (because Mirnyj is situated approximately on the antarctic circle). One gets

Fig. 22a and b. Seasonal variation of the geomagnetic activity at Mirnyj, Antarctica. a Monthly mean of quarter-hourly index Q around 05...06 h UT (upper curve), and around 17...18 h UT (lower curve). b Monthly mean of daily equivalent planetary amplitude A_p (full curve) and Zürich sunspot number R (broken curve).

the impression that the geomagnetic disturbance in the polar-cap region depends mainly on the solar elevation angle at this observatory. Such a characteristic is found more or less at all high-latitude stations inside the auroral zone[5], certainly in contrast with the conditions at stations in or outside the auroral zones. The geomagnetic activity shows a night- or early morning-maximum in or outside the auroral zone, as will be seen in Sect. 43.

23. Coordinates convenient for the study of geomagnetic disturbances. For the analysis of geomagnetic disturbances and associated phenomena, geographic coordinates are not convenient in most cases. The world-wide distribution of disturbance phenomena is better described, when geomagnetic coordinates are used. The ordinary geomagnetic coordinates refer to an axis through the earth's centre,

---

[5] K. LASSEN: Publ. Danske Meteorol. Inst., Comm. Mag., No. 23 (1958).

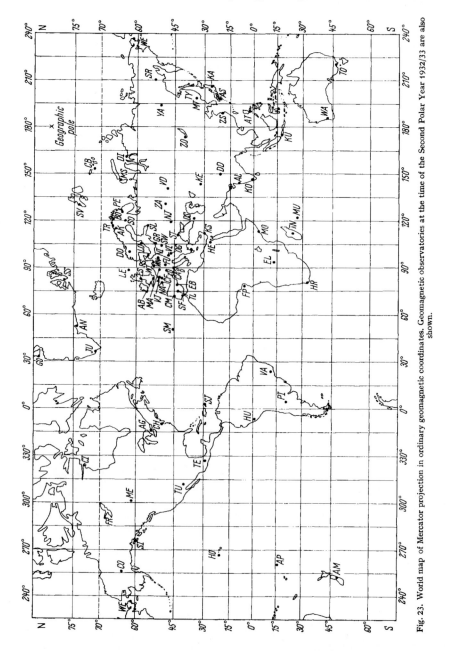

Fig. 23. World map of Mercator projection in ordinary geomagnetic coordinates. Geomagnetic observatories at the time of the Second Polar Year 1932/33 are also shown.

which intersects the earth's surface at 78.5 °N and 69.0 °W, 78.5 °S and 111.0 °E. Though these values have been determined for the epoch of 1922, they are still conventionally used to identify the geomagnetic latitude and longitude of magnetic observatories. The world map in these *ordinary geomagnetic coordinates* is given by Fig. 23.

Geomagnetic disturbances and associated phenomena depend on the real distribution of the magnetic field on and around the earth, which is not of a dipole

Sect. 23. Coordinates convenient for the study of geomagnetic disturbances.

Fig. 24. World map in geographic coordinates, with contours of corrected geomagnetic latitudes and longitudes (in middle and high latitudes), and with ionospheric stations during the International Geophysical Year 1957/58 (after HAKURA).

type. Therefore, a better representation can be obtained, when the higher-order terms of the spherical harmonic representation of the earth's permanent magnetic field are taken into consideration[1,2]. Latitude and longitude calculated in this way will be called "*corrected geomagnetic coordinates*". Fig. 24 gives contours of corrected geomagnetic latitude and longitude calculated by HAKURA [8]. It has been shown that various aeronomical disturbances in high latitudes are much better represented by corrected geomagnetic coordinates than by ordinary ones.

For describing the distribution of trapped energetic particles in the magnetosphere, $L-B$ coordinates have been introduced by McILWAIN[3] and are widely

---

[1] B. HULTQVIST: Arkiv Geofys. **3**, 63 (1958); — Nature **183**, 1478 (1959).

[2] P. N. MAYAUD: Ann. Géophys. **16**, 278 (1960).

[3] C. E. McILWAIN: J. Geophys. Res. **66**, 3681 (1961).; see also contribution by W. N. HESS in Vol. 49/4 of this Encyclopedia.

used. Here B describes the magnetic field intensity, and L is the so-called magnetic shell parameter explained in the following:

With the exeption of particles with extreme momentum, like cosmic-ray particles, the motion of charged particles in an inhomogeneous magnetic field is decomposed into three kinds of motion: gyration around a line of magnetic force, to-and-fro motion of the gyration centre of the particle along the field-line, and a slow drifting motion of the gyration centre perpendicular to the magnetic meridian plane. In the Earth's magnetic field, which is slightly different from the dipole field, the center of gyrational motion along a particular line of force of charged particles with different energy (hence with different mirroring points) will remain on approximately the same *shell*. Therefore, for each shell a unique labelling is possible; it is obtained by a number called the *magnetic shell parameter*, L. For any charged particle moving in the Earth's magnetic field, one finds a corresponding line of force in an ideal magnetic dipole field, along which the particle's second adiabatic invariant I (the integral along the line of force of the component of momentum parallel to the line, divided by the total momentum of the particle, when there is no electric field) is constant. The maximum (equatorial) distance of such a dipole field-line from the Earth's center in Earth radii is L. For example, $L=4$ means the shell which intersects the Earth surface at 60° latitude of the idealized dipole magnetic field.

E. Sudden commencement of magnetic storms.

24. Form of SC on magnetograms. A typical magnetic storm begins with a sudden change in all the geomagnetic elements, over the whole world (see Sect. 20). The sudden commencement of a magnetic storm (**SC** or **ssc**) is very clearly seen at temperate and low latitudes, and a world-wide abrupt increase in the horizontal intensity $\mathcal{H}$ of the geomagnetic field is recorded. The increase in $\mathcal{H}$ amounts to 10 ... 100 γ, and the rise occurs within several minutes; most frequently the rise time is about 3 minutes. A simultaneous sudden change is also recognized in the other components of the geomagnetic field, viz. declination D and vertical intensity Z. The geomagnetic variation of D at the time of an **SC** depends on local time and season, whereas the simultaneous change in Z depends on the location of the observing station; the sign may sometimes be different at two nearby stations. In contrast to D and Z, the variation in $\mathcal{H}$ shows a world-wide similarity, except in polar regions.

In high latitudes, irregular disturbances or pulsative changes of considerable magnitude are so frequently recorded that the sudden commencement of a magnetic storm is not always distinguishable, without referring to simultaneous records from lower latitudes, particularly for small **SC**'s. Moreover, in high latitudes, the sudden commencement is not always characterized by an abrupt increase in $\mathcal{H}$ but sometimes by a small decrease. Cases are frequent where a small negative kick is recorded on the magnetograms just preceding the main impulse of the sudden commencement. Such a storm commencement is called **SC*** or **ssc***.

For convenience the new international nomenclature of the phenomena (IAGA Resolution at Copenhagen Meeting, April 1957) is given in the following:

ssc: A sudden impulse followed by an increase in activity lasting at least one hour. The higher activity of the storm itself may appear immediately, or may be delayed by a few hours.

ssc*: Similar to an **ssc**, except that, on at least one component, the sudden impulse is immediately preceded by one or more small reverse oscillations. When the reverse movement has approximately the same amplitude as the principal movement, it will be reported as **ssc** (not **ssc***). In the case where an **ssc*** is reported, the observer is requested to put a star into that column[1] where the algebraic sign of the principal movement is recorded, so that the star identifies the element(s) in which reverse movements occurred. If during a magnetic storm, the observer sees additional **ssc**'s or **ssc***'s, he should report them as separate phenomena and call them **ssc** or **ssc*** as appropriate.

si: If during a storm, the observer sees an important sudden impulse, but doubts that it represents the beginning of a new storm, he should report it as **si**.

---

[1] This column refers to the table of principal magnetic storms, which appears for example in J. Geophys. Res.

Sect. 24. Form of SC on magnetograms. 43

Fig. 25. Classification of sudden commencements of magnetic storms after their shape. From left to right: **SC, inverted SC, SC*, inverted SC***.

Fig. (caption partially shown) H component only at the beginning of severe magnetic storms. Stations (with geomagnetic College (64.7°, 256.5°), SItka (60.0°, 275.4°), FRedericksburg (49.6°, 349.9°), TUcson (40.0°, ...), (−3.6°, 203.5°), little AMerica (−74.0°, 312.0°), WIlkes (−77.8°, 179.0°), marie BYrd −79.0°, 336.0°). (After Sato.)

Occasionally if a magnetic storm apparently begins with two or more sudden movements, the observer should report each movement as **ssc**, unless he doubts that one or the other is actually the beginning of a storm. In the latter case the clear commencement should be reported as **ssc**, the other as **si**.

Before the above definitions were introduced, various notations had been used to describe the characteristic magnetogram traces of sudden commencements of magnetic storms, particularly those of the $\mathcal{H}$-component. For example,

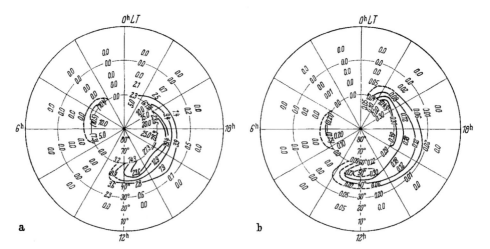

Fig. 27 a and b. (a) Distribution of the average magnitude of the preliminary reverse impulse of sudden commencements, $\Delta\mathcal{H}_k$. (Coordinates: geomagnetic latitude and local time; unit: γ.) (b) Distribution of the ratio of the average preliminary reverse impulse, $\Delta\mathcal{H}_k$, to the average main impulse of sudden commencements, $\Delta\mathcal{H}$. (Coordinates: geomagnetic latitude and local time.)

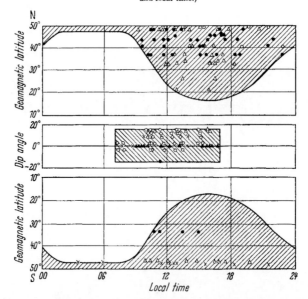

Fig. 28. World-wide occurrence of **SC*** in 1957/58 (IGY), between the geomagnetic latitudes 50° N and 50° S against the local time of occurrence and latitude of the station (dip angle in the magnetic equatorial zone, and geomagnetic latitude at other latitudes). The solid circles, triangles, crosses, and open circles indicate, respectively, the following four longitude zones: Europe and Africa (geomagnetic 50° E ... 140° E); Asia and Australia (geomagnetic 140° E ... 230° E); northwest America and New Zealand (geomagnetic 230° E ... 320° E), and northeast and South America (geomagnetic 320° E ... 50° E). Hatched regions indicate the three zones of predominant occurrence of **SC***. (After MATSUSHITA.)

FERRARO et al.[2] used the symbols **SC**, **SC***, **inverted SC** and **inverted SC***. MATSUSHITA[3] suggested to classify sudden commencements in $\mathcal{H}$ into the following four types: $^-$**SC** (a small negative impulse precedes the main positive impulse), **SC** (a common main positive impulse alone), **SC**$^-$ (an increase lasting from 0 to about 6 minutes, followed by a decrease to a level lower than the initial pre-**SC** level), and $^-$**SC**$^-$ (a combination of $^-$**SC** and **SC**$^-$, which is a negative impulse preceding the main positive impulse, followed by a sharp decrease). According to the international classification, $^-$**SC** and $^-$**SC**$^-$ will usually be reported as **SC***.

The four types of sudden commencement of magnetic storms on ordinary magnetograms are shown in Fig. 25; Fig. 26 shows some quick-run magnetograms of the $\mathcal{H}$-component in high latitudes, at the time of severe magnetic storms[4].

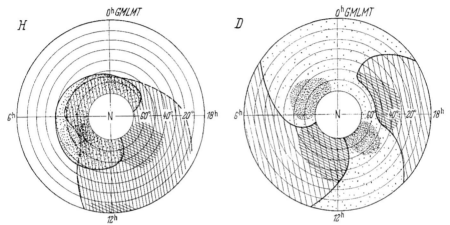

Fig. 29. Zones of predominant occurrence of complex types of in $\mathcal{H}$ or D during **SC** and **SI**. Legend: ▥ **SC*** and **SI***; ▨ **inverted SC*** and **inverted SI***. Area of double hatching or of dense dots indicates the region of more frequent occurrence of the phenomenon. Coordinates in the figure refer to geomagnetic latitude and geomagnetic local mean time). (After SANO.)

25. Local appearance of SC*. The storm sudden commencement with small preliminary impulse appears not all over the world, but only in a restricted region. Fig. 27 shows a statistical distribution of the average magnitude of preliminary reverse impulses, $\Delta \mathcal{H}_k$, of the $\mathcal{H}$-component, and the mean ratio of $\Delta \mathcal{H}_k$ to the following main impulse, $\Delta \mathcal{H}$ [9]. The magnitude of the preliminary reverse impulse shows a great maximum around 16 h, and a smaller one around 9 h local time.

MATSUSHITA [10] showed the presence of a predominant region of occurrence of **SC***, as shown in Fig. 28, which is one result of recent intensified observations in the international network. Observatories located in the hatched region of Fig. 28, report most sudden commencements of magnetic storms as **SC*** or **ssc***, whereas the same commencements are described as ordinary **SC** or **ssc** in other regions of the world. It is worth noting that during sunlit hours, **SC*** does appear sometimes, but not always, also in a narrow belt along the geomagnetic equator.

Little work has been done so far on the change of declination D at the time of **SC**'s. In a recent analysis, SANO[1] classified the declination change into four types, **SC**(D), **SC***(D), **inverted SC**(D), and **inverted SC***(D), just as in the case of

---

[2] V. C. A. FERRARO, and W. C. PARKINSON: Nature **165**, 243 (1950). — V. C. A. FERRARO, W. C. PARKINSON, and W. H. UNTHANK: J. Geophys. Res. **56**, 177 (1951).
[3] S. MATSUSHITA: J. Geophys. Res. **62**, 162 (1957).
[4] T. SATO: Rep. Ionos. Space Res. Japan **15**, 215 (1961); **16**, 295 (1962).
[1] Y. SANO: Mem. Kakioka Mag. Obs. **11**, No. 1, (1963); **11**, No. 2, 5 (1964).

the $\mathcal{H}$-component. Fig. 29 compares the occurrence of the four types of $\mathbf{SC}(\mathcal{H})$ and $\mathbf{SC}(D)$ in a polar map.

The simultaneous change in the vertical component Z at the time of $\mathbf{SC}$'s depends so much on the location of the observatory, that the world-wide systematic distribution of ΔZ can only be examined with careful choice of data[2]. The reason is that electromagnetic induction in the earth plays a fundamental rôle and influences the records considerably.

The dependence of short-period geomagnetic variations on underground electric conductivity is now widely used to study the electric state of the earth's interior, especially anomalies of subterranean structure [11].

26. World-wide simultaneity of sudden commencements.
There have been a number of reports concerning the simultaneity of the sudden commencement of magnetic storms or sudden impulses over the world. The corresponding accuracy is a problem which caught the attention of geomagneticians from the early stage of such investigations. In a number of historical papers it has been reported that the sudden commencement of a magnetic storm takes place simultaneously all over the world within one minute. In order to study this problem in more detail, quick-run records of geomagnetic variations have been made since the 1930s[1]. Even with quick-run magnetograms it is difficult to determine the simultaneity more accurately, because the records of an $\mathbf{SC}$ at different stations are different in shape.

Different authors [12][2-4] suggest that a sudden commencement or sudden impulse (Sect. 32) appears first in high and middle latitudes in the sunlit hemisphere, and propagates around the world within a minute; this refers to the time of maximum deviation. These authors presented an isochronic curve of the time of onset of world-wide geomagnetic changes (see Fig. 30), which indicates the propagation of the main impulse around the world in 60 seconds. The $\mathbf{SC^*}$'s at high latitude appear within the region enveloped by the contour line '20 sec' in Fig. 30. This diagram is valid also for a sudden impulse, which is similar to an **ssc** in form but not followed by magnetic activity.

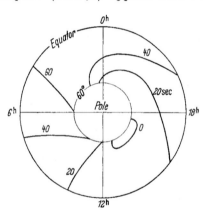

Fig. 30. Isochronic curve of the time of onset of world-wide changes for the phenomena **SI** and **SC**. On each curve the average value of the relative time of onset is given in sec. (After NISHIDA and JACOBS.)

27. Rise-time of SC's.
The time-interval from the beginning to the final stage of an $\mathbf{SC}$ will be called its rise-time. The rise-time for $\mathcal{H}$ in temperate latitudes is several minutes, most frequently around 3 minutes. Considering the rise-time at a fixed station, it seems that it has no definite correlation with the magnitude of the $\mathbf{SC}$ and that of the following main phase[1]. The rise-time for an $\mathbf{SC}$ is slightly different from one station to another; it is generally short in daytime. The de-

---

[2] R. MAEDA, T. RIKITAKE, and T. NAGATA: J. Geomag. Geoelectr. **17**, 69 (1965).
[1] A. TANAKADATE: Compt. Rend. Assemblée de Lisbonne 1933 (A. T. M. E. Bull. No. 9), 149 (1934).
[2] V. B. GERARD: J. Geophys. Res. **64**, 593 (1959).
[3] V. L. WILLIAMS: J. Geophys. Res. **65**, 85 (1960).
[4] M. YAMAMOTO, and H. MAEDA: J. Atmosph. Terr. Phys. **20**, 212 (1961).
[1] H. MAEDA, K. SAKURAI, T. ONDOH, and M. YAMAMOTO: Ann. Géophys. **18**, 305 (1962).

pendence of the **SC** rise-time on local time is given in Fig. 31. Each line in this diagram connects the rise-times of individual **SC**'s recorded on quick-run magnetograms in middle or low latitudes with different longitudes[2]. The rise-time is 200 ... 300 seconds in the night, whereas it is 30 ... 130 seconds around noon.

Fig. 31. Longitudinal (local time) distribution of ssc rise-times at low and middle latitudes from September 1957 to September 1958. (After ONDOH.)

28. Equivalent electric current-systems for SC* and SC. When studying geomagnetic variations, it is very convenient and useful to explain the geomagnetic change, observed on the ground over the whole world, by an electric current-system situated somewhere in the earth's upper atmosphere, as explained in Sect. 7. We introduce, therefore, an equivalent overhead current-system representing the world-wide aspect of the geomagnetic changes during an **SC*** and **SC**.

An equivalent overhead electric current-system for the preliminary reverse impulse of an **SC*** is given by Fig. 32, which was drawn by NAGATA and ABE [9]. This electric current-system describes the behaviour of an **SC*** in middle latitudes very well, not only the variation of the $\mathcal{H}$-component but also the change in declination, at least in the daytime. The observed **SC***(D) in the night hours (see Fig. 29) is, however, not explicitly shown in the current-system of Fig. 32. An additional weak current vortex may be needed to explain **SC***(D) in the dark hemisphere.

The occasional appearance of **SC***$(\mathcal{H})$ at the geomagnetic equator shown in Fig. 28 may be interpreted in the following way: the westward electric current in the daytime sometimes extends to lower latitudes, though it is not intense enough to produce an **SC***$(\mathcal{H})$ in low latitudes. However, the very large electric conductivity in the ionosphere along the geomagnetic equator (see Fig. 2 in Sect. 2) concentrates the westward current so that an **SC*** is observed even at the equator. The duration of the preliminary reverse impulse is only several tens of seconds, before the main impulse of the sudden commencement develops all over the world.

Since the geomagnetic change at the time of preliminary impulse of an **SC*** is observed only in a limited region over the world, the hypothetic 'equivalent current-system' must not be too far from the earth's surface, probably lower than

---

[2] T. ONDOH: J. Geomag. Geoelectr. **14**, 198 (1963).

a few hundred km above the ground[1]. It is interesting to point out that the **SC*** current-system is very similar in form to the S_q^p current-system already shown in Fig. 9, Sect. 10, but that the direction of the electric current is just reversed.

The geomagnetic change at the main stage of an **SC** is shown in Fig. 33, which was drawn by OBAYASHI and JACOBS [13] after examining a large number of **SC**'s observed at various places, all over the world. The electric current distribution is given with geomagnetic latitude and local time as coordinates. The geomagnetic

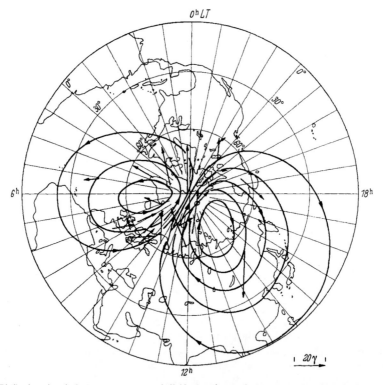

Fig. 32. Distibution of equivalent current vectors at individual stations and average current-system for the preliminary reverse impulse of the ssc* at 06^h 25^m UT on May 29, 1933. (After NAGATA and ABE.)

change at the time of an **SC**, denoted by **D(SC)**, is conveniently decomposed into a longitude-independent component $\mathbf{D}_{st}(\mathbf{SC})$ and the remainder $\mathbf{D}_S(\mathbf{SC})$, as is usually done at the separation of the geomagnetic variations, see Chap. F. The $\mathbf{D}_{st}(\mathbf{SC})$-component is explained by an overhead eastward zonal electric current flow, because it produces a northward magnetic field on the earth's surface.

In the superposition of electric current-systems for $\mathbf{D}_{st}(\mathbf{SC})$ and $\mathbf{D}_S(\mathbf{SC})$, the combined current pattern **D(SC)** in high latitudes is due mainly to the $\mathbf{D}_S$-part of **SC**, which depends on longitude or local time. It is very important that the direction of the electric current flow in the polar vortex in the $\mathbf{D}_S(\mathbf{SC})$ current-system is just opposite to that of the preliminary reverse impulse shown in Fig. 32. In other words, the current-system for $\mathbf{D}_S(\mathbf{SC})$ is very similar to the S_q^p current pattern. Because of the superposition of $\mathbf{D}_{st}(\mathbf{SC})$ and $\mathbf{D}_S(\mathbf{SC})$, the magnetogram traces of an **SC** are quite different at two points of different local times in high

---

[1] T. NAGATA: Rep. Ionos. Res. Japan **6**, 13 (1952); — Nature **169**, 446 (1952).

Sect. 28. Equivalent electric current-systems for SC* and SC. 49

Fig. 33. Electric current-system for the average sudden commencement of many magnetic storms. View from above the geomagnetic North pole; 10^4 A flow between adjacent stream lines. The total current system, $D(SC)$, is shown at right. That part of it which is independent of geomagnetic longitude, $D_{st}(SC)$, is given at left, the remainder, $D_S(SC)$ in the center. See explanation in the text. (After OBAYASHI and JACOBS.)

latitudes, and this explains the complicated appearance of **SC**'s in high latitudes. On the other hand, the increase in H is always observed in lower latitudes due to the predominance of the $\mathbf{D}_{st}$-component, although there remains a local time dependence of the **SC** magnitude, even in middle and low latitudes as will be described in the next section.

29. Diurnal variation of magnitude and frequency of occurrence of SC's. A diurnal variation of the magnitude of **SC's** is noticed from statistical examination of the data at each station; this fact is revealed by the presence of the $\mathbf{D}_S$-component mentioned in the preceding section. Near the geomagnetic equator, the magnitude of the change ΔH during **SC's** is anomalously large in the daytime[1] as is illustrated by Fig. 34. Such an *equatorial anomaly* has been found also for $\mathbf{S}_q$ and **L** variations. It has been shown that the greater is the magnitude of $\mathbf{S}_q$ on a storm day, the greater are the amplitudes of the **SC's**, and of the following

Fig. 34. The diurnal variation of the ratio of amplitudes, ΔH, of **SC's** at HUancayo (Peru) to those at CHeltenham (near Washington, D.C.) on the left hand ordinate (full curve). The broken curve gives the number of **SC's** used in the statistics (right hand ordinate). A total of 183 **SC's** during the period 1922 ... 46 were examined. (After Sugiura.)

initial phase, too[2]. As to the diurnal variation of the amplitude of **SC's**, it depends on latitude as shown in Fig. 35. The hour of maximum amplitude varies with the latitude[3]. At high latitudes the average magnitude of **SC's** is largest during early morning hours, in contrast to the tendency at lower latitudes.

Since the sudden commencement is a world-wide phenomenon, there should be no local-time dependence of the occurrence frequency of **SC's**. However, insofar as the observational result at a single station is concerned, there is a tendency[4] for an early morning minimum of occurrence of **SC's**. This is only apparent and due to the diurnal variation of magnitude and shape of **SC's**.

30. Seasonal influence on SC's in middle and low latitudes. It has been generally considered that the **SC**-vectors lie almost in the geomagnetic meridian plane[1]. This is true insofar as average **SC**-vectors are concerned.

The $\mathbf{D}_{st}(\mathbf{SC})$-field in middle and low latitudes is equivalent to a uniform external magnetic field around the earth; it is directed northward for the average of all seasons. However, the individual $\mathbf{D}_{st}(\mathbf{SC})$-field is not parallel to the geomagnetic axis, especially for **SC's** in the solstitial months. The $\mathbf{D}_{st}(\mathbf{SC})$-axis direction shows a systematic difference against the geomagnetic axis, which remains within 20°, i.e., the $\mathbf{D}_{st}(\mathbf{SC})$-axis is inclined toward the dayside in summer and towards the night side in winter[2], as is shown in Fig. 36.

---

[1] M. Sugiura: J. Geophys. Res. **58**, 558 (1953). — H. Maeda, and M. Yamamoto: J. Geophys. Res. **65**, 2538 (1960).
[2] S. E. Forbush, and E. H. Vestine: J. Geophys. Res. **60**, 299 (1955).
[3] A. J. Shirgaokar, and H. Maeda: J. Geophys. Res. **68**, 2344 (1963).
[4] H. W. Newton: Mon. Not. Roy. Astr. Soc. Geophys. Suppl. **5**, 159 (1948).
[1] A. G. McNish: Compt. Rend. Ass. Lisbonne 1933, A. T. M. E. Bull. No. 9, 238 (1934).
[2] R. Maeda, N. Fukushima, and T. Nagata: J. Geomag. Geoelectr. **16**, 239 (1964).

This effect appears as a declination-change at the time of an SC near the sunrise and sunset hours. For example, at Kakioka, Japan (geomagnetic latitude 26.0°), the declination-change in the early morning hours is eastward in summer and westward in winter, whereas it is eastward in winter and westward in summer around dusk hours[3].

Fig. 35. Amplitude ratio of ΔH between pairs of stations against the local time of the numerator station, for SC's at different geomagnetic latitudes. Pairs of stations with nearly the same latitude but different longitude, in the order of decreasing geomagnetic latitude: Cape WEllen and ESkdalemuir (average geomagnetic latitude 60°), san FErnando and FRedericksburg (average 43°), M'Bour and HOnolulu (average 21°), GUam and BAngui (average 5°), ELisabethville and APia (average −14°), and TOolangi and HErmanus (average −39°). (After SHIRGAOKAR and MAEDA.)

31. Progressive change in the geomagnetic disturbing force vector at and immediately after an SC. During a short time-interval of about one minute, the equivalent overhead current-system for SC* shown in Fig. 32 disappears, and the one shown by Fig. 33 develops over the world. In high latitudes the direction of the geomagnetic disturbing force vector is almost reversed.

The progressive change in the world-wide current-system during and immediately after an SC has not yet thoroughly been analyzed, because the number of stations with quick-run recording is insufficient. Therefore, we have no definite conclusion yet about this problem, though there are suggestions that around the 16^h meridian the high-latitude SC* current-

---

[3] N. FUKUSHIMA: J. Geomag. Geoelectr. **18**, 99 (1966).

Fig. 36a and b. (a) Relation between the latitude, φ_0, defined by the D_{st}(SC) axis and that of the subsolar point, φ_S, in geomagnetic coordinates. (b) Relation between the longitudinal difference, $(\lambda_0 - \lambda_S)$, of the D_{st}(SC) axis and the subsolar point, and the latitude of subsolar point, φ_S (geomagnetic coordinates) (After MAEDA et al.)

system may rotate clockwise[1] during the **SC**, or that it is pushed away by the developing low-latitude current-system[2] of the **SC**.

The development of the horizontal disturbing force corresponding to the **SC** has been recently examined in fair detail combining $\mathcal{H}$- and D-, or $\mathcal{X}$- and $\mathcal{Y}$-traces. According to recent analyses by WILSON and SUGIURA [14]; the following systematic features have been noticed. In high latitudes, near the auroral zone, the horizontal vectors (seen from above) rotate counter-clockwise in the morning hours, whereas in the afternoon hours they rotate clockwise, as can be seen for

---

[1] T. OGUTI: Rep. Ionos. Res. Japan **10**, 81 (1956).
[2] Y. SANO: Mem. Kakioka Mag. Obs. **10**, No. 2, 19 (1962); — J. Geomag. Geoelectr. **14**, 1 (1962).

example from Fig. 37. In temperate latitudes, the vectors are almost parallel to the geomagnetic meridian plane, because the variations in D are small compared with those in $\mathcal{H}$, but the sense of rotation seems still to be as mentioned above, except for the low-latitude region.

A series of damped oscillations of small magnitude often follows after an **SC**. Such oscillations accompanying an **SC** are not synchronous over the world; the oscillations in higher latitudes have longer period and larger amplitude than those in lower latitudes[3]. At low latitudes, it has been noticed that these pulsations have counter-clockwise polarization in the morning and late afternoon hours, while

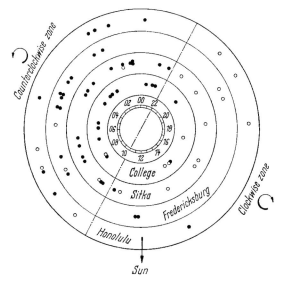

Fig. 37. Local time distribution of **SC**'s with counterclockwise rotation (dots) and with clockwise rotation (open circles) of the magnetic vector, for stations in the northern hemisphere. The zones of opposite senses of rotation are separated by the meridian plane through 1000 and 2200 h local mean time. (After WILSON and SUGIURA.)

it is clockwise in the early afternoon and post-midnight hours[4]. This tendency is readily seen from Fig. 37. The polarization of the horizontal disturbing force will be discussed with the aid of the magneto-gas-dynamic theory of magnetic storms in Chap. K.

32. Sudden impulse SI, and its similarity with SC. The geomagnetic field frequently shows an abrupt change, which for shape and amplitude is very similar to an **SC**, but not followed by considerable magnetic activity as in the case of ordinary magnetic storms. These variations are called sudden impulses, and abbreviated to **SI** or **si** (see the IAGA resolutions described in Sect. 24). Variations of this kind are classified into positive and negative **SI**'s according to whether the horizontal intensity $\mathcal{H}$ shows an increase or decrease. Negative **SI**'s take place less frequently than positive ones.

The sudden impulses are a world-wide phenomenon, and the morphology of positive **SI**'s of large amplitude has been proved to be almost the same as that of **SC**'s[1,2]. In addition, it has been reported that the geomagnetic variation after

---

[3] T. SATO: Rep. Ionos. Space. Res. Japan **15**, 215 (1961); **16**, 295 (1962).
[4] Y. KATO, and T. SAITO: Sci. Rep. Tohoku Univ., Ser. 5 (Geophysics) **9**, 99 (1958).
[1] Y. SANO: Mem. Kakioka Mag. Obs. **12**, No. 1, 55 (1965).
[2] Y. YAMAGUCHI: Mem. Kakioka Mag. Obs. **8**, No. 2, 33 (1958).

large SI's shows a small and short decrease in $\mathcal{H}$ compared with the main phase of a magnetic storm[2,3]. Therefore, one may assume that positive SI's and SC's belong to the same family of events with only a quantitative difference in the geomagnetic activity following the sudden change.

The negative SI is a disturbance, in which the sign of the geomagnetic variation of a positive SI is just reversed. The polarization of the magnetic field vectors during a positive SI is nearly the same as during an SC, while it is almost reversed for a negative SI. The mechanism provoking positive SI's is considered to be nearly the same as that for SC's, namely a compression of the magnetosphere due to an intensified solar wind, as explained in the next section. On the other hand, negative SI's may correspond to an expansion of the magnetosphere[4]. The study of sudden impulses has been extended to deal with those of a few γ magnitude. It resulted that these are also a world-wide phenomenon; they occur very frequently [12], several times a day on the average.

33. Outline of theory for SC. The sudden commencement of a magnetic storm is thought to take place when an intensified solar wind arrives at the magnetosphere boundary and pushes it towards the earth. The compression is transmitted

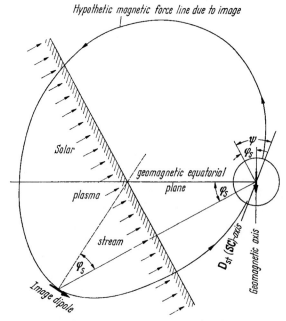

Fig. 38. Image dipole in an idealized solar plasma stream attacking the earth, and its magnetic effect.

to the interior as a gasmagnetic perturbation. The compressional effect is observed in its simplest form as an abrupt increase of the horizontal intensity of the geomagnetic field in middle and low latitudes.

A simplified illustration of the compressional effect of intensified solar wind is given in Fig. 38, which is a modification of Chapman-Ferraro's historical picture[1] for the case of non-perpendicular incidence of solar plasma with respect

---
[3] S. Matsushita: [10] (1962).
[4] A. Nishida, and L. J. Cahill jr.: J. Geophys. Res. **69**, 2243 (1964).
[1] S. Chapman, and V. C. A. Ferraro: Terr. Mag. **36**, 77, 171 (1931); **37**, 147 (1932).

to the geomagnetic dipole. The compressional effect observed near the earth is approximated by the magnetic effect of a mirror image of the geomagnetic dipole, which is symmetric with respect to the front surface of the solar stream. When the solar wind vector is not parallel to the plane of the geomagnetic equator, the magnetic image field is not parallel to the geomagnetic axis in the vicinity of the earth. This model can explain quantitatively the world-wide seasonal variation of the $\mathbf{D}_{st}(\mathbf{SC})$-field, and also the diurnal variation of declination in the solstitial seasons at low latitudes, as described in Sect. 30.

In fact, the solar wind front is not of a plane form as in Fig. 38, but encloses the magnetosphere thus forming the "geomagnetic cavity" immersed in the solar plasma stream. Therefore, at high latitudes, the consequences of the impact of an intensified solar wind are less straightforward. Fig. 39 is a schematic drawing of the equatorial cross-section of the magnetosphere boundary containing the projections of magnetic field-lines originating at high latitudes. During the impact of an intense solar wind, the plasma with the frozen-in magnetic force-lines moves clockwise on the morning side and counter-clockwise on the afternoon side. This rotational motion is transmitted to the earth's surface as Alfvén wave, and results in a systematic polarization of the SC disturbance field vectors in high latitudes [14], [15] as explained in Sect. 31. The propagation time of the SC disturbance over the world, and the rise-time of SC's may be explained by the different delay times of gasmagnetic waves coming from different parts of the magnetospheric boundary [2,3].

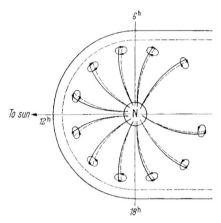

Fig. 39. Schematic drawing of the equatorial cross-section of the magnetosphere boundary and the projections of magnetic field lines originating at high (northern) latitudes; solid lines for the equilibrium state before SC, broken lines for the new equilibrium state after SC. Small loops (with two arrows) indicate the direction of displacement of the field lines during the SC. (After SUGIURA and HEPPNER.)

The frequent occurrence of a reverse kick as SC* in a limited region of the world (described in Sect. 25) is a very important problem.

The current-system shown in Fig. 32 for the preliminary reverse impulse of SC* in high latitudes may be due to a Hall current in the ionosphere, produced by an electric field and/or electric currents, accompanying a magneto-gas-dynamic displacement in the outer magnetosphere, when the magnetosphere is suddenly compressed by an intensified solar wind [4-6].

F. Average storm-time variation and disturbance-daily variation of geomagnetic field.

34. Distinction of S_q, S_D, D_{st} and D_S in the observed magnetic variations. On magnetically disturbed days, the observed variation differs from the daily variation on quiet days, S_q. One may say that, on disturbed days, another kind of solar daily variation is superposed on S_q, and this additional daily variation shows

---

[2] A. J. DESSLER, W. E. FRANCIS, and E. N. PARKER: J. Geophys. Res. **65**, 2715 (1960).
[3] E. J. STEGELMANN, and C. H. VON KENSCHITZKI: J. Geophys. Res. **69**, 139 (1964).
[4] E. H. VESTINE, and J. W. KERN: J Geophys. Res. **67**, 2181 (1962).
[5] T. TAMAO: Rep. Ionos. Space Res. Japan **18**, 16 (1964).
[6] J. H. PIDDINGTON: Planet. Space Sci. **9**, 947 (1962).

a considerable time-change in its intensity. If we denote the average solar daily variation on disturbed days by S_d, and that of all days (in a given period) by S_a, respectively, the differences $S_a - S_q$ and $S_d - S_q$ have similar form, but the amplitude of the latter is greater. The difference $(S_d - S_q)$ is called the *disturbance-daily variation*, and is denoted by S_D. At high latitudes S_D is much greater than at low latitudes, as described later in Sect. 38. We must bear in mind that even the mean daily variation on international quiet days seems to contain very often a contamination with S_D at least to some extent[1] (cf. the definition of international quiet days in Sect. 16).

The magnetic storm variations D at any point P (of colatitude θ and longitude λ), are defined as being added during a storm to the otherwise existing magnetic variation there; they are functions of the storm time t_{st}, which is the time reckoned from the storm commencement. The average of D over λ, (i.e. around the parallel-circle of colatitude θ) at the storm time t_{st} is denoted by $D_{st}(\theta, t_{st})$, and is called the *storm time variation* at colatitude θ. The difference $D(\theta, \lambda, t_{st}) - D_{st}(\theta, t_{st})$ is denoted by $D_S(\theta, \lambda, t_{st})$[2], indicating that it is a function of all three variables, t_{st} and λ and the colatitude θ. The longitude λ can also be replaced by the local time of the observing point. In a paper by CHAPMAN[3], D_S was named the *disturbance local-time variation*, but it now seems preferable to call it *disturbance longitudinal inequality*, because D_S is not uniquely dependent on local time, but changes its phase slightly with time [*16*].

The best method to analyse the morphology of magnetic disturbances is to consider the average disturbance field of magnetic storms, because of the extreme complexity of individual disturbances. In the average disturbance field, irregular features of individual disturbances are almost smoothed out.

In practice, the general world-wide disturbance field of magnetic storms after the sudden commencement is examined by averaging the sequences of hourly mean (or sometimes instantaneous) values for a number of magnetic storms of similar magnitude. (Hourly mean values are the mean of the values in an hour, so that all variations with periods shorter than an hour are almost eliminated.) Even the sequence of hourly mean values does not always show a smooth curve, because of the existence of irregular variations with quasi-periods longer than one hour.

35. Average storm-time variation D_{st} at various latitudes. The storm-time variation at a latitude is the average of the disturbance over the parallel-circle of that latitude. In Fig. 40 the mean storm-time variation is given for the three geomagnetic components, horizontal intensity $\mathcal{H}$, eastward component (referring to geomagnetic coordinates) $\mathcal{E}_{gm}$, and vertical intensity Z, at various geomagnetic latitudes for weak, moderate and severe magnetic storms [*16*].

α) The horizontal intensity, $\mathcal{H}$, shows the most characteristic storm-time variation, particularly in low latitudes. It begins with an initial increase above the preceding undisturbed value, which lasts for a period of several hours. This is the *initial phase* of a magnetic storm.

In Fig. 40 the graphs for high latitude stations (groups 1 and 2; i.e. 80° and 65° geomagnetic latitude) do not show a distinct initial phase, because only daily mean values are plotted; at these latitudes the irregular variations (particularly in daytime in the polar cap, see Sect. 22, and during night hours in the auroral zone) are so great that a reliable trend for D_{st} is hardly obtainable by analysis of hourly or bihourly mean values. The initial increase is most clearly seen in severe storms: there is little systematic difference between the data for weak and moderate storms — except at Huancayo, Peru (−1° geomagnetic latitude). There the increase is abnormally great in all three classes.

---

[1] P. N. MAYAUD: Ann. Géophys. **21**, 121, 219, 369, 514, (1965).

[2] For typographical convenience, S_D, D_S and D_{st} are often printed simply as SD, DS and Dst respectively.

[3] S. CHAPMAN: Ann. Geofis. (Roma) **5**, 481 (1952).

Sect. 35. Average storm-time variation D_{st} at various latitudes. 57

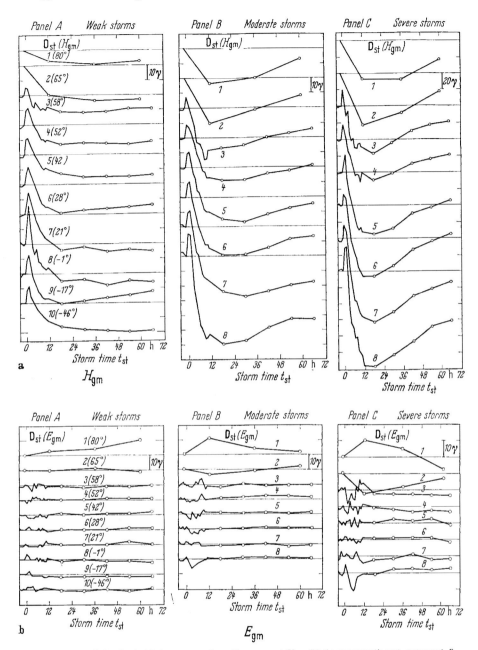

Fig. 40a—c. The variation D_{st} in (a) the geomagnetic north component $\mathcal{H}_{gm}$, (b) the geomagnetic east component $\mathcal{E}_{gm}$, and (c) (see next page) the vertical force $\mathcal{H}$, against storm time (abscissa). Hourly indications from -4 h to $+11$ h storm time, later half-day averages only (circles). For geomagnetic latitudes 80° and 65° (graphs 1 and 2, on top) circles refer to daily averages. In each of the three figures (a), (b), (c), the average results are given separately for weak storms (panel A, at left), moderate storms (panel B, center) and severe storms (panel C, at right). The individual graphs have different scales, as indicated. (After SUGIURA and CHAPMAN.)

A few hours after the sudden commencement, $\mathcal{H}_{gm}$ (geomagnetic north component) returns to its pre-storm value. This occurs during the transition to the main phase of the storm, which is characterized by a large decrease in $\mathcal{H}_{gm}$. It seems to be significant that the interval, during which $\mathcal{H}_{gm}$ is above the pre-storm level, is shorter for the more intense storms.

β) Thereafter, in the *main phase*, $\mathcal{H}_{gm}$ attains a minimum value; the depression in $\mathcal{H}$ is much greater than the rise in the initial phase. The delay required for $\mathcal{H}$ to attain its minimum value depends on the intensity of the magnetic storm;

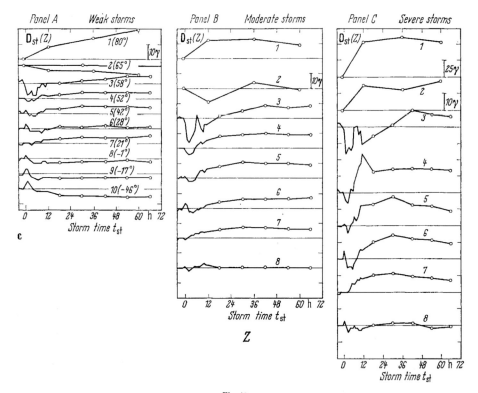

Fig. 40 c.

it is about 18 hours for an average severe storm, 24 ... 30 hours for a moderate one. For an average weak storm the delay is uncertain, because $\mathbf{D}_{st}(\mathcal{H}_{gm})$ continues to be near its minimum level during the second and third days; even at the end of the third day it is not clear whether the recovery phase has begun.

The depression of $\mathcal{H}$ near the equator at the time of a most severe magnetic storm is only of the order of 1% of the earth's permanent magnetic field. The largest equatorial depression ever recorded reached about 1100 γ, amounting to about 4% of the permanent field.

γ) The period during which $\mathcal{H}$ is below its normal value until the turning point where, during the recovery, $\dfrac{d^2 \mathcal{H}}{dt^2} = 0$ approximately, is called the main phase of a magnetic storm, and the subsequent period is named the *last phase* or the *recovery phase*. The names 'initial', 'main' and 'last' phase refer to the average $\mathbf{D}_{st}$-variation. In an individual magnetic storm, the identification of these three phases is rather difficult even when world-wide data are at hand, because of superimposed irregular variations of fairly large amplitude. But for the sake of convenience, each observatory reports the times of beginning of the main and the last phase of every magnetic storm.

The recovery of the depression in $\mathcal{H}$ lasts about 30 hours, but the large $\mathcal{H}$-depression occurring in the early stage of severe magnetic storms decays rapidly, in several hours. This problem is considered below. The recovery continues sometimes many days, the magnetogram traces being quite free from any disturbance effect other than the post-perturbation effect.

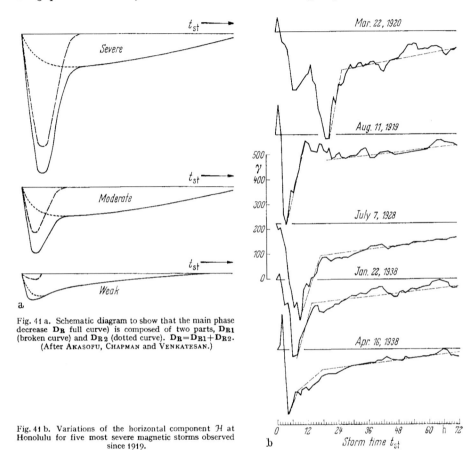

Fig. 41 a. Schematic diagram to show that the main phase decrease D_R full curve) is composed of two parts, D_{R1} (broken curve) and D_{R2} (dotted curve). $D_R = D_{R1} + D_{R2}$. (After AKASOFU, CHAPMAN and VENKATESAN.)

Fig. 41 b. Variations of the horizontal component $\mathcal{H}$ at Honolulu for five most severe magnetic storms observed since 1919.

As described later in Sects. 39 and 40, the major part of the D_{st}-field in middle and low latitudes is called D_R-field, where R stands for the ring-current flowing around the earth, to which the world-wide depression of $\mathcal{H}$ is attributed. The D_R-field sometimes seems to be consisting of two components with different decay times[1], as shown in Fig. 41, although this interpretation must be carefully examined by comparing world-wide data[2]. When D_R is decomposed into D_{R1} and D_{R2}, D_{R1} can be seen only with severe magnetic storms, while D_{R2} is present in all storms[3]. D_{R1} develops within several hours, and decays also very rapidly within several hours. On the other hand, D_{R2} develops slower, in half a day or so, even while D_{R1} is already decaying. The decay-time of D_{R2} is of the order of 30 hours, corresponding to a long post-perturbation of the depression of $\mathcal{H}$ over the whole world. The difference in the decay-times of D_{R1} and D_{R2} may suggest that the current provoking D_{R1} may be nearer to the earth than that of D_{R2}, because the current will probably decay faster at lower altitude.

---

[1] S.-I. AKASOFU, S. CHAPMAN, and D. VENKATESAN: J. Geophys. Res. 68, 3345 (1963).

[2] A. GRAFE, and A. BEST: Pure and Applied Geophysics 64, 59 (1966).

[3] For typographical convenience, D_R, D_{R1} and D_{R2} may simply be printed as DR, $DR1$ nd $DR2$, respectively.

δ) We now consider the storm-time variation of elements other than the horizontal intensity $\mathcal{H}$. The variation $\mathbf{D}_{st}(\mathcal{E}_{gm})$ is small except in high latitudes. Fig. 40 shows irregularities in the corresponding diagrams, indicating that the accidental storm variations are not completely averaged out. A comparison of the diagrams for $\mathbf{D}_{st}(\mathcal{H}_{gm})$ and $\mathbf{D}_{st}(\mathcal{E}_{gm})$ shows that, at least over the greater part of the earth, the $\mathbf{D}_{st}$-field is in the geomagnetic north-south direction. In high latitudes, the $\mathbf{D}_{st}(\mathcal{E}_{gm})$ curves may be unreliable because of insufficient data.

The variation $\mathbf{D}_{st}(Z)$ in Fig. 40 is much less than $\mathbf{D}_{st}(\mathcal{H})$. This is true for all latitudes except at high-latitude stations, and at the low-latitude edge of the auroral zone. At latitudes around 60°, $\mathbf{D}_{st}(Z)$ shows a peculiar change during storm time up to 15 hours. This is related to the equatorward shifting of the auroral electrojet, described later (in Sect. 50). In general at moderate and low latitudes, the vertical force Z undergoes a change nearly opposite in sign to that of the $\mathcal{H}$ variation. The recovery in $\mathbf{D}_{st}(Z)$ seems to be notably slower than in $\mathbf{D}_{st}(\mathcal{H})$.

Fig. 42 a and b. (a) Variation with absolute value of geomagnetic latitude of averages of the $\mathcal{H}$-component of $\mathbf{D}_m$ (disturbed days minus quiet days) for different years. (b) Seasonal dependence for the mean of 1922...33. Stations: SOdankylä, SItka, RUde skov, GReenwich, DE Bilt, VAl Joyeux, CHeltenham, cHRistchurch, WAtheroo, TUcson, HOnolulu, LUkiapang, BOmbay, HUancayo. (After CYNK.)

36. Daily mean storm-time effect over the world. A magnetic storm depresses the daily mean value of the horizontal intensity of the geomagnetic field. This is certainly true one day after the commencement, because the depression in $\mathcal{H}$ exceeds by far the slight increase in the initial phase. The difference of the daily mean value of geomagnetic elements on a disturbed day against that on international quiet days is sometimes used as a numerical measure for geomagnetic activity, representing roughly the daily mean storm-time effect on the disturbed days.

α) Fig. 42 illustrates the distribution of the *mean storm-time effect* $\mathbf{D}_m$ for the horizontal intensity $\mathcal{H}$ of the geomagnetic field as a function of the geomagnetic latitude, in different seasons[1] during the period 1922 to 1933. The magnitude of $\mathbf{D}_m(\mathcal{H})$ is maximum at the equator, diminishing almost proportionally to $\cos \phi_{gm}$ (where ϕ_{gm} denotes the geomagnetic latitude), and becomes minimum at about 55° in the equinoctial season. In the solstitial season, the equatorial maximum is smaller than in the equinoctial one. This suggests that the average magnitude of

Fig. 43. Variation with geomagnetic latitude of X'-, Y'-, and Z-components of $\mathbf{D}_m$. Mean of the years 1922...33: squares. Data of International Polar Year 1932/33 have been multiplied by factor 1.21 (circles). (After VESTINE.)

storms is larger in the equinoctial season than in the solstitial season. It is interesting that in the solstitial season the magnitude of $\mathbf{D}_m(\mathcal{H})$ is smaller in the summer hemisphere than in the winter hemisphere.

In moderate latitudes, $\mathbf{D}_m(Z)$ is much smaller than $\mathbf{D}_m(\mathcal{H})$, and has an opposite sign; a slight increase of the vertical force is observed in moderate northern latitudes [3], as shown in Fig. 43. The effect vanishes at the equator and is reversed in the southern hemisphere. On the other hand, there is no appreciable $\mathbf{D}_m(D)$; this is quite natural because of the absence of any regular storm-time variation in the declination D or in $\mathcal{E}_{gm}$, as already shown by Fig. 40.

---

[1] B. CYNK: Terr. Mag. **44**, 51 (1939).

β) Combining the geomagnetic variation in the three elements, we find that the *disturbing force vector* $\boldsymbol{D}_m$ at temperate latitude lies nearly in the geomagnetic meridian plane, its direction being between the horizontal direction and that of the earth's axis[2]. The distribution of $\boldsymbol{D}_m$ over the world is the same as that of the $\boldsymbol{D}_{st}$-field in the main and last phases. The same conclusion has been derived from the world-wide distribution of the difference between the monthly mean value

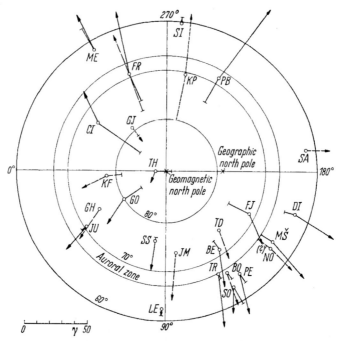

Fig. 44. Daily average vectors $\boldsymbol{D}_m$ in a polar map in geomagnetic coordinates. Horizontal component of disturbance vector as arrow, vertical component as radial line (positive when drawn towards the geomagnetic pole). Dotted: values from Gjöhavn expedition 1904 ... 06; broken: First Polar Year 1882 ... 83 (CHAPMAN's values multiplied by 1.75); full: Second Polar Year 1932 ... 33. Stations with geomagnetic latitude greater than 60°: SItka, SAgastyr, DIkson, Matočkin Šar, NOvaja seml'ja, PEtsamo, BOssekop, SOdankylä, LErwick, MEanook, Point Barrow, TRomsö, Fort Rae, King Point, Franz Josef land, ThorDsen, BEar island, Jan Mayen, Scoresby Sund, JUlianehaab, GodHaab, Kingua Fjord, Chesterfield Inlet, GJöahavn, GOdhavn, THule. (After VESTINE and CHAPMAN.)

of geomagnetic elements and the mean on quiet days. This fact suggests that the form of the disturbance field is the same, regardless of its magnitude, insofar as the daily average effect is considered.

γ) The storm-time variation in high latitudes is not so systematic, as compared with that in temperate latitudes. $\boldsymbol{D}_m$ in high latitudes is fairly large; this is shown in Figs. 42 and 43, revealing a considerable decrease of the horizontal intensity of the geomagnetic field. The distribution of $\boldsymbol{D}_m$ in the northern polar region is given in Fig. 44 on a map in geomagnetic coordinates, where the horizontal vector component of $\boldsymbol{D}_m$ is represented by arrows, and the vertical component by radial butt-ended lines [17]. The horizontal force vectors in Fig. 44 diverge, from the geomagnetic pole rather than from the geographic pole; their magnitude is particularly large between 70° and 65° geomagnetic latitude. The zone of largest $\boldsymbol{D}_m$ coincides with the region of maximum auroral appearance, which is often

---

[2] L. SLAUCITAJS, and A. G. MCNISH: Trans. Edinburgh Meeting 1936, A. T. M. E. Bull. No. 10, 267 (1937).

37. Non-cyclic change after storms. The recovery after a storm is clearly characterized by a steady increase of the value of $\mathcal{H}$, provided the period after the magnetic storm is quiet. An example[1] is given in Fig. 45, from which one clearly sees a general tendency for $\mathcal{H}$ to increase gradually after each magnetic

Fig. 45. Upper curve: Midnight value at the equator of the horizontal disturbance, $D_{st}(\mathcal{H})$, Lower curve: Magnetic activity expressed by the three-hourly equivalent planetary amplitude a_p. Period: Aug. through Oct. 1957. (After SUGIURA.)

Fig. 46. Average non-cyclic change of the horizontal component $\mathcal{H}$ (in γ) for locally quiet days during 1958. Numerical values are given as far as available. (After PRICE and STONE.)

storm. This increase in $\mathcal{H}$ seems to continue as long as no subsequent storm begins. If midnight values of $\mathcal{H}$ (which are free from the S_q-variation) are considered, it appears that they continuously increase from one day to the next one, even in a most quiet period, though the increase is then less than a few γ per day. This systematic change is called the aperiodic or *non-cyclic change* of $\mathcal{H}$ on quiet days. The non-cyclic change of $\mathcal{H}$ seems to be slightly asymmetric around the equator[2,3], as shown in Fig. 46.

---

[1] M. SUGIURA: Annals of the IGY **35**, 9 (1964).
[2] A. T. PRICE: J. Geophys. Res. **68**, 6383 (1963).
[3] A. T. PRICE, and D. J. STONE: Annals of the IGY **35**, 65 (1964).

38. Distribution of the S_D-field over the world.

As already explained in Sect. 34, the geomagnetic variation over the world on disturbed days is considered to be composed of two parts: a universal component $\mathbf{D}_{st}$ and the remainder $\mathbf{D}_S$, i.e.

$$\mathbf{D} = \mathbf{D}_{st} + \mathbf{D}_S. \tag{38.1}$$

α) $\mathbf{D}_{st}$ is the longitudinal average of the disturbance field, whereas $\mathbf{D}_S$ depends also on the longitude or local time of the observing station. $\mathbf{D}_S$, which is called "disturbance longitudinal inequality", varies in magnitude and phase with storm-time, i.e. the time counted from the beginning of the magnetic storm. When the average of a sufficient number of individual $\mathbf{D}_S$ is calculated, we get an *average disturbance-daily variation* which is called $\mathbf{S}_D$. In other words, $\mathbf{S}_D$ is an idealized daily geomagnetic variation, which appears additionally on days of geomagnetic disturbance. On disturbed days, the total solar daily variation is $\mathbf{S}_q + \mathbf{S}_D$, where $\mathbf{S}_q$ is the solar daily variation of the geomagnetic field on quiet days.

In Fig. 47, the $\mathbf{S}_D$-field for the elements $\mathcal{H}, \mathcal{E}, Z$ are compared with $\mathbf{S}_q$, for weak, moderate and severe storms [16]. The $\mathbf{S}_D$ variations on the 1st, 2nd and 3rd days from the storm commencement (denoted respectively by $\mathbf{S}_D^1, \mathbf{S}_D^2$ and $\mathbf{S}_D^3$) are essentially similar for each grade of magnetic activity; they have a constant shape (considerably different from that of $\mathbf{S}_q$) but decreasing intensity. The $\mathbf{S}_D$ variation at night is comparable in magnitude to that during daytime, the diurnal component being clearly predominant. In contrast, $\mathbf{S}_q$ shows a large variation only during the day hours, so that the second harmonic is quite large. The $\mathbf{S}_D$ variation increases considerably towards auroral latitudes, but seems to decrease again from there towards the pole. These are fundamental differences between $\mathbf{S}_q$ and $\mathbf{S}_D$.

β) As functions of the geomagnetic latitude, both $\mathbf{S}_D(\mathcal{H})$ and $\mathbf{S}_q(\mathcal{H})$, are approximately symmetrical with respect to the equator, a reversal appearing at certain "focal" latitudes. For $\mathbf{S}_q(\mathcal{H})$ this latitude is about 35°, but for $\mathbf{S}_D(\mathcal{H})$ it is slightly above 52° for weak and moderate storms; for great storms the focal latitude may be a little lower than 52° during the first two storm days.

At equatorial stations $\mathbf{S}_q(\mathcal{H})$ is abnormally large, but $\mathbf{S}_D(\mathcal{H})$ appears normal, except that its range seems to decrease from the first to the third storm day more rapidly than elsewhere (in weak storms at least). $\mathbf{S}_D(\mathcal{H})$ for weak and moderate storms seems to have a secondary maximum near noon, as if $\mathbf{S}_q$ were enhanced during such storms; if this were so, the enhancement would be superposed on $\mathbf{S}_D$.

As to $\mathbf{S}_D(\mathcal{E})$, its phase is uniform from auroral zone latitudes down to the equator, and it is reversed on crossing the equator. The amplitude of $\mathbf{S}_D(\mathcal{E})$ increases considerably from the equatorial region towards auroral latitudes, and further up. $\mathbf{S}_D(Z)$ preserves a constant shape, with reversal at the equator. Its amplitude increases steeply towards 65°, and is reversed on entering the polar cap. On the first day of great storms, the reversal occurs at a geomagnetic latitude lower than 65°.

γ) When vectograms for the daily geomagnetic variations are presented, the difference between $\mathbf{S}_D$ and $\mathbf{S}_q$ becomes still more clear. Fig. 48 (for moderate magnetic storms) shows the horizontal projection of the vectograms in high and middle latitudes. Comparison of $\mathbf{S}_D^1, \mathbf{S}_D^2$ and $\mathbf{S}_D^3$ shows little change of type, only a decrease of amplitude.

$\mathbf{S}_D$ vectograms in the polar cap (at Godhavn, $\phi_{gm} = 80°$) are nearly circular; the horizontal force vector rotates clockwise. Those for auroral zones (for example 65°) are strikingly different, viz. quite thin, and extended along a direction lying between geomagnetic and geographic north (which is approximately perpendicular to the auroral zone). From this form of the vectogram it is inferred that $\mathbf{S}_D$ is produced by a laterally limited current flowing nearly overhead along the zone, changing with local time in direction and magnitude. Such a narrow current is called an electrojet — in this case, the *auroral electrojet* (see Sect. 39γ).

Near a geomagnetic latitude of 58°, the $\mathbf{S}_D$ vectograms for weak storms are more nearly oval; with increasing storm intensity they become elongated, resembling more those at 65°, at least during the first two storm days. The $\mathbf{S}_D$ vectograms describe clockwise rotation during weak and moderate storms.

Sect. 38. Distribution of the S_D-field over the world.

Fig. 47 a—c. S_q and S_D of horizontal force $\mathcal{H}$ (not the geomagnetic north component $\mathcal{H}_{gm}$), declination force $\mathcal{E}$ (not the geomagnetic east component $\mathcal{E}_{gm}$), and vertical force Z. The very first panel shows S_q; the following three sets (of three panels each) refer to the three intensity classes: weak, moderate, severe. Each set gives S_D^n ($n = 1, 2, 3$) for the first three storm days. The force scale is uniform except where otherwise indicated by a number (the scale contraction factor) enclosed in a circle. The abscissae for S_q refer to local mean time, and those for S_D to geomagnetic local mean time. (After SUGIURA and CHAPMAN.)

Around a geomagnetic latitude of 52°, the S_D vectograms are elongated in the E-W direction, though on the first day of a great storm there is some tendency to come back to the auroral S_D type. The size of the S_D vectograms is much smaller than that for 58°. During

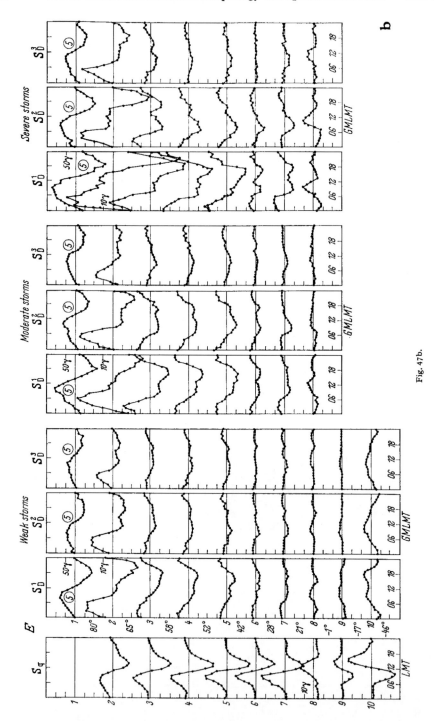

Fig. 47b.

weak and moderate storms, the S_D vectograms are described counter-clockwise. During great storms the vectogram consists of two loops, described in opposite sense. The decline of S_D from the first to the third storm day is most marked for the great storms.

Sect. 38. Distribution of the S_D-field over the world. 67

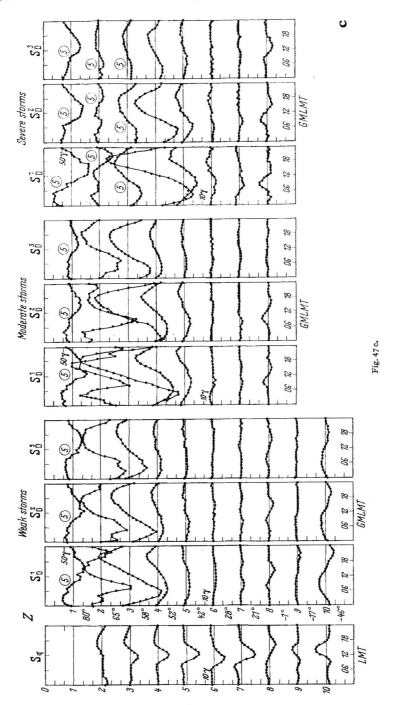

Fig. 47 c.

With decreasing geomagnetic latitude, the S_D diagrams are almost circular, and their sense of polarization remains counter-clockwise. Near the equator, the S_D vectograms do not usually share the equatorial enhancement. The S_D vectograms in the southern hemisphere

5*

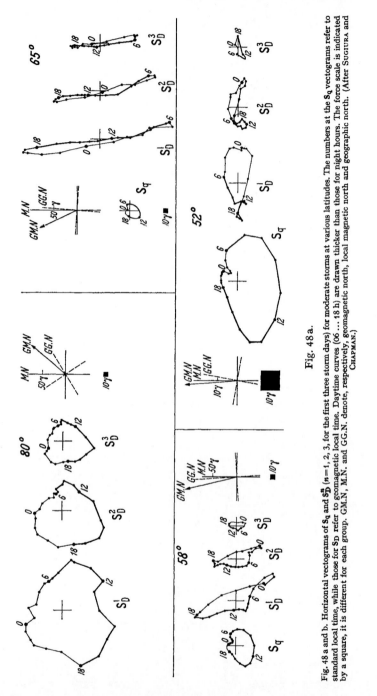

Fig. 48a and b. Horizontal vectograms of S_q and S_D^n ($n=1, 2, 3$, for the first three storm days) for moderate storms at various latitudes. The numbers at the S_q vectograms refer to standard local time, while those for S_D refer to geomagnetic local time. Daytime curves (06 ... 18 h) are drawn thicker than those for night hours. The force scale is indicated by a square, it is different for each group. GM.N, M.N. and GG.N. denote, respectively, geomagnetic north, local magnetic north and geographic north. (After SUGIURA and CHAPMAN.)

resemble those for the northern hemisphere at the same latitude, except that their sense is reversed (clockwise): this is because $S_D(\mathcal{E})$, but not $S_D(\mathcal{H})$, is reversed at the equator crossing.

39. Overhead electric current-systems for D_{st}- and S_D-fields. The spherical harmonic analysis of the geomagnetic disturbance variations over the world sug-

Sect. 39. Overhead electric current-systems for D_{st}- and S_D-fields.

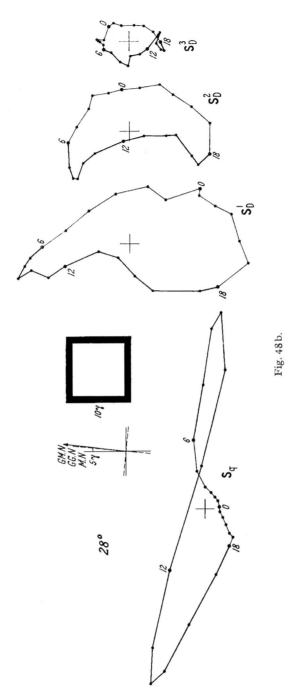

Fig. 48 b.

gests that the principal origin (amounting to about 70%) of the variations on the earth's surface must be situated somewhere in space, above the earth. This part of the **D**-field is attributed to an electric current-system in the upper atmosphere, because no probable origin of this magnetic field is known except for an electric current flowing in the space around the earth.

α) It is, in principle, impossible to determine the *location of this current-system* from observations of magnetic field variation at the earth's surface alone, see Sect. 3. But it seems probable that, at least the currents of the S_D or D_S system are flowing not too far above the earth's surface, because the magnetic disturbance is quite localized in the neighbourhood of the auroral zone. On the other hand, the current corresponding to D_{st} is not necessarily situated very near to the earth.

β) CHAPMAN's famous drawings showing *idealized current-sytems* for D_{st} and S_D, and for the combined field are reproduced in Fig. 49 (the height of the current-system is assumed to be 100 km) [18]. The systems refer to an ideal earth with coincident geomagnetic and geographic axes. Fig. 49 (C) illustrates an idealized current-system for S_D, the left-hand side is a view from the sun, while the right-hand side is from above the north-pole. The current-system for D_{st} is shown by Fig. 49 (B), while the combined system of (B) and (C) is given by Fig. 49 (A).

These figures refer to magnetic storms of weak intensity. The maximum reduction of $\mathcal{H}$ at the equator is 40 γ, and the current-systems are drawn for the epoch of this intensity; at other epochs the current distribution may be similar in shape but of lower intensity, except during the initial phase when $\mathcal{H}$ is increased, and the current direction in (B) must be reversed. An electric current of 10,000 A flows between adjacent stream lines. However, the current flow is so concentrated along the auroral zone that it is impossible to indicate each individual stream-line, so that they appear there merged in a dense mass. In detail the distribution of the magnetic field produced by this idealized current-system does not completely agree with the actual average observations over the world, but it shows fairly well the world-wide geomagnetic variation. Since the asymmetries of the disturbance field with respect to geographic axes almost vanish when one rearranges with respect to geomagnetic coordinates, one may better replace the axis of the earth in Fig. 49 by the geomagnetic one.

γ) The most important feature of the S_D current-system is the presence of an intense current flowing in a narrow auroral zone, which is customarily called the *auroral zone current*, or the *auroral electrojet*. This current flows towards the noon meridian, westward in the morning and eastward in the afternoon, corresponding to a large depression or increase in the horizontal intensity of the geomagnetic field occurring at those times. The greater part of the auroral zone current forms a closed current-system with return current through the polar region, back to the sunlit hemisphere; only a part of the return current flows over middle and low latitudes. The intensity of the temperate-latitude current produces a magnetic field the magnitude of which is comparable with S_q, the daily variation on quiet days. On the other hand, the polar-cap currents of the S_D system produce a much larger magnetic field than does the S_q-current at these high latitudes.

δ) The current-system for the S_D-field in Fig. 49 (C) may appear to be too much idealized, in comparison with the real distribution of the disturbance field in higher latitudes, as given for example by Fig. 47. Detailed examination of the polar disturbance field shows that the phase of the polar-cap current and that of the auroral zone current should be advanced by a few hours [17]. A more *realistic current-system* for the S_D-field is given by Fig. 50 referring to geomagnetic coordinates[1]; even this presentation is idealized to some extent, particularly near the auroral zone around midnight hours (see remarks in Sect. 42). It is worth noting that the current pattern in the polar-cap regions is very similar to that of S_q^p, with much larger current density.

ε) The current-system for D_{st} is, according to the original definition of the D_{st}-field, a zonal current around the earth. The current flows westward every-

---

[1] N. FUKUSHIMA, and T. OGUTI: Rep. Ionos. Res. Japan **7**, 137 (1953).

Sect. 39. Overhead electric current-systems for D_{st}- and S_D-fields.

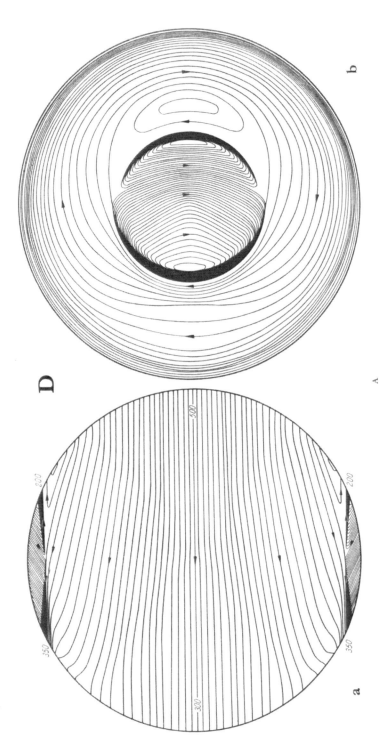

Fig. 49 (A)—(C). Idealized overhead electric current-system for magnetic disturbances, **D**. (A) shows the total disturbance, **D**. In (B) (see page 72) its zonal part, $\mathbf{D}_{st}$ is reproduced, and in (C) (see page 73) its diurnal part, $\mathbf{S}_D$. Left side (a): view from sun; right side (b): seen from above the northern pole. 10^4 A = 10 kA flow between adjacent stream lines. (After CHAPMAN.)

Fig. 49 (B).

Sect. 39. Overhead electric current-systems for D_{st}- and S_D-fields. 73

Fig. 49 (C).

where, reducing the horizontal intensity of the geomagnetic field as is observed during the main phase of storms. Except in high latitudes, the current density varies almost proportionally to cosine of geomagnetic latitude. Such a distribution of current density produces a uniform magnetic field inside the current shell, parallel to the axis of the earth. The westward zonal current has an increased density at the auroral zone; this results from the large $\mathbf{D}_m$ there.

Fig. 50. Improved current-system of the diurnal disturbance field S_D, for the equinoctial season. 15 kA flow between adjacent stream lines. (After FUKUSHIMA and OGUTI.)

40. Location of the current-system for D_S- and D_{st}-fields. α) The electric current-system responsible for S_D or D_S is considered to flow in the ionospheric regions; this conclusion is also supported by the study of the electric conductivity in the ionosphere (see Sect. 2). Measurements of the magnetic field in the upper atmosphere above the auroral zone by means of rockets at the time of magnetic disturbances could give experimental evidence for the location of this current-system; a remarkable change of the magnetic field should be expected when a rocket passes through the current-sheet. Some preliminary results in high latitudes suggest that the intense auroral electrojet is located in the E region of the ionosphere.

β) Before direct measurements of the ionospheric current were carried out, the height of the auroral zone current had been estimated by the following method, assuming that the auroral electrojet can be approximated by a line current[1, 2]. The magnetic potential of the disturbance field $\overline{W}(x, z)$ along a magnetic meridian, consists of an external part $\overline{W}^{(e)}$ and the internal part $\overline{W}^{(i)}$; it can be expressed by a Fourier series as

$$\overline{W}(x, z) = \overline{W}^{(e)} + \overline{W}^{(i)} = \sum_{n=1}^{\infty} e^{nz}[a_n^{(e)}\cos nx + b_n^{(e)}\sin nx] \\ + \sum_{n=1}^{\infty} e^{-nz}[a_n^{(i)}\cos nx + b_n^{(i)}\sin nx] + X_0 x + \overline{W}_0, \quad (40.1)$$

where x- and z-axes are chosen to be southward and upward, and the earth's curvature is neglected (i.e. the plane earth model is adopted). In Eq. (40.1), $\overline{W}_0$ and X_0 are constants and

---

[1] A. G. McNISH: Terr. Mag. **43**, 67 (1938).
[2] T. NAGATA: Rep. Ionos. Res. Japan **4**, 87 (1950).

Sect. 40. Location of the current-system for D_S- and D_{st}-fields. 75

$a_n^{(e)}$, $b_n^{(e)}$, $a_n^{(i)}$ and $b_n^{(i)}$ are the coefficients of the Fourier expansion. Then, the magnetic north component X_m and the vertical downward component Z of the disturbance field on the earth's surface ($z = 0$) are given as follows:

$$X_m = \sum_{n=1}^{\infty} [n (b_n^{(e)} + b_n^{(i)}) \cos nx - n (a_n^{(e)} + a_n^{(i)}) \sin nx] + X_0, \qquad (40.2)$$

$$Z = \sum_{n=1}^{\infty} [n (a_n^{(e)} - a_n^{(i)}) \cos nx + n (b_n^{(e)} - b_n^{(i)}) \sin nx]. \qquad (40.3)$$

The numerical values of $a_n^{(e)}$, $b_n^{(e)}$, $a_n^{(i)}$ and $b_n^{(i)}$, are determined from the Fourier expansion of the observed disturbance field along a magnetic meridian. If we assume further that a plane

Fig. 51 a and b. X_m and Z as functions of geomagnetic latitude (full curves), and their decomposition after external (e) and internal (i) origin (broken and dotted curves); local mean time 22 ... 23 h (a) and 23 ... 24 h (b) on 15 Oct. 1932.

sheet of electric current of intensity* $J(x)$ in y-direction at a height h is responsible for the external part of the disturbance field, and $J(x)$ can be written

$$J(x) = J_0 + \sum_{n=1}^{\infty} (J_n^c \cos nx + J_n^s \sin nx), \qquad (40.4)$$

the following relations should hold;

$$\left.\begin{array}{l} J_n^c = -2n e^{nh} b_n^{(e)}, \quad J_n^s = 2n e^{nh} a_n^{(e)}, \\ J_0 = -2X_0. \end{array}\right\} \qquad (40.5)$$

Fig. 51 gives an example of the distribution along a meridian of X_m and Z, separated after external and internal parts, at the time of a moderate magnetic storm. The mean of the ratio

$$\{(a_n^{(e)})^2 + (b_n^{(e)})^2\}^{\frac{1}{2}} / \{(a_n^{(i)})^2 + (b_n^{(i)})^2\}^{\frac{1}{2}} \quad (n = 1 \ldots 4)$$

is about 2.5 in this case, and this value is in fair agreement with the mean ratio of external to internal part for the geomagnetic variations S_q and L. For various assumed heights of the current sheet, the latitude distribution of the electric current density $J(x)$ is illustrated by Fig. 52. Judging from the convergency of higher order terms of the Fourier expansion of $J(x)$, it may be presumed that the auroral zone current is at the level of the ionospheric E layer, roughly and certainly lower than 150 km above the earth's surface.

---

* J is a "surface current density" to be measured in A/m.

γ) The height of the electric current-system responsible for S_q has also been estimated to be roughly in the ionospheric E layer. By rocket experiments near the geomagnetic equator the existence of an eastward equatorial electrojet has been detected in the E layer at about 100 km height (see Sect. 8). We do not yet

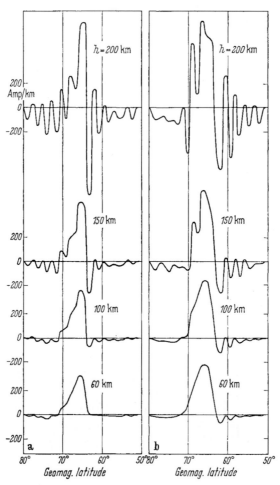

Fig. 52a and b. Latitudinal distribution of auroral zone current, assumed at different heights (as indicated), computed from the ground geomagnetic field given by Fig. 51. (a) 22 ... 23 h; (b) 23 ... 24 h, 15 Oct. 1932.

know whether the heights of the electric currents for S_q and D_S are the same, or slightly different. This should be checked with the aid of rockets by measuring the level of S_q at high latitudes, and that of D_S at lower latitudes.

δ) The D_{st} geomagnetic variation can also be represented by an equivalent overhead electric current-system surrounding the earth, as already shown in Sect. 39. The intensity of the D_{st} current-system in middle and low latitudes is proportional to cosine of the geomagnetic latitude, and this means that the D_{st} field in middle and low latitudes is a uniform magnetic field parallel to the geomagnetic axis. There exist other possibilities than an ionospheric current-system to produce such a uniform magnetic field around the earth. A compression of the earth's magnetosphere, particularly at the time of a sudden commencement of a magnetic storm, produces a northward magnetic field. Charged particles in the magnetosphere, which are trapped in the earth's magnetic field, can also produce some geomagnetic effect at the earth's surface, viz. a northward field due to their own gyro-motion in the geomagnetic field; but a southward field is generated by their slow drifting motion around the earth (see Sect. 23). Since the latter effect is larger than the former one, trapped charged particles produce a southward magnetic field around the earth. The $\mathbf{D}_{st}$ geomagnetic variation, at least that observed in middle and low latitudes, is now considered to be mainly produced by the effects described here*, and not by an electric current flowing in the ionosphere (see Sect. 69). An asymmetry of the $\mathbf{D}_m(\mathcal{H})$ field in the northern and southern hemispheres in the solstitial season has been shown in Fig. 42, Sect. 36. It may be explained by the tilting of the equatorial ring current in the magnetosphere.

41. Polar disturbance field and world-wide disturbance field. During the initial phase of magnetic storms at middle and low latitudes the average disturbance is a uniform northward field, while during the main phase it is southward. If these uniform fields are subtracted from the $\mathbf{D}_{st}$ field, the remaining equivalent overhead current-system is an intense zonal current near the auroral zone. This remaining current may be flowing in the ionosphere as does the current for $\mathbf{D}_S$. Hence it will be a convenient way to separate the whole disturbance into two parts, a uniform magnetic field along the earth's axis and the remaining part. The latter contains, of course, $\mathbf{D}_S$ and the polar part of $\mathbf{D}_{st}$, and may be called the *polar disturbance field*, because of its origin in the polar ionosphere. If we denote the polar disturbance field and the world-wide uniform field by $\mathbf{D}_p$ and $\mathbf{D}_w$ respectively, the following relation holds:

$$\mathbf{D} = \mathbf{D}_{st} + \mathbf{D}_S = \mathbf{D}_p + \mathbf{D}_w \tag{41.1}$$

or

$$\mathbf{D}_p = \mathbf{D}_S + (\mathbf{D}_{st} - \mathbf{D}_w). \tag{41.2}$$

The uniform field, which is here denoted by $\mathbf{D}_w$, gives a large change of the horizontal intensity at lower latitudes, and a small one at high latitudes. In the neighbourhood of the pole, including the auroral zone, the disturbance field can mainly be identified with the polar disturbance field $\mathbf{D}_p$. An idealized current-system for $\mathbf{D}_p$ will be almost the same as that of Fig. 49(A) in high latitudes, and as Fig. 49(C) in middle and low latitudes (see Sect. 39). However, the auroral zone current corresponding to $\mathbf{D}_p$ is much more intense in that part with westward current-flow than in that with eastward current-flow.

The world-wide component of the disturbance comes from the interaction of the solar wind (Sect. 4) with the distant geomagnetic field, and that of the "electric ring current", resulting from the drifting motion of trapped charged particles in the magnetosphere[1] (see Sect. 40). AKASOFU and CHAPMAN [19], [20] propose to use the following notations:

$\mathbf{D}_P$: polar disturbance field,

$\mathbf{D}_{CF}$: magnetic effect of the corpuscular flux from the sun,

$\mathbf{D}_R$: magnetic field produced by the ring current turning around the earth.

---

* This modern interpretation of the ring current producing $\mathbf{D}_{st}$ (see also Sects. 35, 41 and 69) was presented first by S. F. SINGER [Trans. Amer. Geophys. Union **38**, 175 (1957)]. For the motion of charged particles in a magnetic field, the perturbation method developed by H. ALFVÉN [36] is very useful.

[1] See contribution by W. N. HESS in Vol. 49/4 of this Encyclopedia.

Then, we may write[2]

$$D = D_P + D_{CF} + D_R. \tag{41.3}$$

Comparing this expression with (41.2), D_P is common to both, and

$$D_{CF} + D_R = D_w. \tag{41.4}$$

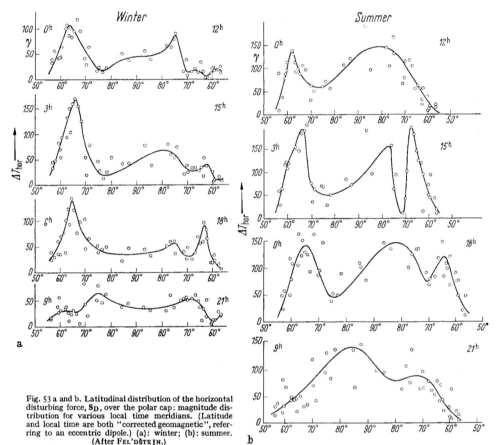

Fig. 53 a and b. Latitudinal distribution of the horizontal disturbing force, S_D, over the polar cap: magnitude distribution for various local time meridians. (Latitude and local time are both "corrected geomagnetic", referring to an eccentric dipole.) (a): winter; (b): summer. (After FEL'DŠTEJN.)

42. Local-time dependence of the position of the auroral zone. In the idealized current-system for S_D or D_S the auroral electrojet is drawn to flow along a circle of constant latitude (in geomagnetic or corrected geomagnetic coordinates). However, careful analysis of geomagnetic disturbances in high latitudes shows that the maximum disturbance does not always appear at the same latitudes; its location changes considerably with local time, even in the statistical average.

Fig. 53 shows the distribution of

$$\Delta T_{\text{hor}} = \{(\Delta \mathcal{H})^2 + (\Delta D)^2\}^{\frac{1}{2}} \tag{42.1}$$

for the disturbance-daily variation during IGY (1957/58) at various meridians in the northern high-latitude zone[1,2]. It is evident that the maximum disturbance

---

[2] For typographical convenience, Eq. (41.3) is usually printed as $D = DP + DCF + DR$.

[1] Ja. I. FEL'DŠTEJN, and A. N. ZAJTSEV: Geomagnetizm i Aeronomija [Translation: Geomagnetism and Aeronomy] **5**, 477 (1965).

[2] Ja. I. FEL'DŠTEJN: Planet. Space Sci. **14**, 121 (1966).

varies considerably with the hour (local time). Fig. 54 shows the local-time dependence of the latitude of maximum disturbance, and it can be concluded from the distribution of horizontal and vertical disturbing forces that the intense westward electric current is most concentrated along the thick lines **1, 2**, and the eastward current along the line **3**. The position of the auroral zone, i.e. the latitude of maximum magnetic disturbance, comes down to 60° ... 62° around midnight, while it is higher than 70° around noon. A similar conclusion was reported earlier by HARANG [21].

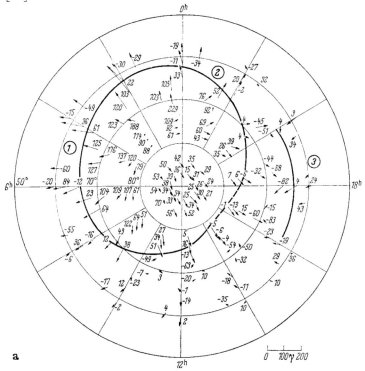

Fig. 54 a and b. Lines of maximum geomagnetic disturbance of the S_D-field in the northern hemisphere, (a) in winter and (b) (see page 80) in summer. Mean observational data are given in the diagram: the horizontal disturbing force is shown by arrows, while the vertical one by its numerical value (both in gamma). (After FEL'DŠTEJN.)

Such a marked local-time dependence of the auroral zone is not shown in the idealized S_D current-systems in Figs. 49 and 50, because the depression of the daily-mean value of H on disturbed days does not affect the S_D current-system, attributing the mean daily effect to D_{st}. Then the auroral zone will be centred at the latitude of maximum daily variation of H. Hence, the intense westward and eastward electrojets must be situated at the same latitude, according to the method of analysis used, unless higher-order terms are taken into account in the spherical harmonic analysis of the disturbance field. The word "auroral zone" means originally the locus of the places of most frequent auroral appearance over the whole globe. As the polar aurora is most frequently seen around midnight and then has maximum equatorward shift, the instantaneous world-wide auroral display is not along the so-called auroral zone (established by statistics), but along an oval belt, the point of which farthest from the geomagnetic pole is just around midnight on the auroral zone. The *auroral oval belt* expands with geomagnetic activity, as described later in Sect. 50.

It is seen in Fig. 54 that in summer the magnetic disturbance is quite large in the sunlit polar cap, its magnitude being nearly equal to that in the auroral zone (see Sect. 22 also). In winter, however, the agitation in the polar cap is small in comparison with the midnight activity in the auroral zone.

43. Spiral pattern of geomagnetic disturbance in high latitudes. In the preceding section, it has been described that the auroral zone is not a circle of constant latitude. The same conclusion has also been obtained by statistical analysis of:

(i) The time of maximum disturbance at various places of different latitudes, studied by NIKOLSKY [22], WHITHAM et al. [23].

(ii) Local-time dependence of the latitude of maximum disturbance along a certain longitude as explained by HARANG [21].

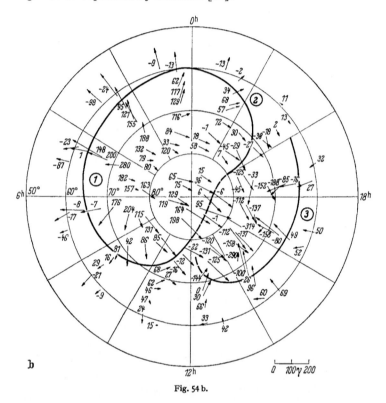

Fig. 54 b.

These results suggest that the geomagnetic activity in the north polar zone is distributed in the form of a spiral.

The spiral pattern of intense geomagnetic activity at high latitudes is now considered as consisting of three spiral branches: the morning branch (M), the afternoon branch (A), and the night branch (N), as shown in Fig. 55. Among these three, the M-spiral is most sharp and conspicuous, while the A-spiral is broad and not very distinct. It has been reported that the activities along the A- and M-spirals are well correlated amongst each other, while the correlation between N and M, or between N and A is much poorer [24], [25], [26]. These spiral curves are well coincident with the curves 1, 2, and 3 in Fig. 54.

Spiral patterns in high latitudes have been found not only for geomagnetic disturbance, but also for auroral activity including hydrogen emission, and ionospheric disturbances such as blackout and appearance of the auroral Es layer[1]. Fig. 56 shows geomagnetic, auroral and ionospheric spiral patterns as reported by different workers. The geomagnetic Morning- and

---

[1] K. RAWER, and K. SUCHY: this Encyclopedia, Vol. 49/2, Sect. 44, Figs. 211 and 212.

Sect. 43.   Spiral pattern of geomagnetic disturbance in high latitudes.   81

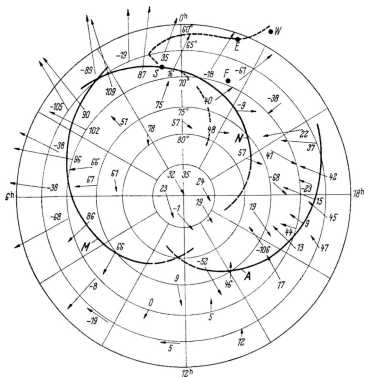

Fig. 55. Loci of morning (M), night (N), and afternoon (A) maxima of ΔH and of geomagnetic agitation in geomagnetic time versus "auroral" latitude. The disturbance vector is given as an arrow for the horizontal component, and as a numerical value for the vertical component (like in Fig. 54). The dotted line represents the locus of night agitation as found by MAYAUD. Locus M is identical with NIKOLSKY's morning spiral of agitation, aurora and ionospheric absorption. (Geomagnetic time as coordinate is approximately equivalent with geomagnetic longitude. "Auroral" latitude is an adjusted geomagnetic latitude: the auroral maximum zone is arbitrarily taken as 70° N and latitudes are measured north and south from this line.)
(After HOPE.)

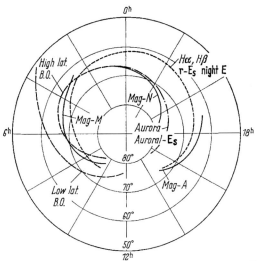

Fig. 56. Loci of maximum geomagnetic agitation (Mag-M for morning, Mag-A for afternoon, Mag-N for night), maximum appearance of aurora, maximum intensity of hydrogen emission (Hα, Hβ), maximum occurrence of ionospheric auroral E_s, type-r E_s, night E, and blackout (B.O.). The polar diagram has geomagnetic latitude for radius and local time for azimuth.
(After OGUTI.)

Handbuch der Physik, Bd. XLIX/3.   6

Afternoon-spirals approximately coincide with the spiral patterns of hydrogen emission, auroral diffuse echoes, 'night-E' layer and the 'r'-type Es ionization ['r' signifies a retardation in the ionogram (h' vs. f curve) indicating that the layer is not so thin as a low-latitude Es]. Auroral H_α and H_β emission is probably caused by proton precipitation. The corresponding spiral seems to form an oval shape surrounding the geomagnetic pole, but shifted towards midnight. On the other hand, the geomagnetic Night-spiral roughly coincides with those of visible aurora, discrete radar echoes, and 'f'-type (meaning flat) and 'a'-type (auroral) Es ionization. It is now believed that the Night-spiral is caused by electron precipitation.

The pattern of ionospheric blackout has been reported to consist of outer and inner branches, as shown in Fig. 56. The inner branch seems to be an extension of the Night-spiral on the evening side and is connected to the lower-latitude side of the Morning-spiral, whereas the outer branch shows little correlation with geomagnetic and ionospheric disturbances and auroral displays. The low-latitude blackout pattern alone appears as a real spiral, which may be caused by direct impinging of high-energy protons from outside the geomagnetic cavity. The spirals for geomagnetic, ionospheric and auroral activities may be a part of two oval-shaped patterns, shifted slightly against the pole. Such oval patterns could be considered as to be caused by impinging of protons and electrons respectively, with orbits deformed by the solar wind pressure [25]. The position of the spiral pattern around the noon meridian is reported to be connected by geomagnetic field-lines with the neutral points on the geomagnetic cavity surface (see Sect. 4, in particular Fig. 5).

It must be stated that the spiral patterns have all been found from statistical studies of phenomena at high latitudes. In individual cases, the instantaneous disturbance has not always its latitudinal maximum on one of the spiral curves.

44. Comparison of magnetic disturbance in the northern and southern hemispheres. It has been shown that geomagnetic disturbances take place simultaneously in the northern and southern hemispheres of the earth[1], and the development is quite similar in both hemispheres[2-5]. Therefore, the typical current-system for a geomagnetic disturbance is almost symmetric with respect to the geomagnetic equator. An example of the D_S current-system in the northern and southern hemispheres for the equinoctial season is shown in Fig. 57; one sees that the current pattern for the southern hemisphere is almost a mirror image of that in the northern hemisphere[6]. On the other hand, we see in Fig. 54a (Sect. 42) considerable seasonal variation of the magnitude of S_D.

It is important to study to which extent magnetic disturbances in the northern and southern hemispheres are conjugated. For large disturbances the problem will be discussed later in Sect. 55. Small disturbances, which may be called "geomagnetic agitation" (or "noise") are definitely stronger in the summer hemisphere[7], and particularly pronounced in daytime in the polar region of the earth. When local measures of geomagnetic activity such as the K-indices are compared in the northern and southern hemispheres, a seasonal dependence of magnetic activity appears clearly; quiet intervals are more frequent in winter, and intervals of small and moderate activity are more frequent in summer. The conjugacy of magnetic disturbance in the northern and southern polar regions becomes poorer with increasing latitude, particularly in daytime[8-10].

A slight asymmetry of the disturbance field in the northern and southern hemispheres is also seen in the magnitude of D_m shown in Fig. 42. D_{st}(SC), a uniform magnetic field near the

---

[1] S. J. AHMED, and W. E. SCOTT: J. Geophys. Res. **60**, 147 (1955).
[2] E. M. WESCOTT: J. Geophys. Res. **66**, 1789 (1961); **67**, 1353 (1962).
[3] G. M. BOYD: J. Geophys. Res. **68**, 1011 (1963).
[4] E. M. WESCOTT, and K. B. MATHER: J. Geophys. Res. **70**, 29, 43, 49 (1965).
[5] T. NAGATA, S. KOKUBUN, and T. IIJIMA: Japanese Antarctic Res. Exp. Sci. Rep., Ser. A, No. 3 (1966).
[6] T. NAGATA, and T. IIJIMA: J. Geomag. Geoelectr. **16**, 210 (1964).
[7] J. M. STAGG: Terr. Mag. **40**, 255 (1935); — Proc. Roy. Soc. London A **149**, 298 (1935).
[8] E. M. WESCOTT, and K. B. MATHER: Planet. Space Sci. **13**, 303 (1965).
[9] L. A. JUDOVIČ: Geomagnetizm i Aeronomija **3**, 723 (1963).
[10] M. S. BOBROV: Geomagnetizm i Aeronomija **3**, 537 (1963).

earth at the time of sudden commencement of a magnetic storm, shows a slight seasonal variation of its axis as already described in Sect. 30. This may be explained by the seasonal change of the angle of incidence of the solar wind with respect to the geomagnetic dipole axis (see Sect. 40).

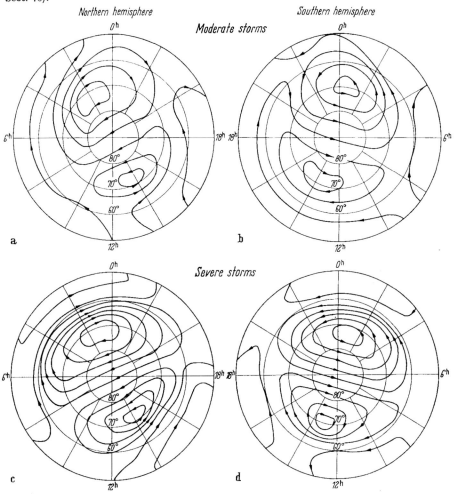

Fig. 57 a—d. Equivalent ionospheric current pattern of the average D_S-field for moderate magnetic storms ($K_p=4.3$; upper) and for severe storms ($K_p=7.2$; lower) during the IGY in the northern (left) and southern (right) polar regions. Between adjacent current lines flow 10^5 A $=100$ kA. (After Iijima.)

G. Development and decay of magnetic storms.

45. Change in amplitude and phase of D_S during the development of magnetic storms. It has been shown in the preceding chapter that development and decay of a disturbance are easily seen in the D_{st}-field, namely the world-wide increase and subsequent depression in the horizontal intensity H of the geomagnetic field, in middle and low latitudes. For that part of a disturbance which depends on longitude, D_S, it was only shown that the average of D_S for the first, second and third days of storms (called S_D^1, S_D^2 and S_D^3) decreases with the order number of the day. In this chapter, it will be illustrated for the average of a number of magnetic storms how the D_S-field changes (in magnitude and phase) with storm time.

Fig. 58. Harmonic dials for the first harmonic component of D_S in $\mathcal{H}, \mathcal{E}, Z$, for weak, moderate and severe storms: mean geomagnetic latitude 30°. The $D_S(\mathcal{E})$ dial points are shown relative to dial axes turned clockwise by 90°, relative to those for $D_S(\mathcal{H})$. The $D_S(Z)$ dial points are shown relative to dial axes rotated by 180°. Arrowed arcs show the rotations needed to bring the dial axes for $D_S(\mathcal{E})$ and $D_S(Z)$ into the usual directions. The dial points 1 ... 8 refer to the first 6-hour intervals, and the dial points 9 ... 11 to the following three 8-hour intervals. (After SUGIURA and CHAPMAN.)

In the harmonic analysis of D_S and therefore also of S_D, the first harmonic component is predominant at all latitudes, in all three elements, $\mathcal{H}, \mathcal{E}$ (or D) and Z, at all storm-times and for all classes of magnetic storms. The phase of this first harmonic component changes as storm-time proceeds. The change of phase implies an earlier time of maximum with progressing storm-time. Each of the functions $D_S(\mathcal{H})$, $D_S(\mathcal{E})$, and $D_S(Z)$ is assumed to have the form

$$D_S = a_n(t_{st}) \cos \lambda_\odot + b_n(t_{st}) \sin \lambda_\odot \tag{45.1}$$

where $\lambda_\odot$ is the longitude relative to the sun (measured eastward from the midnight meridian), thus equivalent to local time. The coefficients a_n and b_n depend on storm-time, t_{st}. In Fig. 58 the change with storm-time of a_n and b_n (for three geomagnetic components: $\mathcal{H}, \mathcal{E}$ and Z) is illustrated for moderate latitudes; amplitude and phase of the $\mathbf{D}_S$ field show a regular variation [16]. The phase changes are essentially similar in all three elements, for all three intensity classes. The change is greatest during the first six hours storm-time. The phase-shift in $\mathbf{D}_S$ amounts to about 6 hours between the early and the later stage of magnetic storms.

46. Mutual relation between $\mathbf{D}_S$- and $\mathbf{D}_{st}$-fields. Fig. 59 illustrates the different rates of evolution of $\mathbf{D}_{st}$ and of the range ($2c_1$) of the first harmonic component of $\mathbf{D}_S$ at various latitudes during the first three days storm-time [16]. For compari-

Fig. 59. Illustration of the different rates of evolution of $\mathbf{D}_{st}$ and of the range ($2c_1$) of the first harmonic component of $\mathbf{D}_S$ during the first three days of weak, moderate and severe storms. $\mathbf{D}_{st}$ graphs are shown by full lines, and $\mathbf{D}_S$ by broken lines (ordinates). For ease of comparison the latter ranges are plotted with negative sign. (In all graphs the height of the white rectangles corresponds to 50 γ and the width to 12 h.) (After SUGIURA and CHAPMAN.)

son, the ranges $2c_1$ are plotted with negative sign, because the main storm-time varriation $\mathbf{D}_{st}(\mathcal{H})$ is nearly always negative. The change in phase of $\mathbf{D}_S$ with storm-time is not shown in this figure.

At 80° geomagnetic latitude, i.e. well inside the auroral zone, and at 65°, in the auroral zone, $\mathbf{D}_S$ largely exceeds $\mathbf{D}_{st}$. At 65° latitude, $\mathbf{D}_S(\mathcal{H})$ exceeds $\mathbf{D}_S(\mathcal{E})$, because of the proximity of the auroral electrojet. At 58°, though $\mathbf{D}_{st}$ is still small compared with $\mathbf{D}_S$, it begins to show the characteristics that mark it throughout middle and low latitudes. At still lower latitudes $\mathbf{D}_{st}(\mathcal{H})$ and $\mathbf{D}_S(\mathcal{H})$ increase and reach a maximum at the equator, while there is a decrease in $\mathcal{E}$ and Z, with reversal of sign at the equator crossing.

The amplitude of $\mathbf{D}_S$ varies with storm-time very differently from $\mathbf{D}_{st}$. The main difference is that $\mathbf{D}_S$ reaches its maximum earlier, and decays more rapidly at high latitudes. Fig. 60 shows the characteristics for temperate latitudes (corresponding to 21° ... 40° in Fig. 59). The maximum range ($2c_1$) of $\mathbf{D}_S(\mathcal{H})$ occurs at nearly the same time when the maximum decrease in $\mathbf{D}_{st}(\mathcal{H})$ occurs.

The time in which $\mathbf{D}_S(\mathcal{H})$ decays to half its maximum amplitude is about 30 hours, 27 hours and 12 hours, for weak, moderate and severe storms, respectively. The corresponding time for $\mathbf{D}_{st}(\mathcal{H})$ for weak and moderate storms is longer than 3 days. By the middle of the third day (storm time $t_{st} = 60$ h) the recovery of

$D_{st}(\mathcal{H})$ is only a small percentage for weak storms, while about 25% and 50% for moderate and severe storms. These remarks apply only to temperate latitudes. At high latitudes, the small magnitude of D_{st} and the greater irregularity of D_S make such estimate rather difficult and doubtful.

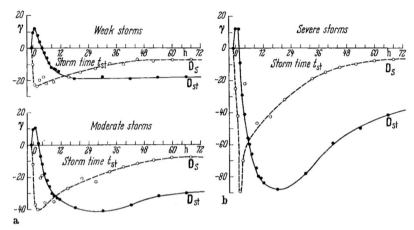

Fig. 60a and b. A comparison of the rates of evolution of $D_{st}(\mathcal{H})$ and the range $(2c_1)$ of the first harmonic component of D_S ($\mathcal{H}$), during the first three days of weak, moderate and severe storms at mean geomagnetic latitude 30°. D_{st} curves are drawn as full lines, those for D_S as broken lines. (After SUGIURA and CHAPMAN.)

47. S_q^p-enhancement and development of the auroral electrojet. In the preceding section, it has been shown that the polar disturbance field generally develops earlier than the world-wide D_{st}-field which is attributed to the ring current in the magnetosphere. Looking at Fig. 59 carefully, we notice that the D_S-fields at various latitudes do not develop synchronously with each other. This means that D_S changes its form while it is developing. The polar disturbance field is composed of the intense auroral electrojet, and the electric current-system within the polar cap. On disturbed days the latter may be called S_q^p-enhancement (see Sect. 4). Fig. 61 shows an example of the changes in the S_q^p-type field, and in the auroral electrojet, at the time of a magnetic storm. We see that in the developing stage of magnetic storms, the enhancement of the S_q^p-type field comes earlier than the development of the auroral electrojet. At middle and low latitudes, the D_S-field is mainly attributable to the return current of the auroral electrojet. In high latitudes, from the polar-cap region down to the auroral zone, the D_S-field consists of the S_q^p-type field and the auroral electrojet. Therefore, the D_S-field in high latitudes develops earlier than that in low latitudes, because the S_q^p-type variation is enhanced before the auroral electrojet develops.

48. Decay of the disturbance field in middle and low latitudes. As explained in the preceding sections, the D_S-field does not develop perfectly parallel with the D_{st}-field, the former developing earlier in the early stage of a magnetic storm. It is also seen that the decay of the D_{st}-field is very slow compared with that of the D_S-field, especially for weak and moderate magnetic storms.

In practice, irregular geomagnetic variations (for example bays) are very often superposed during the recovery stage of magnetic storms, making it rather difficult to examine the idealized decay characteristics of magnetic disturbance over the whole world. Fig. 62 shows mean decay characteristics obtained from 16 magnetic storms during IGY, for which the K_p-indices showed an almost monotonic decrease for more than 24 hours. The equatorial D_{st} field recovers with a rate of approxima-

Decay of the disturbance field in middle and low latitudes.

Fig. 61. Comparison of the intensities of parallel current over the polar cap S_q^p, auroral electrojet, and equatorial D_{st}-field at the time of the magnetic storm which began with a sudden commencement at 08 h 40 min on 3 Sept. 1958. (After IIJIMA.)

Fig. 62. Time-variation of a_p-index (upper diagram) and the equatorial D_{st} (lower diagram) at the decaying stage of magnetic storms (16 storms with regular decay during 1957/58). The decaying rate is 10%/h for the polar disturbance field (represented by a_p), and 3.3%/h for the D_{st}-field.

tely 3.3%/hour, while the disturbance field in moderate latitudes (mainly D_S) recovers quicker, with 10%/hour on the average.

49. Decay of the disturbance field in high latitudes. Records at high-latitude stations do not show a regular decay of the geomagnetic disturbance, even when the disturbance in middle and low latitudes (represented by the K_p-index or by the intensity of the equatorial ring-current) shows a monotonic decrease of activity. Fig. 63 shows the magnitude of the disturbance field at some high-latitude stations

Fig. 63. Examples of the disturbance force versus time relation at high-latitude stations in the decaying stage of magnetic storms (16 storms with regular decay during 1957/58). Thin lines show the variation in the night hours, 18 ... 06 h local mean time, while the thick lines refer to 06 ... 18 h local mean time. Stations are ordered after geomagnetic latitude (given on the left side, together with the stations' abbreviation: THule, REsolute bay, GOdhavn, MUrchison bay, BAker lake, TIhaja bay, DIkson isl., cape WEllen).

for the average of the same 16 storms selected for Fig. 62. For stations at geomagnetic latitudes higher than 75°, the activity seems to increase in general during the day hours, even in the decaying stage of the storm. On the other hand, stations in the auroral zone seem to show an increase of activity in the nighttime.

These observations may be understood, if we take into consideration the following two factors: (i) the presence of a persistent disturbance in the daytime polar-cap region (see Sect. 22), (ii) the local-time dependence of the auroral zone, the latitude of maximum geomagnetic activity of which depends on local time (as explained in Sect. 42). When a polar-cap observatory comes into the daytime region of permanent agitation, the magnetic activity at that station increases, even in the decaying stage of magnetic storms. The geomagnetic activity at an auroral

50. Shifting of the auroral zone with magnetic activity. As a magnetic storm develops, the D_S-field changes not only its intensity and phase, but also its shape.

Fig. 64 a and b. Dependence of the magnitude of geomagnetic disturbance on the corrected geomagnetic latitude along the noon—midnight meridian for cases with solar inclination angles 0°, 11.5° and 23° with respect to the geomagnetic equator. Curves are drawn for $K_p=0, 2, 4, 6$ (parameter). (a, left side) for the Second Polar Year 1932/33. (b, right side) for the International Geophysical Year 1957/58.

An interesting and important deformation of the D_S-field is the shifting of the auroral zone towards the equator during the development of a magnetic storm. Such a tendency has been noticed in early studies of aurorae and magnetic storms. There are many historic descriptions that aurorae could be seen at lower latitudes at the time of severest magnetic storms. On 11 February 1958, an aurora of red colour was photographed even in the central part of Japan, where the geomagnetic latitude is as low as 25°.

The latitudinal dependence of the magnitude of magnetic disturbance along the noon-midnight meridian for different world-wide magnetic activity K_p is shown[1] in Fig. 64. The centre of the midnight auroral zone (the latitude of maximum geomagnetic disturbance) is at the same latitude throughout the year. On the other hand, the position of the daytime auroral zone is asymmetric with respect to the geomagnetic equator, the summer auroral zone being located at a higher latitude than the winter zone. During moderate magnetic activity, the latitude of the daytime auroral zone is higher around noon than at midnight, even in the winter season. The position of the auroral zone shows a considerable equatorward shift with increasing magnetic activity, and this tendency is more pronounced on the day-side than around midnight.

A pronounced equatorward shift of the auroral zone is also seen from individual magnetic storms[2,3]. Fig. 65 shows a mutual comparison of the equatorial shift of the auroral zone in the European region, the world-wide equatorial D_{st}-field, and the intensity of polar-cap currents. The centre of the auroral zone is defined as the point where the horizontal disturbing force is maximum and the vertical disturbing force vanishes. It appears clearly from the figure that the polar distance of the centre of the auroral zone increases in close relation with the development of the storm (see in particular the D_{st}-field). However, in this example[4], the local time dependence of the position of the auroral zone (illustrated in Figs. 54 and 64) must be taken into account, in order to get a precise quantitative relation for the shifting of the auroral zone with magnetic activity. It must also be examined in detail, whether or not the poleward shift of the auroral zone in the decaying stage of magnetic storms is just the reverse process of the equatorward shift during the developing stage of magnetic storms.

Fig. 65. Variations of the intensity of S_D- and D_{st}-fields, and in the position of the auroral zone during the magnetic storm, 1/2 May 1933. I. Magnitude of the field corresponding to the polar-cap (parallel) current at Thule. II. Average equatorial depression ($-\Delta X_m$) at 6 stations, representing the magnitude of the D_{st}-field. III. Polar distance of the centre of the auroral electrojet. IV. ΔX_m under the auroral zone.

H. Instantaneous and average disturbance field.

In the preceding chapters, the characteristics of magnetic storms have been described for the average of a number of magnetic storms. The irregular geomagnetic variations superposed on the general trend were neglected, although their magnitude is not always small but is sometimes even larger compared with the magnitude of the average variations. In this chapter, it is shown that instantaneous polar disturbance fields are often quite different from the field described by the S_D current-system; only their average is still given by the D_S- or S_D-field.

51. Examples of instantaneous disturbance fields. Most examples of hourly means of the disturbance field can be described by a combination of D_{st} and D_S

---

[1] N. Fukushima: Rep. Ionos. Space Res. Japan **19**, 367 (1965).
[2] S.-I. Akasofu, and S. Chapman: J. Atmosph. Terr. Phys. **25**, 9 (1963).
[3] S.-I. Akasofu: J. Atmosph. Terr. Phys. **26**, 1167 (1964).
[4] T. Nagata: J. Geophys. Res. **55**, 127 (1950).

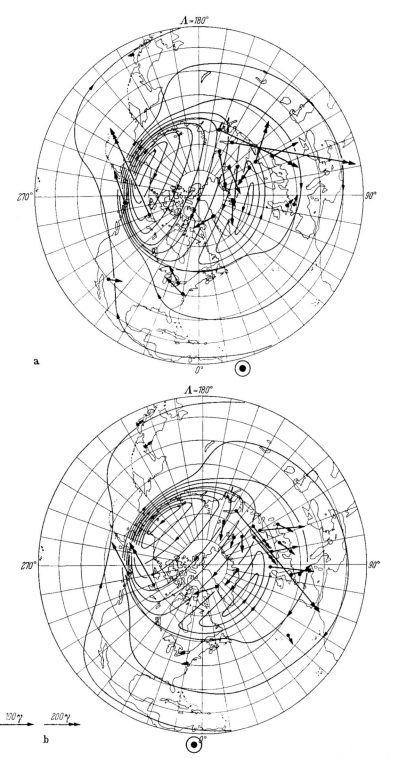

Fig. 66a–d. Geomagnetic disturbance field represented by combination of current-system (closed curves) and current-arrows. Current arrows show the instantaneous disturbance field at the times indicated; each current-system is valid for the hourly-mean of the disturbance field. 1 May 1933; (a) 15 h 43 min; (b) 16 h 50 min; (c) (see next page) 20 h 15 min; (d) (see next page) 22 h 15 min UT. ⊙ Position of the sun.

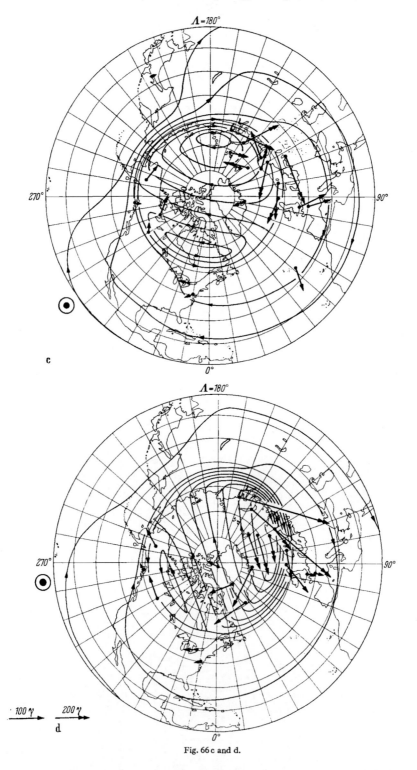

Fig. 66 c and d.

Examples of instantaneous disturbance fields.

Fig. 67a and b. **Examples** of the equivalent overhead current-system of the instantaneous disturbance field in the early stage of the magnetic storm of 30 Apr. 1933. (a) 20 h 52 min; (b) 21 h 45 min UT. ⊙ Position of the sun.

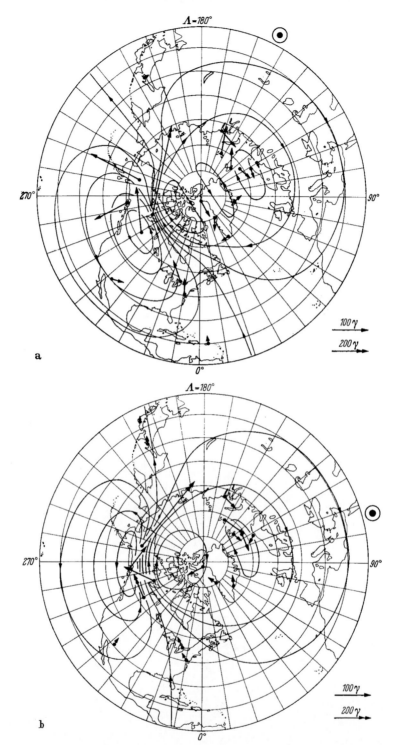

Fig. 68a—d. Examples of the equivalent current-system of the instantaneous disturbance field in the later stage of magnetic storms (D_{st}-field is eliminated). (a), (b) 2 May, (c), (d) 29 May 1933. (a) 06 h 30 min; (b) 09 h 40 min; (c) 18 h 10 min; (d) 21 h 12 min UT. ☉ Position of the sun.

Sect. 51. Examples of instantaneous disturbance fields. 95

Fig. 68 c and d.

current-systems; in Fig. 66 stream lines of overhead electric current have been derived from hourly mean values at stations all over the world [27]. The hourly mean disturbance field is, of course, the average of the varying disturbance field during a certain hour. Low-sensitivity magnetograms show, however, that fluctuations of fairly large amplitude occur intermittently or successively during storms, the duration of each individual disturbance being from several minutes to a few hours. The magnetograms obtained at high-latitude stations also show this ten-

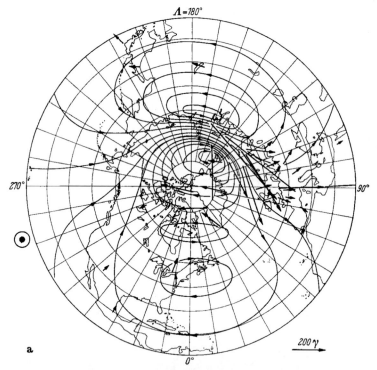

Fig. 69a and b. Examples of the (hourly mean) equivalent current-system of polar magnetic storms. An electric current of 10^5 A = 100 kA flows between adjacent stream lines. (a) 30 Apr. 1933, 21 ... 22 h UT; (b) 19 Feb. 1933, 23 ... 24 h UT. ⊙ Position of the sun.

dency, where S_q is negligibly small compared with the disturbance field. Such individual polar disturbance is not always recorded simultaneously at all stations in high latitudes, but is observed sometimes only in a limited region. The current arrows in Fig. 66 refer to such instantaneous disturbance fields. One sees that these arrows are not always parallel to the stream lines corresponding to the hourly mean disturbance field[1] [28]. This means that the instantaneous disturbance field is sometimes very different from the hourly-mean disturbance field. Fig. 67 shows some typical examples of the instantaneous disturbance field during the main phase of a magnetic storm[2] [28]. These current-systems are very different from the average current-system for magnetic storms as given by Fig. 49 (see Sect. 39). It is also often observed that rather important activity appears during a short time when no other average disturbance field than that of the post-pertur-

---

[1] N. FUKUSHIMA: Rep. Ionos. Res. Japan 6, 185 (1952).
[2] T. NAGATA, and N. FUKUSHIMA: Rep. Ionos. Res. Japan 6, 85 (1952); — Indian J. Meteor. Geophys. 5, Special No., 75 (1954).

bation remains. Fig. 68 gives some examples of the instantaneous aspect of such a polar disturbance.

In Figs. 66 and 67 the westward auroral zone current alone is intense, and the eastward auroral electrojet is almost lacking. In Fig. 68 even the westward auroral electrojet does not extend along the auroral zone circle, but an intense current is concentrated only in a small range of longitude along the zone, and the current seems to be closed over the whole remaining part of the world. In other words,

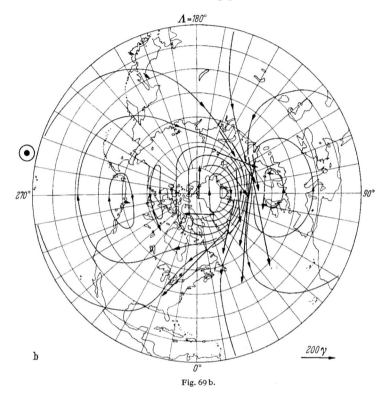

Fig. 69 b.

such a disturbance field can be considered as to be a partial appearance of a D_p current-system; it can hardly be detected in hourly mean maps, rare examples of such cases are given in Fig. 69. The time-resolving power of the "hourly mean values" is not good enough to pick out such simple individual patterns, so that the hourly mean aspect is the mean of varying aspects during an hour.

The illustration and explanation of the instantaneous disturbance field in this section is made with the aid of examples only from the Second Polar Year, 1932—33. Since IGY, 1957—58, and IGC, 1959, we have now much more useful data than those of the Second Polar Year for this kind of analysis, and a number of new papers on the instantaneous disturbance field have been published. According to these new papers, the conclusions reported in this section remain almost unchanged[3].

In most cases of instantaneous individual disturbances, the geomagnetic variation in the polar-cap regions, well inside the auroral zone and near the geomagnetic poles, can be described by assuming a parallel overhead electric current, which flows towards the meridian corresponding to 8 ... 11 h local geomagnetic

---

[3] See for example: D. G. KNAPP: J. Geophys. Res. **66**, 2053 (1961). — S. I. AKASOFU, and S. CHAPMAN: IUGG Monograph No. 7, 93 (1960). See also Refs. [30], [35].

time. The direction of the polar parallel current is sometimes outside this range[4-6], namely when an extremely intense disturbance takes place near the auroral zone.

52. Magnetic field during calm periods on disturbed days in high latitudes. On disturbed days, disturbance fields of various durations and magnitudes are superposed upon the daily variation of the geomagnetic field, intermittently or sometimes successively. In fact, the magnetic field is not always agitated throughout a

Fig. 70. Diurnal variation of $\mathcal{H}$ at TIhaja bay (71.5° geomagnetic latitude), DIkson (63.0°) and cape WEllen (61.8°) for quiet hours of disturbed days (dotted curve), for quiet days (broken curve), and for disturbed days (full curve). (After NIKOLSKY.)

disturbed day, but there appear relatively calm periods even on disturbed days. This character is particularly notable on low-sensitivity records (storm magnetograms) taken at high-latitude stations. Picking up the hourly values in these calm periods (for disturbed days alone) at a high-latitude station, a composite mean daily variation can be obtained. NIKOLSKY found that the composite daily variation does not differ much from the S_q variation at that station [22]. Fig. 70 shows the diurnal variations of the horizontal intensity $\mathcal{H}$ on international quiet days, on disturbed days, and that of calm hours on disturbed days at Tihaja Bay (geomagnetic latitude 71.5°), Dikson Island (63.0°), and Wellen (61.8°). We see in this figure that the dotted curve (calm periods on disturbed days) is almost exactly the same as the broken one (S_q on quiet days). This means that the intensity of the polar disturbance field D_p is extremely small during such calm periods on disturbed days. Polar disturbances usually take place intermittently, or sometimes

---

[4] D. H. FAIRFIELD: J. Geophys. Res. **68**, 3589 (1963).
[5] M. NAGAI: Mem. Kakioka Mag. Obs. **11**, No. 2, 39 (1964).
[6] M. NAGAI, and Y. HAKURA: Mem. Kakioka Mag. Obs. **12**, 15 (1965); — Rep. Ionos. Space Res. Japan **20**, 69 (1966).

successively, on magnetically disturbed days but not throughout a day. This tendency is better seen on high-latitude records.

53. Polar elementary storms or D_P substorms. K. BIRKELAND [29] first examined a number of instantaneous disturbance fields with his data of the Norwegian arctic expedition 1902—1903. He found a systematic pattern of the polar disturbance field, which he named *polar elementary storms*. They were classified into positive and negative perturbations, according to whether in the auroral zone the disturbing force increases or decreases the horizontal intensity of the field.

a) An idealized overhead electric current-system for a negative polar elementary storm is given by Fig. 71. It is simply the distribution of electric current lines on an electrically conductive spherical shell, when an electric dipole is placed with its axis parallel to a latitude circle corresponding to the auroral zone of the earth.

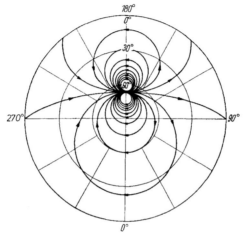

Fig. 71. An idealized electric current-system for a negative polar elementary storm. (After BIRKELAND.)

BIRKELAND reported that though, as a rule, the positive and negative polar elementary storms take place simultaneously but in different areas, they rarely occur alone. Their occurrence was found to depend on the relative position of the sun, i.e. on local time. Negative polar storms are more frequent on the morning and night sides of the globe, and positive ones on the afternoon side (noon to evening). In the case of negative polar elementary storms, the auroral electrojet, as a rule, covers a wider longitude range than in the case of positive ones. In recent years with more comprehensive data of geomagnetic observation, the description given by BIRKELAND has been proved to be really true. Negative polar elementary storms take place very often during night hours, and it is not too much to say that the occurrence of small negative elementary storms is almost a daily event in the auroral zone. During magnetic storms, negative elementary storms of large amplitude (sometimes more than 10 times the S_D variation) take place successively or intermittently from the evening to the morning hours. On the other hand, the occurrence of positive polar elementary storms is much less frequent in the auroral zone; isolated positive storms rarely seem to appear during disturbed periods.

Magnetic storms with large D_{st}-field (as revealed by a large equatorial depression of the horizontal intensity) seem to be accompanied more frequently by polar elementary storms, and a quantitative relation between the magnitude of the D_{st}-field and the occurrence of polar elementary storms[1] is searched for.

---

[1] S.-I. AKASOFU, and S. CHAPMAN: J. Geophys. Res. **68**, 125, 3155 (1963).

From the average characteristics of magnetic disturbances, as shown in Figs. 59 and 60 (Sect. 46), it appears that the $\mathbf{D_s}$-field becomes most intense before the maximum development of the $\mathbf{D_{st}}$-field is reached. This is understandable, provided that polar elementary storms take place most frequently when the equatorial ring current is growing. Fig. 45 (Sect. 37) and Fig. 61 (Sect. 47) show such a tendency in some individual examples of developing magnetic storms.

β) In the history of geomagnetism, the polar elementary storm was once thought to be a mere fluctuation of S_D or D_S fields, while the main interest was to know the average characteristics of the disturbance field. In other words, S_D or D_S was considered to be a really existing disturbance field with occasional fluctuations in intensity and form. However, in recent years, it was proved that this is not so, and the polar elementary storms are now considered to be a

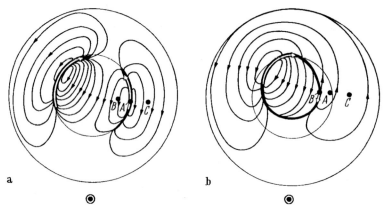

Fig. 72a and b. Schematic illustrations of model current-systems for a polar magnetic substorm viewed from above the (dip-)north pole. ⊙ Direction to the sun. (a) Ordinary case with westward and eastward auroral electrojets; (b) Extreme case without eastward auroral electrojet. (After AKASOFU, CHAPMAN and MENG.)

fundamental disturbance, while now the S_D field is felt to be only the space- and time-average of a number of polar elementary storms all over the world during magnetically disturbed periods [28]. In recent papers, the world "D_P substorm" is often used[2]. This is identical with "polar elementary storm".

γ) Fig. 72 shows the model current-systems for such a polar substorm [30]. Note that in the model current-system (b), the polar electrojet is drawn to be westward at all longitudes, and flows along a closed oval curve, which is not of constant geomagnetic latitude but is shifted towards the midnight side. On the other hand, in picture (a) a weak eastward electrojet was drawn at the same geomagnetic latitude where the intense westward current flows in model (b). Model (b) is an extreme case without an eastward electrojet. In such a case, the positive $\mathcal{H}$-variation observed near the point (A) might be better attributed to the overhead eastward current, which is the return current of an intense westward electrojet flowing at a higher latitude (B), rather than to attribute it to the eastward auroral electrojet.

It must be remembered here that the equivalent overhead current-system of a geomagnetic disturbance could not be accurately determined from ground observation in the present network. If we had some well-distributed chains of magnetic observatories along some specific meridians in high latitudes, the current-system could be determined more precisely. Even current-system (b) shown here may be subject to future correction. It may be reasonable to connect the auroral appearance with the polar electrojet. The instantaneous auroral zone is not a circle of constant geomagnetic latitude, but has an oval form, the centre of which is shifted from the geomagnetic pole towards the midnight side.

---

[2] S.-I. AKASOFU, and S. CHAPMAN: J. Geophys. Res. **66**, 1321 (1961). The word is printed as '*DP* substorm'.

54. The average current-system of polar elementary storms.

In the preceding discussion, it seemed natural to express the simple pattern of a geomagnetic disturbance field by a current-system as given in Fig. 71. On the other hand, the average polar disturbance field should be expressed by the current-system of D_p which is defined as

$$D_p = D_S + (D_{st} - D_w) \equiv D - D_w. \tag{54.1}$$

According to the appearance tendency of polar elementary storms reported by BIRKELAND and recent workers, the direction of the corresponding electric dipoles along the auroral zone is always subjected to the relative position of the sun, as described in the preceding section. If one calculates the average current-

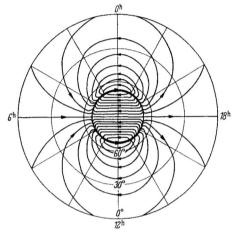

Fig. 73. Resultant electric current-system obtained by superposition of a large number of elementary current-systems (of dipole type each).

system of a large number of polar elementary storms originating along the auroral zone (the intensity of which varies with geomagnetic longitude λ proportionally to cos λ), the resultant current-system becomes very similar to the S_D current-system. This is seen from Fig. 73.

If we take into account the more frequent occurrence of negative polar elementary storms, the resultant average current-system gets the form of the D_P current-system, in which the westward auroral electrojet is much more intense than the eastward one. The current-systems in Fig. 72 are also obtainable by appropriate summation of the elementary storms. In calculating the current-systems of Fig. 73 as well as Fig. 71, the electric dipoles corresponding to the polar elementary storms are distributed only along the northern auroral zone. If they are distributed also along the southern auroral zone to have symmetry with respect to the equator, the resultant electric current-system becomes symmetric with respect to the equator. In this case, the distribution of current flow in the northern high latitudes is not much affected by the presence of electric dipoles along the southern auroral zone.

55. Space- and time-correlation of the magnetic disturbance at different places over the world.

In the preceding sections, it has been explained that the current-system corresponding to an instantaneous disturbance field differs greatly, in general, from that of the average D_S-field. The local appearance of high-latitude geomagnetic disturbance is shown in this section.

Fig. 74 gives the *auto-correlation* values $r(t)$ of geomagnetic Q-indices in a disturbed interval at a few high-latitude stations[1,2] (where the Q-index roughly

---
[1] N. FUKUSHIMA: J. Phys. Soc. Japan 17, Suppl. A-I, 70 (1962).
[2] T. NAGATA, S. KOKUBUN, and N. FUKUSHIMA: J. Phys. Soc. Japan 17, Suppl. A-I, 35 (1962).

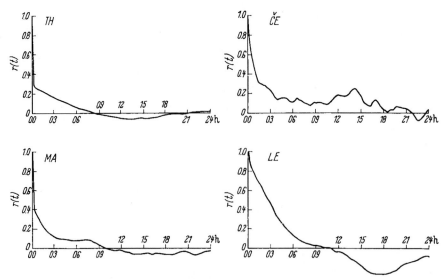

Fig. 74. Auto-correlation function $r(t)$ of the geomagnetic Q-indices for the period 2 ... 6 Sept. 1957 ($A_p=$102, 135, 145, 112, 36). Stations: THule (77.5° N, 69.2° W; $\phi_m=$89.0°, $\Lambda_m=$357.9°), cape ČEljuskin (77.7° N, 104.3° E; $\phi_m=$66.2°, $\Lambda_m=$176.4°), MAwson (67.6° S, 62.9° E; $\phi_m=$−73.1°, $\Lambda_m=$103.0°), LErwick (60.1° N, 1.2° W; $\phi_m=$62.5°, $\Lambda_m=$88.6°). (ϕ_m geomagnetic latitude, Λ_m geomagnetic longitude).

Fig. 75 a—d. Correlation function $r(X, t; Y, (t+\Delta t))$ computed with the geomagnetic Q-indices at two distant stations, X, Y, for the period 2...6 Sept. 1957. ($A_p=$102, 135, 145, 112, 36.) Stations (and their combinations) are (ϕ_m g.m. latitude, Λ_m g.m. longitude);

{ WEllen (66.2° N, 169.8° W; $\phi_m=$61.8°, $\Lambda_m=$237.0°),
  MAcquarie isl. (54.5° S, 158.9° E; $\phi_m=$−61.1°, $\Lambda_m=$243.1°),
  LErwick (60.1° N, 1.2° W; $\phi_m=$62.5°, $\Lambda_m=$88.6°),
{ HAlley bay (75.5° S, 26.6° W; $\phi_m=$−65.8°, $\Lambda_m=$24.3°),
  TIksi bay (71.6° N, 129.0° E; $\phi_m=$60.4°, $\Lambda_m=$191.4°),
{ SOdankylä (67.4° N, 26.6° E; $\phi_m=$63.7°, $\Lambda_m=$120.0°),
  DIkson isl. (73.5° N, 80.4° E; $\phi_m=$63.0°, $\Lambda_m=$161.5°).

represents the intensity of the overhead current at each station). The result shows that a geomagnetic disturbance is of very short duration: the auto-correlation value drops very rapidly with increasing time difference.

Fig. 75 shows four examples of the correlation function $r(X, t; Y, (t+\Delta t))$ for the Q-indices at two stations, X and Y, with a time difference Δt varying from −12 to 12 hours[1,2]. Simultaneous correlation is the best, but the correlation

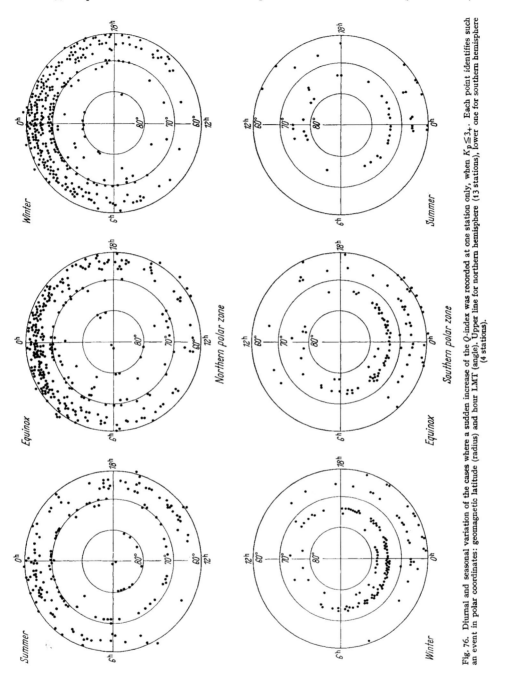

Fig. 76. Diurnal and seasonal variation of the cases where a sudden increase of the Q-index was recorded at one station only, when $K_p \leqq 3_+$. Each point identifies such an event in polar coordinates: geomagnetic latitude (radius) and hour LMT (angle). Upper line for northern hemisphere (13 stations), lower one for southern hemisphere (4 stations).

coefficient is not so high. This indicates that the magnetic disturbance does not develop and decay simultaneously all over the world, but rather locally. However, at a pair of stations, which are conjugate to each other with respect to the geomagnetic equator, the simultaneous correlation is very high; this is the case of Cape Wellen and Macquarie Island (Fig. 75a).

The frequent occurrence of local geomagnetic disturbances at high latitude is shown in the following result. A sudden increase in the quarter-hourly Q-indices is often recorded at high-latitude stations. However, more than 80 per cent of the events are recorded only at a single station [31]. Fig. 76 shows the local time and geomagnetic latitude of the points where such a sudden increase of the Q-index at a single station is observed during one year. We see that the occurrence of local geomagnetic disturbance is concentrated in night hours in winter, while also in the daytime in summer. The concentration of the points in the nighttime in winter is partly due to the very low noise-level of geomagnetic disturbance in winter night in comparison with other seasons.

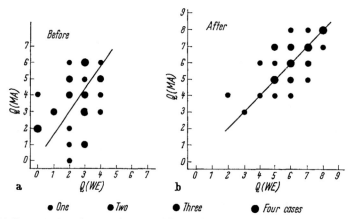

Fig. 77. Q-indices at nearly conjugate stations, cape WEllen and MAcquarie isl., before and after the sudden increase of the Q-index at Wellen (winter at Wellen, summer at Macquarie).

Once a local disturbance takes place, the disturbance at a pair of conjugate stations in the northern and southern hemispheres becomes almost equal in magnitude, as will be seen in Fig. 77. This is one reason why one finds a good correlation of magnetic disturbance at conjugate stations in general, at least for strong perturbations (see also Sect. 44). However, at very high latitudes, well inside the auroral zone, particularly on the sunlit side, conjugate stations show different magnetic variations, because the daytime agitation has considerable magnitude as described in Sect. 22.

I. Geomagnetic bays and other worldwide disturbances of short duration.

56. Geomagnetic bay as the simplest polar disturbance. On magnetically almost quiet days, the magnetogram sometimes shows a small disturbance of a simple character. As illustrated by Fig. 78, the geomagnetic curves are deflected from the normal daily variation; the deviation gradually increases, with superposed minor oscillations, till it attains a maximum or minimum value, from which it gradually returns to its undisturbed value. The duration of this particular variation is generally a few hours. Because of the resemblance of the deviations on magnetograms with typical coast-line forms, these disturbances are called *geomagnetic bays* They are called positive or negative according to increase or decrease of the horizontal intensity H of the geomagnetic field.

Geomagnetic bays are not a local phenomenon, but are the simplest type of a worldwide polar magnetic disturbance. As will be seen later in Sect. 61, the equivalent current-system for individual geomagnetic bays is nearly the same as that of polar elementary storms. Therefore, one may say that geomagnetic bays are polar elementary storms, which are seen when the worldwide magnetic activity is rather small.

At high latitudes especially in the so-called auroral zone, geomagnetic bays are accompanied by auroral displays, and additional ionization in the lower ionosphere often of the "sporadic"

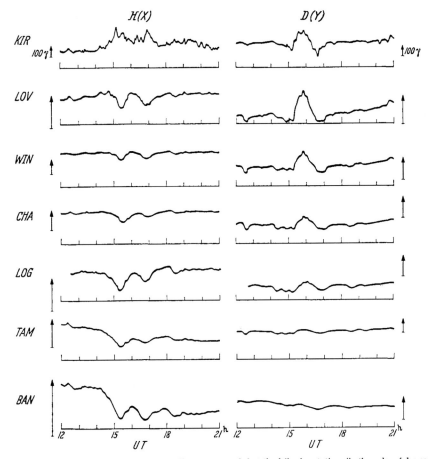

Fig. 78. An example of a geomagnetic bay on 13 Dec. 1957 recorded at the following stations (in the order of decreasing geomagnetic latitude): KIRuna, LOVö (Sweden), WINgst (Germany), CHAmbon la forêt (France), LOGroño (Spain), TAManrasset (Algeria) and BANgui (Equatorial Africa). Different scales, identified by vertical arrow for 100 γ. (After Akasofu and Meng.)

type. Since geomagnetic bays are polar magnetic disturbances of the simplest type, synoptic studies of the relation between geomagnetic bays and the simultaneous phenomena in the ionospheric regions, including auroral displays, will certainly improve the interpretation of the physical mechanism of magnetic disturbances (Chap. K).

57. Statistical tendency of appearance of geomagnetic bays. Geomagnetic bays do not occur equally often throughout the day, but they are most frequently noticeable during night hours. In temperate latitudes, geomagnetic bays observed in night hours are mostly positive ones ($\Delta H > 0$), whereas negative bays are predominant during the daytime. Thus far, many investigators have examined the occurrence frequency of geomagnetic bays in temperate latitudes; positive bays are much more frequently found than negative bays[1-5]. The ratio of occurrence frequency of positive bays to that of negative bays at a middle-latitude station ranges from 2 to 4 or more. This seems to be in apparent contradiction with the statement that a geomagnetic bay is not a local phenomenon but a worldwide one.

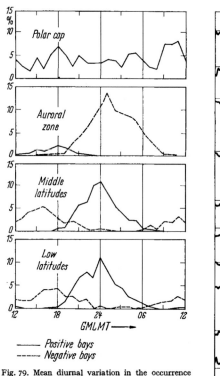

Fig. 79. Mean diurnal variation in the occurrence frequency of geomagnetic bays (in per cent.). (After SILSBEE and VESTINE.)

Fig. 80. Record of the H-component of the geomagnetic field at Fort Churchill (58.8° N, 94.1° W; geomagnetic lat. 68.8°, long. 322.5°), Canada, in successive nights. (Time: local mean time of 90° W.)

SILSBEE and VESTINE [32] made a more careful study of the frequency distribution of positive and negative bays, in which geomagnetic bays are selected by mutual comparison of simultaneous records at many stations with adequate geographic distribution over the world. The mean frequency distribution of positive and negative bays in the polar cap, auroral zone, middle and low latitudes is given in Fig. 79. In this investigation the ratio of occurrence of positive and

[1] L. STEINER: Terr. Mag. 26, 1 (1921).
[2] H. HATAKEYAMA: Geophys. Mag. 12, 16 (1938).
[3] J. MA. PRINCEP CURTO: Memorias del Observatorio del Ebro, No. 10 (1949).
[4] P. ROUGERIE: Ann. Géophys. 10, 47 (1954).
[5] G. ROSTOKER: J. Geophys. Res. 71, 79 (1966).

negative bays in middle and low latitudes is reduced to about 3:2. This means that there are often cases, in which a small positive bay is observed in the night region at moderate latitudes, but no appreciable simultaneous geomagnetic disturbance could be detected on the day-side (see Sect. 61).

In the auroral zone, negative bays take place during night hours and positive ones in the daytime, just opposite to the tendency in middle and low latitudes. Negative geomagnetic bays outnumber positive ones in a proportion of 9:1 in the auroral zone. The definite nighttime predominance of geomagnetic bays in high latitudes is especially worth noting.

As for the seasonal occurrence of geomagnetic bays, there are two maxima in the equinoctial seasons. This tendency is similar to the seasonal variation of the general geomagnetic activity itself.

58. Recurrence tendency of geomagnetic bays on consecutive days. A number of examples of an interesting recurrence tendency of geomagnetic bays have been reported. They are often recorded at nearly the same local time on consecutive days, sometimes more than five days in

Fig. 81. Simultaneous recordings of the horizontal (H) or northward (X) component from five observatories distributed in longitude along or near the auroral zone (indications at left: longitude and northern geomagnetic latitude). A, B, C, and D identify principal negative bays occurring within the 24 hours shown. (After SUGIURA and HEPPNER.)

succession. In the auroral zone, small negative bays are recorded very often around midnight. Sometimes this is an almost regular event occurring day by day, although they do not take place exactly at the same local time on consecutive days. Fig. 80 shows the recurrence of small negative bays around midnight at Fort Churchill (an auroral zone station in Canada).

When magnetograms of several stations situated along the auroral zone are compared with each other, it is noticed that geomagnetic bays are often recorded several times a day, and their magnitude is generally greatest at the station nearest to the midnight meridian [15], see the example shown in Fig. 81.

59. Local-time dependence of the disturbing vector of geomagnetic bays in moderate latitudes. The disturbing force of a geomagnetic bay is usually taken to be the deviation of the geomagnetic elements from an interpolated curve on the magnetogram connecting the undisturbed curves before and after the bay. The direction of the disturbing force vector depends on the local time at the observing station. The projection of the disturbing vector onto a horizontal plane usually describes a loop. In Fig. 82 such a trace is given[1] for the average of a number of geomagnetic bays observed at different local times at Toyohara (46° 58' N, 142° 45' E; now called Južno Sahalinsk) during the four years 1932 ... 35. As far as the middle-latitude region is concerned, the disturbing vector of geomagnetic bays shows a rather similar behaviour[2-5]. The direction of the maximum disturbing force (or that of the major axis of the loop) changes with local time; it is almost northward at about 2h, westward in the early morning, southward in the early afternoon, and eastward in the evening.

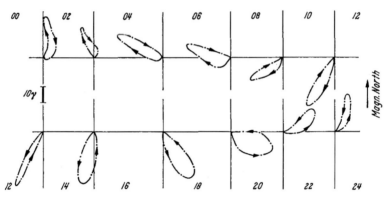

Fig. 82. Progressive change of the mean horizontal disturbing vectors of geomagnetic bays observed at Toyohara (now called Južno Sahalinsk). (After HATAKEYAMA.)

The rotational sense of the loop described by the endpoint of the horizontal disturbing force during a geomagnetic bay is always clockwise in the forenoon (in middle latitudes), and counter-clockwise in the afternoon. This tendency holds for the average of a number of bays. In the case of individual bays, there may be exceptions if irregular geomagnetic variations are superposed. The rotation of the disturbing vector of geomagnetic bays is discussed in Sect. 62 in relation with progressive changes of the equivalent current-system of geomagnetic bays.

In middle and low latitudes the simultaneous change in the vertical component, Z, of the geomagnetic field depends largely on the location. This suggests that the local structure of the earth's crust plays an important rôle for the electromagnetic induction within the ground, also for geomagnetic bays [11]. Geomagnetic bays are variations of short duration, so that electromagnetic induction is very important. Therefore, an inhomogeneous distribution of electric conductivity in the earth's upper mantle has a great influence on the phenomena that are observed on the ground.

60. Equivalent overhead current-system for geomagnetic bays. In general, a geomagnetic bay is not a local disturbance but a worldwide phenomenon. From the worldwide distribution of the disturbance field of geomagnetic bays, an equivalent overhead current-system can be drawn. An average current-system corresponding to the maximum stage of geomagnetic bays depicted by SILSBEE

---

[1] H. HATAKEYAMA: Geophys. Mag. **12**, 16 (1938).
[2] L. STEINER: Terr. Mag. **26**, 1 (1921).
[3] J. MA. PRINCEP CURTO: Memorias des Observatorio del Ebro, No. 10 (1949).
[4] P. ROUGERIE: Ann. Géophys. **10**, 47 (1954).
[5] N. FUKUSHIMA, and H. ŌNO: J. Geomag. Geoelectr. **4**, 57 (1952).

and VESTINE [32] is reproduced as Fig. 83. The current-system is referred to geomagnetic latitude and local time; its height has been assumed to be 150 km above the ground. Other workers have obtained similar current-systems. The height of the electric current has been estimated to be 100 ... 150 km above the earth's

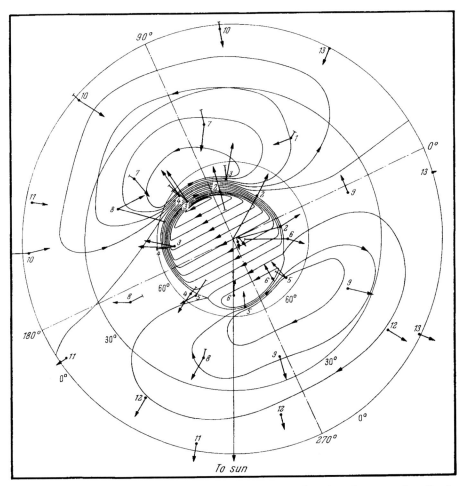

Fig. 83. An equivalent overhead current-system for geomagnetic bays. View from above the geomagnetic north pole; presentation of observed data as in Fig. 44. Values for 00 h UT observed during 12 months 1932/33. (After SILSBEE and VESTINE.)

surface, also from a mathematical analysis of the latitudinal distribution of the disturbing force of bays recorded on the ground[1].

At a glance, we note that this average current-system for geomagnetic bays resembles the current-system for the idealized S_D-field shown in Fig. 49 (C), except for a phase shift of a few hours and an intensity difference between westward and eastward auroral electrojets. Since the real pattern of the polar parallel current of the S_D-field is as described by Fig. 50, one may therefore say that the average equivalent overhead current-system for geomagnetic bays is quite similar to the

---

[1] A. G. McNISH: Terr. Mag. **43**, 67 (1938).

ionospheric current-system for S_D, except for a phase difference of about 4 hours of the current flow in middle and low latitudes.

We may consider, as a first approximation, that geomagnetic bays are caused by the development and decay of such a current-system as given by Fig. 83. But that current-system is drawn using the world-wide distribution of the *average* disturbing force of a number of geomagnetic bays, so that the averaging process must be discussed[1]. An average of the disturbing force is first derived for bays that occurred at a given local time. Then, from the local-time dependence of the average disturbance force at different latitudes, the overhead current-system for geomagnetic bays is drawn, such that the uneven occurrence frequency of bays with respect to local time is not taken account of. Therefore, the current-system in Fig. 83 is not a true average of the disturbance force of individual geomagnetic bays. We must bear in mind that the current-system for an individual geomagnetic bay might differ considerably from it.

61. Current-system for individual geomagnetic bays. If the current-system for individual geomagnetic bays were like that shown in Fig. 83, the occurrence frequency of positive and negative bays at a middle-latitude station should be approximately equal. However, the observations show another statistical distribution of positive and negative bays. There are many cases where negative bays alone are observed along the auroral zone. Thus, the number of geomagnetic bays used for drawing the westward auroral electrojet in Fig. 83 considerably exceeds that for the eastward one. In other words, there should be many bays without noticeable eastward auroral jet current, the current-system of which might be as shown by Fig. 72(b). One may also say that the eastward auroral zone current in the average current-system of Fig. 83 is obtained by averaging the eastward auroral electrojets which appear rather rarely at the time of great bays only. The conclusion is that a real current-system for individual geomagnetic bays has, in most cases, a much weaker eastward auroral electrojet as that of the average current-system shown in Fig. 83. In other words, many bays have no remarkable eastward auroral zone current in the sunlit hemisphere, but only a westward auroral electrojet on the night side[2]. Such a current-system is just the pattern for a polar elementary storm, the ideal form of which was shown by Fig. 71. At lower latitudes, the current density on the sunlit side should be weaker for individual bays than that of Fig. 83, because at lower latitudes negative bays are less frequent than positive ones, at nighttime. The above discussion is based on the natural assumption that small bays occur more frequently than larger ones. For the above reasons, a typical geomagnetic bay is considered to be a polar elementary storm, as described previously in Sect. 53. The occurrence frequency of positive and negative polar elementary storms reported by BIRKELAND [29] is the same as described in this section for geomagnetic bays.

The uneven occurrence of positive and negative bays at temperate latitudes seems to be in an apparent contradiction with the fact that geomagnetic bays are a worldwide phenomenon. In the case of a simple negative polar elementary storm, or in general, when the westward auroral zone current on the night side is much more intense than the eastward flow in the sunlit hemisphere, the electric current at a certain middle latitude should have stronger eastward flow in the dark hemisphere than westward current in the sunlit hemisphere. This means that the magnitude of positive bays is larger than that of negative bays (which occur at the same latitude but with 12 hours difference in local time). Therefore, it may often happen that at lower latitudes negative bays on the sunlit side are overlooked. Furthermore, the solar daily variation makes it more difficult to detect a small negative bay during the daytime. On the other hand, positive bays are easily detected, because the magnetogram trace is almost straight in night hours, which are free from S_q. These two reasons explain why positive bays are more frequent than negative ones at temperate latitudes.

62. Progressive change in the current-system of geomagnetic bays. It might be considered, as a first approximation, that a geomagnetic bay is caused by the

---
[1] N. FUKUSHIMA: J. Geomag. Geoelectr. **10**, 164 (1959).
[2] N. FUKUSHIMA: J. Geomag. Geoelectr. **10**, 164 (1959).

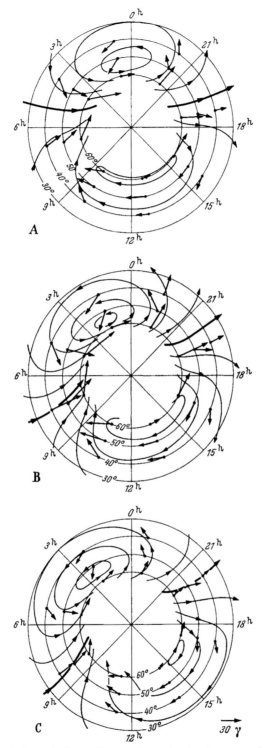

Fig. 84. Progressive change in the overhead current-arrows and current-system for geomagnetic bays in middle and low latitudes. A=developing, B=maximum, C=decaying stage. Coordinates: geomagnetic latitude and local mean time.

development and decay of a current-system in the ionosphere such as shown by Fig. 83, or its modification. However, a progressive change in the current-system must be assumed, in order to explain the systematic rotation of the disturbing force vector during a bay. This has already been shown in Fig. 82, namely clockwise rotation in the forenoon and counter-clockwise one in the afternoon, everywhere at middle and low latitudes. The effect of electromagnetic induction in the earth cannot explain this tendency, at least as long as the external current-system conserves its form with varying intensity[1]. Therefore, we must conclude that the external current-system itself changes with time.

Fig. 85. Progressive change in the geomagnetic latitude of the centre of the auroral electrojet ($\Delta Z=0$, bold line), and of the northern and southern boundaries of the zone of auroral luminosity. Full lines and broken lines are determined independently from the records at BEttles (geomagnetic latitude 65.8°) and KOtzebue (63.7°), in Alaska (31 Oct. 1957). (After KOKUBUN.)

α) In order to account for the characteristic rotational changes of the disturbing forces shown in Fig. 82, the progressive change of the external current-system should be such that the area of positive bays in the dark hemisphere is initially small and becomes wider, whereas the area of negative bays decreases as is illustrated in Fig. 84. Moreover, there is a slight but recognizable tendency that the current-system on the whole rotates eastward[2,3]. From an analysis of several individual geomagnetic bays, it was suggested that the westward auroral electrojet in the dark hemisphere develops quicker and reaches its maximum intensity considerably earlier than the eastward auroral zone current on the sunlit side[4]. The intensity of the eastward electrojet attains its maximum when the westward jet is already in its decaying stage. This conclusion should, however, be carefully examined using more detailed observational data from more stations.

β) These observational results concerning the auroral electrojet explain the broadening of the area of positive bays at middle and low latitudes in the following way. In the initial stage of a geomagnetic bay, the electric current-system would be very similar to that shown by Fig. 71, as only the westward electrojet is predominant in the auroral zone. At that stage, the positive bay area at temperate latitudes is much narrower than that of negative bays. In the later stage, when the westward auroral electrojet is already in its decaying stage, the eastward jet has maximum intensity, and the magnitude of westward and eastward auroral

---

[1] H. J. LIPPMANN: Diss. Univ. Göttingen (1955).
[2] N. FUKUSHIMA: Geophys. Notes, Tokyo Univ. 3, No. 22 (1950).
[3] N. FUKUSHIMA, and H. ŌNO: J. Geomag. Geoelectr. 4, 57 (1952).
[4] N. FUKUSHIMA: J. Geomag. Geoelectr. 3, 59 (1951).

electrojets is nearly equal. At that time, the overhead electric current-system would be as shown in Fig. 83, with nearly equal areas of positive and negative bays at temperate latitudes. Hence, during the development of a geomagnetic bay, the area of positive bays becomes larger with time, and that of negative bays decreases.

During the development of a bay, the auroral electrojet changes not only its intensity but also its position on the earth. Although only a few examples so far have been analyzed[5], the westward auroral electrojet in the dark hemisphere seems to shift poleward in most cases, in the developing as well as in the decaying stage of a bay; this is shown by Fig. 85. The latitudinal width of the electrojet seems to widen as the current intensity increases. On the other hand, the eastward auroral electrojet sometimes tends to move equatorward during its development. But these conclusions are not yet definitive.

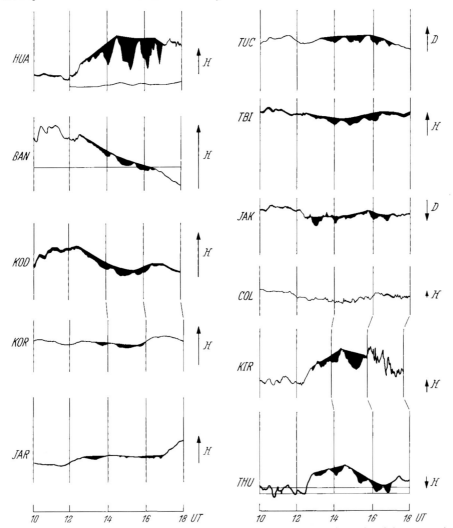

Fig. 86. Example of magnetograms (17 Mar. 1958) showing a world-wide simultaneous fluctuation of the geomagnetic field. Stations: HUAncayo, BANgui, KODaikanal, KORor, JARvis isl., TUCson, TBIlisi, JAKutsk, COLlege, KIRuna, THUle. (The length of the arrow is 100 γ for each station.) (After Nishida, Iwasaki and Nagata.)

63. Fluctuation of the equatorial electrojet in simultaneity with polar geomagnetic variations.
Recently it has been pointed out that a large fluctuation of the

---
[5] S. Kokubun: Rep. Ionos. Space Res. Japan **19**, 177 (1965).

equatorial electrojet in the daytime often accompanies a bay-like geomagnetic variation at high latitudes[1-3]. A typical example of this kind of variation is illustrated by Fig. 86. The equivalent overhead electric current-system for the maximum stage of this world-wide variation is given by Fig. 87. The electric current-system in the southern hemisphere is nearly the mirror image of that in the northern hemisphere. In these cases, in daytime, the magnitude of the variation in the

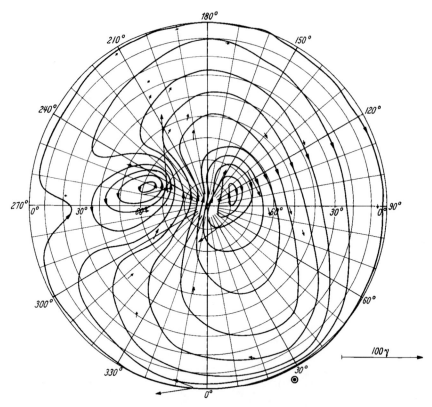

Fig. 87. Equivalent overhead current-system for the northern hemisphere of the world-wide fluctuation of the geomagnetic field shown in Fig. 86 for 17 Mar. 1958, 1450 UT. (After NISHIDA, IWASAKI and NAGATA.)

auroral zone does not exceed very much the simultaneous variation at other stations, including those near the geomagnetic equator. When the record at a single observatory is examined, this kind of variation may be designated as a geomagnetic bay. However, the equivalent electric current-system is not at all like that of Fig. 71, or Fig. 83, so that this variation is not a bay disturbance originating in the auroral zone.

The current-system of Fig. 87 is very similar to that of the S_q^p-field (Sect. 10) at high latitudes. One may say that the system is an extension of the S_q^p-field to middle latitudes in the case of a large fluctuation of the equatorial electrojet; this may be caused by a weak east-west electric field applied to the equatorial

---

[1] A. ONWUMECHILLI, and P. O. OGBUEHI: J. Atmosph. Terr. Phys. **24**, 173 (1962).
[2] S.-I. AKASOFU, and S. CHAPMAN: J. Geophys. Res. **68**, 2375 (1963).
[3] A. NISHIDA, N. IWASAKI, and T. NAGATA: Ann. Géophys. **22**, 478 (1966).

ionosphere where the electric conductivity is very high (see Sect. 3). This additional electric field may accompany a temporary change in the magnetospheric plasma circulation (see Sect. 13 ε), [39].

Fig. 88 a and b. Ordinary magnetogram recorded at Kakioka (36.2° N, 140.2° E) and (b) simultaneous induction magnetogram at Onagawa (38.4° N, 141.5° E), both during the magnetic storm of 23 Jul. 1953.

J. Geomagnetic pulsations.

64. Short period variations on magnetograms. So far we considered geomagnetic variations with fairly large amplitude lasting longer than several minutes. Even when describing polar elementary storms, geomagnetic fluctuations with periods of not more than several minutes are disregarded. Continuous observation of the geomagnetic field shows the presence of many short-period fluctuations, even on quiet days. Generally, the short-period variations appear in series or groups.

Because the period of these variations is not longer than a few minutes, they are not easily detected on ordinary magnetograms (paper speed 20 mm/h, or so), but with quick-run records (paper speed 180 mm/h, or so) and a highly sensitive magnetometer one can detect minor oscillations, down to amplitudes of 0.1 γ and periods as short as several seconds. Observations by means of an induction loop or an induction coil with a core of high permeability enables the recording of transient geomagnetic variations with even smaller amplitudes and periods. An example of an induction magnetogram at the time of a magnetic storm (with sudden commencement) is given by Fig. 88. In recent years, the short-period geomagnetic variations have caught more attention, and the study of geomagnetic pulsations has made great progress[1]. In this chapter, only a brief review of phenomena will be presented. Their importance for space physics is briefly discussed in Chap. K.

65. Morphological classification of geomagnetic pulsations. Various kinds of pulsations, have been observed and a number of different names have been used to describe their characteristics with respect to period and waveform. By a recent international agreement[1] the geomagnetic pulsations are classified into two categories; the type denoted by **pc** has a fairly regular waveform and appears for longer time, while the other one designated by **pi** has irregular waveform and appears only for short time. A subdivision is made as given in Table 1, according to the predominant period or quasi-period of pulsation.

In general, the geomagnetic pulsations with longer period have larger amplitude. The term "giant pulsations" was once used for large-amplitude (**pc5**) pulsations often observed at high-latitude stations[2] (see Sect. 67). So-called "pearls" or "pearl necklace" of short duration[3] belong to **pc1**. The behaviour of geomagnetic pulsations depends on latitude and local-time of the observing point over the world, and shows also seasonal and secular variations. The term "**pt**" was formerly used for trains of magnetic pulsations lasting about 10 minutes or so; these are now denoted by **pi**.

66. Appearance tendency of geomagnetic pulsations. Fig. 89 shows a latitudinal profile of the period spectra of geomagnetic pulsations. The pulsations of longer period occur more often at higher latitudes. Short-period continuous pulsations appear very often at lower latitudes.

Table 1. *Classification of geomagnetic pulsations* **pc** *and* **pt** *according to their predominant period.*

Type		pc 1	pc 2	pc 3	pc 4	pc 5	pi 1	pi 2
Quasi-period	from	0.2	5	10	45	150	1	40 s
	up to	5	10	45	150	600	40	150 s

The diurnal variation of activity is different for each class of pulsations given in Table 1, and it depends on the latitude. At high latitudes, **pc5** shows an early-morning maximum, while in the subauroral zone **pc4** and **pc3** have a daytime maximum (see for example Fig. 90). Fig. 91 illustrates schematically the variation as function of local time of the occurrence of various kinds of pulsations at a low-

---

[1] A detailed description is given in the contribution by E. SELZER in Vol. 49/4 of this Encyclopedia.

[1] At IAGA Assembly at Berkeley in August 1963.

[2] B. ROLF: Terr. Mag. **36**, 9 (1931). — L. HARANG: Terr. Mag. **37**, 57 (1932); **41**, 329 (1936). — E. SUCKSDORFF: Terr. Mag. **44**, 157 (1939).

[3] V. A. TROITSKAYA: J. Geophys. Res. **66**, 5 (1961). — R. R. HEACOCK, and V. P. HESSLER: J. Geophys. Res. **67**, 3985 (1962).

Fig. 89. Relative occurrence frequency (ordinate) of geomagnetic pulsations with different periods (abscissa) observed on LaCour magnetograms at various stations, for March 1958. Stations: Point Barrow (corrected geomagn. latitude $\phi=69.7°$), COLlege ($\phi=64.9°$), SITka ($\phi=59.8°$), LOVö ($\phi=55.9°$), FREdericksburg ($\phi=51.8°$), BOUlder ($\phi=48.8°$), TUCson ($\phi=39.7°$), GUAm ($\phi=3.7°$). (After HIRASAWA.)

latitude station, Kakioka[1] (geomagnetic latitude 26.0°). The observational characteristics for middle and low latitudes are summarized in Table 2.

[1] T. HIRASAWA, T. OGUTI, and T. NAGATA: Rep. Ionos. Space Res. Japan **19**, 452 (1965). — T. HIRASAWA, and T. NAGATA: Pure Appl. Geophys. **65**, 133 (1966).

Continuous pulsations are often observed while large-amplitude variations are not appearing on ordinary magnetograms, whereas **pi** occurs in night hours in association with noticeable disturbance on ordinary magnetograms such as a geomagnetic bay[2]. The best method for studying the characteristics of geomagnetic pulsations is to take their dynamic spectrum[1], an example is given as Fig. 92.

Fig. 90. Contours of average equi-activity (left) and occurrence frequency (right) in 1958 showing distinct diurnal variations in pc at BYRd Station (geomag. lat. −70.4°) and FREdericksburg (49.7°). The activity-scale in the left diagram is proportional to the product of amplitude and duration of geomagnetic pulsations. In the right figure, small circle means one series of pc, while large circle corresponds to ten small circles. (After SAITO.)

As to secular and annual variations of the pulsational periods, they are in a close relation with upper atmospheric condition and solar activity[3], as is seen from Fig. 93.

The spatial extension of the region, where geomagnetic pulsations occur simultaneously, has not yet been studied in detail. Not only the observational methods but also the frequency response of the recording apparatus for short-period geomagnetic pulsations differ from one place to another, so that simultaneous data of widely separated stations cannot readily be compared. It has been pointed out that there are cases where the period of the D-component variation is constant along a middle-latitude meridian, whereas that of the H-component varies with latitude[4]. An equatorial enhancement of **pc5** in daytime is worth noting[5].

[2] O. MEYER: Deut. Hydrograph. Z. **4**, 61 (1951).
[3] Y. KATO, and T. SAITO: Rep. Ionos. Space Res. Japan **18**, 183 (1964).
[4] M. SIEBERT: Planet. Space Sci. **12**, 137 (1964).
[5] N. MATUURA: Rep. Ionos. Space Res. Japan **15**, 192 (1961).

Sect. 66.   Appearance tendency of geomagnetic pulsations.   119

Table 2. *Characteristics of geomagnetic pulsations observed at Kakioka, Japan.*

IAGA notation	Average period sec	Waveform	Duration	Diurnal variation in Activity	Diurnal variation in Period	Other remarks
pc 4	~70	sinusoidal	several hours	At any local time, but most active around 9 h	Shortest around 4...6 h, Longest around 20...21 h	Noticeable frequency-rise after SC
pc 3	~30	sinusoidal	several hours	In the daytime, most active around 6...9 h	Increases from morning to evening	On stormy days, occurrence with **pc 4** around 9 h
pc 2	~3	sub-sinusoidal	almost throughout the day when occurring at all	At any local time, but most active around 23...6 h	Shortest around 4...6 h, Longest around 19...20 h	More frequent occurrence on stormy days. Period becomes shorter on stormy days. Enhanced by longer-period pulsation with irregular waveform
pi 2	~70	irregular with damped train	One train lasts several minutes	Most active around midnight	—	Higher-frequency components superposed
pc 1 or pi 1	1~3	sub-sinusoidal	15 minutes to several hours	Most active in the morning and evening hours, and absent in the midday	Shortest in the evening	Frequency rises or falls during the pulsation. Sometimes intermittent appearance during night hours.
pi 1	0.5~3	irregular	~5 minutes on average	Most active around midnight	—	Like a noise burst without any fine spectral structure. Occurrence with or without **pi 2**.
pc 1	1~2	beating-type	30 minutes to several hours	In night hours, and most frequent in the morning and evening	—	Fan-shaped and stratified spectral structure. Occurrence after the recovery of magnetic storm.

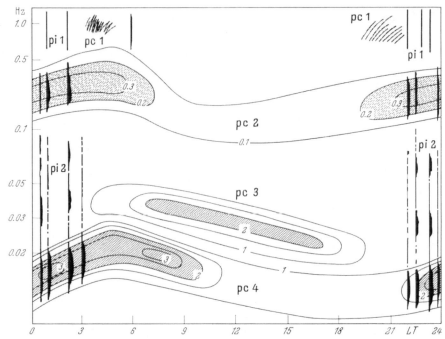

Fig. 91. Average relative intensity in the power spectrum of geomagnetic pulsations at the low-latitude station Kakioka (geomagnetic latitude 26°). (After HIRASAWA.)

Fig. 92. Dynamic spectrum of geomagnetic pulsations during and after a magnetic storm of 15 ... 20 Jun. 1965, at the low-latitude station, Kakioka (geomag. lat. 26.0°). (After HIRASAWA, OGUTI and NAGATA.)

67. Large-amplitude pulsations (pc 5) in high latitudes.

The morphology of large-amplitude and long-period pulsations (**pc 5**) in and near the auroral zone is

Sect. 67. Large-amplitude pulsations (pc 5) in high latitudes.

best known, because these are clearly recorded even on ordinary magnetograms. An example of **pc5** at various stations is given in Fig. 94. Note here that the disturbance field in the D-component is comparable to that of the $\mathcal{H}$-component. This is especially noticeable in the morning hours. (For ordinary geomagnetic disturbances in the auroral zone, the variation of the $\mathcal{H}$-component is usually

Fig. 93. Clear synchronism in the secular and annual variations of various phenomena. 1) period of **pc5** at Byrd Station; 2) period of **pc4** at Fredericksburg, 3) whistler dispersion at Stanford, 4) satellite drag obtained from Explorer I orbit, 5) $foF2$ at Washington, D.C., 6) sunspot relative number. Monthly mean values are given by thin full lines, 3-month running average by thin broken lines, and 12-month running average by thick lines. (After SAITO.)

much more predominant than that of the D-component.) The latitudinal distribution of **pc5** amplitudes is shown by Fig. 95. The polarization of the disturbing force vector (looking downwards in the northern hemisphere) is counter-clockwise in the forenoon and clockwise in the afternoon hours, in general[1]. The forenoon pulsations have larger amplitudes than the afternoon ones, as is seen in Fig. 96. The spatial extent of large-amplitude **pc5** is known to be about 1000 km in longitude[2,3].

---

[1] T. NAGATA, S. KOKUBUN, and T. IIJIMA: Japanese Antarctic Res. Exp. Sci. Rep. Ser. A, No. 3 (1966).
[2] J. OLSEN: ATME/IUGG Assembly, Circular Letter No. 3, 38 (1954).
[3] J. VELDKAMP: J. Atmosph. Terr. Phys. **17**, 320 (1960).

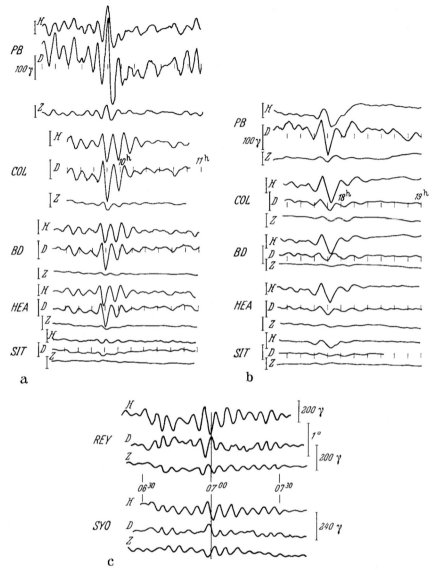

Fig. 94a—c. Examples of **pc5** observed at Alaskan stations (a), (b): Point Barrow, COLlege, Big Delta, HEAly, SITka on 14 March 1958; (c): **pc5** recorded at geomagnetically conjugate stations, REYkjavik and SYOwa Base, on 18 September, 1959.

A large-amplitude **pc5** is observed simultaneously at conjugate stations in the northern and southern hemispheres. This is for example shown by simultaneous records at Syowa Base, Antarctica, and Reykjavik, Iceland, see Fig. 94. It is interesting to point out that auroral radar echoes are often observed[4] in association with **pc5**, see Fig. 97. The period of **pc5** is confined to a very narrow frequency-band in each case[5], and it shows a sudden change when an **SI** is observed in low latitudes. The period becomes shorter or longer, according to whether the **SI** is positive or negative[3,6].

[4] E. KANEDA, S. KOKUBUN, T. OGUTI, and T. NAGATA: Rep. Ionos. Space Res. Japan **18**, 165 (1964).
[5] K. V. PAULSON, A. EGELAND, and F. ELEMAN: J. Atmosph. Terr. Phys. **27**, 943 (1965).
[6] T. HIRASAWA, A. NISHIDA, and T. NAGATA: Rep. Ionos. Space Res. Japan **20**, 51 (1966).

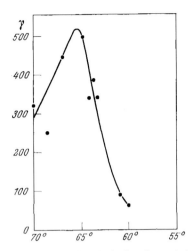

Fig. 95. Geomagnetic latitude dependence of the mean range of the horizontal component of a pc5 observed on 19 March 1958. (After Kokubun.)

Fig. 96. Local mean time dependence of the mean horizontal range of pc5 at College, Alaska (geomagn. latitude 64.6°). Black circle: counter-clockwise pulsation; open circle: clockwise pulsation. (After Kokubun.)

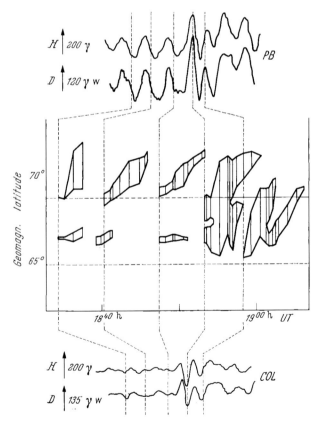

Fig. 97. An example of range variation in the auroral radar echo and corresponding variation on the magnetogram in the event of 8 Dec. 1957. Stations: Point Barrow and COLlege (Alaska). (After Kaneda, et al.)

K. Inter-relation among geomagnetic disturbance and associated phenomena, and outline of theoretical interpretation.

68. Aurora, ionospheric variation, and other geophysical phenomena accompanying geomagnetic disturbance. As already mentioned in the general introduction (Chap. A), geomagnetic disturbance is observed when the earth's outer space is disturbed by the attack of an intense solar plasma stream.

α) We know at present that geomagnetic disturbance takes place together with the following *phenomena* detectable from the ground, or by balloons, rockets and satellites [26], [33], [40]:

Auroral display including polar-cap glow[1];

Blackout or absorption of transmitted radio waves: polar or auroral zone blackout or absorption, short wave fade-out, polar-cap absorption, etc.[2];

Anomalous change in ionospheric conditions: appearance of 'Es-layer', 'spread F', abnormal decrease and increase of electron density and height of the upper ionospheric layers[3];

Cosmic noise absorption: absorption in the ionosphere of natural extraterrestrial high-frequency radio noise[4];

Very low frequency (v.l.f.) radio noise of magnetospheric origin[5];

Intensity variation of primary cosmic rays[6];

Bremsstrahlung X-rays in the lower ionosphere (detected by balloons)[6];

Variation in flux-intensity of charged particles, both precipitated into the ionosphere and trapped in the radiation belts in the magnetosphere, especially in the outer belt (detected by satellites)[7] [40].

Recent advance in space sounding techniques has brought interesting information about the magnetic field distribution and its variation in the magnetosphere[7], and even on the solar stream in interplanetary space[8].

β) The *interrelation* of various simultaneous phenomena has been studied in order to understand the physical processes involved. Fig. 98 gives an example of the interrelation of geomagnetic disturbance, auroral luminosity and ionospheric change at an auroral zone station, Syowa Base, Antarctica [33]. Examination of a number of typical examples of polar elementary storms[9] resulted in the following quantitative relation, valid for auroral zone stations, between the maximum deviation of the horizontal intensity, ΔH, the luminosity of the green auroral line in the zenith*, $I(5577)$, and the maximum electron density in the Es-layer, N:

$$I(5577)/\text{kR} \approx 5 \times 10^{-22} \, (N/\text{m}^{-3})^2, \qquad (68.1)$$

$$I(5577)/\text{kR} \approx 3 \times 10^{-3} \, |\Delta H/\gamma|^2. \qquad (68.2)$$

---

* Measured as photon flux, unit Rayleigh ≡ R = 10^6 cm^{-2} s^{-1}; 1 kilo Rayleigh ≡ kR = 10^9 cm^{-2} s^{-1}.

[1] S.-I. Akasofu, S. Chapman, and A. B. Meinel: This Encyclopedia, vol. 49/1.
[2] K. Rawer, and K. Suchy: This Encyclopedia, Vol. 49/2, Sects. 39, 43, 46, 47.
[3] K. Rawer, and K. Suchy: This Encyclopedia, Vol. 49/2, Sects. 38, 40, 42, 44.
[4] K. Rawer, and K. Suchy: This Encyclopedia, Vol. 49/2, Sects. 46 and 47.
[5] See contribution by R. Gendrin in this volume, p. 461.
[6] See S. E. Forbush: This Encyclopedia, Vol. 49/1.
[7] See contribution by W. N. Hess in Vol. 49/4 of this Encyclopedia.
[8] See contribution by E. N. Parker, and V. C. A. Ferraro in this volume, p. 131.
[9] Similar studies have been made by A. Omholt: J. Atmosph. Terr. Phys. 7, 73 (1955), by J. K. Walker: J. Atmosph. Terr. Phys. 26, 951 (1964), and others.

Fig. 98a and b. Typical examples of simultaneous records of geomagnetic variation, auroral luminosity at zenith, total auroral activity and Es top frequency during upper atmosphere disturbances in the auroral zone (Syowa Base, Antarctica). (a) 7 Apr. 1959. (b) 11 ... 12 Apr. 1959. (After OGUTI.)

Hence the geomagnetic deviation is proportional to the electron density in the Es-layer (for moderate disturbances). It has also been shown that the magnitude of geomagnetic disturbance is also proportional to the increase of cosmic noise absorption at auroral zone stations[4]. The backgrounds of these relations are ionization, excitation and transportation of electric charges caused by the particles precipitating into the ionosphere over the auroral zone. Electrons seem to be mainly responsible for such phenomena. For moderate polar elementary storms in the auroral zone the order of magnitude of the influx is $10^{12} \ldots 10^{13} \, \text{m}^{-2} \, \text{s}^{-1} \, \text{sterad}^{-1}$, and the energy range $1 \ldots 100$ keV.

γ) The geomagnetic variations of simple and typical polar elementary storms observed in the auroral zone seem to be explained by an anomalous increase in the electric conductivity at the level of the E-region. However, the simultaneous increase of ionospheric conductivity seems sometimes to be insufficient to produce an ionospheric current intense enough to explain severe magnetic disturbances. Therefore, some additional electromotive force seems to be necessary to explain the intense auroral zone current.

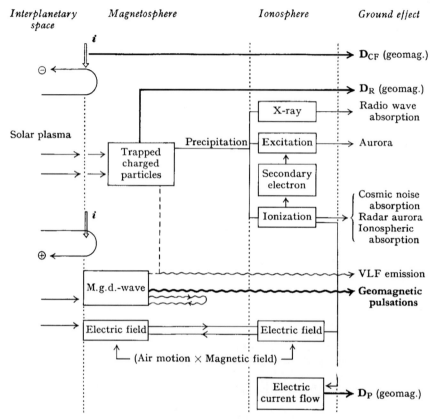

Fig. 99. Schematic diagram for causality and inter-relations of different upper atmosphere disturbance phenomena. (M.g.d.-wave = Magneto-gasdynamic wave.)

In the polar-cap region, the proportionality constant in the I versus ΔH relation is much smaller than the value given in Eq. (68.2). This means that the geomagnetic change in the polar cap is either (i) not only due to the overhead electric current but also influenced by nearby currents such as the auroral zone current, or (ii) due to an external electric field supplied from the magnetosphere. In general, ionospheric changes in the polar-cap zone (such as blackout of radio waves, or the 'polar glow aurora') are produced by deeply penetrating high-energy protons (and helium ions) of solar origin. These reach far down into the lower E- or D-regions, without provoking an intense electric current which could produce a strong geomagnetic disturbance.

Fig. 99 gives a schematic diagram, which summarizes the causality of geomagnetic disturbance and associated phenomena.

69. Outline of theoretical interpretation of magnetic disturbance[1]. A theoretical consideration for the sudden commencement of magnetic storms has already been given briefly in Sect. 33. In this section, an outline of theoretical interpretation of the D_{st}-field and of S_D or D_S after an SC is discussed. Recent review papers by COLE [34], AKASOFU [35], PIDDINGTON [37], HINES [38], OBAYASHI and NISHIDA [39], and others [40] give detailed theoretical discussion.

α) As already mentioned in Sect. 41, the disturbance field **D** can be separated:

$$\mathbf{D} = \mathbf{D}_{CF} + \mathbf{D}_R + \mathbf{D}_P. \tag{41.3}$$

$\mathbf{D}_{CF}$ is the *effect of corpuscular flux*, it is produced by an electric current in the magnetospheric boundary, which confines the geomagnetic field-lines in a cavity immersed in the solar wind. $\mathbf{D}_{CF}$ is always present, but its intensity varies according to the solar wind pressure. A sudden commencement of a magnetic storm is then to be described as a sudden enhancement of $\mathbf{D}_{CF}$ when an intensified solar wind just arrives at the magnetospheric boundary.

β) The $\mathbf{D}_R$-field is attributed to the *ring current* turning around the earth, which itself is a consequence of the motion of charged particles trapped in the geomagnetic field. Let W denote the total kinetic energy of the charged particles trapped in the earth's magnetic field of dipole moment M, the charged particles produce a southward magnetic field near the earth's centre, the intensity of which is $\Delta \mathcal{H} \cong 2W/M$ *. To a first approximation this is independent of the position of the $\mathbf{D}_R$ current; it is now supposed that it would be centred at about 5 or 6 earth radii from the earth's centre. As already described in Sect. 35, $\mathbf{D}_R$ can be decomposed into $\mathbf{D}_{R1}$ with fast decay and $\mathbf{D}_{R2}$ with slow decay rate. The current corresponding to $\mathbf{D}_{R1}$ is supposed to be situated nearer to the earth than that of $\mathbf{D}_{R2}$, because a fast decay is easier explained in a denser atmosphere. The ring current seems to be have a slight day-night asymmetry, because of the unidirectional pressure of the solar wind on the magnetosphere[1].

γ) The *intense $\mathbf{D}_R$-field* observed after SC's is supposed to be due either to acceleration of charged particles in the magnetosphere by gasmagnetic waves, or to diffusion of energetic particles from interplanetary space, or to both mechanisms. The charged particles provoking the field will remain in the magnetosphere, even after the intense solar plasma stream has passed away from the vicinity of the earth. The result is a slow recovery of the $\mathbf{D}_{R2}$ field and the non-cyclic changes of midnight $\mathcal{H}$-value after magnetic storms, described in Sect. 37.

δ) The main origin of the $\mathbf{D}_P$ *field* is the electric current flow in the ionosphere, especially an *intense electrojet along the auroral zone*. The intensification of the current flow is caused either by an increase in the electric conductivity of the ionosphere (for moderate magnetic disturbances in the auroral zone) or/and by an extra electric field, as explained in the preceding section. The additional electromotive force may be produced either within the ionosphere itself (if the air motion is enhanced there), or by transmission of an electric potential from the magnetosphere down to the ionospheric region along the magnetic field-lines. At present the latter possibility has been widely discussed. Since the Hall conductivity greatly exceeds the Pedersen conductivity in the E region, the additional electric field will be nearly perpendicular to the auroral electrojet, i.e., the electric field is equatorward for the westward jet current, and poleward for the eastward jet, as is shown by Fig. 100. Then the problem may be reduced to the mechanism of production of such an electric field in the magnetospheric plasma.

ε) However, we must not neglect the possibility of an *additional ionospheric current* which may be produced without an extra electric field supplied from the magnetosphere. When the electric conductivity in the ionosphere over the auroral zone is locally increased by the ionizing effect of precipitated charged particles, for example in a small elliptic area (with its long-axis along the auroral-zone latitude-circle), and if an electric current has already been flowing in that region of the ionosphere, an additional current is produced without any additional electric field. This additional current is concentrated in the elliptic area; its return current flows in a wide area around. The configuration is as if an electric dipole were placed at the auroral zone; the current-system is thus very similar to that of a polar elementary storm or $\mathbf{D}_P$ substorm[2]. In such a case, an east-west electric field is generally more effective in producing the concentrated electric current in the elliptic area than a north-south electric field. (We deal with a stationary two-dimensional current flow in the ionosphere.) This is due to the effect of the electric polarisation field originating from a charge accumulation at the boundary of the elliptic region of high electric conductivity. When the electric conductivity of the entire auroral-zone ionosphere is increased against the polar-cap and middle-latitude

---

* For example: $\Delta \mathcal{H} = 100\,\gamma$ for $W \cong 4 \cdot 10^{22}$ erg.

[1] A detailed discussion is given in the contribution by E. N. PARKER, and V. C. A. FERRARO: in this volume, p. 131. See also [34] to [40].

[2] L. G. ASAULENKO, and M. I. PUDOVKIN: Geomagnetizm i Aeronomija [Translation: Geomagnetism and Aeronomy] **5**, 322 (1965).

regions, the resultant electric current-system resembles in form that of the average S_D-field[3]. Ionospheric dynamo effects of this kind are worth to be studied in detail. A more accurate approach should take account of the mutual interaction of ionospheric and magnetospheric dynamos. In fact, we must discuss the electric current flow in three dimensions[4]. It is also highly desirable to take account of the height-dependence of the Pedersen and Hall conductivities within the ionosphere, in order to examine the partial current closure in the ionosphere.

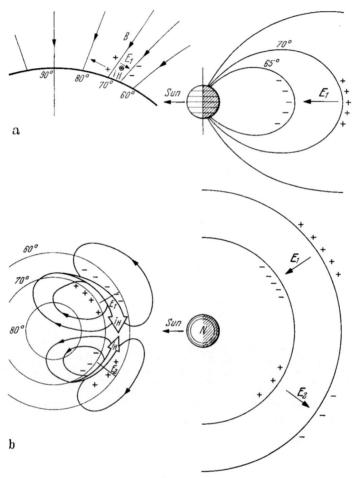

Fig. 100a and b. Illustration of the gross features of the electric field that must exist if auroral electrojet currents are Hall currents, see text. (a): Side view from dusk meridian; (b): polar view. (After SUGIURA and HEPPNER.)

ξ) It has been pointed out that the S_q^p-like variation* is enhanced at the time of magnetic storms (Sect. 47). This should be a consequence of the intensification of the solar wind attacking the magnetosphere, [39].

---

* The term D_{P2} is used because of two current-vortices in the polar rgion, while D_{P1} designates the polar substorm with auroral electrojet.

[3] G. A. LOGINOV: Geomagnetizm i Aeronomija [Translation: Geomagnetism and Aeronomy] **5**, 328 (1965).

[4] R. BOSTRÖM: J. Geophys. Res. **69**, 4983 (1964). J. W. KERN: J. Geomag. Geoelectr. **18**, 125 (1966).

η) One important fact found in local summer[5] near the geomagnetic pole is a noticeable decrease of the vertical intensity Z of the geomagnetic field. This seems to indicate a significant seasonal difference in the number of charged particles coming from the magnetosphere, and impinging or approaching the polar-cap region. The eastward zonal current responsible for the decrease in the vertical component, Z, may be produced by a Hall current in the ionosphere.

70. Geomagnetic pulsations and gasmagnetic oscillation in the outer atmosphere[1]. Geomagnetic pulsations are now considered to be caused by gasmagnetic waves generated in the magnetosphere and propagated to the ionosphere, where they are converted into electromagnetic waves of ultra-low frequency. The predominant period of stronger geomagnetic pulsations (**pc 5** in high latitudes and **pc 4** in temperate latitudes) is shown to be roughly coincident with the eigen-period of a torsional oscillation in the magnetospheric plasma of the lines of geomagnetic force. The daytime **pc 3** might be due to poloidal oscillations excited by solar wind pressure, and **pc 2** might be due to a gasmagnetic resonance in the region between the ionosphere and the level where the Alfvén[2] velocity is maximum (about 3000 km above the earth surface). When **pi 1** or **pi 2** occurs, **pc 4** and **pc 2** are also excited.

Large-amplitude pulsations at high latitudes and simultaneous geomagnetic field-changes at lower latitudes occur after the sudden commencement of a magnetic storm. These pulsations are interpreted as being due to magneto-gas-dynamic waves generated by a sudden compression of the magnetosphere, as explained in Sect. 33. When **pc 5** is already existent, a sudden contraction or expansion of the magnetosphere results in a change of the period of **pc 5**, due to the change in the magnetic field-line configuration. The noteworthy conjugacy of **pc 5** in the northern and southern polar zones also supports the magneto-gas-dynamic explanation of geomagnetic pulsations.

Although this interpretation seems to be very promising, there are a lot of open problems, such as the resonance mechanism of magnetic field-lines, the mutual conversion of gasmagnetic waves of isotropic and anisotropic modes during their propagation in the magnetosphere, the effect of the ionosphere, and so forth. The interaction of gasmagnetic waves with charged particles in the magnetosphere is also a very important problem, not only for geomagnetic pulsations but also for v.l.f. (very low frequency) emission, precipitation of aurora-producing electrons, and the simultaneous polar magnetic disturbance.

General references.

[1] CHAPMAN, S., and J. BARTELS: Geomagnetism, 2 vols. Oxford: Clarendon Press 1940. Reprinted in 1951. This is a monumental book in the field of geomagnetism. A complete list of papers published before 1939 is given in the bibliography of this book.

[2] NAGATA, T., and S. KOKUBUN: A particular geomagnetic daily variation (S_q^p) in the polar regions on geomagnetically quiet days. Nature **195**, 555—557 (1962). An additional geomagnetic daily variation field (S_q^p-field) in the polar region on geomagnetically quiet day. Rep. Ionos. Space Res. Japan **16**, 256—274 (1962).

[3] VESTINE, E. H., L. LAPORTE, I. LANGE, and W. E. SCOTT: The geomagnetic field, its description and analysis. Carnegie Inst. Washington Publ. 580 (1947).

[4] HASEGAWA, M.: Provisional report of the statistical study on the diurnal variations of terrestrial magnetism in the north polar regions. Trans. Washington Meeting 1939, IATME Bull. No. 10, 311-318 (1940).

[5] OHSHIO, M., N. FUKUSHIMA, and T. NAGATA: Solar flare effect on geomagnetic variation. Rep. Ionos. Space Res. Japan **17**, 77—114 (1963). — NAGATA, T.: Solar flare effect on the geomagnetic field. J. Geomag. Geoelectr. **18**, 197—219 (1966).

[6] FUKUSHIMA, N.: Gross character of geomagnetic disturbance during the International Geophysical Year and the Second Polar Year. Rep. Ionos. Space Res. Japan **16**, 37—56 (1962).

[7] MAYAUD, P. N.: Activité magnétique dans les régions polaires. Exp. Polaires Francaises, Resultats Scientifiques N⁰ S: IV. 2 (1955) or Ann. Géophys. **12**, 84—101 (1956).

[8] HAKURA, Y.: Tables and maps of geomagnetic coordinates corrected by the higher order spherical harmonic terms. Rep. Ionos. Space Res. Japan **19**, 121—157 (1965).

[9] NAGATA, T., and S. ABE: Notes on the distribution of **SC*** in high latitudes. Rep. Ionos. Res. Japan **9**, 39—44 (1955).

---

[5] A. NISHIDA, S. KOKUBUN, and N. IWASAKI: Rep. Ionos. Space Res. Japan **20**, 73 (1966).

[1] See contributions by R. GENDRIN in this volume, p. 461, and by E. SELZER in Vol. 49/4 of this Encyclopedia.

[2] This is the propagation velocity $V_A = B/(\mu_0 \varrho)^{\frac{1}{2}}$ of gasmagnetic "shear waves", see [36] and the contribution by V. L. GINZBURG, and A. A. RUHADZE iu Vol. 49/4 of this Encylcopedia.

[10] MATSUSHITA, S.: On geomagnetic sudden commencements, sudden impulses, and storm durations. J. Geophys. Res. **67**, 3753—3777 (1962).
[11] RIKITAKE, T.: Electromagnetism and the earth's interior. Amsterdam: Elsevier Publ. Co. 1966.
[12] NISHIDA, A., and J. A. JACOBS: World-wide changes in the geomagnetic field. J. Geophys. Res. **67**, 525—540 (1962).
[13] OBAYASHI, T., and J. A. JACOBS: Sudden commencements of magnetic storms and atmospheric dynamo action. J. Geophys. Res. **62**, 589—616 (1957).
[14] WILSON, C. R., and M. SUGIURA: Hydromagnetic interpretation of sudden commencements of magnetic storms. J. Geophys. Res. **66**, 4097—4111 (1961).
[15] SUGIURA, M., and J. P. HEPPNER: The earth's magnetic field. Introduction to Space Science, ed. W. N. HESS. New York: Gordon & Beach 1965.
[16] SUGIURA, M., and S. CHAPMAN: The average morphology of geomagnetic storms with sudden commencements. Abh. Akad. Wiss. Göttingen, Math.-phys. Kl., Sonderheft Nr. 4, 53 pp. (1960).
[17] VESTINE, E. H., and S. CHAPMAN: The electric current-system of geomagnetic disturbance. Terr. Mag. **43**, 351—382 (1938).
[18] CHAPMAN, S.: The electric current-system of magnetic storms. Terr. Mag. **40**, 349—370 (1935).
[19] AKASOFU, S.-I., and S. CHAPMAN: A neutral line discharge theory of the aurora polaris. Phil. Trans. Roy. Soc. London A **253**, 359—406 (1961).
[20] AKASOFU, S.-I., and S. CHAPMAN: The ring current, geomagnetic disturbance and the Van Allen belts. J. Geophys. Res. **66**, 1321—1350 (1961).
[21] HARANG, L.: The mean field of disturbance of polar geomagnetic storms. Terr. Mag. **51**, 353—380 (1946).
[22] NIKOLSKY, A. P.: Dual laws of the course of magnetic disturbances and the nature of mean regular variation. Terr. Mag. **52**, 147—173 (1947).
[23] WHITHAM, K., E. I. LOOMER, and E. R. NIBLETT: The latitudinal distribution of magnetic activity in Canada. J. Geophys. Res. **65**, 3961—3974 (1960).
[24] HOPE, E. R.: Low-latitude and high-latitude geomagnetic agitation. J. Geophys. Res. **66**, 747—776 (1961).
[25] OGUTI, T.: Inter-relations among the upper atmosphere disturbance phenomena in the auroral zone — II. Spiral pattern of the polar aeronomical disturbances. Rep. Ionos. Space Res. Japan **16**, 363—386 (1962).
[26] NAGATA, T.: Polar geomagnetic disturbances. Planet. Space Sci. **11**, 1395—1427 (1963).
[27] VESTINE, E. H.: The disturbance-field of magnetic storms. Trans. Washington Meeting, IATME Bull. No. 10, 360—381 (1940).
[28] FUKUSHIMA, N.: Polar magnetic storms and geomagnetic bays. J. Fac. Sci. Univ. Tokyo, Sect. II, **8**, 293—412 (1953).
[29] BIRKELAND, K.: Norwegian aurora polaris expedition, 1902—1903, 2 vols. Christiania, 1908, 1913.
[30] AKASOFU, S.-I., S. CHAPMAN, and C.-I. MENG: The polar electrojet. J. Atmosph. Terr. Phys. **27**, 1275—1305 (1965).
[31] FUKUSHIMA, N., and T. HIRASAWA: Frequent occurrence of local geomagnetic disturbance in high latitudes. J. Geomag. Geoelectr. **15**, 161—171 (1964).
[32] SILSBEE, H. C., and E. H. VESTINE: Geomagnetic bays, their frequency and current-system. Terr. Mag. **47**, 195—208 (1942).
[33] OGUTI, T.: Inter-relations among the upper atmosphere disturbance phenomena in the auroral zone. Japanese Antarctic Res. Exp., Sci. Rep. Series A, No. 1, 82 pp. (1963).
[34] COLE, K. D.: Magnetic storms and associated phenomena. Space Sci. Rev. **5**, 699—770 (1966).
[35] AKASOFU, S.-I.: Electrodynamics of the magnetosphere: geomagnetic storms: Space Sci. Rev. **6**, 21—143 (1966).
[36] ALFVÉN, H.: Cosmical electrodynamics. Oxford: Clarendon Press 1950; 2nd ed. with C.-G. FÄLTHAMMAR, 1963. ALFVÉN, H.: Hydromagnetics of the magnetosphere. Space Sci. Rev. **2**, 862—870 (1963).
[37] PIDDINGTON, J. H.: Geomagnetic storms, aurorae and associated effects. Space Sci. Rev. **3**, 724—780 (1964).
[38] HINES, C. O.: Hydromagnetic motions in the magnetosphere. Space Sci. Rev. **3**, 342—379 (1964).
[39] OBAYASHI, T., and A. NISHIDA: Large-scale electric field in the magnetosphere. Space Sci. Rev. **8**, 3—31 (1968).
[40] MATSUSHITA, S., and W. H. CAMPBELL (Editors): Physics of Geomagnetic Phenomena, 2 vols. (Especially Chap. III. Quiet variation fields; IV. Disturbed variation fields; VI. Recent storm models.) New York: Academic Press.

Theoretical Aspects of the Worldwide Magnetic Storm Phenomenon.

By

E. N. Parker[*] and V. C. A. Ferraro[**].

With 16 Figures.

A. Introduction.

1. Survey description. The geomagnetic field at the surface of Earth is observed to undergo continual fluctuation of small amplitude. The fluctuations cover a broad range of frequencies and intensity, and are collectively called *geomagnetic activity*. The most important of these fluctuations are worldwide and are called *geomagnetic storms*. (We exclude the slow secular changes in the geomagnetic field and the field reversals which seem to occur every 10^6 years or so.) Local magnetic fluctuations during a storm are often much larger than the characteristic worldwide average changes, particularly in the auroral zones. The worldwide storm phenomenon is generally associated with enhanced auroral activity and cosmic ray depression. On good grounds magnetic storms are nowadays attributed to the interaction of the active solar wind with the geomagnetic field. The physical problem presented by the storm is to understand how the forces of the wind lead to the observed magnetic fluctuations.

α) Like most natural phenomena in the real world, the geomagnetic storm is enormously complex and varied. So the first step is to distinguish the central from the peripheral effects. Early analyses of the magnetic storm fluctuations sought to derive their average characteristics and a useful point of view is to treat the *worldwide average* magnetic variations as the major physical effects [7]. The auroral, cosmic ray, ionospheric, and local magnetic effects may then be treated arbitrarily as peripheral, providing important clues where needed, but otherwise not carrying the central thread of the theory.

β) Consider, then, the worldwide average magnetic variations which characterize the storm. The typical storm begins suddenly and almost simultaneously all over the world, this being referred to as its *sudden commencement*.

It is characterized in low and temperate latitudes by a sudden worldwide increase in $\mathcal{H}$, the horizontal component of the geomagnetic field at the surface of Earth. The increase in $\mathcal{H}$ for a typical storm varies from 20 to 30 γ[***] representing an increase of almost one part in 10^3. The increase is largest at equatorial stations and generally has a rise time of the order of 1 to 6 minutes, with 2 or 3 minutes most probable.

The *initial phase* of the storm, which follows the sudden commencement, is the period of 3 to 8 hours during which $\mathcal{H}$ is above its initial undisturbed value; that is, the value preceding the storm. The general increase may be as large as 50 γ or more in very large storms.

---

[*] Principal author of Chapters A through E. Manuscript received June 1966.
[**] Principal author of Chapter F which is a very short résumé of the original paper (written in 1955) received June 1966.
[***] $1\,\gamma = 10^{-5}$ Gs $= 10^{-9}$ Vs m$^{-2} = 10^{-9}$ T.

The *main phase* follows the initial phase and is characterized by a decrease of H below the undisturbed value for greater than the increase in H during the initial phase. The main phase typically endures for 12 to 24 hours, the period being shorter the more intense the disturbance. The main phase is followed by the *recovery phase*, during which the field relaxes back towards its normal prestorm value in a characteristic time of the order of a day. The main phase decrease of H

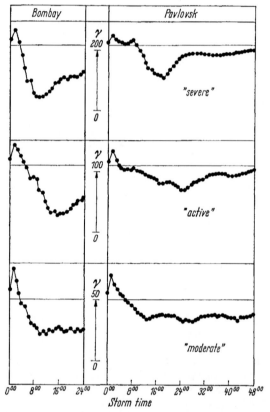

Fig. 1. Averaged magnetic storm variations of the horizontal component of the geomagnetic field as a function of time after the sudden commencement. The magnetic storms are divided into three intensities and the average of each class of storm was taken. Note the difference in scale for each class. The great magnetic storms recover much faster than moderate storms. [After S. Chapman [1].]

below the undisturbed value may be 200 γ in a large storm at middle latitudes, with 50 to 100 γ as a typical figure. The field may be quite 'noisy' during the active main phase, sometimes involving positive and negative local excursions of several hundred γ in times of half an hour. The variations may also be enormously enhanced in the auroral regions (but these are smoothed out in the average characteristics).

γ) Storms differ markedly in the development of their distinctive phases, some storms having no perceptible sudden commencement, or no initial phase, or no perceptible main phase. An individual storm may appear particularly atypical if the record of only one or two observing stations is consulted for that event.

---

[1] S. Chapman: Proc. Roy. Soc. London A **115**, 242 (1927).

The recovery phase or relaxation following the storm proceeds more slowly with time. We mentioned a characteristic period of one day, but in fact the characteristic relaxation time is perhaps only 6—12 hours at the end of the active main phase, increasing rapidly as the recovery proceeds so that some time later the characteristic recovery time may be several days. The recovery thus goes more like an inverse power of time t than an exponential of it. The largest storms go through the initial and main phases more rapidly than moderate storms. There is also a tendency for storms of a given magnitude to recover more rapidly at solar minimum than during maximum.

δ) The average characteristics of magnetic storms was first studied by Moos in 1910 and later more extensively by CHAPMAN in 1919; he analyzed the fluctuations into a part symmetrical with respect to the geomagnetic axis (denoted by $\mathbf{D}_{st}$) and depending on *storm-time*, that is, the time measured from the onset of the storm (usually its sudden commencement) and a part depending on local time but not, of course, necessarily diurnal in character. Fig. 1 shows the storm-time variation $\mathbf{D}_{st}$ for the average of forty storms arranged in low and high latitude for the horizontal force $\mathcal{H}$.

The residual fluctuation, that is, the fluctuations which are left over after subtracting the storm time variation from the actual mean fluctuation is denoted by $\mathbf{S}_D$ or $\mathbf{D}_S$. It is especially intense in the polar regions where it is undoubtedly produced by two polar electrojets, i.e., highly concentrated currents flowing in the ionosphere along the auroral zones, one along the early morning side and the other in the late afternoon side of the earth. For a more detailed description of magnetic storm behavior see [1-7] [4], [5], [6], [7].

ε) Magnetic storms can be classified into two general types, depending upon the conditions of the solar wind which produce them. The *largest storms* usually follow one or two days after a violent local outburst (*flare*) at the sun and are the result of a blast wave propagating in the solar wind from the sun out through interplanetary space[8,9] [19]. They are characterized by a sudden commencement and have no tendency to recur after one or more solar rotations. The other type is the *recurring storm*, which has a tendency to recur after one or more solar rotations (of 27-day duration). These storms arise from the shear, turbulence, contact surfaces, and shock waves [19][10,11] between regions of different wind speed from the sun (see observations[12] and [13]). The rotation of the wind pattern with the sun leads to the 27-day recurrence. The effects produced in the geomagnetic field are much the same under both wind conditions, for reasons that will become clear later.

ζ) The magnetic storm is often accompanied by a cosmic ray decrease first observed by FORBUSH[14]; it is caused by the same blast wave or changing wind

---

[2] See contribution by T. NAGATA, and K. FUKUSHIMA in this volume, p. 5.
[3] M. SUGIURA, and S. CHAPMAN: Abhandl. Akad. Wiss. Göttingen, Math.-Phys. Kl. 1960.
[4] S. MATSUSHITA: J. Geophys. Res. **67**, 3753 (1962).
[5] T. NAGATA: Planet. Space Sci. **11**, 1395 (1963).
[6] S. I. AKASOFU: Planet Space Sci. **12**, 273 (1964); — Space Sci. Rev. **4**, 498 (1965).
[7] See A. B. MEINEL, S. I. AKASOFU, and S. CHAPMAN: this Encyclopedia, Vol. 49/1, contribution on aurorae.
[8] T. GOLD: Nuovo Cimento Suppl. Series 10, **13**, 318 (1959).
[9] E. N. PARKER: Astrophys. J. **133**, 1014 (1961).
[10] V. SARABHAI: J. Geophys. Res. **68**, 1555 (1963).
[11] A. J. DESSLER, and J. A. FEJER: Planet. Space Sci. **11**, 505 (1963).
[12] C. W. SNYDER, and M. NEUGEBAUER: [26] (1964).
[13] N. F. NESS, and J. M. WILCOX: Phys. Rev. Letters **13**, 461 (1964).
[14] S. E. FORBUSH: J. Geophys. Res. **59**, 525 (1954); — This Encyclopedia, Vol. 49/1, pp. 180, 227.

pattern [*19*]⁹, ¹⁰ as the storm. The polar substorm with its large magnetic variations and auroral activity, accompanying the world wide storm, is caused at least in part by the enhanced convection of the magnetosphere[15]. It is evident that the whole storm phenomenon, involving the world wide geomagnetic changes observed on Earth, the polar substorm, the enhanced aurora, the radiation belts detected by van Allen, and the associated ionospheric activity, are all consequences of the impact of the solar wind against the geomagnetic field.

This fact is at once a simplification and a complication. It is a complication because it means that the diverse effects (storm, aurora, etc.) are all intimately related, however distinct the phenomenology may appear. It is not possible to review one without giving some thought to the rôle of all the others.

2. Introductory remarks.

In the past ten years, and particularly in the last five, a widespread interest has developed in the many terrestrial effects of the solar wind. The interest has been given impetus by the direct observation of fields, particles, and plasmas in space, which is responsible for much of the present understanding of the effects. The natural result of the widespread interest, reinforced by automatic data acquisition, is a vast and rapidly increasing literature on the subject. Altogether the observational data is approaching 10^{12} bits of information, the number of published papers is greater than 10^3, and the number of speculations on the meaning of the observational data is probably only a little less than 10^3. Altogether the field of inquiry is complicated by the breadth of the phenomenon being studied, by the vast amount of information available, by the difficulty of communicating among so many workers, and by the other one to bring some extensive ideas and models into agreement with recognized basic principles. It is evident, therefore, that a summary of the physics of the geomagnetic storm involves a critical distillation of a vast and fertile green sea of facts, ideas, and preferences. Such an undertaking cannot avoid reflecting the personal outlook, interests, and biases of the authors in its emphasis on the various facets of the overall problem. The reader is forwarned, therefore, that the present article leans toward the theory of the world wide average variation field; the trend is to relate the variation field to conditions in the wind by the known dynamical properties of plasmas and fields. On the other hand, the authors shall try to give fair abstracts of opposing views.

This article begins, therefore, with an attempt to review the general problem of the geomagnetic field, its internal gases and particles, and the external wind, before taking up the central thread of the world wide magnetic storm. The aurora, the radiation belts, and the ionosphere all come into the construction of the basic processes of the magnetic storm. Sect. B contains a brief survey of the whole interaction of the solar wind with the geomagnetic field, insofar as it is presently understood, with some extended discussion of those aspects bearing directly on the worldwide storm. Sect. C presents the basic properties and principles of the world wide magnetic storm in preparation for Sect. D, which reviews the formal theory of compression of the field by the pressure of the wind (i.e. the sudden commencement and initial phase), and in preparation for Sect. E which treats the formal theory of inflation of the geomagnetic field (i.e. the main phase). Sect. F is a brief review of the history of scientific inquiry into the magnetic storm phenomenon. The long history owes its early origin to the fact that the aurora is visible to the naked eye and the magnetic variations of a large storm can be detected by an experienced and acute observer with a magnetic needle suspended by a thread.

Altogether, it is our impression that the general outline of the magnetic storm and the aurora is now understood and correctly formulated in terms of the principles of Newton, Maxwell, Lorentz, etc. as applied to the behavior of the ionized gases and fields in space. The world wide features of the magnetic storm and aurora appear as straightforward consequences of the solar wind. On the other hand, it is also our impression that there is a considerable lack of understanding about the nature of the local effects and the intermediate

---

[15] W. I. Axford, and C. O. Hines: Can J. Phys. **39**, 1433 (1961).

B. The space environment of earth.
I. General description.
3. The solar wind.

The continued interaction of the solar wind with the geomagnetic field generates some very complicated "weather" in and around the geomagnetic field. The magnetic storm is one manifestation of the most violent of this weather, occurring when the solar wind is strong and more irregular than usual. In turn, the strength and irregularity of the wind are caused by "weather" in the solar corona. The strength and irregularity of the wind increases with solar activity. The reader is referred to such standard works as [7], [4], [6][1] for a presentation of the detailed correlation which exists between solar activity and geomagnetic effects, and to [21] and [2] for a detailed description of aurorae and their correlation with solar activity.

α) The connection between solar activity and geomagnetic effects was established long ago by MAUNDER, CHREE and others; on good grounds geomagnetic disturbance was attributed to a *stream of charged particles* emitted from the Sun. It is now generally recognized that this corpuscular emission is the outward hydrodynamic expansion of the solar corona [19][3,4], called the solar wind, which is a consequence of the very high (10^6 °K) temperature of the corona. The expansion velocity of the wind may reach 300 to 1500 km/sec for coronal temperatures of 0.5 to $4 \cdot 10^6$ °K. The wind is composed of coronal gas, fully ionized and mainly hydrogen, with perhaps one in ten atoms of helium. The wind is now observed directly in space and the velocity measured varies from 350 to 600 km/sec with a density of 2 to 20 hydrogen atoms per cm^3 at the orbit of Earth during the recent solar minimum[5-10]. (The wind velocity and density at solar maximum has yet to be observed.) The temperature corresponding to the wind is observed to fluctuate widely, from 10^4 °K up to 10^6 °K during times of high solar activity[10,11]. The solar wind is believed to be faster, denser, and/or more turbulent during high solar activity; this feeling is based on the stronger geomagnetic, auroral, and cosmic ray effects appearing then.

It was shown many years ago that the delay between a major outburst of light (flare) on the sun and the principal terrestrial effects is typically 1 to 2 days, implying that the wind velocity reached 1 to $2 \cdot 10^3$ km/sec at such times. Sudden heating of the solar corona at the time of a flare is presumed to be responsible for the suddenly enhanced wind which causes the storm. The sudden increase in coronal expansion produced a hydrodynamic blast wave [16] whose outward

---

[1] S. MATSUSHITA: J. Geophys. Res. **67**, 3753 (1962).
[2] A. B. MEINEL, S. I. AKASOFU, and S. CHAPMAN: this Encyclopedia, Vol. 49/1.
[3] E. N. PARKER: Astrophys. J. **128**, 664 (1958); **132**, 1445 (1960); **139**, 72, 93, 690 (1964); **141**, 322 (1965); **142**, 32 (1966); — Space Sci. Rev. **4**, 666 (1965).
[4] L. M. NOBLE, and F. L. SCARF: Astrophys. J. **138**, 1169 (1963).
[5] I. S. ŠKLOVSKIJ, V. I. MOROZ, and V. G. KURT: Astro. Žur. **37**, 931 (1960).
[6] K. I. GRINGAUZ, S. M. BALANDINA, G. A. BORDOVSKIJ, and N. M. ŠUTTE: [25], 432 (1963). — K. I. GRINGAUZ, V. V. BEZRUKIH, S. M. BALANDINA, V. P. OZEROV, and R. E. RIBČINSKIJ: Planet. Space Sci. **12**, 87 (1964). — K. I. GRINGAUZ, V. V. BEZRUKIH, V. D. OZEROV, and R. E. RIBČINSKIJ: Dokl. Akad. Nauk. SSSR **131**, 1301 (1960); (engl. transl.) Soviet Physics Doklady **5**, 361 (1960).
[7] M. NEUGEBAUER, and C. W. SNYDER: Science **138**, 1095 (1963).
[8] A. BONETTI, H. S. BRIDGE, A. J. LAZARUS, E. F. LYON, B. Rossi, and F. SCHERB: [25], 540 (1963).
[9] H. S. BRIDGE, A. EGIDI, A. J. LAZARUS, E. LYON, and L. JACOBSEN: [27] (1965).
[10] M. NEUGEBAUER, and C. W. SNYDER: J. Geophys. Res. **70**, 1587 (1965).
[11] P. A. STURROCK, and R. E. HARTLE: Phys. Rev. Letters **16**, 628 (1966).

velocity is 1 to $2 \cdot 10^3$ km/sec, inferred from the delay times. Judging from the frequency with which the blast waves hit the geomagnetic field, the angular widths of the blast waves are of the order of $90°$, whilst the compression of the geomagnetic field suggests that the density near the earth of the strongest blast waves may occasionally be as high as 50 to 100 atoms per cm^3.

β) The solar wind is filled with the magnetic *lines of force* of the general solar field[12] [*16*] which is observed[13-15] to be of the general order of a few gauss at the solar photosphere. The outward expansion of the solar corona into the solar wind carries the magnetic lines of force out through interplanetary space. The solar wind and the extended solar field evidently fill all of interplanetary space out to a distance of at least ten astronomical units, i.e. $1.5 \cdot 10^9$ km. The magnetic lines of force extending out through interplanetary space maintain a prolonged connection with the sun as a consequence of the high electrical conductivity of the wind. The lines of force would be approximately radial in space were it not for the 27-day rotation of the sun. The result of the rotation is that the lines of force are twisted into a spiral pattern of the form

$$r = v\,\varphi/\Omega \sin \theta, \tag{3.1}$$

where v is the wind velocity, Ω is the angular veloctiy of the sun, r is distance measured from the sun, θ is the polar angle measured from the axis of rotation of the sun, and φ is azimuth measured around the sun.

A quiet day wind velocity of the order of 400 km/sec leads to an angle of about $45°$ between the lines of force and the radial direction at the orbit of Earth. The field strength at the orbit of Earth is about $3 \cdot 10^{-5}$ times that at the photosphere. This theoretical interplanetary magnetic field has been observed recently[16, 17] confirming the predicted overall spiral pattern, and showing that the field strength if usually 5 to $7 \cdot 10^{-5}$ gauss on quiet days, corresponding to 2 gauss at the solar surface. It is the outward convection of the irregularities of the magnetic field in the solar wind which is responsible for the reduced cosmic ray intensity in the solar system[18] [*19*].

γ) There is evidence from observation of "Pioneer V" that the field strength may go as high as $40 \cdot 10^{-5}$ gauss[19] when there has been an *outburst of light radiation* at the sun called a *solar flare*. Theory predicts that the blast waves from the sun should compress the field at such times, giving an asymmetric field pattern which, along with the enhanced turbulence in the field, is responsible for the Forbush-decrease in the cosmic ray intensity[20] [*19*]. The same blast wave is responsible for the large magnetic storm. The Forbush-decrease and the magnetic storm depend upon somewhat different aspects of the blast wave. Hence it is perhaps not surprising to find the relative cosmic ray and magnetic storm effects widely different for different events.

δ) The observations of Ness and Wilcox[17] show that the interplanetary magnetic field had a simple *sector structure*, at sunspot minimum, presumably as a consequence of sectoring of field and corona at the sun. They found that in 1964 the interplanetary field near the equatorial plane of the sun was divided into four sectors, according to whether the field was predominantly toward or away from

---

[12] E. N. Parker: Astrophys. J. **128**, 664 (1958).
[13] H. W. Babcock, and H. D. Babcock: Astrophys. J. **121**, 349 (1955).
[14] H. D. Babcock: Astrophys. J. **130**, 364 (1959).
[15] V. Bumba, and R. Howard: Astrophys. J. **141**, 1502 (1965).
[16] N. F. Ness, C. S. Scearce, and J. B. Seck: J. Geophys. Res. **69**, 3531 (1964).; J. Geophys. Res. **70**, 5793 (1965).
[17] N. F. Ness, and J. M. Wilcox: Phys. Rev. Letters **13**, 461 (1964); J. Geophys. Res. **70**, 5793 (1965).
[18] E. N. Parker: Phys. Rev. **110**, 1445 (1958). See also [*19*].
[19] P. J. Coleman, L. Davis, and C. P. Sonett: Phys. Rev. Letters **5**, 43 (1960).
[20] E. N. Parker: Astrophys. J. **133**, 1014 (1961).

the sun. The sector pattern rotates with the sun, of course. The field reverses abruptly across each sector boundary to give the next sector of field with opposite sense. During the period of observation the widths of the sectors were in the ratio 2:2:2:1, with the field inward toward the sun in the narrow sector. The solar wind tends to be somewhat faster and denser toward the leading edge of each sector. Evidently the sector structure changes slowly with time, there being only two sectors observed a year later. A net residual southward interplanetary field was also observed, which, if correct, is puzzling in view of the enormous net magnetic flux that it represents being carried away from the sun.

Geomagnetic activity is strongly correlated with wind velocity[6, 21, 22]. The wind velocity is now known to correlate with position in the sector, so it can also

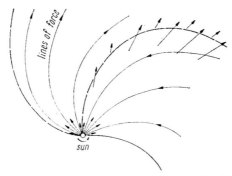

Fig. 2. Sketch of the crowding of the leading edge of one sector into the trailing edge of the next sector. The continuous lines represent magnetic lines of force. The short arrows indicate the direction, and by their length, the relative magnitude of the wind velocity. The broken lines represent sector boundaries. The sketch is drawn for the idealized case that the wind velocity is largest near the sun in the center of the sector. The radial motion of the wind and the spiral boundary soon brings the faster motion to the boundary [19].

be said that geomagnetic activity tends to be higher near the leading edge of each sector[17]. The actual physical connection is not known, but it is interesting to consider the fact that a stronger wind near the leading edge of each sector causes the leading edge to crowd into the trailing edge of the sector ahead, as sketched in Fig. 2. This should lead to turbulence and/or a shock wave toward the leading edge of each sector[23, 24].

SONETT, et al.[25] have given a detailed analysis of a magnetic jump, perhaps caused by the sector observed by "Mariner II" at a distance of $1 \cdot 10^7$ km from Earth and followed 4.7 hours later by a sudden commencement at Earth. They find that the jump was thin ($< 3 \cdot 10^4$ km); it must, evidently, be interpreted as a collisionless shock. Recently the dynamics of the leading edge of a sector has been worked out in an approximate way [26–29], giving a more detailed picture of the possible structure. It is found that there may be both a forward and a backward propagating shock, and of course there may be no shock if the wind velocity differences are not great enough. The sector boundary is followed by enhanced wind velocity and density, enhanced interplanetary field, and/or enhanced turbulence and disorder. Generally speaking, this is also what would be expected with the passage of a blast wave.

---

[21] C. W. SNYDER, M. NEUGEBAUER, and U. R. RAO: J. Geophys. Res. **68**, 6361 (1963).
[22] J. HIRSCHBERG: J. Geophys. Res. **70**, 4159 (1965).
[23] A. J. DESSLER, and J. A. FEJER: Planet Space Sci. **11**, 505 (1963).
[24] V. SARABHAI: J. Geophys. Res. **68**, 1555 (1963).
[25] C. P. SONETT, D. S. COLBURN, L. DAVIS, E. J. SMITH, and P. J. COLEMAN: Phys. Rev. Letters **13**, 153 (1964).
[26] C. P. SONETT, and D. S. COLBURN: Planet. Space Sci. **13**, 675 (1965).
[27] P. A. STURROCK, and J. R. SPREITER: J. Geophys. Res. **70**, 5345 (1965).
[28] J. HIRSCHBERG: J. Geophys. Res. **70**, 5353 (1965).
[29] M. SIMON, and W. I. AXFORD: Planet. Space Sci. **14**, 901 (1966).

So it is now generally accepted that the leading portions of the sectors may be responsible for the recurring magnetic storms[23], in much the same way that the blast wave from an outburst at the sun produces a magnetic storm. The storms produced by blast waves dominate the years of solar maximum. In the absence of solar flares during the less active years, the recurring storms are the major effect.

4. The magnetosphere. We come now to the central problem, the interaction of the solar wind with the geomagnetic field. For purposes of discussion the geomagnetic field and the surrounding wind may be divided into the regions sketched in Fig. 3.

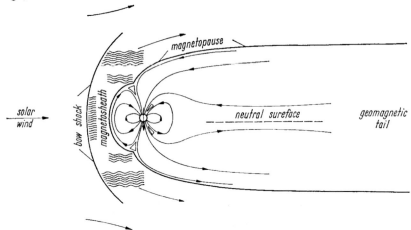

Fig. 3. Sketch of the configuration of the geomagnetic field in the solar wind. The sketch represents a section through the noon-midnight meridian plane, with north at the top. No attempt has been made to sketch the behavior of the interplanetary field, though the geomagnetic lines of force are indicated by the light lines. The length of the tail in the anti-solar direction is unknown. The four wedges indicate the auroral zones.

α) The wind is composed of ionized gas, so it does not penetrate freely into the geomagnetic field. Thus the pressure of the wind confines the geomagnetic field to a finite volume around Earth $[9]^{1-3}$. Hence the geomagnetic field is a blunt obstacle in the path of the solar wind. The Mach number of the solar wind appears to lie in the general range of 3 to 20, so that a standoff shock is formed in the wind upstream from the obstacle[4-11]. The existence of the *bow shock* upstream from the magnetic field is demanded by the fact that there is no strong signal which can propagate upstream in the *un*shocked wind to divert the wind around the obstacle. Only after the gas is shocked, so that the wind is subsonic, can the diversion around the obstacle take place. The shocked gas behind the bow shock forms a thick layer called the *magnetosheath*. The boundary between

---

[1] S. Chapman, and V. C. A. Ferraro: Terr. Mag. **36**, 77, 171 (1931).
[2] S. Chapman, and V. C. A. Ferraro: Terr. Mag. **45**, 245 (1940).
[3] E. N. Parker: Phys. Fluids **1** 171 (1958).
[4] V. N. Žigulev, and E. A. Romiševskij: Dokl. Akad. Nauk. SSSR **127**, 1001 (1960); (engl. transl.) Soviet Physics Doklady **4**, 859 (1960).
[5] T. Gold: Nuovo Cimento Suppl. Series 10, **13**, 318 (1959).
[6] E. N. Parker, and D. A. Tidman: Phys. Fluids **3**, 369 (1960).
[7] W. I. Axford: J. Geophys. Res. **67**, 3791 (1962).
[8] P. J. Kellogg: J. Geophys. Res. **67**, 3805 (1962).
[9] J. R. Spreiter, and W. P. Jones: J. Geophys. Res. **68**, 3555 (1963).
[10] G. K. Walters: J. Geophys. Res. **69**, 1769 (1964).
[11] G. K. Walters: J. Geophys. Res. **71**, 1341 (1966).

the shocked gas and the geomagnetic field is called the *magnetopause* (in analogy to the relatively abrupt transitions between regions in the lower atmosphere). Observations[12-14] place the quiet-day bow shock at about 14 R_E and the quiet-day magnetopause at about 10 R_E in the solar direction*, where they are at their minimum distance from Earth.

β) The region inside the magnetopause is occupied by the geomagnetic field and is normally referred to as the *magnetosphere*. The region is filled with the

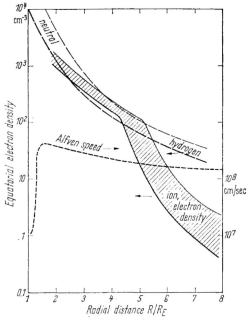

Fig. 4. Plot of the electron density (solid lines), neutral hydrogen density (broken lines) and the approximate Alfvén speed (dotted line) at low latitudes as a function of radial distance from the center of Earth measured in units of the radius of Earth, $R_E = 6.4 \times 10^3$ km. The neutral hydrogen density varies over the eleven year sunspot cycle, reaching the upper curve at sunspot minimum (thin, broken line) and the lower at sunspot maximum (thick, broken line). The electron and/or ion density varies with the level of geomagnetic activity present at any given time, tending to be lower (thick curve) when there is activity.

tenuous gases of the outer terrestrial atmosphere and probably some gas which has diffused or convected inward from the solar wind. The electron and/or ion density is compiled in Fig. 4 from the published literature [13][15-18]. The two solid curves in Fig. 4 represent both the approximate limit of the variations with time and the present uncertainties. The densities are known to vary considerably with time, apparently sometimes dropping to very low values in the outer magnetosphere during magnetic activity[19, 20]. Above about 2000 km altitude the atmosphere is mainly hydrogen. Disturbances in the magnetosphere propagate as gas-magnetic

---

* $R_E = 6370$ km ≈ 6400 km is the radius of Earth which is a convenient unit of length for describing the geomagnetic field.

[12] N. F. NESS, C. S. SCEARCE, and J. B. SECK: J. Geophys. Res. **69**, 3531 (1964).
[13] J. W. FREEMAN, J. A. VAN ALLEN, and L. J. CAHILL: J. Geophys. Res. **68**, 2121 (1963).
[14] J. H. WOLFE, R. W. SILVA, and M. A. MYERS: J. Geophys. Res. **71**, 1319 (1966).
[15] J. H. POPE: J. Geophys. Res. **66**, 67 (1961).
[16] R. L. SMITH: J. Geophys. Res. **66**, 3709 (1961).
[17] H. B. LIEMOHN, and F. L. SCARF: J. Geophys. Res. **69**, 833 (1964).
[18] R. L. DOWDEN, and M. W. EMERY: Planet. Space Sci. **13**, 773 (1965).
[19] D. L. CARPENTER: J. Geophys. Res. **68**, 1675 (1963); J. Geophys. Res. **71**, 693 (1966).
[20] J. J. ANGERAMI, and D. L. CARPENTER: J. Geophys. Res. **71**, 711 (1966).

waves with the *Alfven-speed*,

$$V_A = \frac{B}{\sqrt{u\,\mu_0\,\varrho}}, \tag{4.1}$$

represented approximately by the dotted curve in Fig. 4. The broken curves in Fig. 4 represent the neutral atomic hydrogen density at sunspot maximum and minimum[21].

γ) The magnetosphere stretches out behind Earth to form the *geomagnetic tail*, whose length is very large but unknown and much debated. Estimates range up to several astronomical units*. The magnetic observations carried out by NESS[22] show that at a distance of 30 R_E the magnetic lines of force are extended approximately parallel to the Earth-sun line, with a neutral sheet of no more than a few hundred km thickness between the regions of opposite field. This is illustrated in Fig. 3. There is some evidence that the tail may have an important connection with the aurora[23-26]. It is observed[27,28] that the field in the tail is stronger than usual when magnetic activity is high, but becomes low with the onset of negative bays in the same time sector.

II. Processes and problems.

The overall picture of the geomagnetic field and confining solar wind is constructed from the observational data and the known theoretical properties of gases, particles, and fields. Much, but not all, of the gross picture and the basic physical processes appears to be understood now in terms of the principles of NEWTON, MAXWELL, and LORENTZ. We will try to make clear the questions which are still outstanding as we go over what is presently known about the geomagnetic field and its behavior.

5. The solar wind and the magnetosphere.

α) To begin with the solar wind, there is at present no detailed quantitative understanding of the temperature of the gas, the degree of magnetic disorder, the structure of shock transitions and blast waves, etc., all of which are important parts of the interplanetary "weather". There is very little observational data and theory can proceed only to the extent that observation can point the way. The behavior of the solar wind as it is shocked and passes around the magnetopause is understood only in the gas-dynamic approximation. There must be a bow shock, for the reasons mentioned above, but the shock is collisionless and the structure of collisionless shocks is not fully understood. The situation is complicated by the fact that there appears to be more than one effect contributing to the shock transition[1-14].

* 1 astronomical unit (a.u.) = $1.5 \cdot 10^8$ km.
[21] F. S. JOHNSON: J. Geophys. Res. **65**, 3049 (1960).
[22] N. F. NESS: J. Geophys. Res. **70**, 2999 (1965).
[23] I. B. McDIARMID, and J. R. BURROWS: Can. J. Phys. **42**, 616 (1964).
[24] I. B. McDIARMID, and J. R. BURROWS: J. Geophys. Res. **70**, 3031 (1965).
[25] A. J. DESSLER, and R. D. JUDAY: Planet. Space Sci. **13**, 63 (1965).
[26] W. I. AXFORD, H. E. PETSCHEK, and G. L. SISCOE: J. Geophys. Res. **70**, 1231 (1965).
[27] J. P. HEPPNER: Space Sci. Rev. **7**, 166 (1967).
[28] N. F. NESS, and D. J. WILLIAMS: J. Geophys. Res. **71**, 322 (1966).
[1] J. H. ADLAM, and J. E. ALLEN: Phil. Mag. **3**, 448 (1958).
[2] L. DAVIS, R. LÜST, and A. SCHLÜTER: Z. Naturforsch. **13a**, 916 (1958).
[3] C. S. GARDNER, H. GOERTZEL, H. GRAD, C. S. MORAWETZ, M. H. ROSE, and H. RUBIN: Proc. Second UN intern. Conf. Peaceful Uses of Atomic Energy **31**, 230 (1958).
[4] S. A. COLGATE: Phys. Fluids **2**, 485 (1959).
[5] F. J. FISHMAN, A. R. KANTROWITZ, and H. E. PETSCHEK: Rev. Mod. Phys. **32**, 959 (1960).
[6] P. L. AUER, H. HURWITZ, and R. M. KILB: Phys. Fluids **4**, 1105 (1961).
[7] P. L. AUER, H. HURWITZ, and R. M. KILB: Phys. Fluids **5**, 298 (1962).
[8] P. J. KELLOGG: Phys. Fluids **7**, 1555 (1964).
[9] G. K. WALTERS: J. Geophys. Res. **69**, 1769 (1964).
[10] P. D. NOERDLINGER: J. Geophys. Res. **69**, 369 (1964).
[11] W. BERNSTEIN, R. W. FREDRICKS, and F. L. SCARF: J. Geophys. Res. **69**, 1201 (1964).
[12] F. L. SCARF, W. R. BERNSTEIN, and R. W. FREDRICKS: J. Geophys. Res. **70**, 9 (1965).
[13] D. B. BEARD: J. Geophys. Res. **70**, 4181 (1965).
[14] F. C. MICHEL: Phys. Fluids **8**, 1283 (1965).

The mechanism by which the shock converts the streaming motion into small scale random motions is not understood, nor, at the moment, has the degree of randomization been observed unambiguously. Fast particles are generated somewhere in the region[15] (see also [16]), and it has been suggested that the generation mechanism is Fermi-acceleration[17-19] (see Sect. 9α) and/or resonance with ion acoustic waves [20-22].

β) The shock may alter or produce disorder in the wind on various scales, with the result that the plasma which is actually in contact with the outer boundary of the geomagnetic field is "turbulent" [23-26,11]. DESSLER and FEJER[27] argue that the shock may introduce no new turbulence, but instead serve only to compress the irregularities which are already present in the solar wind, so that a perfectly steady wind would lead to no "turbulence" in the magnetosheath. The relative importance of *turbulence* in the solar wind as it approaches Earth, and the turbulence generated by the bow shock, for agitating the outer boundary of the geomagnetic field is still a subject of debate. The issue is clouded by the question of whether the magnetopause is itself stable, or whether it is unstable and amplifies the agitation in the wind. The issue is complicated by the possibility, suggested by DUNGEY, that the geomagnetic field merges rapidly with southward interplanetary fields (see Subsect. δ). Finally, the issue is complicated by the observational fact that, under both quiet and active conditions, the polar magnetic lines of force are stretched out behind Earth to form a tail, as illustrated in Fig. 3. The tail can be maintained only by strong forces in the anti-solar direction exerted on the extended lines of force. We cannot settle these questions here, so our efforts will be directed toward understanding the different facets and implications of the question.

First of all, it is largely the agitation at the magnetopause which is responsible for the continuing small-amplitude geomagnetic noise observed at ground level, particularly in the polar regions. The agitation is propagated inwards from the magnetopause as gas-magnetic waves. [These are most often called "hydromagnetic" or "magneto-hydrodynamic" in the literature, but as compressibility plays a certain rôle in these phenomena we prefer the terms "gas-magnetic" or "magnetogasdynamic"] [28-30] [*9*]. The characteristic speed of gas-magnetic "shear waves" is the *Alfven speed*

$$V_A = B/(\mathrm{u}\,\mu_0\,\varrho)^{\frac{1}{2}} \qquad [4.1]$$

shown in Fig. 4, Sect. 4β. The propagation of gasmagnetic waves in the magnetosphere is a subject in itself [31-33] and will be discussed briefly in a later section and in [34].

---

[15] C. Y. FAN, G. GLOECKLER, and J. A. SIMPSON: Phys. Rev. Letters **13**, 149 (1964).
[16] L. A. FRANK, J. A. VAN ALLEN, and E. MACAGNO: J. Geophys. Res. **68**, 3543 (1963).
[17] E. SCHATZMAN: Annal. d'Astrophys. **26**, 234 (1963).
[18] J. R. JOKIPII, and L. DAVIS: Phys. Rev. Letters **13**, 739 (1964).
[19] J. R. JOKIPII: Astrophys. J. **143**, 961 (1966).
[20] F. L. SCARF, G. M. CROOK, and R. W. FREDRICKS: J. Geophys. Res. **70**, 3054 (1965).
[21] R. W. FREDRICKS, F. L. SCARF, and W. BERNSTEIN: J. Geophys. Res. **70**, 21 (1965).
— R. W. FREDRICKS, and F. L. SCARF: J. Geophys. Res. **70**, 4765 (1965).
[22] See contribution by V. L. GINZBURG, and A. A. RUHADZE in Vol. 49/4 of this Encyclopedia, Sects. 20 and 26.
[23] A. J. DESSLER: J. Geophys. Res. **66**, 3587 (1961).
[24] A. J. DESSLER: J. Geophys. Res. **67**, 4392 (1962).
[25] W. I. AXFORD: J. Geophys. Res. **67**, 3791 (1962).
[26] P. J. KELLOGG: J. Geophys. Res. **67**, 3805 (1952).
[27] A. J. DESSLER, and J. A. FEJER: Planet. Space Sci. **11**, 505 (1963).
[28] C. O. HINES: J. Geophys. Res. **62**, 491 (1957).
[29] C. O. HINES and R. L. STOREY: J. Geophys. Res. **63**, 671 (1958).
[30] A. J. DESSLER: J. Geophys. Res. **63**, 405 (1958).
[31] A. J. DESSLER, W. B. HANSON, and E. N. PARKER: J. Geophys. Res. **66**, 3631 (1961).
[32] G. J. F. MACDONALD: J. Geophys. Res. **66**, 3639 (1961).
[33] R. L. CAROVILLANO, and J. F. MCCLAY: Phys. Fluids **8**, 2006 (1965).
[34] See contribution by V. L. GINZBURG, and A. A. RUHADZE in Vol. 49/4 of this Encyclopedia Sect. 26.

γ) Consider the stability of the magnetopause. If the boundary is stable, then the only source of agitation of the boundary is the turbulence in the wind. If the boundary is unstable, the boundary is self-agitating. The same question applies to the neutral sheet observed in the tail. Theory does not answer the question of stability in an unambiguous way. Considering the wind as a stream of free particles (which avoids the necessity for a bow shock) leads to a prediction of instability [9][35]. On the other hand, plasma instabilities and the interplanetary magnetic field cause the wind to behave more like a fluid than a stream of independently moving particles [10][36,37]. Treating the wind as a fluid leads to stability unless the field in the wind is very nearly parallel (or anti-parallel) to the geomagnetic field at the boundary[38-41]. DESSLER argues that both the shock and the magnetopause are stable, so that the observed agitation is mainly the result of pressure fluctuations in the wind itself[23,24,27,42,43]. The observed geomagnetic fluctuations would then be a direct measure of the turbulence in the wind approaching Earth. It is suggested also[42] that the large variation of the *Alfvén Mach number*[9] around the bow shock, as a consequence of the obliquity of the general interplanetary field, leads to an asymmetric tail behind Earth. A sudden change in the field direction carried in the wind should lead to a sudden change in the asymmetry of the tail. Altogether, then, in this picture the general level of geomagnetic activity is determined directly by the level of turbulence and/or magnetic variation in the solar wind.

The alternative point of view would be that the magnetopause and the geomagnetic tail have some intrinsic instability so that geomagnetic activity is largely determined by the degree to which the instability is operative. Obviously one may argue too that activity reflects both turbulence and instability.

δ) In addition to the question of turbulence in the wind and magnetopause instability, there is an additional possibility suggested by DUNGEY [9][44-46] that there occurs rapid dissipation and direct connection of the geomagnetic field (which is directed northward at the magnetopause) into southward components of the interplanetary field. The idea is simply that the gas squeezes out from between two regions of oppositely directed field, permitting the fields to move together and annihilate[47] with the fields reconnecting between the two regions at the edges of the region of annihilation, as sketched in Fig. 5. It has not yet been possible to demonstrate the postulated rapid dissipation because the dissipation presumably depends upon nonlinear collisionless processes, perhaps not unlike the bow shock. The proponents of the idea suggest that the rapid dissipation is the basis for understanding the geomagnetic tail and the production of fast particles.

There can be no doubt that oppositely directed fields annihilate much more rapidly when pressed together in such a way that the gas between can escape.

---

[35] E. N. PARKER: Phys. Fluids 1, 171 (1958).
[36] T. GOLD: Nuovo Cimento Suppl. Series 10, 13, 318 (1959).
[37] E. N. PARKER: Astrophys. J. 128, 664 (1958).
[38] J. A. FEJER: Phys. Fluids 7, 499 (1964).
[39] S. P. TALWAR: J. Geophys. Res. 69, 2707 (1964).
[40] A. K. SEN: Planet. Space Sci. 13, 131 (1965).
[41] I. LERCHE: J. Geophys. Res. 71, 2365 (1966).
[42] A. J. DESSLER, and G. K. WALTERS: Planet Space Sci. 12, 227 (1964).
[43] K. MAER, and A. J. DESSLER: J. Geophys. Res. 69, 2846 (1964).
[44] J. W. DUNGEY: Phys. Rev. Letters 6, 47 (1961b).
[45] J. W. DUNGEY: J. Phys. Soc. Japan 17, Suppl. A 2, 15 (1962).
[46] W. I. AXFORD, H. E. PETSCHEK, and G. L. SISCOE: J. Geophys. Res. 70, 1231 (1965).
[47] R. H. LEVY, H. E. PETSCHEK, and G. L. SISCOE: Amer. Inst. Aeronaut. Astronaut. J. 2, 2065 (1964).

On the basis of resistive diffusion alone the *characteristic velocity* of merging and dissipation over a scale L with an electrical conductivity σ would be

$$V_\sigma = \frac{c_0^2 \, \varepsilon_0 \, \mu_0}{u \, \mu_0 \, \sigma \, L}. \qquad (5.1)$$

Combined with the Alfven speed V_A, Eq. [4.1], a *magnetic Reynolds number*[48]

$$\mathcal{R}_m = \frac{V_A}{V_\sigma} = \sqrt{\frac{u \, \mu_0}{\varrho} \, \frac{\sigma \, B \, L}{c_0^2 \, \varepsilon_0 \, \mu_0}} \qquad (5.2)$$

can be defined, and

$$V_\sigma = V_A / \mathcal{R}_m. \qquad (5.3)$$

Since $\mathcal{R}_m$ for the overall system is usually a very large number ($\gg 10^6$), the merging speed is slow. If the fluid is permitted to flow freely out from between the opposite fields, the

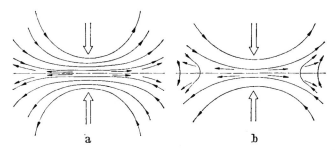

Fig. 5 a and b. A sketch of the rapid annihilation and merging of oppositely directed magnetic fields. The solid lines represent the magnetic lines of force, the broken line represents the plane of symmetry between the opposite fields, the short arrows represent the motion of the fluid, and the open arrows represent the forces pushing the opposite fields together. (a) represents the initial state and (b) represents the configuration after the dissipation and reconnection of the lines of force has progressed somewhat.

merging velocity increases to $V_A/\mathcal{R}_m^{\frac{1}{2}}$. LEVY et al.[47] have shown that the merging may go with $V_A/\ln \mathcal{R}_m$ in some cases, but the dissipation at the neutral point must then evidently arise from something more than simple resistivity. In regions which are sufficiently dense that resistivity has meaning, then perhaps resistive plasma instabilities come into play[49]. In the collisionless magnetosphere one would look to particle-wave interactions for possible dissipation. The basic physical question is whether rapid dissipation of some form plays an important rôle in the behavior of the magnetosphere. The question is still open.

6. The geomagnetic tail. Let us pass on, then, to the problem posed by the geomagnetic tail, the observed structure of which raises serious questions.

α) Note that, if the solar wind were a fluid, free of turbulence, if the magnetopause were stable, if there were no rapid merging of oppositely directed fields, and if there were no internal gas or wave pressure in the geomagnetic field, then there would be no forces on the geomagnetic field tending to stretch the field out behind Earth to form an extended tail. The wind would slide smoothly over the magnetopause, exerting only a pressure normal to the magnetosphere. The geomagnetic field would close off quickly behind Earth as a consequence of the lateral pressure exerted by the solar wind[1], giving the overall tail configuration shown in Fig. 6. The situation can be demonstrated by calculating the idealized geomagnetic field confined inside a circular cylinder with a radius of perhaps $20\,R_E$, representing the geomagnetic tail[2]. The calculations show that the field

---

[48] W. M. ELSASSER: Phys. Rev. **95**, 1 (1954).
[49] R. K. JAGGI: J. Geophys. Res. **68**, 4429 (1963).
[1] F. S. JOHNSON: J. Geophys. Res. **65**, 3049 (1960).
[2] E. N. PARKER: Space Sci. Rev. **1**, 62 (1962).

strength in such a tail declines exponentially with distance along the tail. Hence beyond a distance a of few times the tail diameter, the *magnetic pressure*

$$B^2/2u\mu_0$$

of the geomagnetic field falls below the pressure in the wind outside. The wind pressure squeezes off the tail at the Mach angle (which is essentially the reciprocal of the Mach number). As JOHNSON points out, the tail would then terminate after a distance of the order of perhaps a few hundred R_E.

β) The next point is, that, if the tail extends farther out behind Earth than JOHNSON's model, it can do so only because there is *a force* (external or internal) exerted on the field to stretch out the lines of force into a longer tail. JOHNSON[1,3] pointed out that the pressure of gas-magnetic waves inside the magnetosphere

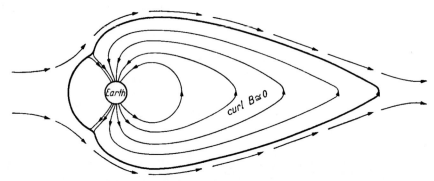

Fig. 6. Sketch of the configuration of the geomagnetic field in the hypothetical circumstance that the solar wind slips smoothly along the magnetopause without ruffling the surface or in any way exerting a tangential drag. The standoff shock has been omitted from the diagram. The arrows outside the magnetopause indicate the direction of motion of the solar wind.

might lengthen the tail. It is known today, largely through the observations of NESS[4,5], that the geomagnetic tail consists of the polar magnetic lines of force extending more or less straight back away from the sun, with a neutral sheet between the north and south polar sides of the tail, as illustrated in Fig. 3, Sect. 4β. The total force of tension in the geomagnetic tail at 40 R_E is of the order of $0.7 \cdot 10^7$ N *, to be compared with the total force of impact of the quiet-day wind against the sunward side which is about $6 \cdot 10^7$ N **. The two obvious questions are, what exerts the tension and how far out behind the Earth does the tail extend?

Consider the tension first. The neutral sheet is observed by NESS[4] to be less than 10^3 km in thickness. The gas pressure p in the neutral sheet must be equal to the magnetic pressure $B^2/2u\mu_0$ of the magnetic field:

$$p = \frac{B^2}{2u\mu_0} \qquad (6.1)$$

on each side of the neutral sheet in order to keep the opposite fields separate. The calculated particle or gas pressure in the neutral sheet is thus of the order of

---

* 1 N = Newton = 10^5 dyn. The value is obtained with a field of $\pm 2 \cdot 10^{-4}$ Gs across a diameter of 40 R_E.
** Obtained with a stream of density 5 H-atoms per cm³ and speed 500 km/sec.
[3] A. J. DESSLER: J. Geophys. Res. **69**, 3913 (1964).
[4] N. F. NESS: J. Geophys. Res. **70**, 2989 (1965).
[5] J. P. HEPPNER, N. F. NESS, T. L. SKILLMAN, and C. S. SCEARCE: J. Geophys. Res. **68**, 1 (1963).

10^{-10} Nm^{-2} (10^{-9} dyn cm^{-2}). But though the neutral sheet can keep the two halves separate, it is too thin for its pressure to draw out the tail.

γ) FEJER[6] has pointed out that the *pressure of the gas* trapped in the two halves of the tail might be responsible for stretching the tail. The pressure requirements of FEJER's idea may be computed from Fig. 7, which is a sketch of the tail with the neutral sheet in the middle and the solar wind on each side. Denote the gas pressure through the tail by p_1. The field in the tail is B. In order to overcome the magnetic tension in the tail, we must have $p_1 = B^2/2u\mu_0$. The total transverse pressure in the tail is $p_1 + B^2/2u\mu_0$ which is then equal to $B^2/u\mu_0$. This total pressure must be balanced by the pressure p_3 in the neutral sheet in order to maintain the separation of the two halves of the tail. This in turn must be balanced by the total pressure p_2 of the external solar wind in order to confine the tail. Hence

$$p_2 = p_3 = 2p_1 = B^2/u\mu_0. \qquad (6.2)$$

Fig. 7. Sketch of the section in the noon-midnight meridian plane through an idealized magnetic tail showing the gas pressures p_1, p_2, p_3 involved in the equilibrium of the model of the geomagnetic tail proposed by FEJER[6].

This requires that the gas pressure within the tail be exactly half what it is in the wind and in the neutral sheet. The required particle densities, and temperatures are perhaps not unreasonable (10 cm^{-3}, at 100 eV per particle). The idea would predict that the tail field density and internal gas pressure vary with the pressure of the solar wind. Observation[7-9] suggest that the total pressure in the wind may vary by a factor of ten or more. The corresponding changes in the tail diameter and field strength should be large.

δ) The alternative to an internal pressure extending the tail is an external force. The external force could be exerted only by the solar wind, which then must have a grip on most of the magnetic lines of force in the tail in order to stretch them back into position.

7. Basic principles and speculations.

It is when we come to inquire into the detailed mechanism by which the solar plasma gets hold of the lines of force in the tail that we again come across the great variety of ideas, mentioned above, that are presently being considered. It is important, therefore, to understand what is theoretically established before trekking out onto the shifting sands of present day speculation.

The two principles to be kept in mind are that (a) the geomagnetic lines of force may be interchanged, as a consequence of the layer of nonconducting atmosphere above the surface of Earth and (b), in a first approximation, the magnetopause is completely covered by lines of force which all come from a small locality on the surface of Earth and represent a very small magnetic flux.

α) The possibility of an *interchange of lines* was first pointed out by GOLD[1]. He stated that the layer of nonconducting atmosphere between the ionosphere and the surface of Earth permits the interchange of two magnetic tubes of force, as illustrated in Fig. 8; there should be no resistance to the motion other than inertia and viscosity. In the absence of viscous forces from gases, etc. it is easily

---

[6] J. A. FEJER: J. Geophys. Res. **70**, 4972 (1965).
[7] M. NEUGEBAUER, and C. W. SNYDER: Science **138**, 1095 (1963).
[8] M. NEUGEBAUER, and C. W. SNYDER: J. Geophys. Res. **70**, 1587 (1965).
[9] N. F. NESS, and J. M. WILCOX: Phys. Rev. Letters **13**, 461 (1964).
[1] T. GOLD: J. Geophys. Res. **64**, 1219 (1959); — Space Sci. Rev. **1**, 100 (1062); — J. Phys. Soc. Japan **17**, Suppl. A **1**, 187 (1962).

shown that the field has neutral stability to such interchange. The inner flux tube in Fig. 8 must be stretched to go into the position of the outer tube, which requires work, but the outer tube shortens upon moving to the position of the inner tube, and gives up as much energy as the other consumes. The work done by the magnetic pressures cancels too. So there is no tendency for the lines to interchange, or to resist interchange. Thus, if the inner of the two tubes were sufficiently inflated with trapped particles, it is evident that instability results. The inner tube becomes inflated and hence buoyant with respect to the surrounding field. Formal calculations may be found in [2] and [3]. The possibility for the interchange of magnetic lines of force* was taken up by AXFORD and HINES[4,5] in connection with the deep convection of the magnetosphere indicated by the D_S

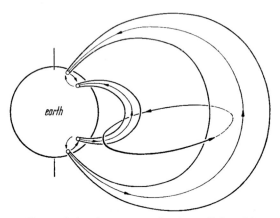

Fig. 8. Schematic diagram to illustrate the interchange or convection of magnetic lines of force permitted by the nonconducting atmosphere at the feet of the lines of force, as pointed out by GOLD[1]. The curved arrows in the equatorial plane indicate the hypothetic motion of two magnetic tubes of force.

current system in the polar ionosphere. They point out that the convection, whatever its cause, is of the general form sketched in Fig. 9, showing the convective motions in the equatorial plane without the complication of diurnal rotation or the neutral sheet. The existence of such convection, with speeds up to the 1 km/sec implied by the polar currents, can account for a number of observed effects, such as the inclination of the regions of auroral activity and the auroral particles themselves. Convection appears as a means for transporting gas and energetic particles rapidly in and out of the geomagnetic field. This is important in the theory of aurorae and the magnetic storm and will be discussed in Sect. 9γ. Presumably the convection is driven in some way by the solar wind, perhaps in cooperation with the diurnal rotation of Earth inside the magnetosphere.

β) Consider what *effect* the interchange of lines of force has *on the formation of the geomagnetic tail*. The geomagnetic lines of force in the magnetopause all

---

* When using this term, here and in the following discussion, it is always understood that it is only an abbreviation for "magnetic tubes of force filled with plasma". In fact magnetic lines of force are "materialized" by "frozen-in" plasma in the whole range which is under discussion here.

[2] B. U. O. SONNERUP, and M. J. LAIRD: J. Geophys. Res. **68**, 131 (1963).
[3] D. B. CHANG, L. D. PEARLSTEIN, and M. N. ROSENBLUTH: J. Geophys. Res. **70**, 3085 (1965).
[4] W. I. AXFORD, and C. O. HINES: Can. J. Phys. **39**, 1433 (1961).
[5] C. O. HINES: Space Sci. Rev. **3**, 342 (1964).

Sect. 7. Basic principles and speculations.

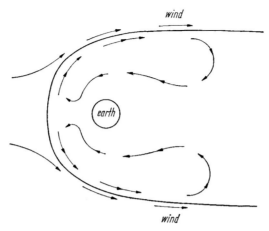

Fig. 9. A highly simplified sketch of possible convective motions of the magnetic lines of force and associated particles in the equatorial plane of Earth. The diurnal rotation has been omitted from the diagram.

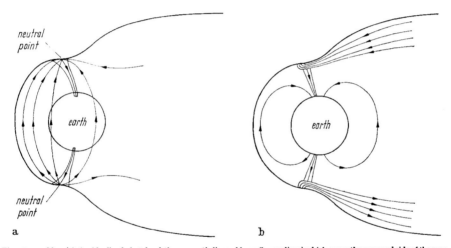

Fig. 10 a and b. (a) An idealized sketch of the magnetic lines of force (heavy lines) which cover the sunward side of the magnetopause. The lines emanate from the two neutral points on the magnetopause and are associated with a bundle of flux lines of vanishing cross section reaching Earth at high latitude on the noon meridian. (b) An idealized sketch of the same lines of force after they have been whisked away into the geomagnetic tail by the wind, uncovering a fresh layer of lines of force underneath (not shown).

come from the near vicinity of the magnetic line of force through the two neutral points at the magnetopause, as illustrated in Fig. 10a. If there is any tangential drag exerted on the magnetopause, it is exerted on these lines of force. It is interesting to note that any drag, no matter how slight, will, if exerted on a sufficiently thin surface layer of field, carry the surface lines back into the geomagnetic tail. This is readily seen considering a strip of surface field exposed to the solar wind. Be the surface drag T_s, then the total drag force on a strip width d exposed to the solar wind for a distance D is $T_s D d$. The drag on the strip is resisted by the magnetic pressure $(B^2/2u\mu_0)$ in the strip, into which the drag might penetrate to a depth ε. The drag overpowers the tension provided only that

$$\left(\frac{B^2}{2u\mu_0}\right)\varepsilon d < T_s D d,$$

or
$$\varepsilon < 2u\,\mu_0\,DT_s/B^2. \qquad (7.1)$$

It is by this basic mechanism that one may imagine the surface lines of force to be whisked back into the tail, as sketched in Fig. 10b, by the drag exerted by the solar wind. But whisking away the surface layer of lines exposes the lines immediately underneath, which are in turn whisked back into the tail, etc. Once a layer of lines is stretched back into the tail, the tension in the lines causes the point where the lines pass through the ionosphere to drift in the direction away from the sun (see, for instance, the discussion of drift by Dungey[6]), Thus the lines of force originally through the neutral point, and covering the magnetopause, may be replaced by a fresh set of lines, which take up the position abandoned by the first set.

If the wind does not lose its grip on the individual lines soon after extending them into the tail, then in time all the polar lines of force become part of the tail. Whether the geomagnetic tail is produced in this way depends upon whether there is a small tangential drag exerted by the wind on the magnetopause. It is interesting to note that the observations of Ness[7] show that most of the polar magnetic lines of force extend into the geomagnetic tail*.

Most of the polar lines of force extend back into the tail regardless of the level of geomagnetic activity (though there are some observed variations of the strength of the field in the tail). So it seems that the existence of the tail is a general thing, which does not require some special condition, such as a southward orientation of the interplanetary field or a high state of turbulence or instability. Altogether it seems plausible to believe that some combination of surface drag and the interchange of lines may be the basis for the extended geomagnetic tail. It was speculated some years ago[8-10] that the solar wind should blow away the exposed magnetic lines of force "like smoke from a chimney" but it required observation to show that a substantial and permanent tail exists.

γ) To digress slightly, the early speculators who suggested that the wind stretched out magnetic lines of force from the magnetopause, went on to suggest that the erosion of lines of force by the wind was intimately associated with the origin of *geomagnetic activity*, and that the main phase of a storm might be the result of the extension of a large amount of magnetic flux into the tail. It was later pointed out by Cowling[11] and Parker[12] that the production of a main phase in this way appeared dubious because of the pressure which is necessarily exerted on the magnetopause by the wind which is eroding away the lines of force into the tail. Recent observations[13-15] indicate that the field in the tail is, in fact, stronger during periods of high magnetic activity. But they also show that the pressure of the wind on the sunward magnetopause exceeds the tension in the tail, so that decreases of the horizontal component at the surface of Earth be interpreted in terms of lines of force being carried away into the tail. In summary then, the observations indicate that the disturbances in the wind producing geomagnetic activity also carry more lines of force into the geomagnetic tail. But this does not lead directly to the main phase.

---

* The total magnetic flux outward through the tail is about 10^9 Vsec (10^{17} Gs cm²), corresponding to all the magnetic lines of force from within about $15°$ of either geomagnetic pole.

[6] J. W. Dungey: J. Geophys. Res. **70**, 1753 (1965).
[7] N. F. Ness: J. Geophys. Res. **70**, 2989 (1965).
[8] E. N. Parker: Phys. Fluids **1**, 171 (1958).
[9] J. H. Piddington: J. Geophys. Res. **65**, 93 (1960); — Geophys. J. Roy. Astron. Soc. (Lond.) **3**, 314 (1960).
[10] E. R. Harrison: Geophys. J. Roy. Astron. Soc. **7**, 479 (1962).
[11] T. G. Cowling: Rep. Progress Phys. **25**, 244 (1962).
[12] E. N. Parker: Space Sci. Rev. **1**, 62 (1962).
[13] J. P. Heppner: Space Sci. Rev. **7**, 166 (1967).
[14] N. F. Ness, and D. J. Williams: J. Geophys. Res. **71**, 322 (1966).
[15] K. W. Behannon, and N. F. Ness: J. Geophys. Res. **71**, 2327 (1966).

δ) Coming back to the question of the wind exerting a tangential drag on the magnetopause, the formation of the geomagnetic tail would seem to require only a rather shallow *drag*. In contrast, the deep convection of the magnetosphere, driven by a drag force which penetrates deeply into the magnetosphere[4], illustrated in Fig. 10, is a similar but perhaps distinct phenomenon and has been discussed by several authors since 1961[16-19]*.

Consider the variety of ideas that are available for producing the surface drag at the magnetopause. As noted earlier, in the absence of any instability or other dissipative effect, a nonturbulent wind should slide smoothly along the magnetopause, exerting only a pressure normal to the surface. Turbulence in the wind as the wind approaches the magnetopause could produce a frictional drag because it ripples the magnetopause with a phase velocity exceeding the Alfven speed, thereby generating waves which propagate into the magnetosphere. The surface drag created in this way may be transmitted to some depth. The situation is explored by Axford[19]. He concludes that, in order of magnitude, the drag which drives the convection may arise in this way. The calculations of Fejer[20] show that the magnetosphere is subject to a Helmholtz-type instability when the field in the solar wind is nearly parallel to the geomagnetic field at the magnetopause. In this way there should be a drag whenever the interplanetary field has the correct orientation. Axford[22] pointed out that a Rayleigh-Taylor instability may occur during the sudden compression of the geomagnetic field at the time of a sudden commencement (**SC**). In all these cases the excitation of the magnetopause, and the resulting geomagnetic fluctuations and drag, depend upon special conditions in the solar wind. Dessler[23] has argued that the geomagnetic observations imply such a dependence, because if the magnetopause were generally unstable, the magnetopause would always be fluttering and the degree of fluttering would increase only as the first or second power of the wind velocity. Hence the observed small variation of wind velocity would not be sufficient to account for the enormous differences in geomagnetic activity that are sometimes observed. But, of course, the geomagnetic tail, which one might expect also to be sensitive to wind conditions, is always present and does not vary in strength by a large factor. The problem is not resolved at the present time and its resolution may lie in some other direction.

ε) As pointed out earlier, Dungey[24,25] has argued that the *interplanetary magnetic field* tends to *merge* with the geomagnetic field at the magnetopause whenever the interplanetary field has a component oppositely directed to the

---

* There exists some confusion in Piddington's later papers concerning the earlier work of Parker[8] and Axford and Hines[4] as compared to Piddington[9]; see [21]. Hence the entire series of papers must be read if one wishes to assess the individual contributions correctly.

[16] J. H. Piddington: Planet. Space Sci. **9**, 947 (1962); — Space Sci. Rev. **3**, 724 (1964); — Planet. Space Sci. **13**, 281, 565 (1965).
[17] E. W. Hones: J. Geophys. Res. **68**, 1209 (1963).
[18] J. A. Fejer: J. Geophys. Res. **69**, 123 (1964).
[19] W. I. Axford: Planet. Space Sci. **12**, 45 (1964).
[20] J. A. Fejer: Phys. Fluids **7**, 499 (1964).
[21] W. I. Axford, and C. O. Hines: Planet. Space Sci. **12**, 660 (1964).
[22] W. I. Axford: J. Geophys. Res. **67**, 3791 (1962).
[23] A. J. Dessler: J. Geophys. Res. **66**, 3587 (1961); **67**, 4392 (1962). — Dessler, A. J., and J. A. Fejer: Planet. Space Sci. **11**, 505 (1963). — Dessler, A. J., and G. K. Walters: Planet. Space Sci. **12**, 227 (1964).
[24] J. W. Dungey: Phys. Rev. Letters **6**, 47 (1961); — J. Phys. Soc. Japan **17** Suppl. A 2, 15 (1962); — Planet. Space Sci. **10**, 233 (1963); — J. Geophys. Res. **70**, 1753 (1965).
[25] R. H. Levy, H. E. Petschek, and G. L. Siscoe: Amer. Inst. Aeronaut. Astronaut. J. **2**, 2065 (1964).

geomagnetic field there. He suggests that the merging takes place on the sunward side, and that the reverse process takes place in the geomagnetic tail, where the temporary connection between the interplanetary and geomagnetic lines of force is broken. The overall magnetic field topology proposed by Dungey is sketched in Fig. 11a. The interplanetary field and plasma close in behind Earth at the thin neutral sheet. The geomagnetic tail terminates at the neutral point, with

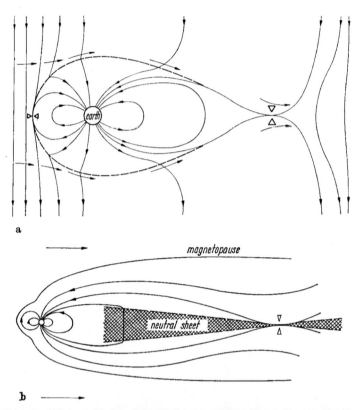

Fig. 11 a and b. (a) An idealized sketch of the interplanetary and geomagnetic lines of force in the noon-midnight meridian plane as envisioned by Dungey [14] in connection with the idea of rapid merging of oppositely directed fields. The short arrows indicate the motion of the wind (without the complication of the standoff shock). The broken line represents the boundary of the magnetosphere. (b) An idealized sketch of the modified picture suggested by Axford [26]. In both sketches the triangles indicate the points of rapid merging.

interplanetary field and solar plasma beyond. Ness observed the neutral sheet at a distance of only 30 to 40 R_E, with indications, from the direction and behavior of the field on both sides of the neutral sheet, that the tail continues beyond, as sketched in Fig. 3. This is not in contradiction with the general topology proposed by Dungey, nor, on the other hand, does it establish the existence of rapid merging.

Axford et al.[26] have recently modified Dungey's idea to the form sketched in Fig. 11b, incorporating some suggestions by Petschek [12][25] on the form of magnetic fields in the process of rapid annihilation at a neutral sheet. In their picture the geomagnetic tail continues for some distance beyond the neutral point, thereby avoiding the objection of observed features to Fig. 11a. Their

[26] W. I. Axford, H. E. Petschek, and G. L. Siscoe: J. Geophys. Res. 70, 1231 (1965).

picture contains the broad, nearly neutral region toward Earth from the neutral point, indicated by shading in Fig. 11b, in which the plasma velocity would be of the order of the characteristic Alfven speed and the field is perpendicular to the plane of the neutral region. They anticipate rapid annihilation of the tail fields.

In contrast, the general quiet-day appearance of the magnetopause and the neutral sheet as revealed by the observations is one of stability and permanence[27] contrary to the idea of universal rapid annihilation of oppositely directed fields[24]. On the other hand, during disturbed times the observations show that the transition between the two halves of the tail is more confused than during quiet times, suggesting that the situation may perhaps be like that pictured by AXFORD[26]. He suggests[28] that the rapid annihilation and broad neutral region is triggered by the disturbed wind in some way, and that the rapid annihilation is responsible for much of the ensuing activity in the magnetosphere.

One problem with the idea of merging the geomagnetic and interplanetary fields as the primary reason for the tail has to do with the question of the balance between the rate of merging at the sunward magnetopause and the rate of merging in the tail. The problem is that the more *rapid* the merging at the sunward side, the more *rapidly* are lines of force picked up by the wind and stretched out behind the Earth into the tail and the denser is the tail. But, on the other hand, the *slower* is the merging in the tail, the longer the lines remain stretched out and the *longer* and *stronger* is the tail. So under steady conditions a well developed tail requires a balance of the merging rates. The balance must exist independently of the direction of the interplanetary field. It was pointed out by PETSCHEK [16], that the merging rate is expected to be proportional to $\sin(\theta/2)$ where θ is the angle between the geomagnetic and interplanetary fields.

It should be borne in mind that the configuration of the geomagnetic tail, as presently observed, was drawn first by PIDDINGTON[9] in connection with his ideas on the cause of the main phase of the storm, and later by DESSLER[29] in connection with the effects of pressure exerted by gas-magnetic waves. Both authors drew pictures essentially identical with Fig. 3., Sect. 4α. It remains to be shown whether rapid merging, if it sometimes occurs, is the result or the cause of the observed tail configuration (see discussion in AXFORD[28] on choking off rapid annihilation). We suggest that convection of lines of force without much reconnection[4] plays the important role in maintaining the tail in its observed form.

ζ) Not unrelated to the question of stability and merging at the magnetopause is the *length of the geomagnetic tail*, which is still unknown. In one extreme DESSLER [8][27,29,30] argues that the spatial and temporal inhomogeneities of the polar cap absorption (PCA) events[31] imply a tail of a length of the order of 1 astronomical unit*. It is said that the energetic solar particles (of 5 to 10 MeV) producing the PCA events are isotropic in space soon after their arrival from the sun, whereas it is observed that the PCA-particles stream into the ionosphere over a very irregular pattern for an hour or more after their arrival outside the magnetosphere. Hence, they argue the tail cannot open into space near Earth or a uniform pattern of particle arrival would be observed, so that the geomagnetic tail must be coherent and closed to low energy solar particles for a large distance, of the order of 0.5 a.u. or more, beyond Earth. Unfortunately, the observed irregular patterns of arrival are not what one would expect from a simple long tail.

DUNGEY's early model, illustrated in Fig. 10, Subsect. β, shows a short tail, of about the same length as JOHNSON's model (Fig. 6, Sect. 6α), though for quite different reasons.

---

* 1 a.u. = $1.5 \cdot 10^8$ km.
[27] F. C. MICHEL, and A. J. DESSLER: J. Geophys. Res. **70**, 4305 (1965).
[28] W. I. AXFORD: Space Sci. Rev. **7**, 149 (1967).
[29] A. J. DESSLER: J. Geophys. Res. **69**, 3913 (1964a).
[30] A. J. DESSLER, and F. C. MICHEL: [23] (1966).
[31] K. RAWER and K. SUCHY: This Encyclopedia, Vol. 49/2, Sect. 47.

DUNGEY recently has modified his earlier arguments about the perpetual necessity for rapid merging of opposite fields and now asserts[6] that the length of the disturbed tail is of the order of 0.04 a.u. based on the rate of migration of lines of force across the polar ionosphere inferred from the D_S current system during polar substorms. VAN ALLEN[32] finds that, whatever the length of the tail, there is no evident flux of energetic particles at a distance of $2 \cdot 10^7$ km $= 0.14$ a.u. in the anti-solar direction, whereas conspicuous fluxes are found in the tail near Earth.

It is not obvious to us that we really understand the origin and nature of the tail yet. The magnetopause is presumably subject to wind erosion[8], but the nature of the drag is not clearly understood. PIDDINGTON[9,33] has considered a number of possible effects caused by the wind erosion, but no final picture has come of it. On the other hand, it should not be concluded that rapid annihilation and merging of oppositely directed fields can be completely rejected because one or two arbitrary applications by its proponents have had to be modified. We must not confuse the validity of a special application with the validity of a general theoretical principle. There is no question that DUNGEY[34] [9], SWEET[35], and PETSCHEK [12] have worked out a magnetic configuration the gas-magnetic behavior of which is extraordinary. But it has not yet been shown that rapid merging of the geomagnetic field with southward interplanetary field plays a rôle in shaping the geomagnetic tail and in producing geomagnetic activity.

8. Energetic particles and the aurora. The aurora is caused by energetic particles penetrating downward along the lines of force of the geomagnetic field through the terrestrial atmosphere to altitudes of 80 to 200 km. Extensive reviews of the visible aurorae may be found in [21], [3][1-7].

α) Most of the auroral electrons appear to be 20 keV or less. Auroral activity is closely correlated with magnetic activity on both a world wide scale, which has been known for decades, and on a very local scale[8]. It is clear from the observations that the *charged particles causing the aurora* are not direct from the sun. Rather the particles have been accelerated in and around the magnetosphere, probably as a consequence of the passage of the solar wind. Presumably the trapped radiation belts originate in much the same way. It is a general rule that fast particles are generated whenever a plasma containing a magnetic field is subject to violent agitation. As noted in Sect. 5α Fermi-acceleration (Sect. 9α) and/or resonance with ion acoustic waves are considered as possible sources of the 50 keV electrons observed in the wind outside the magnetosphere. Within the magnetosphere there is the additional and important possibility of adiabatic compression, see Sect. 9β. It now appears that the general auroral phenomenon can be understood to a large extent as a direct consequence of convection and adiabatic compression of thermal electrons and protons.

β) A brief word should be said on the *observations* before getting into the theory. Recent summaries of the observations of the van Allen belts may be

---

[32] J. A. VAN ALLEN: J. Geophys. Res. **70**, 4731 (1965).
[33] J. H. PIDDINGTON: Planet. Space Sci. **13**, 281, 565 (1965).
[34] J. W. DUNGEY: Phil Mag. **44**, 725 (1953).
[35] P. A. SWEET: [*15*] (1958).
[1] B. J. O'BRIEN: [*26*] (1964).
[2] V. I. KRASSOVSKIJ: Space Sci. Rev. **3**, 232 (1964).
[3] V. I. KRASSOVSKIJ: [*26*], 8 (1964).
[4] C. T. ELVEY: Planet. Space Sci. **12**, 783 (1964).
[5] S. I. AKASOFU: Space Sci. Rev. **4**, 498 (1965).
[6] J. H. PIDDINGTON: Planet. Space Sci. **13**, 565 (1965).
[7] A. B. MEINEL, S. I. AKASOFU, and S. CHAPMAN: This Encyclopedia, Vol. 49/1.
[8] L. J. VICTOR: J. Geophys. Res. **70**, 3123 (1965).

found in [9-22]. The literature on artificially produced radiation belts is summarized in [11][22-24]. It has been calculated that the observed quiet-time proton belt produces about a 10γ decrease of the horizontal component at the surface of Earth[25].

Of particular interest are the observations dealing with time variations of the trapped particles. Marked increases have been found in the energetic electron flux during a magnetic storm, which could be explained only by local acceleration of particles[26]. DAVIS and WILLIAMSON[27] find some increase in proton fluxes during a storm, in contrast to the low altitude observations[19,28,29]. Changes in the electron flux in the outer zone have been observed[30,31]. The evidence is that 50 keV electron fluxes increase at the time of a storm but 1 MeV electron fluxes change little. FREEMAN et al.[32] observed some violent changes in the outer radiation belt and in the magnetosphere at the onset of a magnetic storm, which they interpreted as a compression of the magnetosphere, see also [14,33,34]. The theory of adiabatic motion of particles in the geomagnetic field, and the perturbation of that motion by geomagnetic fluctuations is given in [17][35-46]. Precipitation as a consequence of wave-particle interactions has been pointed out by KENNEL and PETSCHEK[47].

---

[9] V. L. GINZBURG, L. V. KURNOSOVA, L. A. RAZORENOV, and M. FRADKIN: Space Sci. Rev. 2, 778 (1962).
[10] K. I. GRINGAUZ, S. M. BALANDINA, G. A. BORDOVSKJ, and N. M. ŠUTTE: [25] (1963).
[11] K. I. GRINGAUZ: [26] (1964).
[12] S. N. VERNOV, E. V. GORČAKOV, J. I. LOGAČEV, V. E. NESTEROV, N. F. PISARENKO, I. A. SARENKO, A. E. ČUDAKOV, and P. I. ŠAVRIN: [25] (1963).
[13] S. N. VERNOV, I. A. SARENKO, P. I. ŠAVRIN, L. TVERSKAJA: [26] (1964).
[14] I. B. McDIARMID, and J. R. BURROWS: Can. J. Phys. 42, 616 (1964).
[15] I. B. McDIARMID, and J. R. BURROWS: J. Geophys. Res. 70, 3031 (1965).
[16] S. N. VERNOV, V. E. NESTEROV, N. F. PISARENKO, I. A. SARENKO, O. I. SARUN, P. I. ŠAVRIN, and K. N. ŠAVRINA: Planet. Space Sci. 13, 347 (1965).
[17] G. F. PIEPER, C. O. BOSTROM, and A. J. ZMUDA: J. Geophys. Res. 70, 2021 (1965).
[18] C. O. BOSTROM, A. J. ZMUDA, and G. F. PIEPER: J. Geophys. Res. 70, 2035 (1965).
[19] A. J. ZMUDA, G. F. PIEPER, and C. O. BOSTROM: J. Geophys. Res. 70, 2045 (1965).
[20] S. C. FREDEN, J. B. BLAKE, and G. A. PAULIKAS: J. Geophys. Res. 70, 3111 (1965).
[21] M. P. NAKADA, J. W. DUNGEY, and W. N. HESS: J. Geophys. Res. 70, 3529 (1965).
[22] See contribution by W. N. HESS in Vol. 49/4 of this Encyclopedia. Also: W. N. HESS: [23] (1965).
[23] C. R. HAARE, A. J. ZMUDA, and B. W. SHAW: J. Geophys. Res. 70, 4191 (1965).
[24] A. J. ZMUDA, C. R. HAARE, and B. W. SHAW: J. Geophys. Res. 70, 4207 (1965).
[25] R. A. HOFFMAN, and P. A. BRACKEN: J. Geophys. Res. 70, 3541 (1965).
[26] C. Y. FAN, P. MEYER, and J. A. SIMPSON: J. Geophys. Res. 66, 2607 (1961).
[27] L. R. DAVIS, and J. M. WILLIAMSON: Space Res. 3, 365 (1963).
[28] J. W. FREEMAN: J. Geophys. Res. 69, 1671 (1964).
[29] B. J. O'BRIEN: J. Geophys. Res. 69, 13 (1964).
[30] L. A. FRANK: [23] (1965); — J. Geophys. Res. 70, 1593 and 4131 (1965).
[31] D. J. WILLIAMS, and J. W. KOHL: J. Geophys. Res. 70, 4139 (1965).
[32] J. W. FREEMAN, J. A. VAN ALLEN, and L. J. CAHILL: J. Geophys. Res. 68, 2121 (1963).
[33] T. ARMSTRONG: J. Geophys. Res. 70, 2077 (1965).
[34] P. SERLEMITSOS: J. Geophys. Res. 71, 61 (1966).
[35] P. J. KELLOGG: Nature 183, 1295 (1959).
[36] E. N. PARKER: J. Geophys. Res. 65, 3117 (1960).
[37] E. N. PARKER: J. Geophys. Res. 66, 693 (1961).
[38] A. J. DRAGT: J. Geophys. Res. 66, 1641 (1961).
[39] A. J. DESSLER, and R. KARPLUS: J. Geophys. Res. 66, 2289 (1961).
[40] D. G. WENTZEL: J. Geophys. Res. 66, 359, 363 (1961); 67, 485 (1962).
[41] L. DAVIS, and B. CHANG: J. Geophys. Res. 67, 2169 (1962).
[42] B. A. TRERSKOJ: Geomagnetizm i Aeronomija 3, 351 (1964).
[43] C. G. FALTHAMMAR: J. Geophys. Res. 70, 2503 (1965).
[44] J. W. DUNGEY: Space Sci. Rev. 4, 199 (1965b).
[45] D. B. CHANG, and L. D. PEARLSTEIN: J. Geophys. Res. 70, 3075 (1965).
[46] C. E. McILWAIN: J. Geophys. Res. 71, 3623 (1966).
[47] C. F. KENNEL, and H. E. PETSCHEK: J. Geophys. Res. 71, 1 (1966).

γ) The passage of *energetic electrons* downward along the geomagnetic field has been observed and studied[48-54] in connection with aurorae. Some suggestions as to whether the acceleration and/or particle scattering responsible for the precipitation into the atmosphere are at high or low altitudes appear in the analysis of the observations. O'Brien[48] suggests that the electrons may perhaps be accelerated parallel to the magnetic field during the event, rather than merely being scattered from trapped orbits at the time of the precipitation. Mozer[50,51] suggests that some proton precipitation is best explained by low altitude acceleration.

There is widespread precipitation of electrons into the atmosphere in the auroral regions, which has only a loose correlation with the actual visual aurora[55,56,56a]. Auroral X-ray emission has been studied, with attention to conjugate phenomena and short period bursts[57-59], also auroral radio emission [60,61].

δ) At the present time the most exciting observations in the magnetosphere with possible direct connection to the aurora are concerned with the *neutral sheet in the tail* (Fig. 3, Sect. 1α) and the associated high electron fluxes. Observations[15,62] show large fluxes of fast electrons in the geomagnetic tail and neutral sheet. The electrons appear in "islands" or bursts. It is not known what relation the bursts of electrons in the tail bear to the bursts observed in the magnetosheath on the sunward side[63,64]. It appears[65,66] that the outer boundary of the van Allen radiation belt on the night side[67] is determined approximately by the magnetic lines of force which connect inward to the auroral zone and outward into the neutral sheet in the tail. Some possible configurations (see Sect. 9) of auroral particles in space have been discussed[68-70]. A neutral sheet between two opposite fields is considered by many workers in the field to be a likely place for particle acceleration, see, for instance[71]. Hence the magnetic connection of the auroral regions into the neutral sheet suggests that the night time auroral particles may be associated in some way with the neutral sheet.

---

[48] B. J. O'Brien: J. Geophys. Res. **69**, 13 (1964); [*23*] (1965).
[49] J. R. Barthel, and D. H. Sowle: Planet Space Sci. **12**, 209 (1964).
[50] F. S. Mozer, J. F. Crifo, and J. E. Blamont: J. Geophys. Res. **70**, 5699 (1965).
[51] F. S. Mozer: J. Geophys. Res. **70**, 5709, 5717 (1965).
[52] R. D. Sharp, J. E. Evans, W. L. Imhof, R. J. Johnson, J. B. Reagan, and R. V. Smith: J. Geophys. Res. **69**, 2721 (1964).
[53] R. D. Sharp, J. B. Reagan, S. R. Salisbury, and L. F. Smith: J. Geophys. Res. **70** 2119 (1965).
[54] F. S. Johnson: Satellite Environment Handbook. Second edition. Stanford, California: Stanford University Press 1965.
[55] J. A. van Allen: Phys. Rev. **99**, 609 (1955); — Proc. Nat. Acad. Sci. **43**, 57 (1957); — J. Geophys. Res. **70**, 4731 (1965).
[56] J. R. Barcus: J. Geophys. Res. **70**, 2135 (1965).
[56a] O. E. Johansen: Planet. Space Sci. **13**, 225 (1965).
[57] R. R. Brown, J. R. Barcus, and Parsons: J. Geophys. Res. **70**, 2579, 2599 (1965).
[58] G. K. Parks, H. S. Hudson, D. W. Milton, and K. A. Anderson: J. Geophys. Res. **70**, 4976, 4979 (1965).
[59] J. R. Barcus, and A. Christensen: J. Geophys. Res. **70**, 5455 (1965).
[60] B. Hultquist, and A. Egeland: Space Sci. Rev. **3**, 27 (1964).
[61] M. Gadsden: Planet. Space Sci. **12**, 29 (1964).
[62] K. A. Anderson: J. Geophys. Res. **70**, 4741 (1965).
[63] C. Y. Fan, G. Gloeckler, and J. A. Simpson: Phys. Rev. Letters **13**, 149 (1964).
[64] C. Y. Fan, G. Gloeckler, and J. A. Simpson: J. Geophys. Res. **71**, 1837 (1966).
[65] D. J. Williams, and G. D. Mead: J. Geophys. Res. **70**, 3017 (1965).
[66] N. F. Ness, and D. J. Williams: J. Geophys. Res. **71**, 322 (1966).
[67] L. A. Frank, J. A. van Allen, and E. Macagno: J. Geophys. Res. **68**, 3543 (1963).
[68] A. J. Dessler, and R. D. S. Juday: Planet. Space Sci. **13**, 63 (1965).
[69] W. I. Axford, H. E. Petschek, and G. L. Siscoe: J. Geophys. Res. **70**, 1231 (1965).
[70] A. J. Dessler, and F. C. Michel: J. Geophys. Res. **71**, 1421 (1966).
[71] T. W. Speiser: J. Geophys. Res. **70**, 4219 (1965).

ε) A word should be said about the many theoretical ideas that have been developed recently concerning *auroral acceleration*. Considerable attention has been directed to the old idea that the auroral particles are accelerated more or less steadily be an electric field parallel to the magnetic field. Potential differences of 10^5 V are imagined. A number of papers have been written in recent years on this subject [2][72-76]. The papers consider a number of interesting hypothetical circumstances but unfortunately overlook one, or both, of two fundamental facts in the application to the actual aurora. The first fact is that most auroral forms are relatively thin in one direction perpendicular to the magnetic field, whereas a hypothetical electric field cannot be confined to a thin layer without the curl of the electric field being *very* large at the two surfaces of the layer. A large curl leads to an *enormous* time rate of change of the magnetic field according to Maxwell's second equation:

$$\frac{\partial \boldsymbol{B}}{\partial t} = - c_0 \sqrt{\varepsilon_0 \mu_0}\, \nabla \times \boldsymbol{E}. \tag{8.1}$$

Thus electric fields cannot be confined to thin auroral sheets but must be roughly as broad as they are high, so that an electrostatic field cannot produce currents which flow only in the auroral rays, arcs, etc. The second fact follows from the observation that the magnetosphere is filled with a *thermal* plasma with an electron density of 1 cm^{-3} or more. Steady potential differences of 10^5 V along the magnetic lines of force would produce enormous electric currents in this plasma, soon neutralizing the high impedance sources assumed to be responsible for the potential difference. Only if it could be shown, for instance, that under some circumstance the background thermal plasma in the geomagnetic tail were really absent, then perhaps some interesting and remarkable electrostatic acceleration effects could be achieved.

Chamberlain[77] and Kern[78], being aware of the restrictions on electric fields along a magnetic field, pointed out the possibility of *transient* space charge effects as a consequence of the opposite drifts of the electrons and ions in an inhomogeneous plasma. But this process produces no *net* particle acceleration.

9. Particle acceleration.

α) It appears that there are two general directions in which to look for auroral particle acceleration. One is a *Fermi-mechanism* and/or particle resonance with any of various types of plasma wave, as discussed earlier in connection with the fast electrons observed in the magnetosheath.

Fermi acceleration refers to the scattering and increase of individual particle energies as a result of reflection of the particles from moving inhomogeneities (magneto-gasdynamic waves) in the magnetic field[1]. Head-on collisions lead to an energy gain and overtaking collisions to an energy loss. Under most circumstances head-on collisions add more energy and are more frequent, so that there is a net gain of energy. The process was described by Fermi as the trend toward equipartion of energy between the individual particles and the massive gas-magnetic waves in the field. The effect may perhaps be important near the magnetopause where the geomagnetic field and the solar wind plasma and field are particularly active (see, for instance[2]. To take the simplest illustration, consider a particle of mass m with velocity $\pm w$

---

[72] E. T. Karlson: Phys. Fluids **5**, 476 (1962); **6**, 708 (1963).
[73] H. Alfvén: K. Svenska Vet. Akad. Handl. **18** (1939).
[74] H. Alfvén: Space Sci. Rev. **2**, 862 (1963).
[75] D. W. Swift: J. Geophys. Res. **70**, 3061 (1965).
[76] See contribution by W. N. Hess in Vol. 49/4 of this Encyclopedia, Chapt. G, in particular Sect. 45.
[77] J. W. Chamberlain: Astrophys. J. **134**, 401 (1961).
[78] J. W. Kern: J. Geophys. Res. **67**, 2649 (1962).
[1] E. Fermi: Phys. Rev. **75**, 1169 (1949).
[2] J. R. Jokipii: Astrophys. J. **143**, 961 (1966).

in the x-direction. The particle often collides with massive elastic objects with random velocity $\pm v$ in the x-direction. The velocity increase by the particle for a head-on collision is $2v$, so the energy gained is $2mv(w+v)$. For an overtaking collision the energy loss is $2mv(w-v)$. The probability of a collision is proportional to the relative velocity, $\boldsymbol{w}-\boldsymbol{v}$; as $(w-v) \leqq (w+v)$ head-on collision is more likely. The mean kinetic energy gain $\Delta \mathscr{E}_{kin}$ per collision is then just

$$\Delta \mathscr{E}_{kin} = \left(\frac{w+v}{2w}\right) 2mv(w+v) - \left(\frac{w-v}{2w}\right) 2mv(w-v),$$

so that the fractional gain per collision is:

$$\frac{\Delta \mathscr{E}_{kin}}{\mathscr{E}_{kin}} = 8\frac{v^2}{w^2}.$$

Resonance acceleration involves a momentary resonance between the cyclotron frequency of a charged particle in the ambient magnetic field and the frequency of some passing plasma wave (see, for instance [3,4]).

β) Another direction is adiabatic compression. The charged particles move with the magnetic field, so that they are compressed or expanded along with the field. Isotropic compression leads to the usual adiabatic energy increase for a monatomic gas, in which the particle energy increases as the two thirds power of the density. The effective ratio of the specific heats is $\gamma = \frac{5}{3}$.

If the system is compressed only in the direction along the field, then the field, $\boldsymbol{B}$, itself and the particle velocity perpendicular to the field, $w_\perp$, are not affected. The increase in the particle velocity in the direction of the field is directly proportional to the associated density increase, i.e. in one dimension $\gamma = 3$. The kinetic enregy $\mathscr{E}_\|$ of parallel motion then increases with the square of the parallel compression. On the other hand, if the system is compressed only transversely, then the velocity $w_\perp$ of the cyclotron motion of the particle around the lines of force, being a two dimensional motion $(\gamma = 2)$, increases directly with the particle and field densities.

Thus, for instance, if a particle trapped in the geomagnetic field near the magnetopause, or in the tail region, is convected along with the field to a position close to Earth, its energy may be enormously increased. The field at the position of the particle may increase from $3 \cdot 10^{-4}$ gauss to $3 \cdot 10^{-1}$ gauss, so that the energy of the cyclotron motion increases by a factor of 10^3. Very roughly, a thermal electron of 10 eV from the solar wind finds itself with an energy of 10 keV. The energy comes from the forces driving the convection, just as the increased energy of particles compression in a cylinder comes from the forces which produced the compression.

There are electric fields associated with the compression of a magnetic field. According to MAXWELLS' second equation the curl of the electric field is related to the changing magnetic field by

$$c_0 \sqrt{\varepsilon_0 \mu_0}\, \nabla \times \boldsymbol{E} = \frac{\partial \boldsymbol{B}}{\partial t}.$$

The relation is often written differently when treating charged particles and plasmas. The basic fact is that the electric field $\boldsymbol{E}'$ in the frame moving with the particles and/or plasma is essentially zero. In a collision dominated plasma, $\boldsymbol{E}' = 0$ because of the high electrical conductivity. In a collisionless plasma, the component of $\boldsymbol{E}'$ which is parallel to $\boldsymbol{B}$ is again essentially zero because of the usually high mobility of the thermal electrons along the magnetic lines of force. Hence quite generally

$$\boldsymbol{E} \cdot \boldsymbol{B} = 0. \tag{9.1}$$

In the collisionless plasma all charged particles have a drift velocity

$$\boldsymbol{v} = \frac{c_0 \sqrt{\varepsilon_0 \mu_0}}{B^2} \boldsymbol{E} \times \boldsymbol{B}, \tag{9.2}$$

---

[3] E. N. PARKER: J. Geophys. Res. **66**, 2673 (1961).
[4] R. W. FREDRICKS, and F. L. SCARF: J. Geophys. Res. **70**, 4765 (1965).

where v and E are measured in the same frame of reference. It follows that the frame of reference moving with the particles is the frame in which the perpendicular component of E vanishes. Thus, generally speaking $E' = 0$. If now we make a Lorentz transformation from the frame of reference moving with the particles, in which $v' = 0$, to the fixed frame in which the particle drift (convective velocity) is $v(|v| \ll c_0)$, we find that the electric and magnetic fields must be related by

$$E = -\frac{1}{c_0 \sqrt{\varepsilon_0 \mu_0}} v \times B. \qquad (9.3)$$

If we desire, Eq. (9.3) may be used to eliminate E from MAXWELL's equation, giving the usual gasmagnetic relation between the change of the field B and the convective velocity v.

Some authors prefer to deal with the electric field and some with the convective velocity. When using the electric field, the acceleration is not described as an adiabatic compression but rather as a gradient drift down the electric potential, which is a slightly more complicated way of handling the problem for some purposes. But if done correctly, the two descriptions are exactly equivalent.

The first author to point out the possible importance of adiabatic compression for producing energetic particles in the geomagnetic field was KELLOG[5]. He showed that the asymmetric, time-dependent deformations of the magnetosphere (which are part of geomagnetic activity) displace particles in and out through the magnetosphere in a random way. Particles can walk at random all the way in to the top of the atmosphere. The resulting adiabatic compression of these particles, which happen to be pushed deep into the geomagnetic field, leads to particle energies 10^3 times larger close to Earth than at the magnetopause. KELLOG's idea has been developed at some length[6-10]. It now appears[11, 12] that this random displacement and diffusion plays an important role in establishing the observed distribution of energetic trapped protons through the magnetosphere.

γ) AXFORD and HINES[13] in their paper presenting the importance and probable existence of convection of the magnetosphere, pointed out that the convective pattern (Fig. 9, Sect. 7α) continually brings in *particles and plasma from the solar wind*, compressing the collection of particles as it is transported inward to stronger fields. The particle energies are increased by factors of 10^2 to 10^4, depending upon the position of first entrapment, and the depth to which the convection carries them. AXFORD and HINES pointed out that the particles energized in this way could appear at the same places in the magnetosphere where the aurorae are observed. Hence they suggested this unavoidable particle acceleration, as a consequence of the deep convection indicated by the polar D_S current system, for the basic source of auroral particles, and of trapped particles up to energies of 20 to 100 keV.

This fundamental point of AXFORD and HINES[13] has been confirmed in an important calculation by TAYLOR and HONES[14]. These authors set up a working model of the magnetosphere, including the diurnal rotation of Earth, the deep convection of AXFORD and HINES, and an approximation to the tail and neutral sheet. By numerical methods they traced out the magnetic lines of force and the equipotential surfaces along the lines of force. Then they computed the motion of individual electrons and protons after entrapment in the field just inside the magnetopause. The particle motions were traced through the convective cycle by

---

[5] P. J. KELLOGG: Nature **183**, 1295 (1959).
[6] E. N. PARKER: J. Geophys. Res. **65**, 3117 (1960).
[7] L. DAVIS, and B. CHANG: J. Geophys. Res. **67**, 2169 (1962).
[8] K. D. COLE: J. Geophys. Res. **69**, 3595 (1964).
[9] C. G. FALTHAMMAR: J. Geophys. Res. **70**, 2503 (1965).
[10] R. L. KAUFMANN: J. Geophys. Res. **70**, 2181 (1965).
[11] J. W. DUNGEY: Space Sci. Rev. **4**, 199 (1965).
[12] M. P. NAKADA, and G. P. MEAD: J. Geophys. Res. **70**, 4777 (1965).
[13] W. I. AXFORD, and C. O. HINES: Can. J. Phys. **39**, 1433 (1961).
[14] H. E. TAYLOR, and E. W. HONES: J. Geophys. Res. **70**, 3605 (1965).

following the two adiabatic invariants of the particle motion. The end result of their calculations is a graphic display of the particle energies and zones of precipitations, starting with 1 to 10 eV at the magnetopause. The calculations show that particles are delivered into the dense atmosphere with energies of the order of 10 keV. Particles with a given initial magnetic moment are delivered around a thin arc at high latitudes, with a broad initial distribution giving particles over the entire auroral zone. The calculated geographic distribution of the electrons and protons delivered into the atmosphere is so like that of the observed aurora as to support in detail the correctness of Axford and Hines' assertion*. It would appear, then, that the deep convection of the magnetosphere is the primary process responsible for both the aurora and much of the low energy trapped radiation.

δ) It is evident, of course, that a great deal remains yet to be done in exploring and understanding the convection and the aurora, both observationally and theoretically. The *aurora and the trapped radiation* are a very complicated phenomenon, and the convection and compression is only the underlying mechanism. For instance, De Witt and Akasofu[15] suggest that some additional ionospheric motions may contribute to auroral particle acceleration. The computational model of [14] has a number of special assumptions in it whose significance must be assessed, such as a reliance on closed lines of force in the tail out to distances of 50 R_E, and an asymmetric connection of lines of force from the tail into the polar regions of Earth, requiring a severing of 30 percent of the tail lines of force during the daily rotation of Earth[16].

The really fundamental question now lies in other directions, viz. the cause of the deep convection and the detailed behavior of the individual aurorae. Axford and Hines[13] point out, as one possibility, that the tangential drag exerted on the magnetopause by the solar wind would give the convection. Axford[17] finds that gas-magnetic waves generated at the magnetopause could give sufficient drag. On the other hand, Fejer[18] has shown that energetic protons trapped in the geomagnetic field can produce the convection as a consequence of the drift of the protons and the contortions of the geomagnetic field in going through the daily rotation.

Our own feeling is, however, that the solar wind is the prime mover, with most of the trapped particles and the aurora as the consequence, rather than vice versa.

ε) We should remark that *laboratory simulation* of the interaction of the solar wind with the geomagnetic field has been attempted (see for instance[19-21]). The laboratory experiments shoot a burst of ionized gas at a magnetized sphere representing Earth under the conditions that the cyclotron frequency of the ions is large compared to the collision frequency. The experiments demonstrate the compression, and confinement of the geomagnetic field by the wind in a striking manner. But it has not yet been possible to scale the ion cyclotron radius properly, so that it is small compared to the dimensions of the field, nor has it been possible to simulate the ionosphere, where currents flow freely across lines of force, nor has it been

---

* It is surprising in reading the paper of Taylor and Hones[14] to find the statement that the electric fields they use (deduced from the polar D_S currents) are distinct from the convection fields of Axford and Hines[13], when in fact these authors pointed out the immediate relationship that exists between the convective fields and the currents. The electric fields of [14] differ from those of [13] only in precise orientation, not in their nature; possible variations of orientation were recognized, and to some extent discussed, by [13].

[15] R. N. De Witt, and S. I. Akasofu: Planet. Space Sci. **13**, 729 (1965).

[16] As to the influence of this rotation see contribution by H. Poeverlein in Vol. 49/4 of this Encyclopedia.

[17] W. I. Axford: Planet. Space Sci. **12**, 45 (1964).

[18] J. A. Fejer: J. Geophys. Res. **69**, 123 (1965).

[19] F. S. F. Osborne, I. P. Shkarofsky, and J. V. Gore: Can. J. Phys. **41**, 1747 (1963).

[20] J. B. Cladis, T. D. Miller, and Blaskett: J. Geophys. Res. **69**, 2257 (1964).

[21] N. Kawashima, and N. Fukushima: Planet. Space Sci. **12**, 1187 (1964).

possible to begin the experiment with a tenuous ionized gas in the magnetic field, nor has it been possible to get away from an isolated burst of plasma to a broad flow as in the solar wind. Therefore, detailed effects in the laboratory experiment such as particle precipitation cannot generally be compared with the actual magnetosphere.

C. The geomagnetic storm.

10. Basic principles of deformation of the geomagnetic field. Consider the geomagnetic storm and its position in the overall picture of the magnetosphere presented in the preceding section, based on the magneto-gasdynamic behavior of all the fields in space.

α) The basic theoretical principle to be kept in mind is that the large-scale deformations of the magnetosphere are *magneto-gasdynamic* in character[1]. The basis for this assertion is the observed fact[2] that the magnetosphere is filled with a tenuous plasma so that there is little or no electric field in the frame of reference moving with the plasma. Transforming to the fixed frame of reference it can be shown that the electric field is, therefore (see Sect. 9β),

$$\boldsymbol{E} = -\frac{1}{c_0 \sqrt{\varepsilon_0 \mu_0}} \boldsymbol{v} \times \boldsymbol{B}, \qquad [9.3]$$

where $\boldsymbol{v}$ is the plasma velocity. With this expression for $\boldsymbol{E}$, MAXWELL's equations reduce to the magneto-gasdynamic equations [1], which say, in effect, that the plasma and the magnetic lines of force move together as

$$\frac{\partial \boldsymbol{B}}{\partial t} = \nabla \times (\boldsymbol{v} \times \boldsymbol{B}). \qquad (10.1)$$

The common expression is that the magnetic lines of force are "frozen" into the plasma. Thus the gas and field move together, so far as displacements perpendicular to the field are concerned, though the gas may slide more or less freely along the field in many circumstances. From this point of view the magnetosphere is an elastic stringy medium. The *"gas-field medium"* has the elastic properties of the magnetic field. The field has an isotropic pressure

$$p_1 = B^2/2u\,\mu_0, \qquad [6.1]$$

and a tension

$$p_2 = B^2/u\,\mu_0 \qquad [6.2]$$

along the magnetic lines of force. Any deformation from the equilibrium geomagnetic field is resisted by these elastic stresses. The geomagnetic storm constitutes various deformations of the field, so that the theory of the geomagnetic storm consists in discovering the stresses exerted on the geomagnetic field and calculating the consequent distortions. In this sense the theory is a mechanical one. If the forces exerted on the geomagnetic field add up to a new outward force, the result is an expansion of the geomagnetic field, lowering the field density throughout the region of expansion. If the net forces are inward, the result is a compression, raising the field density throughout the region of compression[1,3]. Changes in the strain propagate as magneto-gasdynamic waves[4-6] with

---

[1] E. N. PARKER: J. Geophys. Res. **61**, 625 (1956).
[2] L. R. O. STOREY: Phil. Trans. Roy. Soc. A **246**, 113 (1954).
[3] E. N. PARKER: Phys. Fluids **1**, 171 (1958).
[4] C. O. HINES: J. Geophys. Res. **62**, 491 (1957).
[5] A. J. DESSLER: J. Geophys. Res. **63**, 405 (1958). — W. E. FRANCIS, M. I. GREEN, and A. J. DESSLER: J. Geophys. Res. **64**, 1643 (1959).
[6] J. H. PIDDINGTON: Geophys. J. Roy. Astron. Soc. **2**, 173 (1959).

the Alfven speed

$$V_A = \frac{B}{(u\,\mu_0\,\varrho)^{\frac{1}{2}}}. \qquad [4.1]$$

In the elementary formulation of the problem, the solid Earth is neglected. The geomagnetic field is represented by that of a magnetic dipole containing only a tenuous plasma. The effects of the solid Earth and the ionosphere, into which the lines of force are frozen, do not alter the basic theoretical picture so far as the average world-wide magnetic storm is concerned*. Roughly speaking, the only effect of the inclusion of the solid Earth is to increase the variation of the horizontal component at the surface of Earth at low latitudes by about 50 percent for a given deformation of the magnetosphere.

β) Chapman and Ferraro[7] were the first to appreciate that an external beam of plasma, i.e. a wind, would confine and compress the geomagnetic field into a well defined region surrounding Earth. The *sudden commencement and initial phase* of the geomagnetic storm represent a world wide compression, which is observed as an increase in the horizontal component $\mathcal{H}$. The cause can be only an increased solar wind pressure.

The sudden commencement (**SC**) of the geomagnetic compression is an indication that the wind pressure increases suddenly, as the consequence of a shock preceding the enhanced wind [*10*][8]. Nishida and Cahill[9] and Wolfe and Silva[10] have observed the general compression throughout the magnetosphere at the time of sudden impulses. Pai and Sarabhai[11] have pointed out that the recurring large variations of the geomagnetic field during a storm, with typical periods of 40 minutes, are probably related to variations in the wind with dimensions of the order of 0.02 a.u. ($= 3 \cdot 10^6$ km $\sim 500\ R_E$).

It was pointed out in Sect. 3 that the sudden pressure increase in the wind at Earth may result either from a blast wave from sudden enhancement of the solar corona at the time of a flare, or from the general rotation of a steady pattern of wind with the sun. A significant pressure increase in the former case is probably always preceded by a shock transition. The pressure increase in the latter case is preceded by a shock if the velocity differences in the rotating pattern are large enough. Indeed, there may be a backward propagating shock[12–15] giving what is called a *negative sudden commencement*, which is a sudden release of some of the compression[16].

The sudden compression (or decompression) is applied to the magnetopause over the interval of about 10^2 sec that is required for the shock to engulf the magnetosphere. The compression is transmitted inward from the magnetopause as a magneto-gasdynamic wave, propagating with the Alfven speed V_A [Eq. (4.1)], which is of the order of 500 km/sec (see Fig. 4). The propagation time from various points on the magnetopause (distance $\sim 10\ R_E$) to the surface of Earth varies by a minute or so. Thus after a delay of 2—3 minutes the total compression appears at the surface of Earth. This is the basic explanation for the sudden

---

* Nonsymmetric deformation (Sects. 24, 25) is an exception, but need not concern us here.
[7] S. Chapman, and V. C. A. Ferraro: Terr. Mag. **36**, 77, 171 (1931).
[8] T. Gold: Nuovo Cimento Suppl. Series 10, **13**, 318 (1959).
[9] A. Nishida, and L. J. Cahill: J. Geophys. Res. **69**, 2243 (1964).
[10] J. H. Wolfe, and R. W. Silva: J. Geophys. Res. **70**, 3575 (1965).
[11] G. L. Pai, and V. A. Sarabhai: Planet. Space Sci. **12**, 855 (1964).
[12] C. P. Sonett, and D. S. Colburn: Planet. Space Sci. **13**, 675 (1965).
[13] H. Razdan, D. S. Colburn, and C. R. Sonett: Planet. Space Sci. **13**, 1111 (1965).
[14] P. A. Sturrock, and J. R. Spreiter: J. Geophys. Res. **70**, 5345 (1965).
[15] J. Hirschberg: J. Geophys. Res. **70**, 5353 (1965).
[16] S. I. Akasofu: Planet Space Sci. 573 (1964).

commencement rise time which is observed to be two to four minutes[5,17-19]. The formal methods of ray tracing of hydromagnetic waves have been worked out by BAZER and HURLEY[20]. PIDDINGTON[21] has given an extensive qualitative discussion of possible magneto-gasdynamic waves in the geomagnetic field during a storm. SUGIURA[22] has pointed out recently that there is a further delay of *several* minutes in reaching the final equilibrium of the compressed field because of the time taken for some of the geomagnetic lines of force near the magnetopause to squirm around into their new equilibrium position once the compression is applied throughout the field. The effect contributes to the rise time, and to the partial relaxatoin and readjustment of the compression at the surface of Earth following the onset of the sudden commencement.

The continued enhanced solar wind pressure on the geomagnetic field following the sudden commencement gives the initial phase of the storm, lasting from one to eight hours.

γ) The *main phase of the magnetic storm* follows the initial phase and represents a world wide decrease in the horizontal component below the prestorm level. A decrease of the field density can only be an expansion. A decrease in the solar wind pressure below the prestorm level would permit an expansion, of course, but the amount of the expansion could not exceed the quiet day compression, which is calculated[23,24] to be 14 to 15 γ. However, the main phase of a large storm may be ten times larger. Something more is needed. CHAPMAN and FERRARO, working in the context of an empty and nonconducting space, proposed a westward flowing electric current near the outer boundary of the geomagnetic field (the magnetopause) as the cause of the main phase. Such a current is required by MAXWELL's equations,

$$c_0 \sqrt{\varepsilon_0 \mu_0}\, \nabla \times \boldsymbol{H} = \frac{\partial}{\partial t} \boldsymbol{D} + u\, \boldsymbol{j} \tag{10.2}$$

$$c_0 \sqrt{\varepsilon_0 \mu_0}\, \nabla \times \boldsymbol{E} = -\frac{\partial}{\partial t} \boldsymbol{B} \tag{10.3}$$

the first of which gives in the stationary case:

$$\boldsymbol{j} = \frac{c_0 \sqrt{\varepsilon_0 \mu_0}}{u\, \mu_0}\, \nabla \times \boldsymbol{B} \tag{10.4}$$

so that a distortion $\nabla \times \boldsymbol{B}$ is necessarily associated with a current density $\boldsymbol{j}$. The change in outlook that has developed in the last decade is based on the fact that the geomagnetic field is permeated by a tenuous conducting plasma. The question which arises is whether $\boldsymbol{j}$ should be considered the cause of the distortion $\nabla \times \boldsymbol{B}$, or whether $\nabla \times \boldsymbol{B}$ is the cause of $\boldsymbol{j}$. In the terrestrial laboratory, filled with nonconducting air and containing only small bits of rather poorly conducting metal, we produce a magnetic field $\boldsymbol{B}$ by causing a current $\boldsymbol{j}$ with an applied electromotive force (emf). But in the magnetosphere, and in most astrophysical problems, the situation is reversed. The space is filled with a highly conducting ionized gas

---

[17] E. J. STEGELMANN, and VON KENSCHITZKI: J. Geophys. Res. **69**, 139, 3748 (1964).
[18] E. N. PARKER, and A. J. DESSLER: J. Geophys. Res. **69**, 3745 (1964). — A. J. DESSLER, W. E. FRANCIS, and E. N. PARKER: J. Geophys. Res. **65**, 2715 (1960).
[19] C. E. PRINCE, F. X. BOSTICK, and H. W. SMITH: J. Geophys. Res. **70**, 4901 (1965).
[20] J. BAZER, and J. HURLEY: J. Geophys. Res. **68**, 147 (1963).
[21] J. H. PIDDINGTON: Geophys. J. Roy. Astron. Soc. **2**, 173 (1959); **3**, 314 (1960); — J. Geophys. Res. **65**, 93 (1960); — Planet. Space Sci. **9**, 947 (1962).
[22] M. SUGIURA: J. Geophys. Res. **70**, 4151 (1965).
[23] D. B. BEARD, and E. B. JENKINS: J. Geophys. Res. **67**, 3361 (1962).
[24] G. D. MEAD: J. Geophys. Res. **69**, 1181 (1964).

and there are usually no external emf's applied to the system. The electric current is then automatically induced by the changing magnetic field whenever a force distorts $\boldsymbol{B}$ so that $\nabla \times \boldsymbol{B}$ is nonvanishing. Thus, in the present context it is usually convenient to think of $\boldsymbol{j}$ as a secondary quantity.

This is not to say that electric current is not a useful quantity in some of the theoretical calculations. Rather it is the statement that in thinking about the physics of field distortions, it is the force exerted on the field by the gas that occupies the center of the stage. What is needed, then, is an outward force exerted by the gas on the field to spring the geomagnetic field outward and decrease the horizontal component at the surface of Earth for the main phase of the magnetic storm.

11. The basis for the main phase.

α) It is evident that an *outward force* applied to the outer lines of force of the geomagnetic field (where the field strength is of the order of 40 γ) can have only a limited effect on the field as a whole, some 20 to 40 γ at the most. The force could pull the outer lines away from the magnetopause, relieving only a little of the pressure on the inner field. So the outward force responsible for the main phase must be exerted fairly deep in the field, for a big main phase of 100 to 200 γ not beyond 3 to 5 R_E.

It was suggested[1] that the ionosphere might expand upward some 5 to 10 km to lift the lines of force and cause a decrease in the horizontal component at the surface of Earth. Dessler[2] pointed out, however, that due to collisions the effective resistivity in the ionosphere is so high that the field would slip back through the ionosphere in less than an hour, whereas the relaxation time of the main phase is observed to be several hours or days. Hence it is difficult to believe that the ionosphere is responsible for the world wide main phase.

The fundamental point, first made by Singer[3] is that the main phase expansion is caused by the ionized gases trapped in the field itself.

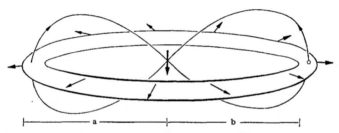

Fig. 12. A sketch of the idealized model as given by a ring of hot gas in a dipole field. The short arrows indicate the direction of the force exerted by the ring on the field.

β) It is possible to demonstrate with a *simple model* how the pressure of gas or particles trapped in a dipole field causes the field to expand outward. To emphasize that the problem and its solution in gasmagnetic, suppose that the dipole field is filled with a tenuous conducting fluid, the field exerting no forces on the field. Then suppose that there is in addition a concentric ring of dense hot gas, with finite pressure, threaded around between the lines of force of a magnetic dipole, as sketched in Fig. 12. In this simple hypothetical case we suppose that the ring is not penetrated by the lines of force of the dipole field, which pass by on each side. Denote the major and minor radii of the torus of gas by a and b respectively. For simplicity assume that the ring is slender, $a \gg b$. Sup-

---

[1] E. N. Parker: J. Geophys. Res. **61**, 625 (1956).
[2] A. J. Dessler: J. Geophys. Res. **64**, 397 (1959).
[3] S. F. Singer: Trans. Amer. Geophys. Union **38**, 175 (1957); — J. Phys. Soc. Japan **17**, Suppl. A **1**, 328 (1962).

pose, for simplicity, that the thermal motions of the gas in the ring are restricted to the azimuthal direction so that the only nonvanishing component of the pressure is azimuthal and of magnitude p. Under the effect of p, the torus will tend to expand its major radius a. The total compression force around the ring is $\pi b^2 p$, so that the outward force exerted by the ring is $\pi b^2 p/a$ per unit length. The expansion of the torus is resisted by the magnetic field in which the gas is embedded acting like an elastic spring. The outward force is exerted on the field of the dipole and obviously causes expansion of the dipole field in the region encircled by the ring. It is obvious, too, that the field of the dipole in the region outside the ring is compressed somewhat. In order to calculate the expansion of the field, suppose that the distortion of the dipole field is not large. When no other forces are present, the equilibrium field satisfies $\nabla \times \boldsymbol{B}=0$ everywhere except at the position of the ring, just as in a vacuum. In that case, the distortion $\delta \boldsymbol{H} = \frac{1}{\mu_0} \delta \boldsymbol{B}$ of the field can be calculated from MAXWELL's first equation, Eq. (10.2), applied to vacuum. This gives a relation between δB and the current I around the ring. Near the center of the ring

$$\delta B = \frac{u \mu_0}{c_0 \sqrt{\varepsilon_0 \mu_0}} \frac{I}{2a}. \tag{11.1}$$

To compute I note that the current flowing westward around the ring must produce a *Lorentz-force* which just balances the outward force per unit length of the ring:

$$\frac{IB}{c_0 \sqrt{\varepsilon_0 \mu_0}} = \frac{\pi b^2}{a} p, \tag{11.2}$$

where B is the unperturbed field at the position of the ring.

γ) The *perturbation field* is related to this induced current I just as in a vacuum, giving a field $\delta \boldsymbol{B}$ with

$$|\delta \boldsymbol{B}| \equiv \delta B = \frac{u \mu_0}{c_0 \sqrt{\varepsilon_0 \mu_0}} \frac{I}{2a} = \frac{u \mu_0}{2} \left(\frac{b}{a}\right)^2 \pi \frac{p}{B} \tag{11.3}$$

in the downward direction near the center of the ring. If the field has a flux density B_0 in the equatorial plane at a radial distance $a > R_E$, then for a *dipole*

$$B = B_0 \left(\frac{R_E}{a}\right)^3 \sqrt{1 + 3 \cos^2 \theta}. \tag{11.4}$$

Making use of Eqs. (11.3), (11.4) the fractional change in field density at the radial distance R_E in the equatorial plane can be written as:

$$\delta B = u \mu_0 \frac{\pi}{2} \frac{a b^2}{R_E^3} \frac{p}{B_0}. \tag{11.5}$$

The relative change*, $\delta B/B_0$, can be expressed as the ratio of two energies, namely:

$$\left|\frac{\delta B}{B_0}\right| = \frac{2 \mathscr{E}_p}{3 \mathscr{E}_m}, \tag{11.6}$$

where

$$\mathscr{E}_m = \frac{4\pi}{3 u \mu_0} B_0^2 R_E^3 \tag{11.7}$$

is the *total magnetic energy* in the field of the dipole external to the solid Earth, i.e. the integral over $(B^2/2 u \mu_0)$ beyond the radial distance R_E. This is about 10^{18} J.

---

* We do not specify the sign of δB here. Later in Sect. 22 it will be specified that in actual storms $\delta \boldsymbol{B}$ is opposed in direction to the field $\boldsymbol{B}_0$ at the surface of Earth, so that δB will be taken as being negative in Eq. (22.14).

On the other side

$$\mathscr{E}_p = \pi^2 a b^2 p \tag{11.8}$$

is the *total particle energy*, equal to $\tfrac{1}{2} p$ multiplied by the volume $2\pi^2 a b^2$ of the torus. Eq. (11.6) tells us that the outward stresses exerted on the dipole field by the ring of gas cause an expansion of the field, giving a fractional decrease of the field density at the surface of Earth which is just two thirds the ratio of the gas energy to the external magnetic energy. $\mathscr{E}_m$ being constant the decrease δB depends only upon the total particle energy $\mathscr{E}_p$. It is independent of the radius of the ring of gas, and, as will be shown later, independent of the particle distribution along the magnetic lines of force.

δ) The reader may inquire at this point why the problem of the expansion of a dipole field by a ring of gas was worked out on the basis of currents rather than from the gasmagnetic equations, which involve only $\boldsymbol{B}$ and p, after the assertion that magneto-gasdynamic theory is the key to understanding the deformation at the dipole field. The answer, as mentioned earliers, is that the algebra for a gasmagnetic problem can be written either in terms of $\nabla \times \boldsymbol{B}$ or in terms of the current density $\boldsymbol{j}$. This duality is made possible by the fact that MAXWELL's first equation requires any field deformation $\nabla \times \boldsymbol{B}$, whatever its origin, to be associated with it an electric current density[4]*

$$\boldsymbol{j} = \frac{c_0 \sqrt{\varepsilon_0 \mu_0}}{u \mu_0} \nabla \times \boldsymbol{B}. \qquad [10.4]$$

Hence one may keep track of a deformation with either $\boldsymbol{j}$ or $\nabla \times \boldsymbol{B}$ in a magneto-gasdynamic problem. Instead of in I, the example just worked out could have been expressed in terms of the total flux $\mathscr{J}$ of the vector $\nabla \times \boldsymbol{B}$ around the torus; after STOKES' theorem

$$\mathscr{J} = \int d\boldsymbol{A} \cdot \nabla \times \boldsymbol{B} = \oint d\boldsymbol{s} \cdot \boldsymbol{B} \tag{11.9}$$

(A being the cross section of the torus and s its circumference). In this case we would have written $\mathscr{J} B/u \mu_0$ for the Lorentz force per unit length, instead of $I B/c_0 \sqrt{\varepsilon_0 \mu_0}$, Eq. (11.2). We chose to use I, however, because the reader probably recognizes Eq. (11.3) as the relation for δB at the center of the ring, whereas he may not be so familiar with $\delta B = \mathscr{J}/2a$. In Chapter C the mathematical formulation will be worked out both ways so that $\mathscr{J}$ and I may be used with equal facility.

12. Specific ideas on main phase inflation.

α) As stated above, the idea that trapped particles and gas are responsible for the main phase is due to SINGER[1]. Modifying an earlier idea by ALFVEN [1][2] he pointed out that the drift of ions and electrons trapped in the dipole field results in a *westward current* which should produce a decrease in the horizontal force, $\mathcal{H}$, at the surface of Earth. He proposed that the particles were injected by the solar wind.

The theoretical formulation of the magnetic effects of gas and particles in the geomagnetic field was given by DESSLER and PARKER[3], together with a reemphasis of the magneto-gasdynamic nature of the problem. Further details of the mathematics may be found in [4,5].

---

* The slowly varying fields of magneto-gasdynamic waves have associated with them a $\partial \boldsymbol{E}/\partial t$ which is smaller than $(c_0 \sqrt{\varepsilon_0 \mu_0/\varepsilon_0 \mu_0}) \nabla \times \boldsymbol{B}$ by v^2/c_0^2.

[4] W. M. ELSASSER: Phys. Rev. **95**, 1 (1954).
[1] S. F. SINGER: Trans. Amer. Geophys. Union **38**, 175 (1957).
[2] H. ALFVEN: Tellus **7**, 50 (1955).
[3] A. J. DESSLER, and E. N. PARKER: J. Geophys. Res. **64**, 2239 (1959).
[4] E. N. PARKER: Space Sci. Rev. **1**, 62 (1962).
[5] J. R. APEL, S. F. SINGER, and R. C. WENTWORTH: [*14*], 131 (1962).

The individual magnetic effects of the motions of a particle in a dipole field were first given by ALFVEN [1], who applied them to a hypothetical geomagnetic field in which there was a large scale electrostatic field of interplanetary origin.

AKASOFU has a number of papers on the magnetic effects of trapped particles. Unfortunately he drops[6] one of the two major terms in the formula for the current density in a collisionless plasma[7], giving the wrong sign for the current at high latitudes. The error is corrected by AKASOFU and CHAPMAN[8] but unfortunately they then apply the definition of a gradient operator in an orthogonal coordinate system to a nonorthogonal system, thereby arriving at an incorrect expression for the current density in a dipole field. The following papers[9] are based on the incorrect formulae in [8]. In [10] a numerical computational error was added to the algebraic error in the formula for the current, giving the erroneous result that the main phase decrease is larger for a given trapped particle energy if the particles are closer to Earth. The error was corrected in [11].

β) DESSLER and PARKER[3] worked out a number of examples to illustrate the *effects of trapped particles* in various distributions of motion parallel and perpendicular to the field. They showed that the field decrease $\delta B/B_0$ at the surface of Earth is in all cases of the order of $\mathscr{E}_p/\mathscr{E}_M$, as in the example worked out above. Thus, a main phase decrease of 100 γ requires trapped plasma with a total internal kinetic energy of about $3 \cdot 10^{15}$ J. They suggested, too, that the particles responsible for the main phase were protons with kilovolt energies located at a distance of about 3 to 5 R_E from the center of Earth. The suggestion was based, among other things, on the requirement that the main phase expansion of the field is observed to relax after the storm with a characteristic time of the order of a day. Charge exchange of kilovolt protons with the neutral hydrogen component of the outer atmosphere of Earth at 3 to 5 R_E gave the correct relaxation time. Charge exchange with the ambient neutral hydrogen converts the trapped kilovolt proton into a kilovolt neutral hydrogen atom, which passes freely out of the geomagnetic field. The charge exchange cross section for protons and neutral hydrogen is known from laboratory work[12]. Over the range of 0.5 to 40 keV the cross section is approximately inversely proportional to the relative velocity. Hence the characteristic time in which the exchange takes place is approximately independent of particle energy in this energy range and is given by

$$\tau = \frac{10^{13}}{N_H/m^{-3}} \sec, \qquad (12.1)$$

where N_H is the number density of the neutral hydrogen atoms (or protons) through which the proton (or neutral atom) is moving. In Fig. 4 (Sect. 4β) the neutral hydrogen density in the outer terrestrial atmosphere has been plotted after JOHNSON[13]. The number density is about 10^8 m^{-3} [10^2 cm^{-3}] at 4 R_E, giving a characteristic relaxation time of the order of 10^5 sec.

The idea that the solar wind injects plasma deep into the geomagnetic field, was adopted in [3,13a]. More recently satellite observations in the outer magnetosphere have failed to find any evidence of *injection of solar wind plasma* deep into the geomagnetic field. So it was suggested instead[14] that the kilovolt protons are the ambient hydrogen ions heated to kilovolt energies by the shock waves in the magne-

---

[6] S. I. AKASOFU: J. Geophys. Res. **65**, 535 (1960).
[7] E. N. PARKER: Phys. Rev. **107**, 924 (1957).
[8] S. I. AKASOFU, and S. CHAPMAN: J. Geophys. Res. **66**, 1321 (1961).
[9] S. I. AKASOFU, J. C. CHAIN, and S. CHAPMAN: J. Geophys. Res. **66**, 4013 (1961); **67**, 2645 (1962). — AKASOFU, S. I., and J. C. CAIN: J. Geophys. Res. **67**, 3537, 4078 (1962).
[10] S. I. AKASOFU, S. CHAPMAN, and D. VENKATESAN: J. Geophys. Res. **68**, 3345 (1963).
[11] S. I. AKASOFU, S. CHAPMAN, and D. VENKATESAN: J. Geophys. Res. **69**, 1025 (1964).
[12] W. L. FITE, R. F. STEBBINGS, D. G. HUMMER, and R. T. BRACKMANN: Phys. Rev. **119**, 663 (1960).
[13] F. S. JOHNSON: Satellite Environment Handbook. Second edition. Stanford, California: Stanford University Press 1965.
[13a] J. R. BARTHEL, and D. H. SOWLE: Planet. Space Sci. **12**, 209 (1964).
[14] A. J. DESSLER, W. B. HANSON, and E. N. PARKER: J. Geophys. Res. **66**, 3631 (1961).

tosphere associated with the general high level of activity. AKASOFU[6,8,15] suggested instead that the van Allen particles are responsible for the main phase. But observations of the radiation-belts show no increases at the time of the main phase which are large enough to account for the effect (see discussion and references in Chapter B). The quiet day belts produce a decrease of only about $10\,\gamma$ [16]. Nor can the quiet decay of the main phase be explained theoretically by particles with lives as long as the quiet day radiation belts.

γ) DESSLER and PARKER[3] pointed out that the observed more rapid decay of the larger main phases can be explained by the charge exchange mechanism from the expectation that the *hot gas* producing the main phase *extends deeper into the geomagnetic field* in the larger storms. The higher decay rate then results from the higher ambient neutral hydrogen density in which the storm protons find themselves. This point may be carried further to explain the observed [7] nonexponential relaxation of a magnetic storm. The innermost portions of the trapped protons will be dissipated first because they lie in the highest ambient atmosphere density. Following their more rapid decay, the remaining trapped protons lie farther out in the geomagnetic field where the atmosphere is less dense. Their decay time is longer, but again it is the innermost remaining protons which decay first, etc. Thus, one would expect that the relaxation of the storm should proceed more and more slowly with time, as is observed.

The increased neutral hydrogen density in the magnetosphere during sunspot minimum [13] predicts[14] that the decay of the main phase from charge exchange proceeds approximately three times more rapidly during sunspot minimum than during sunspot maximum, comparing storms of equal size. An extensive observational study of storm recovery confirms the factor of three[17].

The decay of main phase protons (in the 0.5 to 40 kev energy range) through charge exchange with the ambient neutral hydrogen can be tested directly by looking for the flux of kilovolt neutral hydrogen atoms into which the kilovolt trapped protons are converted[14]. The flux of neutral atoms following a big storm is estimated as $10^9\,\mathrm{m^{-2}\,s^{-1}}\,[10^5\,\mathrm{cm^{-2}\,s^{-1}}]$ at the top of the atmosphere.

δ) AXFORD and HINES[18] have made possible an entirely different view of the growth and decay of the trapped particles producing the main phase. In connection with their general proposal that the magnetosphere is subject to deep convection, they suggest that the energetic plasma producing the main phase is gas which has been convected in close to Earth from regions far out in the field, perhaps originally captured from the solar wind at the magnetopause. The *inward convection* compressed and heats the gas by factors of 10 to 10^3. (See Chapter E, in particular Sect. 20). Thus the total particle energy $\mathscr{E}_p$ may be increased by the convection, producing the main phase field decrease at the surface of Earth. The main phase could then relaxe in the same way that it was formed. The convection, which is observed to slow down after the storm, slowly carries the gas away again to the outer parts of the geomagnetic field. The associated decompression of the gas reduces the total gas energy, and the main phase dies away. Indeed the presence of the gas deep in the field during the main phase may help to drive the convective interchange[19] which permits the relaxation calculations, see [20,21].

---

[15] S. I. AKASOFU: Space Sci. Rev. **2**, 91 (1963).
[16] M. P. NAKADA, and G. P. MEAD: J. Geophys. Res. **70**, 4777 (1965).
[17] S. MATSUSHITA: J. Geophys. Res. **67**, 3753 (1962).
[18] W. I. AXFORD, and C. O. HINES: Can. J. Phys. **39**, 1433 (1961).
[19] T. GOLD: J. Geophys. Res. **64**, 1219 (1959).
[20] B. U. O. SONNERUP, and M. J. LAIRD: J. Geophys. Res. **68**, 131 (1963).
[21] D. B. CHANG, L. D. PEARLSTEIN, and M. N. ROSENBLUTH: J. Geophys. Res. **70**, 3085 (1965).

Finally it has been pointed out that ion and electron cyclotron wave instabilities give particle precipitation[22], which may be important in the decay of the main phase.

The convection does not exclude contributions from other mechanisms for the growth of the main phase, of course. The polar D_S current system makes it clear that the convection, and hence its effects, occur. Hence the convection is responsible for at least some portion of the main phase. The contribution of shock heating, etc. will have to be estimated from observations of hydromagnetic waves in the magnetosphere. Similarly, it is hard to avoid the conclusion that both charge exchange and convection must play a rôle in the decay of the main phase. Observational determination of the neutral hydrogen flux during the decay of the storm would give a measure of the relative importance of the two.

ε) It should be noted that there are other rather *different ideas* on the origin of the main phase. For instance, it was proposed some years ago[23] that the main phase decrease is caused by the outward forces exerted on the equilibrium field when the magnetic lines of force of the outer geomagnetic field are picked up and carried in an enhanced solar wind. PIDDINGTON[24] has proposed the same thing. It is our opinion, however, that such views are ill founded because the overall pressure of solar wind against the magnetopause appears to give a net inward force which overwhelms whatever outward tensions there may be from the line of force picked up in the solar wind[4]. The observed fields and wind pressure confirm that the outward tensions are small compared with the inward pressures exerted by the wind. The tension in the quiet day tail is about $7 \cdot 10^6$ N $[7 \cdot 10^{11}$ dyn$]$ which is rather less than the total pressure force of $6 \cdot 10^7$ N $[6 \cdot 10^{12}$ dyn$]$ exerted by the wind on the sunward magnetopause (see Sects. 6 and 15).

Observations of the tail field during the magnetic storm of 1. April 1964[25]) indicate the total magnetic flux in the tail is increased during the onset of the main phase of a storm, perhaps by as much as 25% in the particular event observed. The tail is also compressed at that time from a radius of about 20 to 22 R_E to about 16 R_E, so that altogether the total tension is tripled, to something of the order of $2 \cdot 10^7$ N $[2 \cdot 10^{12}$ dyn$]$. But this is still small, even when compared only with the quiet day wind pressure.

Pretending that there is not sufficient energy in the solar wind to produce the main phase of a storm the novel idea has been suggested that the main phase is produced instead by *neutral hydrogen ejected from the sun*, which, because its individual atoms are without net charge, moves independently of the fully ionized solar wind and magnetic fields[26-28]. The idea is that some of the neutral atoms exchange charge while passing by the ionized hydrogen component of the outer terrestrial atmosphere. The kilovolt neutral atoms thereby become trapped protons of about the same energy.

It is difficult to believe that neutral hydrogen can pass through the solar corona in the manner proposed by AKASOFU without being ionized by electron impact[29]. It is also difficult to understand how the neutral hydrogen can be accelerated to kilovolt energies at the sun without the temperature exceeding say 10^4 °K at which hydrogen ionizes*. The next difficulty is that the mechanism is extremely inefficient at Earth because there are only 10 to 100 sec for each neutral atom to exchange charge while passing by, and, once converted to a proton, the probability of a second charge exchange, back to a free neutral atom, goes just about as rapidly (because the ambient neutral and ionized hydrogen densities are roughly

---

* The average thermal energy is only 1 eV with 10^4 °K, but there are enough electrons on the high velocity tail of the Maxwellian distribution to maintain essentially complete ionization. Only at rather high densities is recombination so rapid that temperatures higher than 10^4 °K are required.

[22] C. F. KENNEL, and H. E. PETSCHEK: J. Geophys. Res. **71**, 1 (1966).
[23] E. N. PARKER: Phys. Fluids **1**, 171 (1958).
[24] J. H. PIDDINGTON: Geophys. J. Roy. Astron. Soc. (Lond.) **3**, 314 (1960); — J. Geophys. Res. **65**, 93 (1960); Planet. Space Sci. **9**, 947 (1962).
[25] K. W. BEHANNON, and N. F. NESS: J. Geophys. Res. **71**, 2327 (1966).
[26] S. I. AKASOFU, and C. E. MCILWAIN: Trans. Amer. Geophys. Union **44**, 883 (1963).
[27] S. I. AKASOFU: Planet. Space Sci. **12**, 801, 905 (1964); **13**, 297 (1965).
[28] S. I. AKASOFU, and S. YOSHIDA: J. Geophys. Res. **71**, 231 (1966).
[29] P. A. CLOUTIER: Planet. Space Sci. **14**, 809 (1966).

comparable throughout most of the magnetosphere). The neutral hydrogen must continually replenish the trapped protons to maintain the main phase decrease. Hence the neutral hydrogen flux from the sun must continue at a high level throughout the main phase and recovery phase of the storm. The quantity of neutral hydrogen required in space is enormous. BRANDT and HUNTEN[30] have pointed out that the resonant scattering of solar Lyman alpha (1216 Å) by so much neutral hydrogen would be very much higher than observed. So it does not appear that the idea of neutral hydrogen can be taken seriously. The accompanying suggestion that there is not enough energy in the solar wind to produce the main phase was based on the arbitrary assumption that the trapped particles producing the main phase are at a radial distance of $2\,R_E$ or less. AKASOFU and YOSHIDA[28] quote magnetometer observations[31] as evidence that the particles lie at $2\,R_E$. But there is nothing in the observations requiring the trapped particles at $2\,R_E$. At 3 to 5 R_E no such difficulty with rapid decay occurs.

ζ) The true nature of the *main phase inflation* has been brought out clearly by the recent important observations of CAHILL[32]. Briefly, during the great storm of 17 April 1965, the magnetometer on Explorer 26 passed several times from a radial distance of $2\,R_E$ to one of $6\,R_E$, near the equatorial plane. The satellite observations, together with the magnetic records at ground stations, show the inflation responsible for the main phase to be in the region 2.5 to $5.0\,R_E$. At the onset of the main phase there was a strong asymmetry, caused by strong inflation of the late afternoon and evening magnetosphere at about $3\,R_E$. The total energy of this inflation was estimated to be 10^{14} J [10^{21} erg]. The westward drift of the inflation indicates that it consisted of protons. The rate of drift suggests that the protons generally had energies between 15 and 20 keV. The decay of the asymmetry took place with a characteristic time of about two hours. Asymmetric inflation is expected on theoretical grounds to decay rapidly because it is forced to drive currents through the dense ionosphere (see Sects. 24, 25).

Following the decay of the asymmetric inflation there remained inflation with axial symmetry. The inflation was then located around $3.5\,R_E$, with a decay time of about 4 days. It was not possible, obviously, to identify the sign and energy of the particles producing the symmetric inflation, as could be done for the asymmetric inflation. But altogether it appears that much, if not all, of the inflation is low energy protons at 3 to $4\,R_E$, which are dissipated by charge-exchange and/or convection, and in the asymmetric case, by ionospheric resistivity as well.

More recently FRANK[33] has observed the protons and electrons in the energy range 0.2 to 50 keV with instruments aboard the OGO-3 satellite during the main phase decreases of the storm days 25 June and 9 July 1966. Data is available from magnetic shells which cross the equatorial plane from large distances into two Earth's radii. The observations show an enormously enhanced proton flux in this energy interval $3\,R_E$ to $5\,R_E$, between the onset and eventual relaxation of the main phases. The observational values of the energy density show that these particles are indeed the cause of the main phase, giving the observed field decrease of 30 and 50 γ, respectively, at the surface of the Earth. Thus altogether the earlier prediction of DESSLER and PARKER[3], that the main phase is produced largely by kilovolt protons trapped at 3 to 5 R_E, appears to be substantiated by direct observation. The prediction was based upon the several considerations mentioned above, that the particle energy density must be less than $B^2/2u\,\mu_0$ and that charge exchange with the ambient neutral hydrogen is the principal mechanism for the relaxation of the storm. FRANK directly observed the decay of the trapped protons as a function of radial distance and found the protons at

[30] J. C. BRANDT, and D. M. HUNTEN: Planet. Space Sci. **14**, 95 (1966).
[31] L. J. CAHILL, and D. H. BAILEY: Trans. Amer. Geophys. Union **46**, 116 (1965).
[32] L. J. CAHILL: J. Geophys. Res. **71**, 4505 (1966).
[33] L. A. FRANK: J. Geophys. Res. **72**, 3753 (1967).

smaller distances to decay more rapidly, in a manner consistent with the charge exchange hypothesis.

It is evident that these recent observations open up a new era in studies of the geomagnetic storm. The basic theoretical principles, proposed years ago, are evidently sound and we may now progress to direct study of the internal workings of the geomagnetic storm. Not only can the known processes, such as charge exchange, be checked out in detail, but hitherto unanticipated effects are discovered and explored. For instance, FRANK finds that, in addition to the main proton inflation, the enhanced electron fluxes account for about one quarter of the main phase. CAHILL observed the asymmetry of the inflation in space, where it is much larger than the related asymmetry in the field at the surface of Earth. The asymmetric inflation is difficult to study quantitatively from Earth because of the ionospheric currents between the observer and the inflation out in space.

η) In *summary*, then, the geomagnetic storm consists of two principal phases, the initial phase representing a compression of the field by increased solar wind pressure and the main phase representing an inflation of the field by ions trapped in it. The field is an elastic gasmagnetic medium which, in fact, defines the basic dynamical response of the field to external and internal pressures. Once this basic physical principle is understood, the mathematical theory of the geomagnetic storm can be reduced to a number of distinct and well defined formal problems. The principal categories are the confinement and compression of the geomagnetic field by the solar wind, the inflation of the geomagnetic field by internal gases, and the convection of the geomagnetic field. The confinement of the field has been treated mathematically for such simple boundary conditions as a normal pressure. The more complicated problem when there is a tangential drag of the wind on the boundary has not yet been solved, primarily because the drag is not understood (see discussion in Sect. 7). Inflation of the field by internal gases has been treated for a variety of conditions, and is essentially solved so far as the formal mathematics is concerned. The gas is now being observed directly, verifying the earlier general ideas, and providing a wealth of information on the detailed properties and behavior of the gas. But, there is still much work and thought to be given on the questions of how the gas gets there, and how it decays afterward, as noted above. The dynamics of the convection[18] has been treated in a formal mathematical way only for special cases and simplified geometry[34]. The calculations indicate that the forces one might reasonably expect from the solar wind are adequate to give the convective velocities inferred from the polar D_S current system. The driving force is not known; it may be just the tangential drag of the solar wind at the magnetopause but this also is not understood.

The next two chapters survey the formal calculations of the effects. We shall be interested not only in the correct formal equations and their mathematical solution, but in simple illustrations of the physical effects that can be constructed from the formal work.

D. Confinement and compression of the geomagnetic field.

13. The problem of confinement.

α) Confinement of the geomagnetic field by the pressure of solar corpuscular radiation was first pointed out by CHAPMAN and FERRARO[1] (see also [2,3] [*9*]). They noted that the *solar corpuscular radiation* is a *conducting medium* so the geomagnetic lines of force are not able to enter into the corpuscular radiation streaming by in

---

[34] E. WALBRIDGE: J. Geophys. Res. **72**, 5, 213 (1967).
[1] S. CHAPMAN, and V. C. A. FERRARO: Terr. Magn. **36**, 77, 171 (1931); **37**, 147, 241 (1932); **38**, 79 (1933).
[2] V. C. A. FERRARO: J. Geophys. Res. **57**, 15 (1952).
[3] S. CHAPMAN, and P. C. KENDALL: J. Atm. Terr. Phys. **22**, 142 (1961).

space, thus giving a sharp boundary to the geomagnetic field. Ferraro[2] examined the structure of the boundary between the field and wind in some detail. Neglecting the thermal velocities of the particles and assuming normal incidence, he derived the fundamental relation

$$\frac{\partial^2 A(r)}{\partial r^2} = \frac{u \mu_0}{c_0^2 \varepsilon_0 \mu_0} \frac{q^2}{m} N(r) A(r) \equiv \frac{\omega_N^2/c_0^2}{c_0^2 \varepsilon_0 \mu_0} A(r) \tag{13.1}$$

for the vector potential $A(r)$ as a function of distance r into the geomagnetic field, where $N(r)$ is the number of particles per unit volume, each of mass m and charge q, and $\omega_N/2\pi$ the plasma frequency. The relation follows from the fact that the normal derivative of the field density is related by Eq. (10.2) to the tangential current density j

$$\frac{c_0 \sqrt{\varepsilon_0 \mu_0}}{\mu_0} \delta B/\partial r = u\,j \qquad (B = \partial A/\partial r)$$

and the fact that j is equal to $N(r)\,q$ multiplied by the tangential velocity given to the particles by their penetration into the field, which is

$$w_{\tan} = \frac{q}{m\,c_0 \sqrt{\varepsilon_0 \mu_0}} \int d r \cdot B(r), \tag{13.2}$$

where $\int d r \cdot B(r)$ represents the number of lines per unit width penetrated by the particle since encountering the geomagnetic field B.

The ions penetrate farther into the geomagnetic field than the electrons as a consequence of their larger mass to charge ratio. Under stationary conditions the resulting charge separation[4] of the impinging wind particles is neutralized by thermal electrons and ions from the ionosphere, which are free to drift upward along the geomagnetic lines of force. But it takes several seconds or more to achieve neutralization, so that transients in the wind behave differently, with the ions penetrating only a little farther than the electrons while the electrons move tangentially with nearly the initial ion energy.

This case, when the ions and electrons do not separate, was developed by Ferraro[2]. It is readily shown, from Ferraro's basic relation and the fact that the magnetic pressure $B^2/2u\mu_0$, Eq. (6.1) must balance the impact pressure Nmw^2 of the wind, that under stationary conditions, the characteristic thickness of the boundary is of the order of the cyclotron radius of the ions in the field of flux density B. Transients, on the other hand, have a characteristic boundary thickness which is the harmonic mean of the electron and ion cyclotron radii. The ion cyclotron radius is about 100 km, so that the boundary is, under all circumstances, extremely thin compared to the overall dimensions in the magnetosphere, which is of the order of 10^5 km. Hence in looking at the overall confinement of the field by the solar wind it is permissible to approximate the boundary by a mathematical surface[5].

Chapman and Ferraro illustrated the compressive effect on the geomagnetic field by choosing geometrically simple boundaries (plane, cylinder, etc.) between the geomagnetic field and the corpuscular radiation. Solution of the general problem of confinement of the geomagnetic field is rather more difficult. The shape of the boundary is determined only through simultaneous solution of the equations for the geomagnetic field and the equations for the flow of the external wind, subject to the proper conditions at the undetermined boundary. The fields and the boundary must be calculated simultaneously. Even with the simplest boundary conditions, the problem is difficult. The boundary problem has been solved, so far, only for the condition that the wind exerts a normal pressure whose magnitude depends upon the inclination of the boundary to the undeflected wind. The external problem of the deflection of the wind, involving the bow shock, has been solved only by assuming a given shape and size for the magnetopause.

---

[4] T. L. Eckersley: Terr. Magn. Atmosph. Electr. **52**, 305 (1947).
[5] S. Chapman, and V. C. A. Ferraro: Terr. Magn. Atmosph. Electr. **36**, 77, 171 (1931).

14. Compression for simple geometric boundaries. Consider first the compression of the geomagnetic field inside the boundary. The compression by the wind can be studied after the method of CHAPMAN and FERRARO[1] by considering a number of geometrically simple boundaries. The normal component of the field must vanish on the boundary. Neglect the diamagnetic effect of the solid Earth. Neglecting the effects of internal gas pressures, etc. and remembering that the nonconducting atmosphere permits convection of the lines of force, thereby relieving any torsion in the field, it follows that both the parallel and perpendicular components of $\nabla \times \boldsymbol{B}$ are zero under static conditions throughout the geomagnetic field inside the boundary. Then $\boldsymbol{B}$ edrives from a potential Ψ and may be written as $-\nabla \Psi$ with $\nabla^2 \Psi = 0$. For the *unperturbed dipole* after Eq. (11.4)

$$\Psi = -B_0 \left(\frac{R_E^3}{r^2}\right) \cos\theta \tag{14.1}$$

where r is the radial distance from the dipole and θ is the polar angle measured from the axis of the dipole.

α) A *plane boundary* at a distance R in the direction $\theta = \Theta$ distorts the simple case of Eq. (14.1) into the field described by the potential

$$\Psi = -B_0 R_E^3 \left\{ \frac{\cos\theta}{r^2} + \frac{2R\cos\Theta - y\sin 2\Theta - z\cos 2\Theta}{[r^2 + 4R(R - y\sin\Theta - z\cos\Theta)]^{\frac{3}{2}}} \right\} \tag{14.2}$$

where z is distance measured along the dipole axis, y is distance measured perpendicular to the dipole axis and in the direction toward the plane boundary, and x is perpendicular to both the dipole and the direction toward the boundary. In the vicinity of the dipole, $r \ll R$, the change $\delta \boldsymbol{B}$ in the original dipole field is

$$\begin{aligned}
\delta B_x &= 0\left(\frac{R_E^4}{R^4}\right), \\
\delta B_y &= +\frac{B_0}{8}\left(\frac{R_E}{R}\right)^3 \sin\Theta\cos\Theta + 0\left(\frac{R_E^4}{R^4}\right), \\
\delta B_z &= +\frac{B_0}{8}\left(\frac{R_E}{R}\right)^3 (1+\cos^2\Theta) + 0\left(\frac{R_E^4}{R^4}\right).
\end{aligned} \tag{14.3}$$

These formulae show that the compression of the dipole by the plane boundary gives a $\delta \boldsymbol{B}$ which is uniform to first approximation in the vicinity of the dipole. Further, $|\delta B_y|/|\delta B_z|$ is rather less than one for all values of Θ, reaching a maximum value of only 0.35, at $\Theta = 55°$. Thus the compression field $\delta \boldsymbol{B}$ is both uniform and nearly antiparallel to the dipole axis for all values of Θ. This important fact means that compression of the geomagnetic field by *any* distant asymmetric boundary produces a compression in the field near Earth equivalent to adding a uniform field nearly antiparallel to the geomagnetic dipole[2]. The effect of the compression at the surface of Earth is principally to increase the horizontal component at low latitude. The field is compressed slightly more on the side toward the plane boundary than on the side away. For the special case that $\Theta = \pi/2$, so that the *boundary is parallel to the dipole*, the increase in the field has only the z-component

$$\delta B_z \cong \frac{B_0}{8}\left(\frac{R_E}{R}\right)^3 \left(1 + \frac{3y}{2R}\right). \tag{14.4}$$

It is evident that the compression differs by

$$(3/8) B_0 (R_E/R)^4$$

---
[1] S. CHAPMAN, and V. C. A. FERRARO: Terr. Magn. Atmosph. Electr. 36, 77, 171 (1931).
[2] E. N. PARKER: Phys. Fluids 1, 171 (1958).

between the night and day sides of Earth ($y=\pm R_E$). For a plane at $R=10\,R_E$, this difference is 1 γ.

β) Now a plane boundary confines the field only on one side. Confining the field on all sides, as for instance with a *sphere* of radius R produces a much larger compression. The compression in this case has the effect of adding a field which is exactly uniform and antiparallel to the dipole everywhere throughout the confining sphere. The magnitude of the field is $2B_0(R_E/R)^3$, or 16 times as large as for a plane boundary at the same distance R.

In the actual case the geomagnetic field is confined more than by a plane and less than by a sphere (see Fig. 3, Sect. 4α).

γ) A simple boundary resembling the actual confinement of the geomagnetic field a little more closely than the plane or sphere is a *circular cylinder closed at one end*. Consider, then, a circular cylinder of radius b concentric about the z-axis, closed off by a plane at $z=-h$ and extending to $z=+\infty$. It is a straightforward matter to solve the boundary value problem with the normal derivative of the magnetic potential Ψ vanishing on the boundaries to obtain the potential Ψ for a dipole with components $(M_x, 0, M_z)$ at the origin. The x-direction is perpendicular to the axis of the cylinder and the azimuth φ around the z-axis is measured from the x-direction. The potential is

$$\Psi = 2M_x \cos\phi \sum_{m=1}^{\infty} \frac{\beta_m^2 J_1(\beta_m \varrho)\{\exp(-\beta_m|z|)+\exp[-\beta_m(z+2h)]\}}{(\beta_m^2 b^2 - 1) J(\beta_m b)} + \\ + \frac{2M_z}{b^2} \sum_{m=1}^{\infty} \frac{J_0(\alpha_m \varrho)}{J_0^2(\alpha_m b)} \left\{ \frac{z}{|z|} \exp(-\alpha_m|z|) - \exp[-\alpha_m(z+2h)] \right\} \quad (14.5)$$

where α_m and β_m are the m-th roots of

$$J_0'(\alpha_m b) = 0, \quad J_1'(\beta_m b) = 0. \quad (14.5\,\mathrm{a})$$

and ϱ is distance from the z-axis. There are a number of geomagnetic effects illustrated by this example. The cylinder represents the confinement of the field by the solar wind to form the tail. The plane $z=-h$ represents the confinement on the sunward side. For simplicity put* $h=b$ and put $M_z=0$ so that the dipole is perpendicular to the axis of the cylinder. The field in the tail ($z>0$) at the north and south sides of the cylindrical boundary ($\varrho=b$, $\varphi=0, \pi$) is parallel to the axis of the cylinder. The corresponding density is shown in Fig. 13 by the upper broken curve as a function of the distance z/b back in the tail. The lower broken curve represents the total field density of the undistorted dipole at the same position in space. The field along the axis of the cylinder is in the north-south direction, as when the dipole is undistorted. The field density along the axis is represented by the upper solid curve in Fig. 13, the lower solid curve representing the field of the undistorted dipole at the same position in space. It is evident from Fig. 13 that the compression of the field in the tail increases the field density by about a factor of two out to $z\cong 2b$. However, beyond about $z\cong 1.5\,b$ the field in this simple cylindrical model declines as $\exp(-1.84z/b)$ with increasing z, whereas the undistorted dipole field would decline only as z^{-3}. Hence the field in the tail falls below the undistorted dipole field beyond about $z=3b$. This illustrates how rapidly the field in the tail would pinch off behind Earth if the boundaries of the tail were passive, as in the present illustration. In the actual tail the field is parallel to the $x=0$ plane and reverses sign across $x=0$, whereas

---

* This underestimates somewhat the compression on the sunward side compared to the compression in the tail. Presumably $h<b$.

in this passive boundary model the field is perpendicular to the $x=0$ plane and nearly as strong at the plane as anywhere else in the tail at the same distance z.

Further plots of field strength in the passive cylindrical tail can be found in [3], including the case where the tail is closed by a plane at some distance in the antisolar direction. It is readily shown that with $h=10\ R_E$ (as suggested by the observed distance to the boundary in the subsolar direction) the compression increases the field strength by 35 γ at the subsolar point on the boundary ($z=-h$, $\varrho=0$). The field is approximately doubled there by the compression, as with an infinite plane boundary.

The plane $z=-h$ closing off the field in the sunward direction contributes the term involving the exponential of $(z+2h)$ in the expression Eq. (14.5) for the potential. This term is larger on the day than on the night side, giving a diurnal variation. Numerical calculation shows that with $b=h=10\ R_E$ the compression is 7.0 γ at noon on the equator of Earth, compared with 4.8 γ at midnight. The diurnal variation is thus 2.2 γ, to be compared with 1 γ obtained from Eq. (14.4) and the more correct values 3.3 γ[4] and 4 γ[5]. The low value calculated for the cylindrical boundary results from having chosen $h=b$, whereas in fact the effective distance h to the subsolar point of the magnetopause is considerably less than the radius of the tail.

Fig. 13. Plot of the field density of a magnetic dipole confined inside a cylinder of radius b. The coordinate z is measured along the axis of the cylinder. The cylinder is closed off at $z=-b$ by a plane. The field strength B is given in units $B_0\ R_E^3/b^3$. The dipole is perpendicular to the axis and is located at the origin. The (upper) thick solid curve represents the field density along the axis of the cylinder. The (lower) thinner solid curve represents the field density along the axis of the cylinder in the absence of confinement. The (upper) thick broken curve represents the field density as a function of distance z along the intersection of the cylinder wall with the plane defined by the dipole and the cylinder axis. The (lower) thinner broken curve represents the field along the same line in the absence of confinement.

Altogether, then, confinement of the geomagnetic field by the quiet-day solar wind produces strains in the field of 10 to 35 γ throughout. The strains may, to a first approximation, be represented by a uniform northward field of some 10 to 20 γ in the vicinity of Earth*. The calculated diurnal variation, at a point fixed on the surface of Earth, is much smaller than the effects produced by the daily rise and fall of the ionosphere and is probabily unobservable.

The increased compression to form the initial phase of a storm involves boundaries pushed in to 6 to $8\ R_E$ on the sunward side, which is discussed in Sect. 17.

15. Calculation of the boundary.
Now consider the problem of deducing the boundary of the geomagnetic field under the impact of the solar wind. The shape of the boundary depends upon the manner in which the wind interacts with it.

α) The *elementary theory* considers the simplest physical situation in which the individual ions and electrons both move as independent particles as considered by CHAPMAN and FERRARO in their 1931 paper[1]). Then each particle enters the geomagnetic field and undergoes specular reflection after penetrating a distance

---

* 1 gauss $=\Gamma=10^{-4}$ T; T≡Tesla.
[3] E. N. PARKER: Space Sci. Rev. **1**, 62 (1962).
[4] J. E. MIDGLEY, and L. DAVIS: J. Geophys. Res. **68**, 5111 (1963).
[5] G. D. MEAD: J. Geophys. Res. **69**, 1181 (1964).
[1] S. CHAPMAN, and V. C. A. FERRARO: Terr. Mag. **36**, 77, 171 (1931).

comparable to the cyclotron radius. The cyclotron radius for a 1000 km/sec proton (5 keV) in a typical field of 50 γ is 200 km. An electron with the same speed[2] penetrates only about 0.1 km. The penetration is slight and the separation of electrons and protons, as a consequence of their different depths of penetration into the magnetic field, does not lead to significant electrostatic fields under steady conditions. The magnetic lines of force across which they are separated connected into the ionosphere, where the high electrical conductivity permits a flow of thermal particles across the lines of force to neutralize the effect. The elementary theory represents the boundary as a mathematical surface, from which the solar wind particles undergo specular reflection[1,3,4]. Possible instabilities of the surface are neglected. It is evident that with simple specular reflection the wind exerts a pressure normal to the surface and there is no drag force. Thus, the observed drawing out of magnetic lines of force into the tail does not occur within the framework of the elementary theory. But the shape of the boundary is fairly well represented in the general vicinity of Earth.

The boundary conditions are easily written down. The normal component of the magnetic field is put equal to zero, in recognition of the definition of the boundary. The magnetic pressure at the boundary is equal to the impact pressure of the wind incident on the boundary. A collimated stream of particles of particle density N, each with mass m and velocity v, gives a pressure $2Nmv^2\cos^2\vartheta$ for specular reflection from an element of surface whose normal makes an angle ϑ with the wind velocity. Hence the magnetic pressure at the boundary must be[5]

$$\frac{B^2}{2u\mu_0} = 2Nmv^2\cos^2\vartheta. \tag{15.1}$$

ŽIGULEV and ROMIŠEVSKIJ[6] point out that the $\cos^2\vartheta$ dependence of the pressure is also correct for the Newtonian model of extreme hypersonic flow, so the results obtained here may be more generally valid than the special model of independently moving particles (see also [7]). LIGHTHILL[8] gives a general discussion of the validity of the $\cos^2\vartheta$ law. Generally speaking, $\cos^2\vartheta$ is valid to the extent that the Mach number is sufficiently high that the gas behind the shock may be treated as incompressible.

If the reflection were not elastic, the factor 2 in the expression for the pressure [at the right side of Eq. (15.1)] would be correspondingly reduced, reaching the value of one in the limit of no rebound. It is evident from the simple condition Eq. (15.1) that, in order of magnitude, the distance to the subsolar point may be estimated using the field Eq. (14.2) for a simple plane boundary[1]. The result is [9] the *scale of the magnetopause*, R_M:

$$\frac{R_M}{R_E} = \left(\frac{B_0^2}{u\mu_0 Nmv^2}\right)^{\frac{1}{6}}. \tag{15.2}$$

The sixth root renders the numerical value for R_M insensitive to the theoretical approximation for B and to the observational uncertainties in N and v. Putting $B_0 = 0.32$ gauss*, and using the present observations that $v = 300$ km/sec and $N = 5 \cdot 10^6$ m^{-3} [5 cm^{-3}] yields $R_M = 10\ R_E$, in rough agreement with observation.

---

* 1 gauss = Γ = 10^{-4} T; T ≡ Tesla = Vsm^{-2}.
[2] H. GRAD: Phys. Fluids **4**, 1366 (1961).
[3] E. N. PARKER: Phys. Fluids **1**, 171 (1958).
[4] D. B. BEARD: J. Geophys. Res. **65**, 3559 (1960).
[5] V. C. A. FERRARO: J. Geophys. Res. **57**, 15 (1952).
[6] V. N. ŽIGULEV, and E. A. ROMIŠEVSKIJ: Dokl. Akad. Nauk. SSSR **127**, 1001 (1960); (engl. transl.) Soviet Physics Doklady **4**, 859 (1960).
[7] J. E. MIDGLEY, and L. DAVIS: J. Geophys. Res. **67**, 499 (1962).
[8] M. J. LIGHTHILL: J. Fluid Mech. **2**, 1 (1957).
[9] D. F. MARTYN: Nature **167**, 92 (1951).

Calculation of the boundary.

The problem of the boundary and the distorted dipole can be solved exactly in two dimensions by employing a conformal mapping technique[6,10]. Hurley[11] gives both the field inside the boundary and the boundary for arbitrary angle of incidence of the particle stream.

β) The *three dimensional problem* admits no analytic solution in closed form. Three different procedures have been developed to carry out the calculation. Consider first the method developed by Beard[4]. The approximation begins by noting that the tangential component of the field at a plane boundary is precisely double that component in the absence of the boundary. This is obvious from the fact that a plane boundary may be represented by introducing an image dipole. In the first step of the approximation Beard uses the factor two appropriate to the plane. Thus, if the normal to the boundary makes an angle ϑ with the free particle stream, the normal pressure of the particle stream $2Nmv^2 \cos^2 \vartheta$ is equated to $(2B_t)^2/2u\mu_0$ where B_t is the unperturbed component tangent to the boundary. B_t can be calculated as a function of ϑ for a dipole field. If r is radial distance from the dipole, θ the polar angle measured from the dipole axis, and ϕ the azimuth measured around the dipole from the direction of the particle stream, we have

$$B_r \propto 2\cos\theta/r^3, \quad B_\theta \propto \sin\theta/r^3, \quad B_\phi = 0$$

when the unperturbed dipole is at the origin and is directed perpendicular to the particle stream. If ψ is the angle between the radius vector and the boundary $r = r(\theta, \phi)$, then, for instance, in the noon-midnight meridional plane it is readily seen that $\tan\psi = r\, d\theta/dr$. Then for the simple case that the particle stream is perpendicular to the dipole axis, it follows that $\vartheta = \psi - \theta$, and B_t can be expressed in terms of B_θ and B_r, giving

$$B_t \propto \pm(2\cos\theta\cos\psi - \sin\theta\sin\psi)/r^3. \tag{15.3}$$

From the equality of the wind pressure and magnetic pressure, we have $B_t \propto \cos\vartheta$, so that, altogether, the requirement that the magnetic pressure balance the particle pressure can be written

$$\frac{d\tilde{r}}{\tilde{r}} \frac{1 \mp 2/\tilde{r}^3}{1 \mp 1/\tilde{r}^3} = \tan\theta\, d\theta \tag{15.4}$$

in terms of the reduced distance to the boundary, $\tilde{r}(\theta)$, expressed as ratio to a characteristic length which will be determined in the following as being R_M from Eq. (15.2). Thus

$$\tilde{r} \equiv \frac{r}{R_M} = (u\mu_0 Nm)^{\frac{1}{6}} \left(\frac{v}{B_0}\right)^{\frac{1}{3}} \frac{r}{R_E}. \tag{15.4a}$$

It is evident that $\tilde{r} = 1$ is a solution of this equation when the upper sign is taken in Eq. (15.4). When $\tilde{r} \neq 1$ the solution is

$$\frac{\tilde{r}^2}{\tilde{r}^3 \pm 1} = C\cos\theta$$

where C is the integration constant. Obviously $\tilde{r} = 1$ represents the sunlit boundary and

$$\frac{\tilde{r}^2}{\tilde{r}^3 + 1} = C\cos\theta \tag{15.5}$$

represents the night side boundary, with $\tilde{r} \to \infty$ as $\theta \to \pi/2$.

---
[10] J. W. Dungey: J. Geophys. Res. **66**, 1043 (1961a).
[11] J. Hurley: Phys. Fluids **4**, 109, 854 (1961a).

The characteristic length relative to which the reduced distance $\check{r}$ is to be expressed is readily deduced without going through the algebra from the beginning by noting that at the subsolar point ($\theta = \pi/2$) the radius is given exactly by Eq. (15.2) in the present approximation (corresponding to $\check{r} = 1$ in the present solution). The characteristic length to which $\check{r}$ refers is, therefore, the scale of the magnetopause, R_M defined by Eq. (15.2). Thus Eq. (15.4a) is now justified. The integration constant C must be determined by matching to the day side boundary $\check{r} = 1$, which cannot be done across the neutral point in the noon-midnight meridian plane alone. The general shape of the boundary is as sketched in Fig. 3, Sect. 4α.

Elsewhere than the noon-midnight meridian the formulation leads to a non-linear partial differential equation for the boundary, which must be solved by numerical methods. See [4,12] for detailed formulation and solution of the whole boundary. The calculations show that the sunlit boundary is nearly a hemisphere. The simple illustration given here is sufficient to show the basic method.

γ) Beard and Jenkins[13] have calculated the distortion of the *field inside the boundary*. They find a compression of 14 γ at the surface of Earth when the radial distance to the boundary is $R = 10\,R_E$. This is to be compared with the artificial models of the previous subsection and with the other numerical results discussed below.

Beard's method has been applied to deformations of the geomagnetic field under a variety of external particle stream conditions[14] *. A ring current has been added[15], and later a uniform isotropic external pressure (in the fixed frame of reference)[16]. The results for the uniform pressure are interesting but it is not obvious what physical situation they represent. The thermal motions in the supersonic solar wind plasma do not give a uniform pressure in the frame of reference of Earth.

δ) Now a number of *improvements* have been made in Beard's original approximation. The method breaks down seriously in the vicinity of the neutral point, as pointed out by its author. Blum[17] has developed an expansion for treating the boundary near the neutral points in an accurate way. Ferraro[18] has pointed out that the accuracy of Beard's method may be improved if one supposes that the actual field at the boundary is $2f B_t$, where f is a constant of the order of unity but not necessarily exactly equal to one. The constant is maintained throughout the calculations (where it appears *only* as a scale factor) and determined afterward by adjustment of the boundary at some critical point.

Ferraro applied the method to the calculation of the boundary of the field of a straight line current placed across a uniform particle stream. This two dimensional problem can be solved rigorously[19,11] and comparison of the two solutions suggests that the best value of f is 0.68 in this case.

Spreiter and Briggs[14] considered the flow past a two dimensional dipole and obtained $f = 0.913$. These calculations suggest that the correction to the scale as calculated originally by Beard is not large.

Beard[20,13] introduced a second order correction to his method. In the first approximation the field at the boundary is taken as $2 B_t$, appropriate for a plane.

---

* Spreiter and Briggs[14] use the pressure for completely inelastic reflection of particles, contrary to the claim that they have specular reflection. The effect appears only in the scale, of course, which is easily corrected. There is also an error in their first paper in the solution for the boundary on the noon-midnight meridian, involving matching the day and night side boundaries.

[12] L. Davis, and D. B. Beard: J. Geophys. Res. **67**, 4505 (1962).
[13] D. B. Beard, and E. B. Jenkins: J. Geophys. Res. **67**, 4895 (1962).
[14] J. R. Spreiter, and B. R. Briggs: J. Geophys. Res. **67**, 37, 2983 (1962).
[15] J. R. Spreiter, and A. Y. Alksne: J. Geophys. Res. **67**, 2193 (1962).
[16] J. R. Spreiter, and B. J. Hyett: J. Geophys. Res. **68**, 1631 (1963).
[17] R. Blum: J. Geophys. Res. **69**, 1765 (1964).
[18] V. C. A. Ferraro: J. Geophys. Res. **65**, 3951 (1960).
[19] V. N. Žigulev: Dokl. Akad. Nauk. SSSR **126**, 521 (1959); (engl. transl.) Soviet Physics Doklady **4**, 514 (1959).
[20] D. B. Beard: J. Geophys. Res. **67**, 477 (1962).

The calculations show the sunlit boundary to be nearly a hemisphere. So in the next approximation the field at the boundary was chosen appropriate to a sphere instead of a plane. The corrections to the first approximation were found to be small over the sunlit boundary, indicating the general validity of the first approximation there. The polar and nighttime regions were significantly altered, as had been anticipated. Further discussion of the successive approximations may be found in [21,22]. The convergence of the method is rapid. A calculation of the field inside the boundary has been carried out [23], with particular attention to the magnetic lines of force*. See the review [25].

ε) A rather different approach to the solution of the *boundary problem* has been developed [7]. Consider the surface current density $\boldsymbol{J}$ in the boundary instead of the field density $\boldsymbol{B}$ at the boundary. If there is no field outside the boundary, MAXWELL'S equation Eq. (10.2) requires that $u\,\mu_0\,\boldsymbol{J} = c_0\sqrt{\varepsilon,\mu_0}\,\boldsymbol{B}$. Thus if p is the external pressure, the equilibrium requirement $p = B^2/2u\,\mu_0$ is written, instead of Eq. (15.1), as

$$p = \frac{1}{2}\,\frac{u\,\mu_0}{c_0^2\,\varepsilon_0\,\mu_0}\,J^2. \tag{15.6}$$

Starting with a first guess at the shape of the boundary, MIDGLEY and DAVIES [26] calculate p (which is just $2Nmv^2\cos^2\vartheta$ for a stream of particles). The (surface) current density J follows at once. Then they calculate by numerical methods the field outside the surface caused by J. This field should be just equal and opposite to the field of the geomagnetic dipole at the origin since the sum should be identically zero outside the boundary. The calculation proceeds by successive readjustment of the surface until the total external field vanishes to any desired degree of accuracy. The surface for which the external field vanishes is the boundary for whatever form of external pressure was assumed.

They apply the method to the calculation of a dipole surrounded by a static pressure, and in the second paper they calculate the boundary of the field at a dipole lying perpendicular to a stream of free particles. The field inside the boundary is given also. For comparison of the results of the method with BEARD'S calculations see [26, 22, 23, 27]. The agreement is within a few percent, which is essentially the limit of the numerical accuracy with which the computations were carried out.

ζ) A third method for calculating the boundary was developed by SLUTZ [28] based on a method for treating the *free surface of a jet* of liquid [29]. Since no other stresses except those at the boundary surface S and at the dipole at the origin are admitted in the problem, $\nabla \times \boldsymbol{B}$ is zero everywhere between the origin and the surface. Put $\boldsymbol{B} = -\nabla \Psi$. Then from GREEN'S theorem [20] the total potential Ψ at an interior point can be related to the sum of the potential Ψ_i of the dipole at the origin plus a surface integral of Ψ by

$$\Psi(\boldsymbol{r}) = \Psi_i(\boldsymbol{r}) + \frac{1}{4\pi}\int_S d\boldsymbol{S}\cdot\left[\frac{1}{s}\frac{\partial\Psi}{\partial n} - \Psi\frac{\partial}{\partial n}\frac{1}{s}\right] \tag{15.7}$$

where $d\boldsymbol{S}$ is an element of surface, $\partial/\partial n$ represents the outward normal derivative at the surface, and s is the distance between the surface point and the interior point $\boldsymbol{r}$. Now for a dipole field

$$\Psi_i = -B_0(R_E^3/r^2)\cos\theta. \tag{14.1}$$

---

* An idealized but useful mathematical model of the field has been employed when computing particle motions in [24].

[21] J. C. BAKER, D. B. BEARD, and J. G. YOUNG: Phys. Fluids **7**, 504 (1964).
[22] G. D. MEAD, and D. B. BEARD: J. Geophys. Res. **69**, 1169 (1964).
[23] G. D. MEAD: J. Geophys. Res. **69**, 1181 (1964).
[24] H. E. TAYLOR, and E. W. HONES: J. Geophys. Res. **70**, 3605 (1965).
[25] D. B. BEARD: Rev. Geophys. **2**, 335 (1964).
[26] J. E. MIDGLEY, and L. DAVIS: J. Geophys. Res. **68**, 5111 (1963).
[27] J. E. MIDGLEY: J. Geophys. Res. **69**, 1197 (1964).
[28] R. J. SLUTZ: J. Geophys. Res. **67**, 505 (1962).
[29] E. TREFFTZ: Z. Math. u. Phys. **64**, 34 (1916).

Since no field penetrates the boundary, $\partial \Psi/\partial n = 0$. The solid angle $d\Omega$ subtended by dS at the point r is just $-dS(\partial/\partial n)(1/r)$. Further $\int d\Omega = 4\pi$. Thus, altogether, the expression can be written

$$\Psi = \left[4\pi \frac{B_0 R_E^3 \cos\theta}{r^2} + \int_S d\Omega\, \Psi\right] \bigg/ \int_S d\Omega. \tag{15.8}$$

The condition of pressure balance must be introduced to complete the system of equations. Since $\partial \Psi/\partial n = 0$ at the surface, the condition

$$p = B^2/2u\mu_0 \tag{6.1}$$

can be written

$$(\nabla \Psi)_S = (2u\mu_0 p)^{\frac{1}{2}} \tag{15.9}$$

at the surface. The condition Eq. (15.9) determines how rapidly, but not in which direction, Ψ varies at each point on the surface.

Simultaneous numerical solution of Eqs. (15.8) and (15.9) gives a unique determination of the potential. The procedure is to make a first guess at the surface S. Given a surface S the pressure p exerted at each point on the surface is known. Then Eq. (15.8) is solved for Ψ using the result to see if Eq. (15.9) checks. On the basis of the disagreement between the right and left sides of Eq. (15.9) adjustments are made in S and the calculation is repeated.

SLUTZ[28] solved the problem of a dipole field confined by a uniform external pressure p. The results are to be compared with [7]. The boundary has been calculated for a uniform external pressure plus a wind pressure with the $\cos^2\vartheta$ dependence [30]. The results are to be compared with those of [16] based on BEARD's first approximation. The effect of the external pressure is to close off the geomagnetic field in the antisolar direction rather abruptly and without regard for the limiting Mach angle at which the actual thermal pressure in the supersonic wind might pinch off the field.

16. Calculation of the bow shock configuration. Once a first approximation for the magnetopause is achieved, it is possible to consider the shape and standoff distance of the shock wave produced in the supersonic wind by the geomagnetic obstacle in its path. The wind must be considered as behaving something like a fluid because of the weak magnetic field carried in it, as discussed in Chapter B. The Mach number of the wind appears to vary from 2 to something in excess of 20, with a typical value of 8 (corresponding to $10^5\,°K$ temperature in a 400 km/sec wind). The *gas-magnetic Mach number* i.e. the ratio of wind speed to Alfven speed, Eq. (4.1),

$$\sqrt{u\mu_0 \varrho}\,\frac{v}{B}, \tag{16.1}$$

is of the same general order as the Mach number itself.

KELLOGG[1] used the results for the ordinary gas dynamic shock around a spherical obstacle[2] to estimate the shape and standoff distance. An effective Mach number of 2 places the shock at a radial distance of about 1.4 the distance to the magnetopause in the subsolar direction. Thus the shock is at $14\,R_E$ if the magnetopause is at $10\,R_E$. Increasing the Mach number decreases the distance to the shock, giving 1.28 at a Mach number of 4 and 1.23 for infinity. The calculations have been improved by including the fact that the geomagnetic field becomes broader toward the tail rather than closing off as a sphere[3]. The result is an increase in the standoff distance, to 1.62, 1.42 and 1.36 for Mach numbers of 2, 4 and ∞, respectively. Some effects of an interplanetary field on the standoff

---

[30] R. J. SLUTZ, and J. R. WINKELMAN: J. Geophys. Res. **69**, 4933 (1964).
[1] P. J. KELLOGG: J. Geophys. Res. **67**, 3805 (1962).
[2] K. HIDA: J. Phys. Soc. Japan **8**, 740 (1953).
[3] J. R. SPREITER, and W. P. JONES: J. Geophys. Res. **68**, 3555 (1963).

of the shock have been worked out [4]. An oblique interplanetary field may lead to an asymmetric geomagnetic tail[5]. WALTERS[6] has also pointed out the possibility of a second shock associated with the neutral points in the magnetopause.

The major obstacle to further improvement in the calculation of the shock position is the lack of knowledge of the structure of the collisionless shock, and hence the lack of knowledge of the state of the shocked gas in the magnetosheath between the shock and magnetopause. [1] and [3] discuss the state of the shocked gas briefly, based on a collisionless shock structure[7]. The effective ratio of specific heats figures in the calculation of the shock position and may be $\frac{5}{3}$ or larger depending upon the nature of the disorder produced by the shock.

17. Field compression.

α) The "mechanics" of calculating the *first* theoretical *approximation* to the *position* of the *bow shock* and the *magnetopause*, and the resulting compression of the geomagnetic field inside the magnetopause, seems to be fairly well in hand. The calculation of the bow shock is carried out assuming a shape for the magnetopause. The calculation of the magnetopause is carried out more or less ignoring the bow shock and ignoring tangential drag. The assumption in calculating the magnetopause is that the pressure on the magnetopause varies as the square of the cosine of the angle between the normal to the magnetopause and the undisturbed wind direction.

The next order of approximation would be, obviously, to compute the bow shock and magnetopause simultaneously. However, this can not reasonably be done until the collisionless shock phenomenon is better understood. The really interesting questions at the present time have more to do with collisionless shock structure, the state of the shocked gas, fast particle production, and the tangential drag on the magnetopause producing the geomagnetic tail and perhaps the deep convection.

The *first approximation* to the shock and magnetopause are in rather good agreement with the observations[1-3]. Theoretical results of SPREITER and JONES[4] for the *shock* and of BEARD[5] for the sunward *magnetopause* (as explained in Sect. 15) have been plotted by NESS[3], together with observed data reduced to standard conditions, and are reproduced from his paper in Fig. 14. The agreement between theory and observation is good, until the tail region is approached where the simple theory, omitting tangential drag, fails to agree with the observed neutral sheet and the slight broadening of the tail. The breadth of the tail appears to approach a fixed value of about 20 R_E at large distance behind Earth.

β) A summary of the *numerical results* is as follows. With the magnetopause at 10 R_E and a wind of 300 km/sec and atomic density $5 \cdot 10^6$ m^{-3} [5 cm^{-3}] the quiet day compression of the field produces a component parallel to the geomagnetic axis of 23 γ on the midnight side and 21 γ at noon [6]. The resulting 4 γ diurnal variation is negligible compared to the 10^3 γ diurnal ionospheric effects. The mean compression in the vicinity of the dipole is 25 γ. Calculations[6,7] show that the field at the subsolar point of the magnetopause is approximately double

---

[4] C. SOZOU: J. Geophys. Res. **70**, 4165 (1965).
[5] G. K. WALTERS: J. Geophys. Res. **69**, 1769 (1964).
[6] G. K. WALTERS: J. Geophys. Res. **71**, 1341 (1966).
[7] P. L. AUER, H. HURWITZ, and R. M. KILB: Phys. Fluids **4**, 1105 (1961).
[1] W. J. FREEMAN, J. A. VAN ALLEN, and L. J. CAHILL: J. Geophys. Res. **68**, 2121 (1963).
[2] N. F. NESS, and J. M. WILCOX: Phys. Rev. Letters **13**, 461 (1964).
[3] N. F. NESS: J. Geophys. Res. **70**, 2989 (1965).
[4] J. R. SPREITER, and W. P. JONES: J. Geophys. Res. **68**, 3555 (1963). The mechanism applied in this paper involves semi-empirical studies of supersonic flow in wind tunnels.
[5] D. B. BEARD: J. Geophys. Res. **65**, 3559 (1960).
[6] G. D. MEAD: J. Geophys. Res. **69**, 1181 (1964).
[7] J. E. MIDGLEY: J. Geophys. Res. **69**, 1197 (1964).

the unperturbed dipole value of 35 γ. For other wind strengths, recall that δB is proportional to the square root of the wind pressure $2Nmv^2$, or to the reciprocal of the cube of the scale R of the magnetopause.

Consider the implications for the main phase and initial phases of a magnetic storm. A vanishing of the wind would result in the vanishing of δB, representing a decrease of the horizontal component at the surface of Earth at low latitudes of about 25 γ below the normal quiet-day value. Removing the wind cannot

Fig. 14. A comparison of observed shock wave positions (dots) and observed magnetopause positions (plusses) with the theoretical curves (solid lines) of SPREITER and JONES[4], reproduced from[3]. The shock and magnetopause positions are in the equatorial plane. The sun lies to the left and distance is measured in units of Earth radii, R_E. The aberration of the solar wind is supposed to be 5°.

do more than remove the quiet-day compression, as noted in Sect. 10γ. Hence, as noted in Sects. 11 and 12, something else, such as trapped plasma pressure, must be responsible for the usual main phase.

Increasing the dynamic pressure of the wind by a factor of 10 would push the magnetopause in to 6.8 R_E, giving a field increase of 54 γ above the quiet-day value in the vicinity of the geomagnetic dipole, for a total compression of 79 γ. To produce such an increased compression, for the initial phase of a very large magnetic storm, a typical quiet-day wind of 300 km/sec and atomic density 5 cm^{-3} would rise to, say, 600 km/sec and 15 cm^{-3} *. It is clear from the theory[8] that the earlier estimates of the density of the solar wind at the time of a large outburst[9,10] of 300 cm^{-3}, and more, are probably too high. They would produce net increases in the horizontal component of 500 γ and more, whereas the observed compression is typically 20 to 30 γ, and approaches 100 γ only in very rare events.

This, then, is the result of present theory of compression of the geomagnetic field and the initial phase of the geomagnetic storm, based on the original idea

---

* These calculations omit the effect of the solid Earth, so that a somewhat smaller increase in the wind might suffice to produce the temporary increase of the initial phase. The solid Earth is a sufficiently good conductor that the geomagnetic lines of force may be considered frozen into it, so far as periods of several days are concerned. Thus Earth is completely diamagnetic so far as transient δB is concerned. It is well known that the field near the equator of a diamagnetic sphere immersed in a uniform field is 50 percent stronger than the applied field.

[8] E. N. PARKER: Space Sci. Rev. **1**, 62 (1962).
[9] A. UNSÖLD, and S. CHAPMAN: Observatory **69**, 219 (1949).
[10] D. E. BLACKWELL, and M. F. INGHAM: Mon. Not. Roy. Astron. Soc. **122**, 113 (1961).

of CHAPMAN and FERRARO that the "solar corpuscular radiation" confines and compresses the geomagnetic field. The theory is a quasi-equilibrium theory, based on the fact that the initial phase of a storm usually lasts several hours, whereas the gasmagnetic propagation time across the magnetosphere is of the order of a minute.

γ) There remains, then, the *sudden commencement* of the initial phase, which with its rise time of a few minutes, is principally a *magneto-gasdynamic wave phenomenon*[11] (see discussion and references in Chapter C). A quantitative treatment of the problem has not yet been given. The formal methods for computing the rays and wave amplitudes have been given in the limit of short wavelengths[12], with application to some simple illustrative examples. (See [13,14] for a discussion of the ionospheric effects on the sudden commencement.) But the general treatment of waves with scales comparable to the scale of the variation of the Alfven velocity is very difficult (see for instance [15,16]) and has not yet been worked out to the point where the sudden commencement can be investigated quantitatively.

Unfortunately there is still very little quantitative observational information on gasmagnetic waves in the magnetosphere (see references below). Quantitative interpretation of ground based observations of gasmagnetic waves is complicated by the damping, and other effects, in the ionosphere (see, for instance, [17-20]). Significant ionospheric heating may result from gasmagnetic waves during periods of high activity[21]. The general magneto-gasdynamic wave problem may be divided into two parts: The waves generated by a known sudden change in the wind, as in the positive or negative sudden commencement, and the continual presence and propagation of magneto-gasdynamic waves of unknown origin in the magnetosphere. Presumably turbulence in the magnetosheath and the tangential drag mechanism are related to much of the continuing magneto-gasdynamic wave activity observed in the magnetosphere. Of particular interest is the tendency for the wave distributions to be symmetric about a plane significantly different from the noon-midnight or dawn-dusk meridians, compare[22]. The interested reader is referred to the literature for the analysis of ground based observations [23-27] and [28] (review). For comparison with satellite observations see [29-31].

---

[11] A. J. DESSLER: J. Geophys. Res. **63**, 405 (1958).
[12] J. BAZER, and J. HURLEY: J. Geophys. Res. **68**, 147 (1963).
[13] A. A. ASHOUR, and A. T. PRICE: Proc. Roy. Soc. A **195**, 198 (1948).
[14] A. A. ASHOUR, and V. C. A. FERRARO: J. Atmosph. a Terr. Phys. **26**, 509 (1964).
[15] G. J. F. MACDONALD: J. Geophys. Res. **66**, 3639 (1961).
[16] R. L. CAROVILLANO, and J. F. MCCLAY: Phys. Fluids **8**, 2006 (1965).
[17] W. E. FRANCIS, and R. KARPLUS: J. Geophys. Res. **65**, 3593 (1960).
[18] R. KARPLUS, W. E. FRANCIS, and A. J. DRAGT: Planet. Space Sci. **9**, 771 (1962).
[19] E. C. FIELD, and C. GREIFINGER: J. Geophys. Res. **70**, 4885 (1965).
[20] C. GREIFINGER, and P. GREIFINGER: J. Geophys. Res. **70**, 2217 (1965).
[21] A. J. DESSLER: J. Geophys. Res. **64**, 397 (1959); [*27*], 119 (1965).
[22] See contributions by R. GENDRIN in this volume, p. 461, also those by E. SELZER and by H. POEVERLEIN in Vol. 49/4 of this Encyclopedia. The theory of magneto-gasdynamic waves is explained by V. L. GINZBURG and A. A. RUHADZE in Vol. 49/4 of this Encyclopedia. Sects. 24 through 26.
[23] C. R. WILSON, and M. SUGIURA: J. Geophys. Res. **66**, 4097 (1961); **68**, 3314 (1963).
[24] L. R. TEPLEY: J. Geophys. Res. **66**, 1651 (1961).
[25] T. NAGATA, S. KOKUBUN, and T. IIJIMA: J. Geophys. Res. **68**, 4621 (1963).
[26] R. C. WENTWORTH: J. Geophys. Res. **69**, 2689, 2699 (1964). — R. C. WENTWORTH: L. TEPLEY, K. D. AMUNDSEN, and R. R. HEACOCK: J. Geophys. Res. **71**, 1492 (1966).
[27] M. SIEBERT: Planet. Space Sci. **12**, 137 (1964).
[28] V. A. TROITSKAYA: [*18*], 485 (1964).
[29] C. P. SONETT, D. L. JUDGE, A. R. SIMS, and J. M. KELSO: J. Geophys. Res. **65**, 55 (1960).
[30] V. L. PATEL: Planet. Space Sci. **13**, 485 (1965).
[31] R. E. HOLZER, M. G. MCLEOD, and E. J. SMITH: J. Geophys. Res. **71**, 1481 (1966).

E. Inflation of the geomagnetic field.

18. General remarks. The main phase of the geomagnetic storm is the result of energetic plasma trapped in the geomagnetic field[1,2]. The energetic plasma may be the ambient plasma heated by shock waves, etc.[3] and diffusion[4] and/or plasma convected in from the outer magnetosphere and the solar wind[5]. Relaxation of the inflation, leading to the recovery of the main phase, may come about through charge exchange with the ambient neutral hydrogen[2], convection[5], thermal conduction to the ionosphere[6], instabilities[7] etc. (see Chapter C). The source of the inflation, and its subsequent relaxation, are physical questions which have been discussed in Sect. 12. These questions must ultimately be decided by suitable observation, whereas the theoretical mechanics of the magnetic field distortion produced by a given gas distribution is subject to direct mathematical treatment from basic principles and forms the main topic of this Chapter. Much of the descriptive information available on the inflating gas will probably come from inferences based on observations of the magnetic deformations produced at the surface of Earth by the inflating gases in space. This, together with some confusion in the literature on the mechanics of the deformation, requires that a comprehensive derivation and discussion of the theory be given.

The basic theoretical point is, again, that the deformation of the geomagnetic field is magneto-gasdynamic in character[8]. The deformation may be calculated from the gasmagnetic equations for equilibrium of the gas and field stresses*

$$\nabla \cdot \mathbf{M} = \nabla \cdot \mathbf{p} \quad \text{i.e.} \quad \frac{\partial M_{ij}}{\partial x_j} = \frac{\partial p_{ij}}{\partial x_j} \tag{18.1}$$

where both stresses must be described by dyads. If $\mathbf{U}$ is the unit dyad

$$\mathbf{p} \equiv ((p_{ij})) \tag{18.1a}$$

is the pressure tensor for the gas and

$$\mathbf{M} \equiv ((M_{ij})) \tag{18.1b}$$

is the Maxwell stress tensor

$$\mathbf{M} \equiv -\frac{B^2}{2u\,\mu_0}[\mathbf{U} - 2\mathbf{B}^0\mathbf{B}^0]. \tag{18.2}$$

($\mathbf{B}^0 \equiv \mathbf{B}/B$ is the unit vector in the direction of the magnetic field $\mathbf{B}$). Eq. (18.2) reads in components:

$$u\,\mu_0\,M_{ij} = -\frac{1}{2}\delta_{ij}B^2 + B_i B_j. \tag{18.2A}$$

The theoretical connection between the deformation $\delta \mathbf{B}$ and the motions of the particles of the inflating gas was first written down in [2] using the guiding center approximation of ALFVEN [1]. The presentation given here is based on both the

---

* The x_j are cartesian coordinates. $\delta_{ij} = 1$ für $i = j = 0$ for $i \neq j$. The *summation rule* is used such that for any term where two identical indices appear, summation over this index (from 1 through 3) is prescribed.

[1] S. F. SINGER: Trans. Amer. Geophys. Union **38**, 175 (1957).
[2] A. J. DESSLER, and E. N. PARKER: J. Geophys. Res. **64**, 2239 (1959).
[3] A. J. DESSLER, W. B. HANSON, and E. N. PARKER: J. Geophys. Res. **66**, 3631 (1961).
[4] K. D. COLE: J. Geophys. Res. **69**, 3595 (1964).
[5] W. I. AXFORD, and C. O. HINES: Can. J. Phys. **39**, 1433 (1961).
[6] K. D. COLE: J. Geophys. Res. **70**, 1689 (1965).
[7] C. F. KENNEL, and H. E. PETSCHEK: J. Geophys. Res. **71**, 1 (1966).
[8] E. N. PARKER: J. Geophys. Res. **61**, 625 (1956).

formulation given by DESSLER and PARKER and the alternative formulation given by PARKER[9], see also [10].

It is evident that the general solution of Eq. (18.1) for B_i with a given p_{ij} is difficult because the equation is nonlinear in B_i. The general solution has been discussed for the special case of axial symmetry[11,12] but there is no evident elementary application of the results to inflation of a dipole. Fortunately there is a simple but restricted similarity solution applicable to inflation of a dipole[13]. But except for this one solution the theory in its present form is based on the linearized approximation of Eq. (18.1) in which it is assumed that the distortion $|\delta B|$ is small compared to $|B|$ at every point. It is necessary and sufficient to require that the gas pressure is small compared to the magnetic pressure at each point. Altogether, then, the theory gives a good first approximation to the main phase of a storm, but it is not yet possible to study strongly inflated fields in a general way, nor is it possible to give a general treatment of all the instabilities which serve to limit the inflation by plasma[14].

19. Forms of the magneto-gasdynamic equation. It is instructive to examine some of the different forms in which Eq. (18.1) can be written before taking up the various methods of solution. First of all, it is possible to invert the *differential form* Eq. (18.1) solving for $V \times B$ as a function of B and p. Form the vector product of Eq. (18.1) with B. The result

$$V \cdot M \times B = V \cdot p \times B$$

can be written with Eq. (18.2):

$$(U - B^0 B^0) \cdot \mathfrak{E} \cdots \frac{\partial}{\partial r} B = \frac{u\,\mu_0}{B^2} \mathfrak{E} \cdot \cdot B\left(\frac{\partial}{\partial r} \cdot p\right) \quad (19.1)$$

where $\frac{\partial}{\partial r} \equiv V$ is the vector symbol describing local differentiation and $\mathfrak{E} \equiv (((\varepsilon_{jkl})))$ is the usual permutation tensor of third grade*; in components:

$$\left(\delta_{ij} - \frac{B_i B_j}{B^2}\right) \varepsilon_{jkl} \frac{\partial B_l}{\partial x_k} = \frac{u\,\mu_0}{B^2} \varepsilon_{ijk} B_j \frac{\partial p_{kl}}{\partial x_l}. \quad (19.1A)$$

The left hand side of the equation is the component of $V \times B$ perpendicular to B, so that for the special case of a *scalar pressure* p Eq. (19.1) can be written

$$(V \times B)_\perp = \frac{u\,\mu_0}{B^2} B \times V p \quad (19.2)$$

If ionospheric forces do not twist** the tubes of flux making up the geomagnetic field, then the parallel component of $V \times B$ is zero and Eq. (19.1) gives the complete $V \times B$ in terms of the distorting pressure. The magnetic distortion $V \times B$ and the current density j are interchangeable since MAXWELL's equation, Eq. (10.2), requires that

$$u\,\mu_0 j = c_0 \sqrt{\varepsilon_0 \mu_0}\, V \times B, \quad (19.3)$$

---

* ε_{ijk} is equal to ± 1 according as ijk is an even or odd permutation, respectively, on 1 2 3, and zero otherwise.
** The nonconducting atmosphere permits the relief of torsion, as discussed in Chapter B.

[9] E. N. PARKER: Space Sci. Rev. **1**, 62 (1962).
[10] J. R. APEL, S. F. SINGER, and R. C. WENTWORTH: [*14*], 131 (1962).
[11] R. LÜST, and A. SCHLÜTER: Z. Astrophys. **34**, 363 (1954).
[12] S. CHANDRASEKHAR: Astrophys. J. **124**, 232, 240 (1956).
[13] E. N. PARKER, and H. STEWART: J. Geophys. Res. **72**, 5297 (1966).
[14] D. B. CHANG, L. D. PEARLSTEIN, and M. N. ROSENBLUTH: J. Geophys. Res. **70**, 3085 (1965).

so that Eq. (19.1) is also a formula giving the current density induced by the distorting pressure.

We are interested in δB, rather than in $\nabla \times B$ or j, because δB is the observed effect. To accomplish this there is a general theorem which relates $\delta B(r)$ to the associated $\nabla \times B$,

$$\delta B(r) = \frac{1}{4\pi} \int d^3 r' \frac{[\nabla' \times B(r')] \times R}{R^3}. \tag{19.4}$$

The integration symbol $d^3 r'$ means integration over all space (with the assumption that the surface integrals vanish) and $R = r - r'$, $R = |r - r'|$, and $\frac{\partial}{\partial r'} \equiv \nabla'$ denotes the nabla operator with respect to r'. The theorem is readily established by taking the curl of both sides, noting that ∇ operating on any function of R is the negative of ∇' operating on the same function, and then integrating by parts[1].

Using Eq. (19.3) we can rewrite Eq. (19.4) as

$$\delta B(r) = \frac{u}{4\pi} \frac{\mu_0}{c_0 \sqrt{\varepsilon_0 \mu_0}} \int d^3 r' j(r') \times \frac{R}{R^3}. \tag{19.5}$$

This form is recognizable as the familiar Biot-Savart law, and is entirely equivalent to Eq. (19.4). Either j or $\nabla \times B$ may be computed from the particle pressure tensor

$$\mathsf{p} \equiv ((p_{ij})).$$

When there is no torsion in the magnetic field, $\nabla \times B$ (or j) follows directly from Eq. (19.1), permitting Eq. (19.4) to be rewritten as

$$\begin{aligned} \delta B(r) &= \frac{u \mu_0}{4\pi} \int \frac{d^3 r'}{B^2(r') R^3} (\mathfrak{E} \cdot R) \cdot \left[\mathfrak{E} \cdot B \left(\frac{\partial}{\partial r} \cdot \mathsf{p} \right) \right] \\ &= \frac{u \mu_0}{4\pi} \int \frac{d^3 r'}{B^2(r') R^3} \left\{ (B \cdot R) \left(\frac{\partial}{\partial r} \cdot \mathsf{p} \right) - B \left[\left(\frac{\partial}{\partial r} \cdot \mathsf{p} \right) \cdot R \right] \right\} \end{aligned} \tag{19.6}$$

which reads in components:

$$\begin{aligned} \delta B_i(r) &= \frac{u \mu_0}{4\pi} \int \frac{d^3 r'}{B^2(r') R^3} \varepsilon_{ijk} R_k \varepsilon_{jrs} B_r \frac{\partial p_{sl}}{\partial x_l} \\ &= \frac{u \mu_0}{4\pi} \int \frac{d^3 r'}{B^2(r') R^3} \left(\frac{\partial p_{ij}}{\partial x_j} B_k R_k - B_i \frac{\partial p_{jk}}{\partial x_k} R_j \right) \end{aligned} \tag{19.6A}$$

which is the *integral form of* the differential *magneto-gasdynamic equation*, Eq. (18.1)

When torsion is present, so that $(\nabla \times B)_\parallel \neq 0$, then $\nabla \times B$ is not completely determined by Eq. (19.1). The parallel component $(\nabla \times B)_\parallel$ must be determined by other considerations, based on the fact that the divergence of $\nabla \times B$ is identically zero. This more complicated circumstance is taken up in Sect. 24.

Either Eq. (19.5) or (19.6) may be used conveniently to compute the distortion δB in the linear magneto-gasdynamic approximation. Eq. (19.6) gives δB directly in terms of the pressure upon carrying out the integration with the approximation that B under the integral be replaced by the unperturbed field. When the inflating gas distribution lacks rotational symmetry Eq. (19.6) cannot be used. It is then convenient to use Eq. (19.4) or (19.5), which involves an intermediate step in which $\nabla \times B$ (or j) is computed from the motions of the individual gas particles in the unperturbed field. Application of both forms is given in Sects. 22 through 24.

20. Gas pressure tensor.

α) The mathematical formalism representing the gas-field *stress balance* expresses the stress effects of the gas in terms of either the pressure tensor

---

[1] E. N. PARKER: Space Sci. Rev. **1**, 62 (1962).

$p \equiv ((p_{ij}))$, Eq. (18.1a), or the electric current j caused by the pressure. Consider the relation of $p \equiv ((p_{ij}))$ and j to the motions of the gas atoms. The elements of the pressure tensor can be related to the particle velocity distribution function $f(r, w)$, over position r, and velocity, w, by the following integral over the velocity space:

$$p(r) = \int d^3 w f(r, w) m (w-v)(w-v), \qquad (20.1)$$

in components:

$$p_{ij}(r) = \int d^3 w m (w_i - v_i)(w_j - v_j) f(r, w) \qquad (20.1\,\text{A})$$

where m is the mass of the individual particle and r is the bulk velocity given by

$$v = \frac{1}{N} \int d^3 w f(r, w) w, \qquad (20.2)$$

i.e.

$$v_i = \frac{1}{N} \int d^3 w \, w_i f(r, w) \qquad (20.2\,\text{A})$$

with $N = \int d^3 w f(r, w)$ the number of particles per unit volume. The sum is to be carried out over all particles present.

β) Consider the collisionless plasma. The cyclotron radius of the individua ions and electrons is small compared to the scale of the geomagnetic field. Hence for variations of the field which are slow compared to the cyclotron frequency, the well known *guiding center approximation* [1][1] can be used. The individual particle (of mass m, charge q, and speed w) moves with a velocity $w_\perp$ in a circle the instantaneous center of which has a speed along the magnetic lines of force denoted by $w_\parallel$. The radius of the circle can be written

$$\mathscr{R} = c_0 \sqrt{\varepsilon_0 \mu_0} \, \frac{m w_\perp}{q B} \qquad (20.3)$$

where B is the magnetic flux density at the guiding center of the circle. If the magnetic field varies across the circle, the radius of curvature of the particle trajectory is slightly smaller than $\mathscr{R}$ on one side, and slightly larger on the other, so that the actual motion of the particles is a trochoid, i.e. there is a net drift of the guiding center perpendicular to the lines of force. It can be shown [2] that the resulting drift is perpendicular to both B and ∇B with a speed

$$(w_\perp/2) \mathscr{R}/\ell = \mathscr{E}_{\text{kin}\,\perp} \cdot c_0 \sqrt{\varepsilon_0 \mu_0}/q B \ell, \qquad (20.4)$$

ℓ being the scale of the component of the gradient in $|B| \equiv B$ perpendicular to B; $\mathscr{E}_{\text{kin}}$ is the kinetic energy of a particle:

$$\mathscr{E}_{\text{kin}} = \mathscr{E}_{\text{kin}\,\perp} + \mathscr{E}_{\text{kin}\,\parallel}.$$

The motion $w_\parallel$ along the magnetic lines of force produces an additional drift if the lines are curved. The centrifugal force $-m w_\parallel^2 K$ for a curvature K leads to equilibrium with the Lorentz force of the resulting drift. The *drift* is thus

$$-c_0 \sqrt{\varepsilon_0 \mu_0} \, \frac{m w_\parallel^2}{q B} K \qquad (20.5)$$

perpendicular to both B and the local plane of osculation of the line of force. Formally, neglecting the very small gravitational drift

$$m c_0 \sqrt{\varepsilon_0 \mu_0} \, B \times g/q B^2,$$

---

[1] K. M. Watson: Phys. Rev. **102**, 12 (1956).
[2] See contribution by W. N. Hess in Vol. 49/4 (Sect. 2) of this Encyclopedia.

combining the two drifts across the line of force the *total drift velocity* becomes:

$$\boldsymbol{v}_d = c_0 \sqrt{\varepsilon_0 \mu_0}\, \frac{m}{q B^4}\, \boldsymbol{B} \times \left[\frac{w_\perp^2}{4} \nabla(B^2) + w_\parallel^2 (\boldsymbol{B} \cdot \nabla) \boldsymbol{B} \right], \qquad (20.6)$$

$$\boldsymbol{v}_d = \frac{c_0 \sqrt{\varepsilon_0 \mu_0}}{q B}\, \boldsymbol{B}^0 \times \left[\frac{1}{2} \mathscr{E}_{\text{kin}\perp} \nabla(B^{0\,2}) + 2 \mathscr{E}_{\text{kin}\parallel} (\boldsymbol{B}^0 \cdot \nabla) \boldsymbol{B}^0 \right] \qquad (20.6\text{A})$$

when using the unit vector $\boldsymbol{B}^0$ and kinetic particle energy $\mathscr{E}$. Note that both drifts depend upon the sign of the charge of the particle. The motion $\boldsymbol{v}_d$ is the drift which carries *electrons eastward* and *positive ions westward* around the geomagnetic dipole. The drift of all the particles leads to a net current. The drift is of the order of $w\mathscr{R}/l$, which is slow compared to the particle speed w.

γ) There is an important point, now, which comes up in connection with *electric fields*. First of all, note that except under transient conditions, the electric fields parallel to the magnetic lines of force are weak, involving potentials which are of the order of the thermal energy of the coolest plasma component present. For the geomagnetic field this means that the usual potential difference along a magnetic line of force through the whole magnetosphere does not exceed a few V because of the presence of the ionosphere and the tenuous ionized outer atmosphere of Earth. The only significant electric fields are perpendicular to $\boldsymbol{B}$. As was pointed out in Sect. 9β, an electric field perpendicular to $\boldsymbol{B}$ is associated with the bulk motion

$$\boldsymbol{v} = c_0 \sqrt{\varepsilon_0 \mu_0}\, \boldsymbol{E} \times \boldsymbol{B} / B^2 \qquad [9.2]$$

of all charged particles, regardless of sign or energy. Hence the electric drift involves no significant electric currents, but represents *bulk motion of the plasma*. Bulk motion $\boldsymbol{v}$ and a perpendicular electric field $\boldsymbol{E}$ are entirely equivalent. Since both are perpendicular to $\boldsymbol{B}$, the relation between them may be written in either of two ways,

$$\boldsymbol{v} = c_0 \sqrt{\varepsilon_0 \mu_0}\, \boldsymbol{E} \times \boldsymbol{B} / B^2, \qquad \boldsymbol{E} = -\frac{1}{c_0 \sqrt{\varepsilon_0 \mu_0}}\, \boldsymbol{v} \times \boldsymbol{B}. \qquad [9.2];\,[9.3]$$

Hence MAXWELL's second equation

$$\partial \boldsymbol{B} / \partial t = -c_0 \sqrt{\varepsilon_0 \mu_0}\, \nabla \times \boldsymbol{E} \qquad [10.3]$$

can be written

$$\partial \boldsymbol{B} / \partial t = \nabla \times (\boldsymbol{v} \times \boldsymbol{B}), \qquad (20.7)$$

which does not contain $\boldsymbol{E}$, and which tells us that one may think of the magnetic lines of force as being "*frozen*" into the velocity field $\boldsymbol{v}$. The electric field and the bulk velocity are equivalent expressions for the same thing[3-5]. One may therefore state that the magnetic lines of force move with the bulk velocity $\boldsymbol{v}$. There is no significant electric field along the lines of force, so the electric equipotential surfaces lie along the magnetic lines of force. The importance of these facts is illustrated by the deduction of AXFORD and HINES[6], that the magnetosphere is in bulk motion, from the fact that the observed polar $\boldsymbol{D}_S$ current system implies an electric field across the magnetic lines of force.

δ) The various particles each have their own drift $\boldsymbol{v}_d$ relative to the bulk motion $\boldsymbol{v}$, but $|\boldsymbol{v}_d| \ll w$ when the gyroradius is small, and depends upon the sign

---

[3] G. F. CHEW, M. L. GOLDBERGER, and F. E. LOW: Proc. Roy. Soc. London A **236**, 112 (1956).
[4] K. A. BRUECKNER, and K. M. WATSON: Phys. Rev. **102**, 19 (1956).
[5] E. N. PARKER: Phys. Rev. **107**, 924 (1957).
[6] W. I. AXFORD, and C. O. HINES: Can. J. Phys. **39**, 1433 (1961).

of the charge, so v_d is usually neglected when speaking of bulk motion. It is evident from the formula for the drift v_d that to compute v_d we must calculate $w_\perp$ and $w_\parallel$ while the particles are being carried along with the bulk velocity v. The perpendicular velocity $w_\perp$ is readily computed from the fact that the *magnetic moment* of the circular particle motion,

$$\mu = \tfrac{1}{2} m w_\perp^2 / B^2 = \mathscr{E}_{\text{kin}\perp} / B^2, \tag{20.8}$$

is a constant of the motion (unless transient variations occur in times smaller than the cyclotron period). The parallel motion can be computed from the fact that the so called *longitudinal invariant*, represented by the integral

$$J = 2 \int_{r_1}^{r_2} ds\, m\, w_\parallel = 4 \int ds\, \mathscr{E}_{\text{kin}\parallel} / w_\parallel, \tag{20.9}$$

of the parallel velocity along the line of force between mirror points, is constant (unless transient variations occur in times smaller than the transit time between mirror points). It was through the use of the magnetic moment and longitudinal invariant that TAYLOR and HONES[7] were able to trace the individual particle motions through the magnetospheric convective cycle. A comprehensive treatment of particle motions in slowly varying fields may be found in [17] and [8].

ε) It is necessary to sum over the particle motions when computing the *pressure tensor*. Fortunately the magnetic field produces statistical isotropy in the two directions perpendicular to B so that $p \equiv ((p_{ij}))$, Eq. (18.1a), may be expressed simply in terms of the two quantities $p_\parallel$ and $p_\perp$, each of which is defined as a sum over particles per unit volume, $\mathscr{V}_u$, namely:

$$\mathscr{V}_u p_\perp = \tfrac{1}{2} \sum m w_\perp^2 = \sum \mathscr{E}_{\text{kin}\perp}, \tag{20.10a}$$

$$\mathscr{V}_u p_\parallel = \sum m w_\parallel^2 = 2 \sum \mathscr{E}_{\text{kin}\parallel}. \tag{20.10b}$$

The two pressures $p_\parallel$ and $p_\perp$ may vary along a magnetic line of force as a consequence of a nonisotropic particle velocity distribution. For stationary conditions the particle distribution can be computed in a straightforward manner from the Liouville equation. Let $f(w, \vartheta, s) \sin\vartheta\, d\vartheta\, dw$ represent the number of particles per unit volume* with total velocity in the interval $(w, w+dw)$ and pitch angle $\vartheta \equiv \arccos w_\parallel / w$ in the interval $(\vartheta, \vartheta+d\vartheta)$. Then by summing over the particle trajectories it is easily shown[4] that conservation of particles requires

$$\frac{\partial f}{\partial s} \cos\vartheta + \frac{\partial f}{\partial \vartheta} \frac{\sin\vartheta}{2B} \frac{dB}{ds} = 0 \tag{20.11}$$

where s again represents distance measured along the line of force. The general solution is $f(w, \sin^2\vartheta / B)$ along the magnetic line of force, where f is any continuous single valued function with finite velocity moments, etc.

It is instructive to consider the simple case that f is of the form $(\sin^2\vartheta/B)^{\alpha/2}$. Then the particle distribution over pitch angles maintains the invariant form $\sin^\alpha\vartheta$ everywhere along the magnetic lines of force[4]. The pressure and density both vary as $B^{-\alpha/2}$ along the lines of force. For an isotropic particle distribution ($\alpha=0$), the distribution function f as well as the total particle density are independent of $B(s)$, and hence independent of s [20]. If $\alpha > 0$, then the distribution is anisotropic, with fewer particles at small pitch angles (moving along the field) than at large pitch angles (circling around the field). The particle density is lower where B is larger. Conversely, $\alpha < 0$ gives an anisotropy with fewer particles circling the field than moving along it. The particle density is then higher where the field is stronger.

---

* In this definition spherical coordinates are applied in the velocity space such that f has m^{-3} (m/sec)$^{-1}$ for dimension.

[7] H. E. TAYLOR, and E. W. HONES: J. Geophys. Res. 70, 3605 (1965).

[8] See contribution by W. N. HESS in Vol. 49/4 (Chapter B) of this Encyclopedia.

The particle pressures can be calculated from f from the simple relations

$$p_\| = \sum \int dw\, mw^2 \int_0^\pi d\vartheta \sin\vartheta \cos^2\vartheta\, f, \tag{20.12}$$

$$p_\perp = \tfrac{1}{2}\sum \int dw\, mw^2 \int_0^\pi d\vartheta \sin^3\vartheta\, f \tag{20.13}$$

(summed over all masses m and particle speeds w). Note, then, that Eq. (20.11) places a restriction on the spatial variations of $p \equiv ((p_{ij}))$ [Eq. (18.1a)] in a given magnetic field, so that an arbitrary p does not, in general, represent a real distribution of particles. A real p must be compatible with the actual particle motion, as described by Eq. (20.11). This fact prevents us from solving Eq. (18.1) by the reverse process of assuming $\delta \boldsymbol{B}(\boldsymbol{r})$, computing $\nabla \times \boldsymbol{B}$, and integrating Eq. (18.1) to give the p_{ij} responsible for the assumed $\delta \boldsymbol{B}(\boldsymbol{r})$.

21. Current formulation. The considerations given thus far prescribe the behavior of a real collisionless gas distribution in the geomagnetic field. Consider the alternative formulation of the stress relation, Eq. (18.1) in terms of the current density $\boldsymbol{j}$, to be used in Eq. (19.5). Perhaps the simplest approach is to consider the gas as made up of individual particles each with a charge q and velocity $\boldsymbol{w}$. The current density is the sum $\sum q\boldsymbol{w}$ over all particles, per unit volume $\mathscr{V}_u$. In practice, then, the right hand side of Eq. (19.5) becomes an integral over the circular and drift motions of the individual particles everywhere throughout the magnetic field. This was the approach used by DESSLER and PARKER[1]. Alternatively, one may average over the circular and drift motions at a given point, obtaining an explicit expression for $\boldsymbol{j}$ before going on to do the integration in Eq. (19.5). The sum[2] gives with Eqs. (19.2), (19.3):

$$\boldsymbol{j} = \frac{c_0 \sqrt{\varepsilon_0 \mu_0}}{B^2} \boldsymbol{B} \times \left\{ \nabla p_\perp + \frac{p_\| - p_\perp}{B^2}(\boldsymbol{B}\cdot\nabla)\boldsymbol{B} \right\}, \tag{21.1}$$

under quasi-steady conditions when $\boldsymbol{B}\cdot\nabla\times\boldsymbol{B}=0$. This same expression can be obtained from the fact that the diamagnetic moment per unit volume, $\boldsymbol{M}$, of the cyclotron motion of gas particles, is just

$$\boldsymbol{M} = -N|\boldsymbol{\mu}|\frac{\boldsymbol{B}}{B} = -N|\boldsymbol{\mu}|\boldsymbol{B}^0 \tag{21.2}$$

for N particles per unit volume, each with an average magnetic moment $\boldsymbol{\mu}$ *

$$|\boldsymbol{\mu}| = \tfrac{1}{2} m w_\perp^2 / B. \tag{20.8}$$

The current density contributed by $\boldsymbol{M}$ is

$$c_0\sqrt{\varepsilon_0 \mu_0}\,\nabla \times \boldsymbol{M},$$

so that the total current density is

$$\boldsymbol{j} = Nq\boldsymbol{v}_d + c_0\sqrt{\varepsilon_0 \mu_0}\,\nabla \times \boldsymbol{M}, \tag{21.3}$$

where $\boldsymbol{v}_d$ is given by Eq. (20.6).

---

* As the motion perpendicular to $\boldsymbol{B}$ has the magnetic field direction as axis of symmetry, the magnetic moment vector $\boldsymbol{\mu}$ is parallel with $\boldsymbol{B}$, hence the general IUPAP-definition for $\boldsymbol{\mu}$, namely $\boldsymbol{\mu}\cdot\boldsymbol{B}=\mathscr{E}$ reads here: $|\boldsymbol{\mu}|B=\mathscr{E}$. Do not confound the magnetic moment $\boldsymbol{\mu}$ with the permeability (in free space) μ_0.

[1] A. J. DESSLER, and E. N. PARKER: J. Geophys. Res. 64, 2239 (1959).
[2] E. N. PARKER: Phys. Rev. 107, 924 (1957).

In the absence of torsion $\boldsymbol{B} \cdot \nabla \times \boldsymbol{B} = 0$ so that $B^2 \nabla \times \boldsymbol{B} = \boldsymbol{B} \times [(\nabla \times \boldsymbol{B}) \times \boldsymbol{B}]$ and the expression reduces to Eq. (21.1).

22. Perturbation solutions for inflation by axisymmetric gas distributions.

α) Suppose that the inflating gas is distributed symmetrically around the dipole axis. Then $(\nabla \times \boldsymbol{B})_\| = 0$ and Eq. (19.6) is applicable. Use the *linear approximation* in which it is assumed that the gas pressure is small compared to the magnetic pressure $B^2/2u\mu_0$ at each point in the field. Then the field $B_i(\boldsymbol{r})$ under the integral sign in Eq. (19.6) may be approximated by the unperturbed field $\boldsymbol{B}_0(\boldsymbol{r})$ yielding

$$\delta \boldsymbol{B}(\boldsymbol{r}) = \frac{u\mu_0}{4\pi} \int \frac{d^3 \boldsymbol{r}'}{B_0^2(\boldsymbol{r}') R^3} \left\{ (\boldsymbol{B}_0 \cdot \boldsymbol{R}) \left(\frac{\partial}{\partial \boldsymbol{r}} \cdot \boldsymbol{p} \right) - \boldsymbol{B} \left[\left(\frac{\partial}{\partial \boldsymbol{r}} \cdot \boldsymbol{p} \right) \cdot \boldsymbol{R} \right] \right\}, \quad (22.1)$$

which reads in components (summation rule applied):

$$\delta B_i(\boldsymbol{r}) = \frac{u\mu_0}{4\pi} \int \frac{d^3 \boldsymbol{r}'}{B_0^2(\boldsymbol{r}') R^3} \left(\frac{\partial p_{ij}}{\partial x_i} B_{0k} R_k - B_{0i} \frac{\partial p_{jk}}{\partial x_k} R_j \right). \quad (22.1A)$$

The distortion $\delta \boldsymbol{B}$ now follows for any given field $\boldsymbol{B}_0$ and pressure tensor $\boldsymbol{p} \equiv ((p_{ij}))$, Eq. (18.1a). A similar approximation for low gas pressures may be used in Eq. (19.5), computing the current density $\boldsymbol{j}$ from the particle motions in the unperturbed field $\boldsymbol{B}_0$. Two examples are worked out in this section to illustrate the use of Eq. (22.1). The first is simple, involving the magnetic effects of an axially symmetric distribution of particles moving in the equatorial plane of the geomagnetic field, to show the different equivalent ways of doing the calculation. The second example is a general theorem, relating the change in the field in the vicinity of the origin to a general distribution of trapped gas.

β) Consider, then, the *distortion of the geomagnetic dipole field by particles* moving *in the equatorial plane* of the dipole (pitch angle $\vartheta = \pi/2$) and distributed with axial symmetry around the field. Use a cylindrical coordinate system (ϱ, φ, z) with the dipole at the origin and pointing in the negative z-direction. Let ϱ represent radial distance from the z-axis, φ being the azimuth. Then, since there are no particle motions along the field, $p_\| = 0$. As a consequence of the symmetric distribution of the particles, $p_\perp = p(\varrho, z)$ where $p(\varrho, z)$ is the kinetic energy density. With the gas particles moving in the equatorial plane, $p(\varrho, z)$ is a delta function of z. In the vicinity of the equatorial plane, where $\boldsymbol{p} \equiv ((p_{ij}))$, Eq. (18.1a), is nonvanishing, the tensor $\boldsymbol{p}$ is diagonal with components $p, p, 0(pz^2/\varrho^2)$:

$$\boldsymbol{p} = \begin{pmatrix} p & 0 & 0 \\ 0 & p & 0 \\ 0 & 0 & 0(pz^2/\varrho^2) \end{pmatrix}. \quad (22.2)$$

The third component vanishes on the equatorial plane of the dipole and is nonvanishing off the plane only because of the curvature of the magnetic lines of force. The components of $\partial p_{ij}/\partial x_j$ are $\partial p/\partial \varrho$ for $i = 1$ and $0[(\partial/\partial z)pz^2/\varrho^2]$ for $i = 3$. Consider the strain $\delta \boldsymbol{B}(\boldsymbol{r})$ at a general point $(\varrho, 0, z)$ near the origin, with the gas lying well out beyond. Then the distance R' to the point of integration $\boldsymbol{r}' = (\varrho', \varphi', z')$ can be written

$$R' \cong \varrho' - \varrho \cos \varphi',$$

neglecting all terms second order in ϱ/ϱ' and z/z'. Substituting into Eq. (22.1) and neglecting all second order terms leads to first order terms, two of which contain the factors

$$\int_0^{2\pi} d\varphi \cos \varphi, \quad \int_{-\infty}^{+\infty} dz' \frac{\partial}{\partial z'}(zp)$$

which vanish identically. The only nonvanishing first order term is

$$\delta B_z = + \frac{u\mu_0/2}{B_0 R_E^3} \int_{-\infty}^{+\infty} dz' \int_0^\infty d\varrho' \, \varrho'^2 \frac{\partial p}{\partial \varrho'}. \quad (22.3)$$

Integrating by parts, remembering that $\varrho'^2 p$ vanishes at $\varrho' = 0, \infty$, and denoting the total particle energy by $\mathscr{E}$

$$\mathscr{E} \equiv 2\pi \int dz' \int d\varrho' \, \varrho' \, p \tag{22.4}$$

leads to Eq. (11.1). The vanishing of all the terms first order in ϱ and z means that $\delta \boldsymbol{B}(\boldsymbol{r})$ is uniform in the vicinity of the dipole.

γ) Eq. (11.6) can also be obtained from Eqs. (19.5) and (21.1). Compute $\boldsymbol{j}$ from Eq. (21.1) using the unperturbed field, keeping terms of first order in z'. Then since $p_\parallel = 0$ and $\nabla \times \boldsymbol{B}_0 = 0$, Eq. (21.1) reduces to

$$\boldsymbol{j} = c_0 \sqrt{\varepsilon_0 \mu_0} \, \frac{1}{B^2} \, \boldsymbol{B} \times \left[\nabla p - \frac{p}{B} \nabla B \right] = c_0 \sqrt{\varepsilon_0 \mu_0} \, \frac{\boldsymbol{B}}{B} \times \nabla \left(\frac{p}{B} \right). \tag{22.5}$$

In calculating the field in the vicinity of the origin, $\boldsymbol{R} = -\boldsymbol{e}_\varrho \varrho$, it is readily shown that the vector in brackets lies in the equatorial ϱ, z plane (unit vectors $\boldsymbol{e}_\varrho$, $\boldsymbol{e}_z$):

$$\nabla p - \frac{p}{B} \nabla B = \boldsymbol{e}_\varrho \left(\frac{\partial p}{\partial \varrho} + \frac{3p}{\varrho} \right) + \boldsymbol{e}_z \frac{\partial p}{\partial z}, \tag{22.6}$$

so that Eq. (19.5) becomes

$$\delta B_z = + \boldsymbol{e}_z \, \frac{u \mu_0 / 2}{B_0 R_E^3} \int_{-\infty}^{+\infty} dz' \int_0^\infty d\varrho' \, \varrho'^2 \cdot \left(\frac{\partial p}{\partial \varrho'} + \frac{3p}{\varrho'} + \frac{3z'}{\varrho'} \frac{\partial p}{\partial z'} \right). \tag{22.7}$$

Integrating by parts again gives Eq. (11.6).

δ) Finally, it is instructive to compute Eq. (11.6) from Eq. (19.6) using the actual individual particle motions, consisting of the cyclotron motion plus the drift velocity $\boldsymbol{v}_d$ given by Eq. (20.6). It has the advantage of exhibiting the tendency of the diamagnetic effect of the cyclotron motion of the particles to crowd out the field, separately from the tendency for the pressure of the gas to push the field outward from the dipole. The cyclotron motion of an individual particle leads to a net electric current $q\Omega/2\pi$ around a circle of radius $\mathscr{R} = w_\perp/\Omega$ where

$$\Omega = \frac{qB}{m c_0 \sqrt{\varepsilon_0 \mu_0}}. \tag{22.8}$$

$\Omega/2\pi$ is the cyclotron (gyro-) frequency. The resulting magnetic moment $\boldsymbol{\mu}$ is given by the product of the area of the circle and the current, or

$$|\boldsymbol{\mu}| = \tfrac{1}{2} m w_\perp^2 / B, \qquad [20.8]$$

as is well known. The dipole moment is directed opposite to the local field, so that, if the dipole is at a distance ϱ in the equatorial plane of the geomagnetic dipole, $\boldsymbol{\mu}$ produces a field at the geomagnetic dipole of

$$\delta B_z = \frac{u \mu_0}{2} \, m w_\perp^2 / B_0 \cdot R_E^3 \tag{22.9}$$

in the positive z-direction, remembering that the geomagnetic dipole points in the negative z-direction here. Note that this effect is independent of the distance ϱ at which the particle is located in the dipole field. The effect δB_1 of a number of particles with total kinetic energy $\mathscr{E}$ is obtained by summing the δB_z of all the individual particles, giving

$$\delta B_1 / B_0 = + \mathscr{E} / 3 \mathscr{E}_m \tag{22.10}$$

where $\mathscr{E}_m$ is the total magnetic energy of Eq. (11.7). There is, in addition, the field associated with the drift $\boldsymbol{v}_d$. The drift is azimuthal with a westward speed

$$v_d = \frac{3}{2} c_0 \sqrt{\varepsilon_0 \mu_0} \, \frac{m w_\perp^2}{q B \varrho}, \tag{22.11}$$

so that for a total particle energy, $\mathscr{E} = \mathscr{E}_\perp = \tfrac{1}{2} \sum m w_\perp^2$, arranged with axial symmetry, the westward electric current around the circle of radius ϱ is

$$I_w = \frac{3 c_0 \sqrt{\varepsilon_0 \mu_0}}{2\pi \varrho} \, \frac{\mathscr{E}}{B \varrho}. \tag{22.12}$$

In deriving this expression for I_w, remember that $w_\parallel = 0$, and that each particle, with charge q, goes once around the geomagnetic dipole in the time $2\pi \varrho/v_d$. Using Eq. (19.5) or the well known fact that the field B at the center of a circle of current I and radius ϱ is (u $\mu_0/2\,c_0\sqrt{\varepsilon_0\,\mu_0}$) (I/ϱ), it follows that the field in the vicinity of the origin is

$$\delta B_2 = -\frac{\mathrm{u}\,\mu_0}{4\pi}\frac{3\mathscr{E}}{B\varrho^3} = -B_0\frac{\mathscr{E}}{\mathscr{E}_\mathrm{m}}. \qquad (22.13)$$

The total field in the vicinity of the origin is

$$\delta B_1 + \delta B_2 = \delta B$$

giving *

$$\left.\frac{\delta B}{B_0} = -\frac{2}{3}\frac{\mathscr{E}}{\mathscr{E}_\mathrm{m}}\right\}. \qquad (22.14) = [11.6]$$

The calculation shows that the contribution of the diamagnetic pressure of the gas is an increase of the horizontal component at the surface of Earth. The increase is just one third the decrease produced by the tendency for the gas to expand outward from the dipole. Thus the effect as whole is a decrease of $\mathcal{H}$.

In general, a field which in its undistorted state is given by $B = B_0(a/r)^n$ across the equatorial plane has a distortion field $\delta B/B_0 = (n-1)\mathscr{E}r^{n-3}/B_0^2\,a^n$ in the vicinity of the origin as a consequence of a ring of radius r of particles with total kinetic energy $\mathscr{E}$.

A number of examples were worked out [1,2] to illustrate the general fact that $\delta B/B_0$ is of the order of $\mathscr{E}/\mathscr{E}_\mathrm{m}$. The calculations showed that the distortion δB is independent of the sign of the particles, of the energy distribution, and of the scale of the particle distribution in space. The two cases, particles confined to the equatorial plane of the dipole, and particles with an isotropic distribution, gave exactly Eq. (11.6). Particles streaming along the magnetic lines of force (zero pitch angle) gave *

$$\delta B/B_0 = -2\mathscr{E}/\mathscr{E}_\mathrm{m}, \qquad (22.15)$$

when the particles are near the equator, and *

$$\delta B/B_0 = -2\mathscr{E}/3\,\mathscr{E}_\mathrm{m}, \qquad (22.16)$$

when streaming all the way to the pole, but of course particles streaming along the field do not mirror in the field.

23. General theorem of field inflation. Recently Sckopke[1] has discovered and proved the important theorem that *any stationary distribution of particles or plasma trapped and mirroring on the lines of a dipole leads exactly*

$$\frac{\delta B}{B_0} = -\frac{2}{3}\frac{\mathscr{E}}{\mathscr{E}_\mathrm{m}}. \qquad [11.6]$$

The importance of Sckopke's discovery is that one need only add up the total plasma energy to obtain δB in the vicinity of the dipole. No complicated integration over the particle distribution is necessary. The theorem can be proved as follows.

α) Represent the effect of the *trapped plasma* by the *drift* Eq. (20.6) plus the magnetic moment of each particle. Use spherical polar coordinates (r, θ, φ) to represent position in the dipole. Consider the particles in the azimuthal ring between the lines of force $r = r_0 \sin^2\theta$ and $r = (r_0 + dr_0)\sin^2\theta$ and between θ and $\theta + d\theta$. The lines of force cross the equatorial plane at r_0 and $r_0 + dr_0$. The area defined by $dr_0\,d\theta$ is

$$dA = r\,dr\,d\theta = r_0\,dr_0\,d\theta\,\sin^4\theta.$$

---

* The negative sign appearing in Eqs. (22.14) to (22.16) corresponds to the actual conditions during magnetic storms: the drift in azimuthal direction is westward for positive and eastward for negative charges, both producing a westward current and so a perturbation field opposed to the main geomagnetic field B_0.

[1] A. J. Dessler, and E. N. Parker: J. Geophys. Res. **64**, 2239 (1959).
[2] E. N. Parker: Space Sci. Rev. **1**, 62 (1962).
[1] N. Sckopke: J. Geophys. Res. **71**, 1325 (1966).

The circumference of the ring is
$$2\pi r \sin\theta = 2\pi r_0 \sin^3\theta$$
so that the volume is
$$2\pi r_0^2 \, dr_0 \sin^7\theta \, d\theta.$$

Consider from Eq. (20.11) the particle distribution per unit solid angle* in the particular form

$$f_r(r_0, \theta, \vartheta) = \frac{\Gamma\left(\frac{\alpha+3}{2}\right) G(r_0)}{2\pi^{\frac{3}{2}} \Gamma\left(\frac{\alpha+2}{2}\right)} \left[\frac{B(r_0)}{B(r)}\right]^{\alpha/2} \sin^\alpha \vartheta \tag{23.1}$$

over pitch angle ϑ along the line of force $r = r_0 \sin^2\theta$. $B(r_0)$ is the field density $B_0(a/r_0)^3$ at the equatorial plane and, with this normalization, $G(r_0)$ is the particle density** at the equator $r = r_0$. The number of particles per unit volume is

$$2\pi \int_0^\pi d\vartheta \sin\vartheta \, f.$$

To write f_r entirely in terms of r_0, we must express $B(r_0)$ in terms of r_0. For a dipole field, Eq. (11.4), it can be shown that

$$\frac{B(r_0)}{B(r)} = \frac{\sin^6\theta}{(1+3\cos^2\theta)^{\frac{1}{2}}} \tag{23.2}$$

along the line of force. Hence

$$f_r(r_0, \theta, \vartheta) = \frac{\Gamma[(\alpha+3)/2] G(r_0) \sin^{3\alpha}\theta \sin^\alpha \vartheta}{2\pi^{\frac{3}{2}} \Gamma[(\alpha+2)/2] (1+3\cos^2\theta)^{\alpha/4}} \tag{23.3}$$

Now the current dI carried across the area dA is just

$$dI = q \, dA \, 2\pi \int_0^\pi d\vartheta \sin\vartheta \, f_r(r_0, \theta, \vartheta) \, v_d(r_0, \theta, \vartheta), \tag{23.4}$$

where the particle drift velocity $v_d(r_0, \theta, \vartheta)$ is readily shown from Eq. (20.6)***, using the fact that in the undistorted dipole field $(\boldsymbol{B}\cdot\boldsymbol{\nabla})\boldsymbol{B} = \nabla B^2/2$:

$$v_d(r_0, \theta, \vartheta) = c_0 \sqrt{\varepsilon_0 \mu_0} \frac{m}{qB^4}\left(\boldsymbol{B}\times\boldsymbol{\nabla}\frac{B^2}{2}\right)\left(w_\parallel^2 + \frac{1}{2}w_\perp^2\right). \tag{23.5}$$

$$v_d = \boldsymbol{e}_\varphi \frac{3}{2} c_0 \sqrt{\varepsilon_0 \mu_0} \frac{m \, w^2}{qB_0 R_E^2} \frac{r_0^2 \sin^5\theta \, (1+\cos^2\theta)(1+\cos^2\vartheta)}{(1+3\cos^2\theta)^2}. \tag{23.6}$$

The integration over ϑ in Eq. (23.4) is easily carried out for Eqs. (23.3) and (23.6). The current flows in a circle about the dipole axis. The perturbation magnetic field follows from Eq. (19.5). In the neighborhood of the dipole the perturbation field associated with the particle drift current dI is

$$dB_2 = \frac{1}{2} \frac{u\mu_0}{c_0\sqrt{\varepsilon_0\mu_0}} \frac{dI}{r_0},$$

and dI can be obtained from Eq. (23.4), giving, with Eqs. (23.3), (23.6):

$$dB_2 = \frac{3}{4} u \mu_0 \frac{m w^2 (\alpha+4)}{(\alpha+3) B_0 R_E^2} G(r_0) r_0^2 \, dr_0 \frac{\sin^{9+3\alpha}\theta (1+\cos^2\theta) \, d\theta}{(1+3\cos^2\theta)^{2+\alpha/4}} \tag{23.7}$$

in the direction of the dipole.

β) The field in the neighborhood of the dipole produced by the *diamagnetic moment* of a single particle at (r_0, θ) has a component $-\frac{1}{2} m w^2 \sin^2\vartheta / B_0 R_E^3$ in the direction of the dipole so that after integrating over ϑ the contribution of the particles in $dr_0 \, d\theta$ turns out to be

$$dB_1 = -\frac{1}{4} u \mu_0 \frac{m w^2 (\alpha+2)}{(\alpha+3) B_0 R_E^3} G(r_0) r_0^2 \, dr_0 \cdot \frac{\sin^{7+3\alpha}\theta \, d\theta}{(1+3\cos^2\theta)^{\alpha/4}}. \tag{23.8}$$

---

* f_r characterizes spatial distribution only, dimension is m^{-3} (different from f).
** G and f_r are both particle densities dimension m^{-3}, while f in Sect. 20 has m^{-4} sec.
*** e_φ is the unit vector in the direction of the coordinate φ.

γ) The *total perturbation field* in the neighborhood of the dipole as caused by the particles in $dr_0\, d\theta$ is $dB_1 + dB_2$ in the direction of the dipole. The total field change δB as a consequence of all particles trapped in the dipole is the integral of $(dB_1 + dB_2)$ over r_0 and θ. The integral can be rewritten in terms of the total particle energy $\mathscr{E}$ by noting first that the *kinetic energy of the particles* in $dr_0\, d\theta$ is

$$d\mathscr{E} = \frac{\pi m w^2 G(r_0) r_0^2\, dr_0 \sin^{7+3\alpha}\theta\, d\theta}{(1 + 3\cos^2\theta)^{\alpha/4}}. \tag{23.9}$$

Then $\mathscr{E}$ is obtained by integrating over r_0 and θ. Write the total field change δB as the integral of $-(dB_1 + dB_2)$ over r_0 and θ, divide through by B_0, subtract $2\mathscr{E}/3\mathscr{E}_m$ and add $2/3\mathscr{E}_m$ times the integral of $d\mathscr{E}$. The result can be written

$$\frac{\delta B}{B_0} = -\frac{2\mathscr{E}}{3\mathscr{E}_m} + \frac{2\pi m w^2}{3(\alpha+3)\mathscr{E}_m} \int_0^\infty dr_0\, r_0^2 G(r_0) Z(r_0), \tag{23.10}$$

where $Z(r_0)$ is readily shown to be the following (dimensionless) quantity:

$$Z(r_0) = \int_0^\pi \frac{d\theta \sin^{7+3\alpha}\theta}{(1+3\cos^2\theta)^{2+\alpha/4}} [-2 + 3(3\alpha+8)\cos^2\theta + 3(5\alpha+14)\cos^4\theta]. \tag{23.11}$$

The purpose in writing δB in this form is that Z can be shown to vanish. To prove that $Z \equiv 0$ note that one can write

$$-2 + 18\cos^4\theta = 2(3\cos^2\theta + 1)(3\cos^2\theta - 1)$$

which may be separated from the integrand in Eq. (23.11) to give two terms $Z = Z_1 + Z_2$ where

$$Z_1 = 6\int_0^\pi \frac{d\theta \sin^{7+3\alpha}\theta \cos^2\theta}{(1+3\cos^2\theta)^{2+\alpha/4}} [(5\alpha+8)\cos^2\theta + (3\alpha+8)],$$

$$Z_2 = 4\int_0^\pi [d\theta \sin\theta(3\cos^2\theta - 1)] \frac{\sin^{6+3\alpha}\theta}{(1+3\cos^2\theta)^{1+\alpha/4}}.$$

Integration of Z_2 by parts in the manner indicated by the brackets leads to $-Z_1$, plus an integrated term containing the factor $\sin^{7+3\alpha}\theta$ evaluated at both 0 and π. Hence $Z = 0$.

δ) It has been shown in Subsect. γ that Eq. (11.6) is always valid for the field in the neighborhood of the dipole, for any particle distribution of the form Eq. (23.1). The general particle distribution is $f_r\left(r_0, \theta, \sin\vartheta \left(\frac{B_0}{B}\right)^{\frac{1}{2}}\right)$ which can be expanded in powers of $\sin\vartheta \left(\frac{B_0}{B}\right)^{\frac{1}{2}}$. The relation is true for each term in the expansion and hence for the function itself. Hence

$$\frac{\delta B}{B_0} = -\frac{2}{3} \frac{\mathscr{E}}{\mathscr{E}_m} \qquad [22.14] = [11.6]$$

applies to any stationary particle distribution trapped in a dipole field.

The distortion $\delta B(0)$ is independent of the sign of the particles, of the energy distribution, and of the distribution in space. The *distortion* $\delta B(0)$ *produced* in the neighborhood of the dipole *by the inflation* of an *axially symmetric distribution of gas is uniquely determined* in the linear approximation *by the total internal kinetic energy of the gas.*

24. Perturbation solutions for inflation by nonsymmetric gas distributions.

Fejer[1] and Akasofu and Chapman[2] have pointed out that some observed storm

---
[1] J. A. Fejer: Canad. J. Phys. **39**, 1409 (1961).
[2] S. I. Akasofu, and S. Chapman: Planet Space Sci. **12**, 607 (1964).

asymmetries suggest nonsymmetric particle distribution. CAHILL[3] has directly observed asymmetric inflation in space. Consider the inflation of a magnetic dipole by a gas distribution which does not have rotational symmetry about the dipole. The calculation is complicated by the parallel component of $V \times B$, which must be added to the perpendicular component given by Eq. (19.1). Both $(V \times B)_\parallel$ and $(V \times B)_\perp$ must be known in order to calculate δB from Eqs. (19.4) or (19.5). The calculation of $(V \times B)_\parallel$ requires some additional considerations. In particular, the ionosphere and the nonconducting atmosphere play an essential rôle, in contrast to the symmetric inflation in which a passive ionosphere has little or no effect [4].

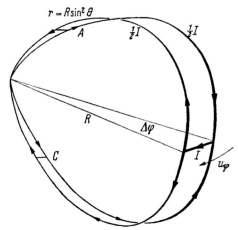

Fig. 15. Sketch of the two dipole lines of force through the ends of the azimuthal segment of particles in the azimutha plane. The drift velocity of the particles is v_φ. The flow of $V \times B$, or the current I, is indicated by the arrows. The crossove segments at $r=a$ and θ_0, $\pi-\theta_0$ are labeled A and B. The geomagnetic dipole is directed downward.

α) To understand the principles involved in nonsymmetric inflation, consider first the nonsymmetric inflation of a dipole pervaded everywhere by a *passive background of tenuous conducting fluid*. Once this case is worked out, it will be a simple matter to put in a nonconducting shell to simulate the terrestrial atmosphere.

Consider the hypothetical situation that the inflating gas consists of a total number of n positively charged particles distributed uniformly along a short azimuthal segment of angular length $\Delta \varphi$ at a radial distance R in the equatorial plane, as sketched in Fig. 15. Let all the particles have the same velocity w perpendicular to the magnetic field, so that the segment drifts around the dipole without dispersing. The azimuthal *drift velocity* u_φ is then

$$u_\varphi = 3 c_0 \sqrt{\varepsilon_0 \mu_0} \frac{|\mu|}{qR}. \tag{24.1}$$

where q is the charge on each particle, m its mass and

$$|\mu| = \tfrac{1}{2} mw^2/B(R) \tag{20.8}$$

is the magnetic moment of each particle.

The distortion of the geomagnetic field by the particles consists of two parts, as in the symmetric case. First, the centrifugal force of the cyclotron motion of each particle tends to push the lines of force out of the region occupied by the

---

[3] L. J. CAHILL: J. Geophys. Res. **71**, 4505 (1966).
[4] E. N. PARKER: J. Geophys. Res. **71**, 4485 (1966).

particles, i.e., the particles are diamagnetic, in the amount indicated by $|\boldsymbol{\mu}|$. The surrounding field is an elastic medium which reacts against the expulsion of lines of force from the region traversed by the cyclotron motion of the particle. Since there is a gradient in the surrounding field, the reaction leads to a net force in the outward direction: Diamagnetic material is always pushed toward regions of weaker field. The particles are, of course, tied to the lines of force which they circle, with the result that the lines of force through the segment are distorted outward, permitting some outward expansion of the field nearer the dipole, just as when the particles are distributed with axial symmetry[5].

Consider the $\nabla \times \boldsymbol{B}$ produced by the particle segment. In the segment itself there is some circulation of $\nabla \times \boldsymbol{B}$ (or $\boldsymbol{j}$) up the outer edge and back along the inner edge as a consequence of the cyclotron motion of the particles. There is, however, a net flow of $\nabla \times \boldsymbol{B}$ (or $\boldsymbol{j}$) forward along the segment, associated with the outward force of the segment on the field. The divergence of $\nabla \times \boldsymbol{B}$ vanishes everywhere, so the net flow of $\nabla \times \boldsymbol{B}$ cannot terminate at the ends of the segment but must flow off into the surrounding space. There are no forces exerted on the field in the space *outside the segment*, so

$$(\nabla \times \boldsymbol{B}) \times \boldsymbol{B} = 0$$

there. It follows that the streaming of $\nabla \times \boldsymbol{B}$ to and from the ends of the segment must be along the magnetic lines of force. Thus the flow of $\nabla \times \boldsymbol{B}$ coming off the front end of the particle segment splits and flows both ways along the magnetic lines of force in to the dipole, as shown in Fig. 15 by the arrows. The net flow of $\nabla \times \boldsymbol{B}$ (or $\boldsymbol{j}$) forward in the segment is $\dfrac{u\,\mu_0}{c_0\sqrt{\varepsilon_0\mu_0}}$ multiplied by the net current

$$I = \frac{n q\, u_\varphi}{R\,\Delta\varphi} \tag{24.2}$$

The flow along each line of force is $\tfrac{1}{2} I$. The entire flow of $\nabla \times \boldsymbol{B}$ must be included in the integration in Eq. (19.4) to obtain the distortion $\delta \boldsymbol{B}$. Obviously $\delta \boldsymbol{B}$ is no longer uniform in the vicinity of the origin, as when the inflating gas is symmetrically distributed around the dipole. But since $\nabla^2 \delta \boldsymbol{B} = 0$ in the vicinity of the origin, the value of $\delta \boldsymbol{B}(0)$ computed at the origin is equal to the mean value of $\delta \boldsymbol{B}(\boldsymbol{r})$ computed over any small sphere centered on the origin. In this sense $\delta \boldsymbol{B}(0)$ computed at the origin is a first approximation to the worldwide average[*] $\delta \boldsymbol{B}(\boldsymbol{r})$. The local variations, which may be very large, are discussed briefly later.

It is a relatively easy matter to compute $\delta \boldsymbol{B}$ at the origin from Eq. (19.4). It is readily shown from symmetry that the flow of $\nabla \times \boldsymbol{B}$ along the dipole lines of force contributes no $\delta \boldsymbol{B}$ at the origin (though the flow along the lines contributes to $\delta \boldsymbol{B}$ elsewhere). There remains, then, only the contribution from the particle segment itself. It can be shown from symmetry that the resulting $\delta \boldsymbol{B}$ is in the direction of the dipole. The magnitude of $\delta \boldsymbol{B}$ is the sum of

$$\delta B_1 = +\frac{u\,\mu_0}{4\pi}\,\frac{n|\boldsymbol{\mu}|}{R^3} \tag{24.3}$$

from the diamagnetic moments of the n particles in the segment and

$$\delta B_2 = -\frac{u\,\mu_0/4\pi}{c_0\sqrt{\varepsilon_0\mu_0}}\,\frac{I\,\Delta\varphi}{R} \tag{24.4}$$

as a consequence of the net forward flow of $\nabla \times \boldsymbol{B}$ along the segment. These may be rewritten as

$$\frac{\delta B_1}{B_0} = \frac{u\,\mu_0}{4\pi}\,\frac{n|\boldsymbol{\mu}|}{B_0\,R^3} = +\frac{\mathscr{E}}{3\mathscr{E}_m},\quad \frac{\delta B_2}{B_0} = -\frac{u\,\mu_0}{4\pi}\,\frac{3n|\boldsymbol{\mu}|}{B_0\,R^3} = -\frac{\mathscr{E}}{\mathscr{E}_m} \tag{24.5}$$

---

[*] Still neglecting the diamagnetic effect of the solid Earth, of course.
[5] A. J. Dessler, and E. N. Parker: J. Geophys. Res. **64**, 2239 (1959).

as when the gas distribution has rotational symmetry. The sum just gives

$$\frac{\delta B}{B_0} = -\frac{2}{3}\frac{\mathscr{E}}{\mathscr{E}_m}.\qquad [11.6]$$

β) Now introduce a *nonconducting shell* at $r = a < R$. Then $\nabla \times \boldsymbol{B}$ cannot flow inward across $r = a$. A nonconductor will not transmit torsion in a magnetic field, or equivalently, a nonconductor will not pass an electric current. So $\nabla \times \boldsymbol{B}$ must flow across the magnetic field somewhere outside $r = a$ in order to follow a closed path back to the trailing end of the particle segment. There is no way to avoid this. The flow of $\nabla \times \boldsymbol{B}$ across the field leads unavoidably to a *force density*

$$\boldsymbol{K} = \frac{1}{u\,\mu_0}(\nabla \times \boldsymbol{B}) \times \boldsymbol{B} \qquad (24.6)$$

on the background medium. It follows that nonsymmetric inflation outside a nonconducting shell necessarily drives *motions in the* background *conducting fluids in* addition to the motion of the inflating gas itself. The crossover of $\nabla \times \boldsymbol{B}$ from one line of force to another adjusts itself of the path of least resistance (including the complication of the motions). Obviously $\nabla \times \boldsymbol{B}$ cannot cross over the lines of force where the background gas is incapable of supporting the unavoidable force density, given by Eq. (24.6). So in the geomagnetic field the crossover is in the ionosphere, where there is enough friction to support the force density $\boldsymbol{K}$. As to the associated electric current, we would say that the electrical conductivity perpendicular to the lines of force is a maximum in the ionosphere (as a consequence of collisions) so the crossover occurs there. The ionospheric crossover is indicated in Fig. 15 by the segments A and C between the dipole lines of force.

It is instructive to compute $\delta \boldsymbol{B}$ at the origin again. The flow of $\nabla \times \boldsymbol{B}$ along the dipole lines of force between the particle segment at $r = R$ and the two crossover segments at $r = a$ again contributes nothing. The particle segment contributes the same as before. In order to calculate the contribution of the crossover segments, denote by θ_0 the polar angle of A. Then C lies at $\pi - \theta_0$. Both crossover segments lie on the line of force $r = R \sin^2 \theta$, so that $a = R \sin^2 \theta_0$. The length of each segment is $a \sin \theta_0 \Delta \varphi$ and the current is $\frac{1}{2} I$. The contribution of both segments is readily shown to add up to

$$\delta B = +\frac{u\,\mu_0}{4\pi}\,3n|\boldsymbol{\mu}|/R,$$

which exactly cancels the contribution from the forward flow of $\nabla \times \boldsymbol{B}$ in the particle segment. The net result is δB_1, given by Eq. (24.5), from the diamagnetic moment of the particles.

In this particular case, then, introduction of a nonconduction shell changes $\delta \boldsymbol{B}$ at the origin from

$$\frac{\delta B}{B_0} = -\frac{2}{3}\frac{\mathscr{E}}{\mathscr{E}_m},\qquad [22.14]$$

representing a worldwide average decrease of the horizontal component, to

$$\frac{\delta B}{B_0} = +\frac{1}{3}\frac{\mathscr{E}}{\mathscr{E}_m},\qquad (24.7)$$

representing a world-wide *increase of the horizontal component*. The forces exerted on the field by the ionosphere exactly cancel the effect on $\delta \boldsymbol{B}(0)$ of the outward force exerted on the field by the current in the particle segment. The large effect of the nonconducting atmosphere is evident.

25. A somewhat more realistic example[*] is illustrated in Fig. 16, where in the flow across the lines of force is permitted to go both ways around $r = a$.

---

[*] This more complicated example neglects the possible crossover from θ_0 to $\pi - \theta_0$ which becomes negligible if $\Delta \varphi_1 + \Delta \varphi_2 \ll \pi/2 - \theta_0$.

A somewhat more realistic example.

α) Consider a sequence of equally spaced segments of particles distributed around the equatorial plane at a radial distance R. Denote the angular length of each by $\Delta\varphi_1$ and the separations between their ends by $\Delta\varphi_2$. Denote by I_1 the ionospheric current at $r=a$ crossing over $\Delta\varphi_1$ and by I_2 the current at $r=a$ crossing over $\Delta\varphi_2$. Conservation of current requires that I_1 and I_2 be related to the current I in the particle segment by

$$I = 2(I_1 + I_2).$$

Denote by Z_1 the electrical impedance across $\Delta\varphi_1$ at $r=a$ and by Z_2 the impedance across $\Delta\varphi_2$. The line integral of the electric field around $r=a$ is zero for steady conditions, so $Z_1 I_1 = Z_2 I_2$. Physically $1/Z_1$ is a relative measure of the force exerted on the medium in $\Delta\varphi_1$ at the crossover $r=a$, compared to $1/Z_2$ for the force in $\Delta\varphi_2$.

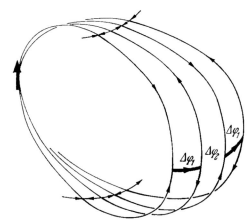

Fig. 16. Sketch of the dipole lines of force through the ends of a sequence of azimuthal segments of particles in the equatorial plane. The arrows indicate the flow of $V \times B$ along and across the magnetic lines of force. The geomagnetic dipole is directed downward.

It is readily shown that the distortion $\delta \boldsymbol{B}$ at the origin is in the direction of the dipole with a magnitude equal to the sum of $\delta \boldsymbol{B}_1$ from the diamagnetic moments, given by Eq. (24.5), and

$$\frac{\delta B_2}{B_0} = + \frac{1 + \Delta\varphi_2/\Delta\varphi_1}{1 + Z_2/Z_1} \frac{\mathscr{E}}{\mathscr{E}_m} \tag{25.1}$$

associated with the flow of $V \times B$ in the particle and ionospheric crossover segments. In the simplest electrical model Z is proportional to the logarithm of the separation, so that

$$\frac{\delta B_2}{B_0} = \frac{1 + \dfrac{\Delta\varphi_2}{\Delta\varphi_1}}{1 + \dfrac{\log(b\,\Delta\varphi_2)}{\log(b\,\Delta\varphi_1)}} \tag{25.2}$$

where $b \equiv (a/s)\sin\theta_0$ with s representing the small radius of the tube of flow of $V \times B$ along the magnetic lines of force into $r=a$. Assume that $b\,\Delta\varphi_1$ is large compared to one as a consequence of the smallness of s. A wide range of possible values for δB_2 results, depending upon the relative magnitudes of $\Delta\varphi_1$ and $\Delta\varphi_2$. If $\Delta\varphi_2/\Delta\varphi_1$ becomes large without limit, then so does $\delta B_2/B_0$. An arbitrarily large average world-wide decrease of the horizontal component results. On the other hand, if $\Delta\varphi_2/\Delta\varphi_1$ becomes small without limit, $\delta B_2/B_0$ lies between 0.5 and 1.0 times $\mathscr{E}/\mathscr{E}_m$. The lower value $0.5\,\mathscr{E}/\mathscr{E}_m$ means that the total δB is small, only $\mathscr{E}/6\,\mathscr{E}_m$, as compared to Eq. (11.6).

β) We should consider briefly the problem of eventually going beyond the illustrative stage to a more complete theory. Consider first the calculation of $(V \times B)_\parallel$ when $(V \times B)_\perp$ is produced by a distribution of particles rather than by a simple line segment. It is sufficient to ignore that part of $(V \times B)_\perp$ associated with the diamagnetic moment of the particles and consider only that *part which is associated with the azimuthal drift* v_d *and given by*

$$\frac{u\,\mu_0}{c_0\sqrt{\mu_0\,\varepsilon_0}}\,Nq\,v_d \tag{25.3}$$

which streams away along the lines of force. (N is the number of particles per unit volume.) The associated $(\nabla \times \boldsymbol{B})_\parallel$ is determined from the differential condition that $\nabla \cdot (\nabla \times \boldsymbol{B}) = 0$. Symmetry about the equatorial plane is a sufficient boundary condition.

To write down the formal conditions, denote the known azimuthal component of $\nabla \times \boldsymbol{B}$ by $\boldsymbol{F}$, which is assumed given. The r- and θ-components F_r and F_θ must form a vector which is parallel to $\boldsymbol{B}$. It follows that in a dipole field [Eq. (11.4)]

$$\frac{F_\theta}{F_r} = \frac{B_\theta}{B_r} = \frac{\sin\theta}{2\cos\theta}. \tag{25.4}$$

Hence $F_r = 2 F_\theta \cot\theta$. The vanishing of the divergence of $\nabla \times \boldsymbol{B}$ leads to

$$\cos\theta \left(2r\,\frac{\partial F_\theta}{\partial r} + 5 F_\theta\right) + \sin\theta\,\frac{\partial F_\theta}{\partial \theta} = -\frac{\partial F}{\partial \varphi} \tag{25.5}$$

(with $F = |\boldsymbol{F}|$). This inhomogeneous linear partial differential equation is readily solved in a variety of circumstances. For instance, if the inflating particles are all confined to the equatorial plane, $\theta = \pi/2$, then the equation reduces to

$$\partial F_\theta/\partial \theta = -\partial F/\partial \varphi$$

and

$$F_\theta(r,\theta,\varphi) = +\int_0^{\pi/2} d\theta\,\frac{\partial F}{\partial \varphi}. \tag{25.6}$$

Or let

$$\partial F/\partial \varphi = \Phi(\varphi)\cos\theta\,\sin^\beta\theta\,R(r)$$

which leads to

$$F_\theta(r,\theta,\varphi) = C\,\Phi(\varphi)\sin^\beta\theta\,Z(r). \tag{25.7}$$

There are a variety of solutions that can be constructed.

The perpendicular component of $\nabla \times \boldsymbol{B}$ in the ionosphere is then to be computed from $(\nabla \times \boldsymbol{B})_\parallel$. But this next step is a difficult task because of fringing fields, inertial effects of the resulting ionospheric motions, the variations in ionospheric conductivity, and the sometimes rapid rate of deformation of the particle distribution itself as a consequence of the dissipation, examples see [1]. All these effects were omitted in the illustrative examples given above, but must be included in any complete theory.

γ) The large *local effects around Earth* are associated mainly with the $(\nabla \times \boldsymbol{B})_\parallel$ flowing into the ionosphere and the $(\nabla \times \boldsymbol{B})_\perp$ flowing through the ionosphere between magnetic lines of force. Only the most rudimentary models can be constructed to illustrate the associated field variations without going into a complete solution of the *ionospheric crossover* of $\nabla \times \boldsymbol{B}$. The magnetic effects of simple crossover segments, as used in Fig. 16, are easily computed and are obviously very large when the crossover segment is directly overhead. The actual crossover fans out into a broad sheet rather than remaining a segment, see example in [1].

We have outlined here the basic principles of magnetic deformation by nonsymmetric inflation. The examples used to illustrate the points have been greatly simplified, but are sufficient to illustrate the main effects of nonsymmetric inflation. The nonconducting atmosphere and the overlying ionosphere enter directly into the picture, whereas they did not in steady symmetric inflation. The ionospheric dissipation of nonsymmetric inflation leads to relaxation times of

---

[1] E. N. PARKER: J. Geophys. Res. **71**, 4485 (1966).

10^2 to 10^4 sec, which is very rapid compared to the observed 10^5 sec relaxation of symmetric inflation. Hence large amounts of nonsymmetric inflation can occur only during the active phase of a storm. The world-wide average decrease of the horizontal component can, in some extreme cases of nonsymmetric inflation, be very large compared to the δB given by Eq. (11.6) for symmetric inflation. It may be the rapid relaxation of nonsymmetric inflation which contributes to the initial rapid recovery of some deep main phases immediately following the active part of the storm. The slower recovery which follows may be the relaxation of the symmetric portion of the inflation.

AKASOFU and CHAPMAN[2] have pointed out that some of the observed asymmetries of the main phase suggest a nonsymmetric particle distribution to produce them. They give an extensive review of the observed asymmetries in their paper, together with their own interpretation of the cause and behavior of the asymmetric particle distribution. The importance of asymmetric inflation can, in the final analysis, be determined only from direct observation. Recent observations[3] show directly the asymmetric inflation during the onset of the main phase of the storm of 17 April 1965, with a rapid decay in about two hours as one might expect. The inflation was mainly in the late afternoon and evening magnetosphere. Following the rapid decay of the nonsymmetric inflation, the remaining symmetric inflation decayed with a characteristic life of the order of 4 days.

F. Historical development.

26. The source of magnetic and auroral activity. The historical development of ideas on magnetic and auroral activity is, first of all, a demonstration of the enormous range of possibilities which confront the investigator when initially faced with a complex natural phenomenon. The historical development shows how widely the naive may fail, and how, in constrast, a shrewd guess, with close attention to the known principles of physics, may point the way for real progress.

α) The *historical development* shows the evolutionary debt owed to the earlier work by the present understanding of magnetic and auroral phenomena. The early ideas are gone and largely forgotten, but in their day they inspired the observational and theoretical analyses which were the next step toward the present general picture.

A brief early history of the discovery of magnetic variations has been given in [6]. The close association of terrestrial magnetic and auroral effects with activity on the sun suggested a hundred years ago that the terrestrial effects are caused by the solar disturbances. The corpuscular hypothesis for aurorae appears to have been proposed first by GOLDSTEIN in 1881. By 1900 such eminent scientists as FITZGERALD[1], BIRKELAND[2], and LODGE[3] were convinced that the magnetic storms were due to "a torrent or flying cloud of charged atoms or ions"; that auroras were caused by cathode rays from the sun; that comet tails could be accounted for by particle radiation projected from the sun with a velocity of about 500 km/sec; that "there seems to be some evidence from auroras and magnetic storms that the earth has a minute tail like that of a comet directed away from the sun". These were shrewd guesses at that time, naive in some respects, but, if quoted selectively, remarkably prophetic. They served to start the inquiry into fruitful channels.

What followed, historically, was an enormous proliferation of ideas on the aurora and magnetic storm, exploring every conceivable possibility. The idea that the individual auroral arcs and rays are a corpuscular phenomenon was popular because of the superficial similarity to the laboratory cathode ray discharge. But how the corpuscles were generated was subject to wide speculation. The view that the auroral particles are particles coming direct from the sun was

---

[2] S. I. AKASOFU, and S. CHAPMAN: Planet Space Sci. **12**, 607 (1964).
[3] L. J. CAHILL: J. Geophys. Res. **71**, 4505 (1966).
[1] G. F. FITZGERALD: The Electrician **30**, 48 (1892).
[2] K. BIRKELAND: Arch. Sci. Phys. Geneve **4**, 497 (1896).
[3] O. LODGE: The Electrician **46**, 249 (1900).

advocated by BIRKELAND, STÖRMER, VEGARD, and others. SCHUSTER[4] attributed the magnetic storm to atmospheric effects caused by corpuscular heating of the terrestrial atmosphere. The alternative hypothesis, that the particles are terrestrial particles activated by ultraviolet light from the sun has been held for many years. It is interesting, in retrospect, to see that both extremes have some elements in them which are now accepted. The ultra-violet light hypothesis is basically incorrect, but it contained in it the correct idea that the auroral particles are locally accelerated. The corpuscular hypothesis is basically correct, but wrong in its belief that the auroral particles are direct from acceleration at the sun.

β) To explore the ideas in a little more detail, consider the *ultraviolet light hypothesis*, which attributes the aurora and the magnetic storm to increased heating and ionization of the ionosphere as the result of enhanced solar ultraviolet emission. The ultraviolet light hypothesis has had a long history. One form of the idea[5,6] is that neutral atmospheric atoms which have acquired high upward velocities and momentarily find themselves at heights of 3 to $4 \cdot 10^4$ km, are ionized by solar ultraviolet radiation. The ions are tied to the geomagnetic field because of their electric charge, and are thus prevented from falling back to the atmosphere far below. To state the effect in simple gasmagnetic terms, the weight of these trapped atoms suspended on the magnetic lines of force compresses the underlying field to give the initial phase of the storm. The main phase was attributed to a general heating of the atmosphere which expanded upward, pushing the field with it.

Other forms of the ultraviolet hypothesis involve atmospheric currents[7-10]. VESTINE[11] has argued in favor of an atmospheric current system as the origin of the magnetic storm.

Without going into the extensive literature on the subject, the fundamental objection to the ultraviolet light hypothesis for magnetic storms produced by a solar flare is the fact that the ultraviolet propagation time from the sun is eight minutes, whereas the magnetic storm is observed to begin sharply 20 to 50 hours, rather than 8 minutes, after the flare. The objection for the 27-day recurring storms, which are not associated with flares, is that they often have a sudden commencement. Their sudden commencement could be explained only by continual emission from the rotating sun of a beam of ultraviolet light with a sharp leading edge. It is difficult to understand how thermal emission from the sun could give rise to a sharp edge.

27. The corpuscular hypothesis. Consider, then, the development of the corpuscular hypothesis. The great storms are observed usually to follow one or two days after a large flare when that flare is usually not farther than about 45° from the center of the solar disk. Hence the solar corpuscular emission was judged to have a characteristic angular width of the order of 45° and a velocity of 1000 to 2000 km/sec[1,2]. MAUNDER[3,4] first suggested that the 27-day recurring storms

---

[4] A. SCHUSTER: Proc. Roy. Soc. London A **85**, 44 (1911).
[5] E. O. HULBURT: Phys. Rev. **31**, 1038 (1928); **34**, 344 (1929); **36**, 1560 (1930).
[6] H. B. MARIS, and E. O. HULBURT: Phys. Rev. **33**, 412, 1046 (1929).
[7] O. R. WULF: Terr. Magn. Atmosph. Electr. **50**, 185, 259 (1945).
[8] O. R. WULF, and S. B. NICHOLSON: Phys. Rev. **73**, 1204 (1948); — Publ. Astron. Soc. Pacific **60**, 37, 259 (1948); **61**, 166 (1949); **62**, 202 (1950).
[9] S. B. NICHOLSON, and O. R. WULF: Pub. Astron. Soc. Pacific **64**, 265 (1952).
[10] O. R. WULF, and L. DAVIS: J. Met. **9**, 82 (1952).
[11] E. H. VESTINE: J. Geophys. Res. **58**, 539 (1953); **59**, 93 (1954).
[1] W. M. H. GREAVES, and H. W. NEWTON: Mon. Not. roy. Astron. Soc. **88**, 556 (1928); — **89**, 84 (1929).
[2] G. ABETTI: Second Rep. Solar and Terr. Relationships, 9 (1929).
[3] E. W. MAUNDER: Mon. Not. Roy. Astron. Soc. **64**, 205 (1904); **65**, 2, 538, 666 (1905); **76**, 63 (1916).
[4] C. CHREE: Proc. roy. Soc. London A **101**, 368 (1922).

were the result of the emission of isolated streams of particles from the sun. Presumably the width of such corpuscular streams was 10 to 15° and the particle speed in the stream was of the order of 500 to 1000 km/sec. ALLEN[5] tentatively identified the corpuscular streams with the long coronal streamers sometimes seen at solar eclipse.

α) The first step in developing the corpuscular hypothesis was taken by BIRKELAND[7]. He showed that *cathode rays projected* in the laboratory against the field of a *magnetized sphere* give a luminosity in the residual gas in the space around the sphere with two bright rings, one over each pole of the sphere, not unlike the terrestrial auroral zones. BIRKELAND, STÖRMER[8], VEGARD[9] and LEMAÎTRE[10] developed ideas for the aurora based on the cathode ray analogy, in which charged particles of one sign are emitted from the sun and, after passing through space, are deflected in the geomagnetic field. STÖRMER's work [21] is probably best known today in connection with its application to the global incidence of cosmic rays. Spectroscopic evidence for the corpuscular nature of the individual auroral arcs and rays was obtained by VEGARD[11] and GARTLEIN[12] and more directly by MEINEL[13] CHAMBERLAIN[14] [3] who observed the H_α (6562 Å) line Doppler shifted by some 3000 km/sec toward the blue.

The most evident difficulty with theories based on charged particles of one sign arises from the observed penetration of auroral particles down to altitudes of 80 to 120 km in the atmosphere. The 10^7 to 10^8 eV particles which STÖRMER needed to explain the latitude of the auroral zone would penetrate too deeply into the atmosphere. Whereas the 10 eV electrons and/or 10^4 eV protons suggested by the transit time from the sun would not penetrate deeply enough. The laboratory data [15,16] show the difficulty very clearly: The atmosphere above 100 km amounts to about 1.5 cm of air at standard temperature and pressure. Thus electrons should have energies of two or three hundred eV, equivalent to 10^5 km/sec and protons should have about 10^6 eV, equivalent to 10^4 km/sec, in rough agreement with the observed Doppler shifts[12].

MARTYN was the first to point out that the only resolution of the conflict is to suppose that the particles are accelerated in the near neighborhood of Earth. Presumably the cause of the acceleration is the passage of corpuscular streams from the sun, but the streams themselves do not flow in to make the aurora directly.

The idea that the aurora is produced by streams of particles of one sign was criticised by SCHUSTER, who pointed out that the electrostatic repulsion of particles of a single sign would disperse the particle streams and clouds. LINDEMANN[17] raised similar objections, noting the enormous charge that would build up

[5] C. W. ALLEN: Mon. Not. roy. Astron. Soc. **104**, 13 (1944).
[6] S. CHAPMAN in: S. MATSUSHITA and W. H. CAMPBELL: Physics of Geomagnetic Phenomena, Vol. I. Chapt. I. New York: Academic Press 1968.
[7] K. BIRKELAND: Arch. Sci. Phys. Geneve **4**, 497 (1896).
[8] C. STÖRMER: Arch. Sci. Phys. Genève **24**, 5, 113, 221, 317 (1907); **32**, 33, 163 (1911); **35**, 483 (1913).
[9] L. VEGARD: Physics of the Earth, Terrestrial Magnetism and Electricity **44**, 611. (1939).
[10] G. LEMAÎTRE, and M. S. VALLARTA: Phys. Rev. **43**, 87 (1933). — G. LEMAÎTRE, and L. BOSSY: Bull. Cl. Sci. Acad. Roy. Belgique **31**, 357 (1946). — BOSSY, L.: Ann. Géophys. **18**, 198 (1962).
[11] L. VEGARD: Nature **144**, 1089 (1939); — Terr. Magn. Atmosph. Electr. **45**, 5 (1940).
[12] C. W. GARTLEIN: Trans. Amer. Geophys. Union **31**, 18 (1950).
[13] A. B. MEINEL: Astrophys. J. **111**, 555 (1950); **113**, 50 (1951).
[14] J. W. CHAMBERLAIN: Astrophys. J. **126**, 245 (1957).
[15] N. N. DAS GUPTA, and S. K. GHOSH: Rev. Mod. Phys. **18**, 225 (1945).
[16] D. R. BATES: Mon. Not. roy. Astron. Soc. **109**, 215 (1949).
[17] F. A. LINDEMANN: Phil. Mag. **38**, 669 (1919).

at the sun. MASSEY and VEGARD[9] suggested that the charged particles might pick up particles of opposite sign in space and avoid the difficulty. But the problem of charge separation is too acute to solve only in this delayed fashion. The important point made by LINDEMANN[17] is that the corpuscular emission must be electrically neutral as it leaves the sun, requiring that the emission is ionized gas consisting of both ions and electrons, with no net charge.

It is interesting to note, then, that the reverse process was suggested by BENNETT and HULBURT[18]. They imagined a slender stream of ionized gas flowing outward from the sun through interplanetary space. They suggested that the electrons in the stream tend to be scattered out to be replaced by nonflowing interplanetary electrons. The gas stream is thus reduced to an ion stream through a stationary background plasma. The ion stream is subject to magnetic self-focussing and upon reaching Earth behaves much as the particles of BIRKELAND and STÖRMER. Unfortunately the theory fails to recognize that electrostatic forces prohibit a net current, whereas if the necessary return current through the background plasma is included, the self-induction prohibits the current[19]. To put the matter differently, the theory overlooks the fact that the field is frozen into the ionized gas, and that the ionized gas is bound into a single fluid by the magnetic field. A jet of gas described by the hydromagnetic equations does not convert itself into a thread of current.

β) The mechanism for *emission of* the *particles from the sun* has been a subject of considerable speculation. LINDEMANN[17] supposed that radiation pressure would eject ions from the sun with sufficient speed. The electrons would be dragged along with the ions by the electrostatic forces. MILNE[20] further developed the general idea of radiative repulsion. But KAHN[21] showed that the L_α radiation (1216 Å) from solar flares is insufficient to produce the corpuscular radiation necessary for even a moderate storm. KIEPENHEUER[22] showed that radiation pressure might produce enough ions to give a moderate storm, but unfortunately the process is effectively inoperative for the principal constituent — hydrogen — of solar streams.

ALFVEN[23] [*1*] suggested that the particles are emitted from the sun because the electrons have temperatures of 10^8 eV. The electrons expand away from the sun, with the ions compelled to accompany them by the electrostatic forces. The final velocity could be as high as 10^5 km/sec if there were no serious impediment to the outflow. ALFVEN expressed the idea in terms of particle motions in magnetic fields, apparently with the view that the magnetic fields played some essential role in the acceleration. SCHLÜTER's "melon seed" mechanism[24] is probably the best known magnetic mechanism for the ejection of gas from the sun, in which a small clump of gas is squeezed out of a strong magnetic field along diverging lines of force. The internal energy of the gas is converted into bulk motion in this way. A somewhat different process was proposed[25], in which the small clump of gas is squeezed ahead of a hydromagnetic wave propagating upward in the solar atmosphere along a field of essentially undiminishing strength. The decreasing atmospheric density leads to increasing Alfven speed, leading to wave velocities of 100 km/sec or more.

28. The nature of the main phase. The early association of the terrestrial magnetic variations with currents in the magnetosphere was based on MAXWELL's

---

[18] W. H. BENNETT, and E. O. HULBURT: Phys. Rev. **95**, 315 (1954); — J. Atmosph. Terr. Phys. **5**, 211 (1954).
[19] V. C. A. FERRARO, and D. M. WILLIS: Astrophys. J. **136**, 288 (1962).
[20] E. A. MILNE: Mon. Not. roy. Astron. Soc. **86**, 459 (1926).
[21] F. D. KAHN: Mon. Not. roy. Astron. Soc. **109**, 324 (1949); **110**, 477 (1950).
[22] K. O. KIEPENHEUER: J. Geophys. Res. **57**, 113 (1952).
[23] H. ALFEN: K. Svenska Vet. Akad. Handl. **18** (1939).
[24] A. SCHLÜTER: Z. Naturforsch. **5a**, 72 (1950).
[25] H. W. BABCOCK, and H. D. BABCOCK: Astrophys. J. **121**, 349 (1955).

equation, see Eq. (10.2), for quasi-stationary fields

$$\boldsymbol{j} = \frac{c_0 \sqrt{\varepsilon_0 \mu_0}}{u \mu_0} \nabla \times \boldsymbol{B}. \tag{28.1}$$

The association appears first in the work of STÖRMER[1] and again in SCHMIDT[2]. SCHMIDT pointed out that the magnetic data required a southward field of as much as 250 γ across Earth during the main phase of a great storm. He postulated that the current was in the form of a ring encircling Earth, as the result of a westward circulation of ions and an eastward circulation of electrons.

α) It is in the work of CHAPMAN and FERRARO[3] that the *dynamical forces of gases and fields* begin to play a central role. They were the first to give explicit recognition to the important fact that the ionized gas from the sun, and the geomagnetic field against which the gas is hurled, do not interpenetrate. They equated the magnetic pressure to the impact pressure of the gas and deduced the resulting compression of the geomagnetic field for the initial phase of a magnetic storm. They abandoned this point of view to some extent in their next step, wherein they assumed that the main phase is the result of ions and electrons moving in large concentric circles around Earth. The centrifugal force of the circular motion leads to a westward current to give the balancing Lorentz force, and the westward ring current is in turn responsible for the main phase decrease. In present day language we would say merely that the centrifugal force flexes the field outward and gives the decrease at Earth. The objection to the theory for the main phase is that ions and electrons do not move in circles around Earth in the manner which they imagine for the main phase. But the ideas on the lack of interpenetration of gas and field, and the importance of magnetic and pressure forces, are correct and form the basis for present theory.

β) The generalization of non-interpenetration of gas and fields to the concept that the magnetic lines of force are "frozen in" to the gas was due to ALFVEN. With this realization came the modern *magneto-gasdynamic theory* [1] in which disturbances are propagated with the Alfven speed, etc. The first overall picture of the individual particle motions in a magnetic field is due also to ALFVEN [1] who demonstrated the curvature and gradient drifts [see Eq. (20.3)] and who emphasized the equivalence of bulk motion and the electric drift, Eq. (20.4). It is with these theoretical tools that the theory of the main phase is now constructed.

Unfortunately ALFVEN's own application of the magneto-gasdynamic theory to corpuscular streams and the geomagnetic field was abortive, as a consequence of his misapplication of the polarization field, Eqs. (9.2), (9.3). ALFVEN [1][4] imagined that the large-scale electric field

$$\boldsymbol{E} = -\boldsymbol{v} \times \boldsymbol{B}/c_0 \sqrt{\varepsilon_0 \mu_0} \qquad [9.3]$$

in an interplanetary corpuscular stream is extrapolated across the geomagnetic field as well. This overlooks the fact that Eq. [9.3] applies to every point in space for the reason that there can be little or no electric field in the frame of reference of the ionized gas (which has velocity $\boldsymbol{v}$). The field $\boldsymbol{E}$ cannot be extrapolated from one region into another. The gas velocity in the magnetosphere is $\boldsymbol{v} \cong 0$ in the frame of reference of the Earth. Hence $\boldsymbol{E} \cong 0$ in the magnetosphere.

γ) It remained for SINGER[5,6] to point out that the particle *motion in the magnetosphere* is the *gradient drift* alone, without the electric drift from the inter-

---

[1] C. STÖRMER: Arch. Sci. Phys. Genève **24**, 5, 113, 221, 317 (1907); **32**, 33, 163 (1911); **35**, 483 (1913).
[2] A. SCHMIDT: Z. Geophys. **1**, 3 (1924).
[3] S. CHAPMAN, and V. C. A. FERRARO: Terr. Magn. Atmosph. Electr. **36**, 77, 171 (1931); **37**, 147, 421 (1932); **38**, 79 (1933); **45**, 245 (1940).
[4] H. ALFVEN: Tellus **7**, 50 (1955).
[5] S. F. SINGER: Trans. Amer. Geophys. Union **38**, 175 (1957).
[6] A. J. DESSLER, and E. N. PARKER: J. Geophys. Res. **64**, 2239 (1959).

planetary polarization field. Axford and Hines[7] then added the *convective motions* of the magnetosphere, to give the present physical picture of the particle motions.

In this connection it is interesting to note that Martyn proposed an idea for the aurora based on the ring current idea of Chapman and Ferraro. The reader will recognize that an electron or a proton cannot by itself circle Earth through the geomagnetic field unless the energy of the particle is 10^9-10^{11} ev (depending upon the radius of the circle). So Martyn suggested that the electrons and ions move together (except for the small velocity difference giving the westward current) as a nearly neutral mass. In this way the charge to mass ratio is sufficiently reduced as to permit the necessary circling. But with both positive and negative particles moving in the same direction, one sign tends to be deflected to the right and the other to the left by the Lorentz force. The positive and negative particles are prevented from separating by the electrostatic forces. Martyn[8] pointed out that the electrostatic field given by Eq. [9.3] becomes as great as 0.1 V/m for a speed of 10^6 m sec^{-1} (and $B = 10^{-3}\ \Gamma$). Across a width of 10^7 m this amounts to 10^6 V = 1 MV. It is evident that a particle can traverse this potential difference by moving down along the lines of force to the ionosphere, passing across through the ionosphere, and up the lines of force to the opposite side of the ring of ionized gas. This electric "discharge" was the aurora. Today we recognize that in the quasi-neutral Chapman-Ferraro whirling ring of ionized gas,

$$\boldsymbol{E} = -\boldsymbol{v} \times \boldsymbol{B}/c_0 \sqrt{\varepsilon_0 \mu_0} \qquad [9.3]$$

would represent a convection of both the particles and the field with velocity $\boldsymbol{v}$, which must extend all the way along the lines of force into the ionosphere. The field Eq. [9.3] gives zero electric field in the frame of the particles. It is this electric field which *prevents* electrical discharges. Of course, the fringing field in the nonconducting nonrotating atmosphere will have the 1 MV potential difference, but only where the atmosphere is nonconducting. It is largely on this basis that the electric discharge theories of the aurora break down.

29. Corpuscular emission from the sun. Finally, we consider briefly the development of ideas on the nature of the solar corpuscular emission is space. The tendency for magnetic and auroral activity to jump to high levels rather suddenly, with extended periods of low activity between, led quite naturally to the idea at the turn of the century that the particles were emitted from the sun in rather narrow collimated beams. This view has persisted to almost the present day.

α) It was a fundamental step, therefore, when Biermann[1] pointed out, first, that the type I (gaseous) comet tails are directed away from the sun as a consequence of *solar corpuscular radiation*, rather than electromagnetic radiation, and second, that the comet fails never fail to point away from the sun, even at high solar latitudes and at minimum solar activity. From this he pointed out that the particle emission from the sun is general, like the electromagnetic emission, and is not confined to streams. It was this point of Biermann's which indicated that the corpuscular radiation from the sun is a continuum flow and therefore has a hydrodynamic origin in the expansion of the solar atmosphere[2]. The name *solar wind* became more appropriate than corpuscular terms[3]. It became evident that Earth was subject to a continual wind from the sun, which accounted, then, for the unceasing magnetic and auroral acrivity at the auroral zones and poleward. The observed bursts of activity are the result of shocks in the wind with turbulence behind, etc. rather than to isolated streams*.

---

* The authors wish to express their gratitude to A. J. Dessler, C. O. Hines, and W. I. Axford for undertaking a reading of the manuscript and offering many valuable comments and suggestions for improving the clarity and accuracy of the presentation. E. N. Parker's part of the present work was supported by US Air Force Office of Scientific Research under contract AF 49 (638)—1642.

[7] W. I. Axford, and C. O. Hines: Can. J. Phys. **39**, 1433 (1961).
[8] D. F. Martin: Nature **167**, 92 (1051)
[1] L. Biermann: Z. Astrophysik **29**, 274 (1951); — Observatory **107**, 109 (1957).
[2] E. N. Parker: Astrophys. J. **128**, 664 (1958).
[3] See contributions by H. Poeverlein and by W. N. Hess in Vol. 49/4 of this Encyclopedia.

General references.

[1] ALFVEN, H.: Cosmical Electrodynamics. Oxford: University Press 1950.
[2] —, and C. G. FALTHAMMAR: Cosmical Electrodynamics, Fundamental Principles. Oxford: University Press 1963.
[3] CHAMBERLAIN, J. W.: Physics of the Aurora and Airglow. New York: Academic Press 1961.
[4] CHAPMANN, S.: The Earth's Magnetism. London: Methuen & Co. 1951.
[5] — The Earth's Environment. New York: Gordon & Breach 1962.
[6] — Solar Plasma, Geomagnetism and Aurora. New York: Gordon & Breach 1964.
[7] —, and J. BARTELS: Geomagnetism. London: Oxford University Press 1940.
[8] DESSLER, A. J.: Radio Sci. (J. Res. NBS) 65 D, 603 (1964).
[9] DUNGEY, J. W.: Cosmic Electrodynamics. Cambridge: Cambridge University Press 1958.
[10] GOLD, T.: Gas Dynamics of Cosmic Clouds. Amsterdam: North Holland Publ. Co. 1955.
[11] HESS, W. N. (ed.): Collected papers on artificial radiation belt. In: J. Geophys. Res. **68**, No. 3. (1963).
[12] HESS, W. N. (ed.): The Physics of Solar Flares. Washington, D. C.: Nat. Aeron. Space Adm. 1963.
[13] JOHNSON, F. S.: Satellite Environment Handbook. Stanford: Stanford University Press 1961.
[14] LANDSBERG, H. E., and J. VAN MIEGHEM: Advances in Geophysics, Vol. 9. New York: Academic Press 1962.
[15] LEHNERT, B. (ed.): Electromagnetic Phenomena in Cosmical Physics. Cambridge: Cambridge University Press 1958.
[16] MACKIN and NEUGEBAUER (eds.): The Solar Wind. New York: Pergamon Press 1966.
[17] NORTHROP, T. G.: The Adiabatic Motion of Charged Particles. New York: John Wiley & Sons 1963.
[18] ODISHAW, H. (ed.): Research in Geophysics, Vol. 1, Chap. 19. Cambridge, Mass.: MIT Press 1964.
[19] PARKER, E. N.: Interplanetary Dynamical Processes. New York: John Wiley & Sons 1963.
[20] SPITZER, L.: Physics of fully ionized Gases. New York: Interscience Publ. 1956.
[21] STÖRMER, C.: The Polar Aurora. Oxford: Clarendon Press 1955.
[22] STRATTON, J. A.: Electromagnetic Theory. New York: McGraw-Hill Book Co. 1941 [see p. 167].
[23] McCORMAC, B. M. (ed.): Radiation Trapped in the Earth's Magnetic Field (Proc. NATO adv. Study Inst., Bergen, Norway, 16. 8.—3. 9. 1965). Dordrecht: de Reidel Publ. Co. 1966.
[24] HULST, H. C. VAN DE, D. DE JAGER, and A. F. MOORE (eds.): Space Research II (Proc. 2nd Intern. Space Sci. Sympos., Florence 1961). Amsterdam: North-Holland Publ. Co. 1963.
[25] PRIESTER, W. (ed.): Space Research III (Proc. 3rd Intern. Space Sci. Sympos. Washington, D. C. 1962). Amsterdam: North-Holland Publ. Co. 1963.
[26] MULLER, P. (ed.): Space Research IV. (Proc. 4th Intern. Space Sci. Sympos., Warsaw 1963). Amsterdam: North-Holland Publ. Co. 1964.
[27] KING-HELE, D. G., P. MULLER, and G. RIGHINI (eds.): Space Research V (Proc. 5th Intern. Space Sci. Sympos., Florence 1964). Amsterdam: North-Holland Publ. Co. 1965.
[28] SMITH-ROSE, R. L. (ed.): Space Research VI (Proc. 6th Intern. Space Sci. Sympos., Mar del Plata 1965). New York: Spartan Books 1966.
[29] SMITH-ROSE, R. L., S. A. BOWHILL, and J. W. KING: (eds.): Space Research VII (Proc. 7th Intern. Space Sci. Sympos., Vienna 1966). Amsterdam: North-Holland Publ., Co. 1966.
[30] (Special edition): Space Sci. Rev. **7**, No. 2/3, 139—395 (1967).

Maßzahlen der erdmagnetischen Aktivität.

Von

MANFRED SIEBERT[*].

Mit 20 Figuren.

A. Einleitung.

1. Statistische Merkmale der erdmagnetischen Aktivität. Das Ausgangsmaterial zur Untersuchung der verschiedenartigen Einzelvariationen und der allgemeinen Unruhe des Magnetfeldes der Erde sind die kontinuierlichen Aufzeichnungen erdmagnetischer Elemente an den Observatorien. Üblicherweise werden registriert: die *Horizontalkomponente* $\mathcal{H}$ oder die *Komponente* X *nach geographisch Nord*, die *Deklination* D oder die *Komponente* Y *nach geographisch Ost* und die *Vertikalkomponente* Z. Jede an einem Observatorium registrierte Komponente erscheint als *Zeitkurve*, digitalisiert als *Zeitreihe*, mit statistischen Merkmalen. Man erkennt dies, wenn man versucht, eine Zeitkurve dieser Art aus möglichst einfachen Grundformen aufzubauen, die auf das Auftreten von physikalischen Vorgängen mit bestimmten Eigenschaften hinweisen. Vier verschiedene Typen können unterschieden werden: 1. Periodische Kurvenformen, herrührend von streng periodischen Vorgängen, von quasi-periodischen Vorgängen und — als schwächstem Ausdruck einer Periodizität — von Vorgängen mit einer gewissen Wiederholungsneigung. 2. Kurven, die aus mehr oder weniger langen monotonen Abschnitten bestehen, entsprechend Vorgängen mit einer vorherrschenden Tendenz oder Erhaltungsneigung. 3. Kurven mit unregelmäßiger Wiederkehr bestimmter, klassifizierbarer Formen, worin sich das unregelmäßige Auftreten spezieller Effekte widerspiegelt. 4. Kurven mit völlig unregelmäßig variierendem Verlauf auf Grund komplizierter und in den Einzelheiten nicht durchschaubarer Störungen mit dem Charakter zufälliger Ereignisse.

Überlagern sich Kurven dieser vier Typen, so können äußerst verworrene Kurvenzüge entstehen, wie es beispielsweise die Registrierungen erdmagnetischer Stürme sind. Darüber hinaus sind die vier Grundformen nicht für alle Zeitintervalle mit gleichbleibenden Anteilen an der resultierenden Kurve beteiligt. Die Intensität eines Typs kann also wiederum zeitabhängig sein. Beim Vergleich der Aufzeichnungen verschiedener Observatorien kommt zur Zeitabhängigkeit die Ortsabhängigkeit hinzu. Alles in allem ist man also beim Studium der erdmagnetischen Aktivität in der für die Geophysik typischen Situation, daß der Beobachter nicht selbst sinnvoll experimentieren kann, sondern den zwar gesetzmäßig, aber doch planlos und in großer Zahl zugleich ablaufenden Vorgängen ohne regulierenden Einfluß gegenübersteht. Diesem Mangel an Einflußnahme auf die Vorgänge selbst wird nachträglich mit statistischen Methoden begegnet.

2. Bemerkungen zur Definition von Maßzahlen der Aktivität. Es ist ein gebräuchliches Verfahren, einen komplizierten Sachverhalt durch Verzicht auf weniger wichtige Informationen zu vereinfachen, um wenigstens seine wesentlichen Züge zu verstehen.

---
[*] Endgültiges Manuskript eingegangen Dez. 1969.

α) In diesem Sinn besteht der erste Schritt einer statistischen Behandlung von Beobachtungen darin, ein umfangreiches Ausgangsmaterial möglichst gruppen- oder abschnittsweise durch *charakteristische Maßzahlen* zu ersetzen. Diese Parameter werden dann zur Auffindung von Zusammenhängen und Gesetzmäßigkeiten benutzt. Die Eigenschaften der erdmagnetischen Zeitkurven, wie sie in Ziff. 1 skizziert sind, legen ein solches Vorgehen auch bei der Behandlung der erdmagnetischen Aktivität nahe. Von diesem Standpunkt aus gesehen, sind bereits die von den Observatorien routinemäßig veröffentlichten Stundenmittel erdmagnetischer Elemente statistische Maßzahlen. Ob man damit aber schon Maßzahlen der Aktivität hat oder wenigstens das Material für die Berechnung solcher Maßzahlen, ist eine Frage, die über die statistische Betrachtungsweise hinausgeht und davon abhängt, was physikalisch mit diesen Zahlen erfaßt und von ihnen ausgesagt werden soll.

Dieser Gesichtspunkt spielte noch keine oder höchstens eine untergeordnete Rolle im ersten Stadium der Einführung und Verwendung von Aktivitätsmaßen. Sie basieren in der Tat auf Stundenmitteln, auf stündlichen oder täglichen Schwankungswerten, Änderungen von Tag zu Tag und anderen Größen, die je nach Vorschrift für ein einzelnes Element oder auch mehrere Elemente zu bestimmen sind. Noch gröber und subjektiver ist das Verfahren, auf dem das älteste Maß für weltweite Aktivität beruht, gegeben durch die täglichen internationalen erdmagnetischen *Charakterzahlen Ci*, die auch die längste vorliegende Beobachtungsreihe dieser Art darstellen (vgl. Ziff. 14).

β) Ein Maß für die Stärke einer Unruhe muß in irgendeiner Weise von dem Unterschied zwischen Unruhe und Ruhe ausgehen, wobei Ruhe nicht zu bedeuten braucht, daß überhaupt nichts geschieht. Vielmehr kann unter Unruhe mit der Nebenbedeutung des Unregelmäßigen auch die *Abweichung vom Regelmäßigen und Normalen* verstanden werden. Vielen Maßzahlen der erdmagnetischen Aktivität, und besonders den neueren, liegt diese Auffassung zugrunde. In formaler Weise kann die Festlegung und Abtrennung eines ungestörten Ganges in den Zeitkurven der Elemente durch ihre Zerlegung in die Kurventypen der Ziff. 1 erfolgen, wobei die streng periodischen Anteile und in gewissem Umfang auch monotone Anteile bei der Definition des ungestörten Verlaufs herangezogen werden müssen.

Nun erlauben diese regelmäßigen Variationen noch am leichtesten eine Deutung durch physikalische Vorgänge. Dann wird es aber möglich, in einem fortgeschrittenen Stadium die formale und nur statistische Festlegung von Ruhe und Unruhe durch die Unterscheidung der Vorgänge zu ersetzen, die den ruhigen und gestörten Kurvenverlauf im wesentlichen verursachen.

Eine solche Entwicklung, den Maßzahlen immer mehr physikalischen Inhalt zu geben, liegt zwangsläufig in der Natur wissenschaftlichen Arbeitens. Das hat aber die Konsequenz, daß neue und detaillierte Erkenntnisse die Gültigkeit der Konzeptionen einschränken, unter denen solche Zahlen ursprünglich eingeführt worden sind. Damit ist ein Anreiz gegeben, immer neue Aktivitätsmaße für immer speziellere Erscheinungen zu definieren und darüber hinaus die bestehenden Maßzahlen durch bessere zu ersetzen. Da es aber bei statistischen Untersuchungen häufig darauf ankommt, über lange homogene Zeitreihen dieser Zahlen verfügen zu können, ist es keine leichte Aufgabe zu entscheiden, wann eine Maßzahl so veraltet ist, daß ihr Gebrauch nicht mehr empfohlen werden kann, und wann neue Erkenntnisse so wichtig sind, daß sie die Einführung einer neuen Maßzahl erforderlich machen[1]. Außerdem muß jede Inflation von Maßzahlen vermieden werden, da sie in erster Linie einfache Hilfsmittel bei der Behandlung anderer Beobachtungen sein sollen.

Aus der Schwierigkeit, wenn nicht sogar Unmöglichkeit, die einzelnen Vorgänge in der Magnetosphäre und Ionosphäre mit ausreichender Genauigkeit in

---

[1] Gegenwärtig ist die *International Association of Geomagnetism and Aeronomy (IAGA)* mit ihrer Kommission IV (magnetische Variationen und Störungen) dafür zuständig.

den täglichen Registrierungen der Observatorien zu erkennen und voneinander zu trennen, folgt gerade bei physikalischer Definition der Maßzahlen, daß die Einzelwerte solchen Definitionen nur näherungsweise gerecht werden und ihre Zuverlässigkeit in diesem Sinn immer beschränkt bleibt. Erst mit vielen Einzelwerten läßt sich diese Ungenauigkeit reduzieren, aber dann eben nur für Aussagen über das durchschnittliche Auftreten von Ereignissen und deren Verlauf. Ein Benutzer dieser Zahlen sollte daher ihren Aussagewert nicht nur nach ihrer formalen Definition beurteilen und möglicherweise überschätzen; wie es andererseits falsch wäre, eine Annäherung an die ideale Maßzahl durch ein immer komplizierteres Verfahren ihrer numerischen Bestimmung zu verlangen.

Auch die ideale Maßzahl, wenn es sie gäbe, müßte die Forderungen nach einfacher Definition und einfacher Anwendbarkeit erfüllen, damit sie für ihre Aufgabe brauchbar ist. Sie kann nie mehr als einen Extrakt des tatsächlichen Geschehens geben. Da, wo ihre Information nicht ausreicht, muß auf das Ausgangsmaterial, also auf die Registrierungen an den Observatorien zurückgegriffen werden.

B. Die BARTELSsche Kennziffer K und daraus abgeleitete Maßzahlen.

I. Überblick und Grundlagen.

3. Historische Bemerkungen. Die *dreistündliche Kennziffer* K (ursprünglich K_1, s. Ziff. 13 δ) wurde von J. BARTELS[1] im Jahre 1938 als lokales Aktivitätsmaß für das Observatorium Niemegk, der Ausweichstation des Magnetischen Observatoriums Potsdam, eingeführt und anhand der Daten dieses Observatoriums praktisch erprobt[2]. Ihr wesentlicher Unterschied gegenüber früheren Maßzahlen liegt in ihrer von vornherein physikalisch festgelegten Definition, wonach sie ein Maß für die Stärke der lokalen Störung des erdmagnetischen Feldes durch *solare Partikelstrahlung* (**P**-*Strahlung*) sein soll. Diese Potsdamer erdmagnetische Kennziffer wurde bald darauf auch für andere Observatorien routinemäßig bestimmt (BARTELS, HECK und JOHNSTON [*3*]) und 1939 von der damaligen *International Association of Terrestrial Magnetism and Electricity (IATME)* versuchsweise als internationale Maßzahl angenommen.

α) Der nächste Schritt, aus den lokalen Kennziffern der Observatorien ein Maß für den *weltweiten Störungsgrad* des erdmagnetischen Feldes abzuleiten, führte zunächst mit der Kennziffer Kw[3] auf eine nicht ganz befriedigende Lösung[4]. Da Kw als einfacher Durchschnitt aus den vorhandenen K-Werten berechnet

---

[1] J. BARTELS: Z. Geophys. **14**, 68—78 (1938).

[2] Der in Fußnote[1] zitierten 1. Mitteilung über die Potsdamer erdmagnetischen Kennziffern sind in fortlaufender Numerierung weitere 13 Mitteilungen in der Zeitschrift für Geophysik gefolgt: J. BARTELS: Z. Geophys. **14**, 230—231, 272—273 (1938); **15**, 214—221, 333—335 (1939). — A. BURGER: Z. Geophys. **16**, 85—86, 185—194, 247—249 (1940); **17**, 67—69, 147—148, 226—229 (1941). — J. BARTELS u. A. BURGER: Z. Geophys. **17**, 317—327 (1942). — J. BARTELS: Z. Geophys. **18**, 76—78, 180—182 (1943).

[3] Für die Jahre 1940—1951 wurde Kw als arithmetisches Mittel aus den K-Werten von 27 bis 39 Observatorien berechnet und in den IATME Bulletins No. 12 veröffentlicht. Ursprünglich war Kw jedoch nach [*3b*] das arithmetische Mittel der reduzierten Kennziffern Kr, die einen Vorläufer der standardisierten Kennziffern Ks darstellen und nur für wenige Observatorien hergeleitet wurden. In dieser letzten Bedeutung liegen Werte von Kw für 1937 bis 1939 vor. Soweit während dieser Jahre die gewöhnlichen arithmetischen Mittelwerte der K berechnet wurden, wurden sie mit Km bezeichnet.

[4] In diesem Beitrag werden indizierte Kennziffern auf Zeile geschrieben, wie es unter Erdmagnetikern, entsprechend den IAGA-Regeln, Brauch ist. In den übrigen Beiträgen dieses Bandes wurden dagegen die Indizes, wie üblich, tief gestellt: K_p, A_p, C_p etc.

wurde, jedoch die geographische Verteilung der verfügbaren Observatorien höchst inhomogen war, enthielt diese Kennziffer noch starke lokale und ortszeitliche Einflüsse der P-Strahlung, herrührend von der unterschiedlichen Bewichtung der verschiedenen Regionen der Erdoberfläche bei diesem Verfahren. Um diese Mängel auszuschalten, entwickelte BARTELS die Methode, die K-Werte ausgewählter Observatorien nach einem in Ziff. 7 erörterten Verfahren in die *standardisierten Kennziffern Ks* umzurechnen. Ihr Durchschnitt für jedes Drei-Stunden-Intervall ist die bekannte *planetarische Kennziffer Kp*. Der K-Index und das Prinzip, aus diesen Maßzahlen Kp zu berechnen, wurden von der IATME 1951 offiziell anerkannt.

Die erste Veröffentlichung[5] von Kp-Werten erfolgte im September 1949 und enthielt die Werte der Monate Januar, Februar und März des gleichen Jahres. In vierteljährlichem Abstand wurde auch der restliche Jahrgang 1949 veröffentlicht. Ebenfalls noch 1949 erschienen im IATME Bulletin No. 12b[6] die Kp-Jahrgänge 1945—1948. Von da an sind Tabellen mit diesen planetarischen Kennziffern ein fester Bestandteil der IATME- und späteren IAGA-Bulletins.

Um 1950 entstanden auch die weiteren, zur Gruppe der BARTELSschen Kennziffern gehörenden Maßzahlen, die hier noch im einzelnen besprochen werden. Von ihnen wurde die *mittlere tägliche Amplitude Ap* von der IATME 1951 versuchsweise als internationales Aktivitätsmaß angenommen. Dreizehn Jahrgänge von Ap erschienen bereits ein Jahr später im Bulletin No. 12f.

β) Die Berechnung aller dieser Kennziffern wurde von da an nicht nur routinemäßig für jedes neue Jahr fortgesetzt, sondern nach und nach auch auf *vergangene Jahre* ausgedehnt. BARTELS tat dies für die Jahre zurück bis 1937 und für das zweite Internationale Polarjahr 1932/33, während die für 1932/33 dann noch fehlenden Kennziffern und die Jahrgänge 1934 bis 1936 im wesentlichen von P. N. MAYAUD berechnet wurden. Diese Arbeit war 1960 abgeschlossen. Eine erste geschlossene Zusammenstellung der Werte der planetarischen Kennziffern Kp, Ap und Cp erschien 1962 im IAGA Bulletin No. 18, das die 30 Jahrgänge 1932 bis 1961 enthält.

Seit 1951 werden im Rahmen des *International Service of Geomagnetic Indices* die bereits endgültigen Werte dieser Kennziffern für jeden Monat in Form von Tabellen und graphischen Darstellungen noch vor Ablauf des folgenden Monats an zur Zeit über 300 Wissenschaftler und Institutionen in allen Teilen der Welt verschickt[7]. Seit dem Beginn des Internationalen Geophysikalischen Jahres im Jahre 1957 werden die Tabellen auch halbmonatlich herausgegeben.

γ) Seit der Einführung der Potsdamer erdmagnetischen Kennziffer K vor mehr als 30 Jahren hat die Entwicklung von Erdmagnetismus und Aeronomie jedes damals vorhersehbare Ausmaß weit übertroffen. Ein Beispiel dafür ist der durch direkte Messungen mit Raumsonden außerhalb der Magnetosphäre erbrachte Nachweis, daß die BARTELSschen planetarischen Kennziffern tatsächlich ein Maß für die Einwirkung der solaren Partikelstrahlung auf das erdmagnetische Feld sind. Dieses Ergebnis beweist aber ebenso die *Qualität der Maßzahlen*, die trotz der Unterschiede zwischen den damaligen und heutigen Vorstellungen über die Natur der erdmagnetischen Aktivität und trotz einiger daraus resultierender Unzu-

---

[5] J. BARTELS and J. VELDKAMP: J. Geophys. Res. **54**, 295—299 (1949).

[6] Dasselbe Bulletin enthält auch die K-Werte von Potsdam-Seddin-Niemegk für die zurückliegenden Jahre 1900 bis 1936; s. auch Abh. Geomagn. Inst. Potsdam Nr. 17 (1957).

[7] Dieser Teil des Kennzifferndienstes wird am Institut für Geophysik der Universität Göttingen, Deutschland, ausgeführt. Die Kennziffern K und Charakterzahlen C der Observatorien werden am Meteorologischen Institut in De Bilt, Niederlande, dem $C + K$ *Centre*, gesammelt und tabellarisiert; dort werden auch die internationalen Charakterzahlen Ci berechnet und die internationalen ruhigen und gestörten Tage (s. Ziff. 11) bestimmt. Schließlich werden am Geophysikalischen Observatorium in Tortosa, Spanien, die Meldungen über die klassifizierbaren kurzzeitigen magnetischen Störungen gesammelt und überprüft. Alle diese Aufgaben gehören zum International Service of Geomagnetic Indices.

länglichkeiten auch heute noch das brauchbarste und zuverlässigste Werkzeug zur einfachen Erfassung und statistischen Behandlung der in ihren Einzelheiten oft nur schwer durchschaubaren Vorgänge im Magnetfeld der Erde sind.

4. K-Variationen und Nicht-K-Variationen.

α) In Ziff. 1 war ausgeführt worden, daß die Zeitkurve einer an einem Observatorium registrierten Komponente des erdmagnetischen Feldes mit statistischen Methoden auf einfachere Kurventypen zurückgeführt werden kann. Die genannten Grundformen der Kurventypen sind dabei aus der Anschauung abstrahiert, so daß eine solche Zerlegung nur formalen Charakter hat, wenn nicht nachträglich eine Interpretation durch tatsächlich stattfindende Vorgänge gelingt.

Bei der Definition von K wird ebenfalls davon ausgegangen, daß die registrierte Zeitkurve aus anderen Kurven zusammengesetzt ist, in die sie zerlegt werden soll. Der grundlegende Unterschied zu dem zuvor genannten Verfahren besteht jedoch darin, daß bereits die Zerlegung der Kurve im Hinblick auf die beabsichtigte physikalische Aussage der Kennziffer vorzunehmen ist. Dieser physikalische Ausgangspunkt ist zweifellos einer der großen Vorzüge von K; er erschwert allerdings auch eine genaue Bestimmung, da die Identifizierung physikalischer Vorgänge in den Registrierungen schwieriger ist als die Unterscheidung formaler Kurvenformen.

Konkret gesprochen, geht BARTELS bei der Definition von K von der Vorstellung aus, daß der zeitliche Verlauf einer Komponente einen normalen Anteil enthält, dem sich Störungen des erdmagnetischen Feldes infolge solarer Partikelstrahlung überlagern. Lediglich diese Störungen, die **P**-*Effekte*, sollen der erdmagnetischen Aktivität zugerechnet werden. Als ein aus 10 Klassen bestehendes Maß dieser Aktivität hat BARTELS die dreistündliche Kennziffer K eingeführt. Der tägliche Verlauf der Aktivität an einem Observatorium ist dann durch acht Werte von K für die acht Drei-Stunden-Intervalle eines Greenwich-Tages gegeben.

Die ursprünglich strenge Identifizierung der *K-Variationen* mit den **P**-*Effekten* und der *Nicht-K-Variationen* mit den **W**-*Effekten*, also den von der *solaren Wellenstrahlung* abhängigen Variationen und Störungen, ist heute nicht mehr aufrechtzuerhalten. Sie kann aber als Richtschnur für eine zeitgemäßere Festlegung dienen:

K-Variationen sind alle von solarer Partikelstrahlung im betrachteten Drei-Stunden-Intervall verursachten unregelmäßigen Störungen des erdmagnetischen Feldes. Alle anderen regelmäßigen und unregelmäßigen Störungen sind Nicht-K-Variationen. Erdmagnetische Aktivität ist das Auftreten von K-Variationen.

Auch diese Unterscheidung läßt sich in der Praxis nicht mit aller Konsequenz vornehmen, worauf bei den betreffenden Erscheinungen noch hingewiesen wird. Es ist jetzt aber möglich, die wichtigsten Nicht-K-Variationen anzugeben, die zur Bestimmung von K bekannt sein und eliminiert werden müssen. Die morphologische Beschreibung der Störungen kann hier unterbleiben, da sie in diesem Band an anderer Stelle erfolgt[1]. Soweit sich die erdmagnetischen Störungen und Variationen überhaupt klassifizieren lassen, gibt es die folgenden Nicht-K-Variationen:

β) *Der sonnen- und mondentägige Gang an magnetisch ruhigen Tagen:* Die Gezeitenkräfte von Sonne (**S**) und Mond (**L**) und vor allem die Wärmestrahlung der Sonne verursachen im Zusammenhang mit der Rotation der Erde gezeitenartige Schwingungen der Erdatmosphäre. Die damit verbundenen Gezeitenwinde führen in der E-Schicht der Ionosphäre infolge unterschiedlicher Bewegung von

---

[1] Siehe den Beitrag von T. NAGATA and N. FUKUSHIMA, in diesem Band, S. 5—130.

Ionen und Elektronen im Magnetfeld der Erde zu großräumigen Stromsystemen, deren Magnetfelder an der Erdoberfläche als überlagerter ($S_q + L$)-Gang beobachtet wird. Dabei ist ein offensichtlicher Einfluß von L auf den Verlauf des Tagesganges nur für äquatornahe Observatorien im Bereich des Elektrojets nachgewiesen.

Nach dieser Erklärung ist sowohl der Anlaß von ($S_q + L$) als auch die Modulation durch die variable elektrische Leitfähigkeit in der Ionosphäre an magnetisch ruhigen Tagen unabhängig von solarer Partikelstrahlung. Insbesondere kann dann S_q fast ausschließlich der Wirkung der solaren Wellenstrahlung W zugeschrieben werden. Von dieser Vorstellung ausgehend, hat BARTELS[2] die Amplitude von S_q direkt zur Definition einer Maßzahl für die relative Änderung von W benutzt.

Die zuvor gegebene klassische Deutung des sonnen- und mondentägigen Ganges, wie sie sich etwa bei CHAPMAN und BARTELS [2] (Kap. VII, VIII u. XXIII) findet, ist nicht ausreichend, um ein S_q-ähnliches, hauptsächlich in den Polargebieten an magnetisch ruhigen Tagen auftretendes ionosphärisches Stromsystem zu erklären. Der damit verbundene S_q^p-Gang kann, wenn überhaupt, dann höchstens zum Teil als W-Effekt angesehen werden. Eine qualitativ brauchbare Erklärung von S_q^p geht aus von der Annahme einer viscosen Wechselwirkung zwischen solarem Wind und magnetosphärischem Plasma und interpretiert so S_q^p wenigstens indirekt als P-Effekt.

Gleiches gilt für einen Effekt, auf den MEAD[3] hingewiesen hat. Danach müßte ein S_q-ähnlicher Tagesgang an einem Ort infolge seiner Rotation mit der Erde im Magnetfeld der in der Magnetopause fließenden Ströme beobachtet werden, sofern der anströmende solare Wind stationär ist und damit auch die Lage der Magnetopause. Dieser Effekt scheint mit wenigen γ Amplitude klein gegen den normalen S_q-Gang zu sein, macht aber ebenfalls deutlich, daß sich P-Effekte nicht unbedingt nur als K-Variationen äußern müssen.

Die hier vorgenommene Gleichsetzung von P-Strahlung mit solarem Wind wird in Ziff. 24 noch kritisch betrachtet. An dieser Stelle genügt die Beachtung der zuvor gegebenen Definition, wonach alle regelmäßigen Störungen Nicht-K-Variationen sind, also auch ($S_q + S_q^p + L$) und möglicherweise weitere Anteile beim sonnentägigen Gang, den MAYAUD [7] daher als Ganzes mit S_R bezeichnet.

γ) *Eruptions- oder Solar-Flare-Effekte* (s.f.e.'s): Die bei intensiven Flares emittierte Ultraviolett- und Röntgen-Strahlung verstärkt die Ionisation in der Ionosphäre auf der Tagseite bis herab zur D-Schicht, wo sich die ionisierende Wirkung der Röntgen-Strahlung besonders bemerkbar macht. Die Winde in der Ionosphäre bewirken dann während der kurzen Lebenszeit eines Flare den Aufbau eines zusätzlichen Stromsystems, das dem S_q-System sehr ähnlich ist, aber vermutlich etwas tiefer liegt. Sein Magnetfeld erscheint in den Registrierungen als s.f.e. und äußert sich wie eine kurzzeitige, selten eine Stunde überschreitende Verstärkung des S_q-Systems. Die s.f.e.'s gehören zwar zu den unregelmäßigen Störungen; als W-Effekte sind sie aber Nicht-K-Variationen. Trotzdem werden sie in die Bestimmung von K mit einbezogen. Ihre eigentlich notwendige Eliminierung wird nur an den Kp-Observatorien vorgenommen und führt dann für Tages-Achtel mit s.f.e. auf die zusätzliche Kennziffer K'. Dieser Handhabung liegen ausschließlich praktische Erwägungen zugrunde (vgl. Ziff. 5).

---

[2] J. BARTELS: Abhandl. Preuß. Akad. Wiss. Berlin, Math.-Nat. Kl. 1941, Nr. 12; — Terr. Magn. Atmosph. Electr. **51**, 181—242 (1946); vgl. auch [4], S. 751; — FIAT Rev. **17**, Geophys. Teil 1, 39—91 (1948).
[3] G. D. MEAD: J. Geophys. Res. **69**, 1181—1195 (1964).

δ) *Die Nachstörung oder Erholungsphase bei magnetischen Stürmen:* Für den Versuch, den bei magnetischen Stürmen auftretenden Magnetfeldschwankungen physikalische Ursachen zuzuordnen, bieten sich folgende Stromsysteme an[4]: Ströme in der Magnetopause und der neutralen Schicht, der magnetosphärische Ringstrom und ein ionosphärisches Stromsystem mit den polaren Elektrojets als starken Konzentrationen längs der Polarlichtzonen und Rückströmen über die Polkappen und die niederen Breiten. Alle diese Ströme und die dazugehörigen magnetischen Störungen werden von solarer Partikelstrahlung verursacht. Während die beobachteten Feldschwankungen in den ersten drei Phasen, dem Sturmbeginn, der Anfangsphase und der Hauptphase ganz überwiegend von den Grenzflächenströmen und besonders dem ionosphärischen **DS**-System herkommen, sind diese kurzzeitigen Schwankungen in der vierten Phase soweit abgeklungen oder ganz verschwunden, daß die Feldänderung jetzt durch den magnetosphärischen Ringstrom bestimmt wird. Durch Energieabgabe der zwischen den Spiegelpunkten oscillierenden Ringstromteilchen an die Neutralgaskomponente der Erdatmosphäre schwächt sich das Ringstromfeld ab und führt so zu einer allmählichen Erholung der $\mathcal{H}$-Komponente nach ihrer starken Depression in der Hauptphase, wenn der vorwiegend westwärts fließende Ringstrom seine größte Stärke hat. So ist zwar der Ringstrom selbst ein **P**-Effekt, jedoch ist die von ihm in der vierten Phase verursachte weltweite Nachstörung kein Maß für eine zu gleicher Zeit erfolgende Anregung durch **P**-Strahlung. Nach der zuvor gegebenen Definition ist damit die Nachstörung, die einige Tage, aber auch einige Wochen andauern kann, eine Nicht-K-Variation.

ε) Systematische Magnetfeldänderungen von noch längerer Dauer als die Nachstörung bis hin zur Säkularvariation, die ihrer Natur nach selbstverständlich eine Nicht-K-Variation ist, brauchen hier nicht weiter betrachtet zu werden, da ihr Beitrag zu den Feldschwankungen innerhalb eines Drei-Stunden-Intervalls vernachlässigt werden kann. Damit sind neben nicht klassifizierbaren unregelmäßigen Störungen als wichtigste K-Variationen zu nennen: *plötzlicher Sturmbeginn* (**ssc**), *plötzlicher Impuls* (**si**) und die stark von der Ortszeit abhängigen **DS**-Variationen mit der *Bay-Störung* und dem *polaren Teilsturm* als auffallendsten Einzelerscheinungen. Die erdmagnetischen Pulsationen erreichen nur in sehr seltenen Fällen Amplituden, die bei der Bestimmung von K von Einfluß sein könnten.

5. Definition und praktische Bestimmung der dreistündlichen lokalen Kennziffer K.

α) Es wird davon ausgegangen, daß die Magnetfeldschwankungen als Registrierungen der Feldkomponenten X, Y und Z oder $\mathcal{H}$, D und Z vorliegen. In diesen Registrierungen müssen zunächst die Nicht-K-Variationen eliminiert werden, wobei ihrem individuellen Verhalten während des betrachteten Drei-Stunden-Intervalls soweit wie möglich Rechnung getragen werden soll. Die restlichen Feldschwankungen, etwa in den Komponenten $\mathcal{H}$, D und Z von einer beliebigen Basis aus gemessen, seien durch $\Delta \mathcal{H}_K$, ΔD_K und ΔZ_K gegeben und sind damit durch den zeitlichen Verlauf des Störungsvektors $\Delta \mathbf{F}_K$ darstellbar. Von dem festen Ursprung des $\Delta \mathcal{H}_K \Delta D_K \Delta Z_K$-Systems aus beschreibt die Spitze des Vektors $\Delta \mathbf{F}_K$ dabei eine Raumkurve. Betrachtet wird der jeweils während eines Tages-Achtels[1] E_ν nach Weltzeit ($\nu = 1: 00 \ldots 03$; $\nu = 2: 03 \ldots 06$; $\ldots \nu = 8: 21 \ldots 24$ UT) zurückgelegte Teil der Raumkurve. Er wird eingefaßt durch einen Quader, dessen Kanten parallel zu $\Delta \mathcal{H}_K$, ΔD_K und ΔZ_K orientiert sind und dessen Flächen

---

[4] S. Chapman: Studia Geophys. Geodet. **5**, 30—50 (1961).
[1] E als Abkürzung für *Eighth*.

gerade noch jeweils einen Punkt des betrachteten Abschnitts der Raumkurve berühren, wie dies in Fig. 1 illustriert ist. Die längste Kante a dieses Quaders, gemessen in γ, führt auf K nach einer nichtlinearen Klasseneinteilung, die zunächst für jedes Observatorium festgelegt werden muß und dann verbindlich ist.

Fig. 1. Verlauf der durch die Spitze des Feldvektors $\Delta \mathbf{F}_K$ beschriebenen Raumkurve während eines Drei-Stunden-Intervalls in dem nach $\Delta \mathcal{H}_K, \Delta D_K, \Delta Z_K$ orientierten kleinsten Quader, der die Raumkurve gerade einschließt.

Auf Grund seiner Untersuchungen zur Einführung der Potsdamer Kennziffer entschied sich BARTELS [3a] für die folgende *Einteilung*, die für viele weitere Observatorien mit geomagnetischen Breiten um 50° Gültigkeit hat:

$a = 0 \ldots 5 \ldots 10 \ldots 20 \ldots 40 \ldots 70 \ldots 120 \ldots 200 \ldots 330 \ldots 500 \, \gamma \ldots$

$K = 0 1 2 3 4 5 6 7 8 9$

Nur im Hinblick auf eine formale Definition von K sei angemerkt, daß entgegen der Festlegung in [3a] die Grenzen dieser Klasseneinteilung jetzt dem jeweils kleineren K-Wert zugerechnet werden (s. BARTELS [5]). Die Grenzen selbst entstehen bis $40\,\gamma$ durch Verdopplung der jedesmal vorhergehenden Grenze. Die höheren Grenzen wachsen langsamer an, da sonst die Werte 8 und 9 für K fast nie erreicht würden. Daraus folgt, daß sich K nicht exakt als Logarithmus von a darstellen läßt, aber doch ein *quasi-logarithmisches Maß* der erdmagnetischen Aktivität ist.

Ein Observatorium mit der oben angegebenen Einteilung wird im Hinblick auf seine Empfänglichkeit für K-Variationen als *Standard-Station* (S) bezeichnet. Sie ist charakterisiert durch die letzte Grenze $a_9(S) = 500\,\gamma$, bei deren Überschreiten der Störungsgrad $K = 9$ erreicht ist.

Wegen der allgemeinen Zunahme der Aktivität vom Äquator zu den beiden Polarlichtzonen muß für jedes Observatorium zunächst die geeignete Skala gefunden werden, die von der Stärke der dort auftretenden K-Variationen bestimmt wird. Dazu wird von dem Prinzip ausgegangen, die Häufigkeitsverteilungen der Kennziffern an den einzelnen Observatorien für bestimmte, ausreichend lange Test-Intervalle möglichst weitgehend anzugleichen. Dabei soll die zu bestimmende Klasseneinteilung proportional zu der oben angeführten bleiben. Als einziger noch freier Parameter wird die untere Grenze a_9 für $K = 9$ gewählt. Ist sie gegeben, so folgen daraus die anderen Grenzen, indem die obige Skala mit dem Faktor $(a_9/a_9(S))$ multipliziert wird und die Werte sinnvoll abgerundet werden. Auf diese Weise wird letzten Endes jedes Observatorium an die Standard-Stationen angeschlossen. Bemerkungen über eine möglichst praktische Ausführung dieses Verfahrens finden sich ebenfalls in [3] wie auch bei MAYAUD [7].

Um eine Vielzahl unterschiedlicher Skalen zu vermeiden, wird empfohlen, für ein neues Observatorium eine der bereits gebräuchlichen unteren Grenzen a_9 zu verwenden. Es sind dies:

$$a_9 = 300,\ 350,\ 500,\ 600,\ 750,\ 1000,\ 1250,\ 1500,\ 1800,\ 2000 \text{ und } 2500\ \gamma.$$

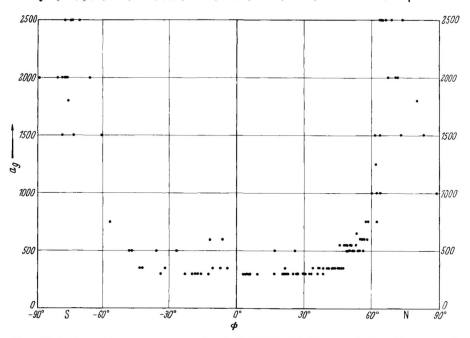

Fig. 2. Die für $K=9$ gültigen unteren Grenzen a_9 in γ der im IAGA Bulletin No, 12 s1 aufgeführten Observatorien in Abhängigkeit von der geomagnetischen Breite.

Die für a_9 auch anzutreffenden Werte 550 und 650 γ sind überflüssig[2]. In Fig. 2 sind die a_9-Grenzen aller im IAGA Bulletin No. 12 s1 aufgeführten Observatorien in Abhängigkeit von der geomagnetischen Breite eingezeichnet. Trotz der Streuung der Punkte tritt deutlich der Anstieg von a_9 zu den beiden Polarlichtzonen hin zutage und darüber hinaus wieder eine Abnahme für Observatorien im Bereich der beiden Polkappen.

Wie die unteren Grenzen für $K=9$ so ändern sich auch die Klassenbreiten von a etwa um den Faktor 8 von den Observatorien in niederen Breiten ($a_9 = 300\ \gamma$) bis zu den Observatorien nahe den Zentren der Polarlichtzonen ($a_9 = 2500\ \gamma$). Diese beiden extremen Einteilungen sollen für einen besseren Vergleich vollständig angegeben werden:

$a = 0 \ldots\ 3 \ldots\ 6 \ldots\ 12 \ldots\ 24 \ldots\ 42 \ldots\ 72 \ldots\ 120 \ldots\ 200 \ldots\ 300\ \gamma \ldots$
$K = 0123456789$
$a = 0 \ldots 25 \ldots 50 \ldots 100 \ldots 200 \ldots 350 \ldots 600 \ldots 1000 \ldots 1650 \ldots 2500\ \gamma \ldots$

β) Da im Durchschnitt die Kennziffern $K = 1$, 2 und 3 am häufigsten vorkommen, ist vor allem in mittleren und niederen Breiten die möglichst voll-

---

[2] Nur die Verwendung eines sehr umfangreichen Ausgangsmaterials würde eine feinere Unterteilung statistisch rechtfertigen.

ständige Eliminierung der Nicht-K-Variationen das eigentliche Problem bei der Bestimmung von K. Die größten Schwierigkeiten macht dabei der regelmäßige sonnentägige Gang $\mathbf{S}_R$ und ebenso der mondentägige Gang $\mathbf{L}$ da, wo er mit genügend großer Amplitude auftritt. Kenntnisse über den Verlauf dieser Variationen müssen aus ihrem Auftreten an ruhigen Tagen gewonnen und auf die gestörteren Tage übertragen werden. Dabei ist allerdings zu beachten, daß neben den systematischen Änderungen mit der Jahreszeit und dem Sonnenfleckencyclus (sowie der Änderung mit dem Mondalter bei $\mathbf{L}$) am Ort einer und derselben Station auch eine deutliche Variabilität in $\mathbf{S}_R$ und $\mathbf{L}$ von Tag zu Tag existiert, wobei diese unregelmäßigen Änderungen ebenso stark sein können wie die systematischen. Eine gewisse Verfälschung der kleinen K-Werte durch unvollständige Elimination von $(\mathbf{S}_R + \mathbf{L})$ ist daher nicht immer zu vermeiden. Es hat sich jedoch als zweckmäßig herausgestellt, in Fällen mit unübersichtlichem Verlauf von $(\mathbf{S}_R + \mathbf{L})$ den wahren Verlauf auf Grund der vorhandenen Erfahrungen und vorliegenden Musterbeispiele für jeden in Frage kommenden Tag individuell abzuschätzen und *nicht* eine aus Beobachtungen über eine lange Zeit hergeleitete „*eiserne Kurve*" dafür zu verwenden. Wegen weiterer Einzelheiten sei auf die äußerst detaillierte Diskussion gerade dieses Problems durch MAYAUD [7] verwiesen.

Bei der Eliminierung der Nachstörung des Ringstroms ist es etwas willkürlich, ihren Beginn festzulegen. Formal wird dafür der Zeitpunkt gewählt, an dem die durch den Ringstrom bewirkte konkave Krümmung im zeitlichen Verlauf der gemittelten $\mathcal{H}$-Komponente während der Hauptphase eines Sturmes in eine konvexe Krümmung gegen das ungestörte Niveau übergeht. Das ungefähre Auftreten dieses Wendepunktes in $\mathcal{H}$ ist auch ein für die Praxis brauchbarer Anhalt. Die durch den Ringstrom dann noch verursachte Feldänderung kann für jedes Drei-Stunden-Intervall linear angesetzt und so als fortschreitender Gang eliminiert werden.

Besondere Schwierigkeiten bereitet die Identifizierung der Solar-Flare-Effekte.

Zusätzlich zu den magnetischen Registrierungen werden im allgemeinen Sonnenbeobachtungen über gleichzeitige starke Eruptionen und Ionosphärenbeobachtungen über gleichzeitig auftretende *Mögel-Dellinger-Effekte* herangezogen werden müssen. Auch bald danach folgende *Polar Blackout's* sind positive Hinweise auf s.f.e.'s. Trotzdem bleibt ihre Identifizierung oft zweifelhaft. Glücklicherweise sind intensive s.f.e.'s, die den Wert von K merklich erhöhen und sich auch noch in der planetarischen Kennziffer Kp auswirken, seltene Ereignisse, die auch bei starker solarer Aktivität nur einige Male im Jahr vorkommen. In der Praxis der K-Bestimmung sind daher die s.f.e.'s in der Vergangenheit recht unterschiedlich behandelt und meistens überhaupt nicht berücksichtigt worden.

Seit Anfang 1968 gilt nun zur Vereinfachung und Vereinheitlichung des Kennzifferndienstes an den Observatorien die Regelung, daß bei der Bestimmung von K mit den s.f.e.'s so zu verfahren ist, als seien sie K-Variationen. Da sie dies aber nach den Ausführungen in Ziff. 4 tatsächlich nicht sind, haben die 12 Kp-Observatorien die Auflage erhalten, für Tages-Achtel mit s.f.e. die zusätzliche Kennziffer K' zu melden, die sich ergibt, wenn der s.f.e. eliminiert wird. Entsprechend werden für die in Frage kommenden Tages-Achtel zwei planetarische Kennziffern Kp und Kp' berechnet.

γ) Der eigentlichen Technik der K-Bestimmung liegt die folgende Vorstellung zugrunde: Für die im Magnetogramm vorliegende Feldkomponente mit den stärksten Schwankungen wird der relative Verlauf der Nicht-K-Variationen während des betrachteten Drei-Stunden-Intervalls oder besser für den ganzen Tag ermittelt und als glatte Kurve dargestellt. Das ist in den meisten Fällen allein der Verlauf des sonnentägigen Ganges. Da K durch den größten Schwankungsbereich festgelegt wird, der in dieser Komponente von den unregelmäßigen Störungen herkommt, brauchen auch die Nicht-K-Variationen nicht absolut bekannt zu sein. Die Kurve der Nicht-K-Variationen denkt man sich nun zweimal so in das Magnetogramm eingezeichnet, daß sie in dem betrachteten Drei-

Stunden-Intervall die registrierte Kurve gerade einmal von unten und einmal von oben berührt. Der vertikale Abstand dieser beiden parallelen Kurven, mit dem Skalenwert der betreffenden Komponente umgerechnet in γ, ergibt den maximalen Schwankungsbereich a, für den aus der Klasseneinteilung von a des betreffenden Observatoriums sofort K folgt. Fig. 3 veranschaulicht dieses Verfahren der Bestimmung von a und macht auch deutlich, daß die beiden tangierenden Kurven an den Grenzen der Tages-Achtel im allgemeinen nicht stetig sein können.

Fig. 3. Vier Tagesachtel mit zur Nicht-K-Variation parallelen Kurven, die in jedem Tagesachtel die Extrema der registrierten Komponente gerade berühren und deren Abstand den maximalen Schwankungsbereich a der K-Variationen im Drei-Stunden-Intervall angibt.

Bei einiger Erfahrung wird es überflüssig, die beiden tangierenden Kurven tatsächlich einzuzeichnen und a zu messen. Es genügt oft, mit einem bereits in K-Bereiche eingeteilten Maßstab dem aus der Registrierung abzulesenden Verlauf der Nicht-K-Variation nachzufahren und dabei den Abstand von absolutem Maximum und absolutem Minimum im Drei-Stunden-Intervall sofort in K zu ermitteln. Ist dabei die erzielte Genauigkeit nicht ausreichend, so trägt man den Verlauf der Nicht-K-Variationen nur einmal innerhalb des Schwankungsbereiches der registrierten Kurve ein und mißt den vertikalen Abstand der beiden Extrema über diese Basis. Weitere Einzelheiten zu diesen technischen Fragen finden sich ebenfalls bei MAYAUD [7].

Es ist im allgemeinen nur nötig, K für eine der drei Komponenten zu bestimmen, da die gestörteste Komponente oft schon ohne Messung mit Sicherheit erkannt werden kann. Jedoch kann das von Tages-Achtel zu Tages-Achtel eine andere Komponente sein. Wegen der elektromagnetischen Induktion der zeitlich variablen Störungsfelder im leitfähigen Erdinnern werden die horizontalen Komponenten der von außen einfallenden Magnetfelder an der Erdoberfläche verstärkt, während die Vertikalkomponente zum Teil kompensiert wird. Ausgenommen die Nachbarschaft der polaren Elektrojets, ist es daher unwahrscheinlich, daß Z als gestörteste Komponente auftritt; und wo dies doch geschieht, deutet diese Erscheinung mit ziemlicher Sicherheit auf das Vorhandensein einer *lokalen Leitfähigkeitsanomalie* im Untergrund hin und nicht etwa auf einen lokalen P-Effekt. Daher ist seit Anfang 1964 die Regelung in Kraft, die Z-Komponente bei der Bestimmung von K generell außer acht zu lassen. Ausgenommen sind jedoch wieder die 12 Kp-Observatorien, um die Homogenität der für die Kp-Berechnung benutzten Ausgangsdaten zu bewahren.

Da sich Leitfähigkeitsanomalien aber auch auf die horizontalen Komponenten auswirken, wenn auch weniger stark als auf Z, kann es vorkommen, daß die Stärke der an einem Observatorium beobachteten magnetischen Störungen von der Richtung der induzierenden Magnetfelder abhängt, ohne daß diese Abhängigkeit in der Natur der einfallenden Störungen liegt. Es ist noch nicht untersucht worden, inwieweit ein solcher Effekt die K-Bestimmung verfälschen kann.

II. Definition und praktische Bedeutung der aus K abgeleiteten lokalen und planetarischen Maßzahlen.

6. Die lokalen Aktivitätsmaße ak, Ak und Ck.

α) Die quasi-logarithmische Klasseneinteilung der Kennziffer K bedeutet eine relativ stärkere Bewichtung der schwachen und mittleren erdmagnetischen Aktivität gegenüber den starken Störungen, wie sie vor allem in der Hauptphase der Stürme auftreten. Die Rechtfertigung dafür liegt in der sehr viel größeren Häufigkeit der kleinen und mittelstarken Störungen, deren zeitlicher Anteil an der Aktivität in einem logarithmischen Aktivitätsmaß besser berücksichtigt wird als in einem linearen Maß. Hinzu kommt, daß die quasi-logarithmische Einteilung von K den Vorteil hat, den ganzen Umfang erdmagnetischer Aktivität bei ausreichendem Auflösungsvermögen mit den zehn einziffrigen Zahlen 0 ... 9 erfassen zu können. Allerdings hat diese Einteilung zur Folge, daß das arithmetische Mittel von K-Werten, also das arithmetische Mittel von Logarithmen der Meßwerte a, gleich dem Logarithmus des *geometrischen Mittels* der Meßwerte ist und nicht etwa gleich dem Logarithmus ihres arithmetischen Mittels. Demzufolge ist die Tagessumme oder das arithmetische Tagesmittel von K (wie übrigens auch von Kp) ein in seinen Aussagen schwer durchschaubares Aktivitätsmaß und daher leicht irreführend.

β) Um mit den lokalen Kennziffern K eines und desselben Observatoriums neben der Aufstellung von Häufigkeitsverteilungen sinnvoll weiterarbeiten zu können, empfiehlt sich daher die Einführung eines *linearen, zu K äquivalenten* lokalen Aktivitätsmaßes. BARTELS[1] hat dafür die Maßzahl[2] ak vorgeschlagen, die er der dreistündlichen Kennziffer K wie folgt zugeordnet hat:

$K =$	0	1	2	3	4	5	6	7	8	9
$ak =$	0	3	7	15	27	48	80	140	240	400

Im Mittelbereich dieser Skala etwa von $K = 2$ bis $K = 6$ ergeben sich die ak so, daß in der Klasseneinteilung von K für eine Standard-Station dem betreffenden K-Wert der *halbe Mittelwert* des zugehörigen a-Bereichs zugeordnet wird; also für $K =$ n ist

$$ak \cong (a_{n+1} + a_n)/4.$$

Für die kleinen K-Werte ist ak kleiner als dieser halbe Mittelwert, für die großen K-Werte größer.

Die Bezeichnung von ak als *äquivalente Amplitude*, wobei Amplitude im Sinne von maximaler Feldschwankung oder Störungsbereich im Drei-Stunden-Intervall gemeint ist, sowie Angaben zur Umrechnung von ak in äquivalente Amplituden (in γ) haben zu Mißverständnissen darüber Anlaß gegeben, was unter ak zu verstehen ist. Tatsächlich hat ak zwei verschiedene Bedeutungen, die voneinander getrennt werden müssen.

ak als Kennziffer: In dieser Eigenschaft soll hier ak als $a\hat{k}$ geschrieben werden. *Die Kennziffer $a\hat{k}$ ist das lineare Analogon zur quasi-logarithmischen Kennziffer K.* Die zuvor gegebene Umrechnung von K in ak gilt *generell* für jedes Observatorium unabhängig davon, welche untere Grenze a_9 für $K = 9$ dort vorliegt. Für ein vorgegebenes K hat $a\hat{k}$ somit stets den Zahlenwert von ak in der vorstehenden Tabelle. Da bei der Definition von K das Anwachsen der erdmagnetischen

---
[1] J. BARTELS: IATME Bull. No. 12e, 132 (1951); ebenso in den nachfolgenden IAGA Bulletins No. 12; s. auch [6].
[2] Ursprünglich nur k genannt.

Aktivität zu den Polarlichtzonen hin im Sinne einer Angleichung von Häufigkeitsverteilungen berücksichtigt ist, gilt das gleiche auch für $a\hat{k}$. Das heißt, statistisch gesicherte Unterschiede in K oder $a\hat{k}$ an zwei Observatorien sollten nicht mehr oder nur noch in geringem Ausmaß von der Breitenabhängigkeit der erdmagnetischen Aktivität herkommen. Im übrigen ist das Auftreten von Unterschieden in diesen Kennziffern an verschiedenen Observatorien selbstverständlich, da K und $a\hat{k}$ ihrer Bestimmung nach lokale Kennziffern sind.

Es ist auch möglich, $a\hat{k}$ mit einer Einheit zu versehen, die dann aber *nicht* 1 γ wäre, sondern $(2\,a_9/a_9(S))\,\gamma$, also speziell für eine Standard-Station 2 γ, und somit von der unteren Grenze a_9 des Observatoriums abhängt. Jedoch lassen sich Verwechslungen leichter vermeiden, wenn $a\hat{k}$ auch in diesem Punkt in Analogie zu K als eine dimensionslose Maßzahl angesehen wird.

ak *als äquivalente Amplitude*[3]: In dieser Eigenschaft soll ak hier als $\hat{a}k$ geschrieben werden. Es wurde bereits zuvor gesagt, was dabei mit Amplitude gemeint ist. *Danach bedeutet $\hat{a}k$ den an einem Observatorium beim Störungsgrad K durchschnittlich auftretenden Störungsbereich der gestörtesten Komponente in γ.* Daher muß in $\hat{a}k$ die Breitenabhängigkeit der erdmagnetischen Aktivität voll berücksichtigt werden. Die Zahlenwerte von $\hat{a}k$ sind dann für ein vorgegebenes K verschieden von denen von ak in der Tabelle und ebenfalls verschieden für Observatorien mit unterschiedlichem a_9. Aus den Definitionen von K und $ak = a\hat{k}$ folgt aber sofort die Vorschrift zur Berechnung von $\hat{a}k$: Da ak beim Störungsgrad K etwa gleich dem halben Störungsbereich in a an einer Standard-Station ist, gilt für ein solches Observatorium mit $a_9(S) = 500\,\gamma$: $\hat{a}k = 2\,a\hat{k}$ in γ. Die Umrechnung auf ein Observatorium mit einer anderen unteren Grenze a_9 erfolgt dann wegen der angesetzten Proportionalität der Skalen durch Multiplikation mit dem Faktor $(a_9/a_9(S))$. Ist also K gegeben, so liefert die Umrechnungstabelle zunächst die lineare Kennziffer $a\hat{k}$, mit der sich dann für ein durch a_9 charakterisiertes Observatorium die dazugehörige äquivalente Amplitude berechnen läßt gemäß

$$\hat{a}k = 2\,a\hat{k}\,a_9/a_9(S)\,\gamma.$$

Da $a_9(S)/2 = 250$, heißt das, man erhält $\hat{a}k$ in γ, indem man $a\hat{k}$ mit $a_9/250$ multipliziert.

Beispiel: An den drei Observatorien Sodankylä SO ($a_9 = 1500\,\gamma$), Fredericksburg FR ($a_9 = 500\,\gamma$) und Kakioka KA ($a_9 = 300\,\gamma$) möge das gleiche Drei-Stunden-Intervall mit $K = 3$ beurteilt werden. An allen drei Observatorien ist dann die lineare lokale Kennziffer $a\hat{k} = 15$. Dagegen ist die äquivalente Amplitude für das der nördlichen Polarlichtzone nahe SO: $\hat{a}k = 90\,\gamma$, für die Standard-Station FR: $\hat{a}k = 30\,\gamma$ und für das in 26° geomagnetischer Breite gelegene KA: $\hat{a}k = 18\,\gamma$.

γ) *Das arithmetische Mittel der acht Werte von ak am Greenwich-Tag wird mit Ak bezeichnet.* Auch bei Ak müssen die beiden Bedeutungen: *tägliche lokale Kennziffer $A\hat{k}$* und *tägliche lokale Amplitude $\hat{A}k$* in γ unterschieden werden. Was über diese Unterscheidung bei ak gesagt wurde, kann sofort auf Ak übertragen werden und macht weitere Erläuterungen dazu überflüssig.

Da ak und Ak lineare Maße der Aktivität sind, kann man aus ihnen *arithmetische Mittel für beliebige Intervalle* berechnen und erhält so Aussagen über die mittlere lokale Aktivität. Gleiche Aussagen folgen aus den K-Werten nur durch mühsames Auszählen von Häufigkeiten. So liegt der Wert der Maßzahlen ak und

---

[3] MAYAUD [7], S. 34, gebraucht als lineare lokale Maßzahl *nur* die äquivalente Amplitude und gibt dementsprechend die Gewichte für eine direkte Umrechnung von K in Amplituden, wobei er außerdem noch von einer Skala mit $a_9 = 300\,\gamma$ ausgeht und sich genau an die Klassenmittelwerte von a hält.

Ak im wesentlichen in ihrer einfachen Benutzung und Interpretation bei statistischen Untersuchungen.

Im übrigen macht auch erst eine große Zahl von Ausgangswerten die Rücktransformation von K in ein Amplitudenmaß zuverlässig, da die Festlegung der Skala durch die Klassenmittelwerte von a reine Konvention ist und besonders bei den breiten, höheren Klassen der Unterschied zwischen einer tatsächlichen Amplitude und dem Mittelwert der Klasse, in die diese Amplitude fällt, beträchtlich sein kann.

δ) In Analogie zu der in Ziff. 10 ausführlich behandelten täglichen planetarischen Charakterzahl Cp läßt sich eine *tägliche lokale Charakterzahl* Ck definieren, die durch die *Tagessumme der* ak in der gleichen Weise bestimmt ist, in der Cp aus der Tagessumme der ap folgt. Das heißt, in der Umrechnungstabelle in Ziff. 10 (Tabelle 4) von der Tagessumme der ap auf Cp braucht nur ap durch ak und Cp durch Ck ersetzt zu werden. Danach kann Ck als ein lokaler Schätzwert für Cp angesehen werden. Allerdings wäre für diesen Zweck die Berechnung einer täglichen *standardisierten Charakterzahl* Cs günstiger, die sich auf die gleiche Weise aus den standardisierten Kennziffern Ks (s. Ziff. 7) über eine lineare Maßzahl as herleiten läßt, sofern die Umrechnung von K in Ks für das betreffende Observatorium bekannt ist.

Die Bedeutung von Ck liegt aber eher in zwei anderen Eigenschaften: 1. Ck ist das auf der Grundlage von K berechnete Gegenstück zur subjektiv geschätzten täglichen Charakterzahl C (s. Ziff. 14) und 2. Ck ist ein quasi-logarithmisches Maß und ergänzt so als tägliche lokale Maßzahl die dreistündliche lokale Kennziffer K mit der gleichen Eigenschaft. Es ist aber noch einmal darauf hinzuweisen, daß Ck zu diesem Zweck über das lineare ak aus K berechnet werden muß.

7. Die dreistündliche standardisierte Kennziffer Ks.

α) Ein planetarisches Aktivitätsmaß basiert naturgemäß auf lokalen Beobachtungsergebnissen und muß daher nach einer Vorschrift berechnet werden können, die es gestattet, den planetarischen Anteil vom lokalen Anteil zu trennen. Auf dem Weg zur planetarischen Kennziffer Kp ist der Schritt von der lokalen Kennziffer K zur standardisierten Kennziffer Ks die Abtrennung des lokalen Anteils in der erdmagnetischen Aktivität.

Fig. 4. Mittlerer Tagesgang der erdmagnetischen Aktivität am Observatorium Wingst in den Jahren 1963/64, ausgedrückt durch die lineare lokale Kennziffer ak.

Nachdem die systematische Zunahme der Aktivität mit Annäherung an die Polarlichtzonen bereits bei der Definition von K berücksichtigt worden ist und damit die K-Werte verschiedener Observatorien in dieser Hinsicht vergleichbar sind, muß noch ein an den meisten Stationen ausgeprägter *Tagesgang der Aktivität nach Ortszeit* als lokaler Anteil eliminiert werden. Eine ebenfalls im Tagesgang vorhandene kleine Komponente nach Weltzeit geht bei dem Standardisierungsverfahren zugunsten einer besseren Homogenität der Ks mit verloren. Gleichzeitig werden dabei aber auch die Fehler infolge einer unvollständigen Eliminierung der Nicht-K-Variationen reduziert. Das Beispiel eines mittleren Tagesganges der Aktivität zeigt Fig. 4. Aufgetragen sind die Mittelwerte der ak am Observatorium Wingst für jedes Tages-Achtel E_ν nach den K-Werten aus den beiden Jahren mäßiger bis schwacher Aktivität 1963/64.

Mit einer großen Zahl gleichmäßig über die Erde verteilter Observatorien würde es wenig Schwierigkeiten machen, eine einwandfreie planetarische Maßzahl zu bestimmen. Die tatsächliche Verteilung der Observatorien, besonders zur Zeit der Einführung von Kp, erwies sich als so unzulänglich, daß Verfahren auf der Grundlage einfacher Mittelwertbildungen wieder aufgegeben wurden. Hinzu kommt, daß der Tagesgang der Aktivität noch von der Jahreszeit und besonders von der geomagnetischen Breite abhängt, derart, daß das Maximum in niedrigen und mittleren Breiten in den Abendstunden bis kurz vor Mitternacht eintritt, während es sich in höheren Breiten auf die Zeit nach Mitternacht verschiebt und bei Annäherung an den Pol sogar bis in den Vormittag hinein. Aus alledem folgt, daß die Abtrennung des lokalen Ganges für jedes Observatorium individuell vorgenommen werden muß. Das Verfahren, dabei zu einer weltweiten Standardisierung der Kennziffern zu kommen, besteht im Prinzip wieder in der Angleichung von Häufigkeitsverteilungen. Bei der praktischen Ausführung kommt noch hinzu, daß sich dieses Ziel mit dem vorhandenen Ausgangsmaterial und einem vertretbaren Rechenaufwand erreichen lassen muß.

β) Ausgehend von den Gegebenheiten um 1948, schlug BARTELS folgenden Weg ein: Die Standardisierung wurde für elf ausgewählte Observatorien vorgenommen. Diese ursprünglichen Kp-Observatorien waren (nach abnehmender geomagnetischer Breite geordnet):

Lerwick, Meanook, Sitka, Eskdalemuir, Rude Skov, Agincourt, Wingst, Witteveen, Abinger, Cheltenham und Amberley.

Für diese Stationen standen die dreistündlichen K-Werte aus den Jahren 1943 bis 1947 und aus der zweiten Hälfte von 1948 zur Verfügung. Diese Standardisierungsjahre enthalten das Sonnenfleckenminimum von 1944 und das Sonnenfleckenmaximum von 1947. Die mittlere erdmagnetische Aktivität der sechs Jahre, ausgedrückt in der linearen planetarischen Maßzahl Ap (s. Ziff. 9), ist $Ap = 15,1$ und stimmt mit dem langjährigen Mittel von Ap überein, das um den Wert 15 schwankt.

Zur Berücksichtigung der jahreszeitlichen Schwankung des Tagesganges der Aktivität wurde die Angleichung der Häufigkeiten gesondert für die Monate Januar, Februar, November, Dezember (**JFND**) um das Dezembersolstitium, für die Monate Mai, Juni, Juli, August (**MJJA**) um das Junisolstitium und für die Monate März, April, September, Oktober (**MASO**) um die Äquinoktien durchgeführt. Dazu wurden aus den Standardisierungsjahren möglichst viele geeignete Tage ausgewählt, z.B. für die Monate **JFND** 424 Tage. Für jedes der elf Observatorien und jedes der Tages-Achtel E_v wurde für jede Monatsgruppe die Häufigkeit der K-Werte ausgezählt. Theoretisch sollten alle 88 Häufigkeitsverteilungen einer Monatsgruppe zur Standardisierung herangezogen werden. Tatsächlich wurden jedoch nur 22 benutzt, nämlich die der *beiden Tages-Achtel um Orts-Mitternacht* an jedem Observatorium. Sie wurden zu einer einzigen Häufigkeitsverteilung vereinigt und daraus die Häufigkeitssummen jeweils bis $K = Ks = 0$, 1, ... 9 berechnet, bezogen auf die Anzahl der verwendeten Tage. Dieses Ergebnis diente nach Einführung von Drittel-Stufen (s. Ziff. 7γ) als Basis, um durch Vergleich mit der entsprechenden individuellen Häufigkeitssumme der K für ein beliebiges Tages-Achtel an einem Observatorium festzustellen, durch welchen Wert, d.h. durch welche standardisierte Kennziffer Ks ein lokales K in diesem Drei-Stunden-Intervall ersetzt werden muß, damit Übereinstimmung mit dem mittleren weltweiten Störungsgrad zu dieser Tageszeit erzielt wird. Weitere Einzelheiten dieses Verfahrens beschreibt BARTELS[1].

---

[1] J. BARTELS: IATME Bull. No. 12b, 97—112 (1949); nachgedruckt in: IAGA Bull. No. 12i, 88—101 (1955).

Für die Monatsgruppe **MJJA** wurde dieses Verfahren in einem Punkt geringfügig abgeändert: Da die Sommermonate einer Halbkugel auf dieser Halbkugel im Durchschnitt gestörter sind als die Wintermonate und zehn der für die Standardisierung verwendeten elf Observatorien auf der Nordhalbkugel liegen, würde die Differenz der Aktivität zwischen Nordsommer und Nordwinter fast vollständig als scheinbar planetarische Erscheinung in die Ks und damit in die Kp eingehen. Aus einer speziellen Betrachtung dieses Effektes während der beiden Tages-Achtel um Ortsmitternacht erwies es sich als ausreichend für seine Eliminierung, in der Standard-Häufigkeitsverteilung der Gruppe **MJJA** die Anzahl der Intervalle mit $K = 0$ um 15% der tatsächlich beobachteten zu erhöhen und diese so hinzugefügte Anzahl von Intervallen zu gleichen Teilen von der Zahl der Intervalle mit $K = 1$ und $K = 2$ zu subtrahieren.

γ) Der Umfang des für die Standardisierung benutzten Ausgangsmaterials, z. B. 9 328 K-Werte für die Monate **JFND**, erlaubte es, für Ks eine feinere Klasseneinteilung zu wählen, als es die zehn Klassen von K sind. Als geeignet erwies sich die Einführung von *Drittel-Stufen*. Zu deren Festlegung wird Ks als kontinuierliche Variable von 0,0 bis 9,0 aufgefaßt und dann z. B. das Intervall von 1,5 bis 2,5 in drei gleiche Teile geteilt, entsprechend $1^2/_3$, 2,0 und $2^1/_3$, die mit 2 −, 2o und 2 + bezeichnet werden. Insgesamt ergibt das eine Einteilung in die 28 Klassen:

$$0o,\ 0+,\ 1-,\ 1o,\ 1+,\ 2-,\ 2o,\ 2+,\ \ldots\ 7+,\ 8-,\ 8o,\ 8+,\ 9-,\ 9o.$$

Lediglich die erste und die letzte Klasse 0o bzw. 9o umfassen nicht ein Drittel, sondern nur ein Sechstel der Klassenbreite von K.

Das praktische Rechnen mit diesen Drittel-Stufen läßt sich dadurch erleichtern, daß die *ganzzahligen Werte* $3\,Ks$ benutzt werden und erst zum Schluß auf Ks übergegangen wird. Dementsprechend wird auch in den *Umrechnungstabellen* jeder Stufe von K ein Wert von $3\,Ks$ zugeordnet, gesondert berechnet für jedes der Observatorien, für jedes Tagesachtel und jede der drei Jahreszeiten. Als Beispiel soll hier die vollständige Umrechnungstabelle von K nach $3\,Ks$ für Wingst angegeben werden.

Will man diese Tabellen zum Tagesgang der Aktivität umdeuten, so ist zu beachten, daß sie ja gerade der Eliminierung dieses Tagesganges dienen und somit bei gleichem K höhere Werte von Ks bedeuten, daß es in diesem Drei-Stunden-Intervall an der betreffenden Station ruhiger ist, als es dem weltweiten Störungsgrad entspricht.

Tabelle 1. *Umrechnungstabelle zur Ermittlung der mit 3 multiplizierten standardisierten Kennziffern Ks aus den lokalen Kennziffern K des Observatoriums Wingst.*

		Jahreszeit		
		JFND	MASO	MJJA
		Tageszeit		
		$E_1\ E_2\ E_3\ E_4\ E_5\ E_6\ E_7\ E_8$	$E_1\ E_2\ E_3\ E_4\ E_5\ E_6\ E_7\ E_8$	$E_1\ E_2\ E_3\ E_4\ E_5\ E_6\ E_7\ E_8$
$K=0$:	$3\,Ks=$	0 0 0 0 0 0 0 0	0 0 0 0 0 0 0 0	0 0 0 0 0 0 0 0
$K=1$:	$3\,Ks=$	3 4 4 3 3 3 2 2	3 4 3 2 2 2 2 2	2 2 3 2 2 1 2 2
$K=2$:	$3\,Ks=$	6 8 8 7 7 6 5 5	7 8 6 6 6 6 5 5	6 6 7 6 5 4 5 6
$K=3$:	$3\,Ks=$	9 11 12 11 10 9 8 8	10 11 11 11 10 9 8 8	9 9 10 10 8 8 9 9
$K=4$:	$3\,Ks=$	13 15 16 15 13 11 10 11	13 15 16 16 13 12 11 12	13 13 14 14 12 12 13 13
$K=5$:	$3\,Ks=$	17 19 20 19 17 13 13 14	16 19 20 20 18 16 15 15	16 17 18 18 17 16 17 17
$K=6$:	$3\,Ks=$	21 22 23 22 20 16 17 18	20 22 23 24 21 19 19 19	21 21 21 22 22 21 21 21
$K=7$:	$3\,Ks=$	23 24 25 25 23 20 21 21	24 25 25 25 24 22 22 22	24 25 25 25 25 25 24 25
$K=8$:	$3\,Ks=$	25 26 26 26 25 23 24 24	26 26 26 26 25 25 25 25	25 26 26 26 26 26 26 26
$K=9$:	$3\,Ks=$	27 27 27 27 27 27 27 27	27 27 27 27 27 27 27 27	27 27 27 27 27 27 27 27

8. Die dreistündliche planetarische Kennziffer Kp.

α) Das in Ziff. 7 beschriebene Standardisierungsverfahren ist weniger gedacht als Grundlage für eine selbständige lokale Kennziffer zur genaueren Abschätzung des weltweiten Störungsgrades durch den lokalen Beobachter, sondern vor allem als Grundlage für die Berechnung einer physikalisch und statistisch fundierten und für den größten Teil der Erdoberfläche repräsentativen Maßzahl der erdmagnetischen Aktivität, der *dreistündlichen planetarischen Kennziffer Kp*.

Wäre das Standardisierungsverfahren vollkommen, so sollten die Ks-Werte der Observatorien für ein und dasselbe Tages-Achtel übereinstimmen, womit zugleich Kp gegeben wäre. In der Praxis ist eine so weitgehende Übereinstimmung nicht zu erwarten, jedoch ist die Streuung der Ks-Werte im Vergleich zu der Streuung der K-Werte derselben Observatorien für dasselbe Tages-Achtel so viel kleiner, daß jetzt Kp als arithmetisches Mittel aus den Ks berechnet werden kann. Die Rechnung selbst läuft über die ganzzahligen Größen $3 Ks$. Durch Abrundung auf *ganzzahlige* $3 Kp$ läßt sich dann auch für die Kp die gleiche *Drittel-Stufen-Skala von* 0o *bis* 9o verwenden wie für die Ks.

β) Werden die standardisierten Kennziffern für die Kp-Observatorien Monat für Monat auf die angegebene Weise bestimmt, so müssen nach Ziff. 7 von Februar auf März, von April auf Mai, von August auf September und von Oktober auf November die Umrechnungstabellen gewechselt werden, um der Änderung des lokalen Tagesganges der Aktivität mit der Jahreszeit bei der Eliminierung des Tagesganges Rechnung zu tragen. Der unmittelbare Übergang von einer zu anderen Umrechnungstabelle würde aber bei der Kp-Berechnung eine willkürliche Unstetigkeit zwischen den angeführten Monaten mit sich bringen. Um diese zu erwartenden Sprünge zu glätten, hat BARTELS zusätzliche *Übergangstabellen* eingeführt.

Eine Übergangstabelle besteht lediglich aus den Differenzen der sich entsprechenden Werte von $3 Ks$ in zwei aufeinanderfolgenden Umrechnungstabellen. Das für die Glättung zu berücksichtigende Übergangsintervall umfaßt 30 Tage, die um den Zeitpunkt 0 h des ersten Tages des neuen Monats zentriert sind, also beispielsweise die Tage vom 14. Februar bis zum 15. März. Zusätzlich zu dem normalen Vorgehen bei der Berechnung der Kp werden für diese Tage auch die in den Übergangstabellen enthaltenen Differenzen je nach Übergang, Observatorium und Tagesachtel berücksichtigt und mit Gewichten für eine stufenweise Glättung versehen. Dabei wird so vorgegangen, daß für die 10 Tage in der Mitte, also etwa die 10 Tage um den 1. März, die Hälfte der auf ein Tagesachtel entfallenden und aufsummierten Differenzen zur Summe der $3 Ks$ der Observatorien mit negativem Vorzeichen für die Februartage und positivem Vorzeichen für die Märztage hinzugefügt wird. Für die ersten 10 Tage und für die letzten 10 Tage des Übergangsintervalls wird nur jeweils ein Viertel dieser aufsummierten Differenzen als Korrektur an $3 Ks$ angebracht und dann Kp berechnet. Die dadurch erzielten Verbesserungen überschreiten selten eine Drittel-Einheit in Kp.

γ) Von den ursprünglichen elf Kp-Observatorien, deren K-Werte auch zur Standardisierung benutzt worden sind, mußten inzwischen Abinger, Cheltenham und Agincourt durch die Ausweichstationen Hartland, Fredericksburg bzw. Ottawa ersetzt werden. Ferner ist Lovö seit 1954 als zwölftes Kp-Observatorium hinzugekommen, wobei die Regelung gilt, daß die 3-Ks-Summen von Lovö und Rude Skov zusammengefaßt werden und ihr arithmetisches Mittel mit einfachem Gewicht in die Kp-Berechnung eingeht[1]. Mit anderen Worten: zehn Observatorien haben je das Gewicht 1 und Lovö und Rude Skov je das Gewicht $1/_2$. Die Namen und Daten der derzeitigen Kp-Observatorien (nach abnehmender geomagnetischer Breite geordnet) gibt die folgende Tabelle[2].

---

[1] Eine gleichartige Regelung ist von 1972 an für die beiden Observatorien Toolangi, Australien, und Amberley vorgesehen mit Toolangi TO als 13. Kp-Observatorium.
[2] Daten nach G. FANSELAU: IAGA Bull. No. 20 (1965).

Tabelle 2. *Namen und Daten der derzeitigen Kp-Observatorien.*

Name	Symbol	Lage	gg. Br.	gg. L.	gm. Br.	gm. L.	a_9
Lerwick	LE	Shetlandinseln	60° 08'	358° 49'	62,5°	88,6°	1000 γ
Meanook	ME	Kanada	54 37	246 40	61,8	301,0	1500
Sitka	SI	Alaska	57 03	224 40	60,0	275,4	1000
Eskdalemuir	ES	Schottland	55 19	356 48	58,5	82,9	750
Lovö	LO	Schweden	59 21	17 50	58,1	105,8	600
Ottawa	OT	Kanada	45 24	284 27	57,0	351,5	750
Rude Skov	RS	Dänemark	55 51	12 27	55,9	98,6	600
Hartland	HA	England	50 58	355 31	54,6	79,0	500
Wingst	WN	Deutschland	53 45	9 04	54,5	94,0	500
Witteveen	WI	Niederlande	52 49	6 40	54,2	91,0	500
Fredericksburg	FR	Virginia/USA	38 12	282 38	49,6	349,9	500
Amberley	AM	Neuseeland	−43 09	172 43	−47,7	252,5	500

gg. Br. = geographische Breite, positiv nach Norden; gg. L. = östliche geographische Länge; gm. Br. = geomagnetische Breite, positiv nach Norden; gm. L. = östliche geomagnetische Länge; a_9 = untere Grenze des Störungsbereiches für $K = 9$ in der gestörtesten Komponente (s. Ziff. 5).

δ) Aus der Art der Berechnung von Kp folgt, daß sich die wichtigsten Eigenschaften der lokalen Kennziffern K auf Kp übertragen. Dazu gehört vor allem die *quasi-logarithmische* Darstellung der Aktivität, durch die das Ausmaß der starken Störungen reduziert wird. Obwohl also in dieser nichtlinearen Skala die schwachen Störungen bereits viel feiner unterschieden werden als die starken, ergibt sich trotzdem noch eine Häufigkeitsverteilung der Kp-Werte, die — wie auch die der lokalen K-Werte — ihr Maximum zwischen 1 und 2 hat. Fig. 5 zeigt die aus allen Kp-Werten der 36 Jahre 1932 bis 1967 berechnete mittlere Häufigkeitsverteilung, normiert auf Häufigkeiten pro Jahr. Danach liegt für diese Zeit das Maximum bei $Kp = 1+$, und die Hälfte aller Werte fällt in die Klassen 1− bis 2+.

Fig. 5 macht auch deutlich, daß die Verteilung der Kp *keine Gauß-Verteilung* ist und eine solche Verteilung auch gar nicht angestrebt werden sollte; denn dazu wäre es erforderlich, die Zahl der Klassen für die schwachen Störungen durch Verkleinerung der Klassenbreite weiter zu erhöhen, also in mittleren und niederen geomagnetischen Breiten etwa mit 1 γ-Klassen zu beginnen. Die Unsicherheit, mit der bereits jetzt die Kennziffern 0 und 1 bestimmt werden, macht aber jeden Versuch illusorisch, die Aktivität eines Drei-Stunden-Intervalls im Bereich der schwachen Störungen noch genauer zu unterscheiden.

ε) Der Umstand, daß sich die Kennziffern Kp und K im wesentlichen nur durch ihren Anwendungsbereich unterscheiden, nämlich planetarisch oder lokal, bringt auch für Kp die gleichen Probleme mit sich, die am Beginn von Ziff. 6 für die Beziehung zwischen K und $\sum K$

Fig. 5. Häufigkeitsverteilung aller von 1932 bis 1967 berechneten planetarischen Kennziffern Kp, für jede Klasse normiert auf die Anzahl der Kennziffern im Jahr.

erörtert wurden. So ist auch das häufig anzutreffende Aktivitätsmaß $\sum Kp$ *nicht proportional dem arithmetischen Mittel der Aktivität* in der betrachteten Zeitspanne. Vielmehr entspricht das arithmetische Mittel der Kp ungefähr dem Logarithmus

Fig. 6. Graphische Darstellung des Kp-Jahrgangs 1967 in der BARTELSschen Notenschrift.

des geometrischen Mittels der Aktivität. Auf diesen Sachverhalt sei noch einmal ausdrücklich hingewiesen[3].

---

[3] Ein Grund für die Benutzung von $\sum Kp$ als selbständiges Aktivitätsmaß könnte in der Veröffentlichung der Tagessummen von Kp in den Göttinger monatlichen Tabellen und den IAGA Bulletins No. 12,1 liegen. Der ursprüngliche Sinn dieser Angaben war aber lediglich, eine Kontrollmöglichkeit zur Überprüfung der veröffentlichten Daten auf Druckfehler mitzugeben.

ζ) Zur anschaulichen Wiedergabe der planetarischen Kennziffern hat BARTELS eine *Notenschrift* eingeführt, die sowohl ein quantitatives Ablesen der Kp-Werte erlaubt als auch eine schnelle qualitative Orientierung dadurch, daß der Störungsgrad durch die Länge und Schwärzung der Noten angezeigt wird. Hinzu kommt die sehr nützliche Anordnung nach Sonnenrotationen, die die Wiederholung von gestörten Zeiten mit diesem Rhythmus hervortreten läßt (s. Ziff. 22). Als Beispiel zeigt Fig. 6 das Jahresdiagramm der Kp für 1967, das auch den Schlüssel für die Notenschrift enthält.

9. Die linearen planetarischen Aktivitätsmaße ap und Ap.

α) Wie für K so empfiehlt sich auch für Kp die Einführung eines äquivalenten *linearen* Aktivitätsmaßes, das dann ebenfalls planetarischen Charakter hat. Das ist die dreistündliche Maßzahl ap (ursprünglich nur als Gewicht g zur Berechnung von Cp gebraucht[1]). Die Zuordnung von ap zu Kp hat BARTELS folgendermaßen vorgenommen:

$Kp=$	0o	0+	1−	1o	1+	2−	2o	2+	3−	3o	3+	4−	4o	4+
$ap=$	0	2	3	4	5	6	7	9	12	15	18	22	27	32
$Kp=$	5−	5o	5+	6−	6o	6+	7−	7o	7+	8−	8o	8+	9−	9o
$ap=$	39	48	56	67	80	94	111	132	154	179	207	236	300	400

Die Überlegungen, die zu dieser Einteilung geführt haben, entsprechen ganz denen zur Aufstellung der Skala für die ak in Ziff. 6. Noch weitgehender als bei den ak, und zwar von $Kp=1$o bis $Kp=8$o, ist hier der einem Kpo zugeordnete ap-Wert gleich dem halben Mittelwert des entsprechenden a-Bereiches in der Klasseneinteilung von K für eine Standard-Station (s. Ziff. 5). Die Zwischenwerte der ap für $Kp\pm$ entstehen dann durch Interpolation mit gleichbleibenden oder nach oben zunehmenden Abständen. Fig. 7 veranschaulicht diesen Zusammenhang zwischen dem linearen Aktivitätsmaß ap und dem quasi-logarithmischen Maß Kp.

Fig. 7. Graphische Darstellung des Zusammenhangs zwischen der quasi-logarithmischen Kennziffer Kp und dem linearen Aktivitätsmaß ap.

Auch die doppelte Bedeutung von ak als Kennziffer und als äquivalente Amplitude ließe sich auf ap übertragen. Jedoch ist eine solche Unterscheidung für ap weniger sinnvoll, da eine planetarische Maßzahl nicht die Breitenabhängigkeit der Aktivität wiedergeben soll. Die eigentliche Bedeutung von ap ist also die einer *Kennziffer* entsprechend $\hat{a}k$ in Ziff. 6. Diese Bedeutung ist in der Praxis auch dann noch gemeint, wenn ap als dreistündliche *äquivalente planetarische*

---
[1] J. BARTELS: IATME Bull. No. 12e, 111 (1951); vgl. auch [*4*].

Amplitude bezeichnet wird; und zwar deshalb, weil ap dabei ausschließlich als *Amplitudenmaß an den Standard-Stationen* interpretiert wird.

Wie bei $\hat{a}k$ ist auch die äquivalente planetarische Amplitude an diesen Stationen durch $2\,ap$ in γ gegeben. Das ist gleichbedeutend mit der Ausdrucksweise, daß, bezogen auf die Standard-Stationen, ap gleich einer äquivalenten Amplitude mit der Einheit $2\,\gamma$ ist. Diese Interpretation ließe sich ohne weiteres auf jedes andere Observatorium mit $a_9 \neq 500\,\gamma$ ausdehnen, wobei ap dann die Amplitude in der Einheit $(2\,a_9/a_9(S))\,\gamma$ angäbe. Jedoch ist davon abzuraten, um unnötige Komplikationen zu vermeiden.

Es wird daher empfohlen, ap in erster Linie als dimensionslose planetarische Kennziffer aufzufassen und, falls erforderlich, sich bei einer Interpretation als äquivalente Amplitude auf die Standard-Stationen zu beziehen, d. h. auf eine geomagnetische Bezugsbreite um 50°. Die Größe $2\,ap$ gibt dann den hier zu erwartenden planetarischen Anteil in γ an der maximalen K-Variation in der gestörtesten Komponente während des betreffenden Drei-Stunden-Intervalls.

Tabelle 3. *Monats- und Jahresmittel des linearen Aktivitätsmaßes Ap, berechnet aus den ap-Werten des betreffenden Monats oder Jahres.*

Jahr	Monat												Jahres-mittel
	I	II	III	IV	V	VI	VII	VIII	IX	X	XI	XII	
1932	11	12	18	17	15	7	7	12	12	10	8	9	11,5
33	10	11	12	12	12	8	7	9	12	10	9	7	10,1
34	6	8	11	6	7	5	6	9	10	6	5	8	7,2
35	9	10	10	8	6	9	7	5	13	12	8	9	8,9
36	9	11	9	15	10	12	11	5	5	9	10	5	9,1
37	7	13	12	20	13	12	12	10	9	20	12	10	12,5
38	28	16	13	18	18	9	13	12	17	16	10	11	15,3
39	7	15	19	28	21	15	19	19	13	22	9	11	16,5
1940	15	12	36	18	13	16	12	11	14	14	16	15	16,1
41	14	18	33	15	11	11	19	16	27	11	16	11	16,8
42	9	12	22	17	8	8	13	13	17	22	15	11	13,8
43	11	9	13	14	14	12	15	31	25	24	20	14	17,0
44	13	12	17	15	9	8	6	9	10	11	6	14	10,9
45	10	10	17	13	9	7	9	7	10	11	8	13	10,4
46	12	22	33	20	18	16	22	11	34	13	12	9	18,6
47	12	12	32	18	14	16	16	25	32	23	14	11	18,8
48	12	13	17	13	19	10	10	20	15	27	16	13	15,4
49	20	14	19	14	18	14	8	14	13	25	15	9	15,4
1950	12	18	14	18	16	14	14	25	22	28	20	16	18,1
51	16	22	21	27	20	17	20	22	40	24	18	20	22,3
52	19	26	33	34	27	18	15	13	23	20	12	15	21,3
53	15	15	21	16	16	13	16	19	21	16	14	7	15,6
54	9	16	16	14	7	6	8	10	17	15	9	6	11,0
55	12	12	14	14	11	9	8	9	13	11	13	8	11,3
56	18	15	20	27	26	17	13	15	18	14	24	10	18,1
57	17	17	26	21	11	22	16	14	49	14	18	18	20,1
58	15	27	26	20	17	24	25	18	20	16	8	15	19,2
59	14	24	24	17	19	15	32	23	28	19	22	19	21,3
1960	15	14	18	42	24	20	20	20	20	36	32	21	23,7
61	12	16	14	14	13	14	28	11	13	16	10	12	14,4
62	7	11	8	14	7	9	12	15	19	20	13	13	12,3
63	11	9	8	10	11	11	12	13	28	15	12	11	12,6
64	12	12	13	13	10	9	9	8	11	10	7	5	10,0
65	6	9	8	8	6	10	8	9	10	7	6	7	7,7
66	7	8	13	7	9	6	9	11	21	11	9	11	10,2
67	11	11	7	9	25	12	8	9	16	10	10	14	12,0
68	11	16	13	13	13	17	10	12	14	16	17	10	13,5
69	8	15	17	14	17	9	8	8	15	9	10	7	11,3

β) *Das arithmetische Mittel der acht Werte von ap am Greenwich-Tag wird mit Ap bezeichnet.* Auch Ap sollte in erster Linie als dimensionslose *planetarische Kennziffer* aufgefaßt werden. Ist analog zu ap eine Interpretation als *tägliche äquivalente Amplitude* in γ angebracht, so sollte auch diese nur in bezug auf die *Standard-Stationen* erfolgen, für die dann Ap ebenfalls die Einheit 2γ hat.

Die bei der Behandlung von ak und Ak ausführlicher dargelegten Vorteile der *linearen* Aktivitätsmaße hinsichtlich ihrer einfachen Interpretation und statistischen Bearbeitung gelten in gleicher Weise für ap und Ap. Insbesondere erhält man durch einfache Mittelwertbildung Monats- und Jahresmittel der Aktivität. Es ist üblich, auch diese wieder mit Ap zu bezeichnen, so daß es ratsam ist, bei Ap immer die Zeitspanne anzugeben, über die gemittelt worden ist. Die Tabelle 3 gibt die Monats- und Jahresmittel von Ap seit 1932 an, d.h. alle Werte, die zur Zeit vorliegen. Jeder Wert dieser Tabelle ist unmittelbar aus den ap-Werten des betreffenden Monats oder Jahres berechnet worden. Die Monatsmittel sind dann auf ganze Zahlen abgerundet worden und die Jahresmittel auf die erste Dezimale.

10. Die täglichen planetarischen Charakterzahlen Cp und $C9$.

α) Ausgehend von ap als linearem Aktivitätsmaß, ist es möglich, eine tägliche planetarische Maßzahl zu definieren, mit der die erdmagnetische Aktivität wieder *quasi-logarithmisch* erfaßt wird. Es liegt nahe, dafür die quasi-logarithmische Skala von Kp zu übernehmen und so als Ergänzung der dreistündlichen Kennziffer Kp eine tägliche Maßzahl mit den sonst gleichen Eigenschaften einzuführen. Dies ist von BARTELS getan worden, der diese Maßzahl Bp nannte. Werte von Bp wurden für das Polarjahr 1932/33 berechnet sowie für die Jahre 1940 bis 1949. Letztere haben hauptsächlich bei der Definition von Cp Verwendung gefunden. Eine weitere Berechnung von Bp ist dann zugunsten von Cp unterblieben.

β) Der Grund für diesen Wechsel muß in einer Empfehlung von IATME aus dem Jahre 1948 gesehen werden, die Möglichkeiten einer Bestimmung der internationalen Charakterzahl Ci (s. Ziff. 14) aus den K-Indizes zu prüfen. Eine Antwort auf diese Empfehlung ist BARTELS' *tägliche planetarische Charakterzahl Cp*[1]. Kernpunkt des Verfahrens ist die Angleichung der Häufigkeitsverteilung der geeignet definierten Cp an die Häufigkeitsverteilung der Ci: Benutzt wurden alle 3653 Tage der 10 Standardisierungsjahre 1940 bis 1949. Da Ci ein 21stufiges, quasi-logarithmisches Maß der Aktivität ist, ergaben sich erste Anhaltspunkte aus einem Vergleich mit den 28 Klassen von Bp, deren Grenzen durch die Tagessumme von ap festgelegt sind. Diese Grenzen mußten nun so verändert werden, daß auf jede der neuen 21 Klassen mit der Numerierung von 0,0 bis 2,0 praktisch die gleiche Anzahl von Tagen fiel, wie sie für die entsprechenden Ci-Klassen gefunden worden war. Damit war auch Cp an die Tagessumme von ap und so an Kp angeschlossen. Um die stark gestörten Zeiten mit $Ci = 2,0$ genauer zu unterscheiden, hat BARTELS die Cp-Skala bis 2,5 erweitert. Die folgende Tabelle 4 gibt den Schlüssel, nach dem Cp aus der Tagessumme der ap ($=\sum ap$) ermittelt wird.

γ) Über die Tagessumme von ap hängt auch Ap mit Cp und damit näherungsweise mit Ci zusammen. Für überschlägige Berechnungen gibt BARTELS [5] die folgende vereinfachte Tabelle an, die es auch gestattet, Ap für die Jahre vor 1932 angenähert aus Ci zu bestimmen:

$\left.\begin{array}{c}Ci\\Cp\end{array}\right\}$ =	0,0	0,1	0,2	0,3	0,4	0,5	0,6	0,7	0,8	0,9	1,0
Ap =	2	4	5	6	8	9	11	12	14	16	19
$\left.\begin{array}{c}Ci\\Cp\end{array}\right\}$ =	1,1	1,2	1,3	1,4	1,5	1,6	1,7	1,8	1,9	2,0	
Ap =	22	26	31	37	44	52	63	80	110	160	

---

[1] J. BARTELS: IATME Bull. No. 12e, 109—137 (1951).

Tabelle 4. *Schlüssel zur Ermittelung von Cp aus der Tagessumme von ap.*

Grenzen von $\sum ap$	Cp	Grenzen von $\sum ap$	Cp	Grenzen von $\sum ap$	Cp
0 ... 22	0,0	121 ... 139	0,9	562 ... 729	1,8
23 ... 34	0,1	140 ... 164	1,0	730 ... 1119	1,9
35 ... 44	0,2	165 ... 190	1,1	1120 ... 1399	2,0
45 ... 55	0,3	191 ... 228	1,2	1400 ... 1699	2,1
56 ... 66	0,4	229 ... 273	1,3	1700 ... 1999	2,2
67 ... 78	0,5	274 ... 320	1,4	2000 ... 2399	2,3
79 ... 90	0,6	321 ... 379	1,5	2400 ... 3199	2,4
91 ... 104	0,7	380 ... 453	1,6	3200*	2,5
105 ... 120	0,8	454 ... 561	1,7		

δ) Die Festlegung von Cp durch Angleichung an die Häufigkeitsverteilung von Ci darf nicht dazu verführen zu übersehen, daß mit diesem Verfahren der Versuch gemacht wird, zwei Maßzahlen mit völlig unterschiedlichen Definitionen aufeinander abzustimmen. Die auch nach den Standardisierungsjahren gefundene gute Übereinstimmung muß als ein bemerkenswertes statistisches Ergebnis gewertet werden, auch wenn die Schätzung von C jetzt objektiver erfolgt.

Als Beispiel seien die Häufigkeiten der absoluten Differenzen $|Ci - Cp|$ für die 4383 Tage der 12 Jahre 1950 bis 1961 angeführt:

$|Ci - Cp| = 0$ an 1638 Tagen $\triangleq$ 37,4%

$|Ci - Cp| = 0,1$ an 2103 Tagen $\triangleq$ 48,0%

$|Ci - Cp| = 0,2$ an 539 Tagen $\triangleq$ 12,3%

$|Ci - Cp| = 0,3$ an 91 Tagen $\triangleq$ 2,1%

$|Ci - Cp| \geq 0,4$ an 12 Tagen $\triangleq$ 0,3%

Da die subjektiven Einflüsse bei der Bestimmung von Cp wesentlich geringer sind als bei Ci, ist Cp die zuverlässigere Maßzahl und dementsprechend Ci vorzuziehen.

ε) Um die Darstellung der erdmagnetischen Aktivität durch die dreistündliche quasi-logarithmische Kennziffer Kp sinngemäß auf die tägliche quasi-logarithmische Charakterzahl Cp zu übertragen, mußte der Weg über das lineare ap-Maß genommen werden, damit bei der mathematischen Operation die physikalische Aussage erhalten blieb. Werden nun von den täglichen Cp-Werten Monats- und Jahresmittel gebildet, so sollten auch diese Schritte wieder über ap oder Ap gehen. In der Praxis werden dagegen die Cp selbst gemittelt, und zwar deshalb, weil seit Anfang an so mit den Ci verfahren worden ist und auch weiter so verfahren wird. Bei der Beurteilung solcher Mittelwerte muß aber berücksichtigt werden, daß durch die implizite *geometrische* Mittelwertbildung die tatsächliche Aktivität zu klein wiedergegeben wird.

---

* Nach der Tabelle in Ziff. 9α wird dem maximalen Störungsgrad $Kp = 90$ eine äquivalente Amplitude $ap = 400$ zugeordnet. An einem maximal gestörten Tag mit $Kp = 90$ für alle acht E_v wird dann $\sum ap = 3200$. Dieser Wert kann also nicht mehr überschritten werden. Es ist unklar, ob bei der Festlegung der Cp-Klassen beabsichtigt worden war, daß die höchste Cp-Klasse nur diesen einen Wert von $\sum ap$ enthalten soll. In der ersten Veröffentlichung[1] wird noch (im Gegensatz etwa zu [4]) $Cp = 2,5$ irrtümlich durch 3200 ... für $\sum ap$ gekennzeichnet. Für die Praxis ist diese Frage fast ohne Bedeutung, da in den bisher vorliegenden 39 Jahrgängen als höchster Störungsgrad nur $Cp = 2,3$ aufgetreten ist, der darüber hinaus sogar nur ein einziges Mal erreicht wurde, nämlich am 13. November 1960. Auch nach der Ap-Skala ist dieser Tag mit $Ap = 280$ der gestörteste Tag seit 1932. Die Festlegung von ap durch die Tabelle in Ziff. 9α hat ferner zur Folge, daß $\sum ap$ nicht alle nach Tabelle 4 möglichen Werte auch tatsächlich annehmen kann. Das gilt auch für die Intervallgrenze 3199.

Um dafür ein Beispiel zu geben, wurden die Monatsmittel von Cp für 1967 einmal auf die übliche Weise berechnet (obere Zeile $\overline{Cp}$) und zum anderen über ap (untere Zeile $Cp(\overline{ap})$):

1967:	I	II	III	IV	V	VI	VII	VIII	IX	X	XI	XII
$\overline{Cp}$ =	0,46	0,46	0,34	0,44	0,78	0,55	0,41	0,45	0,65	0,48	0,53	0,65
$Cp(\overline{ap})$ =	0,65	0,64	0,38	0,49	1,18	0,68	0,42	0,48	0,89	0,55	0,58	0,80

Danach können die beiden Berechnungswege auf beachtliche Unterschiede führen. Vor allem muß aber daraus gefolgert werden, daß man sich eine durch $\overline{Cp}$ gegebene mittlere Aktivität nicht durch die Aktivität an einem einzelnen Tage mit gleichem Cp veranschaulichen darf. Diese Feststellung gilt ebenso für $\overline{Ci}$.

ζ) Für einen ersten Überblick und auch manche statistischen Untersuchungen ist es nützlich, die erdmagnetische Aktivität eines Tages durch eine einstellige Zahl auszudrücken, also wieder die zehn Stufen 0 bis 9 zu verwenden. Wie bei K empfiehlt sich auch dafür eine quasi-logarithmische Skala, die sich leicht durch eine Reduzierung der Cp-Skala auf zehn Klassen erreichen läßt. Das ergibt die *reduzierte planetarische Charakterzahl* $C9$. Die Umrechnung von Cp oder Ci erfolgt nach BARTELS[2] durch:

$C9 =$	0	1	2	3	4	5	6	7	8	9
Cp	0,0	0,2	0,4	0,6	0,8	1,0	1,2	1,5	1,9	2,0
Ci	0,1	0,3	0,5	0,7	0,9	1,1	bis 1,4	bis 1,8		bis 2,5

Die recht unsystematisch aussehende Zuordnung der Cp-Klassen zu $C9$ am Schluß der Tabelle kommt daher, daß $C9$ ursprünglich aus Ci berechnet worden ist und dabei $C9 = 8$ mit $Ci = 1,9$ und $C9 = 9$ mit $Ci = 2,0$ gleichgesetzt wurde. Die Erweiterung der späteren Cp-Skala bis 2,5 wurde dann ausschließlich der letzten $C9$-Klasse zugeschlagen, da ja allen Werten $Cp \geq 2,0$ als internationale Charakterzahl $Ci = 2,0$ entspricht. Diese Problematik läßt sich ganz vermeiden, indem die beiden letzten Klassen von $C9$ zu einer einzigen Klasse zusammengefaßt werden, was auch wegen der Seltenheit der extrem stark gestörten Tage sinnvoll ist.

Auf diese Weise ergibt sich die aus neun Klassen bestehende *Charakterzahl* $C8$ mit der letzten Klasse $C8 = 8$ für $Cp = 1,9$ bis 2,5. Im übrigen stimmt der durch $C9 = 0, 1, \ldots 9$ charakterisierte Störungsgrad nur grob mit dem durch $Kp = 0_0$, $1_0, \ldots 9_0$ bezeichneten Störungsgrad überein, wobei etwa von der Stufe 3 an der gleiche Zahlenwert bei Kp eine stärkere Aktivität bedeutet als bei $C9$.

Die $C9$ werden im Rahmen des internationalen Kennzifferndienstes zusammen mit den Tabellen der Kp, Ap und Cp in Form von Diagrammen veröffentlicht, die eine Kombination von Tabelle und graphischer Darstellung sind und quantitativ wie qualitativ ausgewertet werden können. Die Anordnung der $C9$ nach Sonnenrotationen ermöglicht ein einfaches Erkennen einer Wiederholungsneigung in der erdmagnetischen Aktivität. Unter diesem Gesichtspunkt wird das in Fig. 16 wiedergegebene $C9$-Diagramm in Ziff. 22 β erörtert.

11. Charakterisierung der Aktivität von Tagen, Monaten und Jahren durch planetarische Maßzahlen.

α) Die Vielzahl der in Gebrauch befindlichen unterschiedlichen Maßzahlen ist letzten Endes eine Folge der Unvollkommenheit und Einseitigkeit, mit der die komplexe Natur der erdmagnetischen Aktivität durch eine einzelne Maßzahl wiedergegeben wird. Dieser Umstand ist zu beachten, wenn die Aktivität an einzelnen Tagen oder gar während längerer Zeitspannen zuverlässig charakterisiert werden soll. Dabei wird es im allgemeinen nicht genügen, sich auf nur eine Maßzahl zu beschränken.

---
[2] J. BARTELS: Abhandl. Akad. Wiss. Göttingen, Math.-Phys. Kl., Sonderheft Nr. 1 (1951); vgl. auch [6].

Am Beispiel zweier Tage mit unterschiedlichem Verlauf der Aktivität soll dieser Sachverhalt erläutert werden. Dazu wird von den Kp-Werten der acht Drei-Stunden-Intervalle ausgegangen

E_ν:	$\nu =$	1	2	3	4	5	6	7	8
1. Tag:	$Kp =$	4o	3o	3−	4−	3o	4o	4+	3+
2. Tag:	$Kp =$	1o	1−	0+	1+	1+	2o	5o	6+

Als erstes wird man zur Beurteilung dieser Tage eines der täglichen Aktivitätsmaße berechnen. Die Tagessumme von ap ist an beiden Tagen 168. Damit hat jede der beiden aus $\sum ap$ folgenden Maßzahlen Ap und Cp für beide Tage denselben Wert, nämlich $Ap = 21$ und $Cp = 1{,}1$, woraus noch $C9 = 5$ folgt.

Ob man die mittlere Tagesaktivität mit Ap oder Cp angibt, ist eine Frage der Gewohnheit und der Zweckmäßigkeit im Hinblick auf eine Weiterverarbeitung der Daten, jedoch keine prinzipielle Frage, da sich beide Maßzahlen zwar in ihrer Skala, also zahlenmäßig unterscheiden, aber dadurch noch *nicht in ihrer Aussage über die mittlere Tagesaktivität*.

Betrachtet man nun die Kp-Werte der beiden Tage im einzelnen, so findet man, daß trotz gleicher mittlerer Aktivität der Störungscharakter an beiden Tagen sehr verschieden war. Der erste Tag war durchgehend mäßig stark gestört, während der zweite Tag zu drei Vierteln ausgesprochen ruhig verlief bis zum Ausbruch einer stärkeren Störung in E_7, die in E_8 verstärkt andauert. Will man diesen Unterschied im Störungscharakter andeuten, ohne alle Kp-Werte einzeln aufzuführen, so empfiehlt es sich, wenigstens die Grenzen der Aktivität in Kp mit anzugeben, also die als Beispiel gewählten Tage etwa folgendermaßen zu charakterisieren:

1. Tag: $Cp = 1{,}1$; $3− \leq Kp \leq 4+$
2. Tag: $Cp = 1{,}1$; $0+ \leq Kp \leq 6+$

Fanselau[1] schlägt vor, die Variation der magnetischen Aktivität während eines Tages durch eine zusätzliche Kennziffer anzugeben, die sich aus jeder primären Maßzahl folgendermaßen berechnen läßt: Liegen von dieser Maßzahl n Werte je Tag vor, etwa die acht Werte von Kp oder ap, so bestimmt man die sich aus je zwei aufeinander folgenden Werten ergebenden Differenzen, insgesamt also (n − 1) Differenzen je Tag. Die Differenz vom letzten Wert des Vortages zum ersten Wert des betrachteten Tages und die Differenz vom letzten Wert dieses Tages zum ersten Wert des folgenden Tages werden mit halbem Gewicht hinzugenommen. Die *Summe dieser absolut genommenen Differenzen* ist nach Fanselau ein geeignetes Variabilitätsmaß, das sich ebenso für ein einzelnes Observatorium aus dessen lokalen Kennziffern bestimmen läßt.

β) Ein besonderes Verfahren wird bei der *Auswahl der internationalen ruhigen und gestörten Tage* angewandt. Danach werden für jeden Monat fünf ruhige Tage, zehn ruhige Tage und fünf gestörte Tage festgelegt. Die ersten internationalen Tage wurden für 1906 bestimmt und dann fortlaufend für die folgenden Jahre. Eine nachträgliche Festlegung der fünf ruhigen und der fünf gestörten Tage ist außerdem bis 1884 zurück vorgenommen worden. Die Auswahl erfolgte anfänglich allein auf der Grundlage der internationalen Charakterzahl Ci. Das Verfahren hat dann im Laufe der Zeit Modifikationen erfahren und beruht zur Zeit auf den folgenden drei Größen:

a) auf der *Tagessumme der Kp*,
b) der *Tagessumme der Quadrate von Kp* und
c) dem *größten Kp-Wert des Tages*. Alle Tage eines Monats werden nach jedem dieser drei Kriterien der Größe nach geordnet und je nach ihrer Position in einer solchen Anordnung mit einer Ordnungszahl versehen. Die drei Ordnungszahlen,

---
[1] G. Fanselau: Gerlands Beitr. Geophys. **77**, 42—49 (1968).

die ein Tag auf diese Weise erhält, werden gemittelt und die Tage mit den größten und den kleinsten Mittelwerten als die fünf gestörten Tage bzw. die zehn ruhigen Tage und die fünf ruhigsten Tage ausgewählt[2].

Bei der Benutzung dieser internationalen Tage muß stets ihr relativer Charakter beachtet werden; denn es ist nicht ungewöhnlich, daß die ruhigen Tage von stark gestörten Monaten im Maximum der Sonnenaktivität absolut stärker gestört sind als die gestörten Tage von ruhigen Monaten im Minimum der Sonnenaktivität. Verwendung finden diese Tage vor allem, um den mittleren sonnentägigen Gang in den Komponenten des erdmagnetischen Feldes für gestörte Zeiten (**SD**) und für ruhige Zeiten (**Sq**) zu berechnen. Kommt es darauf an, **Sq** als Nicht-K-Variation zu ermitteln, so genügt es nicht, sich auf die als ruhig bezeichneten Tage zu verlassen, sondern es muß ihr absoluter Störungscharakter herangezogen werden, indem etwa Tage mit $Cp > 0,1$ ausgeschlossen werden. Auch für **SD** ist es sinnvoller, Tage mit annähernd gleichem Störungsgrad zu verwenden. Dagegen gibt die Differenz zwischen dem mittleren Tagesgang der gestörten Tage und dem der ungestörten Tage eines Monats durchaus charakteristische Eigenschaften des Störungsfeldes in jener Zeit wieder.

γ) Die anfangs angestellten Betrachtungen zur Kennzeichnung der Aktivität eines Tages lassen sich unschwer auf längere Zeitspannen ausdehnen. Anders als bei der Behandlung der Tagesaktivität muß aber bei der Angabe der mittleren Aktivität einer längeren Zeitspanne nach Ziff. 10 beachtet werden, daß jetzt das *mittlere Cp einen kleineren Wert* annimmt, als er durch Vergleich mit dem mittleren Ap aus der Umrechnungstabelle der Ziff. 10 gefunden wird. Diese Differenz wird um so größer, je stärker die Tageswerte von Cp in der betrachteten Zeitspanne streuen, also je uneinheitlicher der Charakter der Aktivität ist. Insofern hat also jetzt die Angabe der Mittelwerte von Ap und von Cp einen Sinn, wobei Ap das geeignetere Aktivitätsmaß ist.

Für eine genauere Kennzeichnung reichen aber diese Angaben wieder nicht aus. Auch die Grenzen der Aktivität in Kp sind für eine längere Zeitspanne nicht sehr aufschlußreich. Dagegen bietet sich der *Vergleich der Kp-Häufigkeiten* an, um etwa Monate mit gleicher oder auch ungleicher mittlerer Aktivität gegeneinander abzuwägen. Dabei ist es ratsam, jeweils mehrere Kp-Klassen zusammenzulegen, um die Übersicht zu erleichtern und die Streuung der Häufigkeiten zu reduzieren.

In der beigefügten Zusammenstellung (Tabelle 5) wird der Vergleich von Kp-Häufigkeiten an sechs ausgewählten Monaten veranschaulicht. Die Zahl der Tages-Achtel je Monat ist dabei einheitlich auf 240 Intervalle reduziert worden. Drei Monate (August 1964, September 1954 und Dezember 1964) stammen aus Zeiten minimaler Aktivität, die anderen Monate (November

Tabelle 5. *Häufigkeiten von Kp, bezogen auf 240 Tages-Achtel je Monat, und Monatsmittel von Ap, Cp und R für sechs ausgewählte Monate.*

Kp-Klassen	0 1	2 3	4 5	6 7	8	9	Ap	Cp	R
August 1964	122	102	15	1	0	0	8	0,35	9,3
November 1958	112	115	13	0	0	0	8	0,40	152,3
September 1954	28	144	63	5	0	0	17	0,83	1,5
April 1959	64	120	48	7	1	0	17	0,70	163,3
Dezember 1964	173	71	4	0	0	0	5	0,22	15,1
September 1957	44	86	55	30	18	7	49	1,01	235,8

---

[2] Diese Auswahl erfolgt am $C + K$ Centre De Bilt, Niederlande, die fortlaufende Veröffentlichung in den *Three-Monthly Bulletins* des International Service of Geomagnetic Indices und in den IAGA Bulls. No. 12, 1.

1958, April 1959, September 1957) aus Zeiten maximaler Aktivität, wie aus den hinzugefügten Monatsmitteln der *Sonnenfleckenrelativzahl R* zu ersehen ist. Dennoch haben zwei Monats-Paare aus diesen Gruppen jeweils gleiche mittlere Aktivität nach Ap. Die anomalen Monate sind November 1958 wegen des Ausbleibens der zu dieser Zeit fälligen starken Störungen und September 1954 wegen ungewöhnlich vieler mittelstark gestörter Intervalle auf Kosten der zu dieser Zeit normalen ruhigen Intervalle. Als extreme Monate sind noch der Dezember 1964 und der September 1957 hinzugefügt worden. Der Dezember 1964 gehört zu den sieben Monaten mit dem seit 1932 beobachteten kleinsten Monatsmittel $Ap = 5$, während der September 1957 das höchste bisher aufgetretene Monatsmittel $Ap = 49$ hat. Trotzdem hat der September 1957 mehr ruhige Intervalle mit $Kp = 0$ und 1 als der September 1954.

δ) Auf die gleiche Weise läßt sich auch die Aktivität ganzer Jahre charakterisieren. Neben den Jahresmitteln von Ap und Cp und dem Schwankungsbereich der Monatsmittel sind Häufigkeitsverteilungen der nach Klassen zusammengefaßten Tagesmittel oder besser noch der dreistündlichen Kp-Werte selbst die geeignetste Form der Charakterisierung.

12. Mängel der planetarischen Maßzahlen und Verbesserungsvorschläge.

α) Die Besprechung der BARTELSschen Kennziffern, von denen bisher fast ausschließlich die Rede war, stellt einen wesentlichen Teil dieses Artikels über Maßzahlen der erdmagnetischen Aktivität dar. Diese Bevorzugung entspricht zweifellos ihrer gegenwärtigen Bedeutung, die auch in einer immer noch wachsenden Anwendung zum Ausdruck kommt. Dabei ist es eine natürliche Erscheinung, daß sich 20 bis 30 Jahre nach der Einführung der einzelnen Kennziffern und nachdem fast 40 Jahrgänge mit ihren Werten vorliegen, auch Unzulänglichkeiten bemerkbar machen. Sie betreffen sowohl die Konzeption als auch die Realisierung dieser Konzeption.

Die sich heute anbahnenden neuen Vorstellungen von der Natur der erdmagnetischen Aktivität (s. Ziff. 24) und die beobachteten Verschiedenheiten in der Aktivität der Äquatorzone, der mittleren Breiten, der Polarlichtzonen und besonders der Polkappen lassen es fraglich erscheinen, ob eine wirklich *planetarische Kennziffer der Aktivität* mit einfachen Mitteln bestimmt werden kann. Auf jeden Fall ist es heute schwieriger zu entscheiden, welcher Anteil an der Aktivität lokal und welcher Anteil planetarisch ist. Es scheint auch sicher zu sein, daß eine ausreichende Reduktion der *Breitenabhängigkeit* der Aktivität in der K-Skala allein durch die Wahl von a_9, also der unteren Grenze von $K = 9$, nicht erreicht werden kann. Da inzwischen die Existenz eines *Weltzeitganges* der Aktivität (s. Ziff. 23β) nachgewiesen ist, wäre sein Auftreten in den planetarischen Maßzahlen durchaus wünschenswert, obwohl dies nicht eine Eigenschaft der solaren P-Strahlung ist. Die ebenfalls nachgewiesenen *Unterschiede in der Aktivität zwischen Nord- und Südhalbkugel* (s. Ziff. 22 δ) sollten unbedingt in geeigneter Weise erfaßt werden, wofür die jetzige Berechnung der planetarischen Maßzahlen aus den Daten von elf Observatorien der Nordhalbkugel und nur einem Observatorium der Südhalbkugel keine Möglichkeit läßt.

β) Anhaltspunkte dafür, ob die zugrunde gelegte *statistische Methodik* bei der Kp-Bestimmung die an sie gestellten Forderungen erfüllt hat, ergeben sich, wenn mit dem jetzt vorliegenden Material geprüft wird, inwieweit die planetarischen Kennziffern noch lokale Anteile enthalten. Aus den 30 Jahrgängen von 1932 bis 1961 hat MICHEL[1] den Tagesgang von Kp nach Weltzeit berechnet. Das Ergebnis läßt sich formelmäßig angeben durch

$$Kp = 2{,}345 + 0{,}065 \cos\left(\frac{t - 1{,}5\,\mathrm{h}}{12\,\mathrm{h}}\right),$$

---

[1] F. C. MICHEL: J. Geophys. Res. **69**, 4182—4183 (1964).

wobei t in Stunden nach Weltzeit (UT) einzusetzen ist. Dieser Weltzeitgang ist sehr klein und kaum von praktischer Bedeutung, jedoch statistisch signifikant, soweit sich das bei diesen statistisch nicht unabhängigen Werten abschätzen läßt. Danach ist Kp in den Tages-Achteln E_1, E_2 und E_8 etwas zu groß und in E_4, E_5 und E_6 etwas zu klein.

Eine ähnliche Untersuchung mit der hierfür geeigneteren Maßzahl ap hat MAYAUD[2] durchgeführt. Mit den Werten der Jahre 1932 bis 1960 findet er für die drei Monatsgruppen Tagesgänge, deren Verlauf mit dem für Kp nach MICHEL übereinstimmt. Die Schwankungsbreite dieser Gänge beträgt 2 bis 3 Einheiten von ap und entspricht etwa einer Dritteleinheit von Kp. Wieder ist der Effekt klein, aber wahrscheinlich signifikant.

Obwohl diese Tagesgänge formal nach Weltzeit auftreten, wie es bei planetarischen Maßzahlen nicht anders möglich ist, kann ausgeschlossen werden, daß sie die echte Weltzeitvariation der Aktivität wiedergeben. Ihre Ursache dürfte vielmehr in einer unvollständigen Eliminierung der nach Ortszeit erfolgenden Tagesgänge an den Kp-Observatorien liegen. Damit bietet sich die Möglichkeit, die Zuverlässigkeit der bei der Berechnung von Kp verwendeten Methoden mit dem jetzt verfügbaren Material zu überprüfen, indem aus etwaigen Verkleinerungen der Tagesgänge bei Verwendung modifizierter Methoden auf Unzulänglichkeiten des Verfahrens an entsprechenden Stellen geschlossen wird.

Diesen Weg ist MAYAUD gegangen. Auf Grund seiner Ergebnisse nennt er vier Ursachen für die ortszeitlichen Residuen in Kp: a) Die bei der Aufstellung der Umrechnungstabellen (s. Ziff. 7) zugrunde gelegte *Standardisierungszeit* ist zu kurz bemessen. b) Während dieser Standardisierungszeit wurde der sonnentägige Gang an vielen Observatorien im Gegensatz zu heute noch schematisch mit der „eisernen Kurve" eliminiert. Die daraus resultierende Verfälschung der häufigen niedrigen Kennziffern ist in die Umrechnungstabellen eingegangen. c) Die Benutzung der beiden Tages-Achtel um Orts-Mitternacht für die Standardisierung (anstelle aller acht Drei-Stunden-Intervalle) wirkt sich als *Bezugsniveau der Aktivität* nicht an allen Kp-Observatorien gleich aus, da der Tagesgang der Aktivität und mit ihm die Eintrittszeit des Maximums der Aktivität breitenabhängig ist. d) Die *Häufigkeitsverteilung der K-Werte* ist an den drei nördlichen Observatorien LE, ME und SI mit geomagnetischen Breiten von und über 60° merklich verschieden von der Verteilung der Kennziffern an den anderen Observatorien. Hier garantiert also die Wahl von a_9 keine ausreichende Übereinstimmung der Häufigkeitsverteilungen.

γ) Die angeführten Beanstandungen zeigen auch zugleich, wo und in welcher Weise *Verbesserungen* vorgenommen werden können. Neben der methodisch-statistischen Seite ist dabei nach dem Vorhergehenden die stärkere Berücksichtigung von Observatorien der Südhalbkugel ein besonders wichtiger Punkt. Die Realisierung solcher Verbesserungen verlangt aber eine Entscheidung über die schon in Ziff. 2 aufgeworfene Frage, ob die in Gebrauch befindlichen Kennziffern im Rahmen ihrer Konzeption abgeändert werden sollen und damit ein gewisser Bruch in diesen Zeitreihen in Kauf genommen werden soll, oder ob neue Kennziffern eingeführt und dann mindestens über längere Zeit fast gleichartige Zeitreihen nebeneinander laufen sollen.

MAYAUD [8], [9] empfiehlt die zweite Lösung und bietet dafür neue, zusätzliche Maßzahlen an: Grundlage soll die BARTELSsche Kennziffer K bleiben, deren Definition MAYAUD [7] aber unter Einschränkung des physikalischen Inhalts allein aus dem *morphologischen* Gegensatz von unregelmäßigen Störungen und regelmäßigen Störungen (einschließlich der Nachstörungen) abgeleitet wissen möchte. Alle unregelmäßigen Störungen sind hier K-Variationen. Zur Erfassung der welt-

---

[2] P. N. MAYAUD: Note No. 11, Institut de Physique du Globe, Paris, Avril 1965 und [8].

weiten Aktivität soll eine größere Zahl von *Observatorien beider Hemisphären* herangezogen werden, die jeweils in mittleren Breiten (*subauroral*) liegen und in der magnetischen Länge[3] möglichst gleich verteilt sein sollen. Ihre unteren Grenzen a_0 sind zu korrigieren, falls das auf Grund des jetzt vorliegenden Beobachtungsmaterials erforderlich ist. Angestrebt wird eine Kennziffer Kn für die Nordhalbkugel und eine Kennziffer Ks für die Südhalbkugel (nicht zu verwechseln mit der standardisierten Kennziffer Ks der Ziff. 7), die im Prinzip als *arithmetische Mittel aus den Kennziffern K der Observatorien* jeder Halbkugel berechnet werden sollen. Wegen der nicht zu umgehenden unregelmäßigen Verteilung der Observatorien mit der Länge werden die K-Werte vor der Mittelwertbildung bewichtet. Jedoch werden keine Umrechnungstabellen mit dem Ziel einer Standardisierung der Kennziffern verwendet wie in Ziff. 7. Als Mittelwert aus Kn und Ks folgt als *weltweites Maß der Aktivität Km*. Analog der Transformation von Kp in die linearen Maße ap und Ap werden Kn, Ks und Km die *linearen Maßzahlen an, as, am, An, As und Am* zugeordnet. Ferner werden Maßzahlen σn und σs bestimmt, die die Homogenität der Kennziffern jeder Halbkugel für ein Drei-Stunden-Intervall angeben sollen.

Mit den Daten von zehn Observatorien der Nordhalbkugel und sechs Observatorien der Südhalbkugel hat MAYAUD eine vorläufige Berechnung dieser Kennziffern für drei Jahre vorgenommen[4]. Auf einige seiner Ergebnisse wird noch in Ziff. 23 eingegangen; im übrigen sei auf [8] verwiesen. Soweit sich bisher sehen läßt, enthalten die MAYAUDSCHEN Maßzahlen Informationen, die bei der Berechnung von Kp verlorengehen. So ist neben unterschiedlicher Aktivität der beiden Hemisphären ein echter Tagesgang nach Weltzeit nachweisbar, jedoch tritt dieser zusammen mit zwei anderen gleich großen scheinbaren Weltzeitgängen auf, von denen er erst separiert werden muß. Insgesamt kann noch nicht beurteilt werden, ob der Gewinn an Informationen die fortlaufende Berechnung einer ganzen Reihe sich sehr ähnlicher Aktivitätsmaße rechtfertigt.

δ) Selbst wenn irgendwelche neuen planetarischen Maßzahlen auf der Basis von K alle in sie gesetzten Erwartungen erfüllen sollten, werden sich ihre statistischen Aussagen nur um *Effekte zweiter Ordnung* von den Ergebnissen unterscheiden, die auch aus Kp, ap, und Ap folgen. Das heißt aber auch, daß für die Mehrzahl der statistischen Untersuchungen, bei denen es um den Zusammenhang anderer geophysikalischer Erscheinungen mit der erdmagnetischen Aktivität geht, eine Verbesserung der vorhandenen Maßzahlen nur wenig ins Gewicht fallen würde.

Naturgemäß liegt die größte Unsicherheit bei den einzelnen dreistündlichen Werten und hier wieder stärker bei der linearen Maßzahl ap, die überdies ursprünglich nicht als selbständige Kennziffer gedacht war. Es ist aber fraglich, ob für diese einzelnen Werte Verbesserungen möglich sind, die über ihre natürliche zufällige Streuung hinausgehen. Für die durch Ap gegebene mittlere Tagesaktivität wird der Erfolg eines solchen Vorhabens entsprechend unwahrscheinlicher.

---

[3] P. N. MAYAUD: Ann. Géophys. **19**, 164—179 (1960).
[4] Aufbauend auf den dabei gemachten Erfahrungen, sind einige Observatorien der Nordhalbkugel ausgetauscht worden und ist vor allem auch das Berechnungsverfahren abgeändert worden. Wichtigster Unterschied ist, daß auf dem Weg über die lokalen K-Werte zunächst mittlere äquivalente Amplituden für bestimmte Sektoren auf beiden Hemisphären berechnet werden, aus diesen dann durch einfache Mittelwertbildung an und as und erst daraus Kn, Ks und die planetarischen Maßzahlen. Beginnend mit dem Jahrgang 1964 werden die so berechneten Kennziffern von MAYAUD fortlaufend veröffentlicht. Eine Zusammenstellung der ersten vier Jahrgänge 1964—1967 mit einer ausführlichen Beschreibung des Berechnungsverfahrens enthält [9].

Als Richtlinie bei allen derartigen Überlegungen kann gelten, daß nicht der Selbstzweck des statistisch Möglichen die eigentliche Aufgabe ist, sondern das Ziel, eine *statistisch hinreichend gesicherte Antwort auf eine physikalische Fragestellung* zu geben. Wie eine solche Aufgabe angegriffen und bewältigt werden kann, hat BARTELS zweifellos exemplarisch gezeigt.

C. Andere Aktivitätsmaße.

13. Vorbemerkungen und Übersicht.

α) Die Zahl der frühen Versuche, die erdmagnetische Aktivität durch Maßzahlen zu erfassen, ist heute kaum noch feststellbar. Nur einige dieser Aktivitätsmaße haben sich vorübergehend durchsetzen können, und nur ganz wenige sind heute noch in Gebrauch. Die folgenden historischen Bemerkungen sollen an einigen herausgegriffenen Beispielen lediglich die Bemühungen in den ersten Jahrzehnten dieses Jahrhunderts anschaulich machen.

Obwohl mit dem Beschluß von 1905, die internationale Charakterzahl einzuführen, ein offizielles Aktivitätsmaß vorlag, war gerade die Unbestimmtheit in der begrifflichen Festlegung der mit Ci angegebenen Aktivität ein Anlaß, nach Maßzahlen mit physikalisch begründeter Definition zu suchen. Von großem Einfluß auf die weitere Entwicklung war der von BIDLINGMAIER[1] stammende Gedanke, die *Energiedichte des Störungsfeldes* als Aktivitätsmaß zu benutzen. Das von ihm praktizierte und 1913 veröffentlichte Verfahren wurde von einer Reihe von Observatorien übernommen, an denen diese Maßzahl für jeden Tag des Jahres 1915 berechnet wurde. Als größter Mangel wurde zunächst das äußerst umständliche Verfahren zur Bestimmung der Werte empfunden, bis dann BAUER[2] 1921 nachwies, daß diese Maßzahl im Hinblick auf ihre Konzeption falsch definiert worden war.

In der jetzt gebräuchlichen Darstellung wird dieser Sachverhalt sofort deutlich, wenn man das erdmagnetische Feld F an einem Ort in einen großen ungestörten Anteil F_0 und ein kleines Störungsfeld ΔF zerlegt, so daß gilt

$$F^2 = (F_0 + \Delta F)^2 = F_0^2 + 2F_0 \cdot \Delta F + \Delta F^2. \tag{13.1}$$

Da ΔF^2 klein von zweiter Ordnung ist, erhält man für die Energiedichte ΔE des Störungsfeldes

$$\Delta E \propto F_0 \cdot \Delta F. \tag{13.2}$$

Die BIDLINGSMAIERsche Maßzahl entspricht dagegen dem quadratischen Term ΔF^2, integriert über einen Tag.

Trotz allem hat BIDLINGMAIER damit den Anstoß gegeben, durch Ausmessen der *Feldschwankungen* innerhalb einer gewissen Zeit ein quantitatives Maß der Aktivität herzuleiten. Dieses Prinzip enthält die Möglichkeit zahlreicher unterschiedlicher Definitionen. Beispiele für Maßzahlen dieser Art finden sich bei CHREE[3], BAUER[4], VAN DIJK[5] und anderen. Verschiedene Ansätze diskutiert A. SCHMIDT[6], darunter eine sehr einfach zu bestimmende tägliche Maßzahl, die gleich der Summe der maximalen Unterschiede in den Stundenmitteln der drei Komponenten während eines Tages ist. Bei einem Vorläufer dieser Maßzahl wurden die Stundenmittel eines Jahres zur Berechnung der Schwankungsweite des mittleren täglichen Ganges und des Mittels der täglichen Schwankungsweiten benutzt, wodurch eine Trennung der Anteile des mittleren periodischen und aperiodischen Tagesganges möglich wird[7]. Das Verfahren hat im Prinzip auch später wieder Anwendung gefunden.

Der Gedanke BIDLINGMAIERS wurde dann 1930 doch noch realisiert, indem sich IATME auf Vorschlag von CRICHTON MITCHELL für eine aus Gl. (13.2) abgeleitete Maßzahl aussprach. Unter der Bezeichnung *numerischer magnetischer Charakter* wurde sie durch die folgenden Ausdrücke definiert

$$X_0 R_X + Y_0 R_Y + Z_0 R_Z \quad \text{und} \quad H_0 R_H + Z_0 R_Z.$$

---

[1] F. BIDLINGMAIER: Veröffentl. Kaiserl. Obs. Wilhelmshaven, Neue Folge, H. 2 (Ergebn. magn. Beob. 1911), Berlin (1913).
[2] L. A. BAUER: Terr. Magn. Atmosph. Electr. **26**, 33—68 (1921).
[3] C. CHREE: Terr. Magn. Atmosph. Electr. **22**, 57—83 (1917).
[4] L. A. BAUER: Terr. Magn. Atmosph. Electr. **26**, 53—62 (1921); **27**, 31—34 (1922).
[5] G. VAN DIJK: Kon. Ned. Met. Inst. No. 102, Utrecht (1922).
[6] A. SCHMIDT: Terr. Magn. Atmosph. Electr. **25**, 123—138, (1920).
[7] A. SCHMIDT: Meteorol. Z. **33**, 481—492 (1916).

Die Bestimmung kann auf die eine oder andere Weise vorgenommen werden, je nachdem, ob an einem Observatorium X, Y, Z oder H, D, Z registriert werden. Dabei sind in den beiden Ausdrücken $\mathcal{H}_0$, X_0, $\mathcal{Y}_0$ und Z_0 die Tagesmittel dieser Feldkomponenten und $R_{\mathcal{H}}$, R_X, $R_{\mathcal{Y}}$ und R_Z die dazugehörigen, absolut genommenen maximalen Feldschwankungen, d.h. die täglichen Schwankungsweiten in den betreffenden Komponenten. Zu beachten ist aber, daß die diesen Schwankungsweiten zugrunde liegenden Extrema in den verschiedenen Komponenten im allgemeinen nicht gleichzeitig auftreten, so daß damit die Energiedichte des Störungsfeldes im strengen Sinne ihrer Festlegung durch Gl. (13.2) auch von dieser Charakterzahl nicht angegeben wird. Tägliche Werte wurden an über 30 Observatorien von 1931 bis 1939 berechnet. Ihre Bestimmung wurde dann zugunsten der BARTELSschen Kennziffer K aufgegeben.

β) Ein anderer Weg, Schwankungen in den erdmagnetischen Elementen in ein Aktivitätsmaß umzusetzen, wurde von CHAPMAN[8] 1925 eingeschlagen. Er geht aus von den stündlichen Werten einer Komponente, bestimmt die Unterschiede von Stunde zu Stunde und addiert deren absolute Beträge für einen Tag. Auf diese Weise können auch sehr alte Beobachtungsreihen mit benutzt werden, die üblicherweise als stündliche Terminablesungen vorliegen. CHAPMAN berechnete dieses tägliche Aktivitätsmaß für 63 Jahre aus der D-Komponente von Greenwich und später für 62 Jahre aus der H-Komponente von Greenwich.

In abgewandelter Form ist dieses Verfahren von CHERNOSKY[9,10] aufgegriffen worden, wobei er Stundenmittel anstelle der stündlichen Werte benutzt. Aus den absolut genommenen Differenzen der Stundenmittel wird ein Tagesmittel berechnet; dabei wird als 24. Differenz diejenige zum letzten Stundenmittel des Vortages herangezogen. Diese von CHERNOSKY als *Delta-Indices* bezeichneten Maßzahlen können wieder für jede Komponente bestimmt werden. Mit der Bezeichnung $\Delta \mathcal{H}_1$ für die H-Komponente liegen diese Werte für alle Tage der Jahre 1933 bis 1944 für Huancayo vor[10]. Das Delta-Maß $\Delta \mathcal{H}_1$ läßt sich verallgemeinern zu $\Delta \mathcal{H}_n$ (und entsprechend für andere Komponenten), wenn nicht von Mittelwerten über 1 Std, sondern von Mittelwerten über n Stunden ausgegangen wird. Für äquatoriale Stationen ähneln diese Delta-Indices den Ringstrommaßen (vgl. Ziff. 16).

γ) Die unter α) und β) angeführten Maßzahlen sind Beispiele von Versuchen, das wechselhafte und oft komplizierte Erscheinungsbild der Registrierungen in statistisch vereinfachter Weise als erdmagnetische Aktivität zu charakterisieren, ohne bereits bei der Definition von physikalischen Vorstellungen über die Ursache der auf diese Weise erfaßten Aktivität auszugehen. Zu dieser Gruppe gehört auch die wegen ihrer Bedeutung in Ziff. 14 gesondert behandelte internationale Charakterzahl Ci.

Mit zunehmender Einsicht in die Ursachen entstanden dann aber auch Maßzahlen für spezielle Arten der Aktivität. Dazu gehören vor allem die Angaben über die Stärke des Ringstroms (Ziff. 15 und 16) und spezifische Aktivitätsmaße für Observatorien in hohen Breiten (Ziff. 17). Für die in der Geophysik so wichtigen Erscheinungen der Erhaltungs- und Wiederholungsneigung wurden sekundäre Maßzahlen definiert, deren Berechnung auf Verfahren beruht, die genannten Erscheinungen in den primären Aktivitätsmaßen zum Vorschein zu bringen, sofern sie dort enthalten sind (Ziff. 18).

δ) Eine weitere Möglichkeit der Definition von Kennziffern besteht in der Schematisierung und Charakterisierung von *Kurvenformen*. Bereits in der fünfstufigen Aktivitätsskala von ESCHENHAGEN[11] aus dem Jahre 1892, einer der ältesten Klassifikationen der Aktivität überhaupt, spielt der Verlauf der Registrierkurven eine wesentliche Rolle. Eine spezifische Kennziffer für Kurvenformen wurde dann von BARTELS[12] zusammen mit seiner Kennziffer K für Potsdam eingeführt. Genauer muß dazu gesagt werden, daß die hier im Kapitel B behandelte Kennziffer ursprünglich die Bezeichnung K_1 hatte. Sie wurde ergänzt durch eine *zehnstufige Kennziffer* K_2, ebenfalls mit ganzzahligen Werten von 0 bis 9 für jedes Drei-Stunden-Intervall:

$K_2 = 0$: Glatter Verlauf, Pulsationen unter $0.5\,\gamma$ Amplitude.

    1: Pulsationen bis $5\,\gamma$ Amplitude in mindestens einem Element erkennbar.

    2: Besonders regelmäßige und deutliche Pulsationen mit 2 bis $5\,\gamma$ Amplitude in mindestens einem Element.

    3: Starke Pulsationen mit über $5\,\gamma$ Amplitude in mindestens einem Element.

---

[8] S. CHAPMAN: Phil. Trans. Roy. Soc. London, Ser. A **225**, 49—91 (1925); diskutiert in [*2*], Kap. 11. Terr. Magn. Atmosph. Electr. **46**, 385—400 (1941).

[9] E. J. CHERNOSKY: Trans. Am. Geophys. Union **37**, 339 (1956).

[10] E. J. CHERNOSKY, P. F. FOUGERE, and R. O. HUTCHINSON: The Geomagnetic Field. Handbook Geophys. Space Envir. A. F. Cambridge Res. Lab., Table 11—8 (1965).

[11] M. ESCHENHAGEN: Meteorol. Z. **9**, 450—454 (1892).

[12] J. BARTELS: Z. Geophys. **14**, 68—78 (1938).

4: Bay (oder Bays) von mindestens 10 γ Amplitude, ohne Pulsationen.
5: Bay, außerdem Pulsationen bis 5 γ.
6: Bay, außerdem Pulsationen über 5 γ.
7: Täglicher Störungsgang in Z (mehr als 20 γ), keine starken Pulsationen.
8: Täglicher Störungsgang in Z, gleichzeitig starke Pulsationen über 5 γ Amplitude.
9: Allgemein stürmischer Verlauf.

dazu $K_2 = s$: Plötzlicher Sturmausbruch, Anfangseffekt über 10 γ.

Bei Auftreten mehrerer dieser Formen wird die höchste Ziffer K_2 geschätzt; nur wenn Bay und Störungsgang in Z gleichzeitig auftreten, erhält die Erscheinung mit der größeren Amplitude den Vorrang.

Am Adolf-Schmidt-Observatorium in Niemegk ist auch diese Potsdamer Kennziffer K_2 für die Jahre 1938 bis 1944 bestimmt worden. Sie wurde 1949 von FANSELAU[13] durch Erweiterung auf 16 Klassen neu definiert. Dabei wird zur morphologischen Beschreibung des Kurvenverlaufs von vier Grundformen ausgegangen: Elementarwellen (Pulsationen), Baystörungen, Z-Störungsgang und allgemeiner Störpegel. In dieser Reihenfolge wird jeder Typ nach der Stärke seines Auftretens für jedes Drei-Stunden-Intervall durch eine Kennziffer charakterisiert. Seit 1953 gelten dafür nach JUST[14] die folgenden Schätzintervalle:

$K_2 =$		0	1	2	3
für Elementarwellen	bis:	1 γ	5 γ	10 γ	über 10 γ
für Baystörungen	bis:	20 γ	50 γ	100 γ	über 100 γ
für Z-Störungsgang	bis:	10 γ	30 γ	100 γ	über 100 γ
für Allgemeinen Störpegel	bis:	5 γ	25 γ	100 γ	über 100 γ

In dieser neuen Fassung wird K_2 fortlaufend in den Jahrbüchern des Observatoriums Niemegk veröffentlicht.

Zu dieser Art von Kennziffern gehören auch die Aktivitätszahlen für erdmagnetische Pulsationen. Auf der Grundlage der drei wichtigsten Parameter einer Pulsationserscheinung, nämlich Periode, Doppelamplitude und Anzahl der Schwingungen, lassen sich *Pulsationszahlen* unterschiedlicher Art definieren. Beispiele für derartige Maßzahlen und ihre Anwendung enthalten die Untersuchungen von ANGENHEISTER[15,16] und SAITO[17]. Durch den Einsatz moderner elektronischer Registrier- und Auswerteverfahren ist besonders bei der statistischen Bearbeitung von Pulsationen mit einer schnellen Entwicklung zu rechnen, weshalb auf eine eingehende Darstellung der bisher vorliegenden Methoden und Ergebnisse verzichtet werden soll.

14. Die tägliche internationale Charakterzahl Ci. Sie ist die älteste planetarische Maßzahl, die zur Charakterisierung der Aktivität eines Tages fortlaufend bestimmt wird.

Ausgangspunkt ihrer Einführung waren die von der damaligen internationalen erdmagnetischen Kommission im letzten Jahrzehnt des 19. Jahrhunderts beschlossenen Empfehlungen über die Bildung der Monatsmittel des täglichen Ganges. Dazu war eine Unterscheidung von gestörten und ungestörten Tagen erforderlich. Die Entscheidung, dies mit Hilfe der *Charakterzahl C* zu tun, fiel 1905 auf einem internationalen Kongreß in Innsbruck. Die Charakterzahl selbst wurde von A. SCHMIDT[1] vorgeschlagen, der dazu auf die Erfahrungen mit der in Ziff. 13 erwähnten fünfstufigen Aktivitätsskala von ESCHENHAGEN zurückgriff, die er für den praktischen Gebrauch durch Reduzierung auf drei Schätzstufen und nur eine Schätzung für den ganzen Tag (anstelle der Halbtagsschätzungen bei ESCHENHAGEN) vereinfachte. Mit dem 1. Januar 1906 begann die Bestimmung dieses Aktivitätsmaßes.

Das Verfahren zur Charakterisierung der Aktivität ist äußerst einfach: Anhand des täglichen Magnetogramms klassifiziert der Auswerter den Tag als ruhig ($C = 0$), mäßig gestört ($C = 1$) oder stark gestört ($C = 2$). Diese Beurteilung ist bis zu einem gewissen Grad subjektiv; und die Streuung in den Angaben der

---

[13] G. FANSELAU: Z. Meterol. **3**, 236—240 (1949).
[14] H. JUST: Z. Meteorol. **7**, 95—96 (1953).
[15] G. ANGENHEISTER: Gerlands Beitr. Geophys. **64**, 108—132 (1954).
[16] G. ANGENHEISTER u. C. v. CONSBRUCH: Z. Geophys. **27**, 3—12, 103—111 (1961).
[17] T. SAITO: Rep. Ionosph. Space Res. Japan **18**, 260—274 (1964).
[1] A. SCHMIDT: Meteorol. Z. **33**, 481—492 (1916).

verschiedenen Observatorien ist recht groß. Eine teilweise Objektivierung wird an einer Reihe von Observatorien dadurch erreicht, daß die Klassen von C jetzt an bestimmte Intervalle von K oder $\sum K$ oder direkt an Schwankungsweiten in den Registrierungen angeschlossen werden.

Ein brauchbares Aktivitätsmaß folgt aus den lokalen Charakterzahlen aber erst durch Bildung des Mittelwerts der täglichen Werte einer größeren Anzahl von Observatorien. Das ergibt als weltweites Aktivitätsmaß die *internationale Charakterzahl*, für die jetzt die Bezeichnung Ci gebräuchlich ist. Sie wird auf eine Dezimale berechnet und kann daher die 21 verschiedenen Werte 0,0 ... 2,0 annehmen. Die Anzahl der zur Berechnung von Ci beitragenden Observatorien hat sich im Laufe der Zeit mehrmals geändert. In der ersten Zeit nach Einführung der Charakterzahl wurden die Werte aller sich beteiligenden Observatorien benutzt. Das waren 1906 einige 30, später mehr. Als ihre Zahl weiter anstieg, mußte eine Auswahl getroffen werden. Zur Zeit melden etwa 130 Observatorien ihre C-Werte an das $C+K$ *Centre* in De Bilt, von denen etwa 30 zur Berechnung von Ci herangezogen werden. Tabellen mit den Werten dieser Maßzahl finden sich in den IAGA Bulletins No. 12. Ihre Bestimmung wurde nachträglich von van Dijk und Bartels auch auf die Jahre vor 1906 bis zum Jahr 1884 ausgedehnt, so daß diese Tageswerte der Aktivität jetzt über den Zeitraum von etwa acht Sonnenfleckencyclen vorliegen.

Die heutige Bedeutung dieser Maßzahl liegt in der Länge der Beobachtungsreihe. Schon früh ist aber darauf hingewiesen worden, daß als Folge der subjektiven Schätzung eine Maßstabsänderung vor allem mit dem Sonnenfleckencyclus nicht ausgeschlossen werden kann; denn eine magnetische Störung im Sonnenfleckenminimum wird subjektiv leicht stärker bewertet als die gleiche Störung im Sonnenfleckenmaximum, was sich besonders auf den Vergleich von Jahresmitteln auswirkt. Die ursprüngliche Aufgabe der Bestimmung der internationalen ruhigen und gestörten Tage erfolgt heute mit Hilfe der planetarischen Kennziffer Kp, wie in Ziff. 11 unter β) ausgeführt wurde. Mit der in Ziff. 10 behandelten *planetarischen Charakterzahl Cp* ist außerdem für die Jahre von 1932 an eine Maßzahl gegeben, die bei sonst gleichen Eigenschaften auf Grund ihrer Bestimmung aus Kp und ap ein objektiveres Maß darstellt.

15. Das u-Maß der interdiurnen Veränderlichkeit von $\mathcal{H}_D$ am geomagnetischen Äquator. Zur Berechnung dieser Maßzahl wird von dem Unterschied der Tagesmittel der Horizontalkomponente $\mathcal{H}$ (oder der Nordkomponente X) einer Station P an jeweils aufeinanderfolgenden Tagen ausgegangen. Der Absolutwert dieses Unterschiedes wird die *interdiurne Veränderlichkeit* von $\mathcal{H}$ (oder X) genannt und soll hier mit $u(\text{P})$ bezeichnet werden. Als lokales Aktivitätsmaß ist $u(\text{P})$ als das wohl älteste Aktivitätsmaß bereits 1861 von Broun[1] verwendet worden.

Für $u(\text{P})$ wurde von Moos[2] die historisch erste lange homogene Zeitreihe einer Maßzahl aufgestellt, und zwar für Bombay für die Jahre 1872 bis 1920. Beginnend mit dem Jahr 1920, nahm A. Schmidt[3] seine Bestimmung für Potsdam auf. Mit Werten für u' (s. unten) in den Jahrbüchern des Observatoriums Niemegk wird diese Reihe laufend fortgesetzt.

Der wesentliche Schritt, aus $u(\text{P})$ ein weltweites Maß für die Ringstromstärke abzuleiten, stammt wieder von Bartels[4] und beruht auf der folgenden Überlegung: Für Stationen in nicht zu hohen Breiten mißt $u(\text{P})$ vor allem die Stärke

---

[1] J. A. Broun: Trans. Roy. Soc. Edinburgh **22**, Part 3, 539 (1861).
[2] N. A. F. Moos: Colaba Magn. Data. Magn. Obs. Bombay, 782 pp. (1910); fortgesetzt in den folgenden Bänden.
[3] A. Schmidt: Meteorol. Z. **40**, 186—189 (1923); s. auch: Ergebn. magn. Beob. Potsdam u. Seddin 1921.
[4] J. Bartels: Meteorol. Z. **40**, 301—305 (1923).

und das Abklingen des Ringstroms während der Hauptphase und der Nachstörung eines magnetischen Sturmes (vgl. Ziff. 4δ). Das Ringstromfeld kann im Bereich der Erde näherungsweise als homogen angesetzt werden mit einer zur Dipolachse parallelen Richtung. Der Ablauf der Störung wird dann allein durch einen weltweit gültigen Zeitfaktor bestimmt. Damit können die Werte $u(P)$ verschiedener Stationen durch eine einfache Umrechnung auf vergleichbare Äquatorwerte u' reduziert werden:

$$u' = \frac{u(P)}{\sin \vartheta \cos(\psi - D)}. \tag{15.1}$$

In dieser Beziehung ist ϑ die geomagnetische Poldistanz der Station P[5] und $(\psi - D)$ der Winkel zwischen magnetisch Nord und geomagnetisch Nord in P[6]. Naturgemäß differieren die u'-Werte verschiedener Stationen untereinander, so daß erst ihr Mittelwert u für eine gewisse Zahl von Stationen die Eigenschaft eines weltweiten Aktivitätsmaßes besitzt. Dies ist die interdiurne Veränderlichkeit der Horizontalkomponente $\mathcal{H}_D$ im geomagnetischen oder *Dipol*-Koordinatensystem am geomagnetischen Äquator.

In [1] (Tabelle 2) hat BARTELS monatliche u-Werte (in der Einheit 10 γ) für die Jahre 1872 bis 1930 angegeben und zu detaillierten Untersuchungen über solar-terrestrische Beziehungen herangezogen. Die Berechnung der Werte hat er noch in der Weise modifiziert, daß lediglich die $u(P)$ für Potsdam-Seddin nach Gl. (15.1) auf Äquatorwerte umgerechnet wurden, während für die anderen verwendeten Stationen die Reduktionsfaktoren durch statistischen Vergleich der einzelnen $u(P)$-Reihen mit der Potsdamer Reihe ermittelt wurden. Diese anderen, mit unterschiedlich langen Reihen an dem u-Maß beteiligten Observatorien sind Greenwich, Bombay, Batavia, Honolulu, Porto Rico, Tucson und Watheroo. Ferner enthält [1] Jahresmittel von u in sich überschneidenden Halbjahresabständen von 1835 bis 1872. Diese Werte sind durch andere Einflüsse stärker verfälscht als die Monatsmittel von 1872 an, worauf BARTELS[7] selbst aufmerksam macht. Jedoch gibt es keine andere Maßzahl, für die Angaben so weit in die Vergangenheit zurückreichen. Für die Zeit nach 1930 ist die Reihe der Monatsmittel unter Heranziehung einer wechselnden Zahl von Observatorien bis 1946 fortgesetzt worden. Endgültige Werte für 1872 bis 1936 enthält Tabelle E im Anhang der ersten Auflage von [2]. In der bequemeren Einheit γ findet sich eine Zusammenstellung aller u-Werte in [4] (Tabelle 13). Für die späteren Jahre können die u'-Werte von Niemegk als vorläufige u-Werte benutzt werden.

Seiner Definition nach gehört u zu den linearen Aktivitätsmaßen. Bereits *ein* starker magnetischer Sturm im Monat hebt den Wert von u beträchtlich an. Da dieser Effekt für manche statistischen Untersuchungen unerwünscht ist, hat BARTELS dem linearen u-Maß ein quasi-logarithmisches u_1-Maß zugeordnet. Anstelle der komplizierten Definition, die er für u_1 in [1] angibt, soll hier für einen Überblick die Umrechnungstabelle aus [2] (Kap. 11) genügen:

$u =$	0,3	0,5	0,7	0,9	1,2	1,5	1,8	2,1	2,7	$\geq 3,6$
$u_1 =$	0	20	40	57	79	96	108	118	132	140

Die Einheit von u in der Tabelle ist 10 γ. Die zahlenmäßige Festlegung der u_1-Skala wurde so vorgenommen, daß die Häufigkeitsverteilung der u_1-Werte derjenigen der Sonnenfleckenrelativzahl ähnlich ist. BARTELS benutzt auch das u_1-Maß in [1] ausgiebig für statistische Untersuchungen. Man findet dort sowohl Monatsmittel (Tabelle 4) als auch Jahresmittel in sich überschneidenden Halbjahresabständen von 1872 bis 1930 zahlenmäßig aufgeführt.

---

[5] Das ist der Winkel zwischen dem Durchstoßpunkt der Dipolachse bei Thule in Grönland und P, gemessen im Erdmittelpunkt.

[6] Das ist der Winkel zwischen der Richtung der Kompaßnadel in P und dem durch P und die beiden Durchstoßpunkte der Dipolachse gehenden geomagnetischen Meridian. Denselben Winkel ergibt die Differenz $(\psi - D)$, wobei ψ der Winkel zwischen astronomischem und geomagnetischem Meridian in P (nach Osten positiv) ist und D der Winkel der magnetischen Deklination in P (nach Osten positiv). Wird anstelle von $\mathcal{H}$ von den Tagesmitteln von X ausgegangen, so tritt in Gl. (15.1) der Winkel ψ an die Stelle von $(\psi - D)$.

[7] J. BARTELS: FIAT Rev. **17**, Geophys. Teil 1, S. 71 (1948).

Vom u-Maß gibt es keine kurzfristigeren Werte als monatliche Werte. Die statistischen Eigenschaften dieses Maßes sind außer in [1] ausführlich in [2] (Kap. 11) und in einer Diskussion von Howe[8] und Bartels[9] dargestellt worden. Die physikalische Bedeutung folgt aus seiner Herleitung, wonach u die *Stärke und Häufigkeit der großen Stürme* mißt. Der Beitrag der kleineren Störungen ist gering. In dieser Hinsicht ist u ein echtes Ringstrommaß und damit das Maß einer speziellen erdmagnetischen Aktivität. Für diese Aufgabe ist es aber mit dem Fehler aller von Tagesmitteln ausgehenden Maßzahlen behaftet, daß die Einflüsse der ionosphärischen **DS**[10] und S_q-Stromsysteme im Tagesmittel nicht völlig eliminiert sind. So wird beispielsweise das Tagesmittel von der Stärke des **DS**-Feldes an dem betreffenden Tag mitbestimmt. Dieser Einfluß ist aber von Tag zu Tag verschieden. Auch sind die Tagesmittel nicht unabhängig vom willkürlich festgelegten Zeitpunkt des Tagesbeginns, etwa bei Benutzung des Greenwich-Tages.

Nachdem inzwischen geeignetere Maßzahlen mit besserer Zeitauflösung für die Stärke des magnetosphärischen Ringstroms existieren (s. Ziff. 16), ist die Berechnung von u für diese Aufgabe nicht mehr erforderlich. Andererseits ist die Fortsetzung dieser recht homogenen Reihe schon wegen ihrer ungewöhnlichen Länge wünschenswert, aber auch zur einfachen Charakterisierung der Sturm-Aktivität für längere Zeitabschnitte.

16. Tägliche, dreistündliche und stündliche Maßzahlen für die Stärke des Ringstroms.

α) Ringstromwerte nach Vestine[1]. Bei dem von Vestine angewandten Verfahren wird in noch elementarer Weise von den beiden wichtigsten Aufgaben bei der Berechnung von Ringstromwerten ausgegangen, nämlich der Eliminierung der von ionosphärischen Stromsystemen verursachten täglichen Schwankungen an ruhigen und gestörten Tagen (also im wesentlichen $S_q + $ **DS**) und der Eliminierung der Säkularvariation. Zugrunde gelegt werden mit gleichen Gewichten die Beobachtungen der H-Komponente an den Observatorien San Juan und Cheltenham. Zur Ausschaltung der störenden Tagesgänge werden die Tagesmittel von H benutzt, ein Verfahren, auf dessen Unvollkommenheit schon in Ziff. 15 hingewiesen worden ist. Die Säkularvariation wird bei Vestine dadurch unterdrückt, daß die Abweichungen der Tagesmittel vom Monatsmittel gebildet werden. Dadurch gehen einmal sehr langperiodische Ringstromvariationen verloren, und zum anderen können am Anfang und am Ende eines Monats nichtreelle Ringstromeffekte auftreten als Folge von Sprüngen im Bezugsniveau zwischen zwei Monaten. Trotz dieser Mängel sind die Vestineschen Werte als Tagesmittel der Ringstromvariationen brauchbar. Sie liegen für die Jahre 1905 bis 1942 vor, in denen es sonst keine gleichwertigen Maßzahlen für die Stärke des Ringstroms mit dieser Zeitauflösung gibt.

β) Ringstromwerte U nach Kertz[2]. Bei der aus vier Schritten bestehenden Berechnung der dreistündlichen U-Werte wird zum ersten Male eine möglichst vollständige Eliminierung der störenden Anteile und die Bestimmung eines absoluten Niveaus für die Ringstromstärke angestrebt. Zu diesem Zweck wird in

---

[8] H. H. Howe: J. Geophys. Res. **55**, 153—157 (1950).
[9] J. Bartels: J. Geophys. Res. **55**, 158—160 (1950).
[10] Zur Definition von **DS** s. T. Nagata u. N. Fukushima in diesem Band, Kap. F, S. 55, sowie S. Chapman: Studia Geophys. Geodet. **5**, 30—50 (1961).
[1] E. H. Vestine: Carnegie Inst. Wash. Publ. **580**, 119—128 (1947); **578**, 45—63 (1948).
[2] W. Kertz: Abhandl. Akad. Wiss. Göttingen, Math.-Phys. Kl., Beitr. IGJ, H. 2, 1—83 (1958).

stärkerem Maße von statistischen Methoden Gebrauch gemacht. Grundlage der Berechnung sind dreistündliche Mittel der Nachtwerte von $\mathcal{H}$ an äquatornahen Observatorien. Äquatornähe empfiehlt sich, weil dort der Einfluß von DS am geringsten ist, während der Ringstrom mit voller Stärke in $\mathcal{H}$ zur Geltung kommt. Durch Dreistundenmittel werden außerdem kurzzeitige Schwankungen des DS-Feldes unterdrückt. Die Verwendung von Nachtwerten bewirkt schließlich, daß $(S_q + L)$ ausgeschaltet wird. Um dann aber alle Drei-Stunden-Intervalle eines Tages zu erfassen, müssen solche Observatorien in verschiedenen geographischen Längen benutzt werden, deren Nachtstunden sich für jeweils ein Tages-Achtel überlappen.

Die vier Berechnungsschritte können hier nur skizziert werden: 1. Die Dreistundenmittel werden nach Abzug einer Konstanten von der ungefähren Größe des Hauptfeldes am betreffenden Observatorium vermittels der Transformation Gl. (15.1) auf den geomagnetischen Äquator reduziert. 2. Zur Eliminierung der Säkularvariation wird diese für Huancayo direkt bestimmt. Die anderen Observatorien werden nach einem statistischen Verfahren an diesen Gang angeschlossen. Nach Abzug der Säkularvariation ergeben sich für magnetisch ruhige Zeiten ($Kp \leq 2+$) bereits brauchbare Ringstromwerte für die Mitternachtsintervalle der Observatorien. 3. In den Werten für die dem Mitternachtsintervall benachbarten Tages-Achtel und insbesondere in gestörten Zeiten ist auch noch am Äquator ein Einfluß von DS vorhanden. Um diese Störungen zu eliminieren, wird davon ausgegangen, daß das nächtliche zum DS-Feld gehörige $\mathcal{H}$ an Äquatorstationen nahezu antisymmetrisch zum Mitternachtswert verläuft. Da die sich überlappenden Tages-Achtel benachbarter Observatorien jeweils einen Vormitternachtswert und einen Nachmitternachtswert liefern, ist das Mittel dieser beiden Werte weitgehend von DS befreit und als Ringstromwert brauchbar. 4. Damit fehlen nur noch die Werte für die Mitternachts-Achtel der magnetisch gestörten Zeiten. Für sie erweist sich das arithmetische Mittel aus den beiden nach 3. gegebenen Werten der benachbarten Tages-Achtel und dem nach 2. schon vorliegenden Wert des (noch durch DS gestörten) Mitternachts-Intervalls als geeignete Größe, wobei der Mitternachtswert nach 2. doppeltes Gewicht erhält.

Der Versuch, diese von KERTZ als U-Maß bezeichneten Werte auf ein absolutes Niveau zu beziehen, ist nicht ganz frei von Willkür. Nach KERTZ ist der Fehler bei der Basisbestimmung aber nicht größer als $10 \ldots 20\, \gamma$. Er hat damit etwa die Größe, um die U ohnehin wegen des nur näherungsweise abzuschätzenden Ringstromfeldes während magnetisch ruhiger Zeiten unsicher ist. Aber auch dann ist bei der Interpretation dieses Maßes als Stärke des magnetosphärischen Ringstroms noch zu beachten, daß infolge der elektromagnetischen Induktion des zeitlich variablen äußeren Ringstromfeldes Ströme im leitfähigen Erdinnern induziert werden, deren Magnetfeld bei stärkeren Stürmen nach SCHREIBER[3] etwa ein Viertel zu der beobachteten Horizontalkomponente des Ringstromfeldes (und also auch zu U) beiträgt. Außerdem muß nach den heutigen Vorstellungen ein kleiner Teil des von außen kommenden und sich nach Weltzeit ändernden Störungsfeldes den Grenzflächenströmen in der Magnetopause und den elektrischen Strömen im Schweif der Magnetosphäre zugeschrieben werden (s. Ziff. 4 δ).

Die zuerst berechneten dreistündlichen Werte für die sieben Jahre 1939 bis 1945 basieren auf den Daten der vier Observatorien Huancayo, Apia, Watheroo und Elisabethville. Das ist die kleinste Zahl von Stationen, mit der dieses Verfahren realisierbar ist. Tabellen mit den Werten für diese Jahre finden sich in [2]. Die Einheit der U-Werte ist $3\, \gamma$. Außerdem sind die in den Tabellen aufgeführten Zahlen gegenüber der Definition von U um 10 Einheiten ($=30\, \gamma$) verkleinert, um für ungestörte Zeiten kleinere Zahlen zu erhalten.

Nach dem gleichen Prinzip, aber mit geringfügigen Modifikationen, wurden von KERTZ[4] U-Werte auch für das Internationale Geophysikalische Jahr berechnet, genauer: für die Zeit von Mai 1957 bis Dezember 1958. Die wesentlichste Modifikation des Verfahrens besteht darin, daß diesmal die Daten von 27 Observatorien zugrunde gelegt wurden, die entsprechend ihrer Länge in acht Gruppen zusammengefaßt und für die Gruppenmittel gebildet wurden. Hinsichtlich der Bedeutung der Tabellenwerte gilt das zuvor Gesagte.

---

[3] H. SCHREIBER: Mitt. Max-Planck-Inst. Aeronomie, Lindau, Nr. 35, S. 26 (1968).
[4] W. KERTZ: Ann. Intern. Geophys. Year 35, 49—61 (1964).

γ) Ringstromwerte Dst nach SUGIURA[5]. Auch bei der Herleitung dieser stündlichen Werte wird besondere Mühe darauf verwandt, die störenden Anteile auszuschalten und ein absolutes Niveau zu garantieren. Das Verfahren unterscheidet sich jedoch in wesentlichen Punkten von der Berechnung der U-Werte nach KERTZ. Es wurde in seiner ursprünglichen Form[5] nur für die Bestimmung der Dst-Werte der anderthalb Jahre des Internationalen Geophysikalischen Jahres 1957/58 benutzt. Für die spätere Berechnung der Jahrgänge 1961 bis 1968 wurde das Verfahren in einigen Punkten verbessert. Die folgende kurze Beschreibung schließt sich in diesen Punkten der Darstellung von SUGIURA und HENDRICKS[6] an.

Das Ausgangsmaterial sind die stündlichen Werte der Horizontalkomponente $\mathcal{H}$ einer Reihe äquatornaher Observatorien. Da die $\mathcal{H}$-Werte des ganzen Tages benutzt werden, ist die Eliminierung des S_q-Ganges (einschließlich eines etwaigen Anteils vom äquatorialen Elektrojet) eines der Hauptprobleme. Dazu wird zunächst ein mittleres S_q an jedem Observatorium und für jeden Monat aus den Tagesgängen jener fünf Tage nach Ortszeit bestimmt, die sich mit den fünf internationalen ruhigen Tagen (s. Ziff. 11 β) maximal überschneiden. Nach linearer Anbringung der Mitternachtskorrektur (Lamontsche Korrektur) werden die S_q-Gänge für die zwölf Monate eines Jahres in eine doppelte Fourier-Reihe mit der laufenden Nummer des Monats und der Weltzeit als Variable entwickelt. Diese Darstellung erlaubt es, für jede Stunde des Jahres einen S_q-Wert zu berechnen und als Korrektur an dem dazugehörigen $\mathcal{H}$-Wert anzubringen.

Auch die Säkularvariation wird für jedes Observatorium getrennt ermittelt, indem aus den $\mathcal{H}$-Werten der zehn ruhigen Tage eines jeden Monats ein Jahresmittel bestimmt wird. Aus den Mittelwerten von rund zehn Jahren wird für die Säkularvariation eine Potenzreihendarstellung bis zum quadratischen Glied in der Zeit berechnet. Damit können dann die Korrekturen zur Beseitigung der Säkularvariation bereits an den stündlichen $\mathcal{H}$-Werten vorgenommen werden.

Die so an den verwendeten Observatorien vorliegenden korrigierten $\mathcal{H}$-Werte werden dann durch Bildung des Mittelwertes der zur gleichen Weltzeitstunde eines Tages gehörenden Werte zusammengefaßt. Das ergibt aber gerade den nach Weltzeit ablaufenden Ringstrom-Effekt. Allerdings kann bei dieser Operation noch ein kleines mittleres **DS**-Feld übrigbleiben. Wegen der Äquatornähe und der nahezu gleichförmigen Verteilung der Stationen mit der Länge wird dieser Einfluß von **DS** aber als unbedeutend vorausgesetzt. Dann sind die Mittel der Stationswerte unmittelbarer Ausdruck der Stärke des Ringstroms. Zum Zwecke der Normierung werden sie noch durch Multiplikation mit $\sec \Phi_m$ auf Äquatorwerte umgerechnet, wobei Φ_m die mittlere geomagnetische Breite der verwendeten Observatorien ist. Der auf diese Weise bestimmten stündlichen Maßzahl für die Ringstromstärke hat SUGIURA die gleiche Bezeichnung Dst gegeben, wie sie für die mittlere Sturmzeit-Variation[7] gebräuchlich ist.

Die zunächst nur für die Zeit des Internationalen Geophysikalischen Jahres berechneten Werte[5] basieren auf den $\mathcal{H}$-Werten der Observatorien Hermanus, Alibag, Kakioka, Apia, Honolulu, San Juan, Pilar und M'Bour. Dasselbe gilt für die Dst-Werte des Jahrgangs 1961, während den in der gleichen Veröffentlichung[6] ebenfalls mitgeteilten Werten der Jahrgänge 1962 und 1963 nur noch die Daten von Hermanus, San Juan und Honolulu zugrunde liegen. Daten dieser drei Observatorien wurden von SUGIURA und CAIN[8] auch zur Berechnung der Dst-Werte für die Jahre 1964 bis 1968 benutzt. Beginnend mit dem Jahrgang 1961, tragen alle Werte die Bezeichnung vorläufig. Die Reihe wird fortgesetzt.

---

[5] M. SUGIURA: Ann. Intern. Geophys. Year **35**, 9—45 (1964).
[6] M. SUGIURA and S. HENDRICKS: NASA Techn. Note D-4047, Washington, D.C. (1967).
[7] Zur Definition von **Dst** siehe T. NAGATA u. N. FUKUSHIMA in diesem Band, Kap F, S. 55, sowie S. CHAPMAN: Studia Geophys. Geodet. **5**, 30—50 (1961).
[8] M. SUGIURA and S. J. CAIN: Goddard Space Flight Center X-612-69-20 (1969); X-612-70-3 (1970).

Bei einem Vergleich des Dst-Maßes von SUGIURA mit dem U-Maß von KERTZ ist zu beachten, daß Dst und U von ihrer Definition her entgegengesetztes Vorzeichen haben und die Einheit von Dst: 1 γ und die von U: 3 γ beträgt. Wegen der stündlichen Werte ist die Zeitauflösung beim Dst-Maß größer. Ferner verlaufen die U-Kurven glatter, während die Dst-Kurven unruhiger sind. Im Vergleich mit den anderen in Gebrauch befindlichen planetarischen Aktivitätsmaßen besteht die stärkste Ähnlichkeit mit ap (s. Ziff. 9α). Erwartungsgemäß nimmt aber ap in der Erholungsphase eines magnetischen Sturms schneller ab als die Ringstromwerte.

17. Spezifische Aktivitätsmaße für Observatorien in hohen Breiten.

α) *Storminess S* nach BIRKELAND[1]. Diese Maßzahl geht auf ein Verfahren zurück, das an älteren Observatorien in Indien und Batavia in Gebrauch war. Als Störungsmaß für Observatorien in hohen Breiten wurde S von BIRKELAND eingeführt. Es ist für jede der drei Komponenten H, D und Z definiert als der im Mittel über eine bestimmte Zeit beobachtete Unterschied zwischen dem tatsächlichen Verlauf der Registrierung und dem zu dieser Zeit erwarteten normalen, d.h. ungestörten Verlauf im *absoluten Niveau*. Daraus ergibt sich sofort als Hauptproblem die einwandfreie Bestimmung des normalen Tagesganges für jeden Tag.

BIRKELAND ging dazu von den Stundenmitteln der drei Komponenten aus, wählte die ruhigen Tage oder auch kürzere ruhige Intervalle eines Monats aus und berechnete aus diesen Stundenmitteln den mittleren normalen Tagesgang für den betreffenden Monat. Dieser mittlere Tagesgang dient zur Interpolation während gestörter Zeiten, indem er den ungestörten Stunden eines gestörten Tages angepaßt und so auf das absolute Niveau des Tages bezogen wird. Damit liegen für jeden Tag wenigstens in gewisser Annäherung die Stundenmittel des individuellen normalen Ganges dieses Tages vor und können von den beobachteten Stundenmitteln abgezogen werden, woraus die stündlichen Werte von S in γ resultieren; sie können positiv und negativ sein. Die Summe der absolut genommenen Werte der Storminess für Tage, Monate, Jahre charakterisiert dann den Störungsgrad dieser Zeitabschnitte.

Eine regelmäßige Berechnung von S-Werten ist auf norwegische Observatorien beschränkt geblieben, und zwar liegen in Jahrbüchern veröffentlichte stündliche Werte vor für Dombås von 1916 bis 1951 und für Tromsø seit 1930 fortlaufend. Einzelheiten zur praktischen Handhabung des Verfahrens an diesen Observatorien geben HARANG, KROGNESS und TØNSBERG[2] sowie KROGNESS und WASSERFALL[3]. Eine verbesserte Methode der Bestimmung des normalen Tagesganges haben GJELLESTAD und DALSEIDE[4] entwickelt, indem sie bei ihrem Ansatz von den jetzigen Kenntnissen über Ursache und Veränderlichkeit der magnetischen Variationen an ruhigen Tagen ausgehen und für den wesentlich größeren Rechenaufwand ihrer Methode die modernen Hilfsmittel einsetzen.

β) *Der Q-Index* nach BARTELS und FUKUSHIMA[5]. Anlaß der Einführung dieser Maßzahl waren die im Programm des Internationalen Geophysikalischen Jahres vorgesehenen viertelstündlichen Polarlicht-Beobachtungen an Stationen in hohen Breiten, verbunden mit der Überlegung, daß für einen Vergleich mit der erdmagnetischen Aktivität die an den Observatorien bestimmten dreistündlichen

---

[1] K. BIRKELAND: The Norwegian Aurora Polaris Expedition 1902—1903. Sect. I (1908); Sect. II (1913).
[2] L. HARANG, O. KROGNESS, and E. TØNSBERG: Publ. Norske Inst. Kosmisk Fys. Nr. 2 (1933).
[3] O. KROGNESS and K. F. WASSERFALL: Publ. Norske Inst. Kosmisk Fys. Nr. 9 (1936).
[4] G. GJELLESTAD and H. DALSEIDE: Årbok Univers. Bergen, Mat.-Nat. Ser., No. 7, 1—18 (1963).
[5] J. BARTELS u. N. FUKUSHIMA: Abhandl. Akad. Wiss. Göttingen, Math.-Phys. Kl., Sonderheft Nr. 3, 1—36 (1956).

Kennziffern K eine viel zu geringe Zeitauflösung haben. Der Q-Index ist daher gedacht als ein viertelstündliches Aktivitätsmaß zur Ergänzung der Kennziffer K.

Allerdings kann das Prinzip der K-Messung nicht auf Q übertragen werden. Für ein Drei-Stunden-Intervall sind die in dieser Zeit auftretenden Feldschwankungen durchaus ein Maß für die Störungsintensität und damit für die Definition von K (nach Eliminierung der Nicht-K-Variationen) geeignet; dagegen können innerhalb einer Viertelstunde die Schwankungen des Feldes vergleichsweise klein gegen den Betrag des Störungsfeldes selbst bleiben. Daher muß Q wie zuvor S auf den Verlauf des *absolut* festgelegten normalen Tagesganges bezogen werden. Q wird bestimmt für jedes 15-Minuten-Intervall, das um die Stundenviertel 00, 15, 30 und 45 Minuten nach jeder vollen Stunde zentriert ist. Der Wert von Q beruht bei einseitiger Abweichung der Registrierkurve von der Normalkurve auf dem Maximum Δ dieses Unterschiedes in der betrachteten Viertelstunde. Treten in dieser Zeit Abweichungen nach beiden Seiten auf, so ist Δ die Summe der absolut genommenen maximalen positiven und negativen Abweichung, also in diesem Fall wieder die maximale Feldschwankung. Ferner wird Q als lokales Maß für Stationen in der Nähe der Polarlichtzonen nur aus der größten Komponente des horizontalen Störungsvektors bestimmt, wodurch der weitreichende Einfluß der polaren Elektrojets in der Vertikalkomponente Z ausgeschaltet wird. Den numerischen Zusammenhang zwischen dem Δ der größeren Horizontalkomponente und Q gibt die folgende Tabelle:

$\Delta =$	0 ...	10 ...	20 ...	40 ...	80 ...	140 ...	240 ...	400 ...	660 ...	1000 ...	1500 ...	2200 γ ...
$Q =$	0	1	2	3	4	5	6	7	8	9	10	11

Diese 12stufige *quasi-logarithmische Skala* für Q gilt für alle Observatorien. Zur einziffrigen Bezeichnung der beiden letzten Klassen wird vorgeschlagen, 10 durch den Buchstaben T und 11 durch E zu ersetzen. Die jeweils obere Klassengrenze von Δ gehört noch zu dieser Klasse und damit zu dem kleinen Q-Wert.

Analog dem Zusammenhang zwischen K und a_k in Ziff. 6β ist auch zu Q ein *lineares* Aktivitätsmaß $\Delta(Q)$, die *äquivalente Abweichung*, definiert worden:

$Q =$	0	1	2	3	4	5	6	7	8	9	10	11
$\Delta(Q) =$	0	3	6	12	22	38	64	105	165	250	370	500

Falls diese äquivalente Abweichung wieder als Störungsfeld interpretiert werden soll, ist der betreffende Wert von $\Delta(Q)$ mit 5γ zu multiplizieren.

Hinweise für die Praxis der Schätzung von Q-Werten gibt BARTELS in [5], wobei die praktische Bestimmung von K (s. Ziff. 5γ) in vielen Punkten Vorbild ist. Die eigentliche Aufgabe dabei ist die Konstruktion der Normalkurven, die für die beiden horizontalen Komponenten im Magnetogramm erscheinen würden, wenn in dem betrachteten Intervall keine Störung durch solare Partikelstrahlung aufgetreten wäre. Da nach einer Empfehlung der IAGA die Bestimmung von Q-Indices an Observatorien in höheren geomagnetischen Breiten als 58° Nord und Süd vorgenommen werden soll, braucht hier bei der Bestimmung der Normalkurve nicht der gleiche Aufwand getrieben zu werden, wie dies an äquatornahen Stationen mit ihrem starken S_q-Gang der Fall wäre.

Viertelstündliche Q-Werte wurden von einer Reihe von Observatorien während des Internationalen Geophysikalischen Jahres und zum Teil auch darüber hinaus bestimmt. Eine fortlaufende Reihe liegt für Sodankylä vor.

γ) **Der stündliche R-Index**[6]. Die zuverlässige Bestimmung der zuvor besprochenen Maßzahlen S und Q über Jahre hin erfordert einige Mühe. Daher hat sich an vielen Observatorien in hohen Breiten als wesentlich einfacheres Aktivitätsmaß der R-Index eingebürgert, bei dem die Konzeption der Schwankungsweite

---

[6] Die Verwendung von R für dieses erdmagnetische Schwankungsmaß leitet sich von dem englischen Wort *Range* her. Der hier definierte Index sollte nicht mit der Sonnenfleckenrelativzahl R in Ziff. 20, ff. verwechselt werden.

in dem ursprünglichen Sinn einiger in Ziff. 13α zitierter Maßzahlen wieder aufgegriffen worden ist. Der wichtigste Unterschied zwischen diesen und dem R-Index besteht darin, daß für R stündliche Werte angegeben werden. Genauer gesagt, ist R die *absolut genommene stündliche Schwankungsweite* in jeder der beiden horizontalen Komponenten in der Einheit 10 γ. IAGA hat Observatorien in hohen geomagnetischen Breiten von etwa 65° an aufgefordert, sich ab 1. Januar 1964 an der Bestimmung von R-Werten zu beteiligen. Die schon bei der Besprechung des Q-Index erörterte Frage, inwieweit Feldschwankungen während kurzer Intervalle für die Definition von Aktivitätsmaßen noch brauchbar sind, erhält hier ihre Antwort aus der IAGA-Empfehlung, daß ein stündliches Schwankungsmaß nur noch für Stationen in der Nähe der Polarlichtzonen, also den Zonen stärkster Unruhe, sinnvoll ist.

δ) *Der Auroral Electrojet Activity Index AE nach* Davis *und* Sugiura[7]. Mit dem 1966 eingeführten *Index AE* wird eine globale und fast kontinuierliche Erfassung der von den Strömen in den Polarlichtovals verursachten erdmagnetischen Aktivität angestrebt. Nach dem Prinzip, möglichst Beobachtungen zu benutzen, in denen die zu untersuchende Erscheinung die beherrschende Rolle spielt, werden nur die Registrierungen von Observatorien nahe den beiden Polarlichtzonen verwendet.

Davis und Sugiura gehen von der Vorstellung aus, daß das an diesen Observatorien auftretende Störungsfeld sich zusammensetzt aus dem vorherrschenden Feld der beiden polaren Stromsysteme mit ihren in den Polarlichtovals konzentrierten *Elektrojets*, also im wesentlichen dem **DS**-Feld, ferner einem aus der Magnetosphäre kommenden Störungsfeld und schließlich unregelmäßigen Feldern mit schwer lokalisierbarer Herkunft. Die regelmäßigen Tagesvariationen ($S_q + L$) werden in diesen hohen Breiten vernachlässigt. Der Definition von AE liegt die Beobachtung zugrunde, daß der Polarlichtzonenstrom in der Regel in den frühen Abendstunden ostwärts und sonst westwärts gerichtet ist. Wird nun durch die Auswahl geeigneter Stationen die Horizontalkomponente $\mathcal{H}$ längs des äquatorwärts gelegenen Randes der Polarlichtzone beobachtet, so liegt mit den Abweichungen $\Delta \mathcal{H}$ vom ungestörten Niveau ein Maß für die räumliche und zeitliche Änderung der Stärke des Polarlichtzonenstroms vor. Dabei ist $\Delta \mathcal{H}$ im allgemeinen in den Abendstunden positiv und von etwa Mitternacht bis in die Morgenstunden negativ mit Nulldurchgängen in den Zwischenzeiten. Ein globales, nur von der Weltzeit abhängiges Aktivitätsmaß folgt daraus, indem aus den *gleichzeitigen* $\Delta \mathcal{H}$-Werten längs der Polarlichtzone für jeden Zeitpunkt der maximale Unterschied, also die *Schwankungsweite* von $\Delta \mathcal{H}$, bestimmt wird. Vereinfacht gesehen, ist das ein Maß für die Summe der maximalen Stromstärken der beiden entgegengesetzt gerichteten Anteile des Polarlichtzonenstroms in dem betrachteten Zeitpunkt.

In der Praxis muß von den vorhandenen Observatorien ausgegangen werden. Davis und Sugiura geben an, daß sechs um das Polarlichtzonenoval gleichmäßig in der Länge verteilte Stationen das notwendige Minimum darstellen. Wegen des konjugierten Auftretens magnetischer Störungen an entsprechend gelegenen Observatorien in der Nähe der nördlichen und südlichen Polarlichtzone können Observatorien beider Hemisphären mit dem Blick auf eine möglichst homogene Verteilung zusammengefaßt werden. Observatorien, die eindeutig polwärts der Polarlichtzonen liegen, werden nicht benutzt, um den Einfluß der Rückströme über die Polkappen auszuschalten. Der Einfluß der Rückströme in mittleren Breiten ist von vornherein geringer, so daß es sich sogar empfiehlt, vom Rand der

---

[7] T. N. Davis and M. Sugiura: J. Geophys. Res. **71**, 785—801 (1966).

Polarlichtzone auf der Äquatorseite noch einigen Abstand zu halten, damit der polare Elektrojet die Observatorien nicht bei jedem kleineren magnetischen Sturm überquert.

Die Ausgangsdaten sind Ablesungen der $\mathcal{H}$-Registrierungen im Abstand von 2,5 min. Als Basis für die $\Delta\mathcal{H}$-Werte dient an jedem Observatorium ein aus den Aufzeichnungen an ruhigen Tagen bestimmtes mittleres $\mathcal{H}$, das auf zeitlich benachbarte gestörte Tage übertragen wird und bis auf eine Unsicherheit von 10 γ die Abwesenheit magnetischer Störungsfelder aus der Ionosphäre garantieren soll. Die $\Delta\mathcal{H}$-Werte aller verwendeten Observatorien werden dann nach Weltzeit aufgetragen, und es wird die obere und die untere Hüllkurve dieser Kurvenschar konstruiert. Der Abstand zwischen den beiden Hüllkurven zu einem bestimmten Zeitpunkt gibt in der Einheit γ den dazugehörigen Wert des Index AE an. Entsprechend der Folge der Ablesungen haben die AE-Werte der von DAVIS und SUGIURA[7] untersuchten magnetischen Störungen die hohe Zeitauflösung von 2,5 min. Durch Mittelwertbildung läßt sich auch die Elektrojet-Aktivität längerer Zeitabschnitte durch eine einzige Zahl charakterisieren. Stündliche Werte des AE-Index (AEI) sind für die Zeit vom Internationalen Geophysikalischen Jahr 1957/58 bis 1964 veröffentlicht worden[8].

Das Verfahren zur Bestimmung des Index AE ist naturgemäß auf die Eigenschaften des Polarlichtzonenstroms bezogen. Es muß daher noch betrachtet werden, inwieweit mit dieser Maßzahl auch die anderen, eingangs genannten Anteile des Störungsfeldes erfaßt werden. Sofern Felder unregelmäßiger Stromschwankungen auftreten, sind sie sicher zu einem Teil in AE enthalten. Von den Feldern magnetosphärischer Herkunft überwiegt das Ringstromfeld die Felder der Grenzflächen- und Schweifströme (s. Ziff. 4δ) bei weitem. Da aber das Ringstromfeld in Erdnähe ein nahezu homogenes Feld mit der Dipolachse als Symmetrieachse ist, wird sein Anteil durch die Differenzenmethode, auf der die Bestimmung von AE beruht, praktisch eliminiert. Das Ringstromfeld hat lediglich den Effekt zur Folge, daß das Nullniveau von $\Delta\mathcal{H}$ innerhalb der beiden Hüllkurven gegen die Hüllkurve der positiven $\Delta\mathcal{H}$-Werte verschoben wird. Diese günstige Wirkung der Differenzenmethode in bezug auf eine Beseitigung des Ringstromeinflusses hat aber zugleich die Konsequenz, daß auch jeder axialsymmetrische Anteil der beiden polaren ionosphärischen Stromsysteme ebenfalls eliminiert wird. Der Index AE mißt daher vorwiegend den nichtaxialsymmetrischen Anteil dieser Stromsysteme und damit zweifellos die vorherrschende und wichtigere Komponente.

18. Maßzahlen für Erhaltungs- und Wiederholungsneigung.

α) Wie in anderen Zweigen der Geophysik sind auch bei vielen erdmagnetischen Beobachtungsgrößen, für die längere Zeitreihen vorliegen, die Erscheinungen der Erhaltungs- und Wiederholungsneigung nachweisbar. Ihre gemeinsame Ursache wird mit dem Begriff der *Quasi-Persistenz* beschrieben. Das bedeutet bei der Erhaltungsneigung, daß benachbarte Werte über ein gewisses Intervall ähnlich bleiben und ihre Unterschiede daher nicht rein zufälliger Natur sind. Als Wiederholungsneigung äußert sich Quasi-Persistenz darin, daß nach Ablauf eines Intervalls mit möglicherweise zufälligen Werten in den folgenden gleich langen Intervallen eine ähnliche Anordnung der Werte vorliegt wie im ersten Intervall und diese Ähnlichkeit allmählich aufhört.

---

[8] Scientif. Reports Geophys. Inst. College Alaska, UAG R-194 (AEI 1957 2. Hälfte), R-192 (AEI 1958), R-195 (AEI 1959), R-199 (AEI 1960), R-196 (AEI 1961), R-197 (AEI 1962), R-200 (AEI 1963), R-198 (AEI 1964). Die Reihe wird fortgesetzt.

Maßzahlen für Erhaltungs- und Wiederholungsneigung sind primär keine Aktivitätsmaße. Sie können jedoch zu Aussagen benutzt werden, die mit dem Auftreten der Aktivität in Zusammenhang stehen. Als Ausgangsdaten für ihre Berechnung kommen sowohl Beobachtungswerte als auch daraus abgeleitete Kennziffern in Frage. Daraus folgt, daß wir es mit einer Art von Maßzahlen zu tun haben, die nur indirekt zum Thema dieses Artikels gehören. Wegen ihrer Bedeutung bei der Charakterisierung der morphologischen Eigenschaften erdmagnetischer Zeitreihen sollen sie aber nicht unerwähnt bleiben.

β) Die *Erhaltungsneigung* in den Werten x_i, $(i=1, 2, 3, \ldots, n)$ einer Zeitreihe wird am unmittelbarsten durch die Berechnung des *Autokorrelationskoeffizienten* $r(\nu)$ nachgewiesen, der in bekannter Weise ein Maß für den linearen Zusammenhang zwischen den Zeitreihen x_i und $x_{i+\nu}$ ist. Dabei geht die zweite Zeitreihe aus der ersten durch Verschiebung um ν Schritte gegen diese hervor. Mit $\nu = 0, 1, 2, \ldots \ll n$, läßt sich die Änderung der Erhaltungsneigung innerhalb der Zeitreihe von den unmittelbar benachbarten zu weiter entfernten Werten verfolgen. Für Zeitfunktionen geht $r(\nu)$ in die Autokorrelationsfunktion über oder bei Verzicht auf die Normierung $r(0) = 1$ in die Autokovarianzfunktion[1].

Auf einem ganz anderen Wege ist BARTELS[2] zu einer quantitativen Erfassung der Erhaltungsneigung und einer anschaulichen Interpretation seiner äquivalenten Erhaltungszahl gelangt. Sein Vorgehen beruht auf der Erkenntnis, daß für die Streuung von Mittelwerten nach dem Fehlerfortpflanzungsgesetz nur die Anzahl der voneinander unabhängigen Werte maßgeblich ist und daß also bei der Verwendung korrelierter Werte diesem Sachverhalt im Fehlerfortpflanzungsgesetz Rechnung getragen werden muß, woraus sich ein Maß für die Abhängigkeit aufeinanderfolgender Werte ergibt:

Gegeben sei eine hinreichend lange (theoretisch: unendlich lange) Beobachtungsreihe x_i, deren n Werte nach ihrer natürlichen Anordnung in kleine Gruppen zu je ν Werten unterteilt werden; dabei seien die Gruppen mit $s = 1, 2, 3, \ldots, n/\nu$ numeriert. Ist m^2 die mittlere quadratische Abweichung der x_i, $M_s(\nu)$ der Mittelwert der Gruppe s und $m^2(M_s)$ die mittlere quadratische Abweichung der M_s, so besagt das einfache Fehlerfortpflanzungsgesetz für den Fall, daß die x_i voneinander *unabhängig* sind und aus demselben (normal verteilten) statistischen Kollektiv stammen, daß $m(M_s)$ um den Faktor $\nu^{-\frac{1}{2}}$ kleiner ist als m.

Anhand des einfachen Modells einer Zeitreihe, die aus der ursprünglichen Beobachtungsreihe dadurch hervorgeht, daß jeder der unabhängigen Werte x_i p-mal wiederholt wird, kann BARTELS bereits das Prinzip der Verallgemeinerung des einfachen Fehlerfortpflanzungsgesetzes aufzeigen und erhält daraus die allgemeine Darstellung für *korrelierte* Beobachtungswerte x_i:

$$m(M_s) = m \sqrt{\frac{\varepsilon(\nu)}{\nu}}. \qquad (18.1)$$

Die hier neu auftretende Größe $\varepsilon(\nu)$ wird als *äquivalente Erhaltungszahl*[3] bezeichnet. Für unkorrelierte Beobachtungswerte ergibt sich mit $\varepsilon(\nu) = 1$ die gebräuchliche einfache Form des Fehlerfortpflanzungsgesetzes. Bei der Modellreihe mit jeweils p aufeinanderfolgenden gleichen Werten wird $\varepsilon(\nu) = p$, wobei $p < \nu$ ist. Allgemein

---

[1] Bezüglich der genauen Definition dieser Größen und der Methoden zu ihrer Berechnung sei auf die zahlreichen Lehrbücher der mathematischen Statistik verwiesen, z.B. J. TAUBENHEIM: Statistische Auswertung geophysikalischer und meteorologischer Daten. Leipzig: Akad. Verlagsges. Geest u. Portig 1969.
[2] J. BARTELS: Sitzber. Preuß. Akad. Wiss., Phys.-Math. Kl. 1935, 504—522; — Terr. Magn. Atmosph. Electr. **40**, 1—60 (1935); s. auch [2], Kap. 27.
[3] Bei BARTELS wird nicht immer streng zwischen Erhaltungsneigung und Wiederholungsneigung unterschieden, z.B. in [2], Kap. 27; Naturwissenschaften **31**, 421—435 (1943); FIAT Rev. **7**, Angew. Mathem. Teil 5, 89—99 (1948). Deshalb kommt die hier durch Gl. (18.1) definierte äquivalente Erhaltungszahl bei ihm auch unter dem Namen äquivalente Wiederholungszahl vor und wird mit ω bezeichnet.

sagt $\varepsilon(\nu)$ also aus, daß die ν Werte der Gruppe s die gleiche Abhängigkeit zeigen, die sich ergäbe, wenn diese Gruppe aus *Folgen von je ε gleichen Werten* bestünde bei Unabhängigkeit der Werte der einzelnen Folgen. Das heißt aber nichts anderes, als daß sich unter den ν Werten einer Gruppe effektiv nur

$$\nu' = \frac{\nu}{\varepsilon(\nu)} \qquad (18.2)$$

voneinander unabhängige Werte befinden.

Dabei ist es unerheblich, daß dieses Ersatzmodell nicht realisierbar ist, weil im allgemeinen weder ε noch ν/ε ganzzahlig sind. In diesem Sinne ist $\varepsilon(\nu)$ als „anschauliches" Äquivalent zur tatsächlich vorliegenden Erhaltungsneigung der Zeitreihe zu verstehen. Durch Veränderung des Gruppenumfanges ($\nu = 1, 2, 3, \ldots$) kann dann im ε-Maß auch die Änderung der Erhaltungsneigung von den unmittelbar benachbarten zu den weiter entfernten Werten verfolgt werden. In der Regel ist $\varepsilon(\nu) > 1$ mit $\varepsilon(1) = 1$, jedoch können auch Werte $\varepsilon(\nu) < 1$ vorkommen. Normalerweise strebt $\varepsilon(\nu)$ mit wachsendem ν einem Grenzwert $\varepsilon(\infty)$ zu.

Wird Gl. (18.1) nach $\varepsilon(\nu)$ aufgelöst, so erhält man damit nicht nur die Definitionsgleichung dieser Größe, sondern auch eine in der Praxis benutzbare Rechenvorschrift für ihre zahlenmäßige Bestimmung. Andererseits ergibt die explizite Darstellung von $m^2(M_s)$, daß sich diese Größe bei korrelierten Werten x_i durch deren Autokorrelationskoeffizienten ausdrücken läßt. Das gleiche ist damit auch für $\varepsilon(\nu)$ möglich und liefert zugleich ein bequemeres Bestimmungsverfahren:

$$\varepsilon(\nu) = 1 + 2 \sum_{\tau=1}^{\nu-1} \frac{\nu - \tau}{\nu} r(\tau). \qquad (18.3)$$

In den Autokorrelationskoeffizienten $r(\nu)$ und den äquivalenten Erhaltungszahlen $\varepsilon(\nu)$ kommt daher die Erscheinung der Erhaltungsneigung einer Zeitreihe in verschiedener Weise, aber durchaus gleichwertig zum Ausdruck.

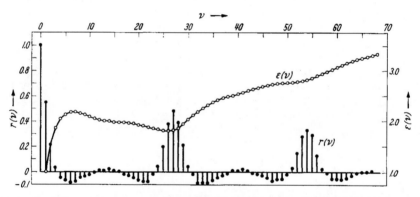

Fig. 8. Autokorrelationskoeffizienten $r(\nu)$ und äquivalente Erhaltungszahlen $\varepsilon(\nu)$ für die täglichen internationalen Charakterzahlen Ci aus 12 Jahren, die den Phasen abnehmender Sonnenaktivität im Fleckencyclus angehören (nach BARTELS[4]); $\nu =$ Anzahl der Tage.

In Fig. 8 wird dieser Sachverhalt an einem von BARTELS[4] stammenden Beispiel veranschaulicht. Als Ausgangsdaten für die Berechnung von $r(\nu)$ und $\varepsilon(\nu)$ sind die täglichen internationalen Charakterzahlen Ci der 12 Jahre 1886 bis 1889, 1896, 1898, 1911, 1922, 1930 bis 1933 benutzt worden, die den Phasen abnehmender Aktivität von sechs 11jährigen Sonnenfleckencyclen angehören. Bei der Beurteilung der Erhaltungsneigung aus den Werten von $r(\nu)$ reicht es nicht aus, nur den Grad der Korrelation zu sehen, den r für ein bestimmtes ν angibt, sondern der mehr oder weniger systematische Verlauf der Kurve $r(\nu)$ enthält eine zusätzliche Information. Beispielsweise findet man in Fig. 8 für $\nu = 3$ die sehr hohe Erhaltungsneigung $\varepsilon = 1,9$, was nach Gl. (18.2) bedeutet, daß nur etwa die Hälfte der Werte in den

---

[4] J. BARTELS: FIAT Rev. 7, Angew. Mathem. Teil 5, 89—99 (1948).

Dreier-Gruppen effektiv unabhängig ist. Dagegen liegt $r(3)$ so nahe bei Null, daß hieraus auf das Fehlen jeglicher Korrelation geschlossen werden müßte. Der Verlauf von $r(\nu)$ weist aber auf einen im Mittel systematischen Wechsel von Tagen mit höherer und geringerer Aktivität hin, wobei der magnetische Charakter des Tages etwa bei $\nu = 3$ zum ersten Mal wechselt und dann eine Zeitlang gegensinnig bleibt ($r < 0$). Trotz der schnellen Abnahme von $r(\nu)$ ist aus dem Verlauf dieser Kurve die weitergehende Abhängigkeit der aufeinanderfolgenden Tageswerte zu verfolgen, die im ε-Maß ein relatives Maximum für $\nu = 6$ erreicht.

Der Vorzug der äquivalenten Erhaltungszahl liegt in ihrer unmittelbaren Interpretierbarkeit für jeden einzelnen Wert von ν, eine Eigenschaft, die letzten Endes aus Gl. (18.3) folgt oder noch klarer aus der analogen Beziehung für Zeitfunktionen, bei denen $\varepsilon(\nu)$ das Integral der mit dem Faktor $2(\nu - \tau)/\nu$ bewichteten Autokorrelationsfunktion $r(\tau)$ mit der oberen Grenze ν ist. Ein weiterer Vorzug von ε ist die nach Gl. (18.2) sofort mögliche Berechnung der Anzahl der effektiv voneinander unabhängigen Ausgangswerte. Damit können statistische Tests, die für nichtautokorrelierte Stichproben aufgestellt worden sind, auch auf Stichproben mit korrelierten Werten (aus stationären Zufallsprozessen) angewendet werden, nachdem der Stichprobenumfang auf die effektive Anzahl unabhängiger Werte reduziert worden ist.

γ) Gegenüber den $\varepsilon(\nu)$ geben die Autokorrelationskoeffizienten einen besseren Einblick in die Struktur einer Zeitreihe mit Erhaltungsneigung und lassen dabei auch das Vorhandensein einer Wiederholungsneigung erkennen. So ist in Fig. 8 als auffallendste Erscheinung eine 27tägige Wiederholung von Tagen mit gleichartiger Aktivität aus den $r(\nu)$ abzulesen. Genauer gesagt, wird unter *Wiederholungsneigung* die quasi-periodische Wiederkehr einer Erscheinung verstanden, wobei aber die Wiederkehr selbst nur quasi-persistent ist, also die Erscheinung hinsichtlich ihrer Wiederholung eine gewisse Erhaltungsneigung zeigt. Die Autokorrelationsfunktion kann also zum qualitativen Nachweis periodischer und quasi-periodischer Anteile in einer Zeitfunktion herangezogen werden, wie dies bei der Berechnung von quadratischen Spektren der Fall ist; sie ist jedoch kein quantitatives Maß der Wiederholungsneigung einer bestimmten Folge von Beobachtungswerten über l gleichlange Abschnitte.

Damit ist gesagt, was von einer *äquivalenten Wiederholungszahl* $\omega_k(l)$ als Maßzahl der Wiederholungsneigung erwartet wird. Zur Lösung dieser Aufgabe braucht das Verfahren der Bestimmung von $\varepsilon(\nu)$ nur geringfügig modifiziert zu werden: Gegeben sei wieder eine hinreichend lange Beobachtungsreihe x_i, deren n Werte nach ihrer natürlichen Anordnung ebenfalls in Gruppen zu je k Werten unterteilt werden. Jedoch werden jetzt diese Gruppen untereinander geschrieben, so daß jede Gruppe eine (waagerechte) Zeile bildet. Das ergibt ein Buys-Ballot-Schema mit k (senkrechten) Spalten und $l = n/k$ Zeilen. Die Spalten seien mit $\varkappa = 1, 2, 3, \ldots k$ numeriert. Das Verfahren von BARTELS läßt sich auf dieses Schema derart anwenden, daß die aus der gesamten Stichprobe der x_i berechnete mittlere quadratische Abweichung m^2 verglichen wird mit der Streuung der Durchschnittszeile. Diese besteht definitionsgemäß aus den k Spaltenmittelwerten $M_\varkappa$. Ihre mittlere quadratische Abweichung $m^2(M_\varkappa)$ ist die Streuung der Durchschnittszeile und sollte bei nichtkorrelierten x_i das einfache Fehlerfortpflanzungsgesetz erfüllen. Besteht nun innerhalb der einzelnen Spalten eine Erhaltungsneigung, was einer Wiederholungsneigung in den aufeinanderfolgenden Zeilen gleichkommt, so lassen sich die zuvor für die quantitative Erfassung der Erhaltungsneigung angestellten Überlegungen ohne weiteres übertragen und führen auf ein verallgemeinertes Fehlerfortpflanzungsgesetz für Zeitreihen mit Wiederholungsneigung:

$$m(M_\varkappa) = m \sqrt{\frac{\omega_k(l)}{l}}. \tag{18.4}$$

Auflösung von Gl. (18.4) nach $\omega_k(l)$ ergibt die Definition und das Berechnungsverfahren dieser äquivalenten Wiederholungszahl. Auch für sie gilt in der Regel $\omega_k(l) > 1$ und besagt, daß die l k-spaltigen Zeilen die gleiche Abhängigkeit zeigen, die sich ergäbe, wenn je ω *aufeinanderfolgende Zeilen exakt übereinstimmen* würden. Mit dieser Modifikation lassen sich die für die äquivalente Erhaltungszahl gegebenen Erläuterungen auf die äquivalente Wiederholungszahl übertragen.

Es sei aber noch darauf hingewiesen, daß die Voraussetzung eines (theoretisch unendlich) großen Beobachtungsmaterials eine Aufteilung der Werte in genügend lange Zeilen ($k \to \infty$) verlangt, wenn die Änderung der Wiederholungsneigung von wenigen zu vielen Zeilen ($l = 1, 2, 3, \ldots$) verfolgt und dazu von der einfachen Beziehung Gl. (18.4) Gebrauch gemacht werden soll. Das Verfahren läßt sich in dieser Form also nicht auf zu kurzzeitige Wiederholungen anwenden. Modifikationen, die sich aus dem endlichen Umfang der Stichprobe und dem Zusammenwirken von Erhaltungs- und Wiederholungsneigung einer Zeitreihe ergeben, hat J. MEYER [10] erörtert. Er gibt auch eine der Beziehung Gl. (18.3) analoge Darstellung der äquivalenten Wiederholungszahl durch geeignet definierte mittlere Korrelationskoeffizienten zwischen den l Zeilen.

Die Arbeit von MEYER enthält aber insbesondere Tabellen mit Werten von $\omega_k(l)$ für $k = 27$ und $l = 2, 4, 8, 16, 32$, berechnet auf der Grundlage der täglichen Charakterzahlen $C8$ (s. Ziff. 10ζ) aus den Jahren 1884 bis 1964. Jede der äquivalenten Wiederholungszahlen $\omega_{27}(l)$ wird als gleitendes Maß mit dem Abstand einer Sonnenrotation für jeweils l aufeinanderfolgende Sonnenrotationen bestimmt. Die Ergebnisse werden in Form einer Notenschrift, die der graphischen Darstellung von Kp (s. Fig. 6) nachgebildet ist, anschaulich gemacht und im Hinblick auf die 27tägliche Wiederholungsneigung der erdmagnetischen Aktivität (s. Ziff. 22) diskutiert. Diese Untersuchung kann als Muster einer quantitativen Behandlung der Wiederholungsneigung angesehen werden.

D. Statistische Aussagen der Maßzahlen über Gesetzmäßigkeiten im Auftreten erdmagnetischer Aktivität und deren Ursache.

19. Vorbemerkungen und Übersicht. Dieses letzte Kapitel muß sogleich mit der Feststellung beginnen, daß aus der kaum übersehbaren Fülle statistischer Ergebnisse hier nur eine ganz spezielle Auswahl angeführt wird. Im Vordergrund steht dabei die Absicht, an diesen Beispielen noch deutlicher zu machen, was erdmagnetische Aktivität ist.

Es wird daher nicht eingegangen auf die zahlreichen nachgewiesenen Korrelationen der hier besprochenen Maßzahlen mit anderen geophysikalischen Parametern, insbesondere mit Beobachtungsgrößen des Polarlichts, der Ionosphäre, der Magnetosphäre und der neutralen Hochatmosphäre. Näheres darüber findet sich in den Artikeln, in denen diese Sachgebiete behandelt werden. Nur hingewiesen werden kann in diesem Zusammenhang auch auf die große Zahl publizierter Ergebnisse, die in ihrer statistischen Signifikanz umstritten sind. Hierhin gehört als ein Musterbeispiel der immer wieder als bewiesen behauptete Einfluß des Mondes auf die erdmagnetische Aktivität. Das Für und Wider dieser Kontroverse hat SCHNEIDER[1] zusammenfassend dargestellt mit dem Ergebnis, daß die gegenwärtig vorliegenden Beobachtungsreihen der verschiedenen Aktivitätsmaße nicht ausreichen, einen etwa vorhandenen kleinen lunaren Effekt von sehr ähnlichen Effekten solaren oder anderen Ursprungs (s. Ziff. 22) abzutrennen und seine Existenz in statistisch überzeugender Weise zu belegen. Zu dieser kritischen Beurteilung hat wesentlich der von BARTELS[2] als Test auf behauptete Mondeinflüsse angewendete Schüttelversuch beigetragen.

Sowohl die indirekten Schlüsse aus einer mehr als 100jährigen Beobachtung der Häufigkeit und Stärke erdmagnetischer Störungen als auch die direkten Messungen der Erdsatelliten und Raumsonden in jüngster Zeit lassen in eindeutiger Weise die Aktivität der Sonne als Ursache der mit den Maßzahlen erfaßten Störungen des Magnetfelds der Erde erkennen. Das kommt besonders in den Schwankungen der erdmagnetischen Aktivität zum Ausdruck, in denen sich die naturgegebenen Perioden und Rhythmen widerspiegeln, die Sonne und Erde anhaften: Sonnenfleckencyclus, Umlauf der Erde um die Sonne, Sonnenrotation und Erdrotation. Ausgehend von diesen Perioden, die sich jeweils um etwa eine Zehnerpotenz unterscheiden, werden in den folgenden Ziffern die dazugehörigen Aussagen einiger Aktivitätsmaße behandelt und zu Überlegungen über die spezielle Ursache jeder dieser Aktivitätsschwankungen herangezogen.

---

[1] O. SCHNEIDER: Space Sci. Rev. 6, 655—704 (1967).
[2] J. BARTELS: Nachr. Akad. Wiss. Göttingen, Math.-Phys. Kl., Nr. 23, 333—365 (1963)

20. Die langjährige Änderung der Aktivität.

α) Die herausragende Erscheinung in diesem Periodenbereich ist die weitgehende Parallelität der erdmagnetischen Aktivität mit dem im Mittel 11jährigen Cyclus der Sonnenaktivität. Sie wurde von SABINE[1] aus den Beobachtungen erdmagnetischer Störungen der Jahre 1841 bis 1848 in Toronto entdeckt und unabhängig davon auch von anderen Observatoren, nachdem zuvor SCHWABE[2] in seinen 1826 beginnenden Beobachtungen der Sonnenflecken auf die quasiperiodische Variation der Fleckenzahl gestoßen war. Diese ersten, wegen der Kürze der Beobachtungszeit noch keineswegs allgemeingültigen Nachweise lenkten die Aufmerksamkeit generell auf die erdmagnetische Wirkung solarer Vorgänge. Das durchaus nicht beste, aber wegen der Länge der Beobachtungsreihe gebräuchlichste Maß der Sonnenaktivität ist die *Sonnenfleckenrelativzahl* R, von R. WOLF 1848 definiert durch*

$$R = k(10g + f).$$

Dabei ist g die Anzahl der Fleckengruppen, f die Anzahl der Einzelflecken und k ein Reduktionsfaktor, mit dem Beobachtungsreihen verschiedener Herkunft an den Züricher Standard anzuschließen sind.

Tägliche Relativzahlen liegen nahezu lückenlos vom Jahre 1818 an vor, Monatsmittel ab 1749 und Jahresmittel ab 1700[3]. Die Maxima der Cyclen und damit die Cyclen selbst werden fortlaufend numeriert, beginnend mit dem Maximum von 1761.

Zur quantitativen Erfassung der Sonnenaktivität werden auch andere Beobachtungsgrößen verwendet, zum Beispiel die von den Fackeln bedeckte Fläche und die Intensität der Sonnenstrahlung in bestimmten Spektralbereichen, insbesondere im Bereich der Radiofrequenzen. Einen sehr engen Zusammenhang mit der Fleckenrelativzahl zeigt die solare Radiostrahlung auf 2800 MHz, entsprechend 10,7 cm Wellenlänge, der sogenannte COVINGTON-Index[4].

β) Die Variation der erdmagnetischen Aktivität mit dem Sonnenfleckencyclus gehört inzwischen zu den selbstverständlichen Vorkommnissen. Sie wird hier in Fig. 9 anhand der Jahresmittel von R und Ap (s. Ziff. 9β) für etwa drei Cyclen demonstriert. Eine etwas genauere Betrachtung der beiden Kurven zeigt, daß eine Phasenverzögerung der erdmagnetischen Aktivität gegenüber der Sonnenaktivität besteht. Sie ist gering zur Zeit der Minima und beträgt für die drei Maxima im Mittel drei Jahre. Wird die Verschiebung der beiden Kurven gegeneinander mit Hilfe des Koeffizienten $r(\tau)$ der Kreuzkorrelation ausgedrückt, so findet man sein Maximum bei $\tau = 1$ Jahr mit $r(1) = 0{,}71$ gegenüber $r(0) = 0{,}62$ und $r(2) = 0{,}63$. Positives τ bedeutet die Verzögerung von Ap gegen R in Jahren. Dabei ist allerdings zu beachten, daß der Zusammenhang zwischen den beiden Aktivitätsmaßen sicher nicht linear ist.

Auf das verspätete Eintreten der maximalen erdmagnetischen Aktivität hat bereits BARTELS[5] 1925 anhand des u-Maßes (s. Ziff. 15) hingewiesen. Diese Beobachtung wiegt um so schwerer, weil von allen planetarischen Maßzahlen die Jahresmittel von u und u_1 die engste Beziehung zu den Jahresmitteln von R aufweisen. Nach [1] ergibt sich mit den Daten der Jahre 1872 bis 1930 ein

---

* Nicht zu verwechseln mit dem in Ziff. 17γ definierten Index R.
[1] E. SABINE: Phil. Trans. London 1851, 123—139; 1852, 103—124.
[2] S. H. SCHWABE: Astron. Nachr. Nr. 495 (1843).
[3] M. WALDMEYER: The Sunspot-Activity in the Years 1610—1960. Zürich: Schulthess & Co. 1961. Die täglichen R-Werte werden auch weiter an der Eidgenössischen Sternwarte Zürich bestimmt und jahrgangsweise veröffentlicht.
[4] Tabellen dieses Index finden sich in diesem Handbuch, Band 49/5, im Beitrag von G. KOCKARTS und P. BANKS.
[5] J. BARTELS: Meteorol. Z. **42**, 147—152 (1925).

Korrelationskoeffizient $r(0) = 0{,}87$ für die Korrelation von u mit R und $r(0) = 0{,}88$ für die von u_1 mit R.

Eine detailliertere Beschreibung des Verhaltens der Aktivität gibt Fig. 10 für den Fleckencyclus Nr. 19, der das Internationale Geophysikalische Jahr 1957/58 enthält. In dieser von BARTELS[6] bevorzugten Form der Darstellung sind die Monatsmittel von R und Ap zusammengestellt sowie die auf 240 Drei-Stunden-Intervalle je Monat reduzierten absoluten und relativen Häufigkeiten von Kp-Werten für ausgewählte Kp-Bereiche. Ein Vergleich von R und Ap bestätigt die

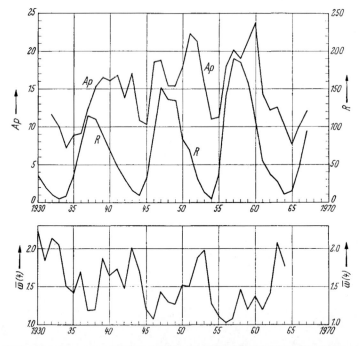

Fig. 9. Langjähriger Verlauf der erdmagnetischen Aktivität, dargestellt durch die Jahresmittel des linearen Aktivitätsmaßes Ap von 1932 bis 1967, im Vergleich zur Sonnenaktivität, dargestellt durch die Jahresmittel der Sonnenfleckenrelativzahl R von 1930 bis 1967. Darunter die Jahresmittel $\bar{w}(4)$ der [10] entnommenen äquivalenten Wiederholungszahl $\omega_k(l)$ für $k = 27$ und $l = 4$ (s. Ziff. 18γ).

allgemeine Erfahrung, daß der Zusammenhang zwischen R und einer Maßzahl der erdmagnetischen Aktivität um so schlechter wird, je höher die Zeitauflösung ist. Die Berechnung von Korrelationskoeffizienten für jährliche, monatliche und tägliche Wertepaare liefert dafür quantitative Belege.

Als ein Beispiel unter anderen sollen in diesem Zusammenhang lediglich die Werte der Monate September und Oktober 1957 betrachtet werden. Der September hat mit $Ap = 49$ das höchste Monatsmittel in der bisher vorliegenden Beobachtungsreihe dieser Maßzahl; das gleiche gilt für den Oktober-Wert der Sonnenfleckenrelativzahl mit $R = 253{,}8$. Die stärkere erdmagnetische Störung geht hier also der höheren Sonnenaktivität voraus. Immerhin ist auch der September mit $R = 235{,}8$ durch starke solare Aktivität ausgezeichnet; jedoch ist für den Oktober nur $Ap = 14$ gefunden worden. Dieser Wert, der noch etwas unter dem langjährigen Mittel liegt, ist viel zu gering, um mit dem Oktober-Wert von R in direkte Verbindung gebracht werden zu können. Der nächste auffallende Wert ist $Ap = 27$ und wird erst im Februar 1958 erreicht.

---

[6] J. BARTELS: Ann. Géophys. 19, 1—20 (1963).

Die Häufigkeit der Kp-Werte ändert sich in jedem der fünf Kp-Bereiche der Fig. 10 systematisch mit dem Sonnenfleckencyclus in der Weise, daß während der Anstiegszeit von R merklich weniger erdmagnetische Störungen vorkommen als während der Abstiegszeit, worin wieder die Phasenverschiebung zwischen beiden Erscheinungen zum Ausdruck kommt. Auch die extrem starken magne-

Fig. 10. Monatliche Aktivitätswerte für die Jahre 1953 bis 1963 nach BARTELS[6]. Obere Zeile: Monatsmittel der Sonnenfleckenrelativzahl. Die fünf mittleren Zeilen: Auf 240 Drei-Stunden-Intervalle reduzierte Häufigkeiten von Kp-Werten für ausgewählte Kp-Bereiche mit Angabe absoluter und relativer Häufigkeiten. Untere Zeile: Monatsmittel des linearen Aktivitätsmaßes Ap.

tischen Stürme treten demzufolge mit größerer Wahrscheinlichkeit in der Abstiegszeit von R auf, im übrigen aber scheint das Eintreffen dieser seltenen Ereignisse zufälliger Natur zu sein.

γ) Bei der Untersuchung weiterer Gesetzmäßigkeiten der langjährigen Änderung der erdmagnetischen Aktivität ist zu beachten, daß der Ablauf des einzelnen Sonnenfleckencyclus nur sehr entfernte Ähnlichkeit mit einem sinusförmigen Vorgang hat und außerdem noch starke Änderungen in der Kurvenform von einem Cyclus zum nächsten vorkommen. In noch unübersichtlicherer Form spiegelt sich dieses Verhalten der Sonnenaktivität dann auch im Auftreten der erdmagnetischen Unruhe wider. Daher ist es nicht verwunderlich, wenn bei der *Periodogrammanalyse des Aktivitätsverlaufes* für einzelne und mehrere Cyclen sowohl benachbarte Perioden zur 11jährigen Periode als auch höhere Harmonische zu diesen Grundperioden gefunden werden. Am häufigsten wird dabei naturgemäß auf eine Periode von 5,5 Jahren hingewiesen. Zur Vermeidung von Fehldeutungen sei daher auf die in der Astrophysik vorherrschende Ansicht aufmerksam gemacht, wonach jeder Fleckencyclus von Minimum zu Minimum ein *nahezu abgeschlossenes Ganzes* darstellt.

Wird die 11jährige Schwankung aus den Zeitreihen der Aktivität herausgefiltert, so wird in den gefilterten Kurven eine Welligkeit sichtbar, die zur Suche nach längeren Perioden anregt. Dabei werden vor allem die doppelte und die siebenfache Sonnenfleckenperiode diskutiert. Die 22jährige Periode wird schon durch den entsprechenden Cyclus in der Polarität der Magnetfelder der bipolaren Fleckengruppen auf der Sonne nahegelegt. CHERNOSKY[7] hat eine solche Periodizität im Verhalten der durch Ci charakterisierten Aktivität und der Halbjahreswelle der Aktivität (s. Ziff. 21) untersucht. Alle derartigen Bemühungen leiden aber darunter, daß die vorliegenden Beobachtungsreihen zur Behandlung dieser Fragen doch noch recht kurz sind. Aus dem gleichen Grunde sind auch noch keine sicheren Angaben über die säkulare Änderung der erdmagnetischen Aktivität zu machen. Der Verlauf der 11jährigen Mittelwerte einer Reihe von Aktivitätsmaßen zum Teil seit 1840 ist von CHERNOSKY, FOUGERE und HUTCHINSON[8] graphisch dargestellt worden. Wieder fällt das Ausmaß der Parallelität zwischen R und der interdiurnen Veränderlichkeit u ins Auge.

Da nach dem zuvor Gesagten die Korrelation mit R um so besser wird, je länger die Zeitabschnitte sind, die die Einzelwerte repräsentieren, ist in Ergänzung zu Fig. 9 und dem Korrelationskoeffizienten $r(0) = 0{,}62$ für die Jahresmittel von R und Ap der r-Wert für die 10jährigen gleitenden Mittel dieser Maßzahlen von Interesse. Mit den Daten seit 1932 ergibt sich in der Tat $r = 0{,}82$. Die Verwendung 10jähriger Mittelwerte trägt der Kürze der drei letzten Cyclen Rechnung.

δ) Die bisher besprochenen Ergebnisse gestatten einige Aussagen, die über die allgemeine Feststellung einer Einwirkung der Sonne auf das Magnetfeld der Erde hinausgehen. Die mit zunehmender Zeitauflösung schlechter werdende Korrelation zwischen R und den erdmagnetischen Aktivitätsmaßen ist nur ein Beweisstück unter anderen für die Folgerung, daß die Flecken nicht selbst für die erdmagnetischen Störungen verantwortlich sind. Das Auftreten solcher Störungen, ohne daß zuvor in angemessener Zeit Flecken den Zentralmeridian der Sonne überquert haben, ist ein weiteres Argument dafür. Da es andererseits Hinweise auf erdmagnetisch wirksame Gebiete auf der Sonne gibt, worauf bei der Behandlung der Wiederholungsneigung von Störungen (s. Ziff. 22) eingegangen wird, hat BARTELS [1] für diese optisch nicht identifizierbaren Gebiete die Bezeichnung *M-Regionen* eingeführt. Gelegentlich kommt es vor, daß sie mit auffälligen Gebieten auf der Sonne zusammenfallen. Aber auch als rein hypothetische Erscheinungen können ihnen auf indirektem Wege bestimmte Eigenschaften zugeschrieben werden.

Durch unterschiedlichste Beobachtungsverfahren bis hin zur Direktmessung mit Raumsonden ist inzwischen nachgewiesen worden, daß die erdmagnetische

---

[7] E. J. CHERNOSKY: J. Geophys. Res. **71**, 965—974 (1966).

[8] E. J. CHERNOSKY, P. F. FOUGERE, and R. O. HUTCHINSON: The Geomagnetic Field. Handbook Geophys. Space Envir. A. F. Cambridge Res. Lab., Fig. 11—31 (1965).

Unruhe durch solare Partikelstrahlung hervorgerufen wird. Wird diese Tatsache mit den Beobachtungsbefunden verknüpft, daß es in der gesamten Ap-Reihe von 1932 bis 1968 nur 20 Tage mit $Ap = 0$ gegeben hat, also völlige magnetische Ruhe extrem selten ist, daß ferner eine rege erdmagnetische Aktivität herrschen kann, auch wenn die Sonne frei von Flecken und Eruptionen ist, so muß daraus gefolgert werden, daß auch ohne erkennbare Störungsquellen eine nahezu kontinuierliche solare Partikelstrahlung auf das Magnetfeld der Erde einwirkt. Diese Vorstellung entspricht voll und ganz der Konzeption vom *solaren Wind*, die, anders als historisch geschehen[9], auch aus der Betrachtung der erdmagnetischen Aktivität hätte entwickelt werden können.

Sofern sich kurzzeitige Anstiege der Aktivität von den statistischen Schwankungen abheben, kann darin die Wirkung von M-Regionen gesehen werden. Das Erscheinungsbild solcher Ereignisse ändert sich im Verlauf eines Fleckencyclus von starken, häufig einmaligen und in ihrer zeitlichen Verteilung zufälligen Störungen im Maximum zu mehr oder weniger regelmäßig wiederkehrender Verstärkung der Unruhe in der Abstiegszeit von R und im Minimum. Die im unteren Teil von Fig. 9 aufgetragenen Jahresmittel der äquivalenten Wiederholungszahl $\omega(4)$ für $k = 27$ (s. Ziff. 18 γ) lassen genau diese Tendenz erkennen. Die einzige Folgerung, die hier daraus gezogen werden soll, besagt, daß M-Regionen offenbar eine größere Lebensdauer haben als Sonnenflecken. Da es naheliegt anzunehmen, daß die Entstehung von Flecken und M-Regionen entweder ursächlich zusammenhängt oder zumindest parallel zum Rhythmus der solaren Aktivität verläuft, muß die *Lebensdauer* der M-Regionen so groß sein, daß sie das Fleckenminimum erreichen. Auch übertrifft die Dauer der von einer M-Region verursachten Störungssequenz die bekannte Lebensdauer der Flecken häufig bei weitem. Damit wird aber zugleich die Phasenverzögerung der erdmagnetischen Aktivität gegenüber der Sonnenaktivität erklärt, wie sich an einem einfachen Modell auch quantitativ zeigen läßt.

In der üblichen Interpretation dieses Effekts wird dagegen von der äquatorwärts gerichteten *Zonenwanderung* der Flecken im Ablauf eines Cyclus ausgegangen: Nimmt man für die M-Regionen ein gleiches Verhalten an, so läßt sich die erhöhte erdmagnetische Aktivität im absteigenden Cyclus rein geometrisch deuten durch eine größere Wahrscheinlichkeit dafür, daß infolge der dann günstigeren äquatornahen Abschußpositionen der M-Regionen die Erde von einem Plasmastrom mit bestimmtem Öffnungswinkel auch getroffen wird[10].

So einleuchtend diese Vorstellung auf den ersten Blick ist, beruht sie im Grunde doch auf dem alten Konzept eines leeren interplanetaren Raumes, durch den die Sonne von Zeit zu Zeit Plasmaströme hindurchspritzt. Sowohl wegen des heute kaum noch zutreffenden Bildes von der M-Region als einer sehr lokalen Emissionsquelle, abgeleitet aus den optisch nachweisbaren Quellen bei starken Flares, als auch wegen der Vernachlässigung der Wechselwirkung zwischen einem solchen Partikelstrahl und dem interplanetaren Plasma kann diese rein geometrische Betrachtungsweise nicht mehr vorbehaltlos akzeptiert werden. In Ziff. 22 ist noch einmal die Rede davon.

21. Der Jahresgang der Aktivität.

α) Neben der langjährigen Änderung im Rhythmus des Sonnenfleckencyclus ist die *Halbjahreswelle* die zweite persistente Variation der erdmagnetischen Aktivität mit der Zeit, auch wenn diese Erscheinung in Einzeljahren von unregelmäßigen Schwankungen verdeckt sein kann. Im Gegensatz zur 11jährigen Variation ist jedoch die Frage nach der Herkunft der auffallenden halbjährigen Kompo-

---

[9] Bemerkungen zur historischen Entwicklung dieser Konzeption finden sich bei A. J. DESSLER: Rev. Geophys. **5**, 1—41 (1967).
[10] A. L. CORTIE: Monthly Notices Roy. Astron. Soc. **76**, 15—18 (1916).

nente der Aktivität Anlaß einer sich seit Jahrzehnten in der Literatur widerspiegelnden Kontroverse und zugleich ein Beispiel dafür, wie wichtig für die Interpretation die statistische Absicherung der Ergebnisse ist. In Fig. 11 (oben) kommt die den Verlauf des mittleren Jahresganges völlig beherrschende Halbjahreswelle unmittelbar zum Ausdruck. Eingezeichnet sind hier die aus den Daten von 35 Jahren berechneten mittleren Monatsmittel von Ap. Auch ihr durch harmonische Analyse gefundener sinusförmiger Verlauf läßt im Vergleich zu dem verbleibenden Rest, d.h. der Summe der übrigen Harmonischen, diesen Sachverhalt in Fig. 11 (unten) klar erkennen.

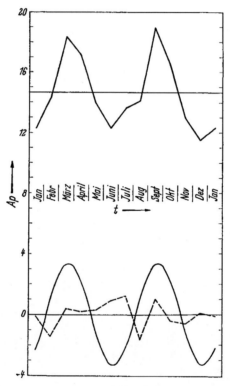

Fig. 11. Der aus den Monatsmitteln von 35 Jahren (1932 bis 1966) abgeleitete *mittlere Jahresgang* des linearen Aktivitätsmaßes Ap (oben), die daraus durch harmonische Analyse berechnete *Halbjahreswelle* von Ap (ausgezogene Kurve, unten) und der nach Abzug der Halbjahreswelle und des 35jährigen Mittelwertes verbleibende *Rest* von Ap (gestrichelte Kurve, unten).

Obwohl es ältere Hinweise auf eine systematische jahreszeitliche Änderung der Aktivität gibt, ist hier die Untersuchung von CORTIE[1] als erste zu nennen, in der das statistische Resultat erstmalig zu einer Antwort auf die Frage nach der Herkunft der Halbjahreswelle benutzt wird. Aus der Analyse eines 23jährigen Datenmaterials eines einzelnen Observatoriums bestimmte CORTIE die Eintrittszeiten der Maxima zu Ende Februar und Ende August. Er schloß daraus auf eine Variation der erdmagnetischen Aktivität mit der Änderung der *heliographischen Breite der Erde* auf ihrer Bahn um die Sonne, verursacht durch den Winkel von 7,25° zwischen der Ebene der Ekliptik und der Äquatorebene der Sonne. Da dieser geometrische Sachverhalt gleichbedeutend ist mit der Aussage, daß die Rotations-

---

[1] A. L. CORTIE: Monthly Notices Roy. Astron. Soc. **73**, 52—60 (1913).

achse der Sonne um 7,25° gegen die Ekliptik geneigt ist, wird die Interpretation von CORTIE auch als *Axial-Hypothese* bezeichnet. Ihr liegt folgende Überlegung zugrunde, die sich auch bei RODÉS[2] findet: Erfahrungsgemäß bleibt eine äquatoriale Zone von etwa $\pm 8°$ Breite auf der Sonne fleckenfrei. Die Erde erreicht am 6. März und 8. September mit 7,25° ihre maximale südliche bzw. nördliche heliographische Breite. Geht man von dem bereits in Ziff. 20 erörterten geometrischen Modell aus, daß erdmagnetische Störungen durch einen annähernd scharf begrenzten Plasmastrom verursacht werden, der mit einem bestimmten Öffnungswinkel von einer lokalen Störungsquelle auf der Sonne emittiert wird, so bestehen zu den angegebenen Zeiten Anfang März und Anfang September die günstigsten Bedingungen für die Einwirkung der Sonne auf das Magnetfeld der Erde.

Diese plausible Deutung der Halbjahreswelle wurde von BARTELS [1] durch subtile statistische Analysen der Eintrittszeiten der Maxima und der Auswirkung von Unterschieden in der Aktivität der beiden Hemisphären der Sonne in Frage gestellt. Harmonische Analyse der Monatsmittel einer 59jährigen Reihe der Maßzahl u_1 (s. Ziff. 15) und einer 25jährigen Reihe der internationalen Charakterzahl Ci (s. Ziff. 14) führte auf eine zeitliche Lage der Maxima, die wohl mit den *Äquinoktien* am 21. März und 23. September verträglich ist, dagegen von den Zeiten maximaler nördlicher und südlicher heliographischer Breite der Erde so weit entfernt ist, daß dieser Unterschied einen nichtzufälligen Anlaß haben muß. Ein erdmagnetischer Effekt des bevorzugten Auftretens von aktiven Gebieten auf nur einer Sonnenhemisphäre war aus einem entsprechend angeordneten Beobachtungsmaterial ebenfalls nicht nachweisbar, obwohl nach der Axial-Hypothese ein positives Ergebnis zu erwarten war. Wegen der zeitlichen Nähe der Eintrittszeiten der Maxima zu den Äquinoktien ist daher der Axial-Hypothese die *Äquinoktial-Hypothese* gegenübergestellt worden, ohne daß damit schon eine Erklärung für die Existenz der Halbjahreswelle gegeben ist. Als ein erster Schritt auf eine solche Erklärung hin kann die von BARTELS[3] schon 1925 aufgestellte Behauptung angesehen werden, daß die Wahrscheinlichkeit für die Auslösung von Störungen mit dem ($\leq 90°$ zu nehmenden) Winkel zwischen der Dipolachse und dem Erdbahnradius wächst. Die Äquinoktien sind dann dadurch ausgezeichnet, daß die Richtung des Radiusvektors zur Sonne senkrecht auf der Richtung der Rotationsachse der Erde steht und damit senkrecht auf der mittleren Richtung der Dipolachse. Der fragliche Winkel erreicht zu diesen Zeiten mit 90° seinen größten Wert.

Die Auseinandersetzung mit diesen beiden Auffassungen dauert noch an. Unter Verzicht auf Einzelheiten ist es hier nur möglich, auf einige der letzten Arbeiten zu diesem Thema hinzuweisen. Im Mittelpunkt steht nach wie vor die Bestimmung und Interpretation der Eintrittszeiten der Maxima der Halbjahreswelle unter Verwendung des inzwischen hinzugekommenen Datenmaterials.

Unter diesem Gesichtspunkt ist die Äquinoktial-Hypothese mit immer größerer Zuverlässigkeit der Ergebnisse bestätigt worden von: MCINTOSH[4], der als Ausgangsdaten 20 Jahrgänge des linearen Aktivitätsmaßes Ap, 52 Jahre der langen Reihe der K-Werte für Potsdam sowie K-Werte weiterer ausgewählter Observatorien verwendet; BARTELS[5], der sich auf die 30jährigen Reihen von 1932 bis 1961 für Kp und Ap stützt; MEYER[6], der anhand der äquivalenten

---

[2] L. RODÉS: Terr. Magn. Atmosph. Electr. **32**, 127—131 (1927).
[3] J. BARTELS: Meteorol. Z. **42**, 147—152 (1925) und Handb. Exp. Phys. **25**, Teil 1, S. 658—665. Leipzig: Akad. Verlagsges. 1928.
[4] D. H. MCINTOSH: Phil. Trans. Roy. Soc. London, Ser. A **251**, 525—552 (1959).
[5] J. BARTELS: Ann. Géophys. **19**, 1—20 (1963).
[6] J. MEYER: J. Geophys. Res. **71**, 2397—2400 (1966) und [10].

Wiederholungszahlen $\omega_{27}(4)$ das Auftreten einer Halbjahreswelle mit Eintrittszeiten der Maxima kurz nach den Äquinoktien auch für die Wiederholungsneigung nachweist; Roosen[7], der ebenfalls die 30jährige Reihe der Ap von 1932 bis 1961 benutzt und den gesamten Verlauf der Halbjahreswelle mit der Variation der absolut genommenen Deklination der Sonne und der absolut genommenen heliographischen Breite der Erde vergleicht. Roosen gibt auch ein Verfahren an, mit dem ein beliebiger Ap- oder Kp-Wert durch Transformation auf das äquinoktiale Störungsniveau vom (statistisch zu erwartenden) Einfluß der halbjährigen Variation der Aktivität befreit werden kann.

Argumente für die Axial-Hypothese haben in den letzten Jahren Priester und Cattani[8] sowie Currie[9] vorgebracht. Als entscheidend muß es aber wohl angesehen werden, daß in keiner der beiden Arbeiten eine quantitativ begründete Erklärung für die etwa drei Wochen Phasendifferenz zwischen einem Extremum der heliographischen Breite der Erde und dem nachfolgenden Maximum der Halbjahreswelle gegeben wird. In beiden Darstellungen wird den Ringstromteilchen die Aufgabe zugeschrieben, die nach der Axial-Hypothese auf das Magnetfeld der Erde einwirkende solare Partikelstrahlung in erdmagnetische Aktivität umzusetzen.

β) Ein einfaches Modell zur quantitativen Erklärung der Amplitude und der Phase der Halbjahreswelle ist vor kurzem von Siebert[10] vorgebracht worden, der die Äquinoktial-Hypothese folgendermaßen modifiziert: *Die Maxima der Halbjahreswelle treten im langjährigen Mittel dann ein, wenn die von der Erde aus bestimmte mittlere Anstromrichtung des solaren Windes in der Ekliptik senkrecht auf die Rotationsachse der Erde zielt.* Ausgehend von der geometrischen Situation zur Zeit der Äquinoktien, sind dann zur Deutung der empirischen Eintrittszeiten der Maxima zwei Korrekturen an der Äquinoktial-Hypothese anzubringen. Die erste folgt aus dem Wesen der harmonischen Analyse, bei der ein gleichförmiger Ablauf des betrachteten periodischen Vorgangs vorausgesetzt wird, und erfordert eine Umrechnung der Daten für das Eintreten der Äquinoktien auf gleichlange Halbjahre. Die zweite, größere Korrektur ergibt sich aus dem Aberrationseffekt beim Anstrom des solaren Windes gegen die sich etwa senkrecht dazu bewegende Erde. Dabei bleibt wegen der Schwankungen der Geschwindigkeit des solaren Windes eine gewisse Unsicherheit. Nach Siebert sollten die beiden Maxima der Halbjahreswelle nach Anbringung dieser Korrekturen mit größter Wahrscheinlichkeit auf den *26. März* und den *25. September* fallen. Die aus den 37 Jahrgängen von 1932 bis 1968 für Ap gefundenen Eintrittszeiten liegen nur jeweils einen Tag später, wie überhaupt die mit den verschiedensten Ausgangsdaten vorgenommenen Analysen ganz vorwiegend auf Maxima kurz nach den Äquinoktien führen. Zeitlich am spätesten liegen die Maxima der halbjährigen Ringstromvariation, was infolge der Dauer der Nachstörungen auch ohne weiteres verständlich ist.

Danach liegt es nahe, die mit dem Jahresablauf erfolgende systematische Änderung der Anstromrichtung des solaren Windes in bezug auf die gegen die Ekliptik geneigte Rotationsachse der Erde als Anlaß der Halbjahreswelle anzusehen. Genauer kommt es dabei auf jene Komponente des solaren Windes an, die senkrecht steht auf der Projektion der Rotationsachse in die durch Anstromrichtung und Normale der Ekliptik aufgespannte *Anstromebene*. Für die quantitative Behandlung wurde noch das auf Messungen der Raumsonde Mariner 2

---
[7] J. Roosen: Bull. Astron. Inst. Neth. 18, 295—305 (1966).
[8] W. Priester and D. Cattani: J. Atmosph. Sci. 19, 121—126 (1962).
[9] R. G. Currie: J. Geophys. Res. 71, 4579—4598 (1966).
[10] M. Siebert: Z. Geophys. 36, 41—56 (1970).

basierende Resultat von MAER und DESSLER[11] herangezogen, wonach der statistische Zusammenhang zwischen der Geschwindigkeit des solaren Windes und Ap durch ein Potenzgesetz beschrieben werden kann. Diese empirischen Grundlagen lassen sich zu folgendem Ansatz für den Jahresgang der Aktivität vereinigen:

$$A(t) = A_0 [1 + \tan^2 \varepsilon \cos^2(t - 173,6°)]^{-\nu/2}. \qquad (21.1)$$

Darin ist $A(t)$ ein lineares Aktivitätsmaß wie Ap, A_0 das Maximum der Aktivität kurz nach den Äquinoktien, ε die Schiefe der Ekliptik, t die während eines Jahres von 0° bis 360° laufende Zeitvariable bei einem Beginn der Zeitzählung am 1. Januar 0 h und ν der aus den Beobachtungen zu bestimmende Exponent des Potenzgesetzes.

Mit $A_0 =$ const, d.h. mit der Annahme, daß der Betrag der Geschwindigkeit des solaren Windes im langjährigen Mittel über das Durchschnittsjahr hin konstant ist, läßt sich Gl. (21.1) in eine Fourier-Reihe entwickeln. Werden die Amplituden der Harmonischen auf das Niveau der mittleren Aktivität reduziert, so hängen sie nur noch von ν ab. Durch Vergleich mit den entsprechend behandelten empirischen Amplituden aus der harmonischen Analyse des Jahresganges ergibt sich ν. In der Praxis kommt für diesen Vergleich nur die halbjährige Komponente in Frage.

Auf der Grundlage der 37 Jahrgänge von Ap hat SIEBERT auf diesem Wege $\nu = 4,7$ gefunden. Der Vergleich mit dem von MAER und DESSLER angegebenen Exponenten 4,6 zeigt einen Grad der Übereinstimmung, der schon wegen der nur vier Monate langen Beobachtungsreihe für die Geschwindigkeit des solaren Windes durch Mariner 2 nicht überschätzt werden darf; dennoch kann dieses Ergebnis als quantitativer Beweis dafür angesehen werden, daß die auffallende Halbjahreswelle der erdmagnetischen Aktivität durch die jahreszeitliche Variation des Anstromwinkels des solaren Windes gegen die in die Anstromebene projizierte Rotationsachse der Erde hervorgerufen wird. Über die zu diesem Effekt führenden physikalischen Vorgänge (s. Ziff. 24) wird dabei nichts ausgesagt. Auch reicht diese einfache geometrische Erklärung sicher nicht aus, um alle Einzelheiten des Jahresganges verständlich zu machen.

γ) Eine Deutung der in dem Rest-Jahresgang von Fig. 11 enthaltenen harmonischen Komponenten macht vom statistischen Standpunkt aus keine Schwierigkeiten.

Die größte dieser Komponenten ist die vierteljährige Variation. Sie enthält als Oberwelle zur Halbjahreswelle noch einen kleinen systematischen Anteil, der sich bei bekanntem ν aus Gl. (21.1) bestimmen läßt. Wird er subtrahiert, so haben die Amplitude der Vierteljahreswelle und die restlichen Amplituden Werte, wie sie auf Grund der zufälligen Schwankungen im Ausgangsmaterial nach den Gesetzen der Statistik zu erwarten sind. Damit sind weitere systematische Anteile, falls sie existieren sollten, aus dem mittleren Jahresgang dieser Ap-Reihe nicht mehr nachweisbar.

Dieses Bild ändert sich, wenn die Ergebnisse der Berechnung quadratischer Spektren herangezogen werden. So zeigen beispielsweise die Analysen von CURRIE[9] und von BANKS und BULLARD[12] neben dem im Spektrum zu erwartenden Maximum bei einer Periode von sechs Monaten ein ebenso ausgeprägtes Maximum bei einer Periode von einem Jahr. Zu dem gleichen Ergebnis gelangt man, wenn die harmonische Analyse der Monatsmittel von Ap für jedes Jahr gesondert vorgenommen wird[13].

Um aus der harmonischen Analyse der Monatsmittel von Einzeljahren brauchbare Ergebnisse zu erhalten, muß der den Jahresgang störende Gang der Aktivität mit dem Sonnenfleckencyclus eliminiert werden. Zu diesem Zweck wurde jeweils von Januar des zu analy-

---

[11] K. MAER and A. J. DESSLER: J. Geophys. Res. 69, 2846 (1964).
[12] R. J. BANKS and E. C. BULLARD: Earth Planet. Sci. Letters 1, 118—120 (1966).
[13] Diese bisher unveröffentlichten Ergebnisse, die auch den Fig. 12—15 zugrunde liegen, wurden von Herrn J.-H. HUFEN, Inst. f. Geophys. Göttingen, zur Verfügung gestellt, wofür ihm auch an dieser Stelle gedankt sei.

sierenden Jahres bis zum Januar des folgenden Jahres der lineare Anteil der unperiodischen Schwankungen mit Hilfe der Lamontschen Korrektur beseitigt. Um zufällige Schwankungen der als Basis dienenden Januarwerte auszuschalten, wurden diese durch mittlere Werte aus den um den betreffenden Januar zentrierten 13 Monaten (bei halber Bewichtung der beiden äußersten Monate) ersetzt.

Fig. 12. Darstellung der durch harmonische Analyse der Monatsmittel von Ap für die Einzeljahre 1933 bis 1966 berechneten *Ganzjahreswellen* Ap_1 durch einen Vektorzug in der Periodenuhr (Summationsuhr). Der störende Einfluß des langsamen Ganges der Aktivität mit dem Sonnenfleckencyclus ist eliminiert.

Fig. 13. Darstellung der durch harmonische Analyse der Monatsmittel von Ap für die Einzeljahre 1933 bis 1966 berechneten *Halbjahreswellen* Ap_2 durch einen Vektorzug in der Periodenuhr (Summationsuhr). Der störende Einfluß des langsamen Ganges der Aktivität mit dem Sonnenfleckencyclus ist eliminiert.

In den Fig. 12 bis 15 sind die Ergebnisse dieser Analysen für die ersten vier Harmonischen zur Grundperiode eines Jahres in der Form von Vektorzügen in Periodenuhren *(Summationsuhren)* dargestellt. Für jedes Jahr von 1933 bis 1966 sind Amplitude und Anfangsphase der betreffenden Komponente Ap_n, $n = 1, 2, 3, 4$, durch eine gerichtete Strecke angegeben. Zur Benutzung der Randbeschriftung für die Bestimmung der Eintrittszeiten der Maxima bringe man den betreffenden Jahresvektor durch Parallelverschiebung in den Ursprung der Periodenuhr.

Bei der Betrachtung der Figuren fällt sofort die ungewöhnliche *Regelmäßigkeit und Stärke der Halbjahreswelle* in Fig. 13 auf, wobei noch zu beachten ist, daß der Maßstab dieser Figur die Länge der Strecken im Vergleich zu den anderen Figuren auf die Hälfte reduziert. Während der Verlauf des Vektorzuges für die *dritteljährige Variation* der Aktivität in Fig. 14 sich kaum von einer nach statistischen Gesetzen erfolgenden Irrfahrt unterscheidet, zeigt die *Vierteljahreswelle* in Fig. 15 das erwartete *teilweise systematische* Verhalten. Ihre mittlere Phase ist in guter Übereinstimmung mit der aus Gl. (21.1) für diese Komponente berechneten Phase[10].

Fig. 14. Darstellung der durch harmonische Analyse der Monatsmittel von Ap für die Einzeljahre 1933 bis 1966 berechneten *Dritteljahreswellen* Ap_3 durch einen Vektorzug in der Periodenuhr (Summationsuhr). Der störende Einfluß des langsamen Ganges der Aktivität mit dem Sonnenfleckencyclus ist eliminiert.

Wird der Grad von Regelmäßigkeit im Vektorzug für die *Ganzjahreswelle* nach Fig. 12 mit der entsprechenden Eigenschaft bei den anderen Harmonischen verglichen, so muß die ganzjährige Variation zwischen der vierteljährigen und der dritteljährigen Variation eingestuft werden. Dabei sind es vor allem die *häufig wechselnden Anfangsphasen*, die die Ganzjahreswelle im langjährigen Mittel zu einer zufälligen Erscheinung reduzieren, während die *Größe und Regelmäßigkeit der Amplituden* mehr auf eine systematische Ursache hindeutet. Dieser Sachverhalt kommt auch zum Ausdruck, wenn man die mittleren Amplituden der Ganzjahreswelle und der Halbjahreswelle miteinander vergleicht; die bei der harmonischen Analyse des Durchschnittsjahres implizierte vektorielle Mittelwertbildung (entsprechend der Resultante in der Periodenuhr) führt mit dem Datenmaterial der Figuren auf das Amplitudenverhältnis $c_1 = 0{,}15\, c_2$, während sich ohne Berücksichtigung der Phasen bei algebraischer Mittelwertbildung der Amplituden der Einzeljahre $c_1' = 0{,}68\, c_2'$ ergibt. Die Amplituden der Ganzjahreswelle sind also in den einzelnen Jahren denen der Halbjahreswelle durchaus vergleichbar und im Mittel sogar noch etwas größer als die der Vierteljahreswelle.

Als eine mögliche physikalische Ursache der Ganzjahreswelle wird im Sinne der Axial-Hypothese eine über längere Zeit (Monate bis Jahre) andauernde stärkere Aktivität der Sonne auf ihrer nördlichen oder südlichen Hemisphäre angesehen. Statistische Untersuchungen auf Grund dieser Vorstellung haben zu widersprüchlichen Ergebnissen geführt. Eine andere Erklärung bietet sich durch eine Erweiterung des Ansatzes Gl. (21.1) in der von SIEBERT gegebenen Deutung der Halbjahreswelle an. Wenn nämlich der anströmende solare Wind ebenfalls über längere Zeit eine vorherrschend positive oder negative Komponente senkrecht zur Ebene der Ekliptik hat, wie dies im Prinzip vom Raumsonden festgestellt worden ist, folgt aus einem anstelle von Gl. (21.1) entsprechend verallgemeinerten Ansatz sofort die Existenz einer Ganzjahreswelle. In ähnlicher

Fig. 15. Darstellung der durch harmonische Analyse der Monatsmittel von Ap für die Einzeljahre 1933 bis 1966 berechneten *Vierteljahreswellen* Ap_4 durch einen Vektorzug in der Periodenuhr (Summationsuhr). Der störende Einfluß des langsamen Ganges der Aktivität mit dem Sonnenfleckencyclus ist eliminiert.

Weise läßt sich aber auch bei einer Verknüpfung des interplanetaren Magnetfelds mit dem Magnetfeld im Schweif der Magnetosphäre auf das Auftreten einer Ganzjahreswelle schließen. Jeder Vorzeichenwechsel der Geschwindigkeitskomponente senkrecht zur Ekliptik oder jedes Umklappen der Richtung des interplanetaren Magnetfelds hat dann in der einen bzw. anderen Deutung einen Phasensprung zur Folge und wäre für den statistischen Charakter der durchschnittlichen Ganzjahreswelle mitverantwortlich. Eine Entscheidung über die Brauchbarkeit einer dieser Interpretationen ist aber nach den bisher vorliegenden Beobachtungsergebnissen (vgl. Fig. 12) noch nicht möglich.

22. Aktivitätsschwankungen und Sonnenrotation.

α) Das Auftreten und Verschwinden aktiver Gebiete auf der Sonne nach nur sehr groben Gesetzmäßigkeiten hat in Verbindung mit der Sonnenrotation eine Erscheinungsform der erdmagnetischen Aktivität zur Folge, die mit dem Begriff *Wiederholungsneigung* beschrieben wird (s. Ziff. 18γ). Die Methoden zu ihrer Behandlung sind ganz wesentlich an diesem Beispiel entwickelt worden.

Ohne die Ursache zu erkennen, scheint BROUN[1] schon 1858 diese Wiederholungsneigung in der Abfolge der erdmagnetischen Störungen bemerkt zu haben. HORNSTEIN[2], LIZNAR[3] u.a. wiesen auf eine Periode von 26 und 27 Tagen hin. MAUNDER[4] benutzte für die Darstellung des Beobachtungsmaterials erstmalig das Schema der untereinander geschriebenen Zeilen mit jeweils 27 Tageswerten und machte auf einen Zusammenhang der Wiederholungsneigung mit dem Auftreten von Sonnenflecken aufmerksam. Den letzten Beweis für das Vorhandensein der Periode der Sonnenrotation im Rhythmus der erdmagnetischen Aktivität erbrachten die statistischen Untersuchungen von CHREE[5] und CHREE und STAGG[5-7] mit der heute als *Synchronisierungsverfahren* bezeichneten Methode. In der nachfolgenden Zeit wurden mit umfangreicherem Beobachtungsmaterial sowie verfeinerten und neuen Methoden zahlreiche Detailergebnisse gewonnen.

Einen Überblick über den Stand der Erkenntnisse bis 1948 gibt BARTELS[8], der durch intensive Anwendung der Methoden des 27tägigen Buys-Ballot-Schemas, der Synchronisierung und der Autokorrelation (s. Fig. 8) und durch die Arbeitshypothese der *M-Regionen* (s. auch Ziff. 20) maßgeblich zu diesen Ergebnissen beigetragen hat. Mit der Entwicklung von Verfahren zur Bestimmung *quadratischer Spektren* wurde auch dieser Weg zum Nachweis der Periode der Sonnenrotation und ihrer Harmonischen im Spektrum der Aktivität eingeschlagen, wofür die Untersuchungen von WARD[9], SHAPIRO und WARD[10] sowie BANKS und BULLARD[11] Beispiele sind. Die umfangreichste und detaillierteste Darstellung der Stärke der Wiederholungsneigung über Zeitabschnitte bis zu 16mal 27 Tagen von 1884 bis 1964 hat MEYER [10] durch Berechnung gleitender äquivalenter Wiederholungszahlen gegeben (s. Ziff. 18γ).

β) Wegen der doppelten Quasi-Persistenz sowohl in der Periode als auch in der Stärke der Wiederholung bietet der Zusammenhang zwischen erdmagnetischen Aktivitätsschwankungen und Sonnenrotation breiten Raum für statistische Aussagen. Ein Überblick, wie er hier allein möglich ist, kann daher gerade bei dieser Erscheinung nur unter Verzicht auf Einzelheiten gegeben werden.

Die Quasi-Persistenz in der Periode resultiert zweifellos aus der differentiellen Rotation der Sonne in Verbindung mit der Zonenwanderung der Flecken, die in gleicher oder ähnlicher Weise auch für die M-Regionen angenommen werden muß. Entsprechend ihrer heliographischen Breite verkürzt sich die synodische Rotationsperiode der Flecken im Verlaufe eines Cyclus von etwa 28,5 Tagen auf etwa 26,5 Tage. Übereinstimmend damit ergeben die Synchronisierungsversuche mit Maßzahlen der erdmagnetischen Aktivität Perioden in der Nähe von 27 Tagen im Fleckenminimum und in den Jahren davor, dagegen Perioden bis zu 28 Tagen für die ersten Jahre eines neuen Cyclus.

Im Gegensatz zur starken Veränderlichkeit der Stärke der Wiederholungsneigung ist die Änderung ihrer Periode wesentlich schwächer und regelmäßiger, so daß für kürzere Zeitabschnitte auch mit einer persistenten Periode gerechnet werden kann.

---

[1] J. A. BROUN: Phil. Mag. **16**, 81—99 (1858). Compt. Rend. **76**, 695—699 (1873).
[2] K. HORNSTEIN: Sitzber. Akad. Wiss. Wien **64**, 62 (1872).
[3] J. LIZNAR: Sitzber. Akad. Wiss., Wien, Math.-Nat. Kl. Abt. II, **91**, 454—475 (1885); **94**, 834—843 (1886); **95**, 394—408 (1887).
[4] E. W. MAUNDER: Monthly Notices Roy. Astron. Soc. **64**, 205—224 (1904); **65**, 2—34, 538—559, 666—681 (1905); **76**, 63—68 (1916).
[5] C. CHREE: Phil. Trans. Roy. Soc. London, Ser. A **212**, 75—116 (1912); **213**, 245—277 (1913); — Proc. Roy. Soc. London, Ser. A **101**, 368—391 (1922).
[6] C. CHREE and J. M. STAGG: Phil. Trans. Roy. Soc. London, Ser. A **227**, 21—62 (1927).
[7] J. M. STAGG: Geophys. Memoirs (London, Meteorol. Office). **4**, No. 40 (1927); **5**, No. 42 (1928).
[8] J. BARTELS: FIAT Rev. **17**, Geophys. Teil 1, 39—91 (1948).
[9] F. W. WARD: J. Geophys. Res. **65**, 2359—2373 (1960).
[10] R. SHAPIRO and F. WARD: J. Geophys. Res. **71**, 2385—2388 (1966).
[11] R. J. BANKS and E. C. BULLARD: Earth Planet. Sci. Letters **1**, 118—120 (1966).

Allerdings hat es wenig Sinn, zu ihrem Nachweis in den Schwankungen der Aktivität die Verfahren der spektralen Analyse von Zeitreihen schematisch anzuwenden und sich allein aus einer Vergrößerung des Beobachtungsmaterials eine höhere Genauigkeit zu erhoffen. Hinzu kommt, daß zumindest die 11jährige Schwankung der Aktivität mit dem Fleckencyclus und die Halbjahreswelle der Aktivität eine Modulation der Wiederholungsneigung bewirken, die sich im Spektrum im Auftreten von Nebenmaxima zum Maximum an der Stelle der Rotationsperiode äußern kann. Dabei unterscheiden sich die von der 11jährigen Modulation herrührenden Perioden nur um etwa ± 0,2 Tage von der Rotationsperiode und konnten bisher im Spektrum nicht aufgelöst werden. Sie sind aber ein anderer Anlaß für eine Verbreiterung des Hauptmaximums. Bei einer ganzjährigen Modulation würden die Abweichungen der Nebenperioden bereits + 2,2 Tage und − 1,9 Tage betragen, wofür trotz der Unregelmäßigkeit der Ganzjahreswelle sogar Analysenergebnisse zu sprechen scheinen[11,12]. Für die Existenz von Nebenperioden mit + 4,8 Tagen und − 3,6 Tagen Unterschied zur Rotationsperiode als Folge einer Modulation durch die Halbjahreswelle liegen seltsamerweise nur erste Hinweise vor[13].

Die Möglichkeit des Auftretens von Nebenperioden zur selbst wieder variablen Rotationsperiode der M-Regionen muß gesehen und geprüft werden, bevor bei entsprechender Lage eines Maximums im Spektrum an einen Einfluß des *Mondes* auf Grund seiner synodischen Umlaufzeit von 29,53 Tagen gedacht werden kann.

Unbeantwortet ist auch noch die Frage, ob die in den Spektren als existent angedeuteten Harmonischen zur Grundperiode der Sonnenrotation mehr als die formale Bedeutung haben, daß eine quasi-persistente, nicht sinusförmige Schwankung der Aktivität über viele Sonnenrotationen erhalten bleiben kann (s. Fig. 18). Allerdings wird auch über Beobachtungen berichtet, nach denen bestimmte Symmetrien in der Anordnung gleichzeitig auftretender Fleckengruppen überzufällig häufig sind.

Soll die Wiederholungsneigung etwa anhand von Kennziffern routinemäßig verfolgt und mitgeteilt werden, so stört dabei die Schwankung der Wiederholungsperiode. Will man außerdem die für die Praxis unbequeme Wahl der nach CARRINGTON in der Astronomie gebräuchlichen mittleren synodischen Rotationsperiode von 27,2753 Tagen vermeiden und statt dessen eine in ganzen Tagen definierte Periode benutzen, so bleibt nur die Wahl zwischen 27 Tagen und 28 Tagen. Von diesen ist die Periode von 27 Tagen vorzuziehen, da sie sowohl näher an der mittleren Periode liegt als auch gerade dann zutrifft, wenn die Wiederholungsneigung am ausgeprägtesten ist. Vor allem durch die Arbeiten von BARTELS hat sich daher bei erdmagnetischen Untersuchungen und Darstellungen nach dem Buys-Ballot-Schema der Gebrauch von 27 Tagen für die Länge der Sonnenrotationsintervalle eingebürgert. Diese auch als *Bartelssche Sonnenrotationen* bezeichneten Intervalle werden mit dem nachträglich in das Jahr 1832 gelegten Beginn fortlaufend numeriert. Die folgenden Daten geben jeweils mit 0 h UT den Beginn der entsprechenden Rotation an:

Rotation Nr.   1:   8. Februar 1832
Rotation Nr. 501: 24. Januar   1869
Rotation Nr. 1001: 11. Januar   1906
Rotation Nr. 1501: 28. Dezember 1942
Rotation Nr. 1751: 21. Juni   1961
Rotation Nr. 2001: 14. Dezember 1979

Für Interpolationen sei an die bequeme Weiterzählung erinnert, die sich wegen: 27 · 27 = 2 · 365 − 1 ergibt.

[12] J. MEYER: Earth Planet. Sci. Letters 1, 392—394 (1966).
[13] R. SHAPIRO: IAGA Bull. No. 26, 203 (1969).

Ziff. 22. Aktivitätsschwankungen und Sonnenrotation. 265

R9	Rot.-Nr.	1st day	C9
665 532 ,22	19	J 23	... ,23 . ,2 . ., 5 , , 5 . , ,35 443 64 , ,2 432
477 643 ,,2		F 19	,,2 432 244 ,22 2,4 62 , ,33 342 ,2 , 243 4 , ,
465 332 2,3	62	M 18	243 4 , , ,12 , ,2 , ,3 243 267 636 52 , ,22 232
655 433 433	1762	A 14	,22 232 356 3,3 322 ,22 ,, ,52 , ,, 2,3 442
322 454 432	63	M 11	2,3 442 , ,3 , ,,,, . ,3 , , ,5 2,, 323 2,5 4,2
333 543 333	64	J 7	2,5 4,2 ,23 , , ,3 34, ,25 444 222 543 23,
222 222 2,,	65	J 4	543 23, 222 32, ,,, 343 ,224 476 52, 363 3,2
,,, ,24 332	66	J 31	363 3,2 556 53, ,,2 555 54, ,64 542 ,, ,4 466
,35 544 422	67	A 27	,,4 466 676 454 433 275 342 2,6 435 4,, 622
444 223 553	68	S 23	4,, 622 547 533 343 665 623 635 246 345 566
333 22, 224	69	O 20	345 566 665 342 244 5,5 42, ,3, ,6 62, ,5
53, 2,3 43,	1770	N 16	62, ,,5 654 5,2 226 2,, 42, ,,, ,63 433 ,67
2,3 2,, 23,	71	D 13	433 ,67 765 3,, ,4,,, ,3,,,,, ,,,, ,,66
,23 2,, 223	19	J 9	,,, ,66 665 45, ,,2 32, ,,, 674 ,, , ,7
43, 222 2,,		F 5	,,,, ,27 556 52, ,,, 2,, ,,, ,,, 4,2 , ,,, ,265
232 2,, 2,,	63	M 4	,, ,265 753 2 ,,, ,2,,, ,3,,,, ,,, ,3,, ,56
224 444 2,,	1775	M 31	,3, ,56 542 2,, ,234 4,, 432 ,32 ,,, 2,, 566
,22 444 553	76	A 27	2,, 566 452 2,3 445 253 ,,, ,,,2 ,,, 2, 245
223 225 642	77	M 24	,2, ,245 323 3,, ,47 32, ,,, ,2, ,253 32,, ,5
,22 22, ,,2	78	J 20	32, ,,5 643 23, ,,4 553 343 ,,, ,2 42, ,63
,22 244 42,	79	J 17	42, ,63 563 44, ,65 442 333 2,22 ,, ,2 ,6
,23 422 232	1780	A 13	,,2 ,26 676 252 224 643 422 2,22 ,4 425 427
236 552 ,,,	81	S 9	425 427 667 464 787 576 675 3,, ,12 ,34 ,25
233 433 434	82	O 6	,34 ,25 666 44, ,,4 2, 742 , ,76 ,2 44, ,36
32, , ,2 232	83	N 2	44, ,36 766 44, ,,4,,, ,,, ,254 ,,, 25, 366
222 22, ,,,	84	N 29	25, 366 654 42 , ,,2 32, ,,, , 533 42, ,34 ,,
,,, ,2, 2,,	1785	D 26	,34,, ,75 52, ,,5 52,,,, 642 ,,,, ,34,,
,,2 ,,, ,,2	19	J 22	,,3 4, ,253 632 ,,33 645 42, ,46 2 232 ,36 454
,33 2,, ,,3		F 18	,36 454 2,4 622 ,,2 764 352 ,24 233 23, ,23
2,3 2,, ,,	64	M 16	23, ,23 665 43, ,,6 276 533 ,34 2,4 ,,, 224
,,, 2,, ,,,	1789	A 12	,,, 224 554 2,, ,33 664 353 ,,2 ,,, ,55 ,56
,,2 ,,, ,,,	1790	M 9	,55 ,56 652 ,,, ,,, ,56, 32,, ,,, ,, ,22 7
,,2 22, ,,,	91	J 5	,,, ,22 763 2, ,,, ,2 ,43, ,,3 , ,, 52, ,5
,,, ,,, ,,,	92	J 2	,52 , , ,5 543 ,,, ,2 564 2,2 ,,, ,,, 442 ,,,
,,, ,, ,22 ,	93	J 29	442 ,, ,,64, 3,2, ,64 ,,, ,,, ,,, ,,, 232 ,,
,,, ,,, ,,,	94	A 25	232 ,, ,35 22, 365 42, ,,, 42,, ,, 722 ,,
,,, ,2 ,,,	95	S 21	,722 ,, ,6, 52, 365 344 4, ,422 ,, ,363 4,,
,,, ,,, ,,,	96	O 18	363 4,,, ,5,,, ,43 ,22, ,253 ,2, ,32 ,,,
,,, ,,, ,,,	97	N 14	,32 ,,, ,,, 5,, 3,,3,32 ,,,, ,3 ,,, , ,2 ,24
,,2 222 222	1798	O 11	,, ,2 ,244 4,2 ,,,, ,,,, ,,, ,32 ,, ,42 , ,3

Symbol	,	2	3	4	5	6	7	8	■
$\bar{R}$ =	0	1-15	16-30	31-45	46-60	61-80	81-100	101-130	131-170	171...
R9, C9 =	0	1	2	3	4	5	6	7	8	9
Cp =	0.0-0.1	0.2-0.3	0.4-0.5	0.6-0.7	0.8-0.9	1.0-1.1	1.2-1.4	1.5-1.8	1.9	2.0-2.5

Fig. 16. Diagramm der aus Cp abgeleiteten täglichen planetarischen Charakterzahlen $C9$ (s. Ziff. 10ζ) für die Jahre 1962 bis 1964, angeordnet nach 27tägigen Sonnenrotationen in der von BARTELS eingeführten Numerierung mit Angabe des Datums für den ersten Tag jeder Rotation; dazu die aus dreitägigen Mittelwerten $\bar{R}$ der täglichen Sonnenfleckenrelativzahl R nach dem beigefügten Schlüssel abgeleitete Maßzahl $R9$ der Sonnenaktivität.

Ein Beispiel für die Anordnung von Maßzahlen nach 27tägigen Sonnenrotationen ist die graphische Darstellung des Kp-Jahrgangs 1967 in Fig. 6. Als weiteres Beispiel zeigt Fig. 16 das Schema für die tägliche planetarische Charakterzahl $C9$ (s. Ziff. 10ζ). Ergänzend sind dort noch aus der Sonnenfleckenrelativzahl R berechnete *dreitägige Maßzahlen $R9$ für die Sonnenaktivität* angegeben. Man erhält die Werte für $R9$, indem aus den täglichen R-Werten die Mittelwerte $\bar{R}$ für die neun dreitägigen Abschnitte einer Rotation bestimmt werden und $\bar{R}$ nach dem beigefügten Schlüssel auf eine 10stufige Skala transformiert wird. Fig. 16 umfaßt zeitlich das Sonnenfleckenminimum von 1964 und die beiden vorhergehenden Jahre. Die durch *Größe und Schwärzung der Werte* unmittelbar mögliche *qualitative Beurteilung* der Aktivität läßt bereits auf den ersten Blick eine mehr als zweijährige Sequenz von Wiederholungen erkennen und weist auf eine entsprechende Lebensdauer der als Störungsquellen anzunehmenden M-Regionen hin.

Auch die Stärke der Wiederholung zu verschiedenen Zeiten im Fleckencyclus ist mit den Verfahren der Synchronisierung und der Autokorrelation untersucht worden. Als quantitatives Maß, zu dem jeder während der l untersuchten Rotationen vorkommende Wert beiträgt, existiert jedoch nur die *äquivalente Wiederholungszahl* $\omega_k(l)$ (s. Ziff. 18γ).

In diesem Zusammenhang sei noch einmal auf den Verlauf der Jahresmittel von $\omega_{27}(4)$ in Fig. 9 hingewiesen, wo am Beispiel der letzten drei Sonnenfleckencyclen abgelesen werden kann, wie die Wiederholungsneigung zum Fleckenminimum hin ansteigt, kurz vor dem Fleckenminimum ihr Maximum erreicht, danach schnell abfällt und ihr Minimum etwa zur Zeit des Fleckenmaximums durchläuft. Dieser Gegenrhythmus zur Sonnenaktivität stimmt auch mit der Beobachtung überein, daß gerade die extrem starken magnetischen Stürme eine äußerst geringe Wiederholungsneigung aufweisen.

γ) Die klassische Deutung dieses Verhaltens der Wiederholungsneigung beruht auf der schon in Ziff. 20δ erläuterten geometrischen Vorstellung der in bezug auf die Erde zunehmenden *Trefferwahrscheinlichkeit* für M-Regionen mit Abnahme ihrer heliographischen Breite, verbunden mit der Annahme, daß der Öffnungswinkel des von einer lokalen Quelle emittierten Plasmastrahls um so größer ist, je stärker die Eruption ist. Die geringe Wiederholungsneigung im aufsteigenden Ast und im Maximum der Sonnenaktivität ist dann eine Folge der ungünstigen Abschußbedingungen in höheren heliographischen Breiten, wobei die Erde lediglich von den weitwinkligen starken Eruptionen getroffen wird. Diese sind aber nach 27 Tagen bereits so weit abgeschwächt, daß sie infolge ihres jetzt kleineren Öffnungswinkels die Erde wieder verfehlen. Nur M-Regionen in niederen Breiten können die Erde auch dann noch erfassen.

Diese Vorstellung erscheint heute in Anbetracht eines mit Plasma erfüllten interplanetaren Raumes, eines ständig von der Sonne etwa radial nach außen strömenden solaren Windes und einer kaum mehr lokal anzunehmenden Ausdehnung der M-Regionen als zu stark vereinfacht. Überhaupt ist die noch immer offene Frage nach der Natur der M-Regionen einer der Schlüssel zum vollen Verständnis der erdmagnetischen Aktivität. Auf die umfangreiche Diskussion, ob die M-Regionen den aktiven Gebieten auf der Sonne angehören, ob sie mit unipolaren magnetischen Gebieten auf der Sonne zusammenfallen, ob sie selbst ein bestimmtes Entwicklungsstadium im Fleckencyclus darstellen, ob es unterschiedliche Arten von erdmagnetisch wirksamen Quellen auf der Sonne gibt[14], und auf dergleichen Fragen mehr kann hier nicht näher eingegangen werden.

δ) Allerdings ist es unumgänglich, in diesem Zusammenhang die in der Zeit vor und während des Fleckenminimums von 1964 gewonnenen Beobachtungen über die *quasi-stationär mit der Sonne rotierende Sektorenstruktur des interplanetaren Magnetfelds* zu erörtern. Besonders deutlich und zuverlässig konnte sie von WILCOX und NESS[15] aus Messungen der Raumsonde Imp 1 für die BARTELSschen Sonnenrotationen 1784/85/86 (Dezember 1963 bis Februar 1964) nachgewiesen werden. Durch den Richtungswechsel des interplanetaren Magnetfeldes, das im übrigen seine normale Spiralform hatte, ergaben sich in der Ebene der Ekliptik vier von der Sonne ausgehende Sektoren. Von ihnen hatten drei Sektoren eine azimutale Ausdehnung entsprechend $^2/_7$ der Rotationsperiode, während der vierte

---

[14] E. R. MUSTEL: Annals IQSY **4**, 67—87, Cambridge, Mass.: M.I.T. Press 1969.
[15] J. M. WILCOX and N. F. NESS: J. Geophys. Res. **70**, 5793—5805 (1965).

Sektor nur die Hälfte dieser Ausdehnung besaß*. Das einfache Schema dieser Struktur besteht also darin, daß das interplanetare Magnetfeld in benachbarten Sektoren entgegengesetzt gerichtet ist, und zwar vorwiegend entweder auf die Sonne hin oder von der Sonne fort. Im kleinen Sektor hatte es zur Zeit der Beobachtungen durch Imp 1 die Richtung auf die Sonne hin. Für die Existenz dieser Struktur auch längere Zeit vor und nach der Beobachtung durch Imp 1 gibt es zuverlässige Anhaltspunkte bis hin zu den Messungen von Imp 2 Ende 1964. Erwartungsgemäß hat sich dieses Bild mit Beginn des neuen Fleckencyclus grundlegend verändert. Sofern es überhaupt zur Ausbildung einer Sektorenstruktur kommt, existiert sie nur für wenige Sonnenrotationen[16].

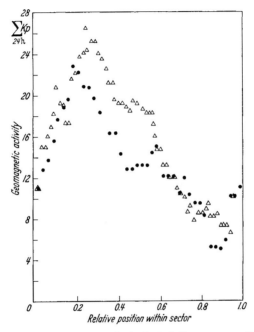

Fig. 17. Variation der erdmagnetischen Aktivität, ausgedrückt durch die Tagessumme von Kp, in Abhängigkeit von der Position der Erde innerhalb eines großen (2/7) Sektors der Sektorenstruktur des interplanetaren Magnetfelds nach Messungen der Raumsonden Imp 1 (offene Dreiecke) und Imp 2 (volle Kreise) (nach NESS und WILCOX[17]).

Innerhalb eines Sektors wurde nicht nur die systematische Änderung der Stärke des interplanetaren Magnetfeldes beobachtet, sondern auch ein ebensolches Verhalten der Dichte und Geschwindigkeit des solaren Windes und damit zusammenhängend ein *systematischer Gang der erdmagnetischen Aktivität*. In Fig. 17 ist diese Schwankung der Aktivität, ausgedrückt durch die Tagessumme von Kp, in Abhängigkeit von der Position der Erde innerhalb eines großen ($^2/_7$) Sektors nach NESS und WILCOX[17] dargestellt. Damit wird aber auch die Wiederholungsneigung der Aktivität von der Sektorenstruktur des interplanetaren Mediums gesteuert. Man braucht sich dazu den Aktivitätsverlauf von Fig. 17 nur nebeneinander und untereinander in Abschnitten von der Länge der Sonnenrotation fortgesetzt zu denken. Wird dann noch der mittlere Verlauf der Aktivität für eine

---

* Siehe Fig. 3 im Beitrag von H. POEVERLEIN in diesem Handbuch, Band 49/4.
[16] J. M. WILCOX and D. S. COLBURN: J. Geophys. Res. **74**, 2388—2392 (1969).
[17] N. F. NESS and J. M. WILCOX: Solar Phys. **2**, 351—359 (1967).

Anzahl von Sonnenrotationen nach der Methode der überlagerten Epochen bestimmt, so tritt die Wiederholungsneigung wie in Fig. 18 als Folge der Sektorenstruktur unmittelbar zutage. In Fig. 18 sind zu diesem Zweck die täglichen Werte von Ap der 41 Sonnenrotationen Nr. 1758—1798 benutzt worden. Deutlich zeichnen sich die Aktivitätsmaxima innerhalb der drei großen Sektoren ab, deren Lage in bezug auf die Tageszählung der Sonnenrotation genau mit den Ergebnissen der durch die Messungen der Raumsonden direkt gewonnenen Stichproben übereinstimmt. Daraus läßt sich folgern, daß diese mit der Sonne rotierende Sektorenstruktur während der drei Jahre 1962 bis 1964 (s. auch Fig. 16) in der Tat quasi-stationär existiert hat.

Fig. 18. Mittlerer Verlauf der erdmagnetischen Aktivität, ausgedrückt durch Ap, während der 41 Sonnenrotationen Nr. 1758 bis 1798 (27. 12. 1961 bis 6. 1. 1965). Die drei Maxima entsprechen dem wiederholten Durchgang der Erde durch die drei großen Sektoren der Sektorenstruktur des interplanetaren Magnetfelds in diesen Jahren.

Diese Entdeckung gewinnt im Hinblick auf die Natur der M-Regionen noch dadurch an Bedeutung, daß es WILCOX und NESS[18] überdies gelungen ist, das mit der Sektorenstruktur verbundene räumliche Schema der Schwankung des interplanetaren Magnetfelds, wie es mit Imp 1 im Abstand der Erde von der Sonne beobachtet worden ist, durch Vergleich mit dem Richtungsverlauf des solaren Magnetfelds in der fraglichen Zeit zwischen etwa 10° und 20° nördlicher heliographischer Breite wiederzufinden. Der Schluß liegt nahe, daß der solare Wind, der in dieser Zeit Ursache der erdmagnetischen Aktivität gewesen ist, in diesem Bereich der Sonne seinen Ursprung gehabt hat. Eine ausführliche Darstellung der sich hier andeutenden Zusammenhänge gibt WILCOX[19].

Diese jüngsten Vorstellungen von der Art der Einwirkung der Sonne auf das Magnetfeld der Erde im Fleckenminimum geben Anlaß zu neuen Fragen und weiteren Untersuchungen. Dazu ist es erforderlich, eine etwa vorhandene Sektorenstruktur des interplanetaren Magnetfeldes auch dann weiterzuverfolgen, wenn Raumsonden für die direkte Beobachtung nicht verfügbar sind. Das sich dann anbietende und auch angewandte indirekte Verfahren, beim Auftreten des in Fig. 17 wiedergegebenen Verlaufs der erdmagnetischen Aktivität auf Vorhandensein, Anzahl und Ausdehnung dazugehöriger Sektoren zu schließen, ist jedoch nicht eindeutig, da ein solcher Störungsablauf auch von lokalen Emissionsquellen der Sonne verursacht werden kann, ohne daß es dabei zur Ausbildung einer Sektorenstruktur zu kommen braucht.

---

[18] J. M. WILCOX and N. F. NESS: Solar Phys. **1**, 437—445 (1967).
[19] J. M. WILCOX: Space Sci. Rev. **8**, 258—328 (1968).

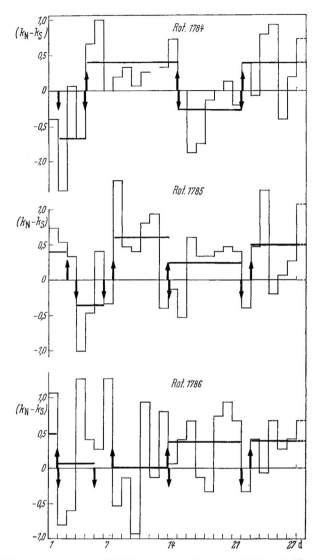

Fig. 19. Unterschiede der erdmagnetischen Aktivität zwischen zwei konjugierten Gebieten der Nord- und Südhalbkugel, ausgedrückt durch die relative Maßzahl $k_N - k_S$ (dünne Linie) für die Verhältnisse um Mitternacht nach Ortszeit (s. Erläuterungen im Text). Die Sektorengrenzen des interplanetaren Magnetfelds sind durch nach oben zeigende Pfeile gekennzeichnet, wenn das Feld vorherrschend von der Sonne weggerichtet ist, und umgekehrt. Auf diese Abschnitte beziehen sich die Sektorenmittel von $k_N - k_S$ (starke Linie), deren Sprünge überwiegend dem Rhythmus der Sektorenstruktur folgen (nach SIEBERT[20]).

Als Hinweis auf das Vorhandensein eines in Sektoren mit der Sonne rotierenden interplanetaren Magnetfelds kann aber ein von SIEBERT[20] gefundener Effekt dienen, wonach die Tendenz besteht, daß sich der *Unterschied in der erdmagnetischen Aktivität zwischen Nord- und Südhalbkugel* während der Tages-Achtel um Mitternacht nach Ortszeit mit dem Wechsel der Sektoren umkehrt.

Fig. 19 zeigt diesen Effekt für die drei Sonnenrotationen Nr. 1784/85/86, für die ein Vergleich mit den Meßergebnissen von Imp 1 möglich war. Auf der Grundlage der lokalen

---

[20] M. SIEBERT: J. Geophys. Res. **73**, 3049—3052 (1968).

Kennziffern K einer regionalen Gruppe von fünf Stationen auf der nördlichen Halbkugel und drei Stationen aus einem magnetisch dazu konjugierten Gebiet auf der südlichen Halbkugel wurde die Maßzahl $(k_N - k_S)$ für den Aktivitätsunterschied um Mitternacht hergeleitet. Da es sich hier im Vergleich zum Gang der Aktivität in Fig. 17 um einen Effekt zweiter Ordnung handelt, ist die Streuung der Einzelwerte nicht verwunderlich, jedoch ändern sich die Sektorenmittel von $(k_N - k_S)$ in auffallender und statistisch signifikanter Weise im Rhythmus der Sektorenstruktur.

Die Anwendung dieses Verfahrens auf die Mittagsaktivität verkleinert den Effekt auf ein Drittel. Auch die Benutzung der von MAYAUD eingeführten Kennziffern für jede Halbkugel (s. Ziff. 12γ) führt nur auf einen geringen Aktivitätsunterschied[21]. Das Hervortreten des Effektes bei den Mitternachtswerten deutet auf Vorgänge im Schweif der Magnetosphäre hin und dürfte mit der Tatsache zusammenhängen, daß die erdmagnetische Aktivität derjenigen Hemisphäre stärker ist, deren Magnetfeld im Schweif antiparallel zum interplanetaren Magnetfeld gerichtet ist. Ausgehend von diesem Effekt für die Mitternachtsaktivität, hat H.-W. BARTELS[22] ein Verfahren entwickelt, daß zusammen mit dem charakteristischen Aktivitätsverlauf nach Fig. 17 bei nicht zu schmalen Sektoren zum Nachweis der Lage der Sektorengrenzen verwendet werden kann.

Mit dem Blick auf die fast 40jährige Geschichte der M-Regionen läßt sich auch nach diesen Erkenntnissen der letzten Jahre nur sagen, daß vermutlich die regionale Struktur des solaren Magnetfelds und ihre zeitliche Variation der beste Indikator für das Auftreten erdmagnetisch wirksamer Gebiete auf der Sonne ist. Die genauere Klärung der Zusammenhänge bleibt die Aufgabe weiterer Untersuchungen.

23. Der Tagesgang der Aktivität.

α) Wird etwa aus Stundenmitteln oder dreistündlichen Kennziffern an einem Observatorium der mittlere Tagesgang der erdmagnetischen Aktivität bestimmt, so findet man eine für die betreffende geomagnetische Breitenzone typische *Variation nach Ortszeit*. Fig. 4 ist ein Beispiel für den Verlauf der täglichen Aktivität an einem Observatorium in mittleren Breiten.

Bei Durchsicht der Literatur nach Angaben dieser Art fällt aber auf, daß anders als bei den zuvor besprochenen Variationen der Aktivität die statistische Untersuchung des Tagesganges mit Hilfe von Maßzahlen zurücktritt gegenüber einer modellmäßigen Beschreibung charakteristischer Störungsabläufe durch Stromsysteme. Der die unregelmäßigen Störungen beherrschende Einfluß der Ortszeit findet sich wieder beim ionosphärischen DS-Stromsystem, von dem der überwiegende Beitrag zur täglichen Schwankung der Aktivität herkommt.

Frühe, statistisch fundierte Untersuchungen über die geographische und jahreszeitliche Veränderlichkeit des Tagesganges, der im wesentlichen durch die Größe und die Eintrittszeiten seiner Extrema charakterisiert ist, stammen von CHREE[1] und STAGG[2]. Auf einige der Ergebnisse ist schon in Ziff. 7 und 12 im Zusammenhang mit der Eliminierung dieses Ganges beim Übergang von der lokalen Kennziffer K zur standardisierten Kennziffer Ks kurz hingewiesen worden. Sie sollen hier nicht wiederholt werden mit Ausnahme des für die Interpretation wichtigen Beobachtungsbefundes, wonach ein deutlicher Bruch zwischen dem Tagesgang in niederen und mittleren Breiten einerseits und dem Tagesgang in hohen Breiten andererseits besteht. Nach STAGG und ebenso nach NIKOLSKY[3]

---

[21] J. M. WILCOX: J. Geophys. Res. **73**, 6835—6836 (1968).
[22] H.-W. BARTELS: Diplomarbeit Math.-Naturw. Fak., Univ. Göttingen, 1970.
[1] C. CHREE: Proc. Phys. Soc. London **27**, 193—207 (1915); **39**, 389—406 (1927).
[2] J. M. STAGG: Proc. Roy. Soc. London, Ser. A, **149**, 298—311 (1935) und Geophys, Memoirs (London, Meteorol. Office) **4**, No. 40 (1927); **5**, No. 42 (1928).
[3] A. P. NIKOLSKY: Terr. Magn. Atmosph. Electr. **52**, 147—173 (1947).

ist der Verlauf der Aktivität in den hohen Breiten in zweifacher Weise ortszeitabhängig. LEBEAU[4] kommt zu dem etwas anderen Schluß, daß die Aktivität in einem weiten Bereich der Polkappen von der Ortszeit und der geomagnetischen Zeit bestimmt wird. Die Grenze zwischen den beiden verschiedenen Typen des Tagesganges innerhalb und außerhalb der Polkappen wird nach MAYAUD[5] durch die Größe der Inklination in 5000 km Höhe charakterisiert.

Auch ohne eine genauere Erörterung der physikalischen Vorgänge läßt sich in diesem Verhalten des Tagesganges die Wechselwirkung zwischen solarem Wind und Magnetosphäre in großen Zügen wiedererkennen. Die beiden wichtigen Punkte sind dabei die auf die Sonne hin orientierte asymmetrische Form der Magnetosphäre und die unterschiedliche Art, mit der der solare Wind auf der Tag- und Nachtseite auf die Magnetosphäre einwirkt. Schließt man die Polkappen aus, so ist die Aktivität von den Morgen- bis zu den Nachmittagsstunden durch kurzzeitige Störungen mit kleinen Amplituden gekennzeichnet, worin die größere Stabilität der Magnetosphäre einschließlich der sie begrenzenden Magnetopause auf der Tagseite zum Ausdruck kommt. Typisch für den Störungscharakter in den Abend- und Nachtstunden sind dagegen Baystörungen, die nach Dauer und Amplitude die Tagesstörungen im allgemeinen weit übertreffen und für das Maximum im Tagesgang der Aktivität verantwortlich sind. Hierin äußert sich die geringere Stabilität und damit die höhere Anfälligkeit des Schweifs der Magnetosphäre gegen Störungen. Es kommt hinzu, daß die für magnetosphärische Vorgänge wichtige Grenze zwischen geschlossenen Feldlinien und in den Schweif laufenden offenen (d.h. nicht in Erdnähe geschlossenen) Feldlinien wegen der Asymmetrie der Magnetosphäre einen auch an der Erdoberfläche asymmetrischen Verlauf hat, der angenähert durch das nördliche und südliche Polarlichtoval wiedergegeben wird. Das bedeutet, daß die polaren Elektrojets ihre größte Stärke zugleich mit ihrer größten Annäherung an die mittleren Breiten erreichen, wodurch die Aktivität in den Abend- und Nachtstunden in verstärktem Maße wirksam wird.

Die Situation in den Polkappen unterscheidet sich dadurch von der der mittleren und niederen Breiten, daß es nur offene Feldlinien gibt. Damit werden diese Gebiete auch auf der Tagseite weniger gut gegen die Einwirkung des solaren Windes abgeschirmt. Außerdem macht sich hier der Einfluß des Unterschiedes zwischen Ortszeit und geomagnetischer Zeit am stärksten bemerkbar, so daß zumindest der andersartige Verlauf des Tagesganges und seine Variabilität von Station zu Station in diesen Gebieten nicht überraschen.

β) Akzeptiert man für die Deutung der Halbjahreswelle der Aktivität die Äquinoktial-Hypothese (s. Ziff. 21α) oder ihre modifizierte Fassung (s. Ziff. 21β), so muß die Halbjahreswelle eine *nach Weltzeit ablaufende Variante* im Tagesgang der Aktivität besitzen. Man hat nur zu beachten, daß die in Ziff. 21 benutzte ausgezeichnete Richtung der Rotationsachse der Erde die mittlere Richtung der Dipolachse repräsentiert. Bei einer tageszeitlichen Variation mit gleicher Ursache muß dann von der mit einem Öffnungswinkel von 11,5° um die Rotationsachse umlaufenden Dipolachse selbst ausgegangen werden. Die Projektion der Dipolachse führt in der Anströmebene des solaren Windes, definiert in Ziff. 21β, einen Tagesgang aus, wodurch sich im gleichen Rhythmus die Empfänglichkeit der Magnetosphäre für Störungen durch den solaren Wind ändert. Die Folge davon sollte eine tageszeitliche Variation der Aktivität nach Weltzeit sein. Dabei hängen Rhythmus und Stärke dieser Variation noch von der Jahreszeit ab. Im Sinne der Senkrecht-

---

[4] A. LEBEAU: Ann. Géophys. **21**, 167—218 (1965).
[5] P. N. MAYAUD: Ann. Géophys. **12**, 84—101 (1956).

bedingung, wonach die Aktivität im statistischen Mittel um so stärker wird, je mehr sich der absolut genommene Winkel zwischen der Projektion der Dipolachse und dem Radiusvektor zur Sonne dem Grenzwert 90° nähert, sollte die Erscheinungsform dieses Weltzeitganges zwischen den folgenden Grenzfällen variieren:

Zur Zeit des Juni-Solstitiums 24stündige Periode mit Maximum gegen 4.30 h UT und Minimum gegen 10.30 h UT;

zur Zeit der Äquinoktien 12stündige Periode mit Maxima gegen 10.30 h und 22.30 h UT und Minima gegen 4.30 h und 16.30 h UT;

zur Zeit des Dezember-Solstitiums 24stündige Periode mit Maximum gegen 16.30 h UT und Minimum gegen 4.30 h UT.

Die große Schwierigkeit beim Nachweis dieses Tagesganges nach Weltzeit besteht darin, daß er an jedem Beobachtungsort von dem zuvor besprochenen, wesentlich größeren Tagesgang nach Ortszeit überlagert wird, der eliminiert werden muß. Die bisher vorliegenden Untersuchungen von McIntosh[6] und Mayaud [8] mit dem Ziel, die oben aufgeführten Grenzfälle dieses Ganges in den Zeitreihen verschiedener Maßzahlen herauszuanalysieren, haben trotz mancher Mängel zu einem insgesamt positiven Ergebnis geführt und untermauern damit indirekt auch die Äquinoktial-Hypothese. Es steht aber auch außer Frage, daß die jetzigen Kenntnisse noch unzureichend sind. Für weitere Untersuchungen bietet sich unter anderem die Aufgabe an, den Ansatz Gl. (21.1) zur quantitativen Behandlung der Halbjahreswelle entsprechend den geometrischen Verhältnissen bei der UT-Variation der Aktivität umzuformen und nach einer quantitativen Verknüpfung beider Erscheinungen zu suchen.

24. Schlußbemerkungen. Die Erörterungen dieses letzten Kapitels machen deutlich, wie die statistische Behandlung der Maßzahlen auf qualitative und quantitative Zusammenhänge führt, ohne daß die physikalischen Vorgänge, die dabei im Spiele sind, im einzelnen durchschaut werden. Sie zu verstehen ist das eigentliche Ziel; die statistischen Aussagen weisen den Weg dorthin.

Eine *physikalische Erklärung* der erdmagnetischen Aktivität muß bei der Wechselwirkung zwischen solarem Wind und Magnetosphäre einsetzen, Aussagen über Art und Ausmaß der Aufnahme von Energie durch die Magnetosphäre liefern und das daraus resultierende Auftreten jener erdmagnetischen Störungen verständlich machen, die mit einer der Maßzahlen entsprechend ihrer Definition erfaßt werden. Es ist augenfällig, daß die dabei ablaufenden physikalischen Prozesse extrem verwickelt und in ihren Einzelheiten noch keineswegs geklärt sind. Aus der Fülle der neueren Arbeiten zu diesem Themenkreis sei auf die Veröffentlichungen von Coleman[1], Obayashi und Nishida[2], Piddington[3], Hirshberg und Colburn[4], Parker[5], Spreiter und Alksne[6], Feldstein[7], Axford[8], Wolfe und Intriligator[9] und deren umfangreiche Literaturangaben hingewiesen.

---

[6] D. H. McIntosh: Phil. Trans. Roy. Soc. London, Ser. A **251**, 525—552 (1959).
[1] P. J. Coleman: J. Geophys. Res. **72**, 5518—5523 (1967).
[2] T. Obayashi and A. Nishida: Space Sci. Rev. **8**, 3—31 (1968).
[3] J. H. Piddington: Geophys. J. **15**, 39—52 (1968).
[4] J. Hirshberg and D. S. Colburn: Planetary Space Sci. **17**, 1183—1206 (1969).
[5] E. N. Parker: Rev. Geophys. **7**, 3—10 (1969).
[6] J. R. Spreiter and A. Y. Alksne: Rev. Geophys. **7**, 11—50 (1969).
[7] Y. I. Feldstein: Rev. Geophys. **7**, 179—218 (1969).
[8] W. I. Axford: Rev. Geophys. **7**, 421—459 (1969).
[9] J. H. Wolfe and D. S. Intriligator: Space Sci. Rev. **10**, 511—596 (1970).

Hier soll lediglich die Frage aufgegriffen werden, ob die auf einer physikalischen Vorstellung beruhende Definition der BARTELSschen Kennziffer K und der daraus abgeleiteten Maßzahlen infolge der neuen Erkenntnisse revidiert werden muß. Ihrer Definition nach sollten diese Kennziffern mit unterschiedlichen Empfindlichkeiten Aussagen über die Stärke der lokalen oder planetarischen Störung des erdmagnetischen Feldes durch solare Partikelstrahlung machen. Abgesehen von den in Ziff. 12 besprochenen Schwierigkeiten, eine solche Konzeption ganz zu realisieren, kann auch die Konzeption selbst nicht ohne Modifikation beibehalten werden. Das heißt konkret, daß es nicht ausreicht, die Bezeichnung „*solare Partikelstrahlung*" durch „*solarer Wind*" zu ersetzen, wenn es um *quantitative* Zusammenhänge geht. Mit der ersten Bezeichnung ist die Vorstellung eines leeren interplanetaren Raumes verbunden, durch den zeitlich und räumlich begrenzte Ströme solaren Plasmas fließen. Dagegen ist der solare Wind mit zeitlich und räumlich wechselnder Stärke kontinuierlich vorhanden und der interplanetare Raum demzufolge ständig mit Plasma angefüllt. Während also früher jede stärkere erdmagnetische Störung als ein Auftreffen solarer Partikelstrahlung auf das Magnetfeld der Erde angesehen wurde und aus dem Fehlen solcher Störungen auf die Abwesenheit dieser Strahlung geschlossen wurde, stellt sich jetzt sinngemäß die Frage, welche Eigenschaften des solaren Windes sich in den erdmagnetischen Störungen widerspiegeln.

Ausdruck der Aktivität, wie sie mit den Kennziffern K, Kp, Ap und anderen erfaßt wird, sind *Feldstärkeschwankungen*. Es liegt daher nahe, auch Schwankungen des solaren Windes als Anlaß dieser Aktivität zu vermuten. Unter dem Blickwinkel dieser Erwartung stellt das Ergebnis von SNYDER, NEUGEBAUER und RAO[10] eine Überraschung dar, wonach der Betrag v der Geschwindigkeit des solaren Windes eine auffallend hohe Korrelation mit der als Aktivitätsmaß benutzten Tagessumme von $Kp (= \sum Kp)$ besitzt, wenn auch für v das Tagesmittel benutzt wird. Aus den Messungen der Raumsonde Mariner 2 in der Zeit vom 29. August 1962 bis zum 3. Januar 1963 wurde der folgende Zusammenhang gefunden:

$$v = (8{,}44 \sum Kp + 330) \; km/sec. \qquad (24.1)$$

Diese Beziehung hat bei einem Korrelationskoeffizienten von $r = 0{,}73$ zwischen v und $\sum Kp$ Gültigkeit für $\sum Kp \leq 40$.

Um diesen unerwarteten Zusammenhang von Kp mit v einer Interpretation zugänglich zu machen, kann Gl. (24.1) durch einfache Zusatzannahmen auf einen Zusammenhang zwischen Kp und der *Dichte der kinetischen Energie* des solaren Windes umgeschrieben werden. Solche und ähnliche Beziehungen, die nach dem Vorbild von Gl. (24.1) auch mit anderen Maßzahlen der Aktivität und mit den Daten anderer Raumsonden aufgestellt worden sind, ändern jedoch nichts an dem Umstand, daß der verwendete Parameter des solaren Windes mit seiner absoluten Größe auftritt und daher die erdmagnetische Aktivität nicht mit Schwankungen des solaren Windes in Verbindung bringt. Das gilt auch, wenn anstelle eines Parameters des solaren Windes die Feldstärke des interplanetaren Magnetfeldes verwendet wird, wie dies WILCOX, SCHATTEN[12] und NESS[11] getan haben.

Erst die Beobachtung und Auswertung der *Querfluktuationen des interplanetaren Magnetfeldes* durch BALLIF, JONES, COLEMAN[14], DAVIS und SMITH[13] erbrachte

---

[10] C. W. SNYDER, M. NEUGEBAUER, and U. R. RAO: J. Geophys. Res. **68**, 6361—6370 (1963).
[11] J. M. WILCOX, K. H. SCHATTEN, and N. F. NESS: J. Geophys. Res. **72**, 19—26 (1967).
[12] K. H. SCHATTEN and J. M. WILCOX: J. Geophys. Res. **72**, 5185—5191 (1967).
[13] J. R. BALLIF, D. E. JONES, P. J. COLEMAN, L. DAVIS, and E. J. SMITH: J. Geophys. Res. **72**, 4357—4364 (1967).
[14] J. R. BALLIF, D. E. JONES, and P. J. COLEMAN: J. Geophys. Res. **74**, 2289—2301 (1969).

einen neuen Aspekt. Die Querfluktuationen wurden dabei aus Magnetfeldmessungen der Raumsonde Mariner 4 auf ihrem Weg zum Mars bestimmt. Ausgewertet wurden die beiden Komponenten B_N normal zur Äquatorebene der Sonne und B_T senkrecht zum Radiusvektor zur Sonne in deren Äquatorebene. Aus Mittelwerten für jede Minute wurden für Drei-Stunden-Intervalle die mittleren Abweichungen $\sigma(B_N)$ und $\sigma(B_T)$ berechnet und daraus $\sigma(B_{T,N})$ als Wurzel aus der Summe der Quadrate dieser beiden Größen. In Fig. 20 werden die 21stündigen gleitenden Mittel von $\sigma(B_{T,N})$ mit den zur gleichen Zeit gehörenden 21stündigen gleitenden Mitteln von Kp verglichen. Der Gleichlauf der beiden Kurven ist

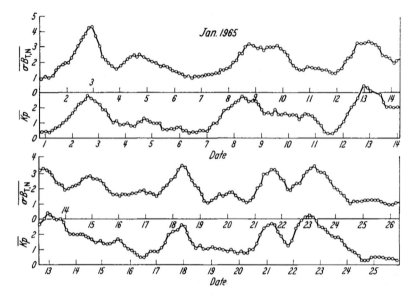

Fig. 20. Stärke der mit der Raumsonde Mariner 4 gemessenen *Querfluktuationen* des interplanetaren Magnetfelds, ausgedrückt durch das 21stündige gleitende Mittel $\overline{\sigma(B_{T,N})}$ der mittleren Abweichungen der transversalen Feldkomponenten in Drei-Stunden-Intervallen (s. Erläuterungen im Text), zum Vergleich mit dem 21stündigen gleitenden Mittel $\overline{Kp}$ zur gleichen Zeit (nach BALLIF et al.[18]).

frappierend, wobei noch die zum Ende jeder Zeile hin anwachsenden Phasenunterschiede eine natürliche Folge der zunehmenden Entfernung der Raumsonde von der Erde und der Sonne sind. Je nach Korrektur dieser Phasenverschiebung liegt der Wert des Korrelationskoeffizienten nach den Daten der Meßperiode vom 29. November 1964 bis zum 12. März 1965 zwischen 0,8 und 0,9.

Die Querfluktuationen des interplanetaren Magnetfeldes deuten auf entsprechende Bewegungsvorgänge im solaren Wind hin, so daß diese letzten Ergebnisse tatsächlich für Schwankungen und Störungen im anströmenden solaren Plasma als Ursache der erdmagnetischen Aktivität sprechen. COLEMAN[1] hat versucht, dies mit einem einfachen Modell auch quantitativ zu belegen. Allerdings werden dadurch die Korrelationen der Aktivität mit den anderen Parametern des solaren Windes und des interplanetaren Magnetfeldes nicht hinfällig. Man steht somit vor einer Fülle statistischer Beziehungen, die sich schwer zu einer einfachen Aussage zusammenfassen lassen.

Die zuvor gestellte Frage, welche Eigenschaften des solaren Windes sich denn nun in den erdmagnetischen Störungen und den sie statistisch erfassenden Kennziffern widerspiegeln, sei daher zum Abschluß Ausgangspunkt der folgenden grundsätzlichen Bemerkung: Zwischen dem solaren Wind als Ursache und der erdmagnetischen Aktivität als Wirkung liegt die Magnetosphäre als Übertragungsmedium. Sowohl die Wechselwirkung zwischen solarem Wind und Magnetosphäre als auch die Übertragungsvorgänge und Übertragungsbedingungen in der Magnetosphäre sind derart vielfältig, daß für die Darstellung der Zusammenhänge beim Ablauf *einer einzelnen Störung* ein entsprechend komplizierter Ansatz erforderlich ist, in dem der Einfluß einer großen Zahl von Parametern berücksichtigt werden muß.

Wird dagegen das *mittlere Verhalten der Aktivität über längere Zeitabschnitte* untersucht, so vereinfachen sich erfahrungsgemäß solche Zusammenhänge. Das bedeutet, daß dann die verschiedenen als wichtig erkannten Parameter mehr oder weniger stark auch *untereinander korreliert* sind und bei Verwendung nur eines Parameters der Einfluß der anderen entsprechend diesen Korrelationen mit im Spiele ist. Es gibt dann zur Herleitung statistischer Gesetzmäßigkeiten Parameter, die brauchbarer sind als andere; aber auch ihre Aussagen betreffen vorzugsweise bestimmte Aspekte des komplizierten Sachverhaltes. Bei zunehmender Einsicht in die Einzelheiten der Vorgänge wird es sich also nicht umgehen lassen, daß allgemeine Aussagen, die zugleich einfach sein sollen, nur durch Unschärfe erkauft werden können.

Literatur.

[1] BARTELS, J.: Terrestrial-magnetic activity and its relations to solar phenomena. Terr. Magn. Atmosph. Electr. **37**, 1—52 (1932).
[2] CHAPMAN, S., and J. BARTELS: Geomagnetism. Vol. 1, pp. 1—542; Vol. 2, pp. 543—1049. Oxford: Clarendon Press 1940; reprinted 1951 and 1962.
[3a] BARTELS, J., N. H. HECK, and H. F. JOHNSTON: The three-hour-range index measuring geomagnetic activity. Terr. Magn. Atmosph. Electr. **44**, 411—454 (1939).
[3b] — — — Geomagnetic three-hour-range indices for the years 1938 and 1939. Terr. Magn. Atmosph. Electr. **45**, 309—337 (1940).
[4] — Erdmagnetische Variationen. In: Landolt-Börnstein, Zahlenwerte und Funktionen, Bd. 3 (Astronomie und Geophysik), S. 728—767. Berlin-Göttingen-Heidelberg: Springer 1952.
[5] — The technique of scaling indices K and Q of geomagnetic activity. Ann. Intern. Geophys. Year **4**, 215—226 (1957).
[6] — The geomagnetic measures for the time-variations of solar corpuscular radiation, described for use in correlation studies in other geophysical fields. Ann. Intern. Geophys. Year **4**, 227—236 (1957).
[7] MAYAUD, P. N.: Atlas of Indices K. IAGA Bull. No. 21; 1. Text 1—113; 2. Figures (1967).
[8] — Calcul préliminaire d'indices Km, Kn et Ks ou am, an et as, mesures de l'activité magnétique à l'échelle mondiale et dans les hémisphères Nord et Sud. Ann. Géophys. **23**, 585—617 (1967).
[9] — Indices Kn, Ks et Km 1964—1967. Paris: Édit. Centre National Rech. Scientif. 1968.
[10] MEYER, J.: Zur 27tägigen Wiederholungsneigung der erdmagnetischen Aktivität, erschlossen aus den täglichen Charakterzahlen $C8$ von 1884—1964. Mitt. Max-Planck-Inst. Aeronomie, Lindau, Nr. 22, 1—80 (1965).
[11] LINCOLN, J. V.: Geomagnetic indices. Physics of Geomagnetic Phenomena **1**, 67—100 New York-London: Academic Press 1967.
[12] CHAPMAN, S., J. BARTELS, and M. SUGIURA: Geomagnetism. Vol. 3 (in Vorbereitung); enthält unter anderem Tabellen der Maßzahlen Kp, ap, Ap, Cp, Ci bis zum Jahrgang 1968.

Classical Methods of Geomagnetic Observations.

By

V. Laursen and J. Olsen*.

With 15 Figures.

A. Geomagnetic observatories and the recording of variations in the geomagnetic elements.

1. General remarks. The geomagnetic field, although extremely weak in comparison with the magnetic fields which can be produced artificially and which can be found for instance near the poles of a strong bar magnet, is readily measurable even by means of very simple instruments, and already a few series of determinations of the field or of one of its components will reveal that at a given place the field is incessantly varying. The variations are due to processes going on partly in the interior of the Earth and partly in the upper atmosphere, and at stations not too far away from each other they may show a high degree of similarity, whereas the variations observed in different parts of the world and particularly at places in different latitudes will as a rule be fundamentally different. In order to give a complete picture of these variations for the entire Earth a number of well distributed observing stations — observatories — have been established[1,2], where the variations in magnitude and direction of the local geomagnetic field vector are continuously recorded[3].

The selection of the site for a geomagnetic observatory and the construction of the buildings require the utmost care. The soil must be of a sufficient rigidity to secure that the pillars, which are often made of concrete, form an absolutely stable support for the sensible instruments. If at all possible the site should be free from local geomagnetic anomalies, and all artificial magnetic disturbances must be avoided as they will disturb the homogeneity of the natural field and may mask the true variations[4]. Special attention should be paid to the disturbing effect of even far-off traffic lines operated on direct current[5]. In the construction of the buildings only non-magnetic material should be used: non-magnetic concrete with non-magnetic brass reinforcement or wood with copper nails. The buildings must be well insulated in order to avoid rapid temperature variations in the instrument rooms.

From what has been said above about the necessity of using non-magnetic material for the construction of the observatory building it is obvious that in the instruments themselves and in their supports even very small magnetic impurities may give rise to considerable systematic errors in the recorded variations or the measured values of the magnetic elements. Apart from the magnet which forms the sensible element of many magnetic instruments all other parts of the equipment should, therefore, be entirely non-magnetic, and it will in general be necessary to test carefully all materials which are to be used in the construction of such equipment. A suspended astatic moving-magnet system provides an adequate and

---

* Original manuscript received Feb. 1956, revised manuscript Jan. 1968.

[1] Description of Geomagnetic Observatories I, II, III. Published by the International Association of Geomagnetism and Aeronomy 1957 and 1959.

[2] G. Fanselau: List of Geomagnetic Observatories. IAGA Bul. **20** (1965).

[3] Annual Mean Values of Geomagnetic Elements since 1900. Report of the Working Group on the Analysis of the Geomagnetic Field, IAGA Commission III (1965).

[4] F. Eleman: Nature **209**, 1120 (1966).

[5] V. V. Sohoni, S. K. Pramanik, S. L. Malurkar, and S. P. Venkiteshwaran: Indian J. Met. Geophys. **4**, 45 (1953).

simple device for performing such testing[6]. (Note that the specimen must be at rest when examined in order to avoid the magnetic effect of eddy currents). Ordinary bronze and brass often contain some magnetic impurities, but special procedures allow to transform these metals into non-magnetic alloys[7]. Aluminium will often be practically non-magnetic.

2. The geomagnetic field and its components. At a given point on the Earth's surface the magnetic meridian is the vertical plane containing the magnetic field vector. The magnetic declination or variation, D, is the angle between the astronomic meridian and the magnetic meridian, reckoned from the true north direction towards east. The inclination or dip, J, is the angle, positive downwards, which the magnetic vector makes with the horizontal plane. The horizontal component of the vector is here denoted by $\mathcal{H}$, the vertical component by Z; X and Y are the components of $\mathcal{H}$ towards true north and east respectively. The following equations connect these quantities, $\mathcal{F}$ being the magnitude of the magnetic vector or the total intensity:

$$X = \mathcal{H} \cos D, \qquad Y = \mathcal{H} \sin D, \qquad \mathcal{H} = \sqrt{X^2 + Y^2},$$

$$\mathcal{F} = \frac{\mathcal{H}}{\cos J} = \frac{Z}{\sin J} \qquad \mathcal{F} = \sqrt{\mathcal{H}^2 + Z^2}, \qquad Z = \mathcal{H} \tan J.$$

Electromagnetic cgs-units are generally used.

The unit for magnetic field strength, $\boldsymbol{H}$ is Oersted (Oe), that for magnetic induction $\boldsymbol{B}$ is called Gauss (Gs or Γ). It depends on the measuring device whether $\boldsymbol{H}$ or $\boldsymbol{B}$ is determined, but in geomagnetism the two quantities will have the same numerical value (in this particular system of units). In geomagnetic measuring practice a smaller unit, gamma (γ), is used, one gamma being equal to 10^{-5} Gs (or Oe).

3. General principles for the recording of geomagnetic variations. At most magnetic observatories the variations of the magnetic vector are recorded as variations of three of its components or elements, and the elements chosen are usually either $D, \mathcal{H}, Z$ or X, Y, Z, since a recording of the dip J meets with technical difficulties.

α) The instrumental equipment used for recording magnetic variations is known as a *magnetograph*, and consists of one or more *variometers* in connection with a *recorder*, generally of the photographic type[1].

The existing types of variometers may be divided into two main groups, magnet variometers, in which the sensitive element is a suspended magnet the movements of which are directly recorded, and electromagnetic variometers, in which the magnetic variations are converted into electromotive force and recorded by means of an appropriate electric device. In all cases, however, the recording technique is almost the same, and it will be sufficient to describe the method in detail in the case of, say, a magnetograph recording variations of D.

β) The *D-variometer* is always of the magnet type. Usually it has a horizontal magnet suspended by a fibre with negligible torsion constant, so that the magnetic axis of the magnet will always be in the magnetic meridian and form the angle D with the astronomic meridian. The suspended magnet will follow all variations of D[2] and so will a mirror, M, attached to the magnet (Fig. 1). The movements are recorded by an optical system similar to the one shown in the figure.

---

[6] H. E. McComb: [9], p. 25.
[7] W. F. Steiner: Terr. Mag. **43**, 47 (1938).
[1] H. Kaiser: Theorie der photographischen Registrierung. Z. Tech. Phys. **16**, 303—314 (1935).
[2] As for the special problems met with in the recording of very rapid variations of the geomagnetic elements see Sect. 13. At the present stage we shall assume the variations to be so slow that these problems can be disregarded.

A light-beam L_1 from the source S, placed at focal distance from the variometer lens L, forms a parallel beam after having passed the lens. The beam is reflected from the magnet mirror M, and having passed the lens once more the reflected beam L_2 is focussed at D on the recorder drum. If the light-source is not in the form of a luminous point it may be necessary to use a cylindrical lens C in front of the drum in order to have the reflected light concentrated in a point on the surface of the drum or on the sheet of photographic paper which is covering the drum when in operation. As long as the magnet mirror does not move the reflected light-spot will produce a straight line on the rotating sheet; movements

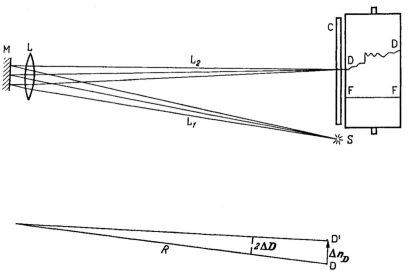

Fig. 1. Declination-variometer and recorder (schematic diagram).

of the mirror, corresponding to variations of D, will be recorded as a more or less irregular curve (DD, at right in the figure). Usually the light from S will be reflected also from a second mirror, placed in the variometer near the moving mirror, but firmly fixed to the housing of the instrument; the corresponding curve on the photographic paper will be a straight reference line (FF, below DD in the figure). The instantaneous distance FD of the moving light-spot from the base line is often called the D-ordinate n_D corresponding to the moment considered.

A change ΔD of the declination will result in a corresponding rotation of the magnet mirror M and the reflected light-beam L_2 will sweep through the angle $2\Delta D$ changing the D-ordinate FD by $DD' = \Delta n_D$ mm (see lower part of Fig. 1). For small deviations there will be proportionality between the variation ΔD and the corresponding displacement Δn_D of the recording light-spot. The ratio

$$S_D = \frac{\Delta D}{\Delta n_D}$$

is called the scale value of the variometer (usually expressed in minutes of arc per mm). The declination D at a given moment will then be

$$D = D_0 + S_D n_D$$

where D_0 is the so-called base-line value of the variometer, that is the value of D which will bring the moving light-spot to coincide with the stationary light-spot producing the base line ($n_D = 0$).

For the ideal D-variometer considered here, set up in a place entirely free from artificial magnetic fields and provided with a suspension fibre with negligible torsion constant, the scale value is depending only on the effective recording distance R, that is the distance from the nearest nodal point of the variometer lens to the recorder drum, reduced by one third of the maximum thickness of the cylindrical lens in front of the drum, the refractive index of the lens set equal to 1.5. The actual scale value S_D will in this ideal case be identical with the so-called optical scale value:

$$\varepsilon = \frac{1 \text{ rad}}{2R} = \frac{3438'}{2R}. \tag{3.1}$$

A more general formula for the scale value of the D-variometer will be given later (Sect. 5).

γ) For the $\mathcal{H}$- and Z-variometers the recording technique is practically identical with the one just described. Each of the variometers will produce a curve showing the variation of the element in question, and a corresponding base-line from where to measure the ordinates. The $\mathcal{H}$-ordinates n_H and the Z-ordinates n_Z are evaluated using the scale values S_H and S_Z, which values are determined experimentally by methods to be described later. It will also be described later in what way the functioning of a magnet variometer is influenced by the *temperature* of the magnet, and how that influence can be allowed for in the evaluation of the records by the introduction of a *temperature coefficient* for the variometer in question, or how the temperature effect can be eliminated by using some special compensating device.

The same remarks are valid for the X- and $\mathcal{Y}$-variometers.

The fundamental question as to how to determine the base-line values D_0 $\mathcal{H}_0$ etc. will be treated in connection with questions relating to the direct determination of the magnetic elements (see Chapt. B).

4. Theory of the unifilar variometer. A unifilar variometer has a horizontal recording magnet, suspended by a fibre (usually quartz) so that it is free to turn about a vertical axis[1]. The movements of the magnet, due to variations of the horizontal geomagnetic field, are recorded photographically as shown in Sect. 3. Sometimes control magnets are placed close to the instrument in order to reduce the temperature effect or to change the scale value.

In Fig. 2 p is the direction of the magnetic axis, reckoned positive towards the north pole; D is easterly declination, $\mathcal{H}$ the horizontal force, and C a vector representing the resulting horizontal component of all foreign magnetic fields originating from control magnets and variometer magnets in the neighbourhood. In the following this foreign field is supposed to be homogeneous in the immediate surroundings of the moving magnet. Reckoning all angles positive in a clockwise direction (in accordance with the adopted positive direction of D), we denote by Θ the angle from true north to p and by ψ the amount of torsion in the fibre (defined as the angle which represents the twisting of the lower end of the suspension fibre relative to the upper end). We indicate by n^0 (see Fig. 2) a horizontal direction normal to p, the angle from p to n^0 being equal to 90°, and denote by ω the angle from $\mathcal{H}$ to n^0 and by ϑ the angle from C to n^0. Let further M be the magnetic moment of the recording magnet and h the torsion constant of the fibre. The suspended magnet will be affected by three torques acting in the horizontal plane of rotation, namely $M\mathcal{H} \cos \omega$ and

---
[1] H. H. HOWE: Terr. Mag. **42**, 29 (1937).

$MC\cos\vartheta$, tending to increase ψ, and $h\psi$, tending to decrease ψ. When the system is in equilibrium we have:

$$\mathcal{H}\cos\omega + C\cos\vartheta = \frac{h}{M}\psi = k\psi \tag{4.1}$$

where $k = h/M$.

If $\mathcal{H}$ and D change infinitesimally to $\mathcal{H}+d\mathcal{H}$ and $D+dD$, the magnet will move slightly from its original position until a new equilibrium is obtained. Differentiating the formula (4.1) above we get:

$$d\mathcal{H}\cos\omega - \mathcal{H}\sin\omega\,d\omega - C\sin\vartheta\,d\vartheta = k\,d\psi.$$

Since

$$d\psi = d\vartheta = d\Theta \text{ and } d\omega = d\Theta - dD$$

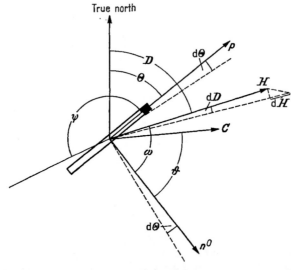

Fig. 2. Theory of the unifilar variometer (diagram defining directions and angles).

this equation may be written:

$$d\mathcal{H}\cos\omega - \mathcal{H}\sin\omega\,(d\Theta - dD) - C\sin\vartheta\,d\Theta = k\,d\Theta$$

or

$$d\mathcal{H}\cos\omega + \mathcal{H}\sin\omega\,dD = (k + \mathcal{H}\sin\omega + C_p)\,d\Theta \tag{4.2}$$

where $\mathcal{H}\sin\omega$ and $C_p = C\sin\vartheta$ are the components of $\mathcal{H}$ and C along the axis of the recording magnet; k may, therefore, also be interpreted as a magnetic field in this direction, expressible in Oersted[2].

The Eq. (4.2) is valid for any unifilar variometer recording variations of a horizontal component of the geomagnetic field. Variometers for D, $\mathcal{H}$, X and Y will be treated separately in the following.

5. Variometer for declination. If the recorded angular movements $d\Theta$ of the suspended magnet are to reflect only the variations dD of the declination, and to be independent of simultaneous variations $d\mathcal{H}$ of the horizontal force, the

---

[2] A. Schmidt: Ergebnisse der magnetischen Beobachtungen in Potsdam und Seddin im Jahre 1908, p. 34.

first term on the left side of Eq. (4.2) must be negligible. This means that $\omega \approx 90°$ (or 270°, see later), or in other words that the magnetic axis of the magnet must be practically in the magnetic meridian. Writing Eq. (4.2) in the form:

$$dD = \left(1 + \frac{k+C_p}{\mathcal{H}\sin\omega}\right)d\Theta - \frac{1}{\mathcal{H}}\cot\omega\, d\mathcal{H} \tag{5.1}$$

and putting ω equal to 90° we get:

$$dD = \left(1 + \frac{k+C_p}{\mathcal{H}}\right)d\Theta \tag{5.2}$$

and if further for $d\Theta$ we substitute $\varepsilon\, dn_D$, where ε is the optical scale value [see Eq. (3.1)] and dn_D the variation of the D-ordinate (usually in mm) we get the scale value of the variometer:

$$S_D = \frac{dD}{dn_D} = \left(1 + \frac{k+C_p}{\mathcal{H}}\right)\varepsilon. \tag{5.3}$$

If in the factor $\left(1 + \frac{k+C_p}{\mathcal{H}}\right)$ the term $\frac{k+C_p}{\mathcal{H}}$ can be neglected, we have the ideal case described above; the suspended magnet will follow exactly all variations of $D(d\Theta = dD)$, and the scale value S_D will be equal to the optical scale value ε. If the term $\frac{k+C_p}{\mathcal{H}}$ cannot be neglected, $d\Theta$ will be different from dD, and after a certain variation of D the magnet may be in a position that is considerably off the magnetic meridian. In this case, as in all other cases where the magnet forms an angle with the magnetic meridian, for instance due to an incorrect setting up of the instrument, or in consequence of the secular variation, the term $-\frac{1}{\mathcal{H}}\cot\omega\, d\mathcal{H}$ in Eq. (5.1) cannot be neglected, and the variations directly computed from the recorded variations of the D-ordinate will need a correction $-\frac{1}{\mathcal{H}}\cot\omega\, d\mathcal{H}$ in order to give the true variation of D. At the same time the base-line value of the D-variometer will be dependent on the *temperature* of the suspended magnet[1].

C_p is the effect of other magnets near the D-variometer and may be computed. The quantity k may be determined by turning the upper end of the fibre until the torsion has produced a certain deviation ν of the recording magnet from its original position in the magnetic meridian. If the upper end of the fibre has been turned through the angle φ and if the deviation ν is small we have:

$$k = \frac{\nu}{\varphi-\nu}(\mathcal{H}+C_p),$$

C_p being considered a constant.

Inserting in Eq. (5.3) we get

$$S_D = \frac{dD}{dn_D} = \frac{\varphi}{\varphi-\nu}\cdot\frac{\mathcal{H}+C_p}{\mathcal{H}}\varepsilon. \tag{5.3a}$$

It appears from the general formula, Eq. (5.3), that a change of scale value of a declinometer can be obtained by changing one or more of the quantities ε, C_p, k, or in other words 1) by altering the optical distance, 2) by the use of control magnets, 3) by using a fibre with a different torsion constant.

Control magnets should preferably be arranged in pairs and symmetrically in order to reduce the inhomogeneities of the field produced[2]. If a scale value smaller than ε is desired and the use of control magnets is impracticable, the magnet may be hung up with its north pole towards magnetic south. It should be remembered, however, that in this position the deviation of the magnet will be in the opposite direction of the variation of D, so that during disturbances the magnet may be considerably off the magnetic meridian.

---

[1] H. H. Howe: Terr. Mag. **42**, 39 (1937).
[2] Ch. Freiburg u. W. Kertz: Z. Geophys. **26**, 227 (1960).

By putting ω equal to 270° in the general formula, Eq. (5.1), we get

$$dD = \left(1 - \frac{k+C_p}{\mathcal{H}}\right) d\Theta. \tag{5.4}$$

In this expression the factor $\left(1 - \dfrac{k+C_p}{\mathcal{H}}\right)$ is negative because the reversed system will only be stable for $k + C_p > \mathcal{H}$.

6. Variometer for horizontal force.

α) If the *unifilar variometer* is intended for the recording of variations of $\mathcal{H}$, the second term $\mathcal{H} \sin \omega \, dD$ in the general formula, Eq. (4.2), should be negligible. That means that ω must be approximately 0° or 180°, or in other words that the magnetic axis of the magnet should be practically in the magnetic prime vertical. Writing Eq. (4.2) in the form:

$$d\mathcal{H} = \left(\mathcal{H} \tan \omega + \frac{k+C_p}{\cos \omega}\right) d\Theta - \mathcal{H} \tan \omega \, dD \tag{6.1}$$

and putting ω equal to 0° we get:

$$d\mathcal{H} = (k + C_p) d\Theta \tag{6.2}$$

and inserting for $d\Theta$ the quantity $\varepsilon \, dn_H$, where ε is the optical scale value (see Sect. 3) and dn_H the variation of the $\mathcal{H}$-trace (ordinate) on the recorder, we get the scale value of the $\mathcal{H}$-variometer:

$$S_H = \frac{d\mathcal{H}}{dn_H} = (k + C_p) \varepsilon. \tag{6.3}$$

The scale value is expressed in γ/mm.

The ideal condition $\omega = 0°$ (or 180°) will generally not be exactly fulfilled. An appreciable deviation of the $\mathcal{H}$-magnet from the magnetic prime vertical may arise from

(1) an incorrect setting up of the instrument
(2) the secular variation of $\mathcal{H}$ and D
(3) accidental great deviations of $\mathcal{H}$ and D from their mean values
(4) the influence of temperature (see Subsect. β through δ).

In all such cases it should be considered whether the above expression, Eq. (6.2), gives the variation of $\mathcal{H}$ with sufficient accuracy, or whether the more complete expression Eq. (6.1) should be used[1]. Using the complete expression the scale value will be

$$S_H = \left(\mathcal{H} \tan \omega + \frac{k+C_p}{\cos \omega}\right) \varepsilon \tag{6.3 a}$$

and to the $d\mathcal{H}$-value obtained by using this scale value a correction $-\mathcal{H} \tan \omega \, dD$ should be added, where dD is the variation of D that may have occurred at the same time as the recorded change of the $\mathcal{H}$-ordinate.

For an $\mathcal{H}$-variometer without control magnets S_H will depend mainly on the quantity $k = h/M$, the ratio between the torsion constant of the fibre and the magnetic moment of the magnet, so that a convenient scale value can be readily obtained by chosing appropriate values of h and/or M. A given scale value can be modified by changing C_p, that is by using control magnets producing an additional horizontal field in the direction of the suspended magnet.

β) The *temperature effect* on an $\mathcal{H}$-variometer is very important, since not only the magnetic moment M of the suspended magnet but also any additional

---
[1] H. H. Howe: Terr. Mag. 42, 36, 38 (1937). See also [9], pp. 95—98.

field C produced by foreign magnets and even the torsion constant of the fibre will be affected by variations of temperature. For bar magnets made of conventional types of steel M will decrease with increasing temperature, the temperature coefficient $\frac{1}{M}\frac{dM}{dT}$ being of the order of $-2 \cdot 10^{-4}$. For quartz fibre suspensions it has been found that the torsion constant is increasing with increasing temperature, the temperature coefficient $\frac{1}{h}\frac{dh}{dT}$ being[2] of the order of 10^{-4}.

For these reasons a variation of the temperature T will in itself produce a deviation of the suspended magnet and will thus be equivalent to a certain variation of $\mathcal{H}$. The field variation $d\mathcal{H}$ that would compensate a certain temperature variation dT, or in other words the temperature correction which should be applied to the $\mathcal{H}$-variations computed directly from the records, can be found by differentiating the general Eq. (4.1) considering ψ as a constant. We get

$$\frac{d\mathcal{H}}{dT}\cos\omega + \frac{dC}{dT}\cos\vartheta = \frac{dk}{dT}\psi = \frac{dk}{dT}\frac{\mathcal{H}\cos\omega + C\cos\vartheta}{k}$$

or

$$d\mathcal{H} = \left(\frac{1}{k}\frac{dk}{dT}\left(\mathcal{H} + \frac{C\cos\vartheta}{\cos\omega}\right) - \frac{dC}{dT}\frac{\cos\vartheta}{\cos\omega}\right)dT.$$

Using for $C\cos\vartheta$ the designation C_n (the component of C normal to the axis of the recording magnet) and indicating by $q_1 = \frac{1}{k}\frac{dk}{dT}$ the temperature coefficient of k and by $q_2 = \frac{1}{C_n}\frac{dC_n}{dT}$ the temperature coefficient of C_n (q_2 is generally negative) we get:

$$d\mathcal{H} = \left(q_1\left(\mathcal{H} + \frac{C_n}{\cos\omega}\right) - q_2\frac{C_n}{\cos\omega}\right)dT. \tag{6.4}$$

For the simple case of a variometer having its magnet in or very near to the prime vertical the temperature coefficient of the instrument will be:

$$\alpha_H = q_1(\mathcal{H} + C_n) - q_2 C_n. \tag{6.5}$$

For negligible values of C we have:

$$\alpha_H = q_1 \mathcal{H}. \tag{6.6}$$

At the same time we have the scale value from Eq. (6.3) $S_H = k\varepsilon$, and since k is dependent on temperature a varying temperature will result in a varying scale value. It is easily seen that the relative variation $\frac{dS_H}{S_H}$ of the scale value for a certain temperature variation dT is equal to the corresponding relative (apparent) variation $d\mathcal{H}/\mathcal{H}$ of the recorded horizontal force.

γ) From Eq. (6.5) it is obvious that the *temperature effect* on the $\mathcal{H}$-variometer can almost be *compensated* by applying an appropriate artificial field C_n, that is by using control magnets producing a field along the magnetic meridian. If the moment of the recording magnet and that of the magnets producing C have approximately the same temperature coefficients, q_1 and q_2 will be of the same order of magnitude but of opposite sign, so that α_H for $C_n = -\frac{1}{2}\mathcal{H}$ will be approximately zero[3,4].

The temperature influence on the suspended system of the variometer, as expressed by the quantity

$$q_1 = \frac{1}{k}\frac{dk}{dT} = \frac{1}{h}\frac{dh}{dT} - \frac{1}{M}\frac{dM}{dT}$$

---

[2] O. REINKOBER: Phys. Z. **38**, 112 (1937).
[3] H. H. HOWE: Terr. Mag. **42**, 35 (1937).
[4] H. E. McCOMB and A. K. LUDY: Terr. Mag. **35**, 29—34 (1930).

can also be eliminated by modifying the system itself. In the expression for q_1, the term $\frac{1}{h}\frac{dh}{dT}$ will be positive, whereas the term $\frac{1}{M}\frac{dM}{dT}$ will be negative, but $\frac{1}{M}\frac{dM}{dT}$ may be made positive by placing — on the magnet itself — an appropriate amount of some special alloy with a high magnetic permeability, provided that the intensity of magnetization of the alloy in question has a sufficiently marked temperature dependence in the temperature interval concerned (see Sect. 22). It will then be possible to adjust q_1 to a zero value. The same result can be obtained by suspending a double magnet consisting of two antiparallel magnets, one strong with a low temperature coefficient, the other weak with a high temperature coefficient, adjusted in such a way that the resulting temperature effect on the magnet system compensates the temperature effect on the fibre[5].

A similar compensating effect has been obtained by using vicalloy magnets, which have been subjected to a special heat treatment around 700°. In addition these magnets have shown a high degree of magnetic stability[6]. Also certain anisotropic magnets (see Sect. 22) the length of which is small as compared with the cross section, have shown properties that would seem to make them fit for the purpose[7].

δ) In the H-variometer designed by LA COUR an *optical temperature compensation* has been used[8]. The light-beam producing the light-spot on the photographic paper is passing a total-reflecting prism suspended in a bimetallic strip (usually silver and platinum), so that the movements of the prism, produced by the effect of temperature variations on the bimetallic strip, will result in a deviation of the light-beam. In a properly adjusted instrument this deviation will just compensate the deviation originating from the effect of the temperature variation on the suspended system.

If the temperature effect is eliminated by means of control magnets, or if the suspended system itself is compensated as described above, the position of the magnet, that is the angle ω, will be unaffected by temperature, and so will the D-correction $-\mathcal{H}\tan\omega\,dD$, Eq. (6.1). For an uncompensated variometer, or for a variometer with optical compensation, the D-term will vary with temperature.

Neglecting the D-correction the complete expression for the horizontal force at a given moment is

$$\mathcal{H} = \mathcal{H}_0 + S_H n_H + \alpha_H T, \tag{6.7}$$

where $\mathcal{H}_0$ is the base-line value, corresponding to $n_H = 0$ and $T = 0$.

7. Variometers for $\mathcal{X}$ and $\mathcal{Y}$.

α) For these variometers, recording the components of the horizontal force towards true north and true east respectively (see Sect. 2), the equilibrium formula, corresponding to Eq. (4.1), may conveniently be written (see Fig. 2, Sect. 4):

$$\mathcal{Y}\cos\Theta - \mathcal{X}\sin\Theta + C\cos\vartheta = k\psi. \tag{7.1}$$

---

[5] H. WIESE: Über die Möglichkeiten zur Beseitigung des Temperatureinflusses auf die Anzeige erdmagnetischer Variometer und Feldwagen. Jahrbuch 1957 des Adolf-Schmidt-Observatoriums in Niemegk, pp. 129—138.

[6] V. N. BOBROV: Magnets with Zero Temperature Coefficients. Geomagn. i Aeronomija **5**, 961—963 (1965) [Engl. translation: Geomagnetism and Aeronomy **5**, 759—761 (1965)].

[7] J. E. GOULD: The Reversible Temperature Coefficient of Anisotropic AL-NI-CO-FE Permanent Magnets, N/T 92, published by The Brit. Electr. and Allied Industry Res. Assoc., Surrey (1962).

[8] D. LA COUR et V. LAURSEN: Le Variomètre de Copenhague. Comm. Mag. **11**, 3, Publ. Inst. Mét. Danois (1930).

Differentiating this formula we get the angular deviation $d\Theta$ corresponding to small variations of dX and dY:

$$\cos\Theta \, dY - \sin\Theta \, dX = (k + X\cos\Theta + Y\sin\Theta + C_p) \, d\Theta \qquad (7.2)$$

where C_p as before stands for $C \sin\vartheta$.

β) For an X-*variometer* the term $\cos\Theta \, dY$ should be negligible, which means that Θ should be approximately 90° or 270°, or in other words that the magnet should be orientated in the true east-west direction. Introducing for $d\Theta$ the quantity $\varepsilon \, dn_X$, where ε is the optical scale value and n_X the recorded X-ordinate, we get the scale value:

$$S_X = \frac{dX}{dn_X} = \left(-X \cot\Theta - Y - \frac{k+C_p}{\sin\Theta}\right)\varepsilon. \qquad (7.3)$$

Putting Θ equal to, say, 270° (north pole towards west) we get:

$$S_X = (-Y + k + C_p)\varepsilon. \qquad (7.4)$$

In case the magnet is not orientated exactly in the east-west direction (for the same reasons as specified in Sect. 6, since X varies with H and D), it should be considered whether the expression Eq. (7.4) gives a sufficient approximation, or whether the complete expression Eq. (7.3) should be used. At the same time due regard should be paid to the correction term $\cot\Theta \, dY$ resulting from Eq. (7.2).

γ) From Eq. (7.2) it is easily seen that the magnet of the Y-*variometer* should be orientated in the astronomic meridian, $\Theta \approx 0°$ or 180°, and that the scale value will be:

$$S_Y = \frac{dY}{dn_Y} = \left(X + Y\tan\Theta + \frac{k+C_p}{\cos\Theta}\right)\varepsilon \qquad (7.5)$$

or (for $\Theta = 0$):

$$S_Y = (X + k + C_p)\varepsilon. \qquad (7.6)$$

In case the magnet is not exactly in the meridian corresponding remarks apply as those given above for the X-variometer.

δ) Using the same procedure and the same designations as used in Sect. 6 for the H-variometer, we find from the above Eq. (7.1) the X- and Y-variations that would compensate a certain *temperature variation* dT. Differentiation gives:

$$\cos\Theta \, dY - \sin\Theta \, dX = [q_1(Y\cos\Theta - X\sin\Theta + C_n) - q_2 C_n] \, dT$$

from which equation we deduce the temperature coefficient for the X-*variometer*:

$$\alpha_X = q_1\left(X - Y\cot\Theta - \frac{C_n}{\sin\Theta}\right) + q_2 \frac{C_n}{\sin\Theta}$$

or for $\Theta = 270°$:

$$\alpha_X = q_1(X + C_n) - q_2 C_n \qquad (7.7)$$

and for the Y-*variometer*:

$$\alpha_Y = q_1\left(Y - X\tan\Theta + \frac{C_n}{\cos\Theta}\right) - q_2 \frac{C_n}{\cos\Theta}$$

or for $\Theta = 0$:

$$\alpha_Y = q_1(Y + C_n) - q_2 C_n. \qquad (7.8)$$

8. Variometer for vertical force.

α) Variometers for the vertical force Z have a *horizontal balancing magnet* free to turn about a horizontal axis normal to the magnetic axis. The rotation axis may be represented by pivots supported by agate planes or knife edges resting on agate cylinders, or it may be a stretched horizontal fibre[1,2]. The movements of the balancing magnet in the vertical plane are recorded by means of a horizontal light-beam, transformed into a vertical beam by passing a prism and reflected from a horizontal mirror attached to the magnet.

Fig. 3, which shows the vertical plane of rotation of the magnet, illustrates the functioning of a variometer with pivot or knife edge suspension. O is the trace of the axis of rotation of the recording magnet, the line $A_S A_N$, normal to this axis, symbolizes the horizontal position of the magnetic axis SN of the magnet. The angle ζ, not shown in the figure, is the angle between the magnetic north direction and OA_N, this angle being reckoned positive towards the east. P is the centre of gravity of the magnet, and β the angle between OP ($=a$) and the magnetic axis. C is a vector representing the component in the rotational plane

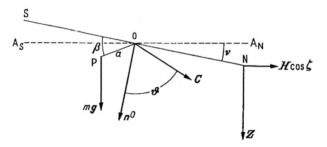

Fig. 3. Theory of the magnetic balance (schematic diagram).

of the magnet of all foreign magnetic fields, and ϑ the angle between C and the direction $n°$, also situated in the rotational plane and perpendicular to the magnetic axis. Let further m be the mass of the magnet, M its magnetic moment and g the gravitational acceleration vector.

When the magnetic axis is deviating the angle ν from the horizontal position, ν reckoned positive from OA_N downwards, the following four torques act on the magnet: $MZ \cos \nu$ and $MC \cos \vartheta$ tending to increase ν, $M\mathcal{H} \cos \zeta \sin \nu$ and $m a g \cos (\beta - \nu)$ tending to decrease ν. If the deviation ν corresponds to the position where the system is in equilibrium we have:

$$MZ \cos \nu + MC \cos \vartheta = m a g \cos (\beta - \nu) + M\mathcal{H} \cos \zeta \sin \nu. \quad (8.1)$$

If $Z, \mathcal{H}$ and D change to $Z+dZ, \mathcal{H}+d\mathcal{H}$, and $D+dD$, ν will change to $\nu+d\nu$, where $d\nu$ can be determined by differentiating Eq. (8.1), putting $d\vartheta$ equal to $d\nu$, and $d\zeta$ equal to $-dD$:

$$M \cos \nu \, dZ - MZ \sin \nu \, d\nu = MC \sin \vartheta \, d\nu + m a g \sin (\beta - \nu) \, d\nu$$
$$+ M\mathcal{H} \cos \zeta \cos \nu \, d\nu + M \cos \zeta \sin \nu \, d\mathcal{H} + M\mathcal{H} \sin \zeta \sin \nu \, dD$$

so that we obtain the following general expression for the change of Z:

$$dZ = \left(Z \tan \nu + \frac{m a g \sin (\beta - \nu)}{M \cos \nu} + \mathcal{H} \cos \zeta + \frac{C_p}{\cos \nu} \right) d\nu \\ + \cos \zeta \tan \nu \, d\mathcal{H} + \mathcal{H} \sin \zeta \tan \nu \, dD \quad (8.2)$$

where $C_p = C \sin \vartheta$ is the component of C along the magnetic axis.

---
[1] G. FANSELAU [6], p. 128.
[2] V. N. BOBROV: Trudy IZMIRAN 18 (28), 55 (1961).

Additional correction terms — depending on the variations of $\mathcal{H}$ and D — will have to be added to the general Eq. (8.2) if the axis of rotation of the balancing magnet is not exactly horizontal[3, 4].

β) If the mean position of the magnetic axis is horizontal the *correction terms* $\cos\zeta \tan\nu\, d\mathcal{H}$ and $\mathcal{H}\sin\zeta \tan\nu\, dD$ will in general be negligible and may be omitted. In this case the recorded variations of ν are independent of $d\mathcal{H}$ and dD. Irrespective of the value of ν the recorded variations will be independent of $d\mathcal{H}$ if the magnet is oriented in the magnetic east-west direction ($\zeta = 90°$ or $270°$), and independent of dD if the magnet is set up in the magnetic meridian ($\zeta = 0°$ or $180°$). When $\cos\zeta \tan\nu$ and $\mathcal{H}\sin\zeta \tan\nu$ have appreciable values the variation dZ derived from the recorded variation of ν must be corrected for superposed variations of $\mathcal{H}$ and D by applying the corrections $\cos\zeta \tan\nu\, d\mathcal{H}$ and $\mathcal{H}\sin\zeta \tan\nu\, dD$. It should be kept in mind that even if the balance has been correctly set up, ν may attain appreciable values due to the secular variation of Z or — if the balance is uncompensated for temperature variations or optically compensated — owing to the effect of extreme temperatures.

γ) We shall assume in the following that ν is sufficiently small to make negligible the correction terms $\cos\zeta \tan\nu\, d\mathcal{H}$ and $\mathcal{H}\sin\zeta \tan\nu\, dD$. The recorded variation of ν will then represent nothing but a variation of Z, and from Eq. (8.2), putting $d\nu$ equal to $\varepsilon\, dn_z$, where ε is the optical scale value and dn_z obtained from the recorded Z-ordinate, we get the *scale value*:

$$S_Z = \frac{dZ}{dn_Z} = \left(Z \tan\nu + \frac{mag\sin(\beta-\nu)}{M\cos\nu} + \mathcal{H}\cos\zeta + \frac{C_p}{\cos\nu}\right)\varepsilon. \qquad (8.3)$$

The variation of S_Z with ν is insignificant. For negligible values of C_p it can be shown that an increase of ν from 0 to ν will bring about an increase of S_Z amounting[5] to only $\nu^2 S_Z$ (ν expressed in radians).

For $\nu = 0$, that is for a horizontal position of the magnetic axis, we have

$$S_Z = \left(\frac{mag\sin\beta}{M} + \mathcal{H}\cos\zeta + C_p\right)\varepsilon. \qquad (8.4)$$

The scale value is largely dependent on the magnetic azimuth ζ of the recording magnet, having a maximum for $\zeta = 0$ (north pole towards magnetic north) and a minimum for $\zeta = 180°$ (north pole towards magnetic south). Only in the magnetic east-west position ($\zeta = 90°$ or $270°$) S_Z will be independent of $\mathcal{H}$, which means that in this position the scale value of a certain magnet will be the same all over the world, neglecting the small variation of g with latitude and the possible influence of C_p.

Since ζ is generally given by the relative position of the variometer and the recorder it is obvious from the expression Eq. (8.4) that a change of the scale value will require a modification of C_p, implying the use of control magnets, or a change of the magnetic moment M or of the quantity $a\sin\beta$ characterizing the position of the centre of gravity relative to the axis of rotation (Fig. 3). This position can be changed by an appropriate grinding of the magnet itself or by changing the position of an adjustable weight attached to the magnet.

δ) Among the quantities entering into the general Eq. (8.1) M, a and C are affected by *temperature variations*. The variation dZ that would compensate a certain temperature variation dT is found in the usual way (see Sect. 6) by differentiating Eq. (8.1). Denoting by q_1 and q_2 the temperature coefficients for

---
[3] Joh. Olsen: Terr. Mag. 39, 173—186 (1934).
[4] J. M. Bruckshaw: Geophys. Prospecting 1, 259—271 (1953).
[5] Joh. Olsen: Terr. Mag. 39, 177 (1934).

a/M and $C_n (= C \cos \vartheta)$ we get:

$$\cos \nu \, dZ = (Z \cos \nu - \mathcal{H} \cos \zeta \sin \nu + C_n) q_1 dT - C_n q_2 dT$$

from which equation we deduce the temperature coefficient for the Z-variometer:

$$\alpha_Z = \left(Z - \mathcal{H} \cos \zeta \tan \nu + \frac{C_n}{\cos \nu} \right) q_1 - \frac{C_n}{\cos \nu} q_2$$

or for $\nu = 0$:

$$\alpha_Z = (Z + C_n) q_1 - C_n q_2. \tag{8.5}$$

Control magnets may be used for compensation of the temperature effect as described above for the $\mathcal{H}$-variometer. Temperature compensation may also be obtained in a mechanical way by attaching to the balancing magnet a small weight which can be adjusted on a spindle of aluminium parallel to the magnetic axis. With reference to the axis of rotation the weight should be placed opposite to the gravity centre, so that the displacement produced by a temperature variation will counteract the temperature effect[6] on M and a. A very simple procedure for the mechanical compensation of a Z-variometer with quartz fibre suspension has been described by BOBROV[7].

The above-mentioned compensation systems will leave the position of the balancing magnet unaffected by temperature variations, whereas the magnet in an optically compensated Z-variometer[8] will change its inclination ν under the influence of such variations.

9. Correct orientation of variometer magnets. The orientation of a variometer magnet may be checked in the following way: Having established, by means of a reliable compass needle or otherwise[1], the direction of the mean magnetic meridian in the recording room, the magnet to be examined is exposed to a fairly strong horizontal field, produced for instance by means of a HELMHOLTZ-GAUGAIN coil (see Sect. 10), or by means of a strong magnet placed at a distance of, say, 100 to 200 cm from the variometer. The direction of the artificial field should be for the D-variometer in the magnetic meridian, for the $\mathcal{H}$-variometer in the corresponding prime vertical, for the X- and Y-variometer in the astronomical east-west and north-south direction respectively, and for the Z-variometer in the rotational plane of the balancing magnet. If the variometer magnet is not in the right position the effect of the artificial field will be a certain deflection, which can be measured on the photographic record or on an appropriate scale, and which may be used to compute the angle through which the magnet should be turned in order to have it correctly orientated. If the intensity of the artificial field is unknown, a few trial adjustments of the magnet will generally bring about a position of the magnet in which it remains unaffected by the field. In the following Section reference will be made to another method of obtaining information as to the orientation of the variometer magnets.

10. Determination of scale values for the variometers.

α) The calculation of the scale value for a D-variometer is described in Sect. 5. The scale value of an $\mathcal{H}$-, X-, Y- or Z-variometer is determined by exposing the variometer magnet to a known artificial magnetic field f, which should be exactly parallel to the geomagnetic component recorded, and which should be as homogeneous as possible in the immediate surroundings of the magnet. The effect of the field will be the equivalent of a certain change of the element in question, and if Δn is the resulting increase or decrease of the recorded ordinate the scale value of the record will be $f/\Delta n$.

The scale value of the D-variometer may be determined or checked in exactly the same way, namely by exposing the D-magnet to a horizontal field f, perpen-

---
[6] H. E. McComb: [9], pp. 109—116.
[7] V. N. Bobrov: Trudy IZMIRAN **18** (28), 55 (1961).
[8] D. la Cour: La Balance de Godhavn. Comm. Mag. **8**, Publ. Inst. Mét. Danois (1930).
[1] H. E. McComb: [9], pp. 141—147.

dicular to $\mathcal{H}$. By vectorial addition it is easily seen that the effect produced will be equivalent to that of a variation $f/\mathcal{H}$ of the declination itself. If Δn_D is the measured change of the D-ordinate on the recorder the scale value will be

$$S_D = \frac{1 \text{ rad } f}{\mathcal{H}\Delta n_D} = \frac{3438' f}{\mathcal{H}\Delta n_D}. \tag{10.1}$$

If S_D is known in advance the method can be used for the determination of an unknown artificial field f, for instance the field produced by a deflecting magnet placed at some well defined distance from the variometer magnet. Once the value of f is determined, such a magnet can be used for the determination of the scale value of other variometers.

β) The simplest way of producing f is to use the *magnetic field of* a HELMHOLTZ-GAUGAIN coil system, consisting of two identical circular and coaxial coils, separated by a *distance b*, equal to the *radius* of either coil. The two coils are connected in series and a direct current flowing in the system will produce a magnetic field of high uniformity along the axis at the centre of the coil system[1]. The intensity of the field at the geometric centre of the system is:

$$f = \frac{4}{25 \cdot \sqrt{5}} \mathfrak{u} \frac{nI}{b} \tag{10.2}$$

$$(f/\text{Oe}) = 0.8997 \frac{n(I/A)}{b/\text{cm}} \tag{10.2a}$$

where n is the number of spires on each coil, b the radius of the coils (or the distance between the two central turns), and I the current.

Since f is proportional to the current I it is a very simple operation to repeat the scale value determination using different values of f, and thereby to examine a possible variation of the scale value with the ordinate (see the remarks made in Sect. 6 when treating the theory of the $\mathcal{H}$-variometer).

γ) The observed scale values and the correct orientation of the variometer magnets may be checked during magnetic disturbances by using a method which is often called the *direct determination* of scale values. By this method the absolute value of the component in question is determined from time to time by appropriate instruments (see Chapt. B) and the corresponding hours are noted. The scale value is then computed by comparing the variations of the absolute value with the corresponding variations of the recorded ordinate. If the variometer magnet is deviating from its standard position, the ordinate variations will to some degree turn out to be dependent also on other magnetic elements (f. inst. an $\mathcal{H}$-ordinate may show dependence on the D-variations occurring during the calibration)[2].

11. Ordinary magnetographs.

α) The principle of *photographic recording* of magnetic variations has already been indicated (Sect. 3). The paper speed with which the recording takes place is usually 15 or 20 mm per hour, and a special device, controlled by a precise chronometer, will produce time marks on the record at appropriate intervals of time. The variations of three magnetic elements, either $D, \mathcal{H}, Z$ or X, Y, Z, are usually recorded on the same magnetogram, and there will often be as a fourth trace a record of the temperature in the variometer room. To make out the effect of a possible shrinkage of the photographic paper two distinct marks at a well

---

[1] S. CHAPMAN and J. BARTELS: [4], p. 85.
[2] D. LA COUR et E. SUCKSDORFF: Exemple d'emploi du QHM pour le contrôle des variomètres pour la déclinaison et pour la force horizontale. Comm. Mag. 16, 1—11, Publ. Inst. Mét. Danois (1936).

defined distance may be applied to the sheet before exposing it to development and drying[1].

β) In order to be able to record both small geomagnetic variations and great geomagnetic disturbances in an adequate manner many observatories operate two magnetographs of the ordinary type, one with sensitive variometers and one with less sensitive ones. The latter equipment is often called a *storm magnetograph*.

As the amplitude of the variations to be expected depends largely on the geomagnetic latitude, due regard should be paid to the latitude when choosing the sensitivity of the variometers. The International Association of Terrestrial Magnetism and Electricity in the International Union of Geodesy and Geophysics, has recommended the scale values given[2] in Table 1.

Table 1.

Geomagnetic latitude	For observatories with sensitive and insensitive variometers						For observatories with only one set of variometers		
	Sensitive			Insensitive			Average sensitivity		
	D '/mm	$\mathcal{H}$ γ/mm	Z γ/mm	D '/mm	$\mathcal{H}$ γ/mm	Z γ/mm	D '/mm	$\mathcal{H}$ γ/mm	Z γ/mm
0°—30°	1	3	4	1	15	4	1	4	4
30°—50°	1	3	4	4	25	15	1	5	6
50°—60°	1	5	5	5	30	25	3	15	15
60°—90°	1	7	7	9	45	45	5	25	25

Several devices have been invented to extend the recording range of a high sensitivity magnetograph beyond the limited width of the paper sheet. Eschenhagen introduced a three-faced variometer mirror giving a central light-spot and two reserve spots so spaced that one of the reserve spots will always have started recording near the border of the paper before the central spot goes out on the opposite side. La Cour obtained a multiple light-beam from a single source, using small prisms appropriately spaced along a rack in front of the recording lamp. Reflected from the variometer mirror this fan-shaped beam will produce a series of equidistant spots so spaced that the distance between two consecutive spots is just slightly smaller than the width of the paper strip available for the record[3]. A. Schmidt[4], using an extra system of inclined variometer mirrors to magnify the scale value, obtained on the same magnetogram not only the normal record, but also a less sensitive record of the same element.

γ) In most magnetographs the *record* is obtained on photographic paper wrapped round a cylinder making one revolution a day. After 24 hours the sheet will have to be changed in order to avoid a mingling of the traces. It is of course possible to modify the recorder so that the record for several days, or even weeks or months, are accumulated on a long strip of photographic paper, moved like a film in a film camera, and if the magnetograph can otherwise work for some time without maintenance or checking, the equipment may be set up for operation at remote places, or at places where no qualified personnel is available.

δ) The German Askania Werke have designed what is called a magnetic variograph or a *portable magnetograph*, which can be set up practically everywhere, and in a very short time, and which is providing automatic records of the variations of three magnetic elements[5].

---

[1] W. Zander: Ausdehnung des photographischen Registrierpapiers in Abhängigkeit von der Lagerung. Jahrbuch 1954 des Adolf-Schmidt-Observatoriums für Erdmagnetismus in Niemegk, pp. 127—129. Berlin 1957.

[2] Int. Ass. Terr. Mag. El., Bul. **13**, 338 (1950).

[3] V. Laursen: Observations faites à Thule, 1ére partie, Magnétisme Terrestre, pp. 21—30. Publ. Inst. Mét. Danois (1943).

[4] A. Schmidt: Z. Instrumentenk. **27**, 137—147 (1907).

[5] G. Fanselau: [*6*], pp. 124—128.

In the Askania equipment the variometers and the recorder are all housed in a well insulated box, 43 × 41 × 23 cm, in which a thermostatically controlled constant temperature can be maintained. The three variometers for D, H and Z have small magnets, fibre suspensions, temperature compensation and three-faced magnet mirrors (see Subsect. β) and the optical path of the recording light-beams has been extended by repeated reflexions from auxiliary mirrors. The recording takes place on a strip of photographic paper, 120 mm wide, which — with a paper speed of 20 mm per hour — will take up 10 days of continuous recording. In a modified arrangement the records are concentrated on a 16 mm film, which may be running continuously for 240 days.

Bobrov's portable magnetograph [6] uses a light-proof box 16 × 32 × 40 cm containing three variometers for recording D, H and Z. The recorder with a 200 mm drum is mounted on the front wall, and the optical path of the recording light-beams has been increased to 100 cm by means of repeated reflexions. The small variometer magnets have a magnetic moment of say 3...10 Oe cm^3 (which is (3...10) 10^{-3} A m^2 in SI-units). They are fixed to light quartz frames provided with quartz mirrors and suspended in quartz fibres which are in turn supported by bigger quartz frames fastened to the portable box. The variometer systems are temperature compensated and each of them is placed in an airtight housing in order to avoid the influence of humidity and dust on the magnets [7].

12. Quick-run magnetographs.
The slow paper speed of an ordinary magnetograph is a serious drawback when a thorough study of rapid geomagnetic variations is intended. Also exact timing of the recorded geomagnetic phenomena is difficult with slow speed. A mere increase of the recording velocity of an ordinary magnetograph would on the other hand involve a considerable increase of the paper costs.

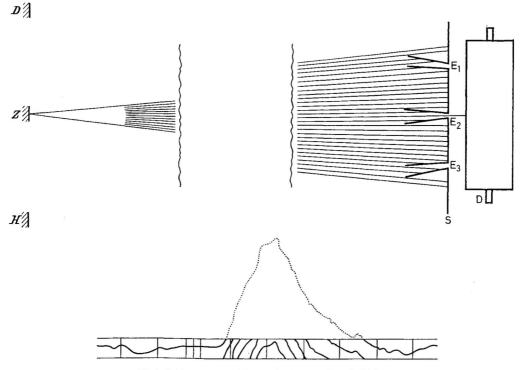

Fig. 4. Quick-run recorder (diagram showing recording principle).

---

[6] V. N. Bobrov: Geomagn. i Aeronomija **5**, 892—894 (1965) [Engl. translation: Geomagnetism and Aeronomy **5**, 695—697 (1965)].

[7] V. N. Bobrov: Trudy NIIZM **11** (21) (1955).

Fig. 5. Magnetograph record obtained with a quick-run recorder in Thule (21. 2. 1949, 0005 h—22. 2. 1949, 0001 h UT).

Fig. 6. Record obtained with an ordinary recorder for the same time interval as in Fig. 5.

Modifying a principle first indicated by AD. SCHMIDT, D. LA COUR has constructed a magnetograph which records the variations of three magnetic elements with a paper speed of 180 mm/hour without mixing the traces and with a consumption of paper not exceeding that of an ordinary recorder[1].

As in the LA COUR magnetograph of the ordinary type (see Sect. 11) a fanshaped bundle of light-beams is sent towards each variometer. The fan, consisting of 51 narrowly spaced beams, is produced by a single lamp in connection with 50 total reflecting prisms, and the light reflected from each magnet mirror would, if no screening was provided for, produce a horizontal row of 51 equidistant light-spots across the photographic paper on the recorder drum. (See Fig. 4, where part of the beams coming from one of the three variometers are indicated.) The distance between two consecutive spots will be slightly under 8 mm, and the effect of a special screen (S in Fig. 4) placed in front of the drum D will be, first, to cut off all the light-beams except those which are accidentally passing through one of three narrow slits cut in the screen about 10 cm apart (E_1, E_2, E_3 in Fig. 4) and, second, to sort out the triple multiplicity of beams so that the beams from a given variometer are allowed to pass through one of the slits only. This selection effect is brought about by appropriate side screens as indicated in the figure.

The width of the slits is 8 mm, so that generally only one light-beam at a time will pass the slit and produce a curve on the photographic paper. When the drum is rotating the records of the three variometers will be concentrated on three narrow strips of paper, one behind each slit. If a magnetic variation causes a certain beam to leave the slit through which it is actually passing, the following beam will in time have started recording near the opposite border of the paper strip, and the whole strip, when developed, may look like the sample record of Fig. 4. The drum is completing one full rotation every two hours and the vertical lines in the record at the bottom of Fig. 4 are time marks, recorded automatically every five minutes with additional marks at 59^m and 01^m. After a full rotation of the drum the special screen will be shifted laterally by 8 mm; thus, recording during the next two hours will continue on three new strips of paper, adjacent to the first ones. The complete 24-hour record will consequently be divided up in 3 · 12 separate strips, each containing a 2-hour record of one element.

If the total variation of an element is required for a certain time interval the discontinuous traces of a strip, and if necessary those of consecutive strips, can of course be combined as indicated by the dotted line in the lower part of Fig. 4, but this operation will rarely, if at all, be of interest at observatories disposing also of ordinary magnetographs.

Fig. 5 is showing a quick-run magnetogram obtained at the Thule Observatory, Greenland, by means of a LA COUR recorder, while Fig. 6 shows the normal magnetogram for the same day Note the registration of a sudden commencement[2] occurring at 1516 UT.

13. Recording of rapid geomagnetic variations.

α) Ordinary and quick-run records of the variations of the geomagnetic field are indicative of the existence of more or less regular oscillations, so-called pulsations[1], having periods of some few seconds or less, but for periods smaller than, say, 10 seconds the *adequate recording* of such pulsations is no longer just a matter of paper speed and of sensitivity. Every recording instrument — in our case the variometer — in which the sensitive element is an oscillating system has its own oscillation period, and when such an instrument is exposed to a pulsation having a period which is not considerably longer than that of the oscillating system itself the amplitude of such a pulsation will not be recorded with the scale value S which we have so far been considering, and which may be called the static scale value, but with a *dynamic* scale value S_d, depending partly on the ratio τ_0/τ between the free (undamped) oscillation period, τ_0, of the oscillating system and the period, τ, of the field pulsation, and partly on the damping applied to the variometer. At the same time there will be a phase difference between the actual pulsations and those appearing in the record.

---

[1] V. LAURSEN: Observations faites à Thule, 1ère partie, Magnétisme Terrestre, pp. 30—39. Publ. Inst. Mét. Danois (1943).

[2] See contribution to this volume by T. NAGATA and N. FUKUSHIMA, pp. 42—55.

[1] See contribution to this volume by T. NAGATA and N. FUKUSHIMA, pp. 115—123, and by E. SELZER, to Vol. 49/4 of this Encyclopedia.

A pulsation of the form $h \sin \frac{2\pi}{\tau} t$, where h is a constant, will therefore be recorded as $a \sin \left(\frac{2\pi}{\tau} t - \varphi\right)$ where a and the phase angle φ are dependent on τ. Let τ_0/τ be denoted by n, and let α be a constant giving the degree of damping (see below) then according to Thellier[2]:

$$\frac{a}{a_0} = [(1-n^2)^2 + 4\alpha^2 n^2]^{-\frac{1}{2}} = \frac{S}{S_d} = f \qquad (13.1)$$

where a_0 would be the response of the variometer to a slow variation with amplitude h of the recorded field component. At the same time we have:

$$\tan \varphi = 2\alpha n/(1-n^2). \qquad (13.2)$$

The damping constant α may be computed from the equation $\delta = \frac{2\pi\alpha}{\sqrt{1-\alpha^2}}$, where δ is the logarithmic decrement of the damped oscillations of the oscillating system[3]. The critical damping is characterized by $\alpha = 1$.

In Table 2 values of the amplification factor f and the phase difference φ are given for increasing values of τ_0/τ and for different values of α.

Table 2.

$n = \tau_0/\tau =$		0	0.1	0.2	0.5	$\frac{1}{2}\sqrt{2}$	1	$\sqrt{2}$	2	5	10	∞
$\alpha = 5 \cdot 10^{-3}$	$f:$	1.00	1.01	1.04	1.33	2.00	100	1	0.33	0.04	0.01	0.00
	$\varphi:$	0°	0°.1	0°.1	0°.4	0°.8	90°	179°.2	179°.6	179°.9	179°.9	180°
$\alpha = \frac{1}{2}$	$f:$	1.00	1.01	1.02	1.11	1.16	1	0.58	0.28	0.04	0.01	0.00
	$\varphi:$	0°	5°.8	11°.8	33°.7	54°.7	90°	125°.3	146°.3	168°.2	174°.2	180°
$\alpha = 1$	$f:$	1.00	0.99	0.96	0.80	0.67	0.50	0.33	0.20	0.04	0.01	0.00
	$\varphi:$	0°	11°.4	22°.6	53°.1	70°.5	90°	109°.5	126°.9	157°.4	168°.6	180°

β) Table 2 shows that for small values of α the variometer will be very sensitive for $\tau = \tau_0$ (effect of resonance), and that the *sensitivity* will be rapidly varying over a certain range of τ near that critical value. For α equal to $\frac{1}{2}$ the factor f will be unity for $\tau_0/\tau = 1$, and it is interesting to note how the value of f remains approximately constant for lower values of τ_0/τ, whereas for $\tau < \tau_0$ the scale value increases so rapidly with decreasing values of τ that the variometer will in practice be incapable of recording the pulsations in an adequate manner.

It should be borne in mind that what we have been treating are variations of a simple sinusoidal type. Other forms of rapid variations will usually be recorded with a certain degree of distortion.

By using oscillating systems of special forms it has proved possible to reduce the dimensions of the systems, and thereby their period of oscillation τ_0, without reducing at the same time the sensitivity of the systems. Variometers with τ_0-values of about 1 sec have been constructed, which are recording the geomagnetic elements with a static scale value of 1 γ per minute of arc or less[2,5]. In a variometer for photographic recording a practical limit to the reduction of the dimensions is set by the fact that the recording mirror must have a certain area. Thiesen[4] has given numerical examples of the response of La Cour variometers to magnetic pulsations.

The sensitivity may be increased by introducing a *photo-electric device* between variometer and recorder. A strong light-beam is reflected from the variometer

---

[2] E. Thellier: Ann. Intern. Geophys. Yr. 4, 255—280 (1957).
[3] Forest K. Harris: Electrical Measurements, p. 623. New York and London 1952.
[4] K. Thiesen: Geophysica (Helsinki) 5 (4), 177 (1958).
[5] V. N. Bobrov: Trudy IZMIRAN 18 (28), 55 (1961).

mirror towards a photo-cell, in which a current is produced that varies with the movement of the mirror. This current may be amplified in order to make it operate a recording milliammeter. A. Graf[6] arranged two identical photo-cells side by side in a symmetrical position relative to the normal direction of the recording light-beam. The difference between the resulting currents can then be used as a measure of the deviation of the beam. In a similar way Dürschner[7] and H. Schmidt[8] used two photo-cells to make a sort of zero-instrument, in which the amplified differential current from the photo-cells feeds a Helmholtz-Gaugain coil placed around the variometer magnet with its axis in the direction of the element to be recorded. When a certain variation of this element causes the light-spot from the variometer mirror to leave its symmetrical position between the photo-cells (zero position) the magnetic field produced by the differential current will tend to bring the magnet back towards its zero position, and an equilibrium is established, in which the current in the coil, which can be recorded by means of a milliammeter, will be practically proportional to the variation of the component observed.

γ) In an equipment designed by G. Grenet for the recording of *rapid geomagnetic pulsations* a rather large variometer magnet is placed in the center of a coil, the axis of which is normal to the rotation axis of the magnet and also to the mean position of the magnet (that means for a D-variometer that the axis is horizontal and orientated in the magnetic east-west direction). By induction the movement of the magnet produces a varying electromotive force in the spires of the coil, which e.m.f. is recorded by means of a galvanometer. For a given frequency of the pulsation the recorded amplitude will be proportional to the amplitude of the pulsation, but the scale value will vary with the frequency for the following reasons: 1. the amplitude of the oscillations of the magnet will be proportional to the amplification factor f defined above, Eq. (13.1); 2. for a given frequency the induced electromotive force will be proportional not only to the amplitude of the enforced oscillations of the magnet, but also to the frequency itself; 3. the e.m.f. will act on a galvanometer which has its own oscillation period and therefore its own amplification factor dependent on the frequency of the variation to be recorded.

G. Grenet has given the theory of the equipment in detail[9]. He finds that the scale value has a minimum for pulsations with a period not far from the oscillation periods of the magnet and of the galvanometer, which periods should preferably be of the same order of magnitude. Towards lower frequencies the scale value increases proportionally with the period, whereas at very high frequencies it is inversely proportional to the cube of the period. It follows that only a limited range of frequencies is recorded, the slow and the very rapid variations being suppressed. Since great variations of the field are generally developing rather slowly, they will not be recorded by the equipment, and the total range of the record will therefore remain small. This gives the possibility of a very compact record, combining a considerable recording speed with a moderate consumption of paper.

It is of interest to note that for magnetic variations having periods longer than the period of the oscillating system (supposed appropriately damped) the amplification factor f will never differ much from unity, see Table 2, so that

---

[6] A. Graf: Beitr. angew. Geophys. 7, 357 (1939).
[7] H. Dürschner: Ann. Géophys. 7, 199 (1951).
[8] G. Fanselau: [6], p. 179.
[9] G. Grenet: Ann. Géophys. 5, 188 (1949).

for such slow variations the effect described under 2. above will be by far the most predominant. In fact, the equipment provides a record of the time derivative of the element in question.

δ) Records of the *time derivatives of magnetic elements* can, however, be obtained in a more direct way by recording directly the electromotive force (e.m.f.) induced in suitably disposed coils by the variations of the geomagnetic field[10, 11].

When a coil with the total winding area F is placed with its axis in the direction of a geomagnetic force component the variation of magnetic induction B with time will produce an electromotive force

$$e = -\frac{1}{c_0 \sqrt{\varepsilon_0 \mu_0}} F \frac{dB}{dt} = -\frac{1}{c_0 \sqrt{\varepsilon_0 \mu_0}} \mu_0 F \frac{dH}{dt} \qquad (13.3)$$

$$\frac{e}{V} = -10^{-13} \frac{F}{cm^2} \frac{d(B/\gamma)}{d(t/s)} \qquad (13.3\,a)$$

in the coil. The e.m.f. may be amplified and recorded by means of an oscillograph or on a tape recorder, but more often the coil is used in connection with an ordinary galvanometer and the scale value of, for instance the dH/dt record, will then — in the same way as the scale value of the magnet variometer treated in Subsect. α — be dependent on the damping factor α and on τ_0/τ, where τ_0 and τ are the oscillation periods for the undamped galvanometer and for the magnetic pulsation respectively. The amplitude recorded will be proportional to the amplification factor f, obtained from Eq. (13.1), and the phase difference will be given by Eq. (13.2). Also for the galvanometer a damping factor $\alpha = \frac{1}{2}$ will result in a fairly uniform scale value for $\tau > \tau_0$.

If the galvanometer used in connection with the coil is of the fluxmeter type[12, 13], that is a heavily overdamped galvanometer with negligible torsion, its deflection will be proportional to the magnetic variations themselves and not to the time derivatives. It is worth noticing that in this special type of variometer the scale value will remain fairly constant over a rather wide range[14] around the resonance period τ_0, although the sensitivity will be rapidly decreasing towards very long periods as well as towards very short ones, so that the equipment to a certain extent will be of the same selective character as the GRENET system described above.

When using coil instruments due regard should be paid to the effect of self-induction. This effect will decrease with increasing dimensions of the coil and may be neglected for coils having a diameter of some meters, whereas for small coils, and quite especially for coils which have been provided with a ferromagnetic core in order to increase the sensitivity, the effect will have to be taken into consideration. For such coil instruments the combined effect of the self-induction and the hysteresis in the core will generally result in a scale value that varies with τ in a rather complicated way, so that special devices are to be used to determine the scale value as a function of τ.

B. Direct determination of the geomagnetic elements.

14. Introduction. The magnetographs described in the preceding sections provide records of the *variations* of the geomagnetic elements, and for many scientific and practical purposes information about the variations is what is

---

[10] E. THELLIER: Ann. Intern. Geophys. Yr. 4, 267 (1957).
[11] H. NAGAOKA and T. IKEBE: Sci. Papers Inst. Phys. Chem. Res. (Tokyo) 36, 183 (1939).
[12] E. J. CHERNOSKY, E. MAPLE, and R. M. COON: Trans. Am. Geophys. Union 35, 711 (1954).
[13] FOREST K. HARRIS: Electrical Measurements, p. 330. New York and London 1952.
[14] G. FANSELAU: [6], p. 357.

really required. For other purposes, however, as for instance studies concerning the main magnetic field of the Earth, or the construction of magnetic charts, it is necessary to know also the *absolute values* of the elements, and most magnetic observatories are therefore equipped not only with magnetographs, but also with instruments for direct determination of the geomagnetic field.

α) In practice the results of such direct determinations made at a magnetic observatory are mainly used to determine the base-line values of the variometers. In order to illustrate the procedure we shall use the *declination variometer* as an example; the other variometers may be treated in exactly the same way.

For the actual value of the declination at a given moment we have found the expression (see Sect. 3):

$$D = D_0 + S_D n_D \qquad (14.1)$$

where n_D is the recorded ordinate and S_D the scale value that may be computed (Sect. 3 and 5) or measured (Sect. 10). D_0 is the so-called base-line value, that is the value of D that will bring the light-spot from the moving magnet mirror to coincide with the stationary light-spot producing the base-line ($n_D = 0$).

If Eq. (14.1) is to give the actual value of the declination, D_0 must be known. This is achieved by making a direct determination of D, see Sect. 15. D_0 can then be computed from Eq. (14.1) as $D_0 = D - S_D n_D$, where n_D is the D-ordinate recorded just at the moment when the direct determination is made. Usually for one determination one makes several successive readings of the instrument in question; n_D should, of course, then be replaced by $\bar{n}_D$, the mean of the D-ordinates at the times when the readings are made.

β) The complete determination of the *geomagnetic vector* with regard to magnitude and direction involves the determination of three geomagnetic elements from which all other elements may afterwards be computed. At many observatories the elements chosen for direct determination are D, H and J; at some observatories the inclination J is replaced by the vertical component Z. A direct determination of Z is certainly preferable at high latitudes, where observational errors in H and J, due to the large value of tan J, will seriously affect values of Z computed from the expression $Z = H \tan J$.

The direct determinations should be made with the instrument on a pillar, not too far from the variometer room, and in a place exempt from all local and artificial magnetic disturbances. Under such conditions it may be safely assumed that the magnetic variations observed on the pillar will be the same as the variations recorded at the same time in the variometer room, but there may easily be an appreciable difference between the absolute values in the two places, and it should be borne in mind, therefore, that the base value found refers to the pillar where the direct determination is made, and that all absolute values which are subsequently derived from the records represent the values at the same pillar, although the ordinates used for the calculation are recorded in the variometer room. If different components are measured on different pillars, a possible field difference between these pillars should be taken into consideration.

Due to instrumental imperfections the observed base value may be slightly variable in the course of time. Direct determinations of the geomagnetic elements must therefore be repeated from time to time. At many observatories such determinations are made once a week.

15. Determination of the declination. Theoretically the determination of the magnetic declination, that is the angle between the astronomic meridian plane and the magnetic meridian plane, is a very simple operation, provided that the astronomic meridian is known from astronomical observations, or otherwise. The magnetic meridian plane is determined as the vertical plane through the magnetic axis of a magnet suspended so as to move freely in the horizontal plane. A rough determination of the magnetic meridian may be made with a good compass, but the high accuracy — one tenth of a minute of arc — required at a magnetic observatory implies the use of a non-magnetic theodolite, f. inst. of the special type which is to be described in Sect. 17γ as part of the equipment for direct determination of the horizontal force.

α) Over the centre of the theodolite a special magnet is suspended in a thin fibre having a comparatively small torsion constant. The upper end of the fibre is attached to a torsion head provided with appropriate divisions.

Fixed to the magnet there may be a plane mirror which, when properly adjusted, should have its normal approximately parallel to the magnetic axis of the magnet. By turning the theodolite table with its telescope, this telescope may be brought to point exactly at the magnet mirror, which means that the mirror normal coincides with the optical axis of the telescope. The suspended magnet may also be formed as a collimator, a hollow cylinder with a frame mounted externally at each end. The frame at one end carries a piece of optical glass on which are engraved two lines at right angles to each other, whereas the frame at the other end serves as setting for a collimating lens. In this case the optical axis of the telescope and that of the collimator are brought to coincide when making the measurement.

In principle the actual value of D is determined as the difference between two readings of the divided circle of the theodolite, one reading with the telescope pointing true north and one reading with the telescope pointing in the actual magnetic meridian plane. Usually both of these readings will have to be *computed* from the set of observations actually made.

β) The setting in the *astronomic meridian* is computed from readings obtained as the result of more pointings at a fixed azimuth mark, situated at a suitable distance from the observatory; the azimuth of the mark is to be determined once for all by astronomic observations.

The setting in the *magnetic meridian* is computed from two sets of readings, one set corresponding to repeated pointings at the magnet mirror or at the collimator as described above, and another set corresponding to a repetition of these pointings after the suspended magnet has been turned exactly 180° about its horizontal geometrical axis. In taking the mean of the readings with magnet erect and magnet reversed the effect of the small unknown angle α, the collimation angle, between the horizontal projections of the mirror normal and of the magnetic axis of the magnet will be eliminated. If the magnet is not reversed exactly 180°, but only $180° - \beta$, where β is a small angle, and if at the same time the magnetic axis and the mirror normal form the small angles γ and δ respectively with the horizontal plane, the mean reading will need a certain numerical correction[1] which is

$$\Delta R = \frac{\beta}{114.6°} (\gamma - \delta). \tag{15.1}$$

γ) Due regard must be paid to the effect of the *residual torsion* in the suspension fibre. It may be done by providing the stirrup at the lower end of the thread suspension with an auxiliary mirror, the angular position of which can be read with sufficient accuracy on a horizontal scale placed at some distance from the instrument. Two readings of this auxiliary mirror are made, one corresponding to the position of the magnet at the moment when the declination reading is made, and one after the magnet has been removed and the stirrup loaded with a non-magnetic bar of the same weight as the magnet. In order to find the torsion correction to be applied to the declination readings the difference between the two readings of the auxiliary mirror should be multiplied by a torsion factor, which has to be determined experimentally in each case from the deviation of the magnet produced by a certain twisting, f. inst. 90°, of the upper end of the suspension fibre. The torsion correction may also be found by repeating the determination of D with a weaker magnet[2]. The ratio between the moments of the two magnets may profitably be chosen as great as two to one.

---

[1] V. H. RYD: private communication.
[2] S. CHAPMAN and J. BARTELS: [4], pp. 31—32.

For a correctly adjusted instrument of the absolute type described above any systematic difference between the observed value and the true value of D should be numerically smaller than the standard error of a single observation[3]. LJUNGDAHL[4] has described a procedure for detecting possible sources of error in the determinations of D.

δ) A rapid and yet fairly accurate determination of the declination may be obtained with the *Transit Magnetometer*[5], in which instrument a compass with a 9 cm long pivoted needle is placed on the table of a non-magnetic theodolite with a telescope used for mark readings. On the north-seeking end of the needle a fine line is engraved along the central line of the magnet, and during a measurement the theodolite is turned until this line is pointing at a reference mark on the adjacent part of the compass limb. The setting is observed by means of a microscope fastened to the telescope barrel, and the corresponding reading of the theodolite will — combined with the azimuth mark reading and a possible instrument correction determined during the standardization at a magnetic observatory — give the declination aimed at. With the telescope adjusted in a vertical position and with a horizontal deflector magnet placed in a heat insolated box at the top of the telescope, the instrument can be used also for a determination of the horizontal force. The deflector magnet is placed first with its north pole towards the east and then with its north pole towards the west, and from the double deflection of the compass-needle the value of H can be computed by means of a calibration formula established as a result of the standardization observations.

In subsequent Sects. other types of instruments will be described which can be used for a determination of D, namely the TSUBOKAWA magnetometer (Sect. 18γ) and the QHM-magnetometer (Sect. 19α).

16. Determination of the inclination.

α) The direction of the total magnetic field vector will theoretically be indicated by the direction of the magnetic axis of a magnet suspended on a horizontal axis so as to move freely in the magnetic meridian plane. This principle has in fact been utilized in the so-called *dip-circle*, where the position of the pointed ends of the suspended magnet is read on a vertical graduated circle, having its zero-division in the horizontal plane through the axis of rotation. The ideal condition — difficult to fulfil — is that the axis of rotation should pass exactly through the centre of gravity of the magnet, but a possible residual effect of gravity, and also the effect of a possible collimation angle (the angle between the projections on the plane of rotation of the magnetic axis and the geometric axis of the magnet) may be largely eliminated by making readings with the vertical circle in two positions, both of them in the meridian, but 180 degrees from one another, and by repeating these readings after the magnet has been magnetized to the same moment in the opposite direction. The inclination J is then computed as the mean of the readings made[1].

β) Owing to the lack of accuracy of the dip-circle this instrument has been replaced at all observatories measuring J by the so-called *earth-inductor*, which may be characterized as a dip-circle in which a rotating coil takes the place of the magnetic needle as indicator of the direction of the field. The rotation takes place about an axis lying along a diameter of the coil, and by placing the bearings of the axis in a frame, which again may be turned about a horizontal axis perpendicular to the axis of rotation, the axis of rotation has been made freely adjustable in a vertical plane. This plane is brought to coincide with the magnetic meridian plane, and before starting the rotation of the coil the axis of rotation is adjusted approximately in the direction of the magnetic field vector. The dip of the axis below the horizontal plane is read by means of microscopes on a vertical divided circle attached to the frame mentioned above. Unless the

---

[3] Int. Ass. Terr. Mag. El., Bul. **11**, 551 (1940).
[4] G. LJUNGDAHL: Terr. Mag. **34**, 73 (1929).
[5] J. W. JOYCE and H. E. McCOMB: Int. Ass. Terr. Mag. El., Bul. **11**, 397 (1940).
[1] D. L. HAZARD: [*8*], pp. 73—77.

axis of rotation is exactly parallel to the field vector, the rotation of the coil in the magnetic field will by induction provoke an alternating e.m.f. in its windings; this e.m.f. is rectified by means of a commutator and indicated by a sensitive galvanometer. The inclination of the axis is then gradually adjusted until the induced e.m.f. is zero, and this final position is read on the divided circle.

The effect of thermo-electric forces at the commutator is eliminated by taking the mean of readings with rotation in opposite directions, and to eliminate the effect of possible inaccuracies in the adjustment of the various axes of the instrument the readings should be repeated after the rotating coil has been turned upside down and combined with a second set of readings, made after the whole instrument has been turned 180° about a vertical axis. As reference reading on the divided circle serves the reading which corresponds to an exact vertical position of the axis of rotation. This position is checked before and after the measurement by means of a very sensitive level.

The general theory of the earth-inductor has been given by DORSEY[2] and the functioning of this type of instrument has been examined experimentally by VENSKE[3-5], JOHNSON[6] and EGEDAL[7]. The observational error may be as small as one tenth of a minute of arc or less.

17. Determination of the horizontal force by the method of Gauss-Lamont.

α) *General remarks.* At most observatories the values of the magnetic force components are based on a direct determination of the horizontal force $\mathcal{H}$. Some observatories are also equipped for a direct determination of the vertical force Z (see Sects. 20 and 21), but this element then takes the place of the inclination, so that the set of elements measured will be $D, Z, \mathcal{H}$ instead of $D, J, \mathcal{H}$.

Up to the present time the method most generally used at magnetic observatories for the determination of $\mathcal{H}$ has been the classical method indicated by C. F. GAUSS and modified by J. LAMONT. The measurement is composed of two operations: 1) a determination of the period of oscillation for a suspended magnet which is oscillating in a horizontal plane about its zero position in the magnetic meridian, and 2) a determination of the angular deflection from the magnetic meridian as provoked by the same magnet when in a horizontal position it is acting on an auxiliary magnet, suspended so as to turn freely in a horizontal plane; the two magnets should be at the same level, and their magnetic axes should be perpendicular to one another. From the result of these two determinations the horizontal force $\mathcal{H}$ and the magnetic moment M of the magnet in question can be computed.

In the following there will be given a few details of each of the two operations involved.

β) *Oscillations.* For a horizontal magnet with the magnetic moment M and the moment of inertia K, suspended in a fibre with the torsion constant h, we have the following expression for the period τ of oscillations which are symmetrical about the geomagnetic meridian, and the mean amplitude of which may be considered to be infinitesimally small:

$$\tau^2 = 4\pi^2 \frac{K}{M\overline{\mathcal{H}} + h}.$$

In this expression $\overline{\mathcal{H}}$ is the mean value of $\mathcal{H}$ during the observation. If the mean angular amplitude α is small, but not infinitesimal, τ is obtained with a

---
[2] N. E. DORSEY: Terr. Mag. **18**, 1—38 (1913).
[3] O. VENSKE: Nachr. kgl. Ges. Wiss. Göttingen, 219 (1909).
[4] O. VENSKE: Z. Geophys. **6**, 248 (1930).
[5] O. VENSKE: Ber. Tätigk. Preuss. Met. Inst. im Jahre 1924, 91 (1925).
[6] E. A. JOHNSON: Terr. Mag. **41**, 251 (1936).
[7] J. EGEDAL: Terr. Mag. **42**, 367 (1937).

good approximation by means of the extended formula:

$$\tau^2 = 4\pi^2 \frac{K}{M\mathcal{H}+h}\left(1+\frac{\alpha^2}{16}\right)^2. \tag{17.1}$$

Methods aiming at an automatic and precise recording of τ have been described[1].

In arctic regions, where the variations of D during the time necessary for making a determination of τ may be rather great, and where a greater oscillation amplitude may therefore be needed, the procedure described by LAMONT[2] may profitably be used for reduction of τ to infinitesimal amplitude. The torsion constant h may be determined by a procedure similar to that mentioned in Sect. 5. Using the same notation we find

$$h = \mathcal{H}M\frac{\nu}{\varphi-\nu}.$$

If the measurement is to include a determination of K, an additional set of oscillations is made, during which the magnet is loaded with an auxiliary non-magnetic weight, the moment of inertia of which, K_1, can be computed from its dimensions and weight.

Indicating by τ_1 the new period of oscillation we have

$$\tau_1^2 = 4\pi^2 \frac{K+K_1}{M\mathcal{H}_1+h}\left(1+\frac{\alpha_1^2}{16}\right)^2. \tag{17.2}$$

From Eqs. (17.1) and (17.2) K can be computed, due regard being paid to the fact that air masses moving with the oscillating systems will tend to increase the values of K and K_1 as shown by BURGER[3,4] and discussed by VELDKAMP[5].

Using an oscillating magnet of a special form K may be determined directly from the dimensions of the magnet[4]. J. A. FLEMING[6] has stressed the importance of a redetermination from time to time of the moment of inertia of the magnet used for oscillation observations.

The temperature of the magnet will affect the period of oscillation in two essentially different ways. By rising temperature the expansion of the magnet steel will result in an increase of K, and at the same time M will decrease, so that the actual values of K and M should correctly be expressed as

$$K = K_0(1+2kT),$$
$$M = M_0(1-qT)$$

where K_0 and M_0 are the values at $T=0$ degree centigrade, k the coefficient of linear thermal expansion of the steel, q (positive) the temperature coefficient of the magnetic moment, and T the actual temperature. As to the determination of q see the end of this Sect. Temperature variations during the measurement should be kept at a minimum, and should in no case exceed half a degree[7].

Finally the induction effect of the horizontal magnetic field should be taken into consideration. Since the oscillating magnet is practically parallel to $\mathcal{H}$, the effect will be an increase of the moment M by $\mu^*\mathcal{H}$, where the induction factor μ^* depends primarily on the composition of the magnet steel and is practically independent of the value of M[8,9].

---

[1] G. FANSELAU: [6], pp. 189—199.
[2] J. LAMONT: [2], pp. 63—74.
[3] A. BURGER: Über die Art und Größe der Fehler die bei Schwingungsbeobachtungen an einem Magneten durch die mitschwingenden Luftmassen entstehen. Diss. Berlin (1935).
[4] G. FANSELAU: [6], pp. 54—56.
[5] J. VELDKAMP: Terr. Mag. 44, 257—258 (1939).
[6] J. A. FLEMING: Int. Ass. Terr. Mag. El., Bul. 10, 243 (1937).
[7] G. FANSELAU: [6], p. 60.
[8] H. E. McCOMB: [9], p. 42.
[9] G. FANSELAU: [6], pp. 63—64.

Using the notations introduced above the complete expression for the period of oscillation will be:

$$\tau^2 = \frac{4\pi^2 K_0 (1+2kT)\left(1+\frac{\alpha^2}{16}\right)^2}{M_0 \overline{\mathcal{H}}(1-qT)\left(1+\mu^* \frac{\mathcal{H}}{M}\right)\left(1+\frac{\nu}{\varphi-\nu}\right)}. \tag{17.3}$$

γ) *Deflections.* AD. SCHMIDT[10] has given the general formula for the interaction of two magnets. For the purpose of determining the torque exerted on the suspended magnet in the deflection experiment the general formula is used in a simplified form, which is based partly on the special position of the two magnets relative to each other (LAMONT's "first position", the two magnets perpendicular to one another, the magnetic axis of the deflecting magnet pointing towards the center of the suspended magnet), and partly on the assumption (to be discussed briefly in the following) that the two magnets can be regarded as *schematic magnets*, each consisting of two equal and opposite poles at a definite distance from one another.

Assuming that the deflecting magnet can be represented with sufficient approximation by a schematic magnet with magnetic moment M and distance $2l$ between the poles, and that the suspended magnet in the same way can be represented by a schematic magnet with moment M′ and distance 2λ between the poles, then the torque exerted on the suspended magnet can be expressed by a series of which the first and significant terms will be:

$$\frac{2MM'}{r^3}\left(1+\frac{2l^2-3\lambda^2}{r^2}+\frac{3l^4-15l^2\lambda^2+\frac{45}{8}\lambda^4}{r^4}\right). \tag{17.4}$$

In this expression r is the distance between the centres of the two schematic magnets; the series will be convergent only if $\frac{2l}{r}$ and $\frac{2\lambda}{r}$ are smaller than $\frac{1}{3}$.

The expression (17.4) is often written in the form

$$\frac{2MM'}{r^3}\left(1+\frac{P}{r^2}+\frac{Q}{r^4}\right) \tag{17.5}$$

where P and Q are the so-called distribution coefficients that may be determined experimentally (see below).

It can be shown[11], that the magnetic field distribution in the space surrounding a schematic magnet gives a good approximation to the field surrounding an ordinary magnet, provided that the latter can be regarded as a so-called *regular magnet*, that means a magnet which produces in the surrounding space a field of complete axial symmetry, and a field which is also symmetrical with regard to a plane through the magnetic centre of the magnet perpendicular to the magnetic axis. Due to inhomogeneities in the magnet steel the two magnets used in the deflection experiment cannot a priori be regarded as regular magnets, but experiment and theory show that if the effect of the deflecting magnet is measured with the magnet in four specified positions, namely two positions where the poles of the magnet are simply reversed, the geometric centre of the magnet keeping its position, and two new positions derived from the first ones by turning the magnet upside down, then the mean effect will represent a good and for our purpose sufficient approximation to the effect of a regular magnet, the magnetic centre of which is coinciding with the geometric centre of the magnet used. This procedure is actually used in the deflection experiment, and in addition the described series of four readings is repeated after the deflecting magnet has been placed in a new position symmetrical with the first one with regard to the suspended magnet. It is easily shown that in the total mean effect of the eight different positions also any lack of symmetry in the suspended magnet will be eliminated, as far as the interaction of the two magnets is concerned. Consequently it is justified to represent the two magnets by schematic magnets as done above.

---

[10] A. SCHMIDT: Terr. Mag. **17**, 181—232 (1912); **18**, 65—70, (1913).
[11] S. CHAPMAN and J. BARTELS: [*4*], pp. 70—79.

In different ways it has been found that for a cylindrical magnet made of one of the classical types of steel the distance between the poles of the equivalent schematic magnet will be between 0.80 and 0.88 of the total length of the magnet itself.

Fig. 7 shows a schematic diagram of the theodolite magnetometer used for the deflection experiment. Many different types have been developed, some of a rather simple design[12] and some of a more elaborate type[13]. Here only the general principle will be given.

The divided circle C is mounted horizontally by means of three footscrews. The circular plate E can be turned around a vertical axis and its position read on C by verniers to $\frac{1}{2}$ minute of arc or by micrometer microscopes to $2''$. Attached to E is the telescope T and two long horizontal arms G_1 and G_2, perpendicular to the optical axis of the telescope. E is turned until the telescope is pointing at a mirror attached to the suspended magnet B, the corresponding reading of the circle will indicate the angular position of the mirror normal, that has been adjusted so as to deviate only some few minutes from the magnetic axis of the magnet.

Under the influence of the deflecting magnet A the suspended magnet takes a position of equilibrium in which its axis forms a certain angle ψ with its zero-position in the magnetic meridian. In this position the torque from the horizontal force, $\mathcal{H}$, will be

$$\mathcal{H} M' \sin \psi.$$

The final value of ψ is to be found from the eight readings which correspond to different positions of the deflecting magnet as described above, namely four with A on the arm G_1 and four with A on the arm G_2. We then find by means of the expression (17.5)

$$\overline{\mathcal{H}} \sin \psi = \frac{2M}{r^3}\left(1+\frac{P}{r^2}+\frac{Q}{r^4}\right)$$

where $\overline{\mathcal{H}}$ is the mean of the eight instantaneous values of $\mathcal{H}$ corresponding to the hours of the eight readings.

Introducing as above the temperature coeffcient q and the induction factor μ^* of M, and denoting by k_1 the coefficient of linear expansion of the distance r, we have, when T_1 is the mean temperature of the deviating magnet during the deflection observations:

$$\overline{\mathcal{H}} \sin \psi = \frac{2M_0(1-qT_1)\left(1-\dfrac{\mu^* \overline{\mathcal{H}} \sin \psi}{M}\right)}{r_0^3(1+3k_1T_1)}\left(1+\frac{P}{r^2}+\frac{Q}{r^4}\right). \tag{17.6}$$

From Eqs. (17.3) and (17.6) $\overline{\mathcal{H}}$ and M_0 can be derived.

When the length of the magnet A is known the distance r can be determined from an exact measurement of the distance between the two stops on G_1 and G_2 (Fig. 7), which define the position of A. A lack of symmetry of the two positions with regard to the suspended magnet will result in a difference between the corresponding circle readings, and the mean deflection ψ will need a slight correction[14].

All readings of the divided circle must, of course, be corrected for the variation of the declination during the measurement. The distribution coefficients P and Q may in principle be computed from Eq. (17.6) by a measurement of ψ for three different values of the distance r.

From the expressions (17.4) and (17.5) we get the following expressions for P and Q:

$$P = 2l^2 - 3\lambda^2; \qquad Q = 3l^4 - 15 l^2 \lambda^2 + \tfrac{45}{8}\lambda^4. \tag{17.7}$$

These expressions can be used to calculate the relative length of two magnets which will make P or Q equal to zero, assuming that the ratio of pole distance to length of magnet is the same in the two magnets. Many observatories are taking advantage of this possibility, some of them using $\lambda/l = 0.817$, which makes $P=0$, others $\lambda/l = 0.467$ or 1.565 corresponding

---

[12] D. L. HAZARD: [8], pp. 62—70.
[13] R. BOCK: Z. Instrumentenk. **48**, 1—14 (1928).
[14] J. LAMONT: [2], p. 31.

Sect. 18. Determination of the horizontal force by electromagnetic methods. 305

to $Q=0$. Ad. SCHMIDT[15] has indicated a method for a more accurate determination of magnet parameters, using a constant value of r, but different azimuths of the deflecting magnet. Using this method on two new cobalt magnets for which λ/l was 0.467, it could be shown that the value of Q was $11 \cdot 10^{-5}$ instead of zero, a departure from the theoretical value that would, undetected, have given rise to a relative error of $6 \cdot 10^{-5}$, in the measured value of $\mathcal{H}$.

In discussing the problem of the constancy of the distribution coefficients, FLEMING[16] summarizes the results of 1300 determinations made by USA scientists during 15 years and concludes that the results on the whole indicate a good constancy of the coefficients in question.

Fig. 7. Theodolite (diagram after CHAPMAN-BARTELS).

The temperature coefficient q may be determined by varying the temperature of the deflecting magnet and by measuring ψ as a function of temperature T [17].

In addition to the value of $\mathcal{H}$ the two Eqs. (17.3) and (17.6) also give the value of the magnetic moment M of the principal magnet. For a good magnet, carefully treated, the magnetic moment will vary very slowly with time, and the successive values of M_0 will therefore, in themselves, give an impression of the accuracy of the $\mathcal{H}$-measurements from which they have been derived.

See [6] for further details[18] concerning the determination of $\mathcal{H}$.

18. Determination of the horizontal force by electromagnetic methods.

α) As we have seen some of the complications encountered in a precise determination of the horizontal force by means of the classical method are due to the inhomogeneity of the field produced by the deflecting magnet. These complications are practically eliminated in the so-called *Sine-galvanometer*[1], where the suspended

---

[15] G. FANSELAU: [6], pp. 34 and 44.
[16] J. A. FLEMING: Int. Ass. Terr. Mag. El., Bul. **10**, 244 (1937).
[17] H. E. McCOMB: [9], pp. 35—42.
[18] G. FANSELAU: [6], pp. 20—70, 437—440 and 452—467.
[1] S. J. BARNETT: Dept. Terr. Mag. Carnegie Inst. Wash. Publ. No. 175, **4**, 373—394 (1921).

magnet is placed in the centre of a Helmholtz-Gaugain coil (Sect. 10) in which a current is producing a homogeneous horizontal magnetic field perpendicular to the magnetic axis of the magnet. The deflection ψ produced in a horizontal geomagnetic field $\mathcal{H}$ will be determined by the expression:

$$\mathcal{H}\sin\psi = CI$$

where I is the current flowing in the coil and C a constant depending on the coil dimensions.

Fig. 8a and b. The SCHUSTER-SMITH horizontal intensity magnetometer. (Courtesy: Royal Greenwich Observatory and National Physical Laboratory).

By appropriate arrangements both $\sin\psi$ and I can be obtained with an accuracy of one part in 50,000, and the principal problem will be, therefore, the construction of a non-magnetic coil, the dimensions of which can be measured with sufficient accuracy, and which does not change with time. In the Sine-galvanometer the coil is wound on a hollow cylinder made of a specially selected Carrara marble, the permeability of which differs only slightly from unity. After having been exposed to a special heat treatment, combined with a paraffin impregnation, the cylinder remains practically unchanged with time, and its thermal changes are small and reversible for small temperature variations. The electric resistance of the marble is found to be very high. The coil consists of two sets of windings with 20 turns in each set. The two sets are separated by a mean distance of 150 mm, and since the outer diameter of the cylinder is 300 mm the windings will in practice represent a Helmholtz-Gaugain coil. The coil windings are bare copperwire, wound under tension in a lathe-cut spiral groove on the cylinder, and by this procedure it has been possible to measure the dimensions of the coil

Sect. 18. Determination of the horizontal force by electromagnetic methods.

with such a precision that the constant C can be computed with an accuracy of one part in 30,000.

In order to make it possible to test the insulation of the coil at any time, each of the two sets of 20 turns is composed of two separate groups of 10 turns, of which the one group is wound in between the windings of the other. The four groups can be combined into two entirely separate Helmholtz-Gaugain coils, and since the winding groups are otherwise quite identical, the two part-coils should have the same deflecting effect. This has in fact been confirmed experimentally within one part in 17,000. By a special arrangement the coil can be adjusted so that its axis is perpendicular to the magnetic axis of the suspended magnet. To eliminate any residual torsion effect of the suspension fibre deflections are made to both sides of the meridian. The current I in the coil is passing through a standard resistance and can thus be measured by means of a potentiometer adjusted by a standard cell. The coil equipment is placed on a horizontal divided circle which can be read to the nearest 10", and a complete measurement of $\mathcal{H}$ takes only a few minutes, whereas a determination by means of the classical method described above takes more than one hour.

To obtain an electromagnetic measurement of high precision it is necessary to secure a highly constant temperature of the standard cells[2], not only during the measurement itself, but also a fairly long time before, and the e.m.f. of the cells must be checked from time to time at a physical laboratory.

Fig. 9. SCHUSTER-SMITH magnetometer (schematic diagram).

β) The SCHUSTER-SMITH *Coil Magnetometer*[3], Fig. 8, for measuring the horizontal force has a coil of the same type as used in the Sine-galvanometer, and the coil is placed in the same way on a horizontal divided circle, but the observational procedure is different, the deflection of the suspended magnet being 90° during the measurement. The procedure may be described as follows:

The suspended magnet is placed in the centre of the coil, with its axis in the magnetic meridian (fibre in torsion-free position). Independently of the magnet suspension the coil is turned until its axis coincides with the magnetic axis of the magnet. This position is verified by producing in the coil an additional horizontal field slightly stronger than $\mathcal{H}$, thereby increasing the total field to a little more than $2\mathcal{H}$. If the position of the coil is correct the additional field will have no deflecting effect on the suspended magnet, and the light-spot from a mirror attached to the magnet will remain in its zero position which is read on an appropriate scale. The current is then reversed and the coil is turned through the small angle α_1, until the deflection of the suspended magnet is exactly 90° from its zero position, that is until the light-spot is coinciding with an experimentally determined mark on an auxiliary scale. At the same time the housing with the suspension fibre is turned 90° in the same direction in order to maintain the torsion-free state of the fibre. The measurement is repeated with a deflection to the opposite side (coil in position $-\alpha_2$). The value of $\mathcal{H}$ is finally calculated from the equation:

$$\mathcal{H} = CI \cos\left(\frac{\alpha_1 + \alpha_2}{2}\right)$$

where C is the coil constant and I the current flowing in the coil (see Fig. 9).

---

[2] FOREST K. HARRIS: Electrical Measurements, pp. 185—190. London and New York 1952
[3] F. E. SMITH: Phil. Trans. Roy. Soc. London, Ser. A **223**, 175—200 (1922) [see also article by A. SCHUSTER: Terr. Mag. **19**, 19—22 (1914)].

The irregular and partly unknown distribution of the magnetism in an ordinary magnet involves some complications when a magnet is used as detector in a coil magnetometer. E. A. Johnson has been the first to use as detector a small rotating coil, utilizing thereby that the mutual influence of the two coils can be calculated on the basis of a precise measurement of their dimensions[4]. A similar detecting method has been used in the magnetometer in which I. Tsubokawa[5] has combined the principle of the earth inductor and that of the Schuster-Smith magnetometer.

γ) In the Tsubokawa *magnetometer* field directions are indicated by a small rotating coil, the centre of which coincides with the centre of a Helmholtz-Gaugain coil. The two coils are built together so that the axis of rotation of the indicator coil is perpendicular to the central axis of the Helmholtz-Gaugain coil. The two axes are lying in the same vertical plane, and the whole system is adjustable in this vertical plane about a horizontal axis. The angular position in the vertical plane is read on a vertical divided circle. The instrument can further be adjusted about a vertical axis and its angular position in the horizontal plane can be read on a horizontal divided circle. When no current is flowing in the Helmholtz-Gaugain coil the instrument acts as an earth-inductor, and the direction of the total geomagnetic force may be determined by adjusting the indicator axis until the output, amplified and detected by a magic eye or a sensible alternating current (a.c.) ammeter, is zero. From the corresponding readings of the horizontal and vertical circles, combined with readings with the indicator axis pointing towards true north and exactly downwards respectively, the values of the declination and of the inclination are derived.

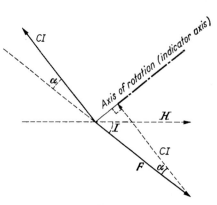

Fig. 10. Theory of the Tsubokawa magnetometer (schematic diagram).

In order to measure the total field intensity $\mathcal{F}$ or one of the field components $\mathcal{H}$ and Z, a current is sent through the Helmholtz-Gaugain coil, producing a magnetic field in the meridian plane, approximately opposite to and of about the same intensity as the field component to be measured. If for instance the total geomagnetic force is to be determined, the Helmholtz-Gaugain coil is directed with its central axis parallel to the direction of **F** (measuring system turned 90 degrees in the vertical plane from the zero position determined above). Having produced in the coil a magnetic field opposite to and slightly weaker than $\mathcal{F}$ the system is turned in the vertical plane until the output of the indicator is zero. If the system has to be turned through the (small) angle α, as shown by Fig. 10, the value of $\mathcal{F}$ is given by the expression:

$$\mathcal{F} = \frac{C I}{\cos \alpha} \qquad (18.1)$$

where I is the current and C the field constant for the Helmholtz-Gaugain coil. The effect of parasitic voltages and of lead-in wires is eliminated by using rotation in opposite directions and by measuring in two azimuths 180° apart.

The Tsubokawa magnetometer is a relative instrument in the sense that before being put into service it has to be calibrated at an observatory where the value of the magnetic elements is otherwise known.

---

[4] E. A. Johnson: Terr. Mag. **44**, 29—42 (1939).
[5] I. Tsubokawa: Bul. Geogr. Surv. Inst. **2**, Part 4, 277 (1951).

δ) A magnetometer of a similar type for *absolute* measurements of the geomagnetic field is the *absolute magnetic theodolite*[6,7]. In this magnetometer the primary coil is not exactly of the Helmholtz-Gaugain type, but has been modified in such a way that higher harmonics of the voltage induced in the rotating secondary coil are partially eliminated. The coil constant of the primary coil is calculated from the dimensions of the coil, which have been measured with an accuracy of one part in 10^5. When the field component to be measured is compensated by the field of the primary coil the secondary coil, driven by a synchronous motor at a speed of 10 rotations per sec, gives zero output as indicated by an appropriate electronic equipment.

19. Determination of the horizontal force by torsion magnetometers.

α) *The QHM-magnetometer.* In the so-called QHM-magnetometer, a relative instrument designed by D. LA COUR for a quick and yet accurate determination of the horizontal force[1], the magnetic field is balanced against the (remarkably constant) elastic force of a quartz fibre. As shown in Fig. 11 the quartz fibre is carrying a small magnet with a mirror attached to it, and the whole instrument with its special telescope is placed in a fixed position on the central table of a non-magnetic theodolite base (Fig. 12), so that it can be turned about a vertical axis, its angular position being read on the divided circle of the base.

If M is the magnetic moment of the magnet and φ the deflection of the magnetic axis from the magnetic meridian, produced by a twisting of the fibre through the angle ϑ, the general expression for the equilibrium of the suspended system will be:

$$\mathsf{H}\,\mathsf{M}\sin\varphi = h\vartheta \tag{19.1}$$

where h is the torsion factor of the fibre.

In the QHM the torque exerted by the horizontal magnetic force is always balanced against one and the same torsional torque, namely the one produced by twisting the upper end of the quartz fibre by exactly one full turn (2π) relative to the lower end. It is easily seen from Eq. (19.1) that in this case the deflection angle φ will be a simple function of $\mathcal{H}$ as long as M and h remain constant. It is a characteristic feature of the QHM that the few, rapidly made settings of the telescope on the magnet mirror that are necessary for a determination of φ give at the same time a very accurate check of the torsion angle producing φ.

The telescope of the QHM is rigidly connected to the housing of the instrument and will always be turned through the same angle as the upper end of the quartz fibre. During such a rotation there will — with an appropriate choice of magnetic moment and torsion factor — be some positions in which the optical axis of the telescope is coinciding with the normal of the magnet mirror, and in Fig. 13 one of these positions of the telescope — in the following called the zero position — has been indicated by T. The normal of the magnet mirror has been indicated by n^0, and m^0 is the magnetic axis of the magnet. Due to a certain torsion in the quartz fibre (torsion angle β, not shown in the figure) m^0 forms an angle δ with the magnetic meridian containing the vector $\mathbf{H}$. (δ defined as shown in the figure and reckoned positive towards the east.) The corresponding reading of the divided circle is called the zero reading.

---

[6] O. LUCKE u. H. SPARRER: Z. Geophys. **24**, 284 (1958).
[7] H. G. REUSCHE: Z. Geophys. **24**, 299 (1958).
[1] D. LA COUR: Le Quartz-Magnétomètre QHM. Comm. Mag. 15, Publ. Inst. Mét. Danois (1936).

From the zero position the theodolite table with the QHM is turned, for instance in a clockwise direction, and as a result of this rotation the magnet, too, will turn in a clockwise direction although at a slower rate. Having turned the QHM through an angle of $(2\pi + \alpha_1)$ the telescope will again be pointing at the magnet mirror, which means that the upper end of the quartz fibre has been turned exactly 2π relative to the lower end, and that the torsion angle is now $(\beta + 2\pi)$. The new position of the telescope, the mirror normal and the magnet has been

Fig. 11.   Fig. 12.

Fig. 11. The QHM-magnetometer (diagram). The Oerstit magnet (dimensions $15 \times 1 \times 1$ mm) with its holder and mirror is seen to the left, suspended by a quartz fibre. The upper support for the quartz fibre is to be screwed into the suspension tube at right, which itself is closed by the closing cap shown in the top right-hand corner. A thermometer is attached to the suspension tube and a clamping mechanism is seen in the lower part of the instrument. The special telescope and its counterweight are shown below.

Fig. 12. QHM-magnetometer placed on a theodolite.

indicated in Fig. 13 by T_1, n_1^0 and m_1^0; the corresponding reading of the divided circle, combined with the zero reading mentioned above, gives the value of α_1.

From the last mentioned position an anti-clockwise rotation of $\alpha_1 + 4\pi + \alpha_2$ will lead to the position indicated in Fig. 13 by T_2, n_2^0 and m_2^0. In this position the telescope will once more be pointing at the mirror, the torsion angle will

Sect. 19. Determination of the horizontal force by torsion magnetometers.

be exactly $(\beta - 2\pi)$ and the angle α_2 will be given by the corresponding reading on the circle and the zero reading.

For the three positions here mentioned the equilibrium is expressed by the above Eq. (19.1):

$$M\mathcal{H}\sin\delta = h\beta, \qquad (19.2)$$

$$M\mathcal{H}\sin(\delta + \alpha_1) = h(\beta + 2\pi), \qquad (19.3)$$

$$M\mathcal{H}\sin(\delta - \alpha_2) = h(\beta - 2\pi). \qquad (19.4)$$

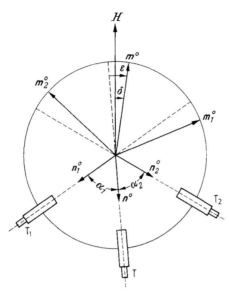

Fig. 13. Theory of the QHM-magnetometer (schematic diagram).

If the sum of Eqs. (19.3) and (19.4) is reduced by means of Eq. (19.2) we get the following expression for the deflection δ in the zero position:

$$\tan\delta = \frac{\sin\alpha_1 - \sin\alpha_2}{2 - (\cos\alpha_1 + \cos\alpha_2)}. \qquad (19.5)$$

During the first calibration of the QHM the instrument has been so adjusted that in the zero position the torsion angle β, and therefore also the deflection δ from the magnetic meridian, is very small, which means that the two additional deflections α_1 and α_2, resulting from the additional torsion of 2π and -2π respectively, will be approximately equal. Under these circumstances Eq. (19.5) may be written in the form:

$$\delta \approx \frac{\cos\dfrac{\alpha_1+\alpha_2}{2}}{2\left(1-\cos\dfrac{\alpha_1+\alpha_2}{2}\right)}(\alpha_1-\alpha_2). \qquad (19.5\,\mathrm{a})$$

The complete expression for $\mathcal{H}$ is found by subtracting Eq. (19.4) from Eq. (19.3):

$$\mathcal{H} = \frac{4\pi h}{M}\,\frac{1}{\sin(\delta+\alpha_1)-\sin(\delta-\alpha_2)}$$

or

$$\mathcal{H} = \frac{2\pi h}{M} \frac{1}{\sin\frac{\alpha_1+\alpha_2}{2} \cos\left(\frac{\alpha_1-\alpha_2}{2}+\delta\right)} . \tag{19.6}$$

As long as $\left(\frac{\alpha_1-\alpha_2}{2}\right) + \delta < 15'$ the omission of the cosine factor in the denominator of Eq. (19.6) will mean an error smaller than one part in 10^5, so that even for fairly great values of δ it may be permissible to use the simple expression

$$\mathcal{H} = \frac{2\pi h}{M \sin\varphi} \tag{19.7}$$

putting φ equal to $\tfrac{1}{2}(\alpha_1+\alpha_2)$. Howe has shown by a numerical example how the critical value of δ may be determined[2].

β) The *working conditions of a QHM* will now be discussed. Among the quantities entering into Eq. (19.7) h will increase[3] and M decrease with rising temperature. M will further be modified through the induction effect of the geomagnetic field. Designating by T the temperature of the QHM, by ν_1 and ν_2 the temperature coefficients for h and M respectively, and by μ^* the induction factor, we may write Eq. (19.7) in the form:

$$\mathcal{H} = \frac{2\pi h_0 (1+\nu_1 T)}{M_0 (1-\nu_2 T) \left(1+\frac{\mu^*}{M}\mathcal{H}\cos\varphi\right)\sin\varphi} \approx \frac{2\pi h_0}{M_0} \frac{(1+(\nu_1+\nu_2)T)\left(1-\frac{\mu^*}{M}\mathcal{H}\cos\varphi\right)}{\sin\varphi}$$

from which we get the logarithmic form*

$$\text{Log }\mathcal{H} = C_0 - \text{Log}\sin\varphi + c_1 T - c_2 \mathcal{H}\cos\varphi \tag{19.8}$$

where $C_0 = \text{Log}\frac{2\pi h_0}{M_0}$, $c_1 = (\nu_1+\nu_2)\text{Log }e$ and $c_2 = \frac{\mu^*}{M}\text{Log }e$.

The constant factors c_1 and c_2 are small quantities that may be determined once for all, c_1 by adjusting the QHM at different temperatures and c_2 by a special method devised for this type of instrument[1]. The principal constant C_0 is then determined from Eq. (19.8) by measuring the deflection angle φ at a place where $\mathcal{H}$ is otherwise known.

When reducing a measurement of the horizontal component by means of Eq. (19.8) an approximate value of $\mathcal{H}$ can be used in the small induction term.

Because of the remarkable constancy of the torsion-factor of the quartz fibres used, and the stability of the magnetic moment of magnets made of modern types of steel the value of C_0 changes in general only very slowly with time[4–6].

During the first minutes after the completion of a rotation of the theodolite, by which a torsion of $+2\pi$ or -2π has been applied to the quartz fibre, there will be a certain lag-motion of the magnet, owing to an elastic after-effect in the fibre[7]. The observed value of φ will consequently be dependent on the time elapsed between the turning of the theodolite and the setting of the telescope on the magnet mirror, so that observations by means of the QHM should preferably be taken at a constant speed (the same as used during the calibration). The after-effect is increasing with temperature, but if the observations are made at a standard speed this temperature dependence simply manifests itself as a slight modification of the temperature coefficient for the QHM in question.

---

* Log means $\log_{10}$

[2] H. H. Howe: Terr. Mag. **43**, 167—168 (1938).
[3] O. Reinkober: Phys. Z. **38**, 112 (1937).
[4] Joh. Olsen: Comm. Mag. **20**, 1—9, Publ. Inst. Mét. Dan. (1942).
[5] W. Zander: Jahrbuch 1960 Adolf-Schmidt-Observatorium, Berlin 1962, p. 183.
[6] C. A. van der Waal: Austr. Bur. of Mineral Resources, Record 1966/207 (1966).
[7] D. la Cour: Int. Ass. Terr. Mag. El., Bul. **11**, 70, 284 (1940).

O. MEYER[8] has first called attention to the fact that the principal constant of a QHM may be slightly dependent on the relative humidity of the surrounding air. When the temperature coefficient c_1 is determined in an artificially heated room the relative humidity will generally vary with temperature, and the humidity effect may thus give rise to an apparent modification of the temperature coefficient. The humidity effect is caused by a coating of the quartz fibre by fungoid colonies, which may often be visible to the naked eye. Clean fibres show no humidity effect at all, and recent experience seems to show that the growth of the fungi may be prevented by treating the surface of the fibre with an alcoholic solution of formalin[9].

Any variation of the declination D, occurring during a determination of $\mathcal{H}$ by means of a QHM, will result in a corresponding variation of the angular position of the suspended magnet, whether in the zero position or in one of the deflected positions. In order to obtain an accurate determination of the deflection φ, corresponding to the actual value of $\mathcal{H}$, all readings (on the divided circle) should, therefore, be reduced to one and the same value of D. Data for the reduction may be obtained from a D-variometer or from an auxiliary QHM, set up in its zero position.

On the other hand a QHM may be used for a direct determination of the declination itself, provided that a suitable azimuth mark is available in or near the horizontal plane through the station, so that the mark can be observed by means of the QHM-telescope. Before the observation of the azimuth mark the magnet system should be clamped in a position where the mirror is parallel to the axis of the telescope.

A determination of D will require a setting, β, on the azimuth mark before and after the ordinary readings are made for the determination of $\mathcal{H}$. If the azimuth, α, of the mark is known or measured, and if the angle, ε, between the normal to the mirror and the magnetic axis of the magnet (ε defined as shown in Fig. 13 and reckoned positive towards the east) is determined at a place where the declination is otherwise known, the easterly declination may be found by the formula:

$$D = \alpha_0 + \varepsilon - \delta - \beta + \alpha \qquad (19.9)$$

α_0 being the ordinary zero-reading of the QHM and δ the deflection angle due to residual torsion in the zero position.

Owing to the elastic after-effect in the quartz fibre the value found for δ may depend on the order in which the standard torsions $+2\pi$ and -2π are applied to the fibre[10], but K. THIESEN[11] has indicated a procedure by which fairly consistent values of δ are obtained, regardless of the order of observations.

L. G. MANSUROVA[12] has described a modified QHM-magnetometer with a temperature compensated magnet system.

γ) The *Universal-Torsion-Magnetometer* described by HAALCK[13] has been designed for a complete determination of the magnetic field vector. A very small magnet, weighing with its mirror and other accessories about 1 g, is suspended by a horizontal fibre, so that it is free to oscillate in a vertical plane, and so that the magnetic torque exerted on the magnet by the geomagnetic component to be measured, can be counterbalanced by applying a certain torsion to the fibre. The torsion angle applied in each case can be read to the nearest tenth of a minute on a divided torsion head. The whole instrument can be turned about a vertical axis and its angular position read on a horizontal divided circle. The inclination of the magnet can be determined by means of a vertical telescope at the top of

---

[8] O. MEYER and D. VOPPEL: Deut. Hydr. Inst. Hamburg, Erdmag. Obs. Wingst, Jahrbuch **10**, 1955—56, p. 143. Hamburg 1959.
[9] E. KRING LAURIDSEN: Private communication (1970).
[10] A. M. VAN WIJK and B. GOTSMAN: S. African J. Sci. **47**, 316—318 (1951).
[11] K. THIESEN: Geophysica (Helsinki) **5**, (2), 57—63 (1956).
[12] L. G. MANSUROVA: Quartz-H-magnetometer. Trudy NIZMIR **12** (22), 183 (1957).
[13] FRITZ HAALCK: Z. Geophys. **19**, Sonderband, 1—7 (1953).

the instrument, provided that the magnetic axis is deviating less than $\pm\frac{1}{2}$ degree from the exact horizontal (or the exact vertical) position.

The horizontal circle-reading which brings the plane of oscillation of the magnet to coincide with the magnetic meridian plane is found by determining those two (symmetrical) positions of the oscillation plane near the magnetic prime vertical where a constant (and small) torsion results in exactly the same (nearly vertical) position of the magnet as read in the telescope. The mean of the corresponding readings of the horizontal circle gives the meridian reading, and since the telescope can be tilted so as to point horizontally at an azimuth mark the determination of the magnetic meridian can be used for a determination of the declination.

For a determination of the horizontal force two readings are made, both with the oscillation plane in the magnetic meridian. Between the two readings the instrument is turned 180° about its vertical axis, and in both azimuths that torsion angle is read that brings the magnetic axis in an exactly vertical position.

The vertical force is measured with the oscillation plane in the magnetic prime vertical. Two torsion angles are read, namely those bringing the magnetic axis in an exactly horizontal position with the north pole pointing east and west respectively.

The instrument is of the relative type and has to be adjusted at a place where the magnetic elements are otherwise known.

δ) The *Combined Field Balance* designed by FANSELAU for the determination of $\mathcal{H}$ and Z follows similar principles as those described in the preceding Subsect. It has a horizontal ribbon suspension carrying a fairly strong magnet, and the measurements of $\mathcal{H}$, as well as of Z, are carried out with the magnet in the magnetic meridian plane. It is a characteristic feature of the balance that within certain limits the readings are practically independent of the accuracy of the levelling of the instrument. This is obtained by choosing for the suspension system appropriate values of the factors entering into the equation of equilibrium. The balance is temperature compensated[14].

20. Determination of the vertical force by electromagnetic methods. As it will be seen from the description given, some of the magnetometers treated in the foregoing sections, namely the TSUBOKAWA magnetometer (Sect. 18γ) and the torsion magnetometers (Sect. 19γ and 19δ) can be used to determine not only the horizontal force, but also the vertical force.

In the magnetometer designed by D. W. DYE[1] the vertical force Z is compensated by the magnetic field produced by a current sent through a Helmholtz-Gaugain coil with its axis exactly vertical. This coil is of the same type as that used in the SCHUSTER-SMITH magnetometer, and a small vibrating coil in the centre of the compensating coil serves as indicator of complete compensation. The determination of Z is then reduced to a measurement of the compensating current.

The small flat indicator coil is suspended on a horizontal axis of vibration in such a way that the windings of the coil are vertical. The axis is lying in the plane of the windings, and the whole detector system can be turned around a vertical axis. As long as the residual vertical field is different from zero the indicator coil will be vibrating around its axis under the combined influence of the residual field and an appropriate alternating current (a.c.) which is sent through the windings of the coil. The a.c. period is in resonance with the oscillation

---

[14] G. FANSELAU: [6], p. 519.
[1] D. W. DYE: Proc. Roy. Soc. (London), Ser. A **117**, 434—458 (1928).

period of the coil itself. The measurement consists in adjusting the current in the Helmholtz-Gaugain coil until the vibration of the indicator is reduced to zero.

If the axis of vibration deviates from the horizontal position, and if the detector coil is not exactly vertical, the result of the measurement may be dependent also on $\mathcal{H}$, but such deviations from the correct position can be detected and adjusted by simple methods. The effect of any residual maladjustment is practically eliminated by combining two readings, both of them with the axis of vibration in the magnetic meridian plane. Between the two readings the detector system is turned 180°, around its vertical axis. A measurement takes a few minutes and the standard error may be as low as one gamma.

21. Determination of the vertical force by balance magnetometers.

α) *The* SCHMIDT *Vertical Magnetometer*. Among the portable magnetometers used in magnetic surveys SCHMIDT's vertical balance is well-known as an easily operated and quickly working instrument[1,2]. From a technical point of view the balance may be characterized as a portable Z-variometer (Sect. 8). The horizontal magnet system, (made of tungsten steel or cobalt steel) is balancing on quartz knife-edges resting on quartz bearings. The system is provided with a horizontal mirror, and by means of an autocollimation telescope, placed vertically on top of the instrument, the angular position of the magnet in its oscillation plane can be read on an appropriate scale in the eyepiece of the telescope. Copper damping serves to reduce the time required for a single reading. The instrument can be turned about a vertical axis, and since the scale value of a Z-variometer is more or less dependent on the horizontal force, unless the oscillation plane of the magnet is coinciding with the magnetic prime vertical (Sect. 8), all readings are made with the magnet normal to the magnetic meridian as indicated by a compass-needle. The corresponding scale value, referred to the telescope scale, is determined in the usual way by exposing the instrument to a known additional field, f. inst. from a Helmholtz-Gaugain coil (Sect. 10). By means of a vertical screw on the magnet system the scale value, which is dependent on the vertical distance of the centre of gravity below the axis of rotation, Eq. (8.4), may be varied within wide limits; it will often be $20\ldots30\,\gamma$ for one scale division.

The instrument housing is temperature-insulated, and by adjusting a special screw on an aluminium spindle along the magnetic axis of the magnet system complete temperature compensation may be achieved. By means of a third screw, adjustable on an invar spindle along the magnetic axis, the gravity centre may be displaced laterally in order to secure that the vertical fields to be measured fall well within the measuring range of the instrument as defined by the length of the scale. The measuring range may be extended by the use of auxiliary magnets placed vertically below the balancing magnet system.

The SCHMIDT balance is used primarily for the determination of local field differences, which differences are simply read on the telescope scale using the known scale value. During a survey the scale value should be redetermined from time to time and the adjustment checked at a base station at regular and frequent intervals. The rather complex construction of the magnet system with its different adjusting screws may imply the risk of small displacements of the gravity centre, especially during transport in the field, and such displacements may appreciably affect the adjustment of the instrument.

β) *The Magnetic Zero Balance*, the BMZ (Fig. 14), is a portable instrument designed by D. LA COUR[3] for relative measurements of the vertical force by a zero method.

---

[1] AD. SCHMIDT: Bericht Tätigk. Preuss. Met. Inst. im Jahre 1914, pp. 109—134; 1915, pp. 87—106. Berlin 1916.
[2] C. A. HEILAND: Geophysical exploration, pp. 318—333. New York: Prentice Hall 1946.
[3] D. LA COUR: The magnetometric zero balance, the BMZ. Comm. Mag. 19, 1—57, Publ. Inst. Mét. Danois (1942).

Fig. 14. The BMZ-magnetometer [T thermometer; A tube containing the tubular magnet (distance between its centre and the balance magnet about 25 cm); B housing for the balance magnet; C clamping device; D telescope; E divided disc and rotating magnet; F supplementary magnet].

The balance-magnet, resting with its knife-edges on agate planes, is of the monad type with its mirror and knives made out of one single piece of steel[4]. By appropriate grinding the magnet has been adjusted once for all so as to be horizontal in a vertical zero field, and during the measurement it serves merely as indicator of the zero field which is produced in the interior of the instrument when the geomagnetic vertical field is compensated by an artificial field due to a system of auxiliary magnets.

The instrument with its auxiliary magnets can be turned about a vertical axis, and to verify that the magnetic axis of the balance-magnet is actually

---

[4] D. LA COUR: La Balance de Godhavn. Comm. Mag. 8, 6—9, Publ. Inst. Mét. Danois (1930).

horizontal, or in other words that the vertical field at the place of the magnet is actually zero, the balance-magnet is observed in two positions, one with the north pole pointing approximately towards magnetic south and one with the north pole pointing approximately towards magnetic north. It has been stated in a previous section, see Eq. (8.4), that the sensitivity of the balance-magnet is varying largely with the orientation of the magnet; it is maximum with the north pole towards magnetic south and minimum in the opposite position. Unless the magnetic axis is exactly horizontal this varying sensitivity will result in a varying inclination of the magnet when the instrument is turned from one position to the other. The inclination of the balance-magnet is observed by means of an autocollimation telescope pointing at the magnet mirror, so that variations of the inclination will change the position of a movable index relative to a scale in the telescope. With the balance-magnet in its sensitive position the adjustment of the compensating magnet system is continued until a telescope reading has been attained that remains unchanged when the instrument is turned 180° from the sensitive to the insensitive position. The corresponding division on the scale is called the neutral division and may remain the same during long series of measurements, although it should be checked at each measurement.

The main part of the geomagnetic vertical field is compensated by means of a large tubular magnet of Oerstit steel, placed vertically and at a well defined distance above the balance-magnet. The residual field is compensated by means of a small and adjustable Oerstit magnet placed below the instrument and mounted on a horizontal axis, so that it can be rotated in a vertical plane that is practically coinciding with the oscillation plane of the balance-magnet. Its angular position can be read on a divided circle. The compensating effect of the rotating magnet will depend on the actual inclination of the magnet, and during the measurement the magnet is adjusted until the movable index is coinciding with the neutral division on the telescope scale. From the corresponding reading of the divided circle and of the thermometer (see below) the total compensating field applied can immediately be derived, using the calibration formula for the instrument. A complete determination of Z takes only a few minutes.

The BMZ with its compensating magnets has to be calibrated in a known vertical field that can be varied over the whole measuring range of the instrument. This field may conveniently be the field of a Helmholtz-Gaugain coil. The rotating magnet alone gives a measuring range of about $2300\,\gamma$, and this range may be extended by means of suitable supplementary magnets placed in a fixed position below the rotating magnet, so that the total measuring range of a standard BMZ will be $7000\,\gamma$, approximately. Greater ranges may be covered by a special device in which the upper, principal, compensating magnet is adjustable along a vertical scale. This type of instrument has been used for survey work extending over large areas or over regions with great magnetic anomalies.

On account of the stability of the monad magnet and the remarkable magnetic qualities of the Oerstit steel the calibration of a BMZ may remain valid for fairly long periods.

As the balance-magnet will always be in a vertical zero field its position will be independent of the magnetic moment of the magnet and therefore also of its temperature. The temperature effect on the compensating magnets will result in a temperature coefficient for the instrument as a whole, but a thermometer placed with its bulb in the middle of the interior of the tubular magnet provides very accurate temperature values for the calculation of the necessary corrections, and reliable results may be obtained even during rapid temperature variations.

The exact horizontal position of the balance-magnet at the moment when the reading is made implies that the result will be independent of the sensitivity of the magnet and consequently also of the orientation of the instrument. The readings are generally made in the most sensitive position in order to obtain maximum accuracy. A different observational method has been suggested by WIESE and GRAFE[5]. The position of the centre of gravity just below the axis of rotation makes the instrument independent of gravity variations.

A modified form of BMZ, with temperature compensated magnet systems, has been described by BOBROV[6].

22. New magnetic materials.

α) In some of the instruments described in the preceding sections two magnets acting on one another form part of the instrument. It is desirable to have these magnets as *compact* as possible. The field to be produced by one magnet at the place of the other is generally of the order of a component of the geomagnetic field, and with the classical types of magnet steel — ordinary steel with small admixtures of tungsten, cobalt and chrome — a magnet of sufficient strength will often have dimensions which are relatively large as compared with the distance between the two magnets, a constellation which gives rise to a considerable inhomogeneity of the field produced. For a given distance between the two magnets the effect of the inhomogeneity will increase with the lengths of the individual magnets. [This is shown, for instance, by expression (17.4)]. Therefore, by using smaller magnets with the same magnetic moments the inhomogeneity would be reduced.

Also in variometers for the recording of magnetic variations it is an advantage to use magnets which combine a great magnetic moment with small dimensions. It is possible to reduce in this way the oscillating period of the moving system, making thereby the variometer more suitable for the registration of rapid variations, see Sect. 13.

β) The possibility of making small magnets with a considerable magnetic moment has been greatly increased by the introduction of *modern magnetic alloys*. Furthermore magnets made from these alloys will generally be highly constant with time.

For a permanent magnet of a given form maximum strength will generally be obtainable with a suitable combination of the following characteristics of the material used:

1) a high remanent magnetic induction, B_r, corresponding to the zero-value of the demagnetizing force,

2) a high numerical value of the coercive force, H_c, corresponding to $B = 0$,

3) a numerically great maximum value of the product BH.

The first real progress with regard to modern types of magnets is due to MISHIMA, who showed in 1931 that a Fe—Ni—Al alloy had a remarkably high numerical value for H_c (515 Oe) combined with a reasonably high value of B_r (6,600 Gs)[1,2].

The addition of cobalt and a few per cent of copper resulted in still higher values of H_c, and later it turned out that the magnetic properties of this so-called Alnico could be further improved by cooling the molten alloy at an optimum rate in a strong magnetic field, thereby obtaining magnets with a certain anisotropic structure. Still better results have been obtained by using a special cooling device which leads to the formation of columnar crystals along the direction of the applied magnetic field (Columax, Alnico VB DG, Ticonal GX)[3].

Theoretical considerations dating back to the late forties showed the possibility of making permanent magnets with a high coercive force by pressing powder containing elongated single domain particles of iron or iron alloys or single domain crystals of bariumferrite. Various magnets of this kind have been developed, for instance type E.S.D. 40C, made from a Fe—Co alloy, or BaO, 6 F_2O_3-magnets with type names Magnadur I, Index I (both isotropic) and Magnadur II, Index V (both anisotropic)[4-7].

Table 3 gives the composition of various magnetic materials and the corresponding values of the magnetic characteristics.

---

[5] H. WIESE u. A. GRAFE: Abhandl. Geom. Inst. Potsdam, Nr. 27 (Sonderheft), 161 (1961).
[6] V. N. BOBROV: Trudy. NIZMIR **15** (25) (1958).
[1] T. MISHIMA: Brit. Pat. Specs. Nos. 378,478 and 392,658.
[2] D. HADFIELD: [*10*], p. 138.
[3] D. HADFIELD: [*10*], pp. 146—148, 156—159.
[4] E. C. STONER and E. P. WOHLFAHRT: Nature **160**, 650 (1947).
[5] E. C. STONER and E. P. WOHLFAHRT: Phil. Trans. Roy. Soc. London, Ser. A **240**, 599 (1948).
[6] E. P. WOHLFAHRT: Phil. Mag., Suppl. **8**, 87 (1959).
[7] D. HADFIELD: [*10*] pp. 178, 182.

Table 3. *Composition and magnetic characteristics of materials used for permanent magnets.* [B_r remanent induction, H_c coercive force, $(BH)_{max}$ maximum numerical value of product induction times force.]

Classical materials.

Composition	C	Admixtures % Cr	Admixtures % Co	W	B_r Gs	H_c Oe	$(BH)_{max}$ MGsOe*
Tungsten	0.7	0.3		6.0	10,500	−65	0.3
Cobalt	0.9	3−6	35	5−6	9,000	−250	0.95

Modern materials.

Composition	Al	Ni	Admixtures % Co	Cu	Ti	Nb	B_r Gs	H_c Oe	$(BH)_{max}$ MGsOe*
Alnico III	12	25	<0.5	<2	0.25		6,700	−575	1.6
Alnico V	8	14	23	3	0.25		12,450	−620	5
Columax	8	13	24	3		0.5	13,400	−840	8.6
Ticonal X	7	15	34		5		9,000	−1,300	4.0
Index V	Bariumferrite = BaO, 6 Fe$_2$O$_3$						3,900	−1,980	3.6

* MGsOe = Mega-Gauss Oersted.

γ) In Fig. 15 *demagnetization curves* for the magnetic materials specified in Table 1 have been drawn. The abscissa is the demagnetizing force H, the ordinate is magnetic induction B. A magnet of, say, Alnico V, which has been magnetized to saturation in an outer magnetic field, will, when the outer field is removed, assume values of H and B that correspond to a certain point R of the Alnico V curve. The position of the point R will depend on the form of the magnet, and the coordinates H_d and B_d of R will represent the strongest magnetization which can possibly be obtained in an Alnico V magnet of the given form. The intensity of

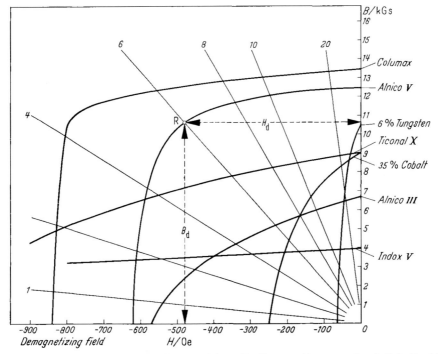

Fig. 15. Demagnetization curves with magnetic characteristics of ellipsoid magnets (electromagnetic units Gs for B and Oe for H, $\mu_0 = 1$). Each of the straight lines corresponds to a certain value of the ratio p of the great axis to the small one. [The intermediate line between those for $p=1$ and $p=4$ corresponds to $p=2.5$.]

magnetization[8] I_d may be calculated from the coordinates by means of the formula

$$B_d = \mu_0 (H_d + u\, I_d). \tag{22.1}$$

Over the greater part of the curve $\mu_0 H$ is small compared with B and in this case we have with a good approximation $B_d = u\mu_0 I_d$.

The position of R on the demagnetization curve can be determined theoretically in the special case that the magnet in question takes the form of an ellipsoid of revolution. In that case the demagnetizing force and the intensity of magnetization are constants throughout the magnet, and the relation connecting H_d and I_d can be expressed by the formula

$$H_d = -u D I_d \tag{22.2}$$

where D is the so-called demagnetizing factor, for which MAXWELL has given the numerical expression (log means log nat)

$$D = \left(\frac{1}{e^2} - 1\right)\left(\frac{1}{2e} \log \frac{1+e}{1-e} - 1\right) \tag{22.3}$$

e being the eccentricity of the ellipsoid. The expression can also be written

$$D = \frac{1}{(p^2 - 1)} \left(\frac{p}{\sqrt{p^2-1}} \log\left(p + \sqrt{p^2-1}\right) - 1\right) \tag{22.3a}$$

where p is the ratio of the axes.

Eliminating I_d from Eqs. (22.1) and (22.2) we get

$$B_d = \mu_0 H_d \frac{D-1}{D}. \tag{22.4}$$

This formula, combined with Eq. (22.3a), determines immediately the working point for a given magnet the shape of which is an ellipsoid of revolution with axes-ratio p. In Fig. 15 the point marked $R(H_d, B_d)$ is the working point for an ellipsoidal Alnico V magnet for which p is equal to 6. R is the point of intersection between the Alnico V curve and a line drawn through origo in a direction determined by

$$\tan \alpha = \mu_0 \frac{D-1}{D} \tag{22.5}$$

where α is the angle between the line and the abscissa[9].

The line through R will intersect also the other demagnetization curves, and the various points of intersection will represent the magnetic characteristics of ellipsoid magnets that all have the same eccentricity, but are made of various kinds of magnetic material.

For the sake of illustration additional straight lines have been drawn in Fig. 15, each of them representing a certain value of the axes-ratio p in the same way as the line through R. The value of p is indicated at the top of each line, and the intersection of the line in question with the various demagnetization curves will give the magnetic characteristics of a group of ellipsoid magnets that all have the same value of p, but are made of different materials.

δ) Comparison of different materials is quite easy with Fig. 15. For all values of $p < 8$ ellipsoidal magnets of the Alnico types will give higher values of B_d than magnets of the cobalt type. For $p = 4$ the B_d value of an Alnico III magnet will be about 55% greater than that of a cobalt magnet; a Columax magnet will be between 3 and 4 times as strong as the cobalt magnet.

In a bar magnet of cylindrical form or with rectangular cross section H and I will vary inside the magnet, and the above Eq. (22.2) will therefore not be strictly valid. For such a magnet the simple procedure Fig. 15 can not be used directly to determine the magnetic characteristics in which we are interested. But since the magnetic characteristics of a bar magnet will depend on the material and on the ratio between length and "diameter" in much the same way as for an ellipsoidal magnet, the curves of Fig. 15 may give a general impression of the great advantages to be obtained from the use of modern magnetic materials when the construction of small and yet strong permanent magnets is aimed at.

The use of "elongated single domain magnets" may be somewhat limited by the fact that their temperature coefficient is extremely high, up to five times that of a normal steel magnet. Such magnets may be profitably utilized, however, for temperature compensation in geomagnetic instruments, a combination of magnets with widely different temperature coefficients being used to reduce or to eliminate the overall influence of temperature.

---

[8] This is the magnetic moment M per unit volume. I_d has dimension Oe in electromagnetic cgs units, and A m^{-1} in international SI-units.

[9] Modified in Fig. 15 because different scales have been used when plotting B and H.

The high coercive force of the magnetic alloys means of course a great resistance against the influence of external magnetic fields. If the magnet is exposed to a strong alternating field (for example in a coil carrying a strong a.c. current) the magnetic moment will generally be somewhat reduced, but the residual magnetization will in return be even more stable. As a general rule the magnetic moment of a magnet to be used in a geomagnetic instrument should always be reduced by some 5 to 15% of its maximum value[10]. In addition to this preliminary reduction of the magnetic moment the magnet should be aged artificially by exposing it to temperature variations within a range that will with certainty cover the range to be expected in observational practice (for instance $-80°\cdots +100°$ C). Treated in this way a magnet made of one of the modern alloys may keep its magnetic moment practically constant for years, the annual variation being less than, say, one part in 10,000. For their application in geomagnetic measuring instruments it is a further advantage of the alloys with great coercive force that they have a relatively small *induction factor*.

ε) For some alloys as *Termoperm* (30% Ni) and *Calmalloy* or *Termalloy* (58—66% Ni, 2% Fe, rest Cu) the Curie-point will be only slightly higher than normal air temperature, so that the variation with temperature of the intensity of magnetization will be very great in the temperature interval in which geomagnetic instruments are generally used (a decrease of 60% between 0° and 50° C for Termoperm). These alloys have been used for temperature compensation of magnets[11,12].

An entirely different type of alloys has come into use in magnetic instruments where small values of coercive force and high values of permeability are required (airborne magnetometers and some variometers recording $d\mathcal{H}/dt$ and dZ/dt. The relative permeability of Mumetal (5% Cu, 75% Ni) has an initial value of 10,000...30,000 and may reach maximum values of 60,000...100,000; for Superpermalloy the maximum value goes up to $8 \cdot 10^5$.

23. International intercomparisons of magnetic instruments.

α) It is the principal aim of any direct determination of a geomagnetic element to obtain an absolute and invariable standard with which data measured or recorded at the same time in *different parts of the world*, could reasonably be compared. Another use is to compare values found at one and the same place at *different epochs*. If values obtained by direct determinations were really absolute in the true sense of the word, they would be directly comparable. However, in many cases there may remain some uncertainty as to the absolute character of the value obtained, and this may even be the case with so-called "absolute methods".

While many of the instruments described above reach a high standard in relative accuracy it is often difficult to avoid systematic errors arising from mechanical imperfections in the instruments, or from traces of magnetism in parts which ought to be non-magnetic. The result of a direct determination of $\mathcal{H}$ by the Gaussian method (Sect. 17) may be slightly erroneous due to unnoticed variations of the moment of inertia or of the distribution coefficients for the magnets used, and in electromagnetic instruments errors may arise from deficiencies in insulation or from a lack of constancy of the standard cells.

β) The *independent check* needed may be obtained from simultaneous measurements of the same element by two different types of instrument, for instance $\mathcal{H}$ measured by the Gaussian method compared with $\mathcal{H}$ measured by an electromagnetic method[1], or Z derived from determinations of $\mathcal{H}$ and $\mathcal{J}$ checked by a direct and independent determination of the vertical force[2].

Even if it may be very difficult to verify the absolute character of determinations made at a magnetic observatory, values from different observatories are readily referred to a common standard provided the local standards are sufficiently

---

[10] K. J. KRONENBERG and M. A. BOHLMANN: WADC Technical Report 58-535; (Astia Document No. AD 203387), p. 32, Wright Air Development Center, Ohio (1958).
[11] B. M. YANOVSKY: Terr. Mag. **43**, 143—147 (1938).
[12] T. KUBOKI: Mem. Kakioka Magn. Obs. **6**, 64—65 (1951).
[1] S. E. FORBUSH and E. A. JOHNSON: Int. Ass. Terr. Mag. El., Bul. **10**, 248—259 (1937).
[2] W. H. M. GREAVES and W. M. MITCHELL: Terr. Mag. **43**, 137—138 (1938).

consistent and the differences between such local standards are known with sufficient accuracy. Comparisons between observatory standards have been initiated long ago[3]. They have been continued on a large scale during the years 1905 through 1948 by the travelling observers of the Department of Terrestrial Magnetism of the Carnegie Institution of Washington, who compared observatory standards all over the world. The portable magnetometers used for these comparisons, mostly instruments for D, J and H, have frequently been compared with each other. They have further been calibrated at Cheltenham Observatory, U.S.A., against the well established standard values maintained at that observatory[4,5].

More recently the Association of Geomagnetism and Aeronomy of the International Union of Geodesy and Geophysics has supported such comparisons by the establishment of a special service on comparisons of magnetic standards. The service has at its disposal a number of QHM-magnetometers, which are being circulated by air freight for intercomparison of observatory standards of the horizontal magnetic force[6,7].

General references.

[1] GAUSS, C. F.: Intensitas vis magneticae terrestris ad mensuram absolutam revocata. Gauss, Werke, herausgeg. von der Kgl. Ges. Wiss., Göttingen, Vol. 5, pp. 79—118 (1877). Ostwalds Klassiker, Nr. 53, The original work was published in 1833. A German translation was published in Pogg. Ann. **28** (1833). In this paper Gauss proposed a system of units of length, mass and time and called these units absolute units.

[2] LAMONT, J.: Handbuch des Erdmagnetismus. VIII + 264 pp. Berlin: Veit & Co. 1849. — Contains a comprehensive theory of geomagnetic observations, especially measurements of H and D.

[3] ANGENHEISTER, G.: Das Magnetfeld der Erde, Instrumente und Meßmethoden. Handbuch der Experimentalphysik, herausgeg. von W. WIEN und F. HARMS, Vol. 25, I, pp. 525—585. Leipzig: Akadem. Verlagsges. 1928.

[4] CHAPMAN, S., and J. BARTELS: Geomagnetism. Oxford: Clarendon Press 1940. — Instrumental questions are treated in chap. II (pp. 29—95) with a great number of bibliographic references on pp. 945—952.

[5] Terrestrial Magnetism and Electricity. Physics of the Earth, Vol. VIII, edit. by J. A. FLEMING. New York and London: McGraw-Hill 1939. — Geomagnetic instruments in general are treated by J. A. FLEMING, H. F. JOHNSTON, and H. E. McCOMB in chap. II (pp. 59—109), and a description of instruments used in magnetic prospecting is given by C. A. HEILAND in chap. III (pp. 118—131). Corresponding bibliographic references are found on pp. 701—706.

[6] FANSELAU, GERHARD, editor: Geomagnetismus und Aeronomie, Bd. II, 648 pp. Berlin: VEB Deutscher Verlag d. Wiss. 1960.

[7] SCHMIDT, ADOLF: Erdmagnetismus, Enzyklopädie der mathematischen Wissenschaften, Bd. VI, 1. B, 10, pp. 265—396. Leipzig 1917.

[8] HAZARD, D. L.: Directions for Magnetic Measurements: U. S. Coast and Geodetic Survey, Serial No. 166, corrected edition, 112 pp. Washington 1947.

[9] McCOMB, H. E.: Magnetic Observatory Manual; U. S. Coast and Geodetic Survey Special Publ. No. 283, 228 pp. Washington 1952.

[10] HADFIELD, D., editor: Permanent Magnets and Magnetism, 556 pp. London: Iliffe 1962.

[11] WIENERT, K. A.: Notes on geomagnetic and observatory practice. Earth Sciences, Vol. 5, 217 pp., published by UNESCO, Paris 1970.

---

[3] E. VAN RIJCKVORSEL: An attempt to compare instruments for absolute magnetic measurements at different observatories. Amsterdam, R. Met. Inst. 1890, 1—15; 1898, 1—8; 1900, 1—4; 1902, 1—6.

[4] Summary of C. I. W. comparisons up to 1922 is given by: L. BAUER, Int. Sect. Terr. Mag. El., Bul. **3**, 93—96 (1923).

[5] Summary up to 1936: J. A. FLEMING, Int. Ass. Terr. Mag. El., Bul. **10**, 241—248 (1937).

[6] J. KERÄNEN: Int. Ass. Terr. Mag. El., Bul. **13**, 335 (1950).

[7] V. LAURSEN: Int. Ass. Terr. Mag. El., Bul. **14**, 267—271 (1954); — IAGA Bul. **15**, 301—306 (1957); Bull. **16**, 310—312 (1960); Bul. **19**, 54—60, 194—197 (1969); Bul. **25**, 29—31 (1968).

Neuere Meßmethoden der Geomagnetik.

Von

H. Schmidt und V. Auster*.

Mit 31 Figuren.

1. Einleitung. Die geomagnetische Forschung ist im Hinblick auf die Zunahme ihrer phänomenologischen Erkenntnisse vielfach durch instrumentelle Entwicklungsstufen gekennzeichnet, die ihre Existenz teils unmittelbar neuen Beobachtungsmöglichkeiten, teils einer Vervollkommnung bekannter Prinzipien verdanken. Überblickt man in diesem Sinne die Entwicklung der neueren geomagnetischen Meß- und Beobachtungstechnik, so lassen sich folgende Wesenszüge erkennen.

α) Die klassische geomagnetische Meßtechnik beruht vorwiegend auf der *Ausnutzung von Drehmomenten*, die auf einen beweglich aufgehängten Magneten wirken und von ponderomotorischen, magnetostatischen und elastisch deformierenden Kräften herrühren[1]. Die magnetostatischen Kräfte werden vom erdmagnetischen Feld, von zusätzlichen Magneten oder von stromdurchflossenen Spulen geliefert. Für die Bestimmung der erdmagnetischen Feldgrößen auf festem Untergrund hat sich seit vielen Jahrzehnten eine in zahlreichen Etappen verfeinerte Präzisionsmeßtechnik entwickelt, die sich in einem weltweiten Einsatz laufend bewährt. Die Notwendigkeit, zur Ausrichtung der Indikatoren das Schwerefeld benutzen zu müssen, schränkte den Anwendungsbereich dieser Methodik auf solche Instrumententräger ein, die eine Horizontierung ermöglichten. Für die Messung von bewegten Trägern aus blieben daher viele Wünsche offen. Messung und Interpretation erfolgten hauptsächlich in Komponenten oder deren Abgeleiteten und die Totalintensität spielte als Interpretationsgröße eine relativ geringe Rolle.

In dieser Situation brachte die Einführung selbstorientierender Nachführsysteme eine entscheidende Änderung, da diese Systeme auf das Schwerefeld verzichten konnten und das erdmagnetische Feld selbst zur Ausrichtung magnetischer Geber benutzten. Die hiermit verbundene Einschränkung der Meßgrößen auf die Totalintensität — anfangs als Nachteil empfunden — erwies sich nach Vorliegen geeigneter Interpretationsmethoden als sehr nützlich. Großräumige Totalintensitätsvermessungen konnten von Flugzeugen, Schiffen, Raketen und Erdsatelliten aus in Angriff genommen werden. Die selbstorientierenden Nachführsysteme enthalten ein Meßorgan, das — mit zwei weiteren ein orthogonales Dreibein bildend — so gesteuert wird, daß seine Meßachse jeweils parallel zum erdmagnetischen Vektor liegt und damit eine Messung der Totalintensität durch Kompensation ermöglicht wird. Obwohl sich Rotations- oder Vibrationsspulen als Meßorgane geeignet hätten, ist der Saturations-Kernsonde der Vorzug gegeben worden. Die Magnetfeldmessung mit derartigen Sonden beruht auf der Bestimmung von Oberwellenanteilen eines auf hochpermeable Kerne wirkenden

---

* Manuskript eingegangen im Februar 1969.
[1] Siehe den vorstehenden Beitrag von V. Laursen u. J. Olsen in diesem Bande.

sinusförmigen Magnetfeldes. Nachdem technologisch gut beherrschbare Sondenkonstruktionen vorlagen und die während des zweiten Weltkrieges vorgenommenen militärischen Entwicklungen geophysikalisch genutzt werden konnten, setzte ab 1947 ein enormer Aufschwung der Aero-Magnetometrie ein. Damit fanden Servomechanismen Eingang in die geomagnetische Instrumenten-Entwicklung und führten zum Einsatz automatisch kompensierender Magnetometer auf bewegten Instrumententrägern. Mit derartigen Apparaturen wurden die ersten großen Erfolge der Raketen-Magnetik erzielt (Nachweis ionosphärischer Stromsysteme, Bestimmung der Größenordnung des Mondmagnetfeldes). Sonden mit hochpermeablen Kernen bewährten sich nicht nur in Nachführsystemen, sondern gestatteten auch als raketenfest eingebaute Dreibeine die Ermittlung der Raketenlage relativ zum (als bekannt vorausgesetzten) Magnetfeld. Kombinationen von Nachführsystemen mit Pendeln oder Kreiselplattformen wurden erstmals zur Dreikomponentenmessung aus der Luft eingesetzt und ermöglichten die erste weltweite Vermessung mit einem gleichbleibenden Instrumentarium[2] (Project Magnet).

Vom Standpunkt der heutigen Datenverarbeitungs-Terminologie aus handelte es sich bei allen bisher genannten Verfahren im wesentlichen um die Anwendung analoger Meßwertwandlung und analoger Aufzeichnung. Alle Meßgrößen wurden in Ströme, Spannungen, Winkel usw. umgesetzt und — falls nicht direkt abgelesen — auf Trägern analoger Daten, z.B. photographischen Papieren, als Kurve geschrieben.

β) Die jetzt überall Eingang findende *Digital-Meßtechnik* kam der Entwicklung von Magnetometern mit *magnetfeldproportionaler Frequenzabgabe* entgegen, die — auf kernphysikalischen und spektroskopisch-optischen Zusammenhängen basierend — zur Zeit den Hauptanteil aller auf bewegten Instrumententrägern eingesetzten Magnetometer darstellen. Besonders auch im Hinblick auf die telemetrische Datenerfassung bieten diese driftfreien Geräte entscheidende Vorteile, da sie die Meßwertwandlung über ein frequenzbestimmtes Signal vornehmen und damit den Einsatz der hochentwickelten Impuls-Zähltechnik gestatten.

Die erste Kategorie der frequenzabgebenden Magnetometer, die der Kernpräzessionsmagnetometer, geht auf die in den ersten Nachkriegsjahren veröffentlichten Arbeiten über Kerninduktion und Kernabsorption zurück. Während die Messung hoher Magnetfelder hiervon unmittelbar profitierte, stellten sich den ersten Erdfeld-Anwendungen (1952) der geringen Signalspannung wegen erhebliche Schwierigkeiten entgegen, die erst von 1955 ab durch das Prinzip der freien Präzession weitgehend überwunden werden konnten. Im letzten Jahrzehnt setzten sich Präzessionsmagnetometer mit Ziffernanzeige der Totalintensität immer mehr durch, lösten die Sonden-Nachführ-Magnetometer zusehends ab und bereicherten die Observatoriums-Meßtechnik dadurch, daß das gyromagnetische Verhältnis des Protons als unveränderliche Atomkonstante zum geomagnetischen Subnormal erklärt wurde.

Die zweite Kategorie der frequenzabgebenden Magnetometer stellen die „optical pumping magnetometer" dar, bei denen eine Resonanzabsorption beispielsweise zum Aufbau selbstoszillierender Systeme ausgenutzt wird. Derartige Magnetometer liefern im Gegensatz zu den rhythmisch abgegebenen Signalen der mit freier Präzession arbeitenden Magnetometer eine kontinuierliche, der Totalintensität jeweils proportionale Frequenz, die bei gleichem Feld etwa 100 bis 700mal höher liegt als die eines Kernpräzessionsmagnetometers. Dieser Vorteil der hohen Frequenzauflösung befähigt solche Geräte, sehr kleine Variationen oder Felddifferenzen sicher zu erfassen.

---

[2] H. P. STOCKARD: Report "Worldwide surveys by Project Magnet". IAGA-Symposium Washington, Oktober 1968, und Beitrag von P. HOOD in diesem Band, S. 422.

Einleitung.

γ) Die präzisen und eleganten Anzeigeverfahren der frequenzabgebenden Magnetometer ließen den Gedanken aufkommen, hiermit unter Anwendung von Zusatzfeldern *Komponentenmessungen* zu versuchen. In den letzten Jahren sind einige Varianten entstanden, die — mit Protonen- oder Rubidium-Meßköpfen ausgerüstet — eine Dreikomponentenanzeige mit digitaler Datenausgabe aufweisen und dem Ziel eines „automatischen Observatoriums" mehr oder weniger nahekommen. Die zur Komponentendefinition nötige Berücksichtigung des Schwerefeldes und azimutaler Richtungen bringt hier allerdings die gleichen Schwierigkeiten wie bei der klassischen Meßtechnik und man sollte sich — zumindest für die strengen Observatoriumsbelange — davor hüten, dem Vorteil der digitalen Datenausgabe die bisher als selbstverständlich geltenden Langzeitstabilitäten der klassischen Methoden zu opfern.

Es hat nicht an weiteren Versuchen gefehlt, die sehr geringe Spin-Besetzungszahl-Unterschiede, die bei der stationären Kernresonanz im Erdfeld kaum nachweisbare Signalspannungen ergeben, künstlich zu erhöhen, z. B. durch Anwendung des OVERHAUSER-Effektes[3], wonach das Kernresonanzsignal durch Sättigung der Elektronenspinresonanz vergrößert werden kann. In Anbetracht der stürmischen Entwicklung der HF (Hochfrequenz)- und NF (Niederfrequenz)-Spektroskopie gilt es als sicher, daß der Weg zum automatischen Observatorium unbedingt über frequenzabgebende Magnetometer führt. Welchem Prinzip hierbei der Vorzug zu geben ist, scheint noch offen und in methodischer Hinsicht sind Überraschungen nicht ausgeschlossen.

Trotz aller Fortschritte der elektronischen Magnetometer werden innerhalb der gesamten erdmagnetischen Meßtechnik immer Aufgaben übrig bleiben, die dem Stabmagnet oder der Rotationsspule ihrer von elektronischen Systemen nur schwer überbietbaren Betriebssicherheit wegen zukommen müssen.

Die folgende Darstellung neuerer Meßmethoden der Geomagnetik ist als Überblick über die gegenwärtig in größerem Maße genutzten Methoden zu verstehen, bei dem versucht wurde, die physikalischen Zusammenhänge im Hinblick auf die geomagnetischen Anwendungen zu betonen. Eine Vollständigkeit konnte des beschränkten Raumes wegen weder in der Aufzählung der Verfahren[4] noch in der Zahl der Anwendungsvarianten angestrebt werden.

δ) Die *Formeln* in diesem Beitrag sind durch Benutzung des Zahlenfaktors u und des Ausdrucks $c_0 \sqrt{\varepsilon_0 \mu_0}$ in jedem der üblichen Maßsysteme gültig[5]. Wünscht man sie beispielsweise im GAUSSschen System zu lesen, das als nichtrationales System zu kennzeichnen ist und für das $\varepsilon_0 = \mu_0 = 1$ vorausgesetzt wird, ist zu setzen:

$$\mathrm{u} = 4\pi; \quad c_0 \sqrt{\varepsilon_0 \mu_0} = c_0. \qquad (1.1)$$

Die Lesart im rational geschriebenen MKSA-System ergibt sich durch

$$\mathrm{u} = 1; \quad c_0 \sqrt{\varepsilon_0 \mu_0} = 1. \qquad (1.2)$$

---

[3] O. OVERHAUSER: Phys. Rev. **91**, 476 (1953).

[4] Es sei auf die ausführlichere, dem Stande von 1960 entsprechende Abhandlung verwiesen: H. SCHMIDT u. O. LUCKE: Neuere Meßmethoden der Geomagnetik. In der Reihe: „Geomagnetismus und Aeronomie", herausgeg. von G. FANSELAU, Bd. II, S. 160—404. Berlin: VEB Deutsch. Verl. d. Wiss. 1960. — Weitere zusammenfassende Publikationen: K. WHITHAM: Measurement of the geomagnetic elements. In: Methods and techniques in geophysics, Vol. 1, pp. 104—167. New York: Interscience Publishers. 1960. — B. M. JANOVSKI: Zemnoi Magnetizm. Izdat. Leningradsk. Univ., Bd. II, 1963 [Russ.]. — K. A. WIENERT: Notes on geomagnetic observatory and survey practice. Earth Sciences UNESCO. München: Oldenbourg 1970.

[5] u ≡ 4π für nichtrationale Systeme, u ≡ 1 für rationale Systeme. Im MKSA-System gelten folgende Werte:

$c_0 = (2{,}997925 \pm 0{,}000003) \cdot 10^8$ m/s;

$\mu_0 = 4\pi \cdot 10^{-7}$ Vs/Am $= 1{,}256637 \cdot 10^{-6}$ Vs/Am,

$\varepsilon_0 = (8{,}85419 \pm 0{,}00002) \cdot 10^{-12}$ As/Vm.

Die Werte für weitere Maßsysteme sind einer Tabelle von RAWER und SUCHY[6] zu entnehmen.

Dem Wunsche, alle Zahlenangaben im MKSA-System vorzunehmen, konnte im Hinblick auf Vergleichsmöglichkeiten mit anderen Publikationen nur teilweise entsprochen werden, da in der Literatur einiger hier zu behandelnder Hauptgebiete das cgs-System immer noch bevorzugt erscheint. Weiterhin sind im Geomagnetismus entsprechend der historischen Entwicklung einige Einheiten in Gebrauch, die — obwohl nicht zu den amtlich empfohlenen zählend — in steigendem Maße angewandt werden. Besonders gilt dies für die altbewährte Einheit γ, die in weltweiten Kartenwerken eine Hauptrolle spielt und die durch die Einführung von „Magnetometern mit direkter γ-Anzeige" erneut an Bedeutung gewinnt. Es gelten folgende Umrechnungsbeziehungen (Gs = Gauß, T = Tesla):

$$1\,\gamma \cong 10^{-5}\,\text{Gs} \cong 10^{-9}\,\frac{\text{Vs}}{\text{m}^2} \cong 10^{-9}\,\text{T}. \qquad (1.3)$$

Für die Schreibweise des Drehmomentes $\boldsymbol{N}$, das ein magnetischer Dipol mit dem Moment $\boldsymbol{\mathsf{M}}$ in einem homogenen Magnetfeld, gekennzeichnet durch die Kraftflußdichte $\boldsymbol{B}$ oder die magnetische Feldstärke $\boldsymbol{H}$, erfährt, gibt es zwei Versionen:

$$\boldsymbol{N} = \boldsymbol{\mathsf{M}} \times \boldsymbol{B} \qquad (1.4\text{a})$$

bzw.

$$\boldsymbol{N} = \boldsymbol{\mathsf{M}} \times \boldsymbol{H} \qquad (1.4\text{b})$$

Wir bevorzugen die erste Version[7] in Übereinstimmung mit der von IUPAP angenommenen Normung.

Zur Wahl der Symbole ist folgendes zu bemerken. Der Vektor des erdmagnetischen Feldes wird mit $\boldsymbol{F}$ bezeichnet, sein Betrag und seine Komponenten sind in Kraftflußdichte-Einheiten anzugeben. Mit den üblichen Komponentenbezeichnungen im geodätisch orientierten xyz-System

$$\mathcal{F}_x = \mathcal{X} \quad \text{(positiv nach geodät. Nord)},$$
$$\mathcal{F}_y = \mathcal{Y} \quad \text{(positiv nach geodät. Ost)},$$
$$\mathcal{F}_z = \mathcal{Z} \quad \text{(positiv nach unten)},$$

ergeben sich die Totalintensität $\mathcal{F}$ und die Horizontalintensität $\mathcal{H}$ zu

$$\mathcal{F} = \sqrt{\mathcal{X}^2 + \mathcal{Y}^2 + \mathcal{Z}^2} \qquad \mathcal{H} = \sqrt{\mathcal{X}^2 + \mathcal{Y}^2} \qquad (1.5)$$

und Deklination D sowie Inklination $\mathcal{J}$ zu

$$D = \arctan\frac{\mathcal{Y}}{\mathcal{X}} \qquad \mathcal{J} = \arctan\frac{\mathcal{Z}}{\mathcal{H}}. \qquad (1.6)$$

Eine eindeutige Wahl aller Symbole wurde zwar angestrebt, konnte jedoch im Hinblick auf die gewohnte Lesbarkeit der Formeln und auf die Vermeidung zu vieler Indices nicht immer getroffen werden. Mißverständnisse dürften dennoch ausgeschlossen sein, da die Symbole ausreichend erklärt wurden.

A. Komponentenmessung mit Hilfe von Stabmagneten als Meßelemente in Verbindung mit elektronischen Anzeige- und Regelorganen.

Der beweglich aufgehängte Magnet stellt das älteste geomagnetische Meßelement dar und ist auch heutzutage in vielen Beobachtungsgeräten vertreten. Als Standard-Magnet in Form eines mit höchster Präzision bearbeiteten Zylinders dient er der Absolut-Meßtechnik, und als „Stabmagnet" bildet er in einer Vielfalt

---

[6] Dieses Handbuch, Bd. 49/2, S. 536. Siehe auch "Introductory Remarks" in diesem Band.

[7] Siehe hierzu: A. SOMMERFELD: Elektrodynamik. Leipzig: Akadem. Verlagsgesellschaft Geest & Portig 1949. — O. LUCKE: Über Größengleichungen und Maßsysteme. Abhandl. Geomagnet. Inst. Potsdam Nr. 22, 17 (1958).

von Formen, vom Quader bis zum dünnen Plättchen variierend, das wesentliche Organ eines klassischen Variometers. Als Meßwertgeber für die photographisch-optische Registrierung beliebig vorgebbarer Komponenten hat er sich seit über 100 Jahren bewährt und wird auch weiterhin dieses Feld behaupten. Allerdings haften diesem photographischen Verfahren die bekannten Nachteile der um die Dauer des Entwicklungsprozesses verzögerten Kurvenwiedergabe an, so daß seit langem Forderungen nach einer direkten Sichtanzeige der geomagnetischen Variationen sowie nach einer Verbesserung des dynamischen Verhaltens erhoben wurden und zu Kombinationen von Stabmagnet und lichtelektrischen Anzeige- und Regelorganen führten. Neuerdings werden diese Forderungen durch Bedingungen der maschinellen Datenverarbeitung im Hinblick auf Anpassung an Analog-Digital (AD)-Umsetzer erweitert.

Zwei Wege haben sich in der Praxis bewährt. Einmal steuert der Magnet mittels Lichtzeiger die Ausleuchtung eines Photoelements und der variationsproportionale Strom wird von einem hochempfindlichen Schreiber im „Ausschlagverfahren" registriert; zum anderen arbeitet der Magnet in einem Regelkreis zur kontinuierlichen Kompensation der Variationen und liefert einen Strom im „Kompensationsverfahren", der auch mit robusten Schreibern aufgezeichnet werden kann. Welcher Methode man den Vorzug gibt, hängt vom Anwendungszweck mit seinen Genauigkeits- und Langzeitstabilitäts-Forderungen ab.

I. Kompensierende Magnetometer.

2. Das Kompensationsverfahren.

α) Der Aufgabe, mit Hilfe eines drehbar aufgehängten Magneten einen variationsproportionalen Kompensationsstrom zu erzeugen, wird man am besten durch den Einsatz lichtelektrischer Hilfsmittel gerecht, obwohl prinzipiell auch andere Methoden, wie z.B. die kapazitive Drehwinkel-Erfassung, geeignet wären. Die Anwendung lichtelektrischer Verfahren verlangt nur die Existenz eines kleinen Spiegels am beweglichen Organ und bietet gegenüber anderen Möglichkeiten Vorteile im Hinblick auf eine günstige Ausnutzung des bereits für den normalen Variometereinsatz getriebenen feinmechanischen Konstruktionsaufwandes und erfordert elektronische Zusätze von vergleichsweise geringem Umfang.

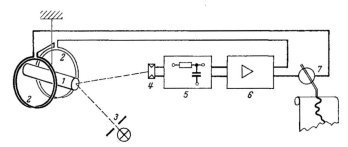

Fig. 1. Prinzipieller Aufbau eines lichtelektrisch kompensierenden Magnetometers mit Magnet (1), Kompensationsspule in HELMHOLTZ-Form (2), Optik (3), lichtelektrischem Schaltelement (4), Zeitglied zur Frequenzgangkorrektur (5), Verstärker (6) und Schreiber (7).

In Fig. 1 wird der prinzipielle Aufbau eines lichtelektrisch kompensierenden Magnetometers angegeben, in dem der Magnet innerhalb eines Regelkreises über einen Lichtzeiger die Beleuchtung eines photoelektrischen Schaltelements (Photo-Zelle, -Widerstand, -Diode, -Transistor, -Vervielfacher oder -Element) ändert.

Ein von diesem Schaltelement über Verstärker oder motorisch betriebene Potentiometer eingestellter Strom erzeugt in der Kompensationsspule ein Magnetfeld, das den Magneten zur Ausführung der Variationen in sehr stark verkleinertem Maße zwingt (proportionale Regelung) oder ihn immer wieder in die gleiche Ausgangslage zurückzuführen sucht (integrale Regelung). Die Meßachse der Spule liegt parallel zur Richtung der aufzuzeigenden Komponente. Spulenmeßachse, Magnetmeßachse und Magnetdrehachse bilden ein orthogonales Dreibein, exakt als Justierbedingung für die Ruhelage gültig und während des Betriebes mit weitgehender Annäherung zutreffend, da die Winkelausschläge des Magneten weit unter 1° liegen. Ein im Regelkreis eingesetztes Zeitglied bestimmt das Einschwingverhalten, die Eigenfrequenz und die regeltechnische Stabilität des gesamten Systems. Durch zweckentsprechende Auslegung des Zeitgliedes gelingt es, proportionale, integrale oder differentielle Anteile der Winkelbewegung im Rückführkreis einzeln oder gemeinsam auszunutzen.

β) Beim kompensierenden Magnetometer gibt es wie bei jeder anderen Regeleinrichtung Betriebszustände, in denen selbsterregte Schwingungen auftreten und damit die Arbeitsweise als automatisch registrierendes Gerät stören. Es ist daher Aufgabe der Entwicklung solcher Magnetometer, eine optimale Verstärkung innerhalb des Regelkreises zu garantieren, die einerseits ein genügend schnelles Einschwingen und weitgehende Eliminierung von Betriebsparameter-Schwankungen gestattet, andererseits aber das Auftreten von Regelschwingungen vermeidet. Der späteren Behandlung nach regeltechnischen Gesichtspunkten geht eine Betrachtung der Gleichgewichtsbeziehungen voraus, mit der man das Verhalten des Magnetometers im eingeschwungenen Zustand charakterisieren und den Skalenwert ableiten kann.

3. Gleichgewichtsbeziehungen an lichtelektrisch kompensierenden Magnetometern.

α) Einer einheitlichen Behandlung aller Registrierfälle wegen wird die *Berechnung der Drehmomente* in einem Koordinatensystem $\xi\eta\zeta$ mit den Einheitsvektoren i^0, j^0, k^0 durchgeführt, das mit dem Gehäuse des Magnetometers als fest verbunden gilt. i^0 kennzeichnet die Richtung der Fadenachse (von Durchbiegungen horizontaler Fäden wird abgesehen) und j^0 gibt die Richtung der Soll-Lage der magnetischen Meßachse an, wobei der Nordpol des Magneten in Richtung positiver η-Werte liegt. Die im erdfesten, geodätisch orientierten xyz-System gegebenen Vektoren werden in Komponenten des $\xi\eta\zeta$-Systems umgerechnet. Während es für die Ableitung von Skalenwerten genügt, die $\xi\eta\zeta$-Komponenten aus Zeichnungen zu ermitteln, erfordert die Diskussion von Orientierungstoleranzen für die Horizontierung des Magnetsystems, die Lage der Spulenachse usw. die Berechnung der Vektorkomponenten über die EULERschen Winkel.

Aus der Forderung, daß die um die Richtung i^0 angreifenden, von der Torsion, dem erdmagnetischen Feld F, dem Spulenfeld C und dem Schwerefeld g herrührenden Drehmomente in ihrer Summe Null ergeben sollen, folgt die Beziehung

$$D^*(\tau - \varphi) + (\mathbf{M} \times F) \cdot i^0 + (\mathbf{M} \times C) \cdot i^0 + m(S \times g) \cdot i^0 = 0 \qquad (3.1)$$

mit D^* als Direktionskraft des Fadens, τ als Torsionswinkel für die Ruhelage, φ als Ausschlagwinkel des Magneten, $\mathbf{M}$ als magnetischem Moment des Magneten, m als seiner Masse und S als seinem Schwerpunktsvektor. Die Beziehung gilt für alle Registrierfälle, gebräuchlich sind bei vertikalem Faden die Aufzeichnungen von X, Y, H und D sowie bei horizontalem Faden X, Y, H, D, Z und F.

β) Als *Beispiel* folgt die Ableitung des Skalenwertes eines Z-Kompensators, der zur Aufzeichnung der Variationen der Vertikalkomponente verwendet wird und dessen Magnetsystem die Fig. 2 zeigt. Auf den in der $\eta\zeta$-Ebene schwingenden Magneten wirke das durch die Komponenten X, Y und Z angedeutete erdmagneti-

Ziff. 3. Gleichgewichtsbeziehungen an kompensierenden Magnetometern.

sche Feld, wobei Z einen Festanteil Z_0 und die Variation Z_v unterteilt werde. In zu Z_v entgegengesetzter Richtung greife das Spulenfeld vom Betrage

$$C = \mu_0 \, k \, I \tag{3.2}$$

mit k als Spulenkonstante und I als Kompensationsstrom an. Die azimutale Lage der Fadenachse ξ gibt der Winkel λ an, der von 0 bis 2π variieren kann und somit

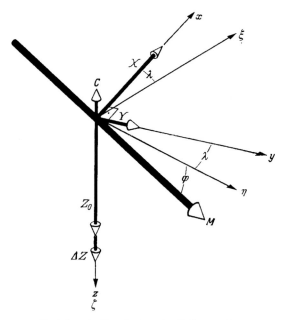

Fig. 2. Lage des Magnetsystems eines Z-Kompensators.

alle möglichen Nordpol-Lagen einschließt. Die Transformation in das $\xi\eta\zeta$-System ergibt mit s_{η_0} und s_{ζ_0} als Schwerpunktskomponenten der Ruhelage:

$$\left.\begin{aligned}
&\mathcal{F}_\xi = X \cos \lambda + Y \sin \lambda, \quad \mathcal{F}_\eta = -X \sin \lambda + Y \cos \lambda, \quad \mathcal{F}_\zeta = Z, \\
&M_\xi = 0, \qquad\qquad\qquad M_\eta = M \cos \varphi, \qquad\qquad M_\zeta = M \sin \varphi, \\
&C_\xi = 0, \qquad\qquad\qquad C_\eta = 0, \qquad\qquad\qquad C_\zeta = -C, \\
&s_\xi = 0, \qquad\qquad\qquad s_\eta = s_{\eta_0} \cos \varphi - s_{\zeta_0} \sin \varphi, \quad s_\zeta = s_{\zeta_0} \cos \varphi + \\
&\qquad\qquad\qquad\qquad\qquad\qquad\qquad\qquad\qquad\qquad\qquad\quad + s_{\eta_0} \sin \varphi, \\
&g_\xi = 0, \qquad\qquad\qquad g_\eta = 0, \qquad\qquad\qquad g_\zeta = g.
\end{aligned}\right\} \tag{3.3}$$

Hiermit folgt

$$\left.\begin{aligned}D^*(\tau - \varphi) + MZ \cos \varphi - M \sin \varphi (Y \cos \lambda - X \sin \eta) - \\ - MC \cos \varphi + mg(s_{\eta_0} \cos \varphi - s_{\zeta_0} \sin \varphi) = 0.\end{aligned}\right\} \tag{3.4}$$

Da der Ausschlagwinkel φ ausnahmslos klein ist, erhält man mit der Zerlegung

$$Z = Z_0 + Z_v \tag{3.5}$$

und der Beziehung

$$I = q\varphi, \tag{3.6}$$

die für eine Proportionalregelung gilt und den Zusammenhang zwischen dem Kompensationsstrom I, dem Ausschlagwinkel φ und dem dimensionsbehafteten Faktor q angibt, die Gleichung

$$\left.\begin{aligned}Z_v = -Z_0 - \frac{mg\,s_{\eta_0}}{M} - \frac{D^*\tau}{M} \\ + I\left[\mu_0 k + \frac{1}{q}\left(\frac{D^*}{M} + \mathcal{Y}\cos\lambda - \mathcal{X}\sin\lambda\right) + \frac{mg\,s_{\zeta_0}}{M}\right].\end{aligned}\right\} \quad (3.7)$$

Der Wert von q wird von der Empfindlichkeit des lichtelektrischen Schaltelementes und von den Verstärkungseigenschaften des Kompensators bestimmt. Für $I=0$ und $Z_v=0$ ist die Justierlage durch

$$Z_0 = -\frac{D^*\tau}{M} - \frac{mg\,s_{\eta_0}}{M} \quad (3.8)$$

gegeben. Der Skalenwert des Z-Kompensators folgt zu

$$\frac{dZ_v}{dI} = \mu_0 k + \frac{1}{q}\left(\frac{D^*}{M} + \mathcal{Y}\cos\lambda - \mathcal{X}\sin\lambda + \frac{mg\,s_{\zeta_0}}{M}\right). \quad (3.9)$$

Je größer q ist, desto mehr gleicht der Skalenwert dem Produkt $\mu_0 k$ und desto weniger gehen Querkomponenteneinflüsse in die Aufzeichnung ein. Einer beliebigen Vergrößerung von q ist allerdings in der Neigung zu Regelschwingungen eine prinzipielle Grenze gesetzt, auf die in den folgenden Abschnitten eingegangen wird.

γ) Zur *Temperaturabhängigkeit* eines kompensierenden Magnetometers ist folgendes zu bemerken. Da die Justierlage vom Kompensationsvorgang nicht beeinflußt wird, gehen die in Gl. (3.8) verzeichneten temperaturvariablen Größen ebenso wie beim nichtkompensierenden Variometer in die Registrierung ein und erfordern die gleichen Maßnahmen zur Beseitigung der Temperaturabhängigkeit wie diese.

Tabelle 1. *Die Kompensator-Skalenwerte für gebräuchliche Registriergrößen.* a_0 *bedeutet das Bogenmaß der Winkeleinheit, in der zu registrieren ist.*

Registriergröße	Faden vertikal	Faden bzw. Band horizontal
$\mathcal{X}$	$k + \frac{1}{q}\left(\frac{D^*}{M} + \mathcal{Y}\right)$	
$\mathcal{Y}$	$k + \frac{1}{q}\left(\frac{D^*}{M} + \mathcal{X}\right)$	$k + \frac{1}{q}\left(\frac{D^*}{M} + Z + \frac{mg\,s_{\eta_0}}{M}\right)$
$\mathcal{H}$	$k + \frac{1}{q}\left(\frac{D^*}{M} + \mathcal{Y}\cos D - \mathcal{X}\sin D\right)$	
D	$\left[k + \frac{1}{q}\left(\frac{D^*}{M} + H\right)\right]\frac{1}{a_0\mathcal{H}}$	$\left[k + \frac{1}{q}\left(\frac{D^*}{M} + Z + \frac{mg\,s_{\eta_0}}{M}\right)\right]\frac{1}{a_0\mathcal{H}}$
$\mathcal{F}$	entfällt	$k + \frac{1}{q}\left(\frac{D^*}{M} + Z\cos J - \mathcal{H}\sin J + \frac{mg}{M}[s_{\zeta_0}\sin J - s_{\eta_0}\cos J]\right)$
Z^*	entfällt	$k + \frac{1}{q}\left(\frac{D^*}{M} - \mathcal{X}\sin\lambda + \mathcal{Y}\cos\lambda + \frac{mg\,s_{\zeta_0}}{M}\right)$

* Die üblichen Lagen des Magnetsystems sind bei geodätischer Orientierung gegeben durch $\lambda = \frac{n\pi}{2}$ mit $n=1,2$ oder 3, desgleichen bei geomagnetischer Orientierung durch $\lambda = \frac{n\pi}{2} + D$, ebenfalls mit $n=1,2$ oder 3.

Demgegenüber ist der Temperatureinfluß des elektrischen Teiles gering, da die in der Klammer von Gl. (3.9) stehenden Größen nur stark reduziert eingehen und die Temperaturvariation der Spulenkonstante k naturgemäß klein ist. Bei hohen Außentemperaturschwankungen ist das gesamte System zu temperieren, für geringe Schwankungen (einige Grad) genügt die Verwendung eines im nichtkompensierenden Zustand temperierten Variometers und auf eine Einbeziehung des elektrischen Teiles kann — wie die Praxis erweist — verzichtet werden.

δ) Die Ableitung der *Skalenwerte* für die anderen üblichen Registrierfälle geschieht auf die gleiche Weise. Das Ergebnis gibt die Tabelle 1 wieder.

4. Regeltechnische Beziehungen.

α) In der Terminologie der Regelungstechnik[1] stellt das automatisch kompensierende Magnetometer ein Folgesystem dar, in dem der erdmagnetischen Variation die Rolle der *Führungsgröße* zufällt und in dem der Magnet als Regler über die *Stellgröße* φ (Lichtzeigerausschlag) die Regelgröße, das Magnetfeld der Kompensationsspule, einstellt.

Betrachtet man wiederum den Z-Kompensator als Beispiel und bezeichnet die Führungsgröße mit Z_v und die Regelgröße mit Z_k, so schreibt sich die Regelabweichung f_Z:

$$f_Z = Z_v - Z_k. \qquad (4.1)$$

Sie wird um so kleiner, je höher die „innere" Verstärkung des Kreises, Schleifenverstärkung oder kurz Regelverstärkung genannt, ist. Die Regelverstärkung, mit V_0 bezeichnet, steht mit der Regelabweichung in folgendem Zusammenhang:

$$f_Z = \frac{Z_v}{1 + V_0}. \qquad (4.2)$$

Mit Gln. (6.2), (4.8) und (3.7) folgt

$$\frac{dZ_v}{dI} = \frac{(1+V_0)\left(\dfrac{D^*}{M} + \mathcal{Y}\cos\lambda - \mathcal{X}\sin\lambda + \dfrac{mg\,s\zeta_0}{M}\right)}{q}. \qquad (4.3)$$

Gleichsetzung mit Gl. (3.9) führt zu

$$V_0 = \frac{\mu_0 k q}{\dfrac{D^*}{M} + \mathcal{Y}\cos\lambda - \mathcal{X}\sin\lambda + \dfrac{mg\,s\zeta_0}{M}} \qquad (4.4)$$

und damit zu

$$\frac{dZ_v}{dI} = \mu_0 k \left(1 + \frac{1}{V_0}\right). \qquad (4.5)$$

Diese Skalenwertbeziehung, die für alle hier erwähnten Registrierfälle zutrifft, enthält nur noch die Spulenkonstante und die den Regelkreis charakterisierende Regelverstärkung V_0. Beide Größen können experimentell einfach bestimmt werden. Sie hängen über die Beziehung

$$V_0 = \frac{\mu_0 k q}{N} \qquad (4.6)$$

voneinander ab, in der N jeweils den Inhalt der neben dem Faktor $1/q$ stehenden Klammer in Tabelle 1 bedeutet.

Gibt man eine zulässige Fehlanzeige εZ_v vor, so darf V_0 zwischen V_{0h} und V_{0n}, gegeben durch

$$\genfrac{}{}{0pt}{}{V_{0h}}{0n} = \frac{V_0 \pm \varepsilon(1+V_0)}{1 \mp \varepsilon(1+V_0)} \qquad (4.7)$$

---

[1] Siehe z.B. W. OPPELT: Kleines Handbuch technischer Regelvorgänge. Weinheim: Verlag Chemie 1956. — J. C. GILLE, M. PELEGRIN u. P. DECAULNE: Lehrgang der Regelungstechnik. München u. Berlin: Oldenbourg 1960.

schwanken. Aus dieser Beziehung geht beispielsweise hervor, daß bei einem gewünschten V_0 von 50 und einem vorgegebenen Fehler von $1^0/_{00}$ die Regelverstärkung zwischen 47,5 und 52,7 schwanken darf, bei 1% Fehler zwischen 32,8 und 103.

β) Zur Untersuchung des *dynamischen Verhaltens* hat die Regelungstechnik mehrere Verfahren entwickelt, von denen sich das *Frequenzkennlinien-Verfahren* wegen seiner leichten rechnerischen Handhabung empfiehlt. Allerdings ist hierbei die Existenz von Geräten[2] zur Prüfsignal-Erzeugung im interessierenden Frequenzbereich und zur Amplituden- und Phasenmessung Bedingung. Man gibt auf den Eingang des aufgeschnittenen Regelkreises ein Signal

$$x_e = x_{e0} \sin \omega t$$

und beobachtet das Ausgangssignal

$$x_a = x_{a0} \sin(\omega t + \psi).$$

Der Frequenzgang $\widetilde{F}(i\omega)$ vereinigt Amplituden- und Phaseninformation in der Form

$$\widetilde{F}(i\omega) = \frac{x_{a0}(\omega)}{x_{e0}} e^{i\psi(\omega)}. \tag{4.8}$$

Bezeichnet man den Frequenzgang des aufgeschnittenen Regelkreises mit $\widetilde{F}_0$ und den des geschlossenen mit $\widetilde{F}_w$, so stehen beide in folgendem Zusammenhang:

$$\widetilde{F}_w = \frac{\widetilde{F}_0}{1 + \widetilde{F}_0}. \tag{4.9}$$

Aufgrund dieser Beziehung läßt sich aus Messungen am offenen Kreis, die relativ leicht vorzunehmen sind, auf das Verhalten des geschlossenen Kreises schließen. In der Praxis hat sich eine graphische Darstellung der Gl. (2.18), das NICHOLS-Diagramm[3] bewährt, in dem in Ordinatenrichtung lg $\widetilde{F}_0$ in Dezibel und in Abszissenrichtung der Phasenwinkel ψ in linearem Maßstab in Grad aufgetragen werden. Liegt $\widetilde{F}_0$ vor, so läßt sich $\widetilde{F}_w$ aus Kurvenscharen für jede Frequenz punktweise ablesen.

Ein Regelkreis arbeitet bei kleinem V_0 sehr stabil, läßt aber ein erhebliches Durchgreifen der Betriebsparameter-Schwankungen zu; mit wachsendem V_0 wird dieser Einfluß reduziert, aber die Gefahr des Auftretens von *Regelschwingungen* wächst. Daher ist mit der Einstellung eines „optimalen" V_0 ein Kompromiß zwischen Stabilität und Unterdrückung von Störeinflüssen zu schließen. Regelungstechnische Überlegungen haben dazu geführt, das in diesem Sinne günstigste Verhalten durch den Zustand zu definieren, in dem der geschlossene Kreis einen Resonanzfaktor von 1,3 aufweist (Verhältnis von Resonanzamplitude zur Amplitude der Frequenz Null).

Das Amplituden-Phasen-Verhalten dieses ausgewählten Zustandes ist im NICHOLS-Diagramm durch eine Normkurve $\widetilde{F}_n$ gegeben. Das Ziel der Regelkreiskorrekturen ist nun darin zu sehen, den Frequenzgang des offenen Kreises durch Einsatz von Korrekturgliedern (meist RC-Gliedern) systematisch so zu verändern, daß eine Verschiebung der Kurve des korrigierten Kreises nach oben bis zur Berührung mit $\widetilde{F}_n$ einen möglichst hohen Betrag in dB ergibt. Dieser Betrag ist äquivalent der höchsten Regelverstärkung V_0, bei der das günstigste Einschwingverhalten bei genügend hoher Stabilitätsreserve gegeben ist.

---

[2] V. Auster: Erdmagnetisches Jahrbuch Niemegk 1965, S. 157—173.
[3] H. James, N. Nichols, R. Phillips: Theory of Servomechanisms. New York: McGraw-Hill 1947.

Mit dieser Methodik lassen sich — wie in Fig. 3 am Beispiel einer Z-Waage demonstriert wird — die optimalen Einstellungen von kompensierenden Magnetometern ohne Kenntnis der Differentialgleichung ermitteln. Untersuchungen dieser Art haben überdies ergeben, daß für das aus einer Kombination von Fadenwaage und lichtelektrischen Wandlern (Photowiderstands-Brücke) bestehende System die Beschreibung durch eine Differentialgleichung 2. Ordnung nicht ausreicht (siehe Fig. 3), da sich F_0 nicht asymptotisch dem Phasenwert $-180°$ nähert, sonder darüber hinausgeht.

Fig. 3. Der Frequenzgang des offenen Regelkreises einer Z-Feldwaage mit Bandaufhängung ohne Korrektur ($\tilde{F}_0$), korrigiert mit einem RC-Glied ($\tilde{F}_{01}$) und einer RC-Kombination ($\tilde{F}_{02}$). Oben der Verlauf der Normkurve $\tilde{F}_n$. Die an den Kurven verzeichneten Zahlen bedeuten nicht Frequenzen, sondern Kreisfrequenzen in s^{-1}.

Hier sei noch auf eine für das Studium schneller Schwankungen wichtige Eigenschaft des kompensierenden Magnetometers hingewiesen. Vergleicht man die Eigenfrequenz des Variometers ohne Kompensationszusatz mit der des kompensierenden Systems, so ergibt sich eine Erhöhung, die etwa bis zu einer Größenordnung getrieben werden kann, d. h. die kompensierende Waage schwingt schneller ein als die nichtkompensierende.

5. Anwendungen kompensierender Magnetometer.

α) Kompensationsmagnetometer finden überall dort *Anwendung* wo es auf folgende Vorteile ankommt, die sie gegenüber den normalen, photographisch registrierenden Variometern haben:

1. Lieferung eines variationsproportionalen Stromes,

2. Betrieb mehrerer Schreiber oder Analog-Digital (AD)-Umsetzer von einem Variometer aus.

3. Änderung der Empfindlichkeit ohne Eingriff in das Variometer.

4. Weitgehende Unterdrückung der Querkomponenteneinflüsse (wichtig bei der Aufzeichnung starker Stürme).

5. Erhöhte Eigenfrequenz gegenüber dem Variometer ohne Kompensationszusatz.

6. Hohe Konstanz des Skalenwertes.

Als Nachteile wären erhöhte Wartungsansprüche und besondere Maßnahmen zur Aufrechterhaltung des Betriebes bei Netzausfall zu nennen.

β) *Komponenten- und Winkelregistrierung.* Für den Frequenzbereich, der auswertetechnisch aus der Normalregistrierung (2 cm/h) folgt, eignen sich kompensierende Magnetometer gut[1], da Abweichungen vom idealen Amplituden- und

Fig. 4. Frequenzgang eines kompensierenden Magnetometers mit Korrekturglied. Abszisse Frequenz bzw. Periode; linke Ordinate zur Amplituden-Kurve, rechte zur Phasen-Kurve gehörig. Genutzt wird der Frequenzbereich unterhalb 10^{-2} Hz.

Phasengang mit Hilfe von Korrekturgliedern weit in den uninteressanten Bereich hinein verschoben werden können (Fig. 4). Der äußerst kleinen Winkelausschläge des Magneten wegen ist auch bei stärksten Stürmen die Aufzeichnung praktisch frei von Querkomponenteneinflüssen. Die Papier-Skalenwerte lassen sich auf rein elektrischem Wege durch Ändern der Schreiber-Stromempfindlichkeiten ohne Eingriff in das Magnetsystem verändern. In Spezialfällen kann das Auflösungsvermögen so stark erhöht werden, daß es fast dem gleichkommt, das die mit optisch gepumpten Gasen arbeitenden Magnetometer aufweisen. Das Anschalten von Grenzkontakten zur Ableitung von Schwankungsmaßen ist ebenso möglich wie der gleichzeitige Betrieb einer empfindlichen, einer unempfindlichen Registrierung und eines Analog-Digital-Umsetzers vom gleichen Variometer aus.

γ) *Pulsationsregistrierungen.* Der Vorteil der Eigenfrequenzerhöhung, den Kompensatoren gegenüber anderen Variometern haben, sichert ihnen auch auf dem Pulsationsgebiet, insbesondere im Hinblick auf magneto-tellurische Auswertungen, eine bleibende Anwendung. Man benutzt kleine, meist flüssigkeitsgedämpfte Magnete und zieht oft den über ein RC-Glied gewonnenen Differential-Anteil der Winkelbewegung unter Benutzung einer zusätzlichen Spule zur Regelung mit heran[2]. Es wurden auch Anordnungen vorgeschlagen, die aus zwei Magnetometern[3] bestehen, von denen eines die langsamen Variationen am Ort

---

[1] H. DÜRSCHNER: Ann. Géophys. 7, 199—207 (1951).
[2] B. E. BRUNELLI, O. M. RASPOPOV i B. M. JANOVSKI: Izv. Akad. Nauk. SSSR, Ser. Geofiz. Nr. 5 (1960).
[3] A. LEBEAU: Dipl.-Arb. Paris, 1956 (siehe E. THELLIER: Instruction Manual Geomagnetism AGI 1957—58, S. 7).

des anderen automatisch kompensiert, so daß auf das letztere nur die schnellen Variationen einwirken und photographisch aufgezeichnet werden können.

δ) *Registrierung fluktuierender Ortsgradienten.* Ein spezielles Anwendungsgebiet automatisch kompensierender Magnetometer ist die Aufzeichnung von Intensitätsdifferenzen zur Erfassung flüchtiger Inhomogenitäten des Variationsfeldes. Zwei oder mehr Magnetometer werden mit einem Subtraktionsnetzwerk (Fig. 5) kombiniert und gestatten unter gewissen Voraussetzungen die Registrierung von $d\mathcal{X}/dx$, $d\mathcal{X}/dy$, $d\mathcal{Y}/dx$, $d\mathcal{Y}/dy$, dZ/dx und dZ/dy als Zeitfunktionen[4]. In einer anderen Methode[5] wird an einer Hauptstation ein Strom

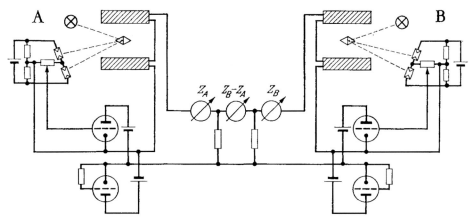

Fig. 5. Prinzipschaltung zur Aufzeichnung zeitlich variabler Magnetfelddifferenzen zwischen den Stationen A und B. In A und B befinden sich identisch aufgebaute, automatisch kompensierende Magnetometer, die jeweils der örtlichen Z-Variation proportionale Ströme liefern. Die mit einem Widerstands-Netzwerk gewonnene Stromdifferenz variiert in der gleichen Weise wie die Magnetfelddifferenz und wird mittels Schreiber aufgezeichnet. Jedes Magnetometer besteht aus dem drehbar aufgehängten Magneten (als Rhombus angedeutet), dem Lichtzeigersystem, einer Photowiderstands-Brückenschaltung, der Kompensationsspule (schraffierte Rechtecke) und dem Widerstand des Subtraktionsnetzwerkes. Der Buchstabe Z ist hier als Z_v im Sinne der Gl. (3.5) zu verstehen.
Die Stationsentfernungen liegen beispielsweise bei einer Nachweisempfindlichkeit von 0,1 γ bei 7 km.

gewonnen, der über Kabel durch Spulen der Nebenstationen fließt. Bei entsprechendem Abgleich der Spulenkonstanten wirkt auf den Magneten der Nebenstation jeweils nur das Differenzfeld zwischen Haupt- und Nebenstation und kann photographisch aufgezeichnet werden.

ε) *Regelung von Magnetfeldern.* Betreibt man ein kompensierendes Magnetometer und schickt den Kompensationsstrom durch eine Spule, deren Meßachse parallel der vom Magnetometer aufgenommenen Komponente liegt, so lassen sich bei zweckentsprechendem Abgleich die Variationen im Spulen-Innern fast völlig kompensieren[6]. Dreikomponenten-Anordnungen gestatten in weitgehender Annäherung die Realisierung eines feldfreien Raumes.

II. Andere Verfahren.

6. Das Ausschlagverfahren. Einen besonders einfachen Weg zur Direktsichtanzeige geomagnetischer Variationen bietet die Kombination eines Variometers mit Differential-Photoelement[1] und Fallbügelschreiber (Fig. 6). Hierbei wird der

---

[4] H. SCHMIDT: Z. Geophys. **23**, 273—286 (1957).
[5] J. H. NELSON: Instruction Manual. Annals of the IGY No. III, Pt. I, 37—39 (1957).
[6] A. GRABNER: Exp. Tech. Physik **11**, 347 (1963).
[1] A. GRAF: Beitr. Geophysik **7**, 357—365 (1939).

Winkel-Ausschlag des Magneten unmittelbar in eine elektrische Größe umgewandelt. Dem Differential-Photoelement gibt man den Vorzug vor anderen Wandlern, weil es eine relativ gute Linearität aufweist und ohne Stromquellen auskommt.

α) *Gleichgewichtsbetrachtungen.* Die Gleichgewichtsbeziehungen werden in der gleichen Weise wie in Ziff. 3 abgeleitet. Bezeichnet man den durch den Fallbügelschreiber fließenden Strom mit I, die lichtelektrische Empfindlichkeit des Photoelementes mit q_{el} und setzt mit φ als Ausschlagwinkel des Magneten

$$I = q_{el}\, \varphi \tag{6.1}$$

an, so ergeben sich die im Gültigkeitsbereich von Gl. (6.1) zutreffenden Skalenwerte aus den Angaben der Tab. 1, falls dort $k = 0$ und $q_{el} = q$ gesetzt werden.

Fig. 6. Im Ausschlagverfahren arbeitendes Variometer mit Direktsichtanzeige. Mit Hilfe eines Lichtzeigers variiert der drehbar aufgehängte Magnet die Lichtintensitätsverteilung eines Differential-Photo-Elementes, so daß dessen Ausgangsspannung als Maß der geomagnetischen Variation aufgezeichnet werden kann (Fallbügelschreiber). Einfache, zuverlässige Methode, der allerdings die Unterdrückung von Betriebsparameter-Schwankungen (wie etwa in Abb. 5 gegeben) fehlt.

β) *Das dynamische Verhalten.* Falls die Trägheit des Differential-Photoelementes außer Betracht bleibt, bestimmen zwei schwingende Systeme, die Drehspule des Schreibers und der am Faden hängende oder auf Schneiden gelagerte Magnet, das dynamische Verhalten. Benutzt man beispielsweise für eine $\mathcal{Y}$-Registrierung einen Magneten am vertikalen Faden, so wird dessen Verhalten beschrieben durch

$$\Theta_M \ddot{\varphi} + p_M \dot{\varphi} + \varphi(D_M^* + \mathsf{M}\mathcal{X}) = D_M^* \tau + \mathsf{M}\mathcal{Y} \tag{6.2}$$

mit Θ als Trägheitsmoment und p als Dämpfungsfaktor. Daraus folgt mit G als dynamischer Galvanometerkonstante für den Ausschlagwinkel α der Galvanometerspule[2] mit Benutzung von Gl. (6.1):

$$\Theta_G \ddot{\alpha} + p_G \dot{\alpha} + D_G^* \alpha = G\, q_{el}\, \varphi. \tag{6.3}$$

Beide Gleichungen bilden ein gekoppeltes System; sie haben deshalb einen durch 2 Resonanzstellen gekennzeichneten Frequenzgang. In der Praxis sind die Eigenfrequenzen von Magnet- und Drehspul-System etwa von gleicher Größenordnung (Sekunden). Durch Auswahl geeigneter Dämpfersysteme und geschickt an das Photoelement angepaßter Galvanometer läßt sich der Frequenzgang dieser Anordnung weitgehend nach speziellen Wünschen auslegen.

γ) *Anwendungen.* Derartige Magnetometer haben für die Komponentenregistrierung vielfach dann Anwendung gefunden, wenn Sofort-Informationen

---

[2] E. MEYER u. C. MOERDER: Spiegelgalvanometer und Lichtzeigerinstrumente. Leipzig: Akadem. Verlagsges. Geest und Portig 1952.

über den momentanen Störungsgrad des geomagnetischen Feldes gebraucht werden oder variationsabhängige Schwellwert-Überschreitungen andere Prozesse auszulösen haben, z.B. Einschalten von Schnellregistrierungen, Apparaturen der Ionosphärenforschung u. a.

Dem Vorteil geringer Wartungsansprüche stehen die Nachteile alterungsabhängiger und von der Lichtintensität der Lampe beeinflußter Skalenwerte gegenüber.

7. Apparaturen zur Schwingzeitbestimmung an Standard-Magneten.

α) Zur GAUSSschen Methode der Horizontalintensitäts-Messung gehört die Bestimmung der Schwingungszeit T_s eines Standard-Magneten mit dem Trägheitsmoment Θ_s und dem magnetischen Moment M_s; man erhält $M_s \mathcal{H}$ nach

$$M_s \mathcal{H} = \frac{4\pi^2 \Theta_s}{T_s^2}. \qquad (7.1)$$

Hieraus resultiert

$$\frac{dT_s}{T_s} = -\frac{1}{2} \cdot \frac{d\mathcal{H}}{\mathcal{H}}; \qquad (7.2)$$

verlangt man ein $d\mathcal{H}/\mathcal{H}$ von 10^{-5} und ein T_s von 10 s, so ergibt sich ein dT_s von $5 \cdot 10^{-5}$ s. Dieser Betrag gibt die *zulässige Meßunsicherheit* bei Benutzung von nur einer Schwingungsperiode an. Eine Reihe von Gründen spricht gegen die Beschränkung des Meßintervalls auf eine Periode, obwohl rein zeitmeßtechnisch die elektronischen Hilfsmittel hierzu ausreichen würden. Der Geomagnetiker steht ja vor der Aufgabe, die Größe $\mathcal{H}$ auf ein $d\mathcal{H}$ genau zu bestimmen, das klein gegen die während der Messung auftretenden Änderungen ist. Eine $\mathcal{H}$-*Registrierung* ist somit unumgänglich, um die Zeiten zu markieren, zwischen denen T_s als Mittelwert aus einer größeren Periodenzahl gewonnen wird.

Für diese Mittelbildung sprechen der reduktionsmäßig schwer zu beherrschende Einfluß von Pulsationen, möglicherweise mikroseismisch verursachte Pendelschwingungen und die durch den Anregungsvorgang bedingten Koppelschwingungen zwischen dem gewünschten Schwingungsvorgang, der $M_s \mathcal{H}$ liefern soll und dem unerwünschten, den der Magnet als Schwerependel am Faden ausführt. Zur Ausschaltung der während der Messung eintretenden Deklinations-Änderungen hat man Links- und Rechts-Durchgänge des Magneten getrennt zu messen und zu mitteln. Methodische Untersuchungen[1] führten zur Berechnung minimaler Meßzeit-Intervalle.

β) *Technik der Schwingzeit-Meßanlagen.* Zur Fixierung der Durchgangszeitmomente des Magneten empfiehlt sich ausschließlich eine lichtelektrische Winkelabnahme, da kein anderes Verfahren eine derart geringe, praktisch unmerkliche Rückwirkung auf den Schwingungsvorgang hat. Der photoelektrische „berührungsfreie Kontakt" fand 1929 Eingang in die magnetische Meßtechnik. Es wurden photoelektrische Verstärker eingesetzt, mit denen über Relais Marken auf Streifenchronographen gewonnen wurden[2]. Eine Auszählung dieser Marken und ein Vergleich mit Normalfrequenzmarken ergab dann die Schwingungszeit T_s. Zur optischen Speicherung von Durchgangszeitmomenten fanden photographische Aufnahmen umlaufender Skalen oder Marken Anwendung. Zuerst wurde der Reflex kurzzeitig aufglühender, vorgeheizter Lampen, von einem umlaufenden Zeiger mit Spiegel wiedergegeben, photographiert[3], später bewährte sich der Einsatz der Blitztechnik beim Erfassen von Zeigerstellungen. Als Zeit-

---
[1] H. SCHMIDT: Abhandl. Geomagnet. Inst. Potsdam-Niemegk **19**, 44—76 (1956).
[2] H. E. McCOMB and C. HUFF: Terr. Magn. **34**, 123—141 (1929).
[3] G. FANSELAU: Z. Geophys. **9**, 93—98 (1933).

skalen dienten auch konzentrische, mit verschiedenen Geschwindigkeiten umlaufende, bezifferte Scheiben oder Trommeln[4]. Neuerdings finden Impulszähler mit Springwagendruckern Verwendung[5] (Fig. 7).

γ) *Praxis der Schwingzeit-Messungen.* Aus den oben geschilderten Gründen ist T_s als Mittelwert über ein Intervall von n Durchgängen zu ermitteln. Automatisch arbeitende Anlagen speichern z.B. für n = 50 die Momente des 1., 51., 101., 151. und 201. Durchganges von links, die des 4., 54., 104., 154. und 204. von rechts und liefern so bei automatischer Markierung des jeweiligen Intervallbeginnes in der Vergleichsregistrierung die Möglichkeit, 4 gleichberechtigte Wertesätze zu ermitteln. Die hieraus berechneten T_s-Werte müssen im Vergleich mit der $\mathcal{H}$-Registrierung der Gl. (7.2) im Rahmen der Meßunsicherheiten genügen. Da bereits ein Intervall zur T_s-Bestimmung ausreicht, ergibt sich durch die Benutzung von 4 Intervallen eine mehrfache Kontrolle, die Rechen- und Ablesefehler mit Sicherheit ausschließt.

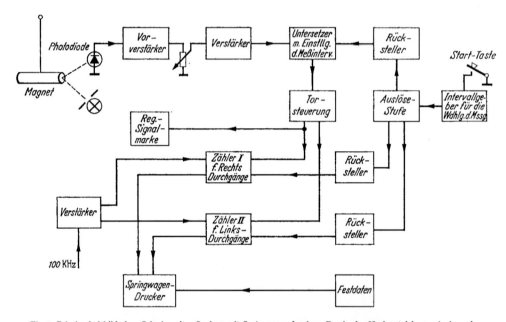

Fig. 7. Prinzipschaltbild einer Schwingzeitmeßanlage mit Springwagendrucker. Der in der Horizontalebene schwingende Magnet verursacht beim Passieren einer bestimmten Winkellage über Lichtzeiger und Photodiode einen Impuls, der nach Untersetzung (z.B. 1 : 50) die Torsteuerung von Impulszählern bedienen. Die von beiden Zählern aufsummierten 100 kHz-Impulse geben ein Maß für die Zeitdauer von 50 Rechts-, bzw. Linksdurchgängen des Magneten an und werden vom Springwagendrucker dokumentiert (mit Datum, Uhrzeit usw.). Zu Beginn der Messung erfolgt automatisch die Auslösung einer Signalmarke auf der Normalregistrierung.

Die vorstehende Methode hat sich für photographische Fixierung der Durchgangszeiten gut bewährt. Wollte man sie auf impulszählende Systeme anwenden, so würde dies einen beträchtlichen Aufwand an Zwischenspeichern erfordern. Man vereinfacht das Verfahren dadurch, daß man nicht ineinandergeschachtelte Intervalle bestimmt, sondern jeweils nur eines mit je einem Zähler für die Links- und einem für die Rechtsdurchgänge. Am Intervallende erfolgt der Druck beider Ergebnisse und mit dem nächsten Magnetdurchgang beginnt die neue Messung.

Die Prüfung der T_s-Werte an den absoluten $\mathcal{H}$-Werten über längere Zeiten ergibt, daß die elektronischen Stoppuhren zwar wesentlich geringere Streuungen innerhalb einzelner Meßsätze liefern als die früher übliche ,,Aug- und Ohr-Methode", daß sich aber die aus beiden Methoden berechneten Jahresmittel von M_s nicht merklich unterscheiden.

---

[4] J. CRUSET: Ann. franc. Chronométrie **24**, 105—115 (1954).
[5] R. KULP u. H. SCHMIDT: Neuerungen auf dem Gebiete der Schwingzeit-Messungen (im Druck).

B. Komponentenmessung mit Hilfe von Induktivitäten.

I. Induktionsspulen als Meßelement.

8. Grundlagen der Meßmethode.

α) Aus dem Induktionsgesetz ergibt sich nach einer einfachen Umformung ein Ausdruck für die *Spannung*, die in einer im erdmagnetischen Feld befindlichen Spule *induziert* wird. Wir schreiben sie als komplexe Wechselspannung der Kreisfrequenz Ω:

$$\mathfrak{U} = \frac{1}{c_0 \sqrt{\varepsilon_0 \mu_0}} n S \mu_{\text{eff}} \frac{d}{dt} \left(\cos \Omega t [\mathcal{F}_0^s + \mathcal{F}_v^s(t)] \right) \tag{8.1}$$

(n = Windungszahl der Spule, S = wirksame Windungsfläche, μ_{eff} effektive Permeabilität des Innenraumes der Spule).

Dabei wird die Komponente $\mathcal{F}^s$ des erdmagnetischen Feldes, die in Richtung der Spulenachse liegt, wie üblich aus einem konstanten Grundwert $\mathcal{F}_0^s$ und einem zeitabhängigen Variationsanteil $\mathcal{F}_v^s$ zusammengesetzt. Ferner ist in Gl. (8.1) noch eine Rotation der Spule um eine zur Spulenachse senkrechte Richtung berücksichtigt, die eine periodische Änderung der Windungsfläche bewirkt. Für die erdmagnetische Meßtechnik ergeben sich aus Gl. (8.1) zwei verschiedene Anwendungsmöglichkeiten:

a) $\Omega \neq 0$. Fall der *Rotationsspule*. Nach (8.1) wären z.B. Momentanwerte der Feldkomponente zu bestimmen, besondere Vorteile bieten Kompensationsverfahren, bei denen die Rotationsspule als Nullindikator wirkt.

b) $\Omega = 0$. Fall der *ruhenden Spule* zur Registrierung der zeitlichen Ableitung der Variation des geomagnetischen Feldes in einer vorgegebenen Richtung. Gl. (8.1) vereinfacht sich dann zu

$$U = \frac{1}{c_0 \sqrt{\varepsilon_0 \mu_0}} n S \mu_{\text{eff}} \frac{d \mathcal{F}_v^s}{dt}. \tag{8.2}$$

Über Meßgeräte, die Rotationsspulen benutzen, wird im folgenden unter β) nur kurz referiert. Geräte mit ruhenden Spulen, die in zunehmendem Maße für das Studium kurzperiodischer Variationen an Bedeutung gewinnen, werden weiter unten in Ziff. 9 ausführlicher behandelt.

β) *Bewegte Spulen*. Die rotierende Spule in Verbindung mit einem mechanischen Gleichrichter (Kollektor) und einem empfindlichen Galvanometer oder mit einem zwischengeschalteten Wechselspannungsverstärker und anschließender phasenempfindlicher Gleichrichtung stellt einen leistungsfähigen Nullindikator für schwache Magnetfelder dar. Verbreitete Anwendung fand dieses Prinzip zur Bestimmung der erdmagnetischen Inklination durch den sog. Erdinduktor[1]. Benutzt man die Rotationsspule als Nullindikator zur (vorzugsweise automatischen) Einstellung von Kompensationsfeldern, so lassen sich prinzipiell beliebig gelegene Komponenten des erdmagnetischen Feldes messen. Derartige Geräte wurden sowohl für den Einsatz zu Vermessungsarbeiten in der angewandten Geophysik wie auch für Forschungszwecke in geomagnetischen Observatorien entwickelt.

Da die Rotationsspule in jeder räumlichen Lage arbeitet, können mit ihr ähnliche Komponentenmeßgeräte wie mit Saturationskernsonden aufgebaut werden (s. Ziff. 14 und 15). Hier wäre z.B. auf Geräte hinzuweisen, die eine

---
[1] Im vorstehenden Beitrag dieses Bandes behandelt von V. LAURSEN und J. OLSEN, S. 300.

Messung der Vertikalkomponente Z in einem Schleppkörper auf See[2] ermöglichen und bei denen die Horizontierung durch eine kardanische Aufhängung von Nullindikator- und Kompensationssystem gegeben ist. Ebenso wurden Geräte entwickelt, die die Totalintensität des erdmagnetischen Feldes anzeigen oder registrieren. Eine von FROWE[3] beschriebene Anordnung benutzt ein Dreibein von oszillierenden Induktionsspulen, die in ähnlicher Weise wie rotierende Spulen eine zur Steuerung von Nachführsystemen geeignete, induzierte Spannung abgeben. Mittels zweier Spulen und entsprechenden Servomechanismen wird die zum erdmagnetischen Vektor senkrecht gelegene Ebene ständig eingehalten und damit das eigentliche Meßsystem parallel zur Erdfeldrichtung orientiert. Je nach dem Verwendungszweck und der möglichen mechanischen Präzision ergeben sich für solche Geräte untere Nachweisgrenzen zwischen 5 und 100 γ.

Außerordentlich kritische Forderungen in mechanischer und elektrischer Hinsicht gelten für Geräte, die zur Absolutmessung der drei Komponenten des erdmagnetischen Feldes entwickelt wurden, wie den elektrodynamischen Theodoliten. Über diese Probleme sind umfangreiche theoretische und experimentelle Arbeiten[4] geschrieben worden. Es scheint jedoch aus heutiger Sicht so zu sein, daß man für die Komponenten-Absolut-Messungen den mit Protonenresonanzen arbeitenden Geräten wegen der digitalen Werteausgabe und dem Verzicht auf bewegte Teile den Vorzug gegenüber den elektrodynamischen Methoden geben wird.

9. Ruhende Spulen.

α) Um eine *Übersicht* über die zu erwartenden Effekte und deren Größenordnung zu gewinnen, wird zunächst eine stark vereinfachte Schaltung (Fig. 8a) berechnet. Das Magnetfeld sei vor Einbringen der Induktionsspule homogen. Die zeitliche Änderung werde als sinusförmig mit der Kreisfrequenz ω und der Amplitude $F_{v_0}^s$ vorausgesetzt. Dabei soll ω so gewählt werden, daß die Ungleichung $\omega L \ll R_i$ gilt (L = Spuleninduktivität, R_i = Innenwiderstand des Meßinstrumentes). Spulen -und Schaltkapazitäten werden vernachlässigt. Die dann für die

Fig. 8a u. b. a) Vereinfachtes Ersatzschaltbild für eine Induktionsspule mit Galvanometer. b) Ersatzschaltbild für eine Induktionsspule mit hochohmigem Verstärker.

Schaltung resultierende Differentialgleichung hat, mit R als Spulenwiderstand die Lösung

$$\mathfrak{U} = \frac{1}{c_0 \sqrt{\varepsilon_0 \mu_0}} n S \mu_{\text{eff}} \omega \mathcal{F}_{v_0}^s \left(1 - e^{-\frac{R+R_1}{L}t}\right) \cos \omega t = U_0 \left(1 - e^{-\frac{R+R_1}{L}t}\right) \cos \omega t. \quad (9.1)$$

Der Amplitudenfaktor wird erheblich durch die Spulenparameter beeinflußt, woraus die eigentliche Problematik dieser Apparaturen, nämlich die Beherrschung des dynamischen Verhaltens in Abhängigkeit von den Spuleneigenschaften, zu erkennen ist.

---
[2] F. ERRULAT u. H. PODSZUS: Ann. Hydrogr. u. marit. Meteorologie **71**, 127—130 (1943).
[3] E. A. FROWE: Geophysics **13**, 209—214 (1948).
[4] Zum Beispiel E. A. JOHNSON: Terr. Magn. and Atm. Electr. **44**, 81 (1939). — J. TSUBOKAWA: J. Geom. and Geoelectr. Kyoto **1**, 10 (1949). — O. LUCKE: Abhandl. Geomagnet. Inst. Potsdam Nr. 16 (1956).

Gl. (9.1) soll zunächst für die Abschätzung der notwendigen Spulendimensionen benutzt werden. Welche Größe man für die die Erdfeldvariationen charakterisierenden Parameter einzusetzen hat, entnimmt man Fig. 9. Hier sind in einem Perioden-Amplituden-Diagramm häufig auftretende Variationstypen dargestellt. Dabei wurden Variationen der Horizontalkomponente in mittleren geomagnetischen Breiten betrachtet. Es ergibt sich überraschend, daß man für ein breites Periodenspektrum einen konstanten Wert für $\omega \mathcal{F}_{v_0}^s$ annehmen kann.

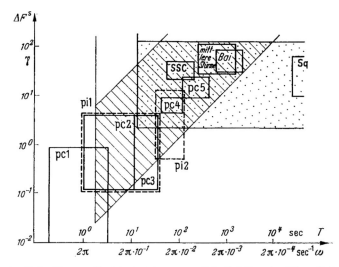

Fig. 9. Perioden-Amplituden-Diagramm für einige typische geomagnetische Variationen in mittleren Breiten. (Skala unter der Abszisse: Kreisfrequenz). Punktiert: Registrierbereich für normales Variometer (Magnet am Faden), schraffiert: Registrierbereich für Induktionsspule mit Galvanometer. Die Bezeichnungen der Variationstypen sind im Beitrag von N. FUKUSHIMA und T. NAGATA angegeben, wo auch ihre Eigenschaften beschrieben werden.

Im vorliegenden Fall gilt etwa

$$\omega \mathcal{F}_{v_0}^s = 2\pi \cdot 10^{-7} \, \Gamma \, s^{-1}.$$

Gibt man ferner als induzierte Spannung für die kleinste, noch nachweisbare Feldänderung $U_{\min} = 10^{-6}$ V vor, so folgt für die Größe $\mu_{\text{eff}} \, n \, S$ ein Wert von $1{,}6 \cdot 10^8$ cm^2. Zwei verschiedene Wege wurden zur Realisierung einer effektiven Windungsfläche dieser Größe beschritten. Eine Reihe von Autoren[1] benutzte große kernlose Flachspulen, während andere[2] lange Spulen geringen Durchmessers mit hochpermeablen Kernen bevorzugten. Zweifellos sind hier Fragen der Anwendung entscheidend, wie weiter unten noch ausführlich behandelt wird. Es wären z.B. zwei spezielle Anordnungen mit folgenden Werten möglich:

a) Kernlose Spule: $\mu_{\text{eff}} = 1$; $S = 3{,}2 \cdot 10^5$ cm^2; $n = 5 \cdot 10^2$.

b) Kernspule: $\mu_{\text{eff}} = 10^3$; $S = 1$ cm^2 ; $n = 1{,}6 \cdot 10^5$.

Diese Werte lassen sich durchaus realisieren.

Eine weitere Klassifizierung der Apparaturen ergibt sich, wenn man die für den Nachweis der Induktionsspannung benutzten Meßgeräte charakterisiert. Man verwendet:

a) elektronische Verstärker mit Schleifenoszillographen oder Direktschreibern als Registriergeräte,

b) Spiegelgalvanometer mit photographischer oder photoelektrischer Registrierung.

---

[1] Zum Beispiel H. WIESE: Erdmagnetisches Jahrbuch Niemegk, S. 1—8, 1953.
[2] Zum Beispiel G. ANGENHEISTER: Gerlands Beitr. Geophys. 64, 108—132 (1955).

Diese Einteilung, die offensichtlich auch einschränkende Bedingungen für die Grenzfrequenzen der Geräte stellt, liegt den folgenden Betrachtungen zugrunde.

β) *Induktionsspule mit Verstärker.* Es werde angenommen, der Verstärker habe bezüglich der Induktionsspule einen hochohmigen Eingang. Interessiert man sich jetzt auch für höhere Frequenzen, so ist im Ersatzschaltbild (Fig. 8b) die Vernachlässigung von C_p, gebildet aus Spuleneigenkapazität und Zuleitungskapazität, nicht mehr erlaubt. Am Verstärkereingang ergibt sich dann für Amplitude und Phase der Induktionsspannung

$$\mathfrak{U} = \frac{U_0}{\sqrt{(1-\omega_n^2)^2 + (2p\omega_n)^2}} \cos(\omega t + \psi); \quad \tan \psi = -\frac{2p\omega_n}{1-\omega_n} \qquad (9.2)$$

wobei die Abkürzungen

$$\omega_0 = \frac{1}{\sqrt{LC}}; \quad 2p = \frac{R}{L\omega_0}; \quad \omega_n = \frac{\omega}{\omega_0} \qquad (9.3)$$

gelten. Die für den stationären Zustand des Systems gültige Gl. (9.2) enthält die bekannten Frequenzgänge der Übertragungsfunktion eines Schwingungskreises (oder, allgemein eines Systems, das durch eine Differentialgleichung 2. Ordnung beschrieben wird) und den Amplitudenausdruck U_0 aus Gl. (9.1). Durch Wahl von ω_0 und $2p$ können einer solchen Schaltung ausgesprochen selektive Eigenschaften gegeben werden. Bei der Untersuchung von kurzperiodischen Variationen haben sich selektive Anordnungen[3] besonders bewährt, weil man damit bereits vor dem Verstärker eine wirkungsvolle Unterdrückung von höherfrequenten Störeinflüssen erreicht. Dazu gehören vor allem Erdströme aus den Wechselstromnetzen, in Europa auf 50 Hz. Allerdings ist die mit einer hohen Selektivität verbundene lange Einschwingdauer des Systems nur dann in Kauf zu nehmen, wenn die gesamte Dauer des periodischen Ereignisses diese Zeit weit übersteigt. Diese Bedingung ist im allgemeinen bei kurzperiodischen Variationen gerätetechnisch zu erfüllen.

Eine weitere Möglichkeit, die Frequenzgänge der gesamten Anordnung zu beeinflussen, ergibt sich durch eine entsprechende Verstärkerauswahl. Im Gebiet von Frequenzen über 1 Hz lassen sich bereits RC-gekoppelte Verstärker anwenden. Die Größe der erzielbaren Verstärkung wird hier durch die Kurzzeitkonstanz der Betriebsspannungen und das Funkelrauschen der Anfangsstufen bestimmt. Der erste Störeinfluß läßt sich durch Stabilisierungen und Anwendung symmetrischer Eingangsschaltungen stark vermindern. Bezüglich des Funkelrauschens ist man auf die Auswahl rauscharmer Verstärkerelemente und die Einengung der Bandbreite angewiesen. Stabilisierungsmaßnahmen (z.B. Gegenkopplung) führen zu Verstärkungsgraden, die um weniger als 1% schwanken.

Größere Schwierigkeiten kann der Einsatz elektronischer Verstärker im Frequenzbereich unterhalb von 1 Hz bereiten. Man wird hier jedenfalls solchen Typen den Vorzug geben, die als Gleichspannungsverstärker mit einem mechanischen oder besser mit einem elektronischen Zerhacker arbeiten. Trotzdem ist mit einer zusätzlichen störenden Drift (im Gegensatz zu RC-gekoppelten Verstärkern) zu rechnen. Maßnahmen gegen diese Effekte sind aus den Gebieten der Technik bekannt, die Gleichspannungsverstärker mit ähnlich hohen Anforderungen an die Empfindlichkeit und Stabilität verwenden, z.B. der medizinischen Elektronik[4].

Bei den in diesem Abschnitt beschriebenen Geräten ergibt sich kein prinzipieller Unterschied, wenn man eine kernlose Induktionsspule oder eine mit hochpermeablem Kern benutzt. Das ist darauf zurückzuführen, daß es sich hier um eine rückwirkungsfreie Messung der Leerlaufspannung an der Induktionsspule bzw. an dem sie enthaltenden Schwingungskreis handelt.

γ) *Induktionsspule mit Galvanometer.* Das Ersatzschaltbild für dieses System (Fig. 10) berücksichtigt die Rückwirkung zwischen Induktionsspule und Meß-

---
[3] B. S. ENENŠTEJN i L. E. ARONOV: Isv. Akad. Nauk SSSR, Ser. Geofiz. 62—70 (1957).
[4] Zum Beispiel: Handbuch medizinischer Elektronik. Berlin: VEB Verlag Technik 1965.

system. Die Anwendung der KIRCHHOFFschen Gesetze und der Drehmomentgleichung für das Galvanometer ergeben die Differentialgleichung

$$U = \varphi \sum_k f_k^{(0)} + \dot{\varphi} \sum_k f_k^{(1)} + \ddot{\varphi} \sum_k f_k^{(2)} + \dddot{\varphi} \sum_k f_k^{(3)} + \ddddot{\varphi} \sum_k f_k^{(4)} \qquad (9.4)$$

mit dem Koeffizientenschema nach Tabelle 2. Dabei wurden die folgenden Bezeichnungen verwendet:

a) Für das Galvanometer: D^* = Direktionsmoment, G = dynamische Galvanometerkonstante, Θ = Trägheitsmoment, δ = Luftdämpfung des Systems, R_i = Innenwiderstand.

b) Für den Stromkreis: C = Parallelkapazität, R_p = Parallelwiderstand, R = Spulenwiderstand.

Fig. 10. Vollständiges Ersatzschaltbild für Induktionsspule mit Galvanometer.

Tabelle 2.

Ableitung	$f_1^{(i)}$	$f_2^{(i)}$	$f_3^{(i)}$	$f_4^{(i)}$
0.	$\dfrac{D^* R_i}{G}$	$+\left(1+\dfrac{R_i}{R_p}\right)\dfrac{D^* R}{G}$		
1.	$G+\dfrac{R_i \delta}{G}$	$+\left(1+\dfrac{R_i}{R_p}\right)\dfrac{D^* L}{G}+\left(1+\dfrac{R_i}{R_p}\right)\dfrac{R\delta}{G}$	$+\dfrac{GR}{R_p}$	$+\dfrac{C_p D^* R R_i}{G}$
2.	$\dfrac{R_i \Theta}{G}$	$+\left(1+\dfrac{R_i}{R_p}\right)\dfrac{D^*\delta}{G}+\left(1+\dfrac{R_i}{R_p}\right)\dfrac{R\Theta}{R_p}$	$+\dfrac{GL}{R_p}$	$+\dfrac{C_p D^* L R_i}{G}+\dfrac{C_p R R_i \delta}{G}-C_p G R$
3.		$\left(1+\dfrac{R_i}{R_p}\right)\dfrac{L\Theta}{G}$		$+\dfrac{C_p L R_i \delta}{G}+\dfrac{C_p R R_i \Theta}{G}-C_p G L$
4.				$+\dfrac{C_p L R_i \Theta}{G}$

Zunächst wird der Spezialfall $R_p \to \infty$ und $C=0$ betrachtet. Praktisch liegt dieser vor, wenn man nur mit den Kapazitäten und den Isolationswiderständen von Induktionsspule und Zuleitung zu rechnen hat. Für den Frequenzgang der Übertragungsfunktion ergibt sich in diesem Fall

$$a(\omega_n) = \dfrac{1}{\sqrt{(1-\omega_n^2)^2 + ([2p+A]\omega_n - A\omega_n^3)^2}}; \quad \tan\psi = -\dfrac{(2p+A)\omega_n - A\omega_n^3}{1-\omega_n^2} \qquad (9.5)$$

wobei folgende Beziehungen gelten:

$$\omega_n = \dfrac{\omega}{\omega_0}, \quad \omega_0^2 = \dfrac{D^*}{\Theta}, \quad 2p = \dfrac{\omega_0}{D^*}\left(\delta + \dfrac{G^2}{R}\right), \quad R_0 = R + R_i, \\ A = \dfrac{\omega_0 L}{R_0}, \quad A\dfrac{\delta\omega_0}{D^*} \ll 1. \qquad (9.6)$$

Neben dem Dämpfungsglied $2p$ tritt hier eine neue (dimensionslose) Größe A auf, die von Spulen- und Galvanometerdaten bestimmt wird. Gelingt es, A durch entsprechende Dimensionierung sehr klein zu machen, so geht Gl. (9.5) in die Lösung einer Differentialgleichung 2. Ordnung über, die aus der Theorie der Galvanometer bekannt ist.

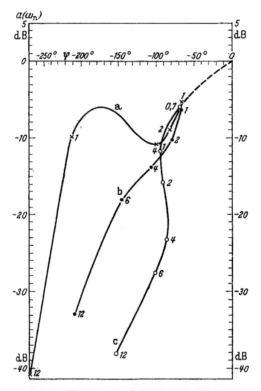

Fig. 1. Experimentell bestimmtes NICHOLS-Diagramm für Induktionsspule mit Galvanometer und lichtelektrischem Wandler. a Ohne Korrektur, b mit Parallelwiderstand, c mit Parallelwiderstand und Korrekturnetzwerk (Integration und Phasenvorhalt). Zahlenangaben wie in Fig. 3.

Verwendet man in diesem Zusammenhang wieder das Beispiel aus Ziff. 9 für zwei verschiedene Spulen gleicher Windungsfläche, so erhält man (mit $R = 200\,\Omega$):

Kernlose Spule: $A = 5 \cdot 10^{-2}$,
Kernspule: $A = 2{,}5 \cdot 10$.

Die Spulenabmessungen für die Induktivitätsberechnung wurden an praktisch ausgeführten Anordnungen orientiert. Es ergibt sich ein entscheidender Unterschied zwischen beiden Spulenarten.

Mit einer *kernlosen Spule* lassen sich Bedingungen realisieren, die der Anpassung von Spannungsquellen mit rein OHMschen Innenwiderstand an Galvanometer gleichkommen. Für eine möglichst phasen- und amplitudentreue Registrierung des zeitlichen Erdfeldgradienten über einen möglichst großen Frequenzbereich wird man daher $p = 0{,}7$ anstreben.

Völlig anders verhalten sich dagegen im allgemeinen die *Kernspulen*. Eine günstige Beeinflussung des Frequenzganges läßt sich hier z. B. durch einen zusätzlichen Parallelwiderstand R_p erreichen.

Auf eine genauere Diskussion dieser Möglichkeit an Hand von Gl. (9.4) sei in diesem Zusammenhang jedoch verzichtet. Günstigere Verhältnisse erzielt man, wenn man über einen photoelektrischen Verstärker eine Rückführung in das System Induktionsspule-Galvanometer einbaut. In diesem Fall kann man zunächst durch eine entsprechende Wahl der Spulenbewicklung für Leistungsanpassung an das verwendete Galvanometer sorgen. Ohne Rücksicht auf den Frequenzgang nehmen zu müssen, erhält man so die maximale Empfindlichkeit der Schaltung. Durch Korrekturnetzwerke im Verstärker oder auch im Rückführungskreis gelingt es dann weiter, die zu einer Linearisierung des Frequenzganges notwendige hohe Regelverstärkung zu realisieren.

Die Wirkung der Korrekturnetzwerke wird in Fig. 11 an einem praktischen Beispiel deutlich. Mit Hilfe des aus der Regeltechnik bekannten NICHOLS-Diagrammes läßt sich die optimale Regelverstärkung besonders einfach graphisch bestimmen (s. auch Ziff. 4).

10. Experimentelle Technik und Anwendungen.

α) Der *praktische Aufbau* der hier beschriebenen Apparaturen kann, soweit er die Verstärker und Registriereinrichtungen betrifft, mit kommerziellen Geräten erfolgen. Die *Induktionsspulen* werden jedoch ausschließlich im Selbstbau hergestellt. Ihre Dimensionierung richtet sich dabei nach dem Anwendungszweck und den Aufstellungsmöglichkeiten. Während bei kernlosen Spulen das Auslegen von horizontalen Schleifen zur Messung der zeitlichen Ableitung der erdmagnetischen Vertikalintensität keine Schwierigkeiten bereitet, ist dies bei Spulen für die horizontalen Komponenten problematisch. Sie werden gewöhnlich an den Wänden von Häusern oder an speziell zu diesem Zweck errichteten Mauern angebracht.

Bei Kernspulen gilt es meist, einen Kompromiß zwischen einer handlichen Abmessung (besonders bei transportablen Geräten) und dem starken Anwachsen des Entmagnetisierungsfaktors bei Verkleinerung des Verhältnisses Länge zu Radius für den Induktionsstab zu finden. Man verwendet im allgemeinen Spulenlängen zwischen 1 und 2 m bei einem Kernquerschnitt in der Größenordnung Quadratzentimeter. Die (relative) Anfangspermeabilität des Materials liegt gewöhnlich zwischen 20000 und 50000. Höhere Werte erweisen sich im Gebrauch besonders im Gelände als zu empfindlich und kommen auch bei den angegebenen Dimensionsverhältnissen noch nicht genügend zur Geltung.

β) Die *Anwendung* von Geräten mit Induktionsspulen zur Aufzeichnung der zeitlichen Ableitung der Komponenten der erdmagnetischen Variationen liegt besonders im Bereich der Pulsationen. Es wurde bereits oben (Fig. 9) darauf hingewiesen, daß wegen der Periodenabhängigkeit der Empfindlichkeit der Apparatur schon ein Registriersystem mit geringer Registrierbreite (z.B. 1:200) genügt, um eine Übersicht über die Variationen innerhalb eines großen Frequenzbereiches zu erhalten. Diese Tatsache wird besonders deutlich, wenn man den analog zu einem normalen Variometer definierten Skalenwert und seine Frequenzabhängigkeit berechnet (Tabelle 3).

Tabelle 3

Variationstyp	Mittlere Periode [s]	Skalenwert [γ/mm]
Bai	$1,8 \cdot 10^3$	4
ssc	$1,5 \cdot 10^2$	$3,4 \cdot 10^{-1}$
pc 1-Pulsation	3	$7 \cdot 10^{-3}$
pc 2-Pulsation	7	$1,6 \cdot 10^{-2}$
pc 3-Pulsation	27	$6 \cdot 10^{-2}$
pc 4-Pulsation	100	$2,3 \cdot 10^{-1}$
pc 5-Pulsation	390	$8,8 \cdot 10^{-1}$
pi 1-Pulsation	20	$4,5 \cdot 10^{-2}$
pi 2-Pulsation	95	$2,2 \cdot 10^{-1}$

Ausgegangen wurde dabei von einer Empfindlichkeit von 70 mm/γ/s. Unter Berücksichtigung der unteren Grenzfrequenz der Gesamtapparatur und der

Grenzen der Maximalaussteuerung ergibt sich der in Fig. 9 schraffierte Registrierbereich. Ein geübter Beobachter kann aus einer solchen Aufzeichnung einen guten Überblick über die geomagnetischen Variationen gewinnen. Es ist jedoch zu beachten, daß bei nichtmonochromatischen Variationen ein relatives Anheben der Oberwellen bewirkt wird und so Verzerrungen entstehen. Zum Vergleich ist in Fig. 9 auch der Registrierbereich eines normalen Variometers angegeben.

In der erdmagnetischen Meßpraxis dienen Geräte mit Induktionsspulen zwei Forschungsrichtungen. Einmal handelt es sich um die bereits erwähnten *Pulsationen*, die an festen Stationen mit kernlosen oder kernhaltigen Spulen erfaßt werden, zum anderen um die *Magneto-Tellurik*, in der Wanderstationen mit Kernspulen Anwendung finden. Beiden Untersuchungskomplexen kommen die Vorteile dieser Methode zugute, daß neben dem Registrierschrieb als analogem Datenträger gleichzeitig maschinenlesbare Datenträger, analog (A) oder digital (D), durch Einsatz von Magnetbandeinheiten, AD-Umsetzern mit Lochstreifenausgabe usw. gewonnen werden können. Für die magnetotellurische Anwendung im Gelände erweist sich die einfache räumliche Justierung der Spulen als sehr günstig und gestattet die Beobachtung der zeitlichen Änderungen des geomagnetischen Feldes in jeder beliebigen Richtung.

Bei der Auswahl magnetotellurischer Meßpunkte und fester Pulsationsstationen muß allerdings berücksichtigt werden, daß nach Gl. (8.1) jede auch geringfügige Erschütterung der Spule vermieden wird. Die sonst mögliche zeitliche Änderung der Kreisfrequenz Ω ruft dann Störungen in der Registrierung hervor. Als Ursachen für die hier in Betracht kommende Bodenunruhe sind neben der Mikroseismik (Meeresbrandung) insbesondere windbewegte Bäume und der Straßenverkehr zu nennen. Oft kann der Einfluß der beiden letztgenannten Störquellen schon durch Eingraben der Spulen stark reduziert werden.

Für besondere Anwendungen läßt sich die Frequenzabhängigkeit der Anordnung in bestimmten Grenzen kompensieren. Beispiele hierfür bieten der Einsatz von Fluxmetern[1] oder die Nutzung eines magnetischen Rückführungszweiges, der direkt auf die Induktionsspule wirkt.

Auf Kombinationen zwischen beweglich aufgehängtem Stabmagnet und ruhender Induktionsspule (GRENETsche Variometer[2]) sei hingewiesen. Auch auf dem Gebiete der Magnetfeldmessung von künstlichen Erdsatelliten aus hat sich die Induktionsspule bewährt. In langsam rotierenden Satelliten benutzt man fest eingebaute Spulen mit hochpermeablem Kern und kann bei konstanter Relativbewegung gegenüber dem Magnetfeld auf dessen Wert schließen[3] oder bei bekanntem Magnetfeld die Lage des Satelliten im Raum ermitteln.

II. Mit magnetischen Wechselfeldern ausgesteuerte hochpermeable Kerne als Meßelemente.

11. Sondenverfahren.

α) Die Fortschritte in der Niederfrequenzmeßtechnik und Regeltechnik sowie die Erfolge in der Herstellung hochpermeabler Legierungen mit stabilen magnetischen Daten beeinflußten die Entwicklung der geomagnetischen Meßtechnik während der letzten 30 Jahre in entscheidendem Maße. Während vorher die Magnetfelderfassung im wesentlichen nur mit beweglichen Gebern (Stabmagnet oder Rotationsspule) möglich war, konnten nunmehr *ruhende Geber* realisiert werden.

---

[1] A. G. KALAŠNIKOV: Fljuksmetr (Fluxmeter). Moskva: Isdat. Akad. Nauk. 1953. — M. E. SELZER: IGY-Instruction manual. Geomagnetism 1—15 (1957/58).
[2] G. GRENET: Ann. Géophys. **4**, 188 (1949).
[3] C. P. SONETT, D. L. JUDGE, A. R. SIMS, and J. M. KELSO: J. Geophys. Res. **65**, 55—68 (1960).

Besondere Vorteile boten sich dann, als es gelang, diese Geber mit Servosystemen zu kombinieren und damit Magnetfeldmessungen auf beweglichen Instrumententrägern wie Schiffen, Flugzeugen und Raketen vorzunehmen[1]. Solche Systeme ermöglichten nach 1945 in vielen Ländern großräumige aeromagnetische Vermessungen und dominierten etwa 20 Jahre lang als Totalintensitäts-Meßgeräte, bis sie von den frequenzabgebenden Magnetometern abgelöst wurden. Für die Komponentenmessung jedoch von Flugzeugen und Schiffen aus, sowie für spezielle Aufgaben der Raketenmagnetik sind sie auch heutzutage ebenso unentbehrlich wie einfachere Geräte dieser Art in der magnetischen Werkstoff-Forschung.

Die Hystereseschleifen hochpermeabler Legierungen wie M 1040, Permalloy, zahlreicher Hyperm-Sorten usw. weisen in ihrem Mittelteil eine nahezu lineare Abhängigkeit zwischen B und H auf, während bei Steigerung von H ein nichtlinearer Übergang, oft in Form eines scharfen Knickes, in den Sättigungszustand erfolgt. Abgesehen von einigen statisch ablenkenden Deflektor-Methoden beruht die Anwendung hochpermeabler Kerne auf einer periodischen Wechselfeld-Aussteuerung der Schleife, wobei gerätetechnisch zwischen linearer und nichtlinearer Aussteuerung unterschieden werden muß. Das hierzu nötige magnetische Wechselfeld kann sowohl von stromdurchflossenen Spulen als auch von Strömen stammen, die den Kern direkt durchfließen.

β) Verfahren mit nichtsättigender Aussteuerung. Induktivitäten mit Eisenkern zeigen bei gegebener Frequenz gewisse Abhängigkeiten des Scheinwiderstandes vom äußeren Magnetfeld. Wählt man für den Kern eine langgestreckte Form und sieht eine geeignete Frequenz vor, so gelingt es, beispielsweise in einer rückgekoppelten Brückenschaltung[2], die Beobachtungsempfindlichkeit in die für die Aufzeichnung erdmagnetischer Variationen nötige Höhe zu treiben.

Befindet sich die Selbstinduktion nicht in Luft, sondern in einem Material der Suszeptibilität χ und der Leitfähigkeit σ, so wird die Brückenspannung von beiden Größen beeinflußt. Durch Anwendung phasenempfindlicher Gleichrichter-Anordnungen ist eine gleichzeitige Anzeige von χ und σ, beispielsweise für Bohrloch-Messungen[3], möglich.

Bereits einzelne Stücke hochpermeablen Drahtes zeigen magnetfeldabhängige Widerstände und lassen sich in Brückenschaltungen zur Komponentenmessung in Drahtlängsrichtung und auch mit Erfolg zur Feldstärkedifferenzmessung (Bombenblindgängersuche)[4] anwenden.

Allen diesen Verfahren mit nichtsättigender Aussteuerung ist eine starke Abhängigkeit der Geber von der Temperatur sowie die Einschränkung auf kleine lineare Anzeigebereiche und daher die Notwendigkeit von Zusatzfeldern eigen. Daher ist der Einsatz dort vorteilhaft, wo es darauf ankommt, die Umgebung des Gebers (etwa im Bohrloch) nicht durch starke Wechselfelder oder impulsartige Magnetfelder zu beeinflussen, wie sie andere, wesentlich weiter verbreitete Methoden aufweisen.

γ) Verfahren mit sättigender Aussteuerung. Die Verfahren mit sättigender Aussteuerung haben eine ungleich größere Verbreitung gefunden als diejenigen, die nur den linearen Teil der Hystereseschleifen ausnutzen. Die Gründe hierfür mögen darin zu suchen sein, daß die nichtlineare $B(H)$-Charakteristik eine hohe zeitliche Konstanz aufweist, eine nur geringe Temperaturabhängigkeit zeigt und daß das pausenlose Ummagnetisieren des Kernmaterials Alterungserscheinungen weitgehend unterdrückt. Weiterhin sind hier die Vorteile zu nennen, die diese Verfahren für den Betrieb von Servomechanismen bieten. Hier wäre auch darauf hinzuweisen, daß sich diese Prinzipien bei geeigneter, gegen Fremdfelder abge-

---

[1] Siehe dazu in diesem Band die Beiträge von P. H. SERSON u. K. WHITHAM, S. 384, von J. R. BALSLEY, S. 395, und von P. HOOD, S. 422.
[2] G. ZICKNER u. E. BLECHSCHMIDT: Feinwerktechnik 54, 171—176 (1950).
[3] R. A. BRODING, C. W. ZIMMERMANN, E. S. SOMERS, E. S. WILHELM, and A. A. STRIPLING: Geophysics 17, 1—26 (1952).
[4] A. BUTTERWORTH: Elec. Eng. 94 (II), 325 (1947); 95 (II), 645—652 (1948).

schirmter Sonde erfolgreich zur Gleichstromverstärkung eignen und in der industriellen Meßpraxis häufig Anwendung finden[5].

Die nichtlinear ausgesteuerte Sonde bezeichnet man im allgemeinen Falle mit *Saturationskernsonde*[6]. Für spezielle Formen sind eigene Bezeichnungen in Gebrauch, z.B. führt die Doppelkernsonde den Namen „FÖRSTER-Sonde"[7].

δ) Die *mathematische Behandlung* der Verfahren mit sättigender Aussteuerung basiert auf einer Berechnung des Spektrums der Sondenausgangsspannung. Für die hierzu nötige analytische Darstellung der Hystereseschleife des Kernmaterials

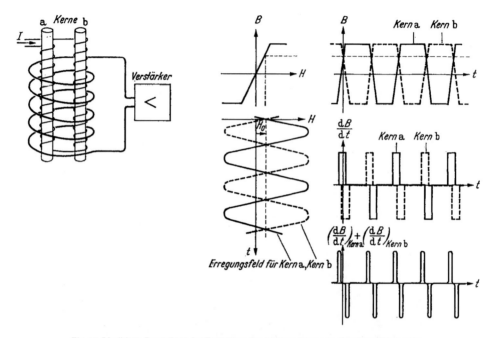

Fig. 12. Idealisierte Darstellung des Entstehens der Ausgangsspannung einer Doppelkernsonde.

existiert kein allen Ansprüchen genügender Ausdruck. Es sind daher zahlreiche Näherungsansätze in Gebrauch, die in der Mehrzahl der Fälle auf eine Berücksichtigung der Verluste verzichten und einen eindeutigen Zusammenhang zwischen B und H vorgeben.

Für eine qualitative Darlegung der Wirkungsweise (und auch weitgehend für eine quantitative Behandlung[8] genügt bereits ein dreiteiliger Linienzug nach Fig. 12. In dieser Abbildung wird die Bildung der Ausgangsspannung einer mit zwei parallelen Kernen (a und b) versehenen Sonde schematisch angegeben, die über zwei gegeneinandergeschaltete Erregerwicklungen und zwei hintereinandergeschaltete Sekundärwicklungen verfügt (Fig. 13 b). In Kernlängsrichtung wirke

---

[5] F. C. WILLIAMS, and S. W. NOBLE: Proc. Inst. Elec. Engrs. (London), **97** (II), 445 (1950).

[6] Die entsprechenden Ausdrücke sind: saturable core device (englisch); sonde éléctromagnétique (französisch); magnetmoduljazionij datšik = magnetmodulierter Geber (russisch). Komplette Magnetometer dieses Prinzips nennt man im Englischen auch „flux gate magnetometer".

[7] F. FÖRSTER: Z. Metallk. **46**, 358—370 (1955).

[8] E. P. FELCH, W. J. MEANS, T. SLONCZEWSKI, L. C. PARRAT, L.-H. RUMBAUGH, and A. J. TICKNER: Elec. Eng. **66**, 641—651, 680—685 (1947).

das Gleichfeld H_0, das gemessen werden soll; falls Felder anderer Richtung existieren, gehen nur ihre Komponenten in Kernlängsrichtung ein. Während die um 180° verschobenen Erregerfelder trapezartige Änderungen von B erzeugen, addieren sich die in den Sekundärwicklungen entstehenden, proportional zu dB/dt verlaufenden Spannungen und ergeben die im unteren Teil der Fig. 12 angegebenen Kurvenformen.

Die Saturationskernsonden basieren alle auf der Tatsache, daß eine Änderung von H_0 zwangsläufig eine reproduzierbare Änderung dieser Kurvenformen nach sich zieht. Bei FOURIERanalytischer Betrachtung folgt unter Beachtung gewisser Symmetriebedingungen, daß die geradzahligen Harmonischen der Erregerfrequenz etwa zu H_0 proportionale Amplituden aufweisen und daß die ungeradzahligen praktisch unabhängig von H_0 sind. Die Feldmessung kann nun selektiv durch Auswahl einer geradzahligen Harmonischen (meist der zweiten) geschehen oder nichtselektiv durch Impulsanzeigeverfahren vorgenommen werden.

12. Das Spektrum der Sondenausgangsspannung. Das Spektrum der Ausgangsspannung ist in zahlreichen Arbeiten berechnet worden, wobei Linienzüge oder transzendente Funktionen sowie Kombinationen beider als Hystereseschleifen-Ersatz benutzt und sinusförmige oder linear auf- und absteigende Erregung vorausgesetzt wurde.

Hier sei eine Berechnung angeführt[1], die eine sehr anpassungsfähige *Schleifendarstellung* durch Linienzüge und Sinusbögen vorgibt und folgendermaßen definiert wird:

$$\left. \begin{array}{l} B = \mu_A \mu_0 H \quad \text{für} \quad -(H^* - H_k) < H < (H^* + H_k), \\[4pt] B = \mu_A \mu_0 \left[\dfrac{H}{2} + \dfrac{H_k}{\pi} \sin (H - H^* + H_k) \dfrac{\pi}{2 H_k} + \dfrac{1}{2}(H^* - H_k) \right], \\[4pt] \qquad \text{für} \quad -(H^* + H_k) < H < -(H^* - H_k), \\[4pt] \qquad \text{und} \quad (H^* - H_k) < H < (H^* + H_k). \end{array} \right\} \quad (12.1)$$

Unter μ_A (dimensionslos) ist die relative Anfangs-Kernpermeabilität zu verstehen. H^* bezeichnet die Feldstärke, die den Schnittpunkt der Linie $B = B_s$ mit $B = \mu_A \mu_0 H$ kennzeichnet*. H_k bestimmt Anfang und Ende der Sinusbögen durch $H^* - H_k$ bzw. $H^* + H_k$. Entsprechendes gilt für die negativen B- und H-Werte.

Da die wesentliche Oberwellenbildung während der linearen Teile der sinusförmigen Erregung (Fig. 12) eintritt, genügt es, eine lineare Erregung in Form einer symmetrischen Sägezahnkurve mit der Amplitude $\dfrac{\pi}{2} \widehat{H}_1$ vorauszusetzen:

$$\left. \begin{array}{l} H_1 = H_0 + \widehat{H}_1\, \omega \left(t + \dfrac{T}{4}\right) \quad \text{für} \quad -\dfrac{T}{2} \leq t \leq 0, \\[6pt] H_1 = H_0 - \widehat{H}_1\, \omega \left(t - \dfrac{T}{4}\right) \quad \text{für} \quad 0 \leq t \leq \dfrac{T}{2}. \end{array} \right\} \quad (12.2)$$

Unter H_0 wird in der einschlägigen Literatur das zu messende, in Kernlängsrichtung wirkende Gleichfeld verstanden[2]. T bezeichnet die Dauer einer Erregungsperiode, t die Zeit und ω die Kreisfrequenz der Erregung. Bildet man abschnittsweise dB/dt und berechnet die Fourierkoeffizienten, so folgen die ungeradzahligen Anteile der (komplexen) Sondenausgangsspannung $\hat{u}$ mit $a = 1, 3, 5, \ldots$ sowie n als Windungszahl und S als Windungsfläche der Sekundär-

---

* Der Buchstabe H kennzeichnet in diesem Abschnitt die magnetische Feldstärke im allgemeinen Sinne; die erdmagnetische Horizontalkomponente wird mit $\mathcal{H}$ bezeichnet.
[1] R. FELDTKELLER: Schriften d. Deutsch. Ak. f. Luftfahrtforschung, Berlin 1943, S. 1—42.
[2] Beispielsweise ist zur Berechnung eines Z-Magnetometers $H_0 = Z/\mu_0$ einzusetzen.

spule zu

mit
$$\hat{u}_a = \frac{1}{c_0 \sqrt{\varepsilon_0 \mu_0}} \mu_A \mu_0 \, n S \omega H^* h_a \cos \frac{a H_0}{\hat{H}_1}$$

$$h_a = \frac{\pi}{4} \cdot \frac{\sin \frac{a H^*}{\hat{H}_1}}{\frac{a H^*}{\hat{H}_1}} \cdot \frac{\cos \frac{a H^*}{\hat{H}_1} \cdot \frac{H_k}{H^*}}{1 - \left(\frac{2}{\pi} \cdot \frac{a H^*}{\hat{H}_1} \cdot \frac{H_k}{H^*}\right)^2}$$

(12.3)

Ebenso folgen die geradzahligen Teilschwingungen mit b = 2, 4, 6 ... zu

mit
$$\hat{u}_b = \frac{1}{c_0 \sqrt{\varepsilon_0 \mu_0}} \mu_A \mu_0 \, n S \omega H^* h_b \frac{\sin \frac{b H^*}{\hat{H}_1}}{\frac{b H_s}{\hat{H}_1}}$$

$$h_b = \frac{\pi}{4} \sin \frac{b H^*}{\hat{H}_1} \cdot \frac{\cos \frac{b H^*}{\hat{H}_1} \cdot \frac{H_k}{H^*}}{1 - \left(\frac{2}{\pi} \cdot \frac{b H^*}{\hat{H}_1} \cdot \frac{H_k}{H^*}\right)^2} \cdot$$

(12.4)

Setzt man das Verhältnis $H_0/\hat{H}_1$ als klein voraus, wie es den Gegebenheiten normalerweise entspricht, so findet man, daß die ungeradzahligen Teilschwingungen praktisch unabhängig vom zu messenden Gleichfeld sind, während die *geradzahligen Harmonischen nahezu proportional* mit H_0 ansteigen oder fallen.

Prinzipiell ist jede geradzahlige Harmonische bis zu einer gewissen oberen Grenze zur Feldmessung geeignet. Man bevorzugt jedoch die zweite, da sie als niedrigste den Betrieb von Servomotoren gestattet.

Wendet man für diese Rechnung eine sinusförmige Erregung[3] an, so ergeben sich ähnliche Resultate, die aber zu den gleichen Folgerungen in technischer Hinsicht wie die aus der Dreiecks-Erregung gewonnenen führen.

Falls dem Gleichfeld H_0 ein Wechselfeld überlagert ist, ergeben sich erweiterte Spektren[4].

13. Sondenformen und ihre Betriebsweise. Die für den geomagnetischen Einsatz gedachten Sonden sollen hohe Empfindlichkeit und geringen Rauschpegel aufweisen sowie ein nachwirkungsfreies, stabiles Verhalten gewährleisten.

Die einfachste Sonde stellt bereits ein hochpermeabler, mit einer Wicklung versehener Kern[1] dar (Fig. 13a), wobei zum Betrieb Generator, 2f-Selektivverstärker und Wicklung parallel zu schalten sind. Diese Anordnung stellt sehr hohe Anforderungen an die Filtertechnik, da sehr schwache geradzahlige Harmonische von sehr starken, benachbarten ungeradzahligen Harmonischen getrennt werden müssen.

α) Wesentlich geringere Filteranforderungen stellt die Kombination zweier Kerne mit getrennten Primär- und Sekundärwicklungen (Fig. 13b), die unter dem Namen Förster-Sonde[2] bekannt ist. In der Schaltung als Feldstärkemesser (Primärwicklungen gegeneinander geschaltet, Sekundärwicklungen hintereinander[3]) und bei Einhaltung gewisser Symmetriebedingungen im Sondenaufbau kompensieren sich die ungeradzahligen Harmonischen und die geradzahligen

---

[3] K. Stambke: Ber. Stuttg. Arbeitsgem. Nr. 16. Stuttgart 1942.
[4] S. Nahrgang: Rech. Aéron. **16**, 11—18 (1950).
[1] D. C. Rose and J. N. Bloom: Can. J. Res. **28** (Sect. A), 153—163 (1950).
[2] F. Förster: Z. Metallkunde **46**, 358—370 (1955).

addieren sich. Diese Anordnung bietet zudem den Vorteil, die Primärwicklungen den Generatoreigenschaften anzupassen und die Sekundärwicklungen den Filterforderungen gemäß auszulegen.

Trennt man räumlich beide Kerne und ordnet sie mit parallelen Meßachsen auf einer Linie an, so eignet sich diese Kombination bei hintereinander geschalteten Primärwicklungen und gegeneinander geschalteten Sekundärwicklungen (Fig. 13c) als (tragbarer) *Feldstärkedifferenzmesser*[3]. Aufgrund dieser Schaltungen addieren

Fig. 13 a—f. Die verschiedenen Formen der Saturationskernsonden. a) Einzelkernsonde, b) Doppelkernsonde als Feldstärkemesser, c) Doppelkernsonde als Feldstärkedifferenzmesser, d) Brückensonde, e) Ringkernsonde und f) kernstromerregte Einzelkernsonde (G Generator, SV Selektivverstärker). Im Falle der Brückensonde (d) wird der Selektivverstärker vorzugsweise durch einen Impulsverstärker (angedeutet durch Stern) ersetzt und es finden nichtselektive Anzeigeverfahren Anwendung.

sich die geradzahligen Harmonischen wiederum, während die ungeradzahligen nicht exakt kompensiert werden und als vom Differenzfeld schwach abhängige Größen gelten. Die Anzeige des Differenzfeldes beruht auf einer Wiedergabe der Amplitude der zweiten Harmonischen.

β) *Brückensonden* enthalten entweder einen einzelnen Kern mit zwei gegensinnigen Wicklungen oder zwei getrennte, parallele Kerne mit gegensinniger Erregung[4]. Sie werden in Schaltungen nach Fig. 13d betrieben, wobei die Brückenspannung zur Anzeige dient. Es sind sowohl symmetrische als auch unsymmetrische Brücken realisiert worden, die unterschiedliche Eigenschaften im Hinblick auf die Störpegel-Unterdrückung aufweisen.

---

[3] Oft vereinigt man die beiden hintereinander geschalteten Sekundärwicklungen zu einer gemeinsamen.
[4] W. E. FROMM: Advanc. Electron. **4**, 257—299 (1952).

Die *Ringkernsonde* (Fig. 13 e) war die erste realisierte Sonde[5]. Sie enthält eine den ringförmigen Kern umschließende Torus-Wicklung sowie eine Sekundärwicklung. Sie findet heute vorwiegend Anwendung in der industriellen Meßtechnik zur Gleichstromverstärkung und zu Regelzwecken.

Bei der *kernstromerregten Sonde* fließt der Erregerstrom direkt durch den Sondenkern (Fig. 13f), wobei die Signalspannung mittels einer Spule abgenommen wird. Mit kernstromerregten Anordnungen lassen sich ähnliche Kombinationen wie oben (Doppelkern- und Brückensysteme) zusammenstellen. Diese Sonden haben allerdings in der Geomagnetik kaum Eingang gefunden, da thermisch bedenkliche Stromstärken Anwendung finden müssen, um in den gewünschten Empfindlichkeitsbereichen messen oder registrieren zu können.

γ) Die *Empfindlichkeit* der Sonden ist in zahlreichen Arbeiten[6] untersucht worden. Als wichtigste Einflußgröße gilt die Amplitude des Erregerfeldes $\hat{H}_1$. Jede Sonde verhält sich generell so, daß die Oberwellenbildung mit wachsender Felderregung H erst bei Erreichen des Knickes in der $B(H)$-Charakteristik einsetzt, dann steil zunimmt, ein Maximum durchläuft und nach großen Feldstärken hin langsam abfällt. $2f$-Sonden betreibt man im allgemeinen in der Nähe dieses Maximums. Als Beispiel werde hier die für eine Doppelkern-Anordnung unter den Bedingungen der Gl. (12.1), (12.2) gültige Beziehung angegeben:

$$\frac{d\hat{u}_2}{dH_0} = \frac{2}{c_0 \sqrt{\varepsilon_0 \mu_0}} \mu_A \mu_0 \, nS \, \omega \, h_2. \tag{13.1}$$

Hierbei wird allerdings vorausgesetzt, daß das Erregerfeld des einen Kernes den anderen nicht beeinflußt. Rückt man bei konstantem $\hat{H}_1$ die Kerne näher zusammen, so nimmt die Empfindlichkeit ab und erreicht ihren geringsten Wert bei unmittelbar benachbarten Kernen[7].

Übliche Werte für $\frac{1}{\mu_0} \cdot \frac{d\hat{u}_2}{dH_0}$ liegen zwischen 10 und 100 μV/γ bzw. zwischen 0,01 und 0,1 V/T (Volt/Tesla).

δ) Obwohl Doppelkernsonden gegenüber den Einzelkernsonden erhebliche Vorteile im Hinblick auf die leichter realisierbaren Filterforderungen bieten, sind sie als exakter Nullindikator in automatischen Magnetometern nur dann brauchbar, wenn ein präziser *Abgleich* der Kerne hinsichtlich ihrer magnetischen und mechanischen Eigenschaften garantiert werden kann[8]. Bei sehr guten Sonden gelingt es, den im Felde Null vorhandenen Spannungsrest so gering zu halten, daß er Magnetfeldern in der Größenordnung $\frac{1}{10}$ γ äquivalent ist. Kohärente Störspannungen, unter denen man Spannungen mit starrer Phasenlage bezüglich der verdoppelten Erregerfrequenz versteht, können Pseudo-Nullpunkte vortäuschen[9]. Die untere Grenze des Einsatzes von Saturationskernsonden bestimmen die BARKHAUSEN-Sprünge im Kernmaterial[10], das elektronische Rauschen in Sonde und Hilfsgeräten sowie andere inkohärente Störspannungen. Magnetische Nachwirkungserscheinungen in den Kernen führen bei unzureichender Materialauswahl zu irreversiblen Nullpunktsänderungen[9].

$2f$-Sonden liefern den für einen Betrieb von Servomechanismen oder phasenempfindlichen Gleichrichtern nötigen Phasensprung der zweiten Harmonischen von 180° und gestatten damit eine Polaritätskennzeichnung des zu messenden Gleichfeldes.

14. Der prinzipielle Aufbau kompensierender Magnetometer.
Während für rohe, auf einige Prozent genaue Feldbestimmungen die Kombination von Generator,

---

[5] H. ASCHENBRENNER u. G. GOUBAU: Hochfrequenztechn. u. Elektroak. **47**, 177—181 (1936).

[6] Eine Übersicht gibt J. GREINER: Nachrichtentechnik **10**, 123—126, 156—162, 495—498 (1960).

[7] M. A. ROSENBLATT: Žurnal techničeskoi Fiziki **24**, 637—661 (1954).

[8] M. WURM: Z. Angew. Phys. **2**, 210—219 (1950).

[9] H. SCHMIDT: Freiberger Forschungsh. C **22**, 20—28 (1956).

[10] F. C. WILLIAMS u. S. W. NOBLE: Proc. Inst. Elec. Engrs. (London) **97** (II), 445 (1950).

Sonde, Selektivverstärker und Gleichrichter-Meßinstrument zur direkten Anzeige der zweiten Harmonischen ausreicht, ist für höhere Ansprüche der Einsatz von Kompensatoren zu empfehlen, insbesondere dann, wenn es sich um die laufende Registrierung von Schwankungen handelt, die sehr klein gegen den mittleren Wert der Meßgröße sind. Dann kommen die Vorteile zur Geltung, die automatische Kompensatoren — hier mit einer Saturationskernsonde als Nullindikator — ganz allgemein auszeichnen und auf die bereits in Ziff. 5 bei der Behandlung des kompensierenden Stabmagnet-Variometers hingewiesen wurde. Aus der Fülle der für stationären oder mobilen Einsatz geschaffenen Magnetometer seien zwei typische Anordnungen erläutert.

Fig. 14. Prinzipieller Aufbau eines kompensierenden 2f-Magnetometers. Der Generator mit der Frequenz f speist die Doppelkernsonde, deren Sekundärwicklung ein Frequenzgemisch abgibt, aus dem der Selektivverstärker den Anteil der Frequenz 2f aussiebt und der Steuerwicklung des FERRARIS-Motors zuführt. Die Erregerwicklung dieses Motors erhält ihre Spannung der Frequenz 2f über Verdoppler, Phasenschieber und Verstärker. Der FERRARIS-Motor verschiebt solange den Schleifer eines Potentiometers, bis der damit eingestellte Strom — durch die Kompensationswicklung der Sonde fließend — gerade ein dem zu messenden Feld entgegengesetztes gleiches erzeugt. Der Potentiometerschleifer kann mit einer Schreibfeder zur Aufzeichnung der Magnetfeldvariationen verbunden werden.

α) *Automatischer Kompensator vom 2f-Typ.* Den prinzipiellen Aufbau dieses wohl am häufigsten eingesetzten Typs[1] zeigt die Fig. 14. Der Generator mit der Frequenz f (einige hundert Hertz) speist die Primärwicklungen der Doppelkernsonde. Aus der in der Sekundärwicklung induzierten Spannung wird mit dem Selektivverstärker der Anteil der zweiten Harmonischen gewonnen und einer Wicklung des Servomotors (FERRARIS-Motor oder Zweiphasen-Induktionsmotor) zugeführt. Die zweite Wicklung erhält eine aus der Verdopplung der Generatorfrequenz hervorgehende Spannung konstanter Amplitude. Der Servomotor erzeugt dann ein maximales Drehmoment, wenn beide Spannungen an den Wicklungen eine Phasenverschiebung von $\pm 90°$ aufweisen. Ein Phasenschieber gestattet, diesen Zustand einzustellen. Über ein Getriebe verstellt der Motor so lange den Schleifer eines Drehwiderstandes, bis der Strom in der Kompensationswicklung der Sonde ein dem zu messenden Felde entgegengesetztes gleiches Feld erzeugt. Die Sondenkerne liegen damit im Felde „Null" und der Anteil der zweiten Harmonischen in der Sondenausgangsspannung ist bis auf einen Rest, den Nullfeldrest, verschwunden. Die Anzeige des Kompensationsstromes geschieht mit Schreibern, wobei meist die Schleiferbewegung direkt zum Transport einer Schreibfeder oder eines Punktdrucker-Wagens ausgenutzt wird. Der Kompensationskreis ist vorwiegend mit einer Brückenschaltung versehen, die eine automatische Umschaltung vieler Registrierbereiche mit gleichbleibenden Skalenwerten gestattet.

---

[1] Ausführliche Schaltbilder s. H. SCHMIDT: „Geomagnetismus und Aeronomie", Bd. II, Abschn. 6.5. Berlin: VEB Deutscher Verlag d. Wiss. 1960.

Fig. 15. Prinzipieller Aufbau eines kompensierenden Brückenmagnetometers. Der Generator mit der Frequenz f erregt die Sonde über einen Transformator. Die Ausgangsspannung der Sonde wird ohne Verformung verstärkt und 2 Gleichrichtern mit Impulsformen zugeführt, die den jeweiligen Anteil als Sägezahnschwingung abgeben. In der Summierstufe werden diese zu einer Sägezahnschwingung der Frequenz $2f$ zusammengesetzt. Ein von der Abgleichbedingung (Grundwellenanteil in der $2f$-Sägezahnschwingung gleich Null) abweichendes, auf die Sonde wirkendes Magnetfeld bewirkt das Auftreten der Frequenz f in der Sägezahnschwingung und das Betätigen des FERRARIS-Motors solange, bis der Strom im Kompensationskreis die Abgleichbedingung wieder erfüllt. Die Benutzung eines Schreibers ergibt sich ebenso wie in Fig. 14.

Wendet man eine Sonde nach Fig. 13c an, so ist mit der gleichen Anordnung eine automatische Anzeige der *Felddifferenz* zwischen beiden Sondenkernen gegeben. Für den Bau transportabler leichter Suchgeräte ersetzt man den Servomotor durch einen phasenempfindlichen Gleichrichter und benutzt zur Anzeige der Felddifferenz ein Drehspulinstrument. Die Generatorfrequenz solcher Geräte liegt bei einigen Kilohertz.

β) *Automatischer Kompensator vom Brücken-Typ*. Neben dem $2f$-Verfahren läßt sich mit Erfolg ein anderes Nachweisprinzip anwenden, das sich impulstechnischer Meßmethoden bedient und in aeromagnetischen Geräten Anwendung gefunden hat. Bei geeigneter Wahl von Kernmaterial und Erregerfrequenz einer Brückensonde (Fig. 15) treten in der Brückenspannung nadelförmige Impulse auf, deren Eigenschaften vom zu messenden Feld H_0 abhängen. Führt man die verstärkten Impulse einer Impulsformerschaltung zu, so werden in der in Fig. 15 angegebenen Weise die Nadelimpulse in zwei, durch Gleichrichtung gekennzeichneten Kanälen zu Sägezahnschwingungen verformt und zu einer Sägezahnschwingung der Frequenz $2f$ zusammengesetzt. Der Abgleich der Anordnung erfolgt so, daß im Feld Null (oder bei einem gewünschten Festwert) beide Sägezahnanteile gleich auftreten, so daß in der zusammengesetzten Sägezahnschwingung nur dann ein Grundwellenanteil vorhanden ist, wenn ein Feld H_0, verschieden von Null bzw. vom Festwert, in Kernlängsrichtung wirkt. Dieser Grundwellenanteil wird mit einem Tiefpaß, der kurz über der erregenden Frequenz f abschneidet herausgesiebt und einem Servomotor zugeführt, der — im Gegensatz zum selektiven $2f$-Verfahren — hier mit der Grundfrequenz f betrieben wird und seine Erregung vom Generator direkt erhält. Der Betrieb von servogetriebenen Schreibern ist ebenso möglich wie beim $2f$-Typ.

Mit unsymmetrischen Brücken (gestrichelter Widerstand in Fig. 15) gewinnt man oft ein besseres Signal-Rausch-Verhältnis beim Nachweis sehr kleiner Feldänderungen als mit symmetrischen Brücken.

15. Anwendungen der Saturationskernsonde. Die Saturationskernsonde hat in zahlreichen Varianten Anwendung gefunden, die besonders die Magnetfeldmessung vom bewegten Instrumententräger aus sehr stark aktivieren konnten. Im folgenden wird ein kurzer Überblick über technisch bedeutsame Wege gegeben.

α) *Selbstorientierende Nachführsysteme*. Ordnet man (unter Verzicht auf den Kompensationskreis) eine Saturationskernsonde mit den in Fig. 14 oder Fig. 15 angegebenen Gerätegruppen so auf der Achse des Servomotors an, daß ihre Empfindlichkeitsrichtung senkrecht zur Achse liegt, so wird der Motor die Sonde so lange drehen, bis er von ihr keine Signalspannung mehr erhält und sie daher die feldfreie Lage einnimmt. Kombiniert man zwei solcher „Orientierungskanäle" so, daß die beiden zueinander senkrecht angeordneten Sonden auf einer Fläche liegen, so stellt sich die Fläche — unabhängig von der Lage des Instrumententrägers (Flugzeug, Schiff, Rakete) so ein, daß die Flächennormale parallel zur Richtung des erdmagnetischen Feldes liegt. Eine dritte Sonde, senkrecht zur Fläche auf dieser befestigt, nimmt somit automatisch die Richtung des geomagnetischen Vektors ein und ermöglicht, mit Anordnungen nach Fig. 14 oder Fig. 15 den Betrag der Totalintensität als automatisch kompensierendes Magnetometer anzuzeigen und die Variationen zu schreiben (Fig. 16a).

Die Bedienung der Orientierungssonden geschieht über Seilzüge von den Servomotoren aus, die — da sie ja ferromagnetische Teile enthalten — etwa einen Meter von der Sondenplattform entfernt sein müssen.

Diese in der Aeromagnetik und Raketenmagnetik besonders verbreiteten Nachführsysteme bildeten lange Zeit das Hauptinstrumentarium, mit dem Totalintensitätsvermessungen, meist kombiniert mit Luftbildaufnahmen sowie geoelektrischen oder radioaktiven Meßvorhaben, vorgenommen wurden[1].

β) *Kombination von Sondensystemen mit Pendeln und Kreiselplattformen.* Wenn auch die nachgesteuerten Sondentripel zu einem enormen Aufschwung der Aeromagnetometrie im Hinblick auf die Totalintensitätsvermessung führten, blieben

Fig. 16a u. b. a) Schematische Darstellung eines Sondendreibeins (a und b sind Hilfssonden, die sich jeweils automatisch in das Feld „Null" ziehen und damit die dritte Sonde c parallel zum erdmagnetischen Vektor ausrichten. b) Prinzipielle Darstellung des Sondenpendels[3]. Das Sondendreibein ist in einem Rahmen (*1*) befestigt, der in einem Kardan (*4*) aufgehängt ist und dessen Bewegung mit Hilfe einer Dämpferplatte (*2*) in Siliconöl (*3*) gedämpft wird. Die Azimut-Einstellung übernimmt der Servomotor (*6*), die der Inklination der Motor (*8*). Die Motoren treiben jeweils Winkelübertragungssysteme (*5*) für das Azimut und (*7*) für die Inklination.

die Forderungen nach einer Komponentenmessung vom bewegten Instrumententräger aus weiterhin erhoben. Daher ging man dazu über, servogesteuerte Sondensysteme mit Anordnungen zu kombinieren, die die Vertikale oder Horizontale im Fahrzeug definieren, verband die Sonden mit Pendeln oder Kreiselplattformen und schuf so die *Sonden-Vektor-Magnetometer*.

Der Meßkopf eines solchen Magnetometers[2], an einer geeigneten Stelle im Flugzeug stationiert, enthält ein flüssigkeitsgedämpftes Pendel (Fig. 16b), an dessen unterem Ende ein Sondentripel befestigt ist, das sich über Servomotoren am Erdfeld orientiert. Eine horizontale Sonde zieht sich selbst immer in das Feld Null und definiert damit eine horizontale Richtung ζ_0 gegen die Flugzeugachse, eine zweite bewegt sich in einer Vertikalebene im geomagnetischen Meridian so

---

[1] Siehe dazu in diesem Band die Beiträge von P. H. Serson u. K. Whitham, S. 384, von J. R. Balsley, S. 395, und von P. Hood, S. 422.

[2] E. O. Schonstedt, and H. R. Irons: Trans. Am. Geophys. Union **36** (1), 25—41 (1955).

lange, bis sie senkrecht zum erdmagnetischen Vektor steht und damit eine gegen die Horizontale zu messende Information für die Inklination J liefert. Die dritte, auf beiden anderen Sonden senkrecht stehende, mißt — in Richtung des geomagnetischen Vektors liegend — laufend die Totalintensität $\mathcal{F}$. Die Winkel werden mit Potentiometern gegen flugzeugfeste Richtungen gemessen und aufgezeichnet, so daß die Komponenten $\mathcal{H}$ und Z sowie die Deklination D aus

$$\left. \begin{array}{l} D = \chi_0 - \zeta_0, \\ \mathcal{H} = \mathcal{F} \cos J, \\ Z = \mathcal{F} \sin J \end{array} \right\} \quad (15.1)$$

folgen. χ_0 bedeutet eine optisch zu bestimmende Hilfsrichtung.

Natürlich birgt diese Methodik eine ganze Reihe von Problemen hinsichtlich der Feststellung der momentanen Flugzeuglage, der Kompensation flugzeugeigener Felder, der Eliminierung von unerwünschten Flugzeugbewegungen sowie der Orts- und Höhen-Bestimmung.

Diese sind innerhalb der technisch und physikalisch gegebenen Grenzen[3] erfolgreich gelöst worden, so daß systematisch angelegte Komponenten-Messungen über den Ozeanen ausgeführt werden konnten[3]. In Anbetracht der auf geomagnetischen Kartierungen basierenden Interpretationsaufgaben ist die weltweite Anwendung eines einheitlichen Instrumentariums als entscheidender Vorteil zu werten.

Eine andere Möglichkeit der Komponentenmessung vom Flugzeug aus beruht auf dem Einsatz kreiselgesteuerter Sondentripel[4]. Hierbei wird ein das Sondendreibein enthaltender Plexiglas-Topf von Kreiseln so gesteuert, daß eine Sonde jeweils vertikal liegt und Z in Kompensation mißt, während die beiden anderen die Horizontalkomponenten in Flugzeuglängsrichtung und quer dazu erfassen. Mit Hilfe von Rechengeräten erfolgt eine Registrierung von D, $\mathcal{H}$ und Z mit Ausgabe gemittelter Werte.

Ähnliche Systeme benutzt auch die Schiffsmagnetik[5] in mehreren Varianten.

Wenn auch die Bedeutung der Sondennachführsysteme für die Totalintensitätsmessung durch die Einführung frequenzabgebender Magnetometer gemindert wurde, so ist sie für die Komponentenmessung vom bewegten Instrumententräger aus offensichtlich unbestritten[6].

γ) *Saturationskernsonden in Raketen, Satelliten und Raumsonden.* Seit etwa 10 Jahren finden Saturationskernsonden Anwendung zur Magnetfeldmessung von Raketen, künstlichen Erdsatelliten sowie von Raumsonden aus, wobei meist dem $2f$-Verfahren der Vorzug vor Impuls-Meßverfahren gegeben wurde. In der ersten Phase benutzte man, beispielsweise zum Nachweis stromführender Schichten in der Ionosphäre[7] oder der Totalintensitäts-Bestimmung im erdnahen Raum[8] Sonden-Nachführsysteme, die weitgehend denen ähnelten, die in der Aeromagnetometrie in Gebrauch waren.

Später verwendete man, vorwiegend zum Studium der Magnetosphäre und des interplanetaren Magnetfeldes Anordnungen von Sonden-Dreibeinen, die fest im Satelliten eingebaut wurden und das Magnetfeld in fahrzeugfesten Komponen-

---

[3] Eine besondere Problematik liegt in der Kompensation flugzeugeigener Magnetfelder. Hierauf wird in [2] und [4] eingegangen.

[4] P. H. SERSON, S. Z. MACK, and K. WHITHAM: Publ. Dominion Observ. Ottawa **19**, 15—97 (1957); s. auch Beitrag von P. H. SERSON und K. WHITHAM in diesem Handbuch, S. 384.

[5] V. AUSTER: Erdmagnet. Jahrbuch Niemegk 1964, S. 102—109.

[6] H. P. STOCKARD: Report "Worldwide surveys by Project Magnet". IAGA-Symposium Washington, Oktober 1968.

[7] S. F. SINGER, E. MAPLE, and W. A. BOWEN: J. Geophys. Res. **56**, 265—281 (1951).

[8] N. W. PUŠKOV, i S. Š. DOLGINOV: Usp. Fiz. Nauk **63**, 645—656 (1957).

ten anzeigten[9]. Im Sinne eines redundanten Instrumentariums setzte man fluxgate-Magnetometer und Rubidium-Dampf-Magnetometer (s. Ziff. 20β) gemeinsam ein.

Die telemetrische Übermittlung der von den Sondenmagnetometern kommenden Spannungs- oder Stromwerte geschieht im allgemeinen so, daß diese in Frequenzen im Niederfrequenz-Bereich umgewandelt und einem hochfrequenten Träger als Modulation angeboten werden. Beispielsweise wird in der Raumsonde IMP 1 der Bereich von $\pm 40\,\gamma$ in eine Niederfrequenz zwischen 333 und 938 Hz umgesetzt und auf 136 MHz zur Bodenstation gegeben, wo durch Frequenzmesser mit Drucker und Schreiber die Dokumentation erfolgt.

Die Unterbringung der Sonden geschieht bei Raketen[10] meist im vorderen, speziell eisenfrei gehaltenen Teil; bei Raumsonden wendet man Sondenhalter von etwa 2 m Länge an, die dann ausgefahren werden, wenn das Raumfahrzeug keinen merklichen Luftwiderstand mehr vorfindet.

Oft benutzt man fest eingebaute Sonden in langsam rotierenden Satelliten und ermittelt damit bei bekanntem Magnetfeld die momentane Lage.

Die vergleichsweise geringe Masse von Sondenmagnetometern (pro Kanal weniger als 0,5 kg) und der gegenüber dem Rb-Magnetometer wesentliche Verzicht auf Temperatur-Regeleinrichtungen werden den Saturationskernsonden einen bleibenden Platz in der Satelliten-Magnetik sichern.

C. Atomphysikalische Feld-Meßmethoden.

In den folgenden Kapiteln sollen Meßverfahren beschrieben werden, die wesentliche Unterschiede zu den klassischen Geräten und den bisher behandelten elektrischen Methoden aufweisen. Während letztere auf Effekten der klassischen Mechanik und Elektrodynamik basieren, nutzt man hier die Aufspaltung der Energiezustände von Atomen oder Atomkernen im Magnetfeld aus, also den ZEEMAN-Effekt im erweiterten Sinne. Die Fortschritte der letzten 30 Jahre in Theorie und Praxis dieses Gebietes ermöglichten schließlich auch die Beobachtung von Energiezuständen, die durch Aufspaltung in einem sehr schwachen Magnetfeld entstehen, wie es das Erdfeld darstellt. Neben allen physikalischen und meßtechnischen Unterschieden der auf diesen Effekten aufbauenden Meßgeräte zeigen sie einige *gemeinsame Merkmale*:

(I) Als Meßgröße tritt eine Frequenz auf, die eine einfache Funktion der Totalintensität des Erdfeldes ist. Im einfachsten Falle besteht zwischen beiden Größen Proportionalität. Die hierbei auftretenden Koeffizienten sind zusammengesetzt aus Konstanten des atom- und kernphysikalischen Bereichs.

(II) Die Grenzen der Empfindlichkeit der Verfahren werden bestimmt durch die Linienbreite der benutzten Übergänge sowie durch das beim experimentellen Nachweis erreichbare Signal-Rausch-Verhältnis.

Hieraus leiten sich für den Einsatz solcher Geräte in der erdmagnetischen Meßpraxis folgende Konsequenzen ab:

1. Die Totalintensität des Erdmagnetfeldes kann im Sinne der klassischen Meßtechnik unmittelbar „absolut" gemessen werden.

2. Eine Messung beliebig gelegener Komponenten des Erdfeldvektors ist nur mit Hilfe von Zusatzfeldern bekannter Orientierung möglich.

3. Da als Meßgröße eine Frequenz auftritt, ist der Einsatz von digitalen Meß- und Registriergeräten zwingend gegeben. Man erreicht damit relativ leicht die

---

[9] Zum Beispiel N. F. Ness, C. S. Scearce, and J. B. Seek: J. Geophys. Res. **69**, 3531—3569 (1964). — J. P. Heppner, N. F. Ness, C. S. Scearce, and T. L. Skillman: J. Geophys. Res. **68**, 1—53 (1963). — L. J. Cahill: J. Geophys. Res. **71**, 4505—4519 (1966).
[10] G. Musmann: Z. Geophys. **33**, 358—361 (1967).

notwendige Meßgenauigkeit (Größenordnung 10^{-5}). Die Daten lassen sich in maschinenlesbarer Form ausgeben, womit die Weiterverarbeitung in Rechenautomaten unmittelbar gegeben ist.

4. Eine Fernübertragung der Meßwerte ist leicht möglich, selbstverständlich auch drahtlos.

Betrachtet man den technischen Aufwand und die Ausfallwahrscheinlichkeit des Dauerbetriebes, so ist vergleichsweise manche klassische Anordnung zur Zeit noch überlegen. Demgegenüber stehen jedoch zahlreiche betriebstechnische Vorteile, von denen beispielsweise die auch bei Messungen von bewegten Trägern aus erzielbare hohe Meßgenauigkeit, die enorme Auflösung bei Beobachtung kleinster Feldänderungen und der geringe Meßzeitaufwand bei nur wenig verzögerter Meßwertwiedergabe zu nennen sind.

Es sollen nun besonders zwei Meßverfahren dieser Gruppe, beruhend auf der *Kernresonanz* und dem *optischen Pumpen*, betrachtet werden, die in vielen Zweigen der erdmagnetischen Forschung Anwendung gefunden haben.

I. Kernpräzessionsmagnetometer zur Bestimmung der Totalintensität.

16. Physikalische Grundlagen der Kernresonanz. Zur Erklärung der Kernresonanzvorgänge wird hier, den zitierten Autoren[1] folgend, ein Weg beschritten, der im wesentlichen anschaulich-klassische Vorstellungen der Kernphysik benutzt und zur quantitativen Behandlung der Resonanzeffekte führt. Allerdings ist hierbei die empirische Einführung leicht meßbarer Materialkonstanten (Relaxationszeiten) nötig; eine exakte Berechnung dieser Werte würde quantenmechanische Betrachtungen erfordern.

α) Zunächst werden einige Beziehungen zwischen atomaren Größen und dem *Kernmoment* zusammengestellt.

Ein analog zum BOHRschen Magneton formal gebildetes „Kernmagneton"

$$|\mu_k| = \frac{1}{c_0 \sqrt{\varepsilon_0 \mu_0}} \cdot \frac{q \hbar}{2 m_P} \tag{16.1}$$

mit q als Elementarladung (Absolutwert) und m_P als Protonenmasse liefert nicht in der erwarteten Art das magnetische Moment des Wasserstoffkerns (Proton).

Es gelten vielmehr die folgenden Beziehungen*:

1. Atomkerne besitzen einen Drehimpuls $\boldsymbol{a}$. Dieser kann durch den dimensionslosen Spinvektor $\mathfrak{J}$ durch folgende Gleichung ausgedrückt werden:

$$\boldsymbol{a} = \hbar \cdot \mathfrak{J}, \tag{16.2}$$

Der Kernspin I ist dann die größtmögliche Komponente von $\mathfrak{J}$ in einer vorgegebenen Richtung.

Der Spinvektor wird allgemein und auch ausnahmslos in der zitierten Literatur mit $\mathfrak{J}$ (gotisch groß i) bezeichnet, entsprechend die Komponente mit I. J wird dagegen als innere Quantenzahl für die Elektronenhülle verwendet [siehe Gln. (19.2), (19.4)].

2. Der Kernspin ist halbzahlig bei ungerader Nukleonenzahl, ganzzahlig bei gerader Nukleonenzahl.

---
[1] A. LÖSCHE: Kerninduktion. Berlin: VEB Deutscher Verl. d. Wiss. 1957. — G. E. PAKE: Amer. J. Phys. **18**, 438—472 (1950). — H. KOPFERMANN: Kernmomente, Akad. Verl.-Ges. Frankfurt 1956. Siehe auch G. LAUKIEN: In diesem Handbuch, Band 38/1, S. 120—376.

* $\hbar = (1{,}05450 \pm 7 \cdot 10^{-5}) \cdot 10^{-34}$ VAs^2; $m_P = (1{,}67252 \pm 8 \cdot 10^{-5}) \cdot 10^{-27}$ kg.

3. Alle Kerne mit nicht verschwindendem Spin haben ein magnetisches Moment

$$\mu_I = \gamma \, \boldsymbol{a} = g_I \, \mu_k \, \mathfrak{J}, \tag{16.3}$$

(γ gyromagnetisches Verhältnis, g_I Kern-g-Faktor) z.B. ergibt sich für das Proton $g_I = 2,8$ und für das Neutron $g_I = -2$. Der zweite Wert ist mit den klassischen Vorstellungen über ein elektrisch neutrales Teilchen nicht vereinbar[2].

4. Die Verteilung der Ladung und deren Bewegung innerhalb der Atomkerne bewirken nicht nur das Auftreten eines magnetischen Dipolmomentes. Vielmehr treten Multipole der Ordnung 2^l auf. Es gilt $l \leq 2I$. Für geradzahlige l existieren nur elektrische und für ungeradzahlige l nur magnetische Multipole. Speziell für das Proton ($I = \tfrac{1}{2}$) ergibt sich ein magnetisches Dipolmoment.

β) Läßt man die Kerne mit einem äußeren Magnetfeld $\boldsymbol{B}$ in *Wechselwirkung* treten, so kann die Wechselwirkungsenergie W_m nur diskrete Werte annehmen. Sie sind anschaulich durch die $2I+1$ Einstellmöglichkeiten des Kernspins I gegeben (Fig. 17). Es gilt

$$W_m = \boldsymbol{\mu}_I \cdot \boldsymbol{B} = |\boldsymbol{\mu}_I| \, B \, \frac{m}{I}. \tag{16.4}$$

Für zwei benachbarte Energiestufen ($\Delta m = 1$) folgt mit Gl. (5.3):

$$\Delta W_m = \frac{|\mu_I|}{I} B = \gamma \hbar B; \quad 2\pi f_L = \omega = \gamma B. \tag{16.5}$$

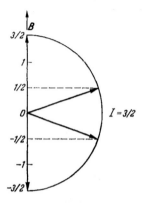

Fig. 17. Einstellmöglichkeit eines Kernspins ($I = \tfrac{3}{2}$) im Magnetfeld $\boldsymbol{B}$.

Für Protonen ergibt sich in einem äußeren Feld von 10^4 Γ die Größenordnung von 10^7 Hz für die Übergangsfrequenz f_L (LARMOR-Frequenz). Dieser Wert liegt unterhalb der natürlichen Linienbreite optischer Spektrallinien. Ein Nachweis des „Kern-ZEEMAN-Effektes" mit Mitteln der optischen Spektroskopie ist daher nicht möglich. Erstmalig gelang 1939 RABI und Mitarbeitern[3] ein experimenteller Nachweis mit der Methode der Molekularstrahlresonanz. Darauf sei in diesem Rahmen nicht näher eingegangen. Man konnte auf diese Weise γ und $|\mu_I|$ direkt bestimmen. Dabei lassen sich die zwischenmolekularen Wechselwirkungen vernachlässigen, da sie wegen der geringen Teilchenkonzentration im Molekularstrahl sehr klein sind. Anders verhält es sich bei der Methode der paramagnetischen Kernresonanz. Die folgende kurze Betrachtung geht von klassisch-physikalischen Vorstellungen aus.

Man betrachtet im folgenden eine Probesubstanz von beliebigem Aggregatzustand, die nur eine Atomsorte mit nicht verschwindendem Kernspin enthält. Alle Moleküle seien im Grundzustand, die Elektronenmomente daher gleich Null. Für viele Moleküle ist diese Bedingung erfüllt, z.B. auch für Protonen in Wassermolekülen. Die Richtungen der Kernmomente seien gleichmäßig über alle Raumwinkel verteilt und unterliegen, durch Zusammenstöße bedingt, statistischen Schwankungen.

γ) Das Einschalten eines äußeren Magnetfeldes $\boldsymbol{B}$ bewirkt, da die Kernmomente mit dem Kerndrehimpuls gekoppelt sind, eine gleichsinnige *Präzession* der magnetischen Momente um die Richtung von $\boldsymbol{B}$ mit der LARMOR-*Frequenz* f_L. Es entsteht eine statische Magnetisierung $\boldsymbol{M}_0$ in Richtung des Feldes $\boldsymbol{B}$ und eine dynamische, die aber von außen nicht wahrgenommen werden kann, da jedes Moment mit einer anderen Phasenlage präzediert.

Läßt man jedoch ein entsprechend zirkular polarisiertes Wechselfeld $\mathfrak{B}_1$ von der Frequenz f_L auf das Spinsystem wirken, so treten Zusatzkräfte auf, die nach

---

[2] G. HERTZ (Herausgeber): Grundlagen und Arbeitsmethoden der Kernphysik, S. 243. Berlin: Akademie-Verlag 1957.
[3] I. I. RABI: Phys. Rev. 55, 526 (1939).

einer Zeit die gleichphasige Präzession erzwingen. Es entsteht so eine rotierende makroskopische Magnetisierung, die sich z. B. als induzierte Spannung in einer die Probe umgebenden Spule nachweisen läßt.

Schaltet man das Wechselfeld $\mathfrak{B}_1$ ab, so läßt die Gleichphasigkeit der präzedierenden Kernmomente allmählich nach. Das bewirken sowohl die Spin-Gitter-Wechselwirkung (Gitter bedeutet hier die Gesamtheit aller Freiheitsgrade der Kerne mit Ausnahme der Spinorientierung) als auch die Spin-Spin-Wechselwirkung. Diese drückt eine Störung der Spins untereinander aus, die ohne Änderung des Gesamtenergiegehaltes des Systems vor sich geht. Die Zeitkonstante dieses Vorgangs wird mit T_2^* (transversale Relaxationszeit)[4] bezeichnet. Schaltet man das Gleichfeld B ab, so verschwindet wegen der Spin-Gitter-Wechselwirkung allmählich die zu B parallele Magnetisierung M_0. Diese Zeitkonstante wird mit T_1 (longitudinale Relaxationszeit) bezeichnet. Mathematisch wird dieser Sachverhalt in den BLOCHschen Gleichungen[5] ausgedrückt:

$$\left.\begin{aligned}\frac{dM_x}{dt} - \gamma(M_y B_z - M_z B_y) + \frac{1}{T_2^*} M_x &= 0, \\ \frac{dM_y}{dt} - \gamma(M_z B_x - M_x B_z) + \frac{1}{T_2^*} M_y &= 0, \\ \frac{dM_z}{dt} - \gamma(M_x B_y - M_y B_x) + \frac{1}{T_1} M_z &= \frac{1}{T_1} M_0, \\ M_0 &= \frac{\chi_I B}{\mu_0}.\end{aligned}\right\} \quad (16.6)$$

Dabei bedeuten M_x, M_y, M_z die (allgemein zeitlich veränderlichen) Komponenten der Kernmagnetisierung, B_x, B_y, B_z die Magnetfeldkomponenten, γ das gyromagnetische Verhältnis nach Gl. (16.3) und χ_I die statische Kernsuszeptibilität, die für Protonen in Wasser bei Zimmertemperatur den Wert $3,3 \cdot 10^{-10}\, 4\pi/\mathrm{u}$ annimmt *.

17. Kernresonanz im Erdfeld.

α) Eine einfache Lösung für das System der BLOCHschen Gleichungen ergibt sich für den sog. „langsamen Durchgang". Man tastet die Resonanzlinie durch Änderung von B oder ω so langsam ab, daß für jeden momentanen Zustand die Nachwirkung vergangener Zustände vernachlässigbar ist. Das Differentialgleichungssystem wird dann durch das Verschwinden der zeitlichen Ableitungen zum linearen Gleichungssystem. Der üblichen Darstellungsweise folgend werden zwei um 90° phasenverschobene Magnetisierungskomponenten eingeführt (u' und v'), die mit der Kreisfrequenz ω in der zu B senkrechten Ebene rotieren:

$$M_x = u' \cos\omega t - v' \sin\omega t; \quad M_y = \mp(u' \sin\omega t + v' \cos\omega t). \quad (17.1)$$

Als Lösungen ergeben sich dann

$$\left.\begin{aligned}u' &= \frac{\omega \chi_I B_1 T_2^*}{\mu_0} \cdot \frac{\Delta\omega T_2^*}{1 + (\Delta\omega)^2 T_2^{*2} + \gamma^2 B_1^2 T_1 T_2^*}, \\ v' &= \frac{\omega \chi_I B_1 T_2^*}{\mu_0} \cdot \frac{1}{1 + (\Delta\omega)^2 T_2^{*2} + \gamma^2 B_1^2 T_1 T_2^*}\end{aligned}\right\} \quad (17.2)$$

---
[4] T_2^* enthält den (verkürzenden) Einfluß von Feldinhomogenitäten. Die transversale Relaxationszeit in streng homogenen Feldern bezeichnet man mit T_2.
[5] F. BLOCH: Phys. Rev. **70**, 460—473 (1946).
* u = 1 in rationalisierten Maßsystemen (z. B. in SI-Einheiten), u = 4π in nicht-rationalisierten Maßsystemen (z. B. in den häufigst benutzten cgs-Einheiten).

als Funktionen von $\Delta\omega = \omega - \omega_0$. Für das Maximum des Absorptionssignals und dessen Linienbreite erhält man bei optimaler Wahl der Größe des Wechselfeldes B_1

$$v'_{\max,\,\text{opt}} = \frac{1}{2}\sqrt{\frac{T_2^*}{T_1}}\,M_0; \quad \Delta\omega_{\frac{1}{2}} = \frac{\sqrt{2}}{T_2^*}. \tag{17.3}$$

Eine typische Versuchsanordnung hierzu zeigt Fig. 18. Mit der Primärspule des Kreuzspulensystems nach BLOCH[1] wird das auf die Probe wirkende Wechselfeld $\mathfrak{B}_1$ erzeugt. Die Sekundärspule ist senkrecht zur Achse der Primärspule innerhalb dieser angebracht. Eine Kopplung beider Spulen erfolgt somit nur durch die Magnetisierungskomponenten u' und v', die im Resonanzfall in der Sekundärspule Spannungen induzieren. Durch besondere, hier nicht

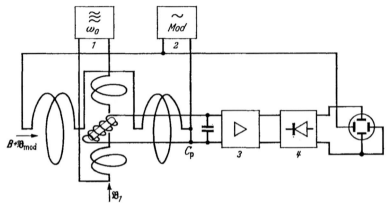

Fig. 18. Schematische Darstellung einer Kernresonanzapparatur mit Kreuzspulanordnung nach BLOCH. Der Generator *1* erzeugt einen Strom mit der Kreisfrequenz ω_0, der durch eine Spule das Wechselfeld $\mathfrak{B}_1$ über der Probe aufbaut. Die dazu um 90° gedrehte Empfängerspule ist normalerweise elektrisch entkoppelt. Am Eingang des Hochfrequenzverstärkers *3* entsteht nur dann eine Spannung, wenn das Gleichfeld B und ω_0 die LARMOR-Beziehung (16.5) erfüllen. Das wird mit Hilfe des Modulationsfeldes $\mathfrak{B}_{\text{mod}}$ periodisch realisiert. Man kann so nach der Gleichrichtung *4* auf dem Oszillographen im Takt der Modulationsfrequenz die beim Durchgang durch die Resonanzstelle auftretenden Kernresonanz-Signale beobachten.

näher beschriebene Maßnahmen (Anwendung phasenempfindlicher Gleichrichter) gelingt es, z.B. nur das v'-(Absorptions-) Signal zu beobachten. Durch eine Modulation des Gleichfeldes B kann man die gesamte Linie periodisch abtasten und nach Verstärkung beispielsweise auf dem Oszillographenschirm darstellen.

β) Wollte man mit einer derartigen Anordnung ein Magnetfeldmeßgerät aufbauen, so würde nicht allein die in der Sekundärspule induzierte Spannung

$$U_B = \frac{\mu_0\,\xi\,n S\omega Q}{c_0\sqrt{\varepsilon_0\,\mu_0}} \cdot \frac{1}{2}\sqrt{\frac{T_2^*}{T_1}} \cdot \mathrm{u}\,M_0 \cos\omega t \tag{17.4}$$

(n = Windungszahl, S = Windungsfläche, Q = Spulengüte, ξ = Füllfaktor für die Probenspule) maßgebend für die Meßgenauigkeit sein, sondern es müßte nach einem möglichst hohen *Signal-Rausch-Verhältnis* S_N getrachtet werden. Berücksichtigt man das Rauschen, das vom Eingangsschwingkreis (gebildet aus Sekundärspule und Abgleichkondensator C_p für die Resonanzfrequenz ω) verursacht wird, so ergibt sich näherungsweise

$$S_N = \frac{\mu_0 \xi \omega \pi r^2 h \sqrt{\mathrm{u}\,\mu_0\,\pi l_s K_1}}{(c_0\sqrt{\varepsilon_0\,\mu_0})^2 8\,\varrho_s\sqrt{0{,}5\,kT}} \cdot \frac{1}{\sqrt{\phi_N}} \cdot \sqrt{\frac{T_2^*}{T_1}} \cdot \mathrm{u}\,M_0 = A_s \cdot A_v \cdot A_p^B. \tag{17.5}$$

(k = BOLTZMANN-Konstante, T = absolute Umgebungstemperatur, r = Spulenradius, h = Wickelhöhe, l_s = Spulenlänge, K_1 = Spulen-Formfaktor, ϱ_s = spezifischer Widerstand des

---

[1] F. BLOCH, W. W. HANSEN, and M. PACKARD: Phys. Rev. **70**, 474—485 (1946).

Drahtes der Spulenwicklung, $\phi_N =$ Rauschzahl des Verstärkers). Die Bandbreite Δf des Verstärkers wurde hierbei gleich der Bandbreite des Eingangsschwingungskreises gesetzt, um eine übersichtliche Formel zu erhalten. Andernfalls ergibt sich für S_N eine Abhängigkeit von $\Delta f^{-\frac{1}{2}}$. Gl. (17.5) läßt sich in drei Faktoren zerlegen, von denen der erste, A_s, nur von Parametern der Probenspule abhängt, A_v den Verstärker kennzeichnet und A_P^B von der Probensubstanz und dem angewandten Magnetfeld bestimmt wird. Zunächst wäre im Hinblick auf einen hohen Wert von A_P^B zu fordern, daß die Probenflüssigkeit eine beträchtliche Relaxationszeit T_2^* aufweist. Ist sie frei von paramagnetischen Ionen (z.B. Cu- oder Fe-Ionen) und ist das Feld B im Probenraume homogen, so ist T_2^* nahezu gleich T_2. Tabelle 4 zeigt die Größenordnung von T_2 für Protonen in verschiedenen Flüssigkeiten. Die daraus folgenden Linienbreiten entsprechen den für verschiedene erdmagnetische Zwecke zu stellenden Forderungen. Große Relaxationszeiten verlangen sehr lange Abtastzeiten. Da im allgemeinen $T_1 \approx T_2$ gilt, wird man $T_2^* \approx T_2$ anstreben. Die generelle Wahl von protonenhaltigen Flüssigkeiten beruht darauf, daß das Proton das höchste gyromagnetische Verhältnis von allen geeigneten Kernen hat.

Tabelle 4 (nach HOCHSTRASSER).

Probe	$\dfrac{T_2}{s}$	$\dfrac{\Delta B_{\frac{1}{2}}}{\gamma}$
Nitromethan	0,29	18
Cis-Dekalin	0,29	18
Wasser	2,7	1,9
Benzol	16	0,33

Im Hinblick auf einen großen Faktor A_v verwendet man schmalbandige, rauscharme Verstärker, vorzugsweise nach dem „lock-in"-Prinzip[2] arbeitende. Dennoch ergeben sich für Kernresonanzversuche im Erdfeld bei Ausnutzung aller Parameter sehr große Probenvolumina[3].

Eine andere Möglichkeit bietet die Verwendung eines stärkeren Zusatzfeldes zur Vergrößerung von M_0. Beispielsweise erhält man das doppelte Erdfeld als Differenz aus zwei Messungen mit verschiedener Polarität des Zusatzfeldes[4] oder man transformiert die Erdfeldvariationen mit Hilfe von Zusatzspulen in einen höheren Feldbereich[5].

Fig. 19. Schematische Darstellung eines mit freier Präzession arbeitenden Magnetometers. Ein Relais verbindet die Spule, die die Kernresonanzprobe enthält, abwechselnd mit einer Gleichspannungsquelle (Aufbau des Polarisationsfeldes) und mit dem Verstärkereingang (Signalempfang). Der gleiche Steuerteil, der diese Vorgänge auslöst, synchronisiert auch entsprechend die Messung der Resonanzfrequenz, die sich im Erdfeld ergibt.

γ) Erst die *Methode der „freien Präzession"* war es, die der Kernresonanz zum erfolgreichen Eingang in die erdmagnetische Meßpraxis verhalf[6]. Sie gestattet bei geringem technischen Aufwand die direkte Messung der Totalintensität. Man läßt

---

[2] R. H. DICKE: Rev. Sci. Instr. **17**, 268 (1946).
[3] G. HOCHSTRASSER: Hochfrequenzspektroskopie, S. 24. Berlin: Akad.-Verl. 1961.
[4] H. PFEIFFER: Nachrichtentechnik **3**, 371 (1953).
[5] H. SCHMIDT: Exptl. Tech. Physik **1**, 121—127 (1953).
[6] M. PACKARD, and R. VARIAN: Phys. Rev. **93**, 941 (1954).

zunächst ein starkes Hilfsfeld B_h (einige hundert Male größer als das Erdfeld) möglichst senkrecht zum Erdfeldvektor auf die Probe wirken. Nach der oben skizzierten Vorstellung ergibt sich eine Präzession der Kernspins um B_h und eine statische Kernmagnetisierung M_h in Richtung von B_h (Fig. 19). Der Gleichgewichtszustand stellt sich mit der Zeitkonstante T_1 ein. Dann wird B_h rasch nicht adiabatisch abgeschaltet, d.h. es muß für die Abschaltzeit Δt die Beziehung

$$\Delta t \ll \frac{2\pi}{\gamma B} \tag{17.6}$$

gelten. Die Magnetisierung M_h bricht nicht sofort zusammen, sondern präzediert nun ihrerseits um das Erdfeld F mit der Kreisfrequenz $\omega = \gamma \mathcal{F}$. Die BLOCHschen Gleichungen reduzieren sich für diesen Fall auf die folgende Form:

$$\left.\begin{aligned}\frac{dM_x}{dt} - \gamma M_y \mathcal{F} + \frac{M_x}{T_2^*} &= 0, \\ \frac{dM_y}{dt} - \gamma M_x \mathcal{F} + \frac{M_y}{T_2^*} &= 0, \\ \frac{dM_z}{dt} + \frac{M_z}{T_1} &= \frac{M_\mathrm{h}}{T_1}; \quad M_\mathrm{h} = \frac{\chi_\mathrm{I} B_\mathrm{h}}{\mu_0}.\end{aligned}\right\} \tag{17.7}$$

Die Lösungen für die Magnetisierungskomponenten haben dann z.B. die Form

$$M_x = M_\mathrm{h} \sin \Theta_\mathrm{h}\, e^{-\frac{t}{T_2^*}} \sin \omega t. \tag{17.8}$$

(Θ_h = Winkel zwischen Spulenachse und Feldvektor, t = Zeit-Variable). Wird die Spule für die Kreisfrequenz auf ω abgeglichen, so ergibt sich für das Signal-Rausch-Verhältnis

$$S_N = \frac{\mu_0\, \xi\, \omega\, \pi\, r^2\, h \sqrt{u\, \mu_0\, \pi\, l_s\, k_1}}{(c_0\sqrt{\varepsilon_0\,\mu_0})^2\, 8\varrho\sqrt{0,5\, kT}} \cdot \frac{1}{\sqrt{\varphi_N}} \cdot \sqrt{\frac{T_2^*}{T_1}}\, u M_\mathrm{h} \sin_\zeta \Theta_\mathrm{h}\, e^{-\frac{t}{T_2^*}} = A_s\, A_v\, A_P^F. \tag{17.9}$$

Dieser Ausdruck zeigt einen ähnlichen Aufbau wie Gl. (17.5). Die von den Spulenparametern und den Eigenschaften des Eingangsverstärkers bestimmten Faktoren A_s und A_v sind dieselben. Entscheidend ist die Vergrößerung von S_N etwa um den Faktor $B_\mathrm{h}/\mathcal{F}$. Diesem Vorteil steht die in A_P^F enthaltene exponentielle Abnahme der Signalamplitude gegenüber. Immerhin erhält man für die Dauer eines Meßzeitintervalles, das in der Größenordnung von T_2^* liegt, auch für relativ kleine Spulenvolumina (ca. 100—250 cm³) Werte für S_N von etwa 50. Nach Tabelle 4 sind somit Meßzeiten von bestenfalls einigen Sekunden realisierbar.

Eine kontinuierliche Frequenzabgabe ist mit der freien Präzession nicht zu realisieren, vielmehr muß ein ständiger Wechsel von Polarisation und Signal erfolgen. Die Signalfrequenz ist nach Gl. (16.5) direkt proportional der Totalintensität $\mathcal{F}$ des Erdfeldes.

18. Technik der Protonenmagnetometer.

α) Die wesentlichen Teile eines Magnetometers, das nach dem Prinzip der freien Präzession arbeitet, sind in Fig. 19 dargestellt. Einige Besonderheiten der einzelnen *Baugruppen* werden im folgenden beschrieben.

(I) *Spule*. Gewöhnlich wird zur Erzeugung des Hilfsfeldes B_h, des sog. Polarisationsfeldes, und zum Empfang des Kernresonanzsignals die gleiche Spule verwendet. Durch richtige Wahl der Spulenabmessungen und des Drahtes kann man erreichen, daß sowohl die notwendigen Polarisationsfeldstärken wie auch ein günstiges Signal-Rausch-Verhältnis realisiert werden.

Die Spulendaten können beispielsweise so gewählt werden:
$l_s = 10$ cm; $r = 3$ cm; $h = 2$ cm; $n = 800$; $d = 1,2$ mm; $V_s = 300$ cm³ (V_s = Spulenvolumen, n = Windungszahl, d = Drahtdurchmesser).

Bei 1,5 A Polarisationsstrom folgt ein Signal-Rausch-Verhältnis von 30 bis 50. Der Draht zum Bewickeln der Spule ist sorgfältig auszusuchen, Metall und Isolation sollen frei von ferromagnetischen Verunreinigungen sein. Ferner zeigen die Spulen der hier gebräuchlichen Abmessungen, sofern Volldraht benutzt wird, bei den Erdfeld-Signalfrequenzen (Größenordnung 2 kHz) einen starken Skin-Effekt[1]. Die hierdurch verursachte Verringerung der Spulengüte umgeht man durch Verwendung von vieladriger Hochfrequenz-Litze.

Die durch den Polarisationsstrom erzeugte Erwärmung der Spule muß besonders für den Dauerbetrieb in zulässigen Grenzen gehalten werden, weswegen das Polarisationsfeld nicht beliebig groß gewählt werden kann. Die Spulendaten bestimmen über A_s in Gl. (17.9) wesentlich das Signal-Rausch-Verhältnis und damit auch, wie weiter unten gezeigt wird, die Empfindlichkeit des gesamten Magnetometers.

(II) *Steuerteil*. Die beiden Arbeitszustände des Magnetometers, Polarisation und Signalempfang, werden im einfachsten Fall durch ein Relais gesteuert. Dieses verbindet die Spule in der einen Kontaktlage mit der Gleichspannungsquelle und in der anderen mit dem Verstärkereingang. Die Schaltzeiten des Relais, vorgegeben durch Multivibratoren oder durch Handbedienung, sind für die Polarisationsintervalle nach T_1 und für die Meßzeiten nach T_2^* zu bemessen.

Die Funktion eines Kernresonanzmagnetometers hängt entscheidend davon ab, wie weit die Abschaltbedingung nach Gl. (17.6) eingehalten wird.

Einen raschen Abbau der magnetischen Energie des Polarisationsfeldes erreicht man z.B. durch stufenweises Abschalten und zusätzliches Bedämpfen der Spule durch einen OHMschen Widerstand im Abschaltmoment[2] oder durch Ausnutzen des Öffnungsfunkens bei genügend langer Ankerflugzeit des Relais. Die dabei auftretenden hohen Spannungen begrenzt man durch Glimmstrecken oder Varistoren. Die Funktionen von Glimmstrecke und Relais werden auch von Schalttransistoren übernommen, die durch von Intervallgebern stammende Impulsfolgen gesteuert werden.

(III) Als *Verstärker* können prinzipiell normale Niederfrequenz-Verstärker verwendet werden, die einen Eingangswiderstand haben, der groß gegen den Widerstand des Eingangskreises (Größenordnung 10 kΩ) ist. Das Rauschen der Verstärkereingangsstufe läßt sich bei entsprechender Wahl der Verstärkerelemente klein halten gegenüber dem, das der Eingangsschwingkreis erzeugt[3].

Die Bandbreite des Verstärkers kann man in diesem Frequenzbereich sowohl durch LC-Schwingkreise wie auch durch RC-Netzwerke einstellen. Um die Forderungen nach einer geringen Bandbreite und der Anwendbarkeit des Gerätes für einen großen Bereich von $\mathcal{F}$ (meist in der Größenordnung von $10^4 \gamma$) zu erfüllen, wird gewöhnlich mit umschaltbaren Frequenzbereichen gearbeitet. Die Umschaltung muß dann auch auf die Abgleichkondensator des Eingangsschwingungskreises ausgedehnt werden.

(IV) *Frequenzmessung*. Die Messung der dem Erdfeldbetrag proportionalen Kernpräzessionsfrequenz kann mit jedem kommerziellen Frequenzmeßgerät erfolgen, das für Niederfrequenzen geeignet ist und die Meßgenauigkeit von 10^{-5} erreicht.

Am Anfang der Kernresonanzmeßtechnik verwendete man Schwebungsverfahren mit einer der Meßfrequenz dicht benachbarten Normalfrequenz oder mit einer durchstimmbaren Normalfrequenz[4]. Die Abstimmung wurde mit Hilfe von LISSAJOUS-Figuren durchgeführt. Gegenwärtig benutzt man ausschließlich digitale Frequenzmeßgeräte mit fester Normalfrequenz, die die Periodendauer oder die Frequenz des Signals als Ziffernfolge anzeigen oder ausdrucken. Die Meßwerte lassen sich auf Lochstreifen, Lochkarte oder Magnetband speichern. Die Frequenzmeßgeräte können so konstruiert werden, daß die Anzeige in magnetischen Einheiten (etwa in γ) erfolgt. Wegen des einfachen Zusammenhanges zwischen Magnetfeld und Kernpräzessionsfrequenz nach Gl. (16.5) ist das leicht möglich. Kernpräzessionsmagnetometer und Frequenzmeßgerät brauchen selbstverständlich keine Einheit zu bilden. Die

---
[1] A. SOMMERFELD: Vorlesungen über theoretische Physik III, S. 175—182, 1954.
[2] G. KLOSE: Z. Angew. Phys. **10**, 495—497 (1958).
[3] H. PFEIFFER: Elektronisches Rauschen. Leipzig: Teubner 1959.
[4] H. SCHMIDT: Erdmagn. Jahrbuch Niemegk 1959, S. 170—172.

Übertragung der Meßfrequenz zwischen beiden Geräten kann, je nach Eigenart der Anwendung, über Kabel oder drahtlos über große Entfernungen erfolgen[5].

β) *Genauigkeitsbetrachtungen.* Die Meßgenauigkeit der Totalintensitätswerte des geomagnetischen Feldes, mit Protonenmagnetometern bestimmt, wird nach Gl. (16.5) durch die absolute Genauigkeit des gyromagnetischen Verhältnisses des Protons γ_P und die Frequenzmeßgenauigkeit begrenzt.

Als Mittelwert zahlreicher, von staatlichen metrologischen Institutionen vorgenommener Messungen ist anzugeben:

$$\gamma_P = 2{,}67513 \cdot 10^4 \, \Gamma^{-1} \, s^{-1} \, [6]; \quad \gamma_P = 2{,}67513 \cdot 10^8 \, \frac{m^2}{V s^2} \, [7]. \tag{18.1}$$

Der kleinste Fehler dieser Absolutbestimmungen liegt bei $2 \cdot 10^{-5}$. In die Genauigkeit der Frequenzmessung gehen zusätzlich die Parameter der gesamten Apparatur ein (Fehler des elektronischen Frequenzmeßteils bei abklingender Signalamplitude, Fehler durch ferromagnetische Unreinheiten des Spulenmaterials sowie unzureichende Isolation der Relaiskontakte usw.[8].

Hier sei nur der an einem Protonenmagnetometer mit Periodendauermessung auftretende Fehler der abklingenden Signalamplitude erläutert. Gl. (16.5) nimmt die Form an:

$$\mathcal{F} = \frac{2\pi \, m \, f_Q}{\gamma_P \, z} \quad \text{mit} \quad z = \frac{m \, f_Q}{f} \quad \text{und} \quad \tau = \frac{m}{f}. \tag{18.2}$$

($\mathcal{F}$ = Totalintensität, f = Signalfrequenz, m = Zahl der gemessenen Perioden, f_Q = Normalfrequenz, z = abgelesene Impulszahl, τ = Meßzeit). Setzt man Konstanz von $\mathcal{F}$ während der Messung voraus, so folgt als Fehler der Impulszahl bei gleichbleibender Signalamplitude ± 1.

Durch das exponentielle Abklingen des Signals aber wird eine zusätzliche Meßzeit $\Delta \tau$ verursacht:

$$\Delta \tau = \tau_E - \tau_A = \frac{1}{\omega} \left(e^{-\frac{\tau}{T_2^*}} \arcsin p - \arcsin p \right). \tag{18.3}$$

Dabei bezeichnet $p A_0$ die Höhe der Spannung, bei deren Erreichen die Steuerimpulse ausgelöst werden (Fig. 20). Als Fehlergleichung erhält man somit

$$\frac{\Delta \mathcal{F}}{\mathcal{F}} \approx \frac{\Delta \gamma_P}{\gamma_P} + \frac{\Delta f_Q}{f_Q} + \frac{1}{c \, f_Q} + \frac{p}{\omega \, T_2^*}, \tag{18.4}$$

woraus nach Einsetzen von Zahlenwerten einige 10^{-5} folgen. Hält man die Amplitude während der Meßzeit durch Regel- oder Begrenzer-Schaltungen konstant, so verursacht das bei der Ableitung der Gl. (18.3) vernachlässigte Rauschen auf sinngemäß ähnliche Weise eine Unsicherheit von τ. Nach Fig. 21 erhält man für den letzten Summanden in Gl. (18.4) näherungsweise

$$\frac{\Delta \tau_N}{\tau} \approx \pm \frac{e^{\tau/T_2^*}}{S_{N_0} \, \omega \, \cos(\arcsin p)}, \tag{18.5}$$

wobei S_{N_0} das Signal-Rausch-Verhältnis zu Beginn des Signals bedeutet.

Zusammenfassend kann man feststellen, daß absolute[9] Messungen der erdmagnetischen Totalintensität mit einem Fehler in der Größenordnung von $0{,}1 \, \gamma$ mit Protonenmagnetometern möglich sind.

γ) *Anwendungshinweise.* Die Einführung von Protonenmagnetometern hat viele Spezialgebiete der geomagnetischen Forschung entscheidend bereichert. Der Einsatz in Observatorien führte nicht nur zu einer Genauigkeitssteigerung der Totalintensitätsbestimmung, sondern ergab auch eine Verbesserung des Niveaus

---
[5] L. R. ALLDREDGE, and I. SALDUKAS: Coast and Geodetic Survey Technical Bull. No. 31 (1966). — D. LENNERS u. H. SCHMIDT: Gerlands Beitr. Geophys. 72, 253—265 (1963).

[6] Dieser Wert wurde international zur Vereinheitlichung aller Erdfeld-Protonenmagnetometer-Messungen mit H_2O als Probe-Material für verbindlich erklärt (IAGA-Resolution Nr. 66 des 12. IUGG-Kongresses, Helsinki 1960).

[7] Siehe hierzu: J. H. NELSON: Rep. of Comm. on Magnetic Instr. Nr. 8 (13. IUGG-Kongr. Berkeley 1963).

[8] H. SCHMIDT: Erdmagnet. Jahrbuch Niemegk 1962, S. 141—144.

[9] Im Sinne der geomagnetischen Meßpraxis.

für die Vertikalintensität Z. Diese konnte nun mit größerer Genauigkeit aus $\mathcal{F}$ und dem klassisch bestimmten $\mathcal{H}$ als aus $\mathcal{H}$- und $\mathcal{J}$-Messungen ermittelt werden. Als weitere Vorzüge muß man die Einfachheit des Meßvorganges und den Verzicht auf Temperaturkorrekturen ansehen.

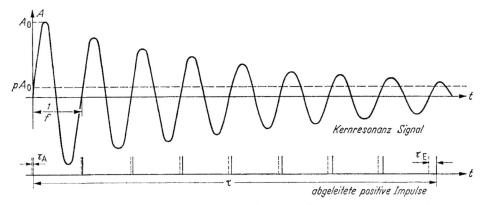

Fig. 20. Wirkung des Impulsformers im Hinblick auf amplitudenabhängige Verschiebung der Auslöse-Impulse für die Torsteuerung.

Fig. 21. Schematische Darstellung des Fehlers, den ein verrauschtes Kernresonanzsignal verursacht (A_0 Amplitude ohne Rauschen, N mittlere Rauschamplitude).

Transportable Geräte eignen sich ihres geringen Gewichtes wegen vorzüglich für Geländemessungen. Hierfür ist als Vorteil zu vermerken, daß die Orientierung der Spule im Erdfeld weitgehend unkritisch ist. Das ermöglicht auch den Einsatz von Protonenmagnetometern in Schleppkörpern für magnetische Seevermessungen und in der Aeromagnetometrie. Auch für geomagnetische Messungen von Ballons, Raketen und künstlichen Erdstatelliten aus wurden Protonenmagnetometer mit Erfolg eingesetzt, wobei man besonders gut die Möglichkeiten der drahtlosen Meßwertübertragung über große Entfernungen nutzen konnte.

Neben der Bestimmung von Magnetfeldern und Magnetfeld-Differenzen (z.B. auch bei archäologischen Erkundungsvorhaben) sind Anwendungen im Hinblick auf die Untersuchung von Materialeigenschaften in der Gesteinsphysik und in der Bohrloch-Meßtechnik zu verzeichnen. Auch die Strukturaufklärung mit Kernresonanz-Spektrometern ist geophysikalisch genutzt worden.

Die Kernresonanz im Erdfeld hat nicht nur geomagnetische Aspekte, sondern auch rein physikalische[10].

---

[10] P. M. BORODIN, A. B. MELNIKOV, A. A. MOROSOV i J. S. ČERNIČEV: Jaderni resonanz v zemnom polje (Kernresonanz im Erdfeld). Leningrad: Izdatelstvo Leningradskovo Universiteta 1967.

II. Magnetometer mit optisch gepumpten Gasen und Dämpfen zur Bestimmung der Totalintensität.

Im vorhergehenden Abschnitt wurden Magnetfeldmeßmethoden beschrieben, die die Aufspaltung der Energiezustände des Atomkerns im äußeren Magnetfeld ausnutzen. In ähnlicher Weise kann man auch die Übergänge zwischen benachbarten ZEEMAN-Termen der Elektronenhülle benutzen, um Rückschlüsse auf die Größe des äußeren Magnetfeldes zu ziehen. Mit Mitteln der optischen Interferenzspektroskopie lassen sich diese Aufspaltungen nicht mehr nachweisen, da die durch den DOPPLER-Effekt bedingten Linienbreiten im sichtbaren Spektralgebiet einige Größenordnungen über den zu erwartenden Übergangsfrequenzen liegen. Ihr Nachweis gelingt jedoch mit Mitteln der Hochfrequenzspektroskopie. Bei der Verwendung von verdünnten Gasen und Dämpfen als Proben und schwachen Magnetfeldern ist leider auch hier eine direkte Beobachtung nicht möglich. Die Besetzung der ZEEMAN-Terme weicht in diesem Falle kaum von der sog. „statistischen Besetzung" ab. Zum sicheren Nachweis der Übergangsfrequenz muß erst künstlich eine Veränderung der Besetzungszahlverhältnisse herbeigeführt werden. Ein Verfahren, das hier zum Ziel führt, ist das erstmalig von KASTLER vorgeschlagene „optische Umpumpverfahren".

Für Magnetometer, die auf diesem Effekt aufbauen, gelten die am Anfang von Kapitel C zusammengestellten allgemeinen Merkmale.

19. Physikalische Grundlagen.

α) Das Verfahren des optischen Pumpens[1] soll hier zunächst unter *Vernachlässigung* der durch den *Kernspin* verursachten Hyperfeinstruktur an der D_1-Linie des Na-Atoms erläutert werden[2]. Die Aufspaltung der Terme der D_1-Linie in einem äußeren Magnetfeld zeigt Fig. 22. Die möglichen Übergänge sind ebenfalls

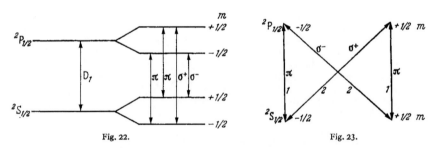

Fig. 22.   Fig. 23.

Fig. 22. Idealisiertes Termschema zur Verdeutlichung des ZEEMAN-Effektes.

Fig. 23. Schema zur Erläuterung des optischen Pumpens (nach KOPFERMANN). Hier sind die Terme als Punkte dargestellt. Die Pfeile bedeuten die möglichen spontanen Übergänge, die Zahlen relative Übergangswahrscheinlichkeiten. Ferner ist der Polarisationszustand für die emittierte bzw. absorbierte Strahlung angegeben.

eingezeichnet. Die dabei emittierten oder absorbierten Wellenzüge haben verschiedene Polarisationszustände, die sich nach der Änderung der Quantenzahl m richten. Für $\Delta m = 0$ erhält man eine Polarisation in Richtung des Magnetfeldes (π), für $\Delta m = +1$ rechtszirkulare (σ^+) und für $\Delta m = -1$ linkszirkulare (σ^-) Polarisation. Man strahlt nun Resonanzlicht mit einem bestimmten Polarisationszustand ein. Dadurch wird eine starke Abweichung von der statistischen

---
[1] A. KASTLER: J. Phys. Radium, **11**, 255 (1950).
[2] Unserer Darstellung legen wir folgende Arbeit zugrunde: H. KOPFERMANN: Sitzungsber. d. Heidelb. Akad. d. Wiss., Heidelberg, 1960.

Besetzungszahlverteilung der ZEEMAN-Terme erreicht. Die dazugehörigen Vorgänge lassen sich leicht an dem Schema in Fig. 23 erkennen, in dem noch die relativen Übergangswahrscheinlichkeiten vermerkt sind. Strahlt man z.B. σ^+-Licht ein, so ist nur der Übergang $^2S_{\frac{1}{2}}$, $m = -\frac{1}{2}$ nach $^2P_{\frac{1}{2}}$, $m = +\frac{1}{2}$ möglich. Dieser Zustand zerfällt in $^2S_{\frac{1}{2}}$, $m = \pm\frac{1}{2}$, wobei dann zwei Drittel der Atome den Zustand $^2S_{\frac{1}{2}}$, $m = -\frac{1}{2}$ und ein Drittel den Zustand $^2S_{\frac{1}{2}}$, $m = +\frac{1}{2}$ bevölkern. Der Zustand $^2S_{\frac{1}{2}}$, $m = +\frac{1}{2}$ reichert sich so auf Kosten von $^2S_{\frac{1}{2}}$, $m = -\frac{1}{2}$, an, da der spontane Übergang innerhalb des $^2S_{\frac{1}{2}}$-Zustandes „verboten" ist. Die Übergänge zwischen $m = +\frac{1}{2}$ und $m = -\frac{1}{2}$ können jedoch erzwungen werden, wenn man ein

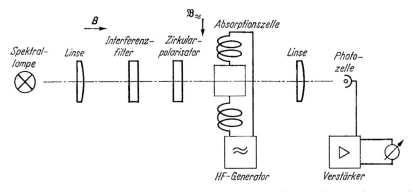

Fig. 24. Prinzipielle Darstellung einer Apparatur zum Nachweis des optischen Pumpens. Licht von der zum optischen Pumpen notwendigen Wellenlänge und des entsprechenden Polarisationszustandes durchsetzt die Absorptionszelle. Hat das senkrecht zum magnetischen Gleichfeld B erzeugte Hochfrequenz-Feld $\mathfrak{B}_{\approx}$ die richtige, durch die Linienaufspaltung im Magnetfeld gegebene Frequenz, dann kann man eine Intensitätsänderung des durch die Absorptionszelle gelangten Lichtes nachweisen.

magnetisches Wechselfeld mit der Übergangsfrequenz $f = f_L$ für den betreffenden ZEEMAN-Übergang mit genügender Intensität einstrahlt. Es ergibt sich so die Möglichkeit, die LARMOR-Frequenz f_L zu bestimmen.

Die Versuchsanordnung hierzu zeigt schematisch Fig. 24[3]. Aus dem von einer Spektrallampe ausgehenden Licht wird der Anteil der gewünschten optischen Resonanzlinie herausgefiltert, durchläuft nach Passieren eines Zirkularpolarisators (z.B. für σ^+) die Probenzelle und gelangt auf eine Photozelle zur Intensitätsbestimmung. Die Frequenz des Hochfrequenz-Feldes, das durch Spulen über der Probenzelle senkrecht zu B erzeugt wird, sei zunächst verschieden von f_L. Während des optischen Pumpens nimmt wegen der geringeren Besetzung von $^2S_{\frac{1}{2}}$, $m = -\frac{1}{2}$ die Absorptionswahrscheinlichkeit ab und die Intensität des durchgelassenen Lichtes zu. Verändert man die Frequenz des Hochfrequenz-Feldes, so werden für $f = f_L$ Übergänge von $^2S_{\frac{1}{2}}$, $m = +\frac{1}{2}$ nach $^2S_{\frac{1}{2}}$, $m = -\frac{1}{2}$ induziert, wodurch die Absorption des durchgelassenen Lichtes wieder zunimmt. f_L läßt sich somit durch *Einstellen des Absorptionsmaximums* bestimmen. Es gilt[4]

$$f_L = \frac{1}{2\pi} \cdot \frac{q}{c_0\sqrt{\varepsilon_0 \mu_0}\, 2m_e} \cdot g_I \cdot B = \frac{|\mu_B|}{h} \cdot g_I\, B. \qquad (19.1)$$

(g_I = LANDEscher Aufspaltungsfaktor, q = Elementarladung, $|\mu_B| = (9{,}2732 \pm 6 \cdot 10^{-4}) \cdot 10^{-24}$ Am^2 = BOHRsches Magneton, m_e = Elektronenmasse, $h = (6{,}6256 \pm 5 \cdot 10^{-4}) \cdot 10^{-34}$ VAs^2 = PLANCKsches Wirkungsquantum). Für den LANDE-Faktor

---
[3] H. G. DEHMELT: Phys. Rev. **105**, 1487 (1957). — W. E. BELL, and A. L. BLOOM: Phys. Rev. **107**, 1559 (1957).
[4] Eine entsprechende Formel für die LARMOR-Frequenz folgt aus Gl. (16.1) und (16.5).

läßt sich folgende Formel angeben:

$$g_J = 1 + \frac{J(J+1) + S(S+1) - L(L+1)}{2J(J+1)}, \qquad (19.2)$$

wobei J, S und L die Quantenzahlen des $^2S_\frac{1}{2}$-Zustandes sind. Die für die erreichbare Meßgenauigkeit wichtige Linienbreite Δf wird durch die mittlere Lebensdauer τ_L eines Atoms im gepumpten Zustand bestimmt:

$$\Delta f = \frac{1}{\pi \tau_L} \qquad (19.3)$$

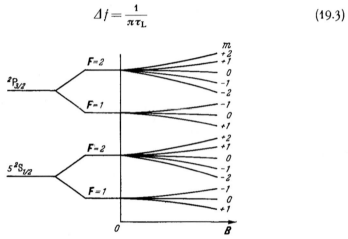

Fig. 25. Schematische Darstellung der Aufspaltung der Terme von Rb[87] im Magnetfeld.

Die Größe τ_L unterliegt Einflüssen durch depolarisierende Stöße der Atome untereinander und Stöße an der Gefäßwandung der Probe. Dieser Effekt kann durch die Anwendung eines neutralen Puffergases verringert werden. Man erhält je nach Gefäßgröße und Gasdruck Linienbreiten der Größenordnungen 10 bis 100 Hz.

β) Weniger übersichtlich gestalten sich die Zusammenhänge, wenn auch der *Kernspin* mit *berücksichtigt* wird. Die dann auftretende Hyperfeinstruktur führt zu einer größeren Anzahl von ZEEMAN-Niveaus, da bereits ohne äußeres Magnetfeld eine Aufspaltung z.B. des Grundzustandes, vorhanden ist. Ein entsprechendes Termschema ist in Fig. 25 dargestellt. Der Effekt des optischen Pumpens ist selbstverständlich auch hier möglich. So werden in dem gezeigten Beispiel bei Einstrahlung von σ^+-Licht die Zustände mit positivem m auf Kosten der mit negativem m angereichert. Bereits bei geringen Feldintensitäten beginnt aber die Entkopplung des Kernmoments von dem magnetischen Moment der Elektronenhülle (PASCHEN-BACK-Effekt)[5]. Man erhält mit der Versuchsanordnung nach Fig. 24 dann bei Frequenzmodulation des HF (Hochfrequenz)-Feldes ein Spektrum dicht benachbarter Linien. Die Lage der Linien beschreibt die BREIT-RABI-Formel[6]. Eine weitere nachteilige Wirkung des Kernspins ist in einer Verringerung des Aufspaltungsfaktors

$$g_F = g_J \frac{F(F+1) + J(J+1) - I(I+1)}{2F(F+1)} < g_J \qquad (19.4)$$

zu sehen. Dabei ist F (Feinquantenzahl)[7] die Resultierende aus dem Kernspin I und der inneren Quantenzahl J der Elektronenhülle.

---

[5] Siehe z.B. Handbuch der Experimentalphysik, Bd. 16 (I). Leipzig 1936.
[6] J. M. B. KELLOGG, and S. MILLMAN: Rev. Mod. Phys. **18**, 323 (1946).
[7] Das Symbol F wurde nur in der Formel (19.4) nicht für die erdmagnetische Totalintensität verwendet, um die in der Literatur übliche Schreibweise wahren zu können.

20. Optisches Pumpen im Erdfeld. Sollen die Zeeman-Übergänge an einem optisch gepumpten Gas oder Dampf zur Messung des relativ schwachen Erdfeldes benutzt werden, so hat man die Substanz nach Gesichtspunkten auszuwählen, die denen bei der Kernresonanz im Erdfeld ähneln. Zunächst wünscht man einen großen Aufspaltungsfaktor g_J. Dann sind Linien geringer Breite für eine hohe Meßgenauigkeit günstig, die wiederum durch das Signal-Rausch-Verhältnis und damit von Parametern der Apparatur mit bestimmt wird. Von den hierfür besonders geeigneten Proben, mit denen der Effekt des optischen Pumpens im Erdfeld realisiert wurde, seien hier zwei typische Vertreter herausgegriffen: Helium und Rubidium.

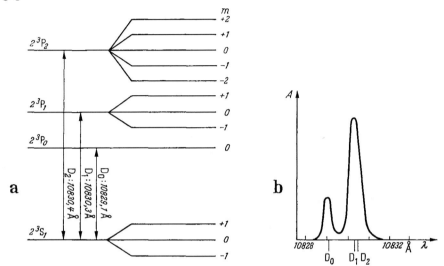

Fig. 26a u. b. a) Schematische Darstellung der Aufspaltung der Terme von He im Magnetfeld. b) Ausschnitt aus dem He-Spektrum (nach COLEGROVE und FRANKEN).

α) Beim *Helium*[1] liegen die Verhältnisse einerseits günstig, da das stabile Heliumisotop He⁴ kein Kernmoment besitzt. Jedoch ist der bei Zimmertemperatur fast ausschließlich besetzte Grundzustand $^1S_0$ im Magnetfeld nicht aufgespalten. Daher wird zum optischen Pumpen der metastabile Zustand $^3S_1$ benutzt, der wegen der Übergangsverbote zum Grundzustand eine genügend große Lebensdauer besitzt. Praktisch erreicht man die Besetzung dieses Zustandes dadurch, daß in der Absorptionszelle eine schwache Gasentladung durch ein außen angelegtes Hochfrequenz-Feld gezündet wird. Die schematische Darstellung für diesen Fall entnimmt man aus dem Termschema Fig. 26a[1]. Ein Ausschnitt aus dem resultierenden Spektrum ist in Fig. 26b qualitativ angegeben. Demnach liegen die D_1- und die D_2-Linie so dicht nebeneinander, daß eine Trennung durch optische Interferenzfilter nicht möglich ist. Dies beeinflußt den Prozeß des optischen Pumpens nicht wesentlich, da der $^3P_2$, $m=+2$-Zustand nur nach $^3S_1$, $m=+1$ zerfallen kann. Allerdings wird dadurch die mittlere Lebensdauer des gepumpten Zustandes verkürzt. τ_L ergibt sich zu etwa 10^{-3} sec. Demgegenüber führt der relativ große Aufspaltungsfaktor für He zu der Beziehung

$$\frac{f_L}{\text{Hz}} \approx 2{,}8 \cdot 10^6 \frac{B}{\Gamma} \tag{20.1}$$

---
[1] Zum Beispiel P. A. FRANKEN and F. D. COLEGROVE: Phys. Rev. **119**, Nr. 2, 680 (1960).

für die LARMOR-Frequenz f_L. Es ist noch zu bemerken, daß sich wegen des Fehlens der Hyperfeinstruktur die ZEEMAN-Terme exakt äuquidistant ergeben. Für beide Übergänge $m = +1 \to m = 0$ und $m = 0 \to m = -1$ resultieren die gleichen Frequenzen.

β) Bei *Rubidium*[2] liegen die Verhältnisse anders, da beide Isotope Rb^{85} und Rb^{87} den Kernspin $\frac{5}{2}$ bzw. $\frac{3}{2}$ besitzen. Da nach Gl. (19.4) der Aufspaltungsfaktor mit größer werdendem Kernspin kleiner wird, bevorzugt man das Isotop Rb^{87}. Das Termschema dafür ist in Fig. 25 angegeben. Die Wirksamkeit des optischen Pumpens verringert sich durch die größere Anzahl von Absorptions- und Emissionsschritten, bedingt durch die größere Zahl von Termen, als sie z. B. unter idealisierten Voraussetzungen in Fig. 23 dargestellt wurden. Aus der Zahl der ZEEMAN-Übergänge zwischen den einzelnen Termen, die eine Anzahl von Linien unterschiedlicher Intensität über einen Bereich von etwa 1,7 kHz verursachen, sei der niedrigste Übergang herausgegriffen. Für ihn ergibt die BREIT-RABI-Formel

$$\left.\begin{aligned} f_{L(-2,-1)} &= -g_{87}\frac{|\mu_B|}{h}B + \frac{1}{2}f_B[(1-C+C^2)^{\frac{1}{2}}-(1-C)]; \\ C &= \frac{1}{f_B}(g_{Rb}+g_{87})\frac{|\mu_B|B}{h} \end{aligned}\right\} \quad (20.2)$$

g_{Rb} = LANDE-Faktor, g_{87} = Aufspaltungsfaktor für Kernmoment; Hyperfeinaufspaltung für den Grundzustand von Rb^{87}: $f_B = (6834682614 \pm 1)$ Hz.

Somit folgt

$$\frac{f_L}{\text{Hz}} \approx 6{,}99623 \cdot 10^5 \left(\frac{B}{\Gamma}\right) - 2{,}16 \cdot 10^2 \left(\frac{B}{\Gamma}\right)^2. \quad (20.2\text{a})$$

Für diese Linie wurden in der Literatur Linienbreiten zwischen 20 und 30 Hz bei einem Signal-Rausch-Verhältnis von 50 angegeben.

21. Experimentelle Technik. Im wesentlichen sind zwei Anordnungen zur Messung des erdmagnetischen Feldes bekannt geworden, die auf dem Effekt des optischen Pumpens basieren[1].

α) *Abtasten der Resonanzlinie.* Bei dieser Methode wird die Linie des ZEEMAN-Übergangs bzw. das Spektrum der möglichen Übergänge abgetastet, ähnlich dem bei der Kernresonanz beschriebenen Verfahren (Fig. 18). Die experimentelle Anordnung gibt schematisch die Fig. 27 wieder. Eine periodische Modulation erlaubt ein vollständiges oder teilweises Abtasten der gewünschten Linie. Mit einer geeigneten Regeleinrichtung läßt sich die Frequenz des Hochfrequenz-Generators so einstellen, daß stets maximale Absorption herrscht. Dann folgt die Frequenz des Generators, abgesehen von regeltechnisch bedingten, geringen Abweichungen, den Variationen des erdmagnetischen Feldes. Man kann selbstverständlich auch durch einmaliges Abtasten der Resonanzlinie die Absolutmessung eines Momentanwertes der Erdmagnetfeldes vornehmen.

β) *Selbstschwingende Anordnung.* Durch die Anordnung des „selbstschwingenden" Magnetometers ist gegenwärtig die unter α) beschriebene Methode weitgehend verdrängt worden. Zum Verständnis der Vorgänge bedient man sich ebenso wieder der klassischen Vorstellungen von präzedierenden Spins wie zur Ableitung der BLOCHschen Gleichungen[2] der Kernresonanz. Man betrachtet dazu

---

[2] Zum Beispiel P. L. BENDER and T. L. SKILLMAN: J. Geophys. Res. **63**, Nr. 3, 513 (1958).
[1] Zum Beispiel P. L. BENDER: Measurements of weak magnetic fields by optical pumprug methods. Vortrags-Niederschrift URSI-Kongreß, München 1960.

eine Versuchsanordnung mit zwei gekreuzten Strahlengängen, wie sie Fig. 28 darstellt. Dabei ist Strahl I parallel und Strahl II senkrecht zum Magnetfeld orientiert. Im Strahl I wird dann eine Präzession der Spins um die Magnetfeldrichtung mit der LARMOR-Frequenz erfolgen, wobei durch Wechselwirkung mit

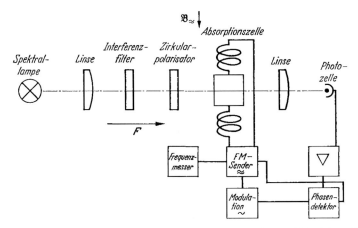

Fig. 27. Prinzipdarstellung eines Magnetometers, das mit teilweiser Abtastung der Resonanzlinie arbeitet. Die Linie wird durch periodische Modulation der Frequenz des angelegten Hochfrequenz-Feldes abgetastet. Das Ausgangssignal des lichtelektrischen Wandlers kann dann nach phasenempfindlicher Gleichrichtung zum automatischen Nachstimmen des Generators auf Linienmitte ausgenutzt werden.

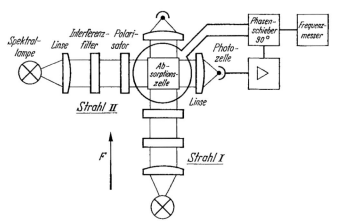

Fig. 28. Selbstschwingende Anordnung nach dem Kreuzstrahlprinzip. Strahl I bewirkt in der Absorptionszelle den Effekt des optischen Pumpens. Bei Strahl II (senkrecht zu F) tritt daher nach Durchlaufen der Zelle eine Helligkeitsmodulation mit der LARMOR-Frequenz auf. Diese benutzt man nach lichtelektrischer Umwandlung, Verstärkung und 90°-Phasendrehung zur Erzeugung des Hochfrequenzfeldes.

dem Hochfrequenz-Feld im Resonanzfall Gleichphasigkeit erreicht wird. Dadurch wird eine Helligkeitsmodulation für Strahl II mit der LARMOR-Frequenz erreicht, da das Spin-Ensemble während einer LARMOR-Periode einmal parallel und einmal antiparallel zu seiner optischen Achse „gesehen" wird. Den Ausgang des lichtelektrischen Wandlers von Strahl II kann man daher nach Verstärkung und 90°-Phasendrehung (wegen der um 90° versetzten räumlichen Anordnung beider

[2] Zum Beispiel L. ENGELHARD: Dissertation, TU Braunschweig, 1967.

Strahlen) als Hochfrequenz-Führungsfeld für Strahl I benutzen. Man erhält einen „rückgekoppelten" Oszillator, dessen Frequenz die LARMOR-Frequenz ist und der Feldänderungen mit einer Zeitkonstante folgt, die von der Größenordnung einer LARMOR-Periode ist.

Die Kreuzstrahlenanordnung läßt sich ersetzen durch eine Einstrahlenanordnung, die Fig. 29 zeigt. Die Orientierung zum äußeren Feld ergibt für den Strahl sowohl eine Komponente in Feldrichtung wie auch eine senkrecht dazu. Das gilt ebenso für das Hochfrequenz-Feld. Ein solches Gerät arbeitet in einem Winkelbereich von ca. $\pm 30°$ um die günstigste Richtung ($\psi = 45°$). Dabei tritt, wenn man z.B. Rb als Probensubstanz verwendet, eine Verschiebung der Oszillatorfrequenz auf, die umgerechnet bis $\pm 5 \gamma$ erreichen kann. Dieser Effekt beruht darauf, daß sich die asymmetrische Amplitudenverteilung auf die einzelnen ZEEMAN-Linien, die in ihrer Gesamtheit die Oszillatorfrequenz bestimmen, geringfügig mit der Orientierung im Magnetfeld ändert.

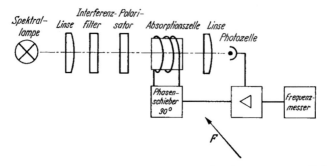

Fig. 29. Selbstschwingende Einzellenanordnung. Man erreicht durch die Orientierung von 45° zum Erdfeldvektor F, daß bei einer Anordnung zum optischen Pumpen mit einem Strahl eine Komponente parallel zum Feld (wie Strahl I in Fig. 28) und eine senkrecht dazu (wie Strahl II in Fig. 28) auftritt. So ergibt sich auch hier die Möglichkeit, eine Rückkopplungsschaltung aufzubauen.

Dieser Einfluß wird bei der sog. Doppelzellen-Anordnung beseitigt. Hier arbeiten zwei Einstrahlenanordnungen gegeneinander, d.h. die Achse des einen Systems ist gegenüber der des anderen um 180° gedreht. Das Modulationssignal des einen Systems wird jeweils als Führungsfeld für das andere benutzt. Die antiparallele Anordnung bewirkt eine insgesamt symmetrische Verteilung der ZEEMAN-Linien und damit die Richtungsunabhängigkeit der Oszillatorfrequenz.

Unabhängig von der Wahl der Meßanordnung ergeben sich bei der Herstellung geeigneter Spektrallampen[3] und Resonanzgefäße einige Schwierigkeiten. Sie beginnen bei der Auswahl geeigneter Gläser für die Gefäße, die z.B. keine Diffusion des verwendeten Dampfes erlauben dürfen. Eine weitere Schwierigkeit bildet die Auswahl des Puffergases. Als Spektrallampen dienen kleine Resonanzgefäße, in denen durch ein kräftiges Hochfrequenz-Feld auf etwa 25 ... 80 MHz eine elektrodenlose Entladung aufrecht erhalten wird. Dafür, wie für die Absorptionszellen werden von verschiedenen Autoren unterschiedliche Angaben gemacht, denen stets umfangreiche Meßreihen zugrunde liegen.

γ) *Genauigkeitsbetrachtungen.* Für eine Meßanordnung wie unter α) beschrieben, läßt sich die Meßgenauigkeit leicht abschätzen. Man setzt z.B. für die vereinfachte Form der Resonanzlinie

$$A = \frac{A_0}{1 + \left(\frac{\Delta f}{\Delta f_{\frac{1}{2}}}\right)^2} \quad \text{mit} \quad \Delta f = f - f_L \qquad (21.1)$$

an ($\Delta f_{\frac{1}{2}}$ = Linienbreite, A_0 = Maximalamplitude). Dann ergibt sich für das differenzierte Signal, das z.B. für die Genauigkeit der unter α) beschriebenen

---
[3] Zum Beispiel F. A. FRANZ: Rev. Sci. Instr. **34**, Nr. 5, 589 (1963).

Anordnung maßgeblich ist, näherungsweise in der Nähe des Nulldurchgangs:

$$\frac{dA}{df} \approx -A_0 \frac{2\Delta f}{\Delta f_{\frac{1}{2}}}. \tag{21.2}$$

Unter Berücksichtigung des Signal-Rausch-Verhältnisses S_N kann man näherungsweise für den Fehler der Meßfrequenz schreiben:

$$\Delta f \approx \frac{1}{2} \frac{\Delta f_{\frac{1}{2}}}{S_N}. \tag{21.3}$$

Benutzt man die Angaben für ein spezielles Helium-Magnetometer[4] ($f = 3 \cdot 10^2$ Hz, $S_N = 2 \cdot 10^2$) und berücksichtigt noch Gl. (20.1), so erhält man

$$\Delta f \approx 0.8 \text{ Hz} \quad \text{und} \quad \Delta \mathscr{F} \approx 0.025 \gamma. \tag{21.4a}$$

Für Rubidium-Magnetometer mit $\Delta f = 30$ Hz und $S_N = 5 \cdot 10$ ergeben sich

$$\Delta f \approx 0.3 \text{ Hz} \quad \text{und} \quad \Delta \mathscr{F} \approx 0.04 \gamma. \tag{21.4b}$$

Im Vergleich zum Kernresonanzmagnetometer liegt die relative Meßgenauigkeit hier also etwa eine Größenordnung höher. Auch fällt die bei dem Kernresonanzeffekt der freien Präzession so störende zeitliche Änderung des Signal-Rausch-Verhältnisses weg, die noch einen zusätzlichen Fehler in der Frequenzmessung verursacht.

Eine weitere Steigerung der relativen Meßgenauigkeit wurde mit den selbstschwingenden Anordnungen erreicht. Für die Absolutbestimmung der Totalintensität des geomagnetischen Feldes ist das allerdings nicht der Fall. Das wird nicht etwa durch eine Ungenauigkeit in der Bestimmung der atomaren Konstanten [vgl. z.B. Gl. (20.2a)] verursacht, sondern durch apparative Einflüsse. Diese wurden bereits oben erwähnt. Sie bewirken eine geringe Abweichung der Oszillatorfrequenz von der LARMOR-Frequenz, die sich gewöhnlich schwer erfassen und eliminieren läßt. Man bestimmt diese Abweichung häufig durch Vergleichsmessung mit Protonenmagnetometern, so daß schließlich höchstens die gleiche absolute Meßgenauigkeit wie bei diesen Geräten erreicht wird. Es liegen auch Präzisionsmessungen mit entsprechend hohem Aufwand[5] vor, bei denen z.B. für das Verhältnis der Aufspaltungsfaktoren für Rubidium 87 (Elektronenhülle) und Protonen (Atomkern) eine Genauigkeit von 10^{-6} erreicht werden konnte.

Es sei auch noch auf äußere Einflüsse hingewiesen, die die Meßgenauigkeit stark beeinträchtigen können. Hierher gehören neben den überall bei dem Bau geomagnetischer Geräte auszuschließenden ferromagnetisch verunreinigten Materialien auch elektromagnetische Störfelder. Weist z.B. das Hochfrequenz-Feld eine geringe Modulation auf der Netzfrequenz (in Europa 50 Hz) auf, so führt das zu Seitenbändern im System der Resonanzlinien, das dadurch zu einer einzigen breiten Linie verwischt werden kann.

δ) *Anwendungen.* Die Anwendung der Magnetometer, die auf dem optischen Umpumpeffekt beruhen, ist zunächst prinzipiell überall möglich, wo man auch Protonenmagnetometer mit Erfolg zur Messung der Totalintensität des geomagnetischen Feldes eingesetzt hat. Es gibt bereits eine Reihe von leistungsfähigen, industriell gefertigten Geräten, die für die verschiedensten Meßaufgaben, die vom Observatoriumseinsatz bis zur Magnetfeldmessung im interplanetaren Raum reichen, eingesetzt wurden. Man kann die Geräte nicht nur zur Absolutmessung im klassischen Sinn benutzen. Besonders die selbstschwingende Version fand auch verbreitet Anwendung zur Messung und Registrierung von Variationen

---

[4] L. D. SHEARER, F. D. COLEGROVE u. G. K. WALTERS: Rev. Sci. Instr. **34**, 1363 (1963).
[5] P. L. BENDER: Phys. Rev. **128**, Nr. 5, 2218 (1962).

des geomagnetischen Feldes. Hierbei wird die hohe relative Empfindlichkeit und die geringe Eigenperiode der Geräte voll ausgenutzt. So registrierte man z.B. Mikropulsationen mit Perioden von 0,1 s und 0,01 γ Amplitude. Diese Eigenschaften führten zum Einsatz derartiger Geräte in der Magnetotellurik. Ein besonders wichtiges Anwendungsgebiet ist ferner die Messung und Registrierung von Komponenten des geomagnetischen Feldes mit Hilfe von Zusatzfeldern (s. auch Ziff. 22).

III. Komponentenmessung mit Totalintensitäts-Magnetometern.

Bereits in Abschnitt II wurde auf einen Nachteil hingewiesen, der allen auf dem ZEEMAN-Effekt basierenden Meßgeräten gemeinsam ist: Man kann sie direkt nur zur Messung des Betrages des erdmagnetischen Feldvektors, der Totalintensität, anwenden. Die vorhandene Richtungsabhängigkeit des Effektes beeinflußt im allgemeinen nur die Signalgröße und dadurch die Meßgenauigkeit. Ebenso läßt sich die geringe Richtungsabhängigkeit der Frequenz der Einzellenanordnung [s. Ziff. 21 α) und β)] des selbstschwingenden Magnetometers nicht derart beherrschen, um darauf eine Komponentenmessung aufbauen zu können. Bisher wurden nur solche Meßverfahren in der geomagnetischen Meßpraxis mit Erfolg eingesetzt, die sich der Hilfe von Zusatzfeldern bekannter Richtung bedienen.

22. Physikalische Grundlagen.

α) Die folgenden Ausführungen beschränken sich auf die *Komponentenmessung mit Zusatzfeldern*. Dazu betrachtet man zunächst die Verhältnisse in der Ebene, die durch den Erdfeldvektor und den Einheitsvektor in Zusatzfeldrichtung aufgespannt wird. Aus Fig. 30 liest man für den Betrag des resultierenden

Abb. 30. Ebener Fall einer Komponentenmessung mit Zusatzfeldern.

Feldes F_g ab, wenn der Betrag des Zusatzfeldes F_1 und der des Erdfeldes $\mathcal{F}$ sei:

$$F_g^2 = \mathcal{F}^2 + F_1^2 + 2\mathcal{F}F_1 \cos \alpha = \mathcal{F}^2 + F_1 + 2F_1^2 \mathcal{F}_\alpha. \tag{22.1}$$

Mißt man demnach $\mathcal{F}$ und F_g und setzt den Betrag von F_1 als bekannt voraus, so ist $\mathcal{F}_\alpha$, die Komponente des Erdmagnetfeldes in Zusatzfeldrichtung, berechenbar. Man versucht jedoch meist, durch eine weitere Messung zu Gl. (22.1) noch eine weitere Gleichung zu erhalten, mit deren Hilfe sich F_1 eliminieren läßt. Von den bekannten Methoden seien zwei angeführt:

(i) Man sucht das *Minimum des resultierenden* Feldes F_g durch Variieren des Zusatzfeldes und erhält ausgehend von Gl. (22.1)

$$F_g^2 = \mathcal{F}^2 + F_1^2 + 2F_1 \mathcal{F}_\alpha, \qquad \frac{\partial(F_g^2)}{\partial F_1} = 0. \tag{22.2}$$

Dieses Gleichungssystem liefert

$$\mathcal{F}_\alpha = (\mathcal{F}^2 - F_{g\,\text{min}}^2)^{\frac{1}{2}}.\tag{22.3}$$

Das Ergebnis enthält nicht mehr die Größe des Zusatzfeldes. Wie man aus Fig. 30 ersieht, ist in diesem Spezialfall gerade $F_1 = -\mathcal{F}_\alpha$. Man bezeichnet diese Art der Messung daher auch als *Kompensationsmethode*[1].

2. Man benutzt zwei Messungen mit dem Zusatzfeld des Betrages F_1, jedoch mit *unterschiedlicher Polarität*. Es gilt

$$\left.\begin{aligned} F_{g1}^2 &= \mathcal{F}^2 + F_1^2 + 2F_1\,\mathcal{F}_\alpha, \\ F_{g2}^2 &= \mathcal{F}^2 + F_1^2 - 2F_1\,\mathcal{F}_\alpha. \end{aligned}\right\}\tag{22.4}$$

Dieses Gleichungssystem ergibt

$$\mathcal{F}_\alpha = \frac{F_{g1}^2 - F_{g2}^2}{2{,}83\,(F_{g1}^2 + F_{g2}^2 - 2\mathcal{F}^2)^{\frac{1}{2}}}.\tag{22.5}$$

Gegenüber dem Kompensationsverfahren hat man hier einige Vorzüge[2]. Man kann F_1 z.B. stets so wählen, daß $F_{g1}, F_{g2} \geq \mathcal{F}$ erreicht wird. Das bedeutet besonders beim Kernresonanzverfahren eine höhere erreichbare Meßgenauigkeit.

β) Zur eindeutigen *Bestimmung* des gesamten *Feldvektors* muß man, wie sich zeigen läßt, im allgemeinen drei Zusatzfelder benutzen, die nicht in einer Ebene liegen dürfen. Bei geschickter Wahl von zwei Zusatzfeldern kann man allerdings die verbleibende Mehrdeutigkeit praktisch bedeutungslos machen. Betrachtet man z.B. die von BACON[3] eingeführte Anordnung, Fig. 31, bei der sich ein Zusatzfeld

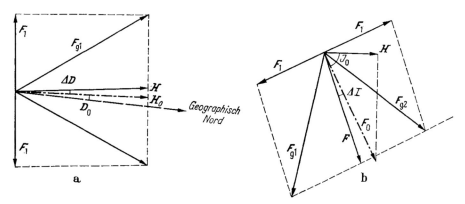

Fig. 31 a u. b. Messung von D und J mit Zusatzfeldern. a) Darstellung in der horizontalen Ebene. b) Darstellung in der Meridianebene.

in der horizontalen Ebene in magnetischer Ost-West-Richtung befindet zur Bestimmung der Deklination D, das andere in der magnetischen Meridianebene senkrecht zu $\mathcal{F}$ zur Bestimmung der Inklination J, so wird offenbar die Richtung des Erdfeldvektors bestimmt, wobei das Vorzeichen unbestimmt bleibt. Das läßt sich aber bereits mit einem Nadelinklinatorium bestimmen, soweit es nicht ohnehin aus magnetischen Karten bekannt ist.

---

[1] L. HURWITZ and J. H. NELSON: J. Geophys. Res. 65, Nr. 6, 1759 (1960).
[2] L. R. ALLDREDGE: J. Geophys. Res. 65, Nr. 11, 3777 (1960).
[3] F. BACON: Master's Thesis. U.S. Naval Postgrad. School, Monterey, Calif. 1955.

Man führt also die Messung der Magnetfeldkomponenten zurück auf die Bestimmung der Richtung künstlicher Magnetfelder. Das erfordert aber, wie im folgenden noch gezeigt wird, ein hohes Maß an mechanischer Präzision für die verwendeten Spulensysteme usw., durchaus vergleichbar mit dem mechanischen Aufwand für die Geräte, die den klassischen geomagnetischen Absolutmessungen dienen.

23. Praktische Anwendung.

α) *Genauigkeitsbetrachtungen.* Der Fehler für die Komponentenbestimmung nach den Gln. (22.2) und (22.4) läßt sich bei entsprechenden Anforderungen an die Magnetfeldmeßapparatur und an die Kurzzeitkonstanz der Zusatzfelder während der Messung auf die Größenordnung von 0,1 γ reduzieren. Schwieriger ist der Einfluß von Fehljustierungen zu beherrschen. Will man z.B. mit einem horizontalen Zusatzfeld in magnetischer Nord-Süd-Richtung die Horizontalintensität $\mathcal{H}$ bestimmen, so ergibt sich wenn $\varDelta D$ die Fehljustierung des Zusatzfeldes im Meridian und $\varDelta J$ die horizontale Fehljustierung bezeichnen:

$$\mathcal{H}+\varDelta \mathcal{H} \approx \mathcal{H}\left(1-\frac{\varDelta J^2}{2}-\frac{\varDelta D^2}{2}+\frac{Z}{\mathcal{H}}\varDelta J\right). \tag{23.1}$$

Möchte man in mittleren geomagnetischen Breiten ($\mathcal{H}=2\cdot 10^4 \gamma$ und $Z=5\cdot 10^4 \gamma$) eine Meßgenauigkeit von $\varDelta \mathcal{H}=0,5\gamma$ realisieren, so liefert Gl. (23.1) als maximal zulässige Größen für $\varDelta D$ ca. 25' und für $\varDelta J$ ca. 2". Eine derartig genaue Justierung des Zusatzfeldes stößt auf Schwierigkeiten. Man kann aber die Spule, die das Zusatzfeld erzeugt, um 180° in der horizontalen Ebene herumdrehen und durch eine zweite Messung mit dieser Orientierung und Mittelbildung aus beiden Meßergebnissen den durch $\varDelta J$ verursachten Fehler eliminieren. Voraussetzung ist dabei allerdings, daß die Achse, um die der 180°-Umschlag erfolgt, auf 2" genau vertikal ist. Das läßt sich aber erreichen und durch Libellen jederzeit nachprüfen.

Schwieriger noch gestaltet sich die absolute Messung von D und J. Das ist darauf zurückzuführen, daß der Schluß auf die Richtung der Zusatzfelder aus den geometrischen Daten der erzeugenden Spulen problematisch ist.

β) *Experimentelle Probleme.* Bei der Berechnung der Spulensysteme zur Erzeugung der Zusatzfelder für Komponentenmeßgeräte muß man zwischen Stabilität, leichter Justierbarkeit und nicht zu großen Abmessungen einerseits und einer möglichst guten Homogenität des Zusatzfeldes über die gesamte Probe des Magnetometers andererseits einen Kompromiß finden. Es ergeben sich je nach Art der verwendeten Magnetometer und der angestrebten Meßgenauigkeit für HELMHOLTZ-Spulensysteme Durchmesser zwischen 0,5 und 1 m. Spulen und Untersatz müssen mit höchster mechanischer Präzision gefertigt sein. Die Ströme in den Spulen werden jetzt allgemein mit einer Genauigkeit von 10^{-5} elektronisch stabilisiert. Da man gewöhnlich ein Magnetometer mit zwei Spulensystemen für die Messung von zwei Komponenten kombiniert, müssen die Zusatzfelder nach einer bestimmten Reihenfolge von Hand für Einzelmessungen oder automatisch geschaltet werden. Für einen kontinuierlichen Betrieb wird das Gerät mit einem Prozeßrechner verbunden[1], der z.B. nach Gl. (22.2) die gewünschten Komponenten berechnet, anzeigt, ausdruckt, oder speichert[2]. Meßgeräte und Prozeßrechner können dabei weit voneinander getrennt betrieben werden. Die Verbindung kann über Kabel oder drahtlos erfolgen[3].

γ) *Anwendung.* Während die Einführung z.B. des Kernresonanzmagnetometers in die erdmagnetische Meßtechnik eine Steigerung der Meßgenauigkeit für $\mathcal{F}$ und Z zur Folge hatte, so ist dies für die anderen Komponenten durch die hier beschriebenen Geräte nicht ohne weiteres der Fall. Ihr Vorzug ist vielmehr in einer Vereinfachung und zeitlichen Verkürzung des eigentlichen Meßprozesses zu sehen. Ferner ist die Ausgabe digitalisierter Daten, ohne daß der Umweg über einen Analog-Digital-Umsetzer notwendig wird, von großem Vorteil. Auf diese Weise wird eine enge Verbindung zwischen der Gewinnung der Meßwerte und der Auswertung mit Hilfe der modernen Datenverarbeitungsmethoden gefördert.

---

[1] I. R. SHAPIRO, J. D. STOLARIK, and J. P. HEPPNER: J. Geophys. Res. 65, Nr. 6, 1759 (1960). — L. R. ALLDREDGE: J. Geophys. Res. 65, Nr. 11, 3777 (1960).

[2] Anwendung mit Programmbeispielen s. A. P. DE VUYST: Inst. roy. mét. de Belgique. Publ. sér. A (1968).

[3] L. R. ALLDREDGE: Coast and Geod. Survey, Techn. Bull. No. 31 (1966).

IV. Andere Methoden.

24. Hinweise auf weitere Methoden. In den Abschnitten I und II wurden die beiden bekanntesten Magnetometertypen beschrieben, bei denen der ZEEMAN-Effekt ausgenutzt wird. In beiden Fällen war ein physikalischer Kunstgriff notwendig, um den im Erdfeld normalerweise schwachen Effekt bedeutend zu verstärken und zum Ausgangspunkt einer empfindlichen Meßmethode zu machen. Auf zwei weitere Möglichkeiten, die in der Literatur erwähnt werden, sei hier noch kurz hingewiesen. Es handelt sich um die paramagnetische Resonanz freier Elektronen und um den OVERHAUSER-Effekt. Daß beide Verfahren zur Zeit noch keine verbreitete Anwendung gefunden haben, kann nicht als Aussage über ihre mögliche Eignung für Erdfeldmessung im Routinebetrieb gewertet werden.

α) *Paramagnetische Elektronenresonanz.* Das für die Meßgenauigkeit entscheidende Signal-Rausch-Verhältnis für Experimente, bei denen die Resonanzlinie langsam abgetastet wird, ist oben unter Gl. (17.5) angegeben. Im Falle der Kernresonanz trug man der Abhängigkeit vom gyromagnetischen Verhältnis durch die bevorzugte Auswahl von wasserstoffhaltigen Proben (Protonenresonanz) Rechnung. In dieser Hinsicht wären ungleich bessere Verhältnisse bei der paramagnetischen Elektronenresonanz zu erwarten, weil das gyromagnetische Verhältnis des Elektrons das des Protons um das $6{,}58 \cdot 10^2$-fache übertrifft. Dieser Faktor erklärt sich ohne weiteres aus den Formeln (16.1) und (16.3), wenn man dort statt m_P die Ruhemasse des Elektrons $m_e = (9{,}1091 \pm 4 \cdot 10^{-4}) \cdot 10^{-31}$ kg und den Aufspaltungsfaktor $g = 1$ einsetzt.

Der experimentellen Nutzung des Effektes stehen allerdings auch einige Schwierigkeiten entgegen. Schon die Auswahl einer geeigneten Probensubstanz ist schwierig. Die Moleküle müssen ein freies Elektron haben, das heißt man kann nur Radikale verwenden. Es gibt nur wenige Substanzen, die so beständig sind, daß man sie für eine Magnetfeld-Meßanordnung verwenden könnte. Als Beispiel sei Diphenylpicryl-hydrazyl (DPPH) angeführt. Weiterhin liegen die bei der Elektronenresonanz auftretenden Linienbreiten in der Größenordnung des Erdfeldes, so daß die in Ziff. 17 beschriebenen Kernresonanz-Meßverfahren hier nicht sinngemäß übernommen werden können.

Für eine von JUNG und CAKENBERGHE[1] beschriebene Meßanordnung werden trotz dieser erschwerenden Momente Meßgenauigkeiten von 5γ angegeben. Als interessante Möglichkeit ergibt sich hier z.B. noch die simultane Verwendung des Modulationsfeldes zur Linienabtastung und als Zusatzfeld zur Komponentenmessung.

β) *OVERHAUSER-Effekt*[2]. Dieser Effekt beruht auf der Wechselwirkung zwischen den magnetischen Momenten der Elektronen und den Kernmomenten, auf die wir in Ziff. 19β bereits in anderem Zusammenhang hinwiesen. Zum qualitativen Verständnis geht man nicht von der BLOCHschen Theorie (Ziff. 16γ) aus, vielmehr von den Gln. (16.4) und (16.5), die die Terme des Kernspins in einem äußeren Magnetfeld beschreiben. Die Intensität der absorbierten oder emittierten Energie bei Übergängen hängt von den Unterschieden der Besetzungszahlen der einzelnen Energiestufen ab. Eine Sättigung der Elektronenresonanz im Magnetfeld durch starke Einstrahlung der entsprechenden Resonanzfrequenz bewirkt nun bei einem bestimmten Wechselwirkungsmechanismus Besetzungszahlverhältnisse der Kernspinniveaus, die stark von der normalen Verteilung abweichen. Die Auswirkungen ergeben hier, analog zu dem Effekt des optischen Pumpens, eine starke Zunahme der Intensität des Kernresonanzsignals. Man spricht in diesem Zusammenhang auch (in ähnlicher Ausdrucksweise wie bei der freien Kernpräzession s. Ziff. 17γ), von einer Art „dynamischer Polarisation" der Kerne. Spe-

---
[1] P. JUNG u. J. VAN CAKENBERGHE: Vortrag Colloque Ampere, Leipzig, 1961.
[2] O. OVERHAUSER: Phys. Rev. **91**, 476 (1953).

ziell für Messungen im Erdmagnetfeld wurde dieses Problem z.B. von BONNET und THOMAS[3] behandelt.

Hierbei benutzte man eine Protonenprobe, die in geringer Konzentration ein paramagnetisches Radikal enthält. Ferner sättigte man nicht die Resonanz der freien Elektronen, sondern einen Übergang der Hyperfeinaufspaltung bei ca. 56 MHz, der erst bei größeren Magnetfeldstärken in ZEEMAN-Terme aufspaltet. Die Signalvergrößerung ist in diesem Falle so hoch, daß man mit ihrer Hilfe einen sog. Spin-Generator[4] im Erdfeld betreiben kann. Es handelt sich hierbei um eine selbstschwingende Schaltung, die aus der Anordnung nach Fig. 18 bei Schaffung eines Rückkopplungsweges vom Signalverstärkerausgang zu den Generatorspulen entsteht, wobei der in Fig. 18 angegebene Generator entfällt. Mit einem kommerziellen Meßgerät nach diesem Prinzip ergeben sich Meßgenauigkeiten von ca. 1 γ.

D. Informationsverarbeitung im Geomagnetismus.

25. Grundlagen.

α) *Anpassung der Methoden und Apparaturen* an die Eigenschaften der maschinenlesbaren Datenträger und Verarbeitungsgeräte ist die bestimmende Forderung, die die elektronische Informationsverarbeitung, wie in vielen Wissenschafts- und Industriezweigen, auch auf geomagnetischem Gebiet stellt. Dabei wird die früher oft erkennbare thematische Trennung von Messung und Auswertung immer weniger wirksam und Tendenzen zur Vereinheitlichung dieser Aufgaben in Gerätekomplexen sind erkennbar. Hier sollen daher einige Betrachtungen über die Wechselbeziehungen zwischen der geomagnetischen Meßtechnik und der Informationsverarbeitung angeschlossen werden.

Die vielfältigen Aufgaben, die aus der Beobachtung des geomagnetischen Feldes an festen Punkten sowie an See-, Luft- und Raumstationen folgen[1], führen zu einer entsprechenden Vielfalt von Digitalisierungsproblemen und Problemen der Anwendung von Analogie-Rechenmethoden unterschiedlicher Zielstellungen. So steht der Geomagnetiker bei der Disposition der Verarbeitungssysteme vor zahlreichen problemorientierten und technischen Alternativen. Für viele Zwecke der angewandten Geophysik liegen eng begrenzte Aufgabenstellungen vor und demzufolge sind klare Konsequenzen beispielsweise für die Wahl der Digitalisierungskenngrößen zu ziehen. Dagegen ist die Lage für die geomagnetische Grundlagenforschung nicht derart eindeutig. Hier verlangt man einerseits eine möglichst detaillierte Aufnahme der Variationen und erreicht aus ökonomischen Gründen bald gewisse technische Grenzen. Andererseits erfordern manche hochspezialisierten Auswerteaufgaben ein Überschreiten dieser Grenzen, meist nur auf einem schmalen Sektor, aber dennoch in einer Weise, daß eine generelle Berücksichtigung solcher Wünsche zu unerfüllbaren Bedingungen führt. Jedes Datenverarbeitungsverfahren ist eben nur innerhalb seiner (oft nicht ohne weiteres erkennbaren) Grenzen aussagefähig.

β) *Analoge oder digitale Verarbeitungsverfahren.* Die Wahl eines Verfahrens der Datenverarbeitung wird durch die Aufgabe und durch die Eigenschaften der verfügbaren Rechengeräte bestimmt. Die meisten Aufgaben dürften sich prinzipiell durch analogrechentechnische Verfahren lösen lassen, doch zeigt ein Vergleich der Analogie- mit der Digital-Rechentechnik sofort eine erheblich größere Flexibilität der letzteren, vereint mit höherer Genauigkeit, größerer Rechengeschwindigkeit, besserer Speichermöglichkeiten und günstigerer Ein- und Ausgabe-Verfahren. Digital-Rechenautomaten sind bequemer programmierbar und

---

[3] J. THOMAS: Vortrag 25. EAEG-Meeting, Liège, Juni 1964.
[4] J. FREYCENON, I. SOLOMON: Onde électr. **40**, 590 (1960).
[1] Siehe die folgenden Beiträge von P. H. SERSON und K. WHITHAM (S. 384), J. R. BALSLEY (S. 395) und P. J. HOOD (S. 422) in diesem Band.

verfügen über einen leichter nutzbaren Fundus an vorbereiteten Programmierhilfen als Analogie-Rechenautomaten. Dennoch werden zahlreiche Probleme weiterhin dem Analogbereich vorbehalten bleiben und es ist in jedem Falle zu prüfen, ob die Vorteile der digitalen Rechentechnik das Opfer an Informationsverlust (Stichprobenentnahme) aufwiegen. Der gegenwärtig zu beobachtende Trend geht allerdings ziemlich eindeutig in Richtung zunehmender Digitalisierungsvorhaben, so daß dem Kurvenblatt als klassischem Datenträger ernsthafte Konkurrenten in Gestalt von Lochkarten, Lochstreifen und Magnetbändern erwachsen sind[2]. Ein genereller Verzicht auf den Kurvenschrieb ist aber nicht zu erwarten, da kein digitaler Datenträger eine unmittelbare visuelle Beurteilung des aufgezeichneten Geschehens gestattet.

26. Probleme der Digitalisierung. Die Digitalisierung von Wertefolgen ist methodisch bedingte Voraussetzung zum Gebrauch ziffernmäßig arbeitender Rechenautomaten. Sie kann als Echtzeit-Digitalisierung (on line-Methodik) oder nachträglich vom Kurvenblatt her (off line-Methodik) geschehen. Im wesentlichen handelt es sich bei solchen Vorhaben um die Zerlegung zeitabhängiger Wertefolgen in äquidistanten Abständen.

α) Eine entscheidende Rolle kommt hierbei der *Wahl des Digitalisierungs-Intervalls* Δt zu. Einerseits sollte Δt sehr klein sein, damit der Verlust an Information durch Ausfall der Werte in den Lücken möglichst gering bleibt.

Das führt zu sehr hohen Werteraten. Andererseits ist aus arbeitsökonomischen Gründen ein möglichst großes Δt erwünscht. Im Hinblick auf Frequenzanalysen kommt bei der Diskussion der Δt-Wahl der NYQUIST-Frequenz

$$f_N = \frac{1}{2\Delta t} \tag{26.1}$$

eine wesentliche Rolle zu. Sie ist die höchste, den diskreten Werten entnehmbare Frequenz und dient zur Kennzeichnung der „Alias-Frequenzen", die das Vorkommen von Signalen der Frequenz f vortäuschen können. Diese „*Alias-Frequenz-Reihe*" ist bei gegebener Frequenz $f < f_N$ durch

$$2f_N - f; \quad 2f_N + f; \quad 4f_N - f; \quad 4f_N + f; \quad \ldots$$

zu beschreiben. Auch wenn eine Frequenzanalyse der digitalisierten Werte die Existenz der Frequenz f anzuzeigen scheint, so kann es möglich sein, daß das Ausgangsmaterial f gar nicht enthält und daß der f-Anteil dem Auftreten von „Alias-Frequenzen" zugeschrieben werden muß. Man vermeidet diese Mehrdeutigkeiten, indem man das Analog-Daten-Material vor der Digitalisierung einer Tiefpaß-Filterung (elektronisch oder mathematisch) unterwirft, wobei das Filter alle Frequenzen oberhalb f_N unterdrückt. In der Praxis geht man sicherheitshalber oft zu etwas tieferen Werten ($f_N/2$) für die maximale Durchlaß-Frequenz über. Die Beachtung dieser Zusammenhänge gestattet auch aus gefiltertem und „zu dicht" digitalisiertem Zahlenvorrat Werte in Abständen von Vielfachen von Δt zu entnehmen und damit für Frequenzanalysen den Eingangsdatenaufwand stark zu verringern.

Jeder Digitalisierung ist ein durch die Ziffernstruktur der Informationsdarstellung bedingter *Quantisierungsfehler* eigen, der jeweils nur die Bestimmung auf ± 1 Quant zuläßt. Im allgemeinen sind die elektronischen Hilfsmittel so leistungsfähig, daß dieser Fehler klein gegenüber den sonstigen Fehlern des Registrierprozesses gehalten werden kann.

---

[2] Es existieren bereits internationale Empfehlungen über die Anordnung geomagnetischer Daten auf Lochkarten, Lochstreifen und Magnetbändern: IAGA News No. 8, 22—26 (1969).

β) Ein besonderes Problem stellt die *Digitalisierung von Kurvenschrieben*[1] dar, die mit semiautomatischen[2] und vollautomatischen[3] Anlagen vorgenommen wird. Die hohen Genauigkeitsforderungen, die bei der visuellen Ablesung mittels einer Glasskala erfüllt werden ($1^0/_{00}$ vom Endwert), bieten für die Realisierung automatischer Systeme erhebliche Schwierigkeiten. Diese verlangen den Einsatz präziser feinmechanisch-optischer Hilfsmittel und lassen z. Z. die Anwendung fernsehtechnischer Abtast-Anordnungen wegen ungenügender Linearität nicht zu. Eine zweite Schwierigkeit liegt darin, daß sich die erdmagnetischen Kurven auf dem gleichen Blatt oft überschneiden. Ein Weg zur Lösung dieses Problems ist durch die Anwendung programmierter Informationszuordnungen vor und nach dem Schnittpunkt unter Einsatz von Ringzählern realisiert worden[3].

Für den Betrieb solcher Anlagen ist die Existenz von Einrichtungen zur Wiedergabe der Kurvenverläufe aus den digitalisierten Daten für Prüfzwecke vorteilhaft. Diese als Digital-Analog-Umsetzer zu bezeichnenden Apparaturen enthalten z. B. Stromgeber, die mit Hilfe einer Summierstufe einen dem digital gegebenen Wert proportionalen Strom liefern. Mit Registriergeräten (vorzugsweise Kompensationsschreiber in Punktdrucker-Ausführung) wird dieser Strom aufgezeichnet. Für die so erhaltene Punktfolge läßt sich bei geeigneter Wahl der Proportionalitätsfaktoren Deckungsgleichheit mit der Original-Kurve erzielen.

γ) Zur *Wahl der Datenträger* (Lochkarte, Lochstreifen, Magnetband) wäre zu bemerken, daß hierfür letzten Endes die Ökonomie der Rechenautomaten den Ausschlag gibt. Für die kontinuierliche Datenerfassung genießt der Lochstreifen bei beschränkten Datenmengen gewisse Vorteile, während für die Speicherung sehr großer Datenmengen das Magnetband vorzuziehen ist.

27. Bemerkungen zum Datenfluß geomagnetischer Observatorien[1].

α) In der *Auswertung und Aufbereitung verdichteter Daten* existiert seit langem in der Geomagnetik eine international nahezu einheitliche Methodik. Besonders gilt dies für die absoluten Stundenmittel und die daraus folgenden Tages-, Monats- und Jahresmittel und für die verschiedenen Aktivitätsschätzungen durch quasilogarithmisch oder linear gestufte Maßzahlen[2]. Aus den absoluten Stundenmitteln lassen sich durch reine Datenverdichtungsprozesse beispielsweise die folgenden Informationen gewinnen:

Mittlere tägliche Gänge an Tagen weitgehend gleichen Störungscharakters (5 Gruppen),

Täglicher Gang an 5 ruhigen (gestörten) Tagen eines Monats, dargestellt durch Stundenmittel,

Monatsmittel des täglichen Ganges, dargestellt durch Stundenmittel,

Abweichungen der fortlaufend gebildeten Tagesmittel vom Normalwert.

Die Schwierigkeiten kleinerer Observatorien, aus Personalmangel auf diese Informationen verzichten zu müssen, bestünde im Prinzip nicht mehr, falls international einheitliche Datenflußpläne und Programmabläufe vereinbart würden. Der Einsatz von Rechenautomaten zur Routine-Auswertung sollte so

---

[1] K. L. SVENDSEN: Transformation of analog records to digital form (Vortragsveröff.). IAGA Assembly Madrid 1969.
[2] Zum Beispiel B. CANER u. K. WHITHAM: J. Geophys. Res. **67**, 5362 (1962).
[3] Zum Beispiel D. LENNERS: Abh. d. Geom. Inst. Potsdam Nr. 38 (1966).
[1] Für eine ausführlichere Darstellung siehe z. B. H. SCHMIDT und D. LENNERS: Gerl. Beitr. z. Geophys. **78**, 180—189 (1969).
[2] Siehe dazu den Beitrag von M. SIEBERT (S. 206) in diesem Band.

disponiert werden, daß ein möglichst hoher Informationsgehalt jährlich publiziert werden kann und daß die Digital-Datenträger — archivarisch abgelegt — ebenso für einen internationalen Austausch benutzt werden können wie bisher die Kurvenblätter[3].

β) *Das „automatische Observatorium"*. Ein erklärtes Ziel der Automatisierungsbestrebungen im Geomagnetismus ist das „automatische Observatorium". Darunter versteht man bislang eine Einrichtung, die die 3 Komponenten in gewissen Abständen (z.B. 1 min) bestimmt und in digitaler Form (auf Magnetband) ausgibt. Als Geber dienen Spulen von Protonenmagnetometern oder Meßköpfe von Magnetometern, die nach dem Prinzip des optischen Pumpens arbeiten. Es finden die im Abschnitt C III geschilderten Methoden Anwendung; mittels Spulenzusatzfeldern werden Meßbedingungen geschaffen, die die Komponentenmessungen jeweils auf „Totalintensitäts-Bestimmungen" zurückführen. Die Komponentenermittlung erfolgt dann durch Rechenautomaten im on line- oder off-line-Betrieb. Während man gegenwärtig mit solchen 3-Komponenten-Digital-Variometern im bereits klassisch weitgehend erforschten Spektralbereich der geomagnetischen Variationen (Periodendauer von Stundenbruchteilen an aufwärts) arbeitet[4], dürfte die Zukunft solcher Anlagen besonders in der äquivalenten Erforschung anschließender Spektralbereiche (bis zu Sekundenbruchteilen hin) liegen. Allerdings sollte man den Begriff „automatisches Observatorium" dann Gerätekomplexen vorbehalten, die nicht nur Komponenten digitalisieren, sondern durch den Einsatz von Prozeßrechnern die dem Observatoriumsbegriff bisher innewohnende Datensicherheit und Stabilität garantieren. Man wird dann wohl mit redundantem Instrumentarium arbeiten, d. h. mehr Meßgeräte als nötig einsetzen, um gegenseitige Kontrollbeziehungen zu erhalten.

Zur Effektivitätssteigerung der derzeitigen Observatorien wird das „automatische Observatorium" eine unumgängliche Erweiterung des Instrumentariums werden. Dabei dürften sich weitreichende Möglichkeiten zur maschinenmäßig orientierten, neuartigen Beurteilung des geomagnetischen Geschehens ergeben.

---

[3] Das würde allerdings auch eine internationale Festlegung der „Formate" dieser Träger bedeuten. Auf dem Gebiet der Magnetbänder besteht eine solche bisher (1970) noch nicht.
[4] A. P. deVuyst: Different aspects of geomagnetic data in computer-form. Vortrags-Veröff. IAGA Assembly Madrid 1969.

Three-Component Airborne Magnetometers.

By

PAUL HORNE SERSON and KENNETH WHITHAM*.

With 5 Figures.

1. Need for three-component instruments. For the construction of world magnetic charts, and for making spherical harmonic analyses of the geomagnetic field, magnetic observations from all parts of the world are necessary. In 1945, the geographical distribution of recent data was quite unsatisfactory[1]. Practically no measurements had been made over the oceans since 1929, and many land areas were sparsely covered, particularly in the Arctic and Antarctic. Since 1950, there has been a great improvement in the distribution of observations due to the work of the Russian non-magnetic ship "Zarja", and to the development in the United States and Canada of airborne magnetometers capable of measuring both the intensity and direction of the magnetic field. This article describes the principles and application of three-component magnetometers with the Canadian three-component airborne magnetometer as example, and discusses the accuracy of surveys made with this instrument.

To a country, such as Canada, with large unpopulated areas where travel on the ground is difficult, the prospect of making observations in an aircraft of the geomagnetic field is particularly attractive. With an aircraft observations can be made with one instrument and one technique over land, sea or ice. Surveys of large regions can be carried out in the air more quickly and in greater detail than would be feasible with ground measurements. Charts prepared from the continuous profiles produced by an airborne magnetometer are certain to represent the average magnetic field more accurately than charts based on observations at widely spaced points, which are likely to contain errors of several hundred gammas due to local anomalies. Furthermore, continuous profiles of the magnetic components reveal interesting geological and geophysical features[2], and are an aid in the selection of magnetically normal values for such purposes as an accurate separation of the internal and external parts of the field[3].

In 1948, development was undertaken in Canada of an airborne instrument capable of measuring the direction of the geomagnetic vector as well as its intensity. The particular instrument described here was designed and built between 1951 and 1953 at the Dominion Observatory, Ottawa[4].

Two basic considerations governed the over-all design of the magnetometer. First, the instrument was to present magnetic results in a form as close as possible to that required for the preparation of magnetic charts, in order to avoid the necessity of a large staff for the processing of data. It was decided that the instrument should indicate the declination, the horizontal component and the vertical component, and that the indicators should show the three quantities

---

* Original manuscript received 1957, modified February 1966.

[1] H. SPENCER-JONES and P. J. MELOTTE: Mon. Not. Roy. Astr. Soc. Geophys. Suppl., 6, 7 (1953).

[2] P. H. SERSON and W. L. W. HANNAFORD: J. Geophys. Res. 62, 1 (1957).

[3] K. WHITHAM: IAGA Bull. 16, 18, Int. Assoc. Geomag. and Aeronomy 1960.

[4] See also the contribution by J. R. BALSLEY in this volume, p. 395.

directly in the usual units, viz. degree and Oersted (Oe). The second aim was that the accuracy of measurement should be 0.1° in declination, 10 γ^* in the horizontal component and 10 γ in the vertical component. To measure the direction of the magnetic vector, it is obviously necessary to establish in the aircraft a direction reference system. From the first requirement stated above, it is apparent that the direction reference system should be that defined by the vertical and geographical north (rather than, for instance, one defined by the directions of two stars). The second aim requires that the direction reference system be accurate to approximately 0.1° in azimuth and one minute of arc in the determination of the vertical.

The accuracy specified for the measurement of the magnetic components may at first seem low. To obtain the classical accuracy of one gamma, however, the direction reference system would have to be accurate to a few seconds of arc. When this project began, the accuracy of the best gyroscope systems in determining the direction of the vertical in an aircraft was of the order of one degree. Ten gamma, requiring the vertical to about one minute of arc, was chosen as a more realistic goal.

In practice, the airborne magnetometer probably falls short of the specified accuracy by a factor of about three. It is shown in the last part of this article that a typical survey observation is subject to errors of an operational nature much larger than the instrumental errors; a single observation of the magnetic vector as plotted on the charts has a probable error of about 100 γ in any component, principally due to errors in navigation and plotting. It can be shown from a statistical consideration of the geographical variations in the geomagnetic field, however, that for surveys of the type made at present with flight-lines at least 100 km apart, a reduction in this 100 γ probable error would not result in a significant increase in the accuracy of the magnetic charts[2]. Also, in computing and plotting airborne observations, considerable time is saved by accepting the lower accuracy. To summarize, the three-component airborne magnetometer is more accurate than the technique with which the surveys are made, and this technique is sufficiently accurate for surveys with widely spaced flight-lines.

Three different instruments have been developed to measure in an aircraft the direction and intensity of the earth's magnetic field by groups working in the United States, the United Kingdom, and Canada[5] respectively. The United States instrument, the Vector Airborne Magnetometer[6], is in regular use by the U.S. Navy Hydrographic Office. Here the total intensity is measured by a detector automatically aligned in the total field, and the angles between the axis of the detector and a reference system, defined by a damped pendulum and the astronomically determined heading of the aircraft are continuously recorded. These angles are automatically averaged over an interval of about two minutes, to reduce the effect of periodic accelerations of the aircraft, giving the average declination and inclination over that interval. The British development[7] proposes to measure the total field and the component of the field in the direction of the sun. A second flight, when the sun has changed position, supplies a third component. In the Canadian instrument[5], three components of the earth's field are measured by detectors mounted on a horizontal gyro-stabilized platform. A directional gyroscope which is checked periodically by astronomical observations supplies the azimuth reference.

---

* Here gamma $\equiv \gamma = 10^{-5}$ Oe; 10 $\gamma = 10^{-4}$ Oe.
[5] P. H. SERSON, S. Z. MACK, and K. WHITHAM: Publ. Dom. Obs. **19**, 15 (1957).
[6] E. O. SCHONSTEDT, and H. R. IRONS: Trans. Am. Geophys. Union **36**, 25 (1955).
[7] C. A. JARMAN: Scien. and Tech. Memorandum TPA 3, British Ministry of Supply (1949).

2. The direction reference system.

Fig. 1 shows the gyro-stabilized platform and the magnetometer head of the Canadian instrument. Explanations are given in the caption. The pitch and roll stabilization systems are shown schematically in Fig. 2. A small disturbance of the platform, due to a sudden pitching of the aircraft for example, will cause the pitch gyroscope mounted on the platform to precess in azimuth. The pitch gyroscope pick-off immediately gives an electrical signal which is amplified and fed to the pitch servomotor, returning the platform to its original attitude. Similarly the roll servomotor, controlled by the roll gyroscope, holds the platform steady when the aircraft rolls. These servo-systems are extremely fast, with a natural frequency of 30 Hz, and will correct for angular velocities of the aircraft up to 44°/sec. Great care is necessary in the design of the gimbals, the gear trains, and the amplifiers to obtain stable operation at this frequency.

The system so far described will hold the platform steady for a short period of time, but will not keep it horizontal. Torques must be applied to the gyroscopes to counteract their tendency to drift, and to rotate the platform slowly as the earth rotates and the aircraft moves over the earth. The necessary torques are derived primarily by integrating the outputs of two accelerometers mounted on the platform. The pitch accelerometer responds to fore-and-aft accelerations in the plane of the platform; the roll accelerometer to transverse accelerations in the plane of the platform. The constants of the system are chosen to make the platform behave approximately like a damped pendulum with a period of 6 min. (A simple pendulum with a 6-min period would be 32 km long.)

A pendulum or a level bubble in an aircraft will give an incorrect indication of the vertical because of the horizontal accelerations of the aircraft. Preliminary measurements in transport aircraft in straight and level flight revealed quasi-periodic accelerations which produce deflections of the apparent vertical from the true vertical with amplitudes of 1° or 2° and periods of 1 or 2 min. Superimposed on these were accelerations with a much shorter period and comparable amplitude. A period of 6 min was chosen for the gyro-stabilized platform to make it act like a low pass filter, attenuating the effect of the long-period accelerations to a few minutes of arc. A longer natural period would have resulted in better filtering, but would have increased the accuracy demanded of the gyroscopes.

In order to reduce the amplitude of forced oscillations of the platform due to long-period accelerations of the aircraft, automatically computed signals proportional to these accelerations are subtracted from the accelerometer signals at the input to the integrators. The computed fore-and-aft signal is the derivative of the output of a true airspeed meter, and the computed transverse acceleration signal is the product of airspeed and the rate of change of heading of the aircraft. The heading is supplied by the directional gyroscope. A small correction for Coriolis acceleration is also applied to the roll accelerometer output.

The introduction of the computed acceleration to the pitch system reduces periodic disturbances in it by a factor of two or three. The roll acceleration computer gives an even more obvious improvement in platform accuracy. The track of the aircraft can depart considerably from a great circle, with frequent minor changes in heading, without disturbing the platform. During major alterations of heading, the accelerations exceed the range which can be handled accurately by the accelerometers and computing circuits, and the input to the integrators is automatically switched off to avoid the introduction of false information.

The directional gyroscope K (Fig. 1) supplies the azimuth reference to the magnetometer system as well as to the roll acceleration computer described above. It is connected in a servo-loop similar to those of the pitch and roll gyroscopes, so that its azimuth remains fixed as the aircraft changes heading, but it

Sect. 2. The direction reference system.

Fig. 1. The gyro-stabilized platform and magnetometer head of the Dominion Observatory three-component airborne magnetometer. The inner gimbal C can rotate about the longitudinal axis of the aircraft (the roll axis) in bearings set in gimbal E. The gyroscope platform A is supported by bearings in gimbal C which allow it to rotate about the pitch axis — an axis perpendicular to the roll axis. The magnetometer yoke N is rigidly connected to the inner gimbal C by an aluminum pipe H. The magnetometer head G is mounted on a small platform which hangs from yoke N in bearings parallel to the pitch axis. The plane of the magnetometer platform is maintained accurately parallel to the plane of the gyroscope platform A by the mechanical connection through pipe H and a parallelogram linkage, the lower part of which can be seen at J. The gyroscope platform A is maintained steady in space, when the aircraft rolls or pitches, by the roll servomotor D and pitch servomotor B, geared to the gimbal axes. A third servomotor is geared to the turntable L which carries the directional gyroscope K.

Fig. 2. Schematic diagram of the vertical reference system.

is not provided with an automatic rate control. It is allowed to wander, and the corrections to be applied to its readings are determined from astronomical measurements made at 10 min intervals with a Kollsman periscopic sextant. This instrument has been modified by the addition of servomechanisms stabilizing it in pitch, roll, and azimuth according to signals transmitted by the gyro-stabilized platform. When the observer looks through the eyepiece he can read directly on a graduated circle the angle in azimuth between the directional gyroscope and the sun or a star. The stabilization, besides increasing the accuracy of observation, often allows observations to be made through clouds when the sun is invisible to the unaided eye.

The most critical components of the direction reference system are of course the gyroscopes. Gyroscopes of the single degree of freedom, floated type manufactured by Minneapolis-Honeywell are used — originally the type HIG-5 and later the type HIG-4. The present system of determination of the vertical was chosen out of the many possible ways of combining gyroscope and accelerometers because it made the smallest demands on the performance of the gyroscopes[8]. Constant drift rates produce no platform error, but the drift rates of the gyroscope should be constant to within a few °/h when the units are subjected to the small accelerations normally encountered in survey flying. In spite of this specification, modest by modern standards, it has proved difficult to maintain satisfactory gyroscope performance for more than 100 hours of operation without overhaul.

It is difficult to measure directly the accuracy of the gyro-stabilized platform in the air, because no standard of comparable accuracy is available. However, continuous records of the outputs of the accelerometers and computing circuits are made during surveys, and typical accelerations can be simulated in laboratory tests by introducing electrical signals into the system. From such experiments, it is concluded that the gyro-stabilized platform remains horizontal to within two or three minutes of arc under normal survey operations. When Doppler navigation equipment becomes available, giving continuous information on wind velocity and direction, it should be possible to reduce errors to one minute of arc. In the case of the azimuth reference system, examination of the scatter of the astronomical observations indicates that the probable error of a single observation is about 0.2°. The smoothed curve of corrections applied to the directional gyroscope is probably accurate to within 0.2°.

3. The magnetometer. The field-sensitive head of the magnetometer is mounted inside the cabin of the aircraft on the small platform shown in Fig. 1 which is stabilized in pitch and roll by mechanical connection to the gyro-stabilized platform A. The head contains three mutually perpendicular magnetic detectors of the saturated transformer type, each of which is continuously maintained at the null by a direct current flowing in a solenoid surrounding it. The direct current is thus proportional to the magnetic field component along the axis of sensitivity of the detector. One of these units, mounted with its sensitive axis vertical, measures the vertical component; the others measure the fore-and-aft and transverse horizontal components respectively.

The direct current proportional to the vertical component is measured by an automatic potentiometer, to which is geared a counter indicating the vertical component in units of 10γ. The range of the potentiometer corresponds to $-10{,}000$ to $+70{,}000\,\gamma$ for the northern hemisphere, and $-70{,}000$ to $+10{,}000\,\gamma$ for the southern hemisphere.

The outputs of the two horizontal field-measuring units are combined in an electrical resolver to give the intensity of the horizontal component and the magnetic heading of the aircraft. The horizontal intensity is displayed in units

---

[8] S. Z. MACK: Report 4-1200-1 Defence Research Board, Ottawa, Canada 1953.

of 10 γ by a counter geared to an automatic potentiometer. The range is 0 to 48,000 γ. The magnetic heading is continuously subtracted from the gyro heading of the aircraft (supplied by the directional gyroscope), and the difference is displayed on a dual indicator dial in degrees. The dial reading, when corrected for the difference between gyro north and true north, is the magnetic declination.

During normal survey flights, a switch controlling the operation of the indicators is thrown to "Average". The declination dial and the two counters, instead of following the changes in the magnetic field now remain fixed at the last average values $\overline{D}$, $\overline{H}$, and $\overline{Z}$

Fig. 3. Schematic diagram of the three-component magnetometer.

determined for declination D, horizontal component H, and vertical component Z respectively. At the end of a 5 min period, each indicator turns automatically to a new value representing the average value of its component over the preceding 5 min interval. While the instrument is being used in this manner, three recording milliammeters record the difference between the instantaneous value of each component and the value appearing on the corresponding indicator. Thus continuous values of any component can be read from the meter records using the last determined average as a baseline.

Fig. 3 shows a block diagram of the magnetometer system. Each magnetic detector contains two parallel strips of annealed Mumetal which are excited in opposite phase by a 1000 Hz alternating current passing continuously through primary windings. As the Mumetal strips become saturated and unsaturated 2000 times per second, they modulate the magnetic induction through a common secondary coil due to the component of the earth's field along the axis of the unit. A signal composed of even harmonics of the excitation frequency[4] is thus induced in the secondary winding, with an amplitude and phase depending on the magnitude and sense of the magnetic field component. The 2000 Hz component of the signal is greatly enhanced without using band-pass filters, which introduce phase-shifts difficult to control, by tuning the secondary winding with a capacitor, and operating the detector according to certain conditions[9].

The signal is amplified, detected by a phase-sensitive detector, and integrated by a d.c. (direct current) amplifier with capacitive feedback. The output of the integrator is passed through a solenoid surrounding the magnetic detector, producing a magnetic field which reduces the field at the magnetic detector to zero. The feedback system behaves like a simple servomechanism with a natural frequency of 20 Hz. Three component station recording magnetometers have been constructed using the same basic design[10].

The currents through the solenoids pass through manganin resistors, producing voltages proportional to the three components of the magnetic field at the head. The voltages representing the two horizontal components are accurately converted from direct to alternating current and combined in a resolver of the synchro type. The angular position of the resolver rotor, the magnetic heading of the aircraft, is subtracted from the gyro heading and the difference is displayed on a dial which can be read to $0.1°$. The electrical output signal of the resolver is measured by a motor-driven potentiometer, to which is geared the counter indicating $\mathcal{H}$. The voltage representing Z is measured directly by a similar potentiometer. The potentiometer circuits are standardized periodically in flight against a standard cell.

When the automatic averaging system is in operation, the $\bar{D}$, $\bar{\mathcal{H}}$, and $\bar{Z}$ indicators remain fixed for 5 min. Each of the three averaging devices contains a motor-generator, whose shaft is made to revolve at a speed proportional to the difference between the actual value of the component and the value shown by the corresponding indicator. At the end of the 5 min averaging period, the indicator reading is changed by an amount proportional to the angle through which the motor-generator shaft has turned. With proper adjustment of the constants of the system, the indicator shows the average value of the component for the preceding 5 min interval. The three recording meters record on charts moving at 20 mm/min the instantaneous values of D, $\mathcal{H}$, and Z, with the center lines of the charts corresponding to the last determined averages. The normal range of the meters, $\pm 2.5°$ in D and $\pm 500\,\gamma$ in $\mathcal{H}$ and Z, can be doubled by the operator when large anomalies are encountered.

The sensitive axes of the three field-measuring units must be mutually perpendicular and must be horizontal or vertical when the gyro-stabilized platform is level. To assist in making these adjustments, the magnetometer head is fitted with level bubbles and mounted on a vertical axis, like a theodolite. The head is removed from the stabilized platform, and the magnetometer readings are plotted as the head is revolved slowly about its vertical axis in a uniform magnetic field. From the curves obtained, errors in the orientation of the axes of the field-measuring units relative to each other and to the vertical axis can be identified, and the appropriate adjustments are made. This procedure also compensates for errors in the electrical resolver. The magnetic components at the head are then measured with a portable magnetometer[9] and the potentiometer circuits are adjusted to indicate the correct values. The head is then replaced on the stabilized platform and levelled as a unit.

The chief source of error in the magnetometer system, considered separately from the direction reference system, is in the non-linearity of the automatic potentiometers, which can produce maximum errors of $20\,\gamma$. Temperature effects are avoided by thermostating the magnetometer head. To ensure that the magnetic detectors give zero signal in zero field, provision is made for balancing the wave-form of the 1000 Hz excitation current. Permanent magnetization of the Mumetal cores is believed to produce zero offsets of less than $5\,\gamma$. It has been found that mechanical creeping occurs in the detector mounting after adjustment. By allowing a day to reach equilibrium before the final measurements are made, this effect is reduced to less than $10\,\gamma$.

It is concluded that the accuracy of the magnetometer in measuring the field at the magnetometer head with respect to the axes supplied by the direction reference system is $0.1°$ in declination (in southern Canada), and $20\,\gamma$ in the horizontal and vertical components.

---

[9] P. H. SERSON, and W. L. W. HANNAFORD: Can. J. Tech. **34**, 232 (1956).
[10] P. H. SERSON: Can. J. Phys. **35**, 1387 (1957).

Fig. 4. Profiles of D, H and Z, Sept. 11, 1953, for a flight over Hudson Bay at an altitude of 3 km. The position is indicated in the insert.

Fig. 5. Isomagnetic chart for the prairie provinces of Canada (epoch 1955.5) showing lines of equal vertical intensity in Oe. The heavy lines show contour intervals of 500 γ and the lighter lines intervals of 250 γ.

4. Reduction and plotting of observations. Since its completion in 1953, the three-component airborne magnetometer has been used in surveys of about a month's duration, at intervals of approximately one year. It is installed in a four-engined aircraft chartered for the period of the survey. Because the aircraft is usually available for only a limited time, it is not practical to compensate the instrument for the magnetic field of the aircraft, as a ship's compass is compensated. Instead, the aircraft field is measured at intervals of about a week, before, during, and after the survey, by swinging the aircraft over a region where the geomagnetic field is accurately known. After the completion of the survey, corrections for the effect of the aircraft field are applied to the observations before they are plotted.

Ideally, the magnetic field of the aircraft can be considered as consisting of two parts: a permanent field and a field induced by the earth's magnetic field. These fields can be described by 12 parameters in Poisson's equations for three orthogonal magnetic components. In Canada, the vertical component of the earth's field is constant to ± 10 per cent over the whole country, and it is difficult — and also unnecessary — to separate the fields induced by the vertical component from the permanent fields of the aircraft. Thus, in practice the number of parameters to be determined is reduced from 12 to 9. A station is chosen where the gradients in the magnetic field are small, and careful measurements are made on the ground, so that the field at the station can be deduced at any time from the records of a nearby magnetic observatory. Eight 20 min flights passing over the station on four orthogonal headings are made at an altitude of about 3 km. From the difference between the magnetic components at the ground and the components indicated by the airborne magnetometer, the parameters representing the aircraft field are computed in a least squares solution, with the aid of a small analog computer. In the unmodified transport aircraft used, a substantially vertical permanent field of about 600 γ is usually found at the magnetometer head, with the induced fields an order of magnitude smaller.

During survey operations, the field of the aircraft is redetermined at intervals of about a week. The results of the separate determinations are usually combined to give a single set of parameters of aircraft magnetism for the whole survey. The corrections to be applied to the readings of the magnetometer are calculated, using the adopted parameters, for several values of the apparent horizontal field intensity, and plotted as functions of the apparent magnetic heading of the aircraft. In reducing the survey results the corrections can be quickly read from these curves to an accuracy of 10 γ.

In flight, the magnetometer operator records the 5 min averages of the field components and Universal Time on an observation sheet. After the survey, the correction to the measured declination for the drift of the directional gyroscope and the corrections to all three components for the magnetic field of the aircraft are entered on the same sheet. Small corrections reducing the observations to ground level may also be applied but these are usually included in the correction for the aircraft field. The corrections are added to the observed values, and the observations are ready to be plotted. One man in a day can correct and plot the results from an 8 hour flight comprising nearly 100 observations of the magnetic vector, each observation representing the average of the components over a 30 km segment of the flight path.

Fig. 4. shows magnetic profile from a flight made over Hudson Bay in 1953, drawn by joining the 5 min averages of the components. The dotted lines represent the instantaneous values of the components over a large anomaly as read from the continuous traces of the recording meters. The smoothing effect of the automatic averaging is apparent. Where airport facilities permit, surveys are flown in a systematic pattern of parallel lines. The corrected average values are then written directly on the charts. Fig. 5 shows an isomagnetic chart drawn by contouring the average values of vertical intensity obtained in a survey made in three weeks in 1955. Measurements were made at an altitude of about 3 km along parallels of latitude one degree apart.

5. The accuracy of survey results. The error in an observation made by the three-component airborne magnetometer as plotted on the charts may be considered as the sum of five independent errors of the following types:

a) Errors in measuring the magnetic field at the magnetometer with respect to the direction reference system.

b) Errors in the direction reference system.

c) Errors due to changes in the magnetic field of the aircraft.

d) Errors due to magnetic disturbances, and

e) Errors in geographical position.

It was shown in Sect. 3 that the accuracy of the magnetic measurements, with respect to the direction reference system, was 0.1° in declination and 20 γ in the horizontal and vertical components. The effect of errors in the direction reference system will vary with magnetic latitude. It is easily shown that the stabilized platform, with maximum errors of 3 minutes of arc, will contribute maximum errors of 0.2° in declination D, 50 γ in the horizontal component $\mathcal{H}$ and 20 γ in the vertical component Z (in southern Canada).

If the parameters representing the magnetic field of the aircraft were constant, the aircraft field would not produce an independent source of error in the survey results, since the appropriate corrections could be determined to within the instrumental accuracy. It has been found, however, that the aircraft field changes during the survey, introducing an uncertainty into the corrections to be applied at any time.

An analysis of successive determinations of the field of the aircraft in 1955 showed that this uncertainty will contribute 60 γ in the horizontal vector, and 30 γ in vertical field intensity to the probable error of an observation.

The effect of magnetic disturbances on airborne magnetic observations will vary widely, depending on the particular region of the survey and the degree of disturbance at the time. In moderate magnetic latitudes, corrections can often be made from the records of magnetic observatories, but near the zone of maximum aurora a dense network of recording stations on the ground would be necessary to make such corrections with any degree of confidence. While this poses a serious difficulty in detailed mineral surveys requiring an accuracy[11] of a few γ, the problem is much less acute for surveys of the type considered here. Even at sunspot maximum, the probability of making magnetic observations during magnetic disturbances of more than 100 γ is rather small, and it is usually possible to suspend survey operations during major world-wide storms.

The effect of magnetic disturbances and diurnal variation on the results of 1955 survey was estimated from an examination of the magnetograms of Meanook Observatory. Although much of the survey lay in the auroral zone and no attempt was made at applying corrections, the probable uncertainty was only about 20 γ in any component of the field.

Experience in airborne magnetic surveys has shown that errors in fixing the geographical position of the observations have a greater effect than was at first anticipated. An examination of existing magnetic charts indicated that the gradients in the magnetic components would be of the order of 10 γ/km. It was thought that the usual methods of aerial navigation (a combination of dead reckoning with occasional fixes by map-reading, astronomical observations and radio bearings) would be sufficient to produce reproducible results to the order of accuracy considered here. It was soon found, however, that the actual magnetic gradients at any point were likely to be many times larger than the regional gradients as shown by the magnetic charts, and errors of a few km in navigation could produce large differences in the magnetic field.

The survey of Western Canada in 1955 was planned to test the over-all operational accuracy of the airborne method[5]. The flightlines were laid out in a grid, so that they intersected at 36 points. Navigation was by map-reading, with 4 or 5 fixes per hour. After the results were plotted, the measurements made at the points of intersection on different flights were compared. Discrepancies were found more than twice as large as would be expected from the determinations of the aircraft field and the contribution of magnetic disturbances, as shown in Table 1. The local gradients at the points of intersection were measured, and it was found that the discrepancies could be accounted for statistically by assuming a probable error in geographical position of about 6 km. Geographical errors of this magnitude are considered reasonable when it is remembered that they include errors in plotting the flight-lines and reading the times of intersection from a chart at a scale of 1:3,000,000. This interpretation is confirmed by the fact that the largest discrepancies occur at intersections with large local gradients.

The effect of geographical errors on magnetic observations is closely related to the more general problem of predicting the accuracy of magnetic charts as a function of the geographical distribution of the observations on which they are based. A statistical analysis of the geomagnetic field as a function of horizontal distance has shown[2] that the probable error of charts based on flight-lines 100 km

---

[11] K. WHITHAM, and E. I. LOOMER: Geophysics **22**, 646 (1957).

Table 1. *Results of an analysis of the discrepancies observed at 36 points of intersection of flight-lines during a survey in 1955.*

	$D/°$	H/γ	Z/γ
Probable error of one instantaneous observation due to			
a) changes in the aircraft field	0.3	70	30
b) magnetic disturbances	0.24	22	21
c) navigation errors of ± 6 km	0.62	95	95
Total probable error of one instantaneous observation due to (a), (b), and (c)	0.73	120	100
Expected probable discrepancy at intersections estimated from from total probable error of one instantaneous observation	1.0	170	140
Measured probable discrepancy at intersections	1.4	140	140

apart will be of the order of 100 γ in any component. This is basically the error resulting from interpolation of the magnetic field between the flight-lines, and is quite distinct from the errors of observation. A reduction of the errors in navigation, by the use of aerial photography, or by the use of special radio aids to navigation which are available over certain ocean areas, would increase the accuracy of knowledge of the magnetic field directly beneath the path of the aircraft, but would not significantly increase the accuracy of the charts as a whole unless accompanied by a great increase in the density of flight lines[2].

The conclusions regarding the accuracy of surveys with this three-component airborne magnetometer can now be summarized. Using flight techniques which are economic and do not require the use of specialized aids to navigation, and therefore allow very quick and easy reduction of observations, an overall accuracy of 100 γ is attained which is sufficient for the production of large scale magnetic charts. Furthermore, major secular change foci should be identified on repeating surveys five or more years later. The errors arise as follows. Experimental results obtained swinging the aircraft at suitable locations show that the probable error of a single observation, after correction for the magnetic field of the aircraft, is 40 γ in declination D, 40 γ in the horizontal component H and 10 γ in the vertical component Z. This indicates that the probable error of the stabilized platform is 2 to 3 minutes of arc under conditions of frequent manoeuvering. The magnetic field of the aircraft changes over periods of a few weeks, producing an uncertainty in the corrections for the aircraft field which increases the probable error of survey observations to 60 γ in the horizontal vector and 30 γ in the vertical component. Errors in navigation and plotting the flight-lines and the effect of magnetic disturbances result in a further increase in the probable error of an observation, as plotted on the charts, to 100 γ in any component.

Aeromagnetic Surveying with the Fluxgate Magnetometer.

By

J. R. BALSLEY*.

With 8 Figures.

I. Fluxgate magnetometers.

1. Introduction. The great speed and broad applicability of airborne magnetometers have been recognized for many years[1], and a few crude measurements had been attempted[2-7], but it was not until 1944 that equipment was produced that was accurate enough to make useful measurements. The basic element of most of the modern airborne magnetometers is the fluxgate or saturable core inductor, a well known device that has been used to measure magnetic fields since 1936[8-10]. However, the full utilization of the fluxgate and its incorporation in a useful functioning geophysical field instrument required cooperative efforts of physicists, geophysicists, geologists, and engineers of several different governmental agencies and private organizations that produced a device that could not have been designed or perfected otherwise[11-19].

---

* Publication authorized by the Director, U. S. Geological Survey, Washington D, C., USA. Manuscript received October 1955.

[1] C. A. HEILAND: Geophysical mapping from the air: its possibilities and advantages. Eng. and Min. J. **136**, 609—610 (1935).

[2] M. GROTEWAHL: Bericht über die Versuchsfahrt des Bidlingsmaierschen Doppelkompasses mit dem Luftschiff Graf Zeppelin. Terr. Magn. **35**, 226—229 (1930).

[3] H. LUNDBERG, and K. SUNDBERG: Discussion of paper by A. S. EVE, A magnetic method of estimating the height of some buried magnetic bodies. Am. Inst. Min. Met. Eng. Geophysical Prospecting 209 (1932).

[4] A. A. LOGACHEV: The development and applications of airborne magnetometers in the U.S.S.R. Geophysics **11**, No. 2, 135—147 (1946); — Petrol. Eng. **17**, No. 10, 211—218 (1946).

[5] A. A. LOGACHEV: Aeromagnetic surveys in prospecting for iron ore deposits (in Russian). Vses. Nauch.-Issled. Geol. Inst. Mater., Geofiz., No. 11, 3—7 (1947).

[6] K. RAMSAYER: Die Änderung magnetischer Störgebiete mit der Höhe und ihr Einfluß auf die Flugnavigation. (Variation of magnetic anomalies with altitude and its influence on aerial navigation.) Beitr. angew. Geophys. **9**, 65—97 (1941).

[7] R. E. GEBHARDT: Investigation of height of local magnetic anomaly at Port Snettisham, southeastern Alaska. Terr. Magn. **47**, No. 2, 165—170 (1942).

[8] HAIG ANTRANIKIAN: Magnetic field direction and intensity finder. U. S. Pat. 2,047,609 (July 14, 1936).

[9] H. ASCHENBRENNER, and G. GOUBAU: Eine Anordnung zur Registrierung rascher magnetischer Störungen. Hochfrequenztech. u. Electroakust. **47**, 177—181 (1936).

[10] G. W. ELMEN: Magnetic alloys of iron, nickel, and cobalt. Bell Systems Tech. J. **15**, 113—135 (1936).

[11] R. BAILEY: Canadian aerial magnetic surveys. Canadian J. Res. F **26**, 523—539 (1948).

[12] J. R. BALSLEY: The airborne magnetometer. U. S. Geol. Surv. Geophys. Invest., Prelim. Rept. 3, 8 pp. Petrol. Eng. **17**, No. 11, 77—87; No. 12, 104—130 (1946).

[13] E. A. ECKHARDT: Airborne magnetometer. Oil and Gas J. **45**, No. 5, 78—79, 91—92 (1946).

[14] H. JENSEN: Geophysical surveying with the magnetic airborne detector AN/ASQ-3A. U. S. Naval Ordnance Lab. Rept. 937, May, 1945, 63 pp., Washington, D. C. (1945).

[15] A. W. KNOERR: The airborne magnetometer, a new aid to geophysics. Eng. and Min. J. **147**, No. 6, 70—75 (1946).

[16] G. MUFFLY: The airborne magnetometer. Geophysics **11**, No. 3, 321—334 (1946).

[17] J. E. QUAILE: Airborne submarine detector. Military Eng. **39**, No. 258, 166—167 (1947).

[18] J. R. BALSLEY: Aeromagnetic surveying in Advances in Geophysics **1**, 313—349 (1952). New York: Academic Press.

[19] W. E. FROMM: The magnetic airborne detector in Advances in Electronics **4**, 257—299 (1952). New York: Academic Press.

The Earth's magnetic field and its variations have been measured for many centuries by a variety of instruments. The most popular is the magnetic balance in which a carefully suspended and calibrated magnet is used to measure the magnetic field in relation to the force of gravity. Obviously such instruments must be used statically and cannot be used in moving vehicles. Magnetometers have also been based on the principle that an electric current is induced in a coil of wire moving in the earth's magnetic field. These instruments generally are too cumbersome or too insensitive to be applicable in aerial measurements. The fluxgate magnetometers makes use of the effect of the earth's magnetic field on the variation of permeability with intensity of magnetization in ferromagnetic material used in the detecting element. The magnetometers, being sensitive, small, and easily stabilized and oriented, are ideally suited for aerial operations.

Because it is difficult to obtain an accurate frame of reference in an aircraft, most airborne magnetometers measure only the total intensity and are useful only for that type of geophysical prospecting in which the objective is to measure the magnetic characteristics of rock units. Recently airborne systems have been developed which measure the three components of the earth's field. Although the two types of equipment are widely different in their application and use they are both fundamentally fluxgate magnetometers and both will be discussed in this article. Many reports and articles have been published that deal with one or more phases of the development and use of airborne magnetometers. The purpose of this chapter is to review and summarize the literature dealing with the instrumentation, use of the equipment, and the interpretation of the data, and to illustrate by a few examples the results that have been obtained.

a) Basic fluxgate magnetometer.

An airborne magnetometer consists of a detecting mechanism either self oriented with respect to the earth's field or oriented against some other frame of reference; electronic oscillators and amplifiers required for its operation; and the equipment used to measure and record its output, the variations of the earth's magnetic total field.

2. Detector element.

α) The *fluxgate magnetometer* consists of a core of easily saturable ferromagnetic material of high permeability (usually permalloy) surrounded by a coil to which is applied an alternating voltage. The alternating current in the coil produces within it an alternating magnetic field, h, equal to $h_0 \cos \omega t$ which acts upon the core (Fig. 1c). If in addition a small external magnetic field, H_0 is acting along the axis of the core, the total field, H, acting upon the core is increased during one-half the cycle and decreased during the other half. If we neglect hysteresis in the core and assume that it has a permeability, μ, and is saturated in a field higher than (H_0+h_0), the flux density B' would be μH or $\mu(H_0+h_0 \cos \omega t)$, shown as the dashed lines in Figs. 1a and b, respectively. The output e.m.f. (electromagnetomotive force), e', of the coil with NA effective area turns would be (dashed curve, Fig. 1d):

$$e' = -NA \frac{dB'}{dt} \quad \text{(without saturation)} . \tag{2.1}$$

$$\frac{e'}{V} = -10^{-8} \frac{NA}{cm^2} \frac{dB'/\Gamma}{dt/s} . \tag{2.1a}$$

Sect. 2.   Detector element.   397

If, on the other hand, the alternating magnetic field in the presence of a small external field is more than sufficient to saturate the core, the core will be forced through the hysteresis cycle shown as the solid curve in Fig. 1a. If its permeability is high the flux density, μH_0, produced in it by the external field will be an appreciable part of the flux density at saturation, B_M, so that the core material will be saturated earlier during one-half cycle and later in the other. The flux density, B, will be a function, f, of μ and $H_0 + h_0 \cos \omega t$, shown as the

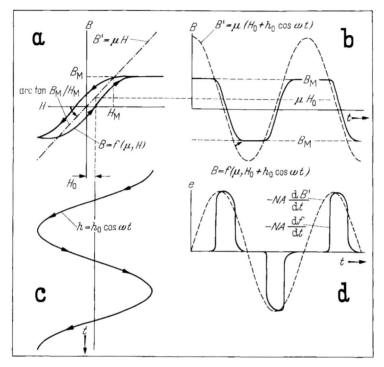

Fig. 1a—d. Schematic diagram of operation of saturable inductor. [Courtesy: American Geophysical Union [20].]

solid line in Fig. 1b, and the output e.m.f. of the coil, e, will have the form shown as the solid curve in Fig. 1d:

$$e = -NA \frac{df}{dt} \quad \text{(with saturation)}, \tag{2.2}$$

$$\frac{e}{V} = -10^{-8} \frac{NA}{\text{cm}^2} \frac{df/\Gamma}{dt/s}. \tag{2.2a}$$

β) This distorted and asymmetrical wave form contains both *even and odd harmonics of* the driving frequency, as has been shown by RUMBAUGH and ALLDREDGE[20] in the following excerpt.

"By neglecting hysteresis for the moment, the even and odd harmonic content of e may be demonstrated analytically in a simple manner. It is evident from the symmetry of an idealized B vs. H curve that the flux density B may be

---

[20] L. H. RUMBAUGH, and L. R. ALLDREDGE: Airborne equipment for geomagnetic measurement. Trans. Am. Geophys. Un. **30**, 836—848 (1949).

expressed as an odd power series in H:

$$B = \sum_{n=0}^{\infty} b_{2n+1} H^{2n+1} \tag{2.3}$$

where coefficients b are fixed by the geometry and material of the core and H is the sum of the external field, H_0, and the driving field $h_0 \cos \omega t$

$$H = H_0 + h_0 \cos \omega t. \tag{2.4}$$

"By the examination of the general term $(H_0 + h_0 \cos \omega t)^{2n+1}$ it is seen that the binomial expansion will give even powers of $h_0 \cos \omega t$ multiplied by odd powers of H_0 in addition to odd powers of $h_0 \cos \omega t$ multiplied by even powers of H_0.

"All even (odd) powers of the cosine may be expressed as the sum of the cosines of even (odd) multiples of ωt. The corresponding coefficients are even (odd) function of h_0 and odd (even) functions of H_0 and

$$B = \sum_{n=0}^{\infty} B_n (H_0, h_0) \cos n \omega t$$

where the coefficients B_{2n} of the even harmonics are even functions of h_0 but odd functions of H_0, while the coefficients B_{2n+1} of the odd harmonics are odd functions of h_0 and even functions of H_0.

"The output e.m.f. is given by

$$\frac{e}{V} = -10^{-8} \frac{NA}{\mathrm{cm}^2} \mathrm{d} \frac{B}{\Gamma} \Big/ \mathrm{d} \frac{t}{\mathrm{s}}, \tag{2.1a}$$

where NA means the effective area turns in the output coil. It follows immediately that $\frac{e}{V}$ is $\left(-10^{-8} \frac{NA}{\mathrm{cm}^2}\right)$ times the sum of even harmonic terms

$$2n \frac{\omega}{\mathrm{Hz}} \frac{B_{2n}}{\Gamma} \sin 2n \omega t,$$

and odd harmonic terms

$$(2n+1) \frac{\omega}{\mathrm{Hz}} \frac{B_{2n+1}}{\Gamma} \sin (2n+1) \omega t.$$

The even harmonic terms vanish for H_0 equal to zero, while the odd harmonics do not vanish" because they contain a term involving H_0^{2n} which is independent of H_0 for $n = 0$. "Likewise, only the even harmonics are sensitive to the sign of H_0." Measurement of one or more of the even harmonics of the output e.m.f. provides therefore the means of measuring the external field [20-22].

γ) A variety of *methods* have been used *to make such measurements;* one method involves the use of a single inductor coil and the second harmonic output is measured [23-24]. Others involve two inductors wound in opposite directions and when balanced the output of all the even harmonics is measured [25-27]. For some

---

[21] E. P. FELCH, W. J. MEANS, T. SLONCZEWSKI, L. G. PARRATT, L. H. RUMBAUGH, and A. J. TICKNER: Airborne magnetometers. Elect. Eng. **66**, 680—685 (1947).

[22] U. S. Naval Ordnance Laboratory. Physical principles of the fluxgate and other inductor magnetometers. Mine Unit Rept. 263, 44 pp. (1941).

[23] A. G. LAIRD, and T. SLONCZEWSKI: Magnetic field detector. U. S. Pat. 2,426,622. (Sept. 2, 1947) (1947).

[24] F. G. MERRILL: Magnetic detector. U. S. Pat. 2,448,613. (Sept. 7, 1948) (1948).

[25] V. V. VACQUIER: The Gulf absolute magnetometer. Terr. Magn. **50**, No. 2, 91—104 (1945).

[26] V. V. VACQUIER: Apparatus for responding to magnetic fields. U. S. Pat. 2,406,870. (Sept. 3, 1946) (1946).

[27] R. W. D. WYCKOFF: The Gulf airborne magnetometer. Geophysics **13**, No. 2, 182—208 (1948).

applications two inductors in an unbalanced circuit are used, and the width of complex pulses is measured. The single harmonic system has the advantage of using a simple and trouble-free electronic circuit, but does not have the large source of energy of the balanced inductor type. The unbalanced inductor circuit has the advantage of producing a signal even with zero external field[19].

δ) The *single inductor system* is illustrated in Fig. 2a. The inductor is excited through a bandpass filter designed to eliminate all but the fundamental 1,000-Hz frequency of the driving oscillator. The second harmonic output of the coil is measured by feeding the signal through a narrow bandpass filter of 2,000 Hz

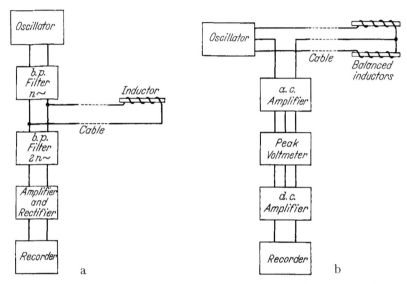

Fig. 2a and b. Typical fluxgate magnetometer circuits. (b. p. = bandpass; a. c. = alternating current; d. c. = direct current.) [Courtesy: American Geophysical Union [20].]

and then through amplifiers and a rectifer to a recorder. The bandpass filter must be of high quality and the oscillator stable. This basic circuit had wide application in submarine detectors, and has been used for many years in geophysical prospecting during which time it has required minimum maintenance.

ε) Several circuits have been devised to utilize the *balanced inductor* which consists of two identical parallel permalloy cores, each surrounded by identical coils so arranged that the driving fields are equal in magnitude but opposite in direction. Thus, when the external field opposes the driving field in one coil it adds to the driving field in the other. In one type of circuit, as shown in Fig. 2b, the coils are electrically connected so that the even harmonics of the output e.m.f. add, and the fundamental odd harmonics oppose. Obviously the critical feature of this design is that the inductors have identical mechanical, magnetic, and electrical characteristics. In another type of circuit the inductors form two of the arms of a balanced bridge. As the core of the coil in which the external field adds to the driving field reaches saturation earlier than the core of the other coil, the bridge is temporarily unbalanced and a pulse produced whose width and amplitude are functions of the magnetic field. Because one core leads the other during the part of the cycle from saturation in one direction to saturation in the other direction, and lags during the other half cycle when the e.m.f. is reversed, the pulses have the same polarity.

ζ) In the *unbalanced bridge circuits* a small resistance is introduced whose function is to reduce the alternating magnetic field in one of the inductors. Thus even with no external magnetic field one coil reaches saturation sooner than the other and a signal similar to that described above is produced except that alternate pulses have opposite polarity because the weakly driven inductor lags as the alternating field increases in either direction from zero and leads as the alternating field decreases. If an external field is applied to this bridge the relative width of positive and negative pulses is changed because the lag time is increased and the lead time is decreased or vice versa depending upon the direction of the applied field. This is usually referred to as the "Peak Type Magnetometer".

η) Most systems make use of compensation: the greater part of the Earth's field acting on the detector coil is nullified by a magnetic field produced by means of a steady current flowing through the coil or through a coaxial secondary winding around it. Thus instead of measuring one γ (gamma) in a field of 60,000 γ, which requires considerable precision, the equipment measures one γ in a field of 5,000 γ or less. In one variation of the basic circuit the output of the detector coil is used to adjust this nullifying current so that it is made exactly equal and opposite to the earth's field. Proper measurement of this current constitutes the measurement of the earth's field.

3. Orientation of the detector element.

The saturable inductor measures the external field parallel to its axis and can therefore be used to measure any component of the magnetic field of the earth by orienting the detector in the right direction. Several instruments using this principle have been constructed for making measurements on the ground where the vertical and geographic frame of reference can be determined and the different magnetic components measured.

α) Several attempts have been made to use a damped *pendulum or* a vertical *gyroscope* to orient the detector in a vertical position so that the vertical magnetic component could be measured in airplanes or other moving vehicles. Neither the damped pendulum nor gyroscope determines the true vertical in a moving vehicle; the most precise system now available for use in aircraft is accurate to about 10' of arc, but only when the aircraft is flown at high altitude where the air is free of turbulence.

The error in the measurement of total magnetic intensity produced by this angular displacement is given by

$$\text{Error} = T[\cos(V-I) - \cos(V'-I)] \tag{3.1}$$

Where T is the total intensity of the earth's magnetic field; I, V, V' are angles: I is the magnetic inclination; V the true vertical of 90°, and V' the erroneous vertical established by the orienting mechanism. An error of 10' of arc in the orienting mechanism gives a magnetic error of 75 γ if the magnetic inclination is 65° and the total magnetic intensity 55,000 γ.

A more accurate orientation system must be applied if the equipment is to be used in exploratory aeromagnetic surveys which usually must be accurate to within a few γ. These surveys are conducted close to the surface of the ground where turbulent air and frequent maneuvers of the aircraft introduce violent motion that materially increases the error in the damped pendulum or gyroscope orienting system, and therefore, in the magnetic measurements. To produce an instrument of sufficient accuracy for aeromagnetic surveying a method of orientation that does not require geodetic reference is necessary.

β) The simplest method is to use the *magnetic field itself as reference* although a system with such a reference can measure only the scalar magnetic intensity

and not the direction of the magnetic vector. If the detector element is oriented parallel to the earth's field:
$$\text{Error} = T(1 - \cos \varphi) \tag{3.2}$$
where φ is the angle between the Earth's field and the axis of the coil. If, as assumed above, this angle is 10' and the total magnetic intensity 55,000 γ, the error is 0.23 γ. Obviously by orienting the detector in the direction of the earth's field the errors produced by misalignment have been greatly decreased, and such a system of orientation has the highest inherent accuracy. Several different systems have been developed to accomplish this orientation. In one system that uses the single-inductor second-harmonic circuit for detecting the field, three inductors are used[28-34]. Because the output of an inductor is zero if there is no field parallel to its axis, and because an inductor is sensitive to the direction of the applied field its output can be used to operate a servo mechanism which orients the inductor to the null position. This single inductor can be oriented and maintained in the direction at right angles to the field and two inductors with an angular separation, usually 90°, can be used to define automatically the plane at right angles to the external, or earth's field. If the axis of the third inductor is at right angles to the plane determined by the axis of the other two it is, therefore, automatically oriented and maintained parallel to the earth's field. Thus a detecting unit of three inductors at mutually right angles can be mounted in gimbals and the output of two of its inductors used to operate two servo motors which orient the detecting unit so that its third inductor can be used to measure the total earth's magnetic field[35, 36]. To correct for the effect of small errors of orientation produced by any time lag in the servo mechanisms, a technique is used based on the principle that the total magnetic intensity is equal to the square root of the sum of the squares of three values of the intensity measured at mutually right angles. As the orienting inductors measure only the small intensities produced by minor misorientations their output is squared and used to increase by the proper amount the output of the detecting inductor. This squaring method can be used without orientation to measure the total field, and special magnetometers of this type have been designed to measure high fields, but the strict requirements placed on the mechanical and electrical characteristics of the system limit its application and it is used only where insensitive measurements are required. The system described can be considered either as an oriented detector with a squaring correction, or as a partially oriented squaring detector. The squaring correction has been employed in conjunction

---

[28] E. P. FELCH jr., and T. SLONCZEWSKI: Magnetic field strength indicator. U. S. Pat. 2,468,968. (May 3, 1949) (1949).

[29] T. SLONCZEWSKI: Magnetic field strength indicator. U. S. Pat. 2,485,931. (Oct. 25, 1949) (1949).

[30] T. SLONCZEWSKI: Detection system. U. S. Pat. 2,488,341. (Nov. 15, 1949) (1949).

[31] E. P. FELCH jr., F. G. MERRILL, and Z. SLONCZEWSKI: Detection system. U. S. Pat. 2,488,389. (Nov. 15, 1949) (1949).

[32] Bell Telephone Laboratories. Magnetic airborne detector, development of a magnetic orienting system. U. S. Office Sci. Res. and Devel. Rept. **1309**, 72 pp. (Jan. 1943) (1943).

[33] U. S. Navy Department. AN/ASQ-3A equipment. Navy Aeronaut. Spec. M 597, 7 pp. (Nov. 1943.) (1943).

[34] U. S. War Department, U. S. Navy Department, and Air Council of the United Kingdom. Handbook of Maintenance Instructions for AN/ASQ-3A Equipment. Army-Navy 08-29-14, 193 pp. (June 1944.) (1944).

[35] M. S. RICHARDSON, and A. C. WEID: Operation of the universal magnetometer head with linear non-ideal control mechanism. U. S. Office Sci. Res. and Devel. Rept. 1930, 13 pp. (Sept. 1943.) (1943).

[36] Columbia University. Airborne Instrument Laboratory. Magnetometer suspensions. U. S. Office Sci. Res. and Devel. Rept. 5051, 28 pp. (May 1945.) (1945).

with the second harmonic single inductor system, but has not been reported as being used with the other system.

γ) Two methods of *orientation* have been reported *for* the systems that use the *balanced inductors* which measure all the even harmonics. In one an orienting inductor pair is mounted on a rotating disk with the detecting pair mounted along the axis of rotation[16, 25]. If the rotating disk is not normal to the Earth's field an alternating field is produced by the orienting inductor. The phase relationship of this signal is established by an electronic reference, and the alternating current is used to drive two servo motors that orient the disk to a position in which no signal is produced by the orienting inductor, that is, normal to the Earth's field. The detector element is then parallel to the Earth's field. No squaring correction is described for this system but the correction can undoubtedly be made (probably with some difficulty).

The second method of orientation of the paired-inductor systems is similar to that described for the single-inductor system. Three pairs of balanced or unbalanced inductors are mounted at mutually right angles and, as in the previously described system, two pairs are used to define the plane at right angles to the Earth's field. As the third pair is also at right angles to this plane it is parallel to the direction of the total field. It was pointed out earlier that the unbalanced inductors used in the peak-type magnetometer produce a signal even when a zero field is applied along their axes. This circuit is less complicated electronically and gives satisfactory performance, and for this reason has been used for the orienting circuits with either balanced or unbalanced inductors in the detecting circuit[19]. No squaring correction is described for this system, but the errors due to misalignment are kept low by the use of high-performance servo motors.

4. Installation of detector mechanism. Any magnetometer installed in a moving vehicle may be subjected to a variety of false magnetic fields or signals. All of the parts of the vehicle made of ferromagnetic material will have induced moments and are likely to have strong permanent moments which are subject to rather rapid change. Any wiring carrying direct current has associated magnetic fields, and some instruments include strong permanent magnets. In addition, eddy currents are frequently set up in the conducting parts of the vehicle whether they are ferrromagnetic or not and often produce a large magnetic field. The best way of overcoming these difficulties is to remove the detector mechanism from the vehicle and tow it in some non-magnetic holder. In an aircraft this is done by installing the detecting head and its orienting servo motors in a streamlined shell that is towed behind the aircraft on a cable consisting of the necessary electrical conductors and a stress-bearing member of non-magnetic material, usually phosphor bronze[37]. The tow cable and nacelle, or "bird", as it is called, are shown in Fig. 3. A similar water tight nacelle has been towed beneath the surface of water by ships for making magnetic surveys in bodies of water[38]. On land vehicles the detecting mechanism has been mounted on a nonmagnetic boom which removes the detecting mechanism not only from the magnetic effects of the vehicle but also from the small magnetic effect produced by fences, road materials, and other superficial foreign material[39].

---

[37] P. M. SMITH: The two bird, mechanical details. Columbia Univ. Airborne Instrument Lab. Rept., 32 pp. (Feb. 1944.) (1944).

[38] B. C. HEEZEN, M. EWING, and E. T. MILLER: Trans-Atlantic profile of total magnetic intensity and topography Dakar to Barbados. Deep Sea Res. **1**, No. 1, 25—33 (1953).

[39] N. S. MORRISEY: This mobile magnetometer accurately maps these contacts. Oil and Gas J. **53**, No. 19, 120 (1954).

Where it is not possible or feasible to remove the detecting mechanism from the vehicle, inboard installations have been made, but they require elaborate compensation for the various magnetic fields discussed above. This is an extremely difficult procedure and requires considerable experimentation[40]. The detector is usually mounted in a place where the effects are least, for example, in the wing tip or in the tail section of an aircraft. The first step is to remove all unnecessary sources of trouble particularly all magnetic materials, replacing these if necessary with nonmagnetic materials. Where possible all electrical wiring is

Fig. 3. Airborne magnetometer. [U. S. Geological Survey.]

rerouted; where not possible it is made twin lead and the conductors twisted. Next, the effect of the permanent magnetism in the various parts of the aircraft is compensated at the location of the detector by small permanent magnets oriented properly. The effect of the induced fields of the magnetic parts is neutralized at the detector by small strips of high permeability material properly situated and oriented. Two methods have been used to neutralize fields due to eddy currents; in one the eddy currents are detected electrically by a pickup coil, and are amplified and fed to an output coil which is placed near the detecting mechanism and adjusted to produce a magnetic field equal and opposite to that produced by the eddy current; in another system disks are properly oriented so that an eddy current is developed which neutralizes that in the conducting parts of the vehicle. All of these compensation mechanisms must be checked frequently because the electrical and magnetic characteristics of the vehicle are subject to frequent change.

5. Measurement and recording of detector output. The various components of a complete magnetometer system required to produce, measure, and record the

---

[40] W. E. TOLLES: Compensation of induced magnetic field in MAD-equipped aircraft. U. S. Office Sci. Res. and Devel. Rept. 1386, 19 pp. (April 1943.) (1943). — W. E. TOLLES, and V. V. VACQUIER: Compensation of magnetic fields in MAD-equipped aircraft. U. S. Office Sci. Res. and Devel. Rept. 4187, 87 pp. (July 1944.) (1944).

output of the detector are indicated in the previous sections. There is the detector mechanism consisting of a unit containing two or three detectors mounted on a two or three gimbal support and the two orienting servo motors. This head mechanism is operated by and supplies the input for the electronic unit, which consists of a driving oscillator, usually 1,000 Hz, and a linear rectifier, amplifier, or differential amplifier depending upon the system employed. There is usually a power unit consisting of a dynamotor and a stabilization unit to convert the vehicle's electrical power to the voltages necessary for the operation of the electronic unit. The output of the electronic unit is fed to a control box and recorder. The operation of the control box depends upon the type of fluxgate circuit employed, but it is usually a precise currentcontrol system whose function is to provide adjustment of the sensitivity and level of the measurement of the equipment.

In all systems the control box is provided with a sensitivity-selector switch that makes it possible to change instantly the full-scale sensitivity of the recorder. Five or six sensitivities are available from 1 to 40 γ/mm (20 to 1000 γ per inch) on the recording chart. The magnetic value of the zero line or base of the recorder chart is changed instantly, either by hand or automatically, in fixed increments, the size of which may be selected by an operator. Thus, a anomaly greater than the set range of the recorder can be recorded either by decreasing the sensitivity of the recorder with the sensitivity-selector switch or by changing the range of the recorder by adding an increment to the magnetic value of the zero line. The usual procedure is to set the sensitivity-selector switch to provide reasonable recording of the desired anomalies, for example, 100 γ full-scale on the recorder chart, and then to change the range of the recording by adding or subtracting increments of 100 γ. By making these adjustments on the control box the recorder will record all intensities between, let us say, 50,000 γ and 50,200 γ. If an intensity greater than this is to be measured the range is instantly changed to record between 50,200 and 50,300 γ. If, however, an exceptionally large anomaly is found, by adjusting the control box the recorder can be set to record from 50,100 to 50,600 γ or perhaps from 50,100 to 51,100 γ. This decision is made by the operator from his observations on the steepness of the magnetic gradient.

The recorder, usually a milliammeter, draws a continuous magnetic profile on a roll of graph paper with rectilinear or curvilinear coordinates. Several chart speeds are available and under most conditions one is chosen that gives a horizontal scale approximately equal to the scale at which the data are to be compiled. The recorder is equipped with a device that stamps the chart with the settings of the sensitivity-selector switch and the magnetic value of the zero line of the chart, and with edge-marking devices whose function will be described in Sect. 10.

b) Special modifications.

The airborne magnetometer as described measures only the intensity of the Earth's total magnetic field. Several major modifications have been made to broaden the use of this instrument, and in particular to measure the components of the Earth's field and to measure the gradient of the Earth's total field[41].

6. Vector airborne magnetometer. Attempts have been made to orient the airborne magnetic detector in the vertical by the use of pendulum or gyroscope, but as discussed above the errors introduced into the magnetic field measurement by misalignments have made these systems impractical. To overcome this difficulty a system has been developed which uses a self-orienting airborne magnetometer with means provided for measuring the angular relationship between the detecting element and a vertical reference and the heading of the aircraft. In the first of these instruments, the detecting mechanism was rigidly fixed to the aircraft and the two angles defining the orientation of the detector with respect to

---

[41] A three component magnetometer is described in this volume in the contribution by P. H. SERSON and K. WHITHAM, p. 384.

the aircraft were measured and recorded. The heading of the aircraft was measured and recorded and a damped pendulum used to establish the vertical against which the orientation of the aircraft could be measured and recorded [42].

In order to reduce the number of angular measurements required, a new system was developed in which the entire detecting mechanism and associated servo motors were pendulously supported and damped [43]. Synchro-transmitters geared to the two orienting servo motors are used to measure directly the inclination and declination of the Earth's field referred to the heading of the aircraft. The heading of the aircraft relative to a celestial body is determined with a sextant which is similarly geared to a synchro-transmitter.

The normal oscillation of an aircraft in motion even in still air at high altitude will produce considerable fluctuations in the angular measurements. This can be averaged from the records, but to speed this process an averaging circuit was introduced which automatically sums the observed values over any predetermined period. An analysis of aerodynamic characteristics of the aircraft in which the equipment has been installed indicates that it has a period of about 50 sec, so the angular measurements obtained are averaged over an interval of 100 sec each.

Because this equipment is used for extended flights where ties to magnetic bases are not possible every effort has been made to reduce electronic and thermal drift by redesigning the mechanical and electronic parts of the basic airborne magnetometer equipment which contribute to these effects. At the present time this equipment has been used in a region where the angle of dip is about 60° and total field $\mathcal{F}$, horizontal component $\mathcal{H}$, vertical component Z, inclination J, and declination D have been measured under average conditions with an estimated probable error of 15, 40, and 30 γ and 3 and 5 minutes of arc, respectively.

7. Airborne magnetic gradiometer.

It is recognized that by observing the gradient of the magnetic field it is possible to determine important magnetic features which are not as easily detected in the recording of the variations of the total intensity. The gradient can be measured by using two magnetometers spaced at some known distance, and each measuring the same component of the Earth's field.

Gradiometers of this type have been constructed which use two matched fluxgate detector elements and their measuring circuits. These will be subject to the same errors of orientation as single elements discussed in Sect. 3, and therefore are not suitable for use in aircraft. Two ordinary, matched self-orienting magnetometers could be used, but as the actual magnetic difference to be detected between the two instruments is very small, a precise and rapid orientation would be required, and in general such a system would not be practical.

A novel means of overcoming these difficulties has been developed [44]. The airborne magnetometer is constantly moving at a speed which can be determined, so it can be used as a gradiometer that measures the magnetic intensity at two different points by measuring at two slightly different times. This principle has been used to modify an airborne magnetometer of the type in which the measurements of the magnetic intensity is determined by measuring a current that is properly adjusted to nullify the earth's magnetic field at the detecting element. This is a peak-type magnetometer in which the output of the detector element is used to produce a voltage suitable for running a motor which in turn drives a potentiometer that controls the nullifying d.c. (direct current) current through a compensating coil on the detector element. An induction type generator is coupled to the shaft of the motor that drives the potentiometer, and as the

---

[42] E. O. SCHONSTEDT, and H. R. IRONS: Airborne magnetometer for determining all magnetic components. Trans. Am. Geophys. Un. **34**, No. 3, 363—378 (1953).

[43] E. O. SCHONSTEDT, and H. R. IRONS: N. O. L. Vector airborne magnetometer type 2A. Trans. Am. Geophys. Un. **36**, No. 1, 25—43 (1955).

[44] W. E. WICKERHAM: The Gulf airborne magnetic gradiometer. Geophysics **19**, No. 1, 116—123 (1954).

voltage output of this generator is directly proportional to the rate of motion of the motor, it is directly proportional to the rate of change of magnetic intensity. By recording the output of this generator and correcting if for the ground speed of the aircraft it is possible to obtain the horizontal gradient of the earth's total magnetic field in the direction of flight.

c) Associated equipment.

To be useful the position of any measurement of the earth's magnetic field must be determined in space. The accuracy required in this determination depends upon that desired for the survey and to some extent upon the purpose for which the data are to be used. In general the precision of location, particularly for geophysical prospecting applications, is considerably higher than is usually possible in aircraft. In order to attain the proper location of the aircraft a variety of methods have been employed[11-15, 18, 45, 46].

8. Aerial cameras. Probably the simplest, least expensive method for determining the position of aircraft is by using an aerial camera which photographs the surface beneath the aircraft. By comparing the photographs with a planimetric map it is possible to find the point on the ground over which the airplane is located. Because one of the big advantages of the airborne magnetometer is that it is continuously recording, it is usually necessary to record similarly and continuously the position of tha aircraft. Using cameras, this has been accomplished by photographing the flight path of the aircraft by taking a series of intermittent pictures with a modified motion picture camera or by making a continuous photograph with a strip camera. In the strip camera the photograph is produced by moving film past a slit at a speed equal to that of the movement of the photographic images of the ground points beneath the airplane. This produces one long narrow photograph of the flight path of the aircraft for the length of time the film moves through the camera.

In any photographic method it is important that the camera photograph the spot vertically beneath the aircraft, and to accomplish this the camera must be gyroscopically controlled. Because the strip camera at any instant takes only a one-dimensional photograph at right angles to the travel of the aircraft, it can be stabilized optically by a gyroscopically controlled mirror which bends the beam of light so that the point directly beneath the airplane always appears in the center of the photograph. The accuracy of this method of location depends upon many factors and in particular upon the accuracy of the base map on which the photographic data must be plotted, but under average favorable conditions the position of aircraft can be determined within 30 m.

9. Navigation aids.

α) *Electronic aids*. In unmapped areas or in regions of water, desert, or featureless terrain the photographic method of location is useless and electronic navigation aids must be used. A variety of these methods are available with a wide range of accuracy and applicability.

One of the best systems that has received wide application is "SHORAN". In this system a high frequency pulse is produced in the aircraft, is received at a ground station at a known geographic location and retransmitted to the aircraft where the transit time of the pulse is determined and converted into distance

---

[45] H. JENSEN, and J. R. BALSLEY: Controlling plane position in aerial magnetic surveying. Eng. and Min. J. **147**, No. 8, 94—95, 1953-154 (1946).
[46] J. M. KLAASE, L. H. RUMBAUGH, and H. JENSEN: Applications of Shoran and photographic techniques to aerial magnetic surveys. Oil and Gas J. **45**, No. 47, 123 (1947).

from the ground station. If two pulse systems at different frequencies are used with two ground stations at known geographic positions the position of the aircraft can be determined generally to within 30 m.

Information supplied by this equipment can be used in many ways, but the most useful is probably the straigth-line computer. With this the pilot is directed along a predetermined straight line by an indicating dial, and the track of the aircraft is plotted automatically on a tracking board which supplies a map of the aircraft's flight path relative to the known geographic position of the two or more ground stations. The chief disadvantage of the "SHORAN" method is that it requires such high frequency energy that the transmission is essentially by "line-of-sight" and, therefore, at least two ground stations must be "visible" from the aircraft at all times. This requirement restricts the use of the method to those areas in which there is level terrain or over which low-level surveys are not required.

β) The "*RAKDIST*" *and* "*DECCA*" systems use lower frequencies and are not hampered by the line-of-sight requirement as "SHORAN". These systems by means of two ground stations establish interference patterns of radio waves, and the vehicle locates itself from a study of the phase relationships of the two sets of waves. A variety of new and precise modifications of the basic equipment are becoming available.

γ) *Celestial or dead reckoning navigation.* Obviously where the location requirements are not strict any of the established celestial or dead reckoning location methods may be used, and many of the newly developed instruments for continuously and automatically recording this position information can be used.

δ) *Altimeters.* The third dimension of the position of the aircraft is determined by one or more altimeters, either barometric or radio. The former measures the height above sea-level and the latter the height above the ground by measuring the transit time of radio waves sent from the aircraft and reflected from the earth's surface. The measurements of any of these altimeters can be recorded either photographically or continuously on recording meters.

10. Correlating circuit. The magnetic and position records obtained in the aircraft must be correlated frequently. This correlation is usually done by an electronic circuit triggered automatically or by an observer. The circuit actuates pens that mark the various recorder charts, shutters of cameras which photographs the various nonrecording instruments, number counters, and solenoids which stamp recorder charts and actuate the edge-marking and numbering devices on the cameras. This is illustrated in Fig. 4. An observer in a position in the aircraft where he has a good view of the terrain the path of the aircraft on the map and triggers the electric circuit every time he passes over an easily recognized landmark. This circuit changes an electric number counter before him and he records on the map at his position the number showing. At the same time the circuit actuates a pen which makes a tick on the edge of the magnetic chart and on the altimeter record and momentarily moves the shutter across the slit of the continuous strip camera therby producing an edge-mark on the photograph. Simultaneously electrical number counters are changes throughout the aircraft, and the same number used by the observer is stamped on the magnetic record, the altimeter record, and is photographed along the edge of the photograph.

Thus, all the records are correlated and it is thereby possible to obtain the magnetic intensity at any point along the path of the aircraft. These records and their correlation are shown in Fig. 4, from left to right; the observer's map with handwritten numbers 26, 27, 28, 29, 30 at the approximate locations of the aircraft at the time the correlating circuit was triggered, the magnetic record showing the total intensity profile with automatic edge-marks 26, 27, 28, 29, 30 and stamped values of the sensitivity and magnetic base level, the height-above-ground record with automatic edge-marks produced by the radar altimeter, and the strip photograph of the flight track with automatic edge-marks and correlating numbers.

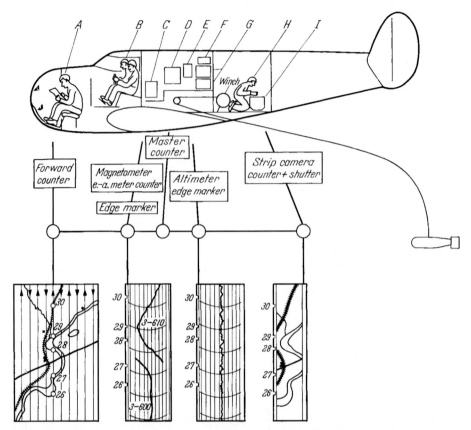

Fig. 4. Sketch of airborne magnetometer and associated equipment installed in AT-11 aircraft showing records obtained and system of correlation. (*A* geophysicist with flight lines on base map; *B* navigator with flight lines on aerial photographs; *C* meter recording magnetic profile; *D* magnetometer control box; *E* meter recording altitude; *F* radar altimeter; *G* magnetometer, oscillator and converter units; *H* geophysicist operating magnetometer; *I* continuous strip camera.) [Courtesy: U. S. Geological Survey.]

II. Field survey technique.

It is not possible to review in detail all the various factors which must be considered to assure that the magnetometer is used most efficiently for the variety of purposes for which it is useful. Many of these factors are the same that affect the use of any geophysical equipment, and are therefore discussed in detail in geophysical texts, but a few of the general features which are unique with the airborne magnetometer are considered here.

a) Operation of airborne magnetometers.

11. Operational conditions.

α) *Aircraft.* The basic requirements for the aircraft carrying the magnetometer are that it be safe and economical to operate, that it have range sufficient for efficient operation, that it have capacity for a crew of three or four and for equipment weighing 150 to 200 kp. Small twin-engine training aircraft have been used, but although they are economical to fly they generally have a range of less than 1200 km, and therefore are not economical to operate unless the project is small or close to an airport. The aircraft most widely used in this work are the twin-engine transports which have ample load-carrying space capacity and a normal range of about 2,000 km, permitting economical operation even when the area to be surveyed is at a great distance from the nearest airport.

In order to make three-component magnetic surveys it is necessary that the airplane fly high above the surface disturbances, both magnetic and meteorological, and for this aircraft similar to the B-29 or P2-V have been used. These have the advantage of extremely long range flying, thus permitting flights of 12 hours or more which lead to the most efficient use of the infrequent times when both magnetic and meteorological conditions are favorable for this work.

Helicopters have been used because of their ability to fly safely very close to the ground[47]. Although their cost of operation and maintenance is high, their range short, and their load-carrying and space capacity small, they have been used to make small, detailed, low-level aeromagnetic surveys where the surveys cannot be satisfactorily made on the ground.

β) *Crew.* The basic crew consists of a pilot, copilot, magnetometer operator, and observer-navigator or "SHORAN" operator, but this is modified depending upon the type of survey and upon the space and load restrictions.

The pilot should be experienced in flying straight and parallel lines as in photo mapping, and in low-level flying as in crop dusting, if a geophysical prospecting survey is planned. In all airborne survey work it is important that the pilots be trained in line flying rather than in the point to point flying used in transport and most commercial operations. If automatic navigational aids are not used the pilots should be able to interpret readily the maps and photographs used to guide them on the predetermined flight lines. The magnetometer operator should be an experienced geophysicist capable not only of operating the magnetometer and making minor repairs on it in flight, but of making "rough" interpretations of the data as they are obtained so that he may change the flight pattern to obtain most economically useful results.

12. Operational procedures. The several factors which must be considered when planning the most efficient use of the airborne magnetometer depend upon the use to which the results are to be put, in particular, whether it is to be used for geophysical prospecting or for mapping the three components of the earth's field at high elevation. It is not possible therefore, to discuss them in great detail, but a few general guiding principles are outlined here[12, 15, 18, 48, 49].

α) *Flight pattern.* When the airborne magnetometer is used in geophysical prospecting, the most efficient pattern is usually a series of parallel traverses flown at right angles to the trend of the major magnetic anomalies. Because this trend often is not known until the survey is well underway, the traverses are generally planned to be flown at right angles to the geologic grain, and if necessary the direction is modified after the first stages of the work have begun. The spacing of flight lines is generally made equal to the supposed depth of the anomaly-producing masses beneath the aircraft. By observing the data as it is recorded, the geophysicist and flight crew are able to change the flight spacing so that the anomalies are adequately mapped with the minimum number of traverses.

The pilots are guided on the flight pattern by photographs or maps on which the desired lines have been plotted. If an electronic navigation aid is used the course is preset by an operator, and the pilot is guided along the course by an electronic indicator that shows the position of the plane with respect to the preset course.

---

[47] H. LUNDBERG: Magnetic surveys with helicopters. Inst. Min. Met. Bull. **488**, 21—28 (July); — Canadian Min. Met. Bull. **423**, 392—400 (July); — Rhodesian Min. J. **19**, No. 243, 233—237 (1947).

[48] H. JENSEN: Operational procedure for the airborne magnetometer. Oil and Gas J. **45**, No. 10, 80—83 (1946).

[49] F. KELLER, J. R. BALSLEY, and W. J. DEMPSEY: Field operations and compilation procedure incidental to the preparation of isomagnetic maps. Photogrammetric Eng. **13**, No. 4, 644—647 (1947).

When the airborne magnetometer is used for preparing the magnetic charts of the three components of the earth's field, the requirements on the pattern are less strict, and in general it consists of a few long flights at rather wide spacing, generally chosen on the basis of experience to provide adequate mapping of the broad features of the earth's main field.

β) *Flight elevation.* The intensity of an anomaly produced by an mass of magnetic material decreases rapidly with increasing distance from the source, and the effects of two masses tend to merge at a distance roughly equal to the distance between them[50]. The effect of nongeologic magnetic materials are also subject to the same laws, but as most of these masses are relatively small, a flight elevation can be chosen that is low enough to give the necessary magnetic detail and is high enough to reduce the effect of nongeologic material to a negligible amount.

In the choice of altitudes, consideration is given to safety and ease of operation in the particular terrain involved. At low altitude the manipulation of the aircraft becomes considerably more tiring and hazardous; because of reduced visibility, the aircraft is more difficult to keep on the prescribed flight path. Experience has shown that a flight elevation of about 300 m above the ground is the most reasonable, although an elevation of 150 m is customary where the fine magnetic detail produced by shallow magnetic materials must be recorded. Where the magnetic materials are deeply buried, the flights are generally made at a constant barometric elevation.

As the elevation for mapping the three components of the Earth's field is less critical it is generally chosen at the maximum elevation for efficient operation of the aircraft and crew.

γ) *Magnetic control pattern.* In ground magnetic surveys the effect of time variations of the Earth's field and instrument drift can be determined by reoccupying a base or by recording the time variations with a second instrument. These methods are sometimes used in aeromagnetic surveys, particularly when long flights are made to measure the three components of the Earth's main field, but because of the large areas involved a very elaborate system of base stations is required and considerable computation necessary to project the ground observations to the flight elevation. In most geophysical prospecting surveys the same effect is accomplished more simply by flying a series of interconnecting base loops or a series of base lines at right angles to the traverse lines.

This base network is internally adjusted and provides the datum to which the traverse data can be adjusted because each time the traverse crosses a base line the result is the same as though a base station were reoccupied. The network is laid out so that traverses cross a base line every 5 to 20 minutes, depending, upon the accuracy required and upon the rate of the time variations of the field. In areas of large magnetic features, an attempt is made to fly the base lines where the magnetic relief is low, and in areas where maps are poor or where the terrain is featureless, the base lines are flown along easily recognizable linear landmarks such as rivers, railroads, or highways.

b) Compilation of field data.

The magnetic intensities measured during the field survey must be plotted their proper position on a map. The techniques used in the office compilation of an aeromagnetic survey are determined by the type of survey, whether it is for the components of the Earth's field in which a series of intermittent measurements and locations are made, or is for geophysical prospecting in which the continuous measurements of the magnetic intensity and position of the airplane are recorded. In both types of surveys the track of the airplane must be determined

---

[50] K. L. WIER: Comparisons of some aeromagnetic profiles with ground-magnetic profiles. Trans. Am. Geophys. Un. **31**, No. 2, pt. 1, 191—195 (1950).

and the magnetic records plotted to map distances as profiles and adjusted to a common magnetic datum. Because the magnetic intensity is recorded on a meter at a constant rate but the ground speed of the aircraft varies widely, the horizontal scale of the recorder chart has no constant relationship on the ground.

13. Compilation of geophysical prospecting survey data.

α) The first step in compilation is plotting the *flight track* of the aircraft. If the photographic method has been used, the position of the point indicated by the correlation marks shown by the intermittent or continuous-strip photographs are plotted on aerial photographs, and are transferred from them to the best base map available. If an electronic navigation aid has been used, the positions of the points indicated by the correlation marks are plotted by their known distances from the geographic positions of the two base stations.

β) The second step in compilation is *plotting* the continuously recorded *magnetic data* at a constant horizontal scale by adjusting the correlation marks on the magnetic record to fit the distance between the correlation marks plotted on the map. This "rectification" may be accomplished on a machine that semiautomatically sets up the proportion between the correlation-mark distance on the magnetic record and the map's correlation-mark distance plotted on graph paper. The operator then traces the magnetic profile on the field magnetic record and a new rectified profile is scribed on rectilinear graph paper at the desired horizontal and vertical scale.

γ) The third step is *adjusting* the rectified profiles *to a common magnetic datum* free from the effects of instrument drift and time variation of the Earth's field. The magnetic datum of the traverses is determined by the network of base lines or loops, discussed in Sect. 12 γ. Each time a traverse crosses a base line, the magnetic datum of the traverse is established, and because the flight pattern is generally chosen so that such intersections occur every ten minutes or less, only the very slight effects of instrument drift and time variations of the Earth's field, which are nonlinear during that period, are not removed from the magnetic record.

Because the position of the airplane on the base line and on the traverses are not known exactly, the magnetic intensities measured at the points of intersection are not real and some averaging technique must be used. This is done by treating statistically the difference in the magnetic intensity obtained on the base line and on the traverse for each point of intersection, and from this data constructing an averaged curve of the correction to be applied to the magnetic intensity measured at any given time.

δ) The fourth step in office compilation is *magnetic intensity contouring*. The contour interval is chosen and intersections of the contour levels and the magnetic profiles determined. The position of these intersections are transferred to the base map, and contours are drawn by connecting points of equal magnetic intensity. This procedure has the obvious advantage of requiring no interpolation along the traverse lines. The magnetic intensity contour map prepared by these procedures represents the observed variations in the total intensity on a surface described by the flight level of the plane.

ε) Various *simplifications* of these procedures are possible under favorable circumstances. Where roads or other culture make surveyed points easily identifiable, the observer, using a gyro stabilized sighting device can plot accurately on the map the check points and thereby eliminate that part of the office compilation usually required to determined the flight path of the aircraft. When the magnetic gradient is low and the anomalies broad, a reasonably accurate contour map can sometimes be made by plotting the magnetic intensities measured only at check points. This eliminates the compilation steps required in "rectification", but at the same time destroys the chief advantage of the airborne over the ground method, viz that of using the continuous profile to eliminate the interpolation required when individual measurements are plotted. Such simplifications increase the speed and reduce the cost of compilation, factors which must always be balanced against the factor of accuracy.

14. Compilation of three component surveys. The compilation of the results of the aeromagnetic surveys of the three magnetic components, X, Y, Z, of the Earth's magnetic field is done by the same techniques used in compiling the results of ground magnetic surveys. As the magnetic gradient is low at the elevations at which these measurements are made, and because the positional and angular measurements must be summed over the 100 sec interval, it is not necessary or possible to make the continuous measurements and compilations that are used in the geophysical prospecting surveys.

In this work, intensities are taken at regular intervals, usually 5 minutes; the various positional, angular, and magnetic measurements of the 100 sec interval centered at this time are determined; the plane's position is plotted; and the various components of the field computed. These intensities are corrected for instrumental drift by determining its rate from measurements from a base station at the beginning and end of the flight and for the time variations of the field by computing it from observations made at ground stations. The intensities thus determined at 5 minute intervals are plotted on a map and are used to prepare by usual means the standard magnetic charts.

c) Interpretation of results.

15. Magnetic prospection. The data collected in an aeromagnetic survey are the same as those obtained by ground methods, the only difference being that in ground magnetic surveys the variations in either the vertical or horizontal component of the Earth's field are measured but in most aeromagnetic surveys the variations of the total field are measured. Therefore the interpretation of the maps and profiles resulting from these data involve the same fundamental theories that for so many years have been applied to the results of ground magnetic surveys, and require only that the original techniques in interpretation be modified so that they are applicable to the study of the total magnetic field[51-53].

This modification is done easily by assuming that the local distortions of the Earth's field are negligibly small. In this manner, the Earth's field and the total field can be assumed to be codirectional and consequently the anomalous field behaves like a potential and satifies the mathematical laws that have been developed for potential functions. Most of the recent developments in the techniques and interpretations, therefore, are applicable to all types of magnetic surveysand in many instances to gravity surveys. They are briefly discussed in this report because they apply to the interpretation of aeromagnetic surveys and in many instances have been developed to assist in the study of the magnetic maps of large areas which, through the use of the airborne magnetometer, are now more common than theretofore.

A magnetic map is a representation of the magnetic field over the surface on which the measurements were made, and the basic aim of the interpretation is to separate the field into its separate components and determine the source of each. It is a mathematical fact, however, that a unique distribution of magnetization which will account for any magnetic potential field cannot be determined, but only that a particular configuration of the field can be produced by an infinite number of distributions of magnetization at distances equal to or less

---

[51] R. G. HENDERSON, and I. ZIETZ: Analysis of total magnetic-intensity anomalies produced by point and line sources. Geophysics **13**, No. 3, 428—436 (1948).

[52] D. S. HUGHES, and W. L. PONDROM: Computation of vertical magnetic anomalies from total magnetic field measurements. Trans. Am. Geophys. Un. **28**, No. 2, 193—197 (1947).

[53] I. ROMAN: The resolving power of magnetic observations. Min. Technol. **10**, No. 6, Tech. Paper 2097, 18 pp. (1946).

than a determinable maximum figure. Thus it may be possible in spite of the ambiguity inherent in the data to use it to provide maximum depths for masses producing particular anomalies. If other information is available, either geophysical or geologic, it can be used with the magnetic data to provide limiting conditions on the size, shape, depth, and composition of the mass producing the anomaly.

16. The use of residuals and vertical derivatives. Any magnetic map is a summation of the magnetic fields produced by sources both shallow and deep and to interpret it properly anomalies of interest must be separated from those that do not relate to the particular problem. This can be done by several means, the simplest of which is by removing from the map an average magnetic gradient which is drawn by smoothing the contours of the map[54]. The residual magnetic anomalies are reflections of the shallower magnetic structures and in this way can be separated from the regional features produced by the deep magnetic masses.

If the magnetic map is complicated the "smoothing" process is difficult to apply. To overcome this difficulty and to arrive at a more objective method, systems have been developed which make use of a grid which is laid over the magnetic map so that the magnetic intensities can be read at various intersections. These intensities are weighted in various ways and the value so obtained substracted from the intensity at the central point of the grid[54-56].

α) The *residual map* produced by this process tends to accentuate anomalies of certain breadth and to reduce comparatively the amplitudes of anomalies of greater or less breadth. This "filtering effect" is most closely related to the grid spacing chosen, so it is possible to isolate in a general way the anomalies produced by magnetic material at different depths. This can best be accomplished by using a grid whose spacing is equal to the depth of burial of the anomalous mass.

β) The *second vertical derivative*, like the residual field brings into greater prominence those geologic structures which have only slight expression in the magnetic field. Unlike the residual, however, the second derivative is used in the downward analytical calculation of magnetic fields and subsequent depth calculations. To compute the derivative, a grid whose spacing is approximately equal to the average depth of the magnetic sources is placed over the magnetic map and the values of the field recorded at the intersections. HENDERSON and ZIETZ[56], among others, have calculated the second vertical derivative and found it to be

$$\frac{\partial^2 (\Delta r)}{\partial z^2} = 2\left(3\Delta T_0 - 4\overline{\Delta T_1} + \overline{\Delta T_2}\right) \tag{16.1}$$

where $\frac{\partial^2 (\Delta T)}{\partial z^2}$ is the second vertical derivative of the total intensity anomaly; ΔT_0 the value at the center of the grid; $\overline{\Delta T_1}$ the average value of the four grid intersections at a radial distance of 1 unit from the center and $\overline{\Delta T_2}$ the average value of the four intersections at a radial distance equal to the square root of 2 times the grid spacing.

---

[54] L. L. NETTLETON: Regionals residuals, and structures. Geophysics **19**, No. 1, 1—22 (1954).

[55] L. J. PETERS: The direct approach to magnetic interpretation and its practical application. Geophysics **14**, No. 3, 290—320 (1949).

[56] R. G. HENDERSON, and I. ZIETZ: The computation of second vertical derivatives of geomagnetic fields. Geophysics **14**, No. 4, 508—516 (1949).

17. Upward and downward continuation.

α) *Upward continuation of total intensity anomalies.* With the introduction of the airborne magnetometer, it became possible to make magnetic surveys at more than one level. It was believed by many geophysicists that this would provide additional information which could be used in interpreting the magnetic data. Various investigators have shown, however, that this additional information could have been obtained more quickly and at less expense by using mathematical techniques. HENDERSON and ZIETZ have shown[57] that the value of the total intensity anomaly, $\Delta T(z)$ at a point at height z above the point at which the anomaly has been measured is given by

$$\Delta T(z) = \int_0^\infty \frac{\overline{\Delta T}(r)\, z\, r\, dr}{(r^2 + z^2)^{\frac{3}{2}}} \tag{17.1}$$

where $\overline{\Delta T}(r)$ is the average value of the anomaly ΔT on a circle of radius r about the point which is to be projected. This formula can be reduced to

$$\Delta T(z) \approx \sum_{i=1}^{n-1} \overline{\Delta T}(r_i) K(r_i, z) \quad + \quad \overline{\Delta T}(r_n) K(r_n, z) \tag{17.2}$$

where $\overline{\Delta T}(r_i)$ is the average value of ΔT around a ring of radius r_i and $K(r_i, z)$ is the coefficient to be applied to the average for that ring. The average for each ring is determined by placing a square grid over the magnetic map and reading the intensities at appropriate intersections. These intersections are chosen so that they fall upon particular circles about the point to be projected; the proper coefficient can then be applied to the averages of various values.

By comparing magnetic maps prepared using these methods with those prepared by aeromagnetic surveys it has been shown that the differences lie within the limits of experimental errors of measurement[57].

β) *Downward continuation of vertical anomalies of magnetic field intensity.* Essentially the same techniques is used in extending a magnetic field downward, but here the objective is to project the data to the surface over which the magnetic material is distributed, and thereby to determine the depth and nature of the geologic features which produce magnetic anomalies observed at the level of measurement.

PETERS[55] has shown that

$$F(0, 0, z) \cong 2[b_0 - 2 b_2 z^2 + \tfrac{8}{3} b_4 z^4] - F(0, 0, -z) \tag{17.3}$$

where $F(0, 0, z)$ is the value of the magnetic field at depth z and b_0, b_2, b_4 are constants which are obtained by least square solutions of

$$\overline{F}(r) \cong b_0 + b_2 r^2 + b_4 r^4 \tag{17.4}$$

$\overline{F}(r)$ is the average value of the magnetic field on a circle r, about the point to be continued downward. As in the case above a grid system is used to evaluate the various $\overline{F}(r)$.

---

[57] R. G. HENDERSON, and I. ZIETZ: The upward continuation of anomalies in total magnetic intensity fields. Geophysics **14**, No. 4, 517—534 (1949).

[58] I. ZIETZ, and R. G. HENDERSON: Magnetic anomalies at high altitudes. Trans. Am. Geophys. Un. **32**, No. 3, 397—404 (1951).

γ) *Computations of depth to source of anomalies.* Most techniques used in the interpretation of magnetic maps have as an important function the computation of the depth to the mass which has produced the magnetic anomaly[59-61].

A new method that is perhaps more accurate and simple to apply than others has been described by Vacquier and others[62]. Depth calculations can be made by assuming fixed geometrical shapes for the anomalous magnetic material. These are right verticle prisms extending indefinitely downward and having horizontal rectangular surfaces. In addition it is assumed that the mass has a uniform susceptibility and that the polarization is in the direction of the Earth's field.

The authors[62] state: "With these assumptions calculations were made of the magnetic field for rectangular prismatic models with various cross sections and orientations and for different inclinations of the Earth's field. The field was contoured on a map with a horizontal scale expressed in terms of the depth to the top of the buried model. As anaid to interpretation, second vertical derivative maps of the field were also computed. The depth determination procedure consists of computing a second vertical derivative map of the observed anomaly and comparing certain diagnostic features for both the observed and second derivative maps with similar features of the corresponding model. When the magnetic fields are studied, the horizontal extent of the steepest alopes are compared. When analyzing the derivative anomalies one compares the horizontal distance between maximum, minimum, and zero curvature contours[60]."

Fig. 7 (Sect. 19) illustrates a contour map of the basement surface prepared by these techniques.

18. Experimental model studies. In addition to mathematical model studies, one of which is described above, various investigators have studied the anomalies produced by relatively thin physical models and have compared them with those measured in the field. This investigation makes the calculation of magnetic structure of arbitrary shape possible. In discussing such a study I. Zietz and R. G. Henderson[63] state that "Any irregularly shaped magnetic body may be approximated by the proper arrangement of prismatic rectangular slabs of constant thickness and varying horizontal dimensions. The contoured total-intensity magnetic fields of such slabs buried at different depths and subjected to including fields of varying inclinations may be experimentally determined".

They prepared models of constant thickness and uniform susceptibility with various lateral dimensions. The magnetic anomaly produced by these models was mapped at various elevations above the model and for various inclinations of the inducing field and "All the experimentally derived field maps will be made available in a 'normalized' form, i.e., one which is independent of the susceptibility of the model and the strength of the inducing field and dependent only on the geometry of the prism. In this manner, complicated bodies may be built up by the proper combination of prismatic blocks and the total 'normalized' field computed by arithmetically summing up each point the 'normalized' fields due to each of the prismatic slabs. Multiplication with the susceptibility of the

---

[59] E. H. Vestine, and N. Davids: Analysis and interpretations of geomagnetic anomalies. Terr. Magn. **50**, No. 1, 1—36 (1945).

[60] I. Zietz, and R. G. Henderson: The Sudbury aeromagnetic map as a test interpretation method. Geophysics **20**, No. 2, 307—317 (1955).

[61] K. L. Cook: Quantitative interpretation of vertical magnetic anomalies over veins. Geophysics **15**, No. 4, 667—686 (1950).

[62] V. V. Vacquier, N. Steenland, R. G. Henderson, and I. Zietz: Interpretation of aeromagnetic maps. Mem. 47, Geological Society of America (November 1951).

[63] I. Zietz, and R. G. Henderson: Total-intensity magnetic anomalies of three-dimensional distribution by means of experimentally derived double layer model fields. Science **119**, No. 3088, 329 (1954).

known rock and the inducing field strength will result finally in the desired anomalous magnetic field".

Another similar study is reported by L. R. ALLDREDGE and W. J. DICHTEL[64] who constructed a magnetic clay model of the basement rocks which produced the same anomaly as that measured over Bikini Atoll. A uniform susceptibility and no permanent magnetization of the basement rocks was assumed. As an infinity of such models is possible it was necessary to choose a susceptibility consistent with that of the basalt which is assumed to be the basement rock and to use depth measurements consistent with those deduced from seismic surveys.

Fig. 5. Geologic map of part of Fairfax Quadrangle Virginia. (Rm sedimentary rocks; Rdb diabase; mb metabasalt; mbc metabasalt complex; p Phyllite; pcs Peters Creek schist; gr granite.)
[Data from U. S. Geological Survey.]

19. Examples of aeromagnetic surveys. It is probable a few million km of detailed and accurate aeromagnetic traverses have been surveyed by commercial and government agencies since the development of the airborne magnetometer in 1944. Because much of this work has been done by commercial organizations for mining and oil companies the results are not generally available to the public; in fact their existence is not generally known. It is known that extensive surveys have been made in most of the United States, Alaska, Canada, Mexico, Venezuela, Bahamas, Mozambique, North and South Africa, the USSR, the Philippines, Australia, Sweden, Liberia, Siam, Great Britain, Germany, and France. Most of this work

---

[64] L. R. ALLDREDGE, and W. J. DICHTEL: Interpretation of Bikini magnetic data. Trans. Am. Geophys. Un. **30**, No. 6, 831—835 (1949).

Sect. 19. Examples of aeromagnetic surveys. 417

has not been published, but enough is available to demonstrate the applicability of the method to specific geologic problems.

It is not possible here to present all the different types of surveys and their results, but two examples are presented to demonstrate some of the projects in which this technique has been used. These examples are of the geophysical prospecting type because the results of three-component aeromagnetic surveys have not yet been published.

Fig. 6. Aeromagnetic map of part of Fairfax Quadrangle, Virginia. (Contour intervall=10 γ.) [Data from U.S. Geological Survey.]

α) *Fairfax quadrangle, Virginia.* The results of geologic and aeromagnetic surveys of Fairfax quadrangle, Virginia by the U. S. Geological Survey are shown in Figs. 5 and 6. This area is underlain by Triassic sedimentary rocks and diabase and by metamorphosed Paleozoic or Precambrian sedimentary and igneous rocks. The Triassic diabase has a high magnetic susceptibility and gives rise to pronounced anomalies. This is shown by the striking correlation between the magnetic pattern produced by it and its outcrop shown on the geologic map. The exposures are poor in the area, and it is difficult to determine from these alone the exact structure of the large masses of Triassic diabase whose outcrops are shown to be U-shaped on the geologic map. Several different structures are possible; a partially complete conical or cylindrical sheet with a vertical axis similar to the characteristic ring dike, a southward-plunging anticlinal sheet, a northward-plunging synclinal sheet, or a spoon-shaped sheet. An analyses of the magnetic data shown in Fig. 7 indicates the mass has a distorted U-shape with a vertical western limb and a gently dipping eastern limb. Subsequent geologic studies of similar features in southeastern Pennsylvania have verified these conclusions. To analyse the

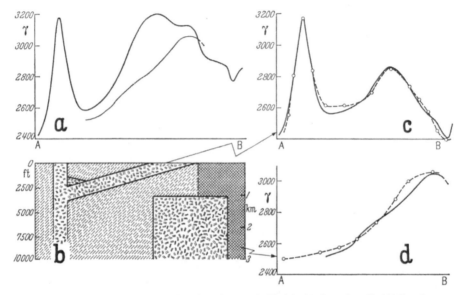

Fig. 7a—d. Analysis of aeromagnetic profile in Fairfax Quadrangle, Virginia. (a) observed profile (thick) and average magnetic ridge (thin); (b) geological section: ▨ triassic diabase; ▨ triassic sedimentary rocks; ▨ metamorphic rocks; (c) observed (thick) and computed (broken) profile in the central part of (b) (arrow), (d) observed (thick) and computed (broken) profile at right (arrow). [Computed by R. Henderson and I. Zietz.] [Courtesy: U.S. Geological Survey.]

observed field it was necessary to remove from the data the magnetic ridge apparently produced by a large mass of deeply buried magnetic diagramatically shown in the lower right hand structure section of Fig. 7.

The Triassic sedimentary rocks are essentially free of magnetic material, have a very low magnetic susceptibility and therefore, essentially no magnetic expression. The older phyllite and schist have varying but generally low susceptibility, which produces small and irregular anomalies. The average susceptibility of the schist seems to be slightly higher than that of the phyllite.

The metabasalt produces a distinctive pattern of large anomalies that can be correlated easily with the geologic map. The large anomalies in the upper right hand quarter of the magnetic map are caused by a swarm of dikes of serpentized periodotite. The dike in the northeastern corner produces a negative anomaly. This is apparently caused by reverse permanent magnetization of the serpentine in this dike, which in turn is apparently related to the fact that the magnetic minerals in this dike have a different composition that those in the periodotite dikes producing positive anomalies.

β) *Indiana.* An aeromagnetic survey consisting of more than 56,000 km of traverses flown 1.5 km apart has been completed by the U. S. Geological Survey of the State of Indiana. The results have been compiled as magnetic contour maps and analyzed by the methods set forth in [62] and a generalized contour map of the basement surface prepared, part of which as shown in Fig. 8. The map is presented here as an example of the results of mathematical analysis of magnetic data.

Fig. 8. Generalized relief map of surface of basement rock in part of Indiana, computed from aeromagnetic data.

III. Discussion of advantages and limitations.

The airborne magnetometer is an instrument that can be used with great rapidity and little cost to obtain accurate magnetic results of a reconnaissance type, but as indicated in the foregoing sections it has certain limitations which restrict its use to certain types of surveys [65, 66].

---

[65] J. R. BALSLEY: Techniques and results of aeromagnetic surveying. Proc. U. N. Sci. Conf. Conservation and Utilization of Resources 3, 8—10 (1951).

[66] C. J. DEEGAN: Economics of aerial magnetics. Oil and Gas J. 47, No. 12, 67—71 (1948).

20. Speed, ease, and economy of operation. One of the greatest advantages of the airborne magnetometer is its speed, for a three-or four-man field crew can generally survey more than 150 km of useful traverse per flying hour and 10,000 to 15,000 km per month. This excludes the time lost in turns, establishing base lines, and flights to and from the airport. This rate is affected slightly by the length of the flight lines and by the size of the area and its distance from the airport, but the rate is practically independent of the type of terrain, the density of vegetation, and other factors which affect greatly the speed and cost of ground magnetic surveys. The rate of office compilation depends upon the type of result desired and to some extent upon the quality of the base maps and the intensity of the magnetic anomalies, but to produce a contour map generally requires about one-man hour of office compilation per traverse mile. Obviously the airborne magnetometer is the easiest means of conducting magnetic surveys of both the geophysical prospecting type and the 3-component type in areas which are not easily accessible on foot. At its present early stage of development the vector airborne magnetometer can be used to obtain rapidly, inexpensively, and accurately the 3-component data necessary to prepare isogonic charts of the ocean areas.

The cost of operation is primarily a function of the speed at which the work is done and is, therefore, affected by the same factors. Commercial organizations quote cost that vary considerably with the company, the project, and the region in which the work is to be made The latter item not only includes the cost of transporting the necessary equipments and personnel to the area of operation, but also the general climatic conditions as they affect the flying conditions. In the United States most companies will make an average survey and prepare a contour map for $ 10.00 to $ 15.00 per traverse mile. If electronic navigation aids are used the cost is generally increased by 50 to 100%.

21. Quality of results. The over-all accuracy, that is, the ability to make and duplicate a true magnetic map, is dependent upon both the precision of the magnetic measurement and the accuracy of the position measurement. The airborne magnetometer can measure the difference between two magnetic fields with an precision of $\pm 2\gamma$, or $\pm 1\gamma$ if particular care is exercised. Because it is possible to make frequent crossings of a base line it is possible to make a more accurate correction for diurnal variation in an aeromagnetic survey than is usually possible in a normal ground survey. For these reasons, the precision of magnetic measurement is somewhat higher with the airborne magnetometer than with the ground magnetometer.

On the other hand, the accuracy of position measurement in most aeromagnetic surveys can be no better than that of the best available map, and even by using a highly accurate map at a scale of at least 1:25,000 or by using Shoran, the accuracy of position measurement is generally no better than 30 m. This accuracy is considerably less than that obtained by most ground magnetic surveys.

Although the precision of magnetic measurement and accuracy of location are the primary factors affecting the over-all accuracy of a magnetic map, two other factors must be considered when comparing air and ground surveys. The usual ground survey consists of a series of measurement at points with interpolation between them but the air survey consists of a series of line measurements with interpolation in only one direction. This permits a more realistic contouring of the magnetic features along a flight line in an aeromagnetic map. However, because any discrepancies in the magnetic measurement exist between lines rather than between points, as in a ground survey, there is a tendency to develop a false "herringbone" in the magnetic contours, a vague linearity parallel to the direction of flight. These discrepancies have a random distribution and are of the same order of magnitude as those found in ground surveys, but because they are expressed along the length of a line rather than at a point they are more apparent in aeromagnetic contour maps than in the maps obtained by ground methods.

Measurements along a line guarantees that no magnetic feature along a traverse are missed as between point measurements, and for this reason airborne surveys for some purposes may be more useful than ground surveys. On the other hand, magnetic anomalies attenuate rapidly with distance and tend to merge, so complex magnetic features detected by a detailed ground magnetic survey are not always apparent in an airborne survey made 150 m above the surface. For the same reason, however, the airborne survey is free of the spurious magnetic features created by most of the nongeologic structures and by making measurements at a high elevation the effect of local geologic features can be essentially eliminated.

The two methods are not directly comparable in quality of results and are not suited to the same type of survey. In general, the quality of their results is similar but the aeromagnetic method is best suited to accurate reconnaissance work and the ground method to accurate detailed work.

22. Instrumentation. The greatest disadvantage of the aeromagnetic method is the large capital expediture required to purchase the necessary equipment and airplane, and the high overhead necessary for their operation. For these reasons it is generally not feasible for an individual oil or mining company to conduct its own aeromagnetic surveys unless their holdings are of sufficient extent that their aerial survey will require several years during which the first cost of the equipment can be amortized.

These capital expenditures and maintenance items are particularly high for the large aircraft required for 3-component surveys but they must be compared with similar large items for purchasing, outfitting, and maintaining a non-magnetic vessel of the type previously used for making these surveys.

23. Applicability. The airborne magnetometer is ideally suited to some magnetic surveys but cannot be used for others. It can be used most productively in relatively flat areas not easily accessible on foot for which good maps and photos are available or in which electronic navigation aids can be used easily. It is ideally suited to nearly all projects that require an accurate reconnaissance map of a large area or whose primary function is prospecting for magnetic anomalies. As the development of the 3-component or vector modification of the equipment continues, it will probably become the best and least expensive means of collecting the data for preparing the isogonic charts of the three-components of the earth's magnetic field.

The airborne magnetometer is of limited usefulness in mountainous regions for which a detailed and accurate magnetic map is required; in such areas the difficulty of aircraft operation introduces errors into the results. It must be used with caution on any project that involves complex and detailed anomalies produced by shallow magnetic deposits.

Unless a helicopter is used, the magnetometer can seldom be applied economically to projects covering less than 60 km^2 or projects requiring accuracy of location to better than ± 15 m.

In summary, the airborne magnetometer can be used to provide a low-cost accurate survey of large areas, which can then be used to delineate localities for more expensive ground work, both geologic and geophysical. It does not eliminate the need for ground magnetic surveys, but rather relieves the ground magnetometer of the load of reconnaissance work and enables it to be used more productively on detailed work. The development of the airborne magnetometer has not broadened the fundamental science of geophysics, but has placed in the hands of the geophysicist an instrument that can provide him with magnetic maps of tremendously greater coverage in much less time, at less cost, and in some instances of greater accuracy than has heretofore been possible.

Geophysical Applications of High Resolution Magnetometers.

By

Peter J. Hood*.

With 30 Figures.

Abstract.

The review begins with a description of the various types of nuclear magnetometer. Then the various ground, airborne, and satellite applications of high resolution magnetometers are discussed. The main applications of high resolution magnetometers may be summarized as follows:

a) In magnetic observatories to record the temporal variations of the Earth's magnetic field;

b) In archaeological investigations of buried sites of historical interest to delineate such features as kilns, ditches, walls etc. Sunken iron ships are a relatively easy target to detect using high sensitivity magnetometers providing the general location is known and the depth of water is of continental shelf dimensions;

c) In satellite investigations of the magnetic fields of planets such as Earth and Venus, and

d) In mineral prospecting surveys both ground and airborne. Mineral exploration is potentially the most important application of high resolution magnetometers because magnetic methods have been successfully used to locate mineral deposits for several centuries. A number of countries have been carrying out systematic aeromagnetic surveys of their territory and the results have indicated the value that these surveys have in geological mapping programs.

Of great interest to the petroleum industry is the evidence that experiments recently carried out with high resolution nuclear magnetometers have shown that detectable magnetic effects seem to be present in at least some sedimentary formations which give rise to "fine structure" in the recorded magnetic profiles. The development of high resolution magnetometers has also made feasible the measurement of the vertical gradient of the earth's magnetic field by using two magnetometers separated vertically by a short distance. The effect of the diurnal variation of the Earth's magnetic field is thus eliminated in the resultant differential output, which is an especially desirable feature in diurnally-active areas. However the use of high resolution airborne magnetometers necessitates a much better figure of merit for the magnetic compensation of the survey aircraft, and the navigational requirements are much more stringent. Moreover the use of high resolution magnetometers has also necessitated that the aeromagnetic survey data be digitally recorded in order to make full use of the high resolution available.

---

* Manuscript received May 1968.

This procedure will therefore permit the compilation of the resultant magnetic maps to be automated as much as possible.

1. Introduction. As the title of this article is the geophysical applications of high resolution magnetometers, it is perhaps as well to begin by defining what is meant in this article by the term high resolution (or high sensitivity) magnetometer. The magnetometer which has had greatest use in commercial aeromagnetic surveys up to the present time, namely the fluxgate variety, is usually considered to be a 1 γ (gamma) — or perhaps a ½ γ — sensitivity instrument. Thus it would appear reasonable to consider magnetometers having a sensitivity at least one order of magnitude better than the foregoing to be high resolution instruments. Thus for the purposes of this discussion we will define a high sensitivity magnetometer as one having a sensitivity of 0.1 γ or better.

The only magnetometers capable of measuring the Earth's magnetic field presently available which have a sensitivity of 0.1 γ or better are the nuclear variety, namely the proton precession and optical absorption varieties. The basic principles of these instruments will therefore be described together with some actual ground and airborne equipment even though these are not all high resolution instruments.

A. Nuclear magnetometers

2. Proton precession magnetometers.

α) It was observed by PACKARD and VARIAN[1] in 1953 that after a strong magnetic field is removed from a sample of water, an audio-frequency signal may be detected for a second or so. The explanation for this free precession is that the hydrogen proton has a magnetic moment due to its spin so that the magnetic moment vectors will align themselves parallel to an applied magnetic field. When the field is removed and the protons come under the influence of the Earth's magnetic field, which will in general have a different orientation to the applied field, the protons precess like gyroscopes. The precession signal will last for several seconds, its amplitude decaying exponentially with time, and during this decay time the precessing protons induce a small alternating voltage at the precession frequency in the polarizing coils. The rate of decrease of amplitude is dependent on the gradient of the ambient field across the sample which causes loss of phase coherence because protons in different parts of the sample precess at slightly different frequencies.

The frequency of precession, f, is directly proportional to the ambient field T so that it is only necessary to ascertain the former in order to calculate the latter[2]. Thus

$$T = Kf \quad \text{and} \quad K = 23.4874 \, \gamma/\text{Hz}. \tag{2.1}$$

Actually

$$K = 2\pi/\text{gyromagnetic ratio of the proton } \gamma_P$$

so that K is an immutable atomic constant, which means that proton precession magnetometers are absolute instruments. An accurate determination of the gyromagnetic ratio of the proton (γ_P) by DRISCOLL and BENDER[3] yielded a value of $\gamma_P = (2.67513 \pm 0.00002) \cdot 10^4 \, \text{Oe}^{-1} \, \text{Hz} \equiv (\ldots) \cdot 10^{-1} \, \text{Hz}/\gamma$.

If the axis of the polarizing/detecting coil is at angle θ to the total field T, the amplitude of the precession signal is proportional to $\sin^2\theta$. Thus the amplitude for $\theta = 45°$ is only half

---

[1] M. PACKARD, and R. VARIAN: Phys. Rev. **93**, 941 (1954).
[2] See contribution by H. SCHMIDT and V. AUSTER, p. 359, in this volume.
[3] R. L. DRISCOLL, and P. L. BENDER: Phys. Rev. Letters **1**, 413—414 (1958).

that for $\theta = 90°$, and it will fall to very low values as θ approaches zero, so that there is a dead zone along the axis of the coil. To avoid this difficulty, two mutually perpendicular coils are commonly used. With this configuration, if the total field T makes an angle θ with the normal to the plane defined by the axes of the coils, then the signal amplitude is proportional to [4]

$$\tfrac{1}{2}(1 + \cos^2 \theta).$$

Thus between $\theta = 0°$ and $\theta = 90°$, the signal amplitude is only reduced by a factor of 2. Alternatively a toroidal coil may be used to avoid the possibility of low precession signal amplitudes; a toroid also picks up less external interference due to nearby electrical systems, because it is the electrical equivalent of a single turn with respect to an external source.

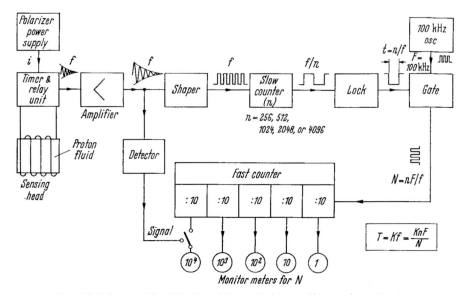

Fig. 1. Block diagram of Elsec 592 reciprocal-type proton free-precession ground magnetometer.

There are two common ways of determining the proton precession frequencies which results in the two types of proton magnetometers, namely the reciprocal and direct-reading types.

β) The *reciprocal-type* proton *free-precession* magnetometer counts the number of oscillations of a high-frequency oscillator (often 100 kHz) which occur during a given binary number (for example 1024) of cycles of the proton precession frequency. Thus the higher the proton precession frequency the less will be the total number of high-frequency oscillations counted, which is the reason for the term reciprocal-type.

This method is used in the *Elsec ground magnetometer* manufactured by Littlemore Scientific Engineering Co. in the U.K. A block diagram of the Elsec Type 592 proton magnetometer is shown in Fig. 1. The length of the polarizing period is controlled by the timer from 0.6 to 4.5 sec. The precession signal of frequency f is amplified, squared, and fed to an 8 to 12 binary chain for frequency division by a factor n.

For a ten stage divider $n = 2^{10} = 1024$, so that the frequency after the tenth stage is around 2 Hz. The signal from this Slow Counter is then applied, via a locking unit, to an electronic gate. The first positive-going edge of a pulse opens the gate and the next positive-

---

[4] K. WHITHAM: Measurement of the geomagnetic elements. Methods and Techniques in Geophysics, pp. 104—167. London: Interscience Publishers Ltd. 1960.

going edge closes the gate which remains closed until the unit is reset. While the gate is open, pulses from the high frequency oscillator of frequency $F = 100$ kHz pass through the gate into the fast counter. The corresponding meter for each decade unit indicates the number of counts in each decade unit, so that the 5-digit number indicated by the five meters gives the number N of 100 kHz pulses which pass through the gate during the time t that it is open. Thus $t = n/f = N/F$ so that

$$T = Kf = \frac{KnF}{N}.$$

Thus T is proportional to the reciprocal of N. The sensitivity S of the system is ± 1 count of N which therefore represents $T/N\ \gamma$. Hence

$$S = \frac{T}{N} = \frac{T^2}{KnF}. \tag{2.2}$$

Thus the sensitivity is better at the lower values of the Earth's magnetic field.

The Elsec Type 592 proton magnetometer has a total mass of 12 kg, with the standard 12 volt battery pack and detector bottle. The range of the transistorized instrument is 24,000 to 70,000 γ with a sensitivity of $\pm 0.5\ \gamma$ or better. The accuracy is 1 part in 50000 over the operating temperature range of $-10°$ C to $+40°$ C. The Slow Counter may be set for $n = 256$, 512, 1024, 2048, and 4096 using a front panel switch. Five automatic polarize/count rates are selectable by a front panel switch between about 1 per sec and 1 per min, and there is also a manual position of this switch so that readings may be initiated by push button. The tolerable magnetic gradient quoted for the detector bottle is 200 γ/m for 0.5 γ accuracy, 400 γ/m for 1 γ accuracy, and 800 γ/m for 2 γ accuracy.

The Elsec-Wisconsin proton precession airborne magnetometer consists of an Elsec ground instrument, which obtains the basic magnetic field values, and a digital-recording system designed and constructed at the University of Wisconsin[5]. The reciprocal magnetic field values and times are outputed in binary coded decimal format on punched paper tape for direct input into a computer. An analog output is also obtained on a suitable recorder.

The PM-1 reciprocal-type proton free precession magnetometer developed by the Institute of Terrestial Magnetism, Ionosphere and Radio Wave Propagation (Izmiran) in the USSR has been described by Kudrevskij[6]. Because it is rather bulky for field use it has been replaced by the transistorized reciprocal-type PM-5 instrument[7]. This has a 100 kHz fast counter and the slow counter is set to $n = 1024$ giving the PM-5 a one-gamma sensitivity.

γ) In *direct-reading* proton precession *instruments*, such as the *Barringer AM-101* (Fig. 2), the induced signal from the sensing head, which occurs after the polarizing field has been switched off, is amplified, squared, and fed to a phase detector. A voltage-controlled oscillator, which is governed by the output of the phase detector, generates a frequency approximately m times the precession frequency f. A portion of this multiplied frequency is fed back to the phase detector through a $1/m$ frequency divider to bring it down to almost f. If a frequency difference exists between the original precession frequency input to the phase detector and the output frequency of the frequency divider, the phase detector will change the output direct-current voltage to the voltage-controlled oscillator to make the voltage-controlled oscillator frequency exactly mf. The multiplied frequency mf is then counted for an exact period t which has been determined so that the resultant digital number gives the total field reading directly in γ. Thus we have the total field [compare Eq. (2.1)]:

$$T/\gamma = mf \cdot t \tag{2.3}$$

and sensitivity $S = \dfrac{K}{mt}$, S being given in γ.

---

[5] R. J. Wold: The Elsec Wisconsin digital recording proton precession magnetometer system. Geophys. and Polar Res. Centre, U. of Wisconsin, Res. Rept. Ser. 64—4, 83 (1964).

[6] A. I. Kudrevskij: Geomagnetizm i Aeronomija 1, 436—440 (1961); [Engl. translation: Geomag. and Aeronomy 1, 390—393 (1961)].

[7] V. I. Nalivajko, A. V. Tjurmin, and U. V. Fastovskij: Geomagnetizm i Aeronomija 2, 343—347 (1962); [Engl. translation: Geomag. and Aeronomy 2, 288—292 (1962)].

Using a digital-to-analog converter, the output of the counter can be displayed on a strip-chart recorder for monitoring purposes.

The Barringer Research AM-101 B airborne nuclear magnetometer consists of three units: the bird, the magnetometer console, and the pre-amplifier for the precession signal from the bird. The instrument is fully transistorized using printed circuitry throughout and a direct readout of the absolute value of the earth's total field is available in one-or five-γ intervals. The magnetic reading may also be read off a NIXIE-tube visual display on the front of the unit. A standard single-potentiometer recorder output is provided, and as an option voltages proportional to each of the last three digits of the total field may be supplied for analog recording, or alternatively outputs can be provided for the operation of digital-recording equipment. The operating range of the 27 kg. AM-101 B is from 20,000 to 110,000 γ in eight

Fig. 2. Block diagram of Barringer AM-101 direct-reading proton free-precession airborne magnetometer.

overlapping ranges of 14,000 γ selectable by a switch. Automatic tuning of the unit to the precession signal frequency is provided over the complete operating range. The polarize/count time of the instrument for a one-gamma sensitivity is one sec, and for a five-gamma sensitivity is $\frac{1}{5}$ sec. The count time t is accurately maintained by a crystal-controlled clock (see Fig. 2). The output from the 136.243 kHz crystal oscillator is first squared by a SCHMIDT trigger and then fed to four series-connected decade counters. The output of the fourth decade is used as a source of 73.3983 msec clock pulses which are used to control the programmer and the counter so that a 5 γ sensitivity is obtained using a frequency multiplier $m = 64$. The sensitivity of the instrument may be increased to 1 γ by increasing the count time five times to 366.99 msec using the : 5 divider incorporated in the circuit.

Transistorized direct-reading proton-precession airborne magnetometers have also been produced by Varian Associates at Palo Alto (USA), the V-4937, and by Prakla at Hannover (Fed. Rep. Germany), the PM-22 and PM-24.

Development of the DC V-4937 airborne proton magnetometer was begun in 1965 to convert the Varian direct-reading marine proton magnetometer for use in small aircraft. This instrument offers a ± 1 γ sensitivity at a 1 sec sampling rate, or ± 2 γ at a 0.5 sec sampling rate. The range of this instrument is 20,000 to 100,000 γ, and it requires about 200 W for operation.

A direct-reading proton-precession airborne magnetometer, the AYaAM-6, has been built in the USSR. The Larmor precession frequency is multiplied 24 times and counted for 0.97867 sec to obtain a one-gamma sensitivity[8].

---

[8] B. M. JANOVSKIJ: Zemnoj magnetizm (Earth's magnetism). Leningrad University 2, 461 (1963).

Barringer Research Ltd. of Toronto have recently introduced a one-gamma direct-reading proton free-precession magnetometer, designated Model SM-103, which has an operating range of 20,000 to 100,000 γ in 16 overlapping ranges. The tuning of the unit to the precession signal is automatically maintained within a preset range of 5,000 γ. The magnetometer has two outlets so that the output may be recorded in analog form (0—99 γ in 1 γ steps and 0—900 γ in 100 γ steps) on a suitable chart recorder. There is also a digital output in binary coded decimal 1—2—4—8 format. The unit has an automatic 2 sec cycling rate. It may also be manually operated or, for airborne surveys, controlled by an external programmer. The electronic unit weighs only 5.7 kg and operates from a 22...28 volt direct current power supply. Actually this versatile instrument may be used either as (1) an airborne magnetometer (2) a station instrument or (3) a ground survey instrument.

3. Spin-precession or OVERHAUSER type.

α) In the spin-precession or OVERHAUSER type of magnetometer the liquid sample in the magnetometer sensor contains a parametric substance which possesses the property that the spin energy of its electrons is transferred by coupling to the proton spin of the solvent. This mechanism is called the OVERHAUSER[1] effect. The spin energy of the electrons is maintained by the application of a suitable high-frequency polarizing field. Thus the polarized protons of the solvent precess continuously at the Larmor frequency which is governed by the ambient magnetic field and which is detected by a suitable pickup coil.

Sud Aviation of France have developed an airborne magnetometer[2] designated the MP 121 under licence from the French Atomic Energy Commission. It operates over the range 27,000 to 75,000 γ.

The sampling rate is one reading per second, and a sensitivity of 0.1 γ is possible with digital recording on punched paper tape. An analog record is also made with a maximum sensitivity of about 0.5 γ. The analog sub-ranges are 100, 200, 500 and 1000 γ. Total mass of the airborne installation with towed bird is 127 kg.

β) One of the difficulties of OVERHAUSER magnetometers hitherto has been that the precession characteristics of the liquid used in the detector head deteriorated over a relatively short period of time. However, several *semi-permanent hydrogenated solvents* have been perfected by workers at the Grenoble Laboratory of the French Atomic Energy Commission[3].

These are a class of organic free radicals consisting of

1) ditertiary butylnitroxide $(NO[C(CH_3)_2])$ and its derivatives such as triacetoneanime nitroxide or tetramethyl-2,2,6,6-aza-1-cyclohexanone-4-nitroxide-1, often called TANO for brevity;

2) those having a five-numbered heterocyclic ring such as tetramethyl-2,2,5,5-aza-1-cyclopentene-3-carbonamide-3-oxide-1 (TANAM-5), tetramethyl-2,2,5,5-aza-1-cyclopentane-3-oxide-1 (TANANE-5), tetramethyl-2,2,5,5-aza-1-cyclopentanol-3-oxide-1 (TANOL-5), and tetramethyl-2,2,5,5-aza-1-cyclopentanone-3-oxide-1 (TANO-5);

3) those having a six-membered heterocyclic ring, such as tetramethyl-2,2,6,6-aza-1 cyclo-hexanol-4-oxide-1 (TANOL-6), tetramethyl-2,2,6,6-aza-1-cyclohexane-oxide-1 (TANANE-6), and tetramethyl-2,2,6,6-aza-1-cyclohexanoneoxime-4-oxide-1 (TANOXIME-6). Triacetoneamine nitroxide is a crystalline paramagnetic substance melting at 36° C, and soluble in water and ether to form red-coloured solutions[4].

The foregoing organic radicals have electron resonance lines which vary from about 64 MHz to 75 MHz in pure water.

γ) A *ground* OVERHAUSER *magnetometer*, the MP 102, has also been developed by Sud Aviation. It consists of a sensing head and pole, a 10 m connecting cable, an electronic

---

[1] A. W. OVERHAUSER: Phys. Rev. **89**, 689—700 (1953).
[2] J. THOMAS: Geophys. Prospecting **13**, 22—36 (1965).
[3] H. LEMAIRE, A. RASSAT, R. BRIERE, R. M. DUPEYRE, and A. SALVI: Magnetometers for measuring the earth's magnetic field and its variations. Can. Patent 764,597,40 (1967).
[4] M. B. NEIMAN, E. G. ROZANTZEV, and YU. B. MAMEDOVA: Nature **196**, 472—474 (1962).

measuring module, and storage batteries. The toal mass of the equipment is 23 kg, and the equipment is designed to operate at temperatures between $-15°$ C and $+50°$ C. This continuously-reading instrument has a range of 11,100 γ and a series of plug-in modules allows fields between 20,000 and 81,000 γ to be measured. Magnetic field variations of 1 γ can be detected using the galvanometer on the electronic module over a working range of 100 γ By using a suitable external recorder changes of 0.1 γ can be detected.

4. Optical absorption magnetometers.

α) The high sensitivity atomic-resonance magnetometers are the latest type to be used in airborne geophysical surveys and are currently (1968) under intensive development. Three varieties have been produced, namely the metastable helium, rubidium and cesium magnetometers. All these magnetometers make use of the *optical-pumping technique* originally devised by KASTLER[1]. The basic physics of these magnetometers is complicated[2] but some idea of the principles involved may be gained from the following simplified explanation which has been modified from the description given by BLOOM[3] and BREINER[4]: —

Atoms can emit and absorb electromagnetic radiation ranging from radio waves, whose wavelength is measured in hundreds of metres, through the visible light spectrum to X-rays. Electromagnetic radiation energy is divided into discrete packets or photons and the energy of a photon is directly proportional to its electromagnetic wave frequency. When photons are absorbed or emitted by an atom, the atom gains or looses the energy they contain, changing its physical state in some way. The photons of light, with their comparatively high energy, cause the transition of an electron from one orbit to another. The photons of radio waves merely shift the axes of spinning electrons within an orbit. Since the electrons behave like tiny magnets with fields aligned along their axes, such a shift produces a small change in the magnetic energy of an atom. The transition between energy states, or levels, is reversible: atoms at a higher level tend to fall spontaneously into the lower one; those at the lower level will jump to the higher one if the requisite quanta of energy are available. In jumping up, the atoms extract photons from a transmitted beam of radiation, producing an absorption spectrum; in jumping down, they send out photons, producing an emission, or "bright line", spectrum.

To understand the principle of optical pumping, consider a simplified atom with only three energy levels, which we shall call A, B and C in increasing order of energy. Levels A and B are low-lying and very close together; the energy difference between them corresponds to a radio-frequency spectrum line, and initially all the atoms are distributed equally between them. Level C is much higher; the transition A—C and B—C correspond to lines in the optical part of the spectrum. Suppose a sample of these atoms is irradiated with a light beam from which the spectral line BC has been filtered out. The beam contains photons that can excite atoms in level A but not in level B. Atoms excited out of A absorb energy and rise to C. They will remain there for a short time and then emit energy, dropping back either to the A or B state. The proportion going to each state depends on the structure of the atoms, but the important thing is that occasionally an atom drops into B. When it does, it can no longer be excited by the incident light. If it returns to A, the light will raise it to the C state again, and again it will

---

[1] A. KASTLER: Quelques suggestions concernant la production optique et la détection optique d'une inegalité de population des niveaux de quantification spatiale des atomes. J. Phys. Radium **11**, 255—263 (1950).
[2] See contribution by H. SCHMIDT and V. AUSTER, p. 368, in this volume.
[3] A. BLOOM: Optical pumping. Sci. Am. **203**, 72—80 (1960).
[4] S. BREINER: Science **150**, 185—193 (1965).

have some probability of dropping to B. Given enough time, every atom must end up in the B state, and the material is then completely 'optically pumped'. Once this condition has been reached the cell becomes transparent, and photons can pass freely through the vapour cell because they are not absorbed.

Sweeping the vapour cell with a weak alternating magnetic field disrupts this state and allows pumping to begin again. If one varies the frequency of the applied field and observes the light transmitted through the vapour cell, a sharp absorption is seen to occur when the applied field has a certain frequency which is the LARMOR precession frequency of the electrons. The electrons undergoing the transition between energy sublevels precess about the magnetic field at a LARMOR frequency determined by the ZEEMAN splitting of the levels.

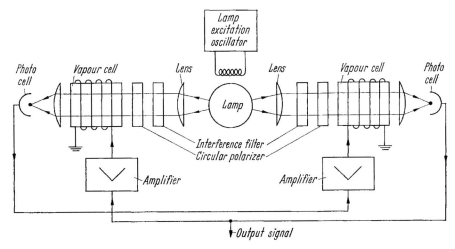

Fig. 3. Block diagram of dual-cell self-oscillating optical absorption magnetometer.

β) In the actual instrument (see Fig. 3), the modulation of the transmitted light is detected by a photocell. The current from the photocell is amplified, shifted in phase by 90°, and used as a *feedback* signal to drive the alternating magnetic field about the vapour cell. This arrangement thus constitutes an *oscillator* whose frequency is proportional to the total magnetic field intensity at a rate fixed by an atomic constant, 4.667370 ± 0.000035 Hz/γ for rubidium 85 and 3.4987 Hz/γ in the case of cesium 133. Actually any of the alkaline metals can be used but the vapours of rubidium and cesium possess the correct vapour pressure without an appreciable heating[5].

There is a second basic method of monitoring the LARMOR frequency by *sweeping* either the magnetic field or the frequency of the radio-frequency (r.f.) oscillator through resonance using a servo-loop to correct the r.f. oscillator frequency. PARSONS and WIATR[6] have described such a swept-field magnetometer.

There is a difference in energy between pairs of ZEEMAN sub-levels proportional to T^2 because of the BACK-GOUDSMIT effect. As the angle between the optical axis and the magnetic field is changed the relative population of the various sub-levels changes producing an appreciable asymmetry in the local pattern when the average line is used for detection. For a typical value of Earth's field of 50,000 γ, this orientation effect may amount to as much as

---

[5] R. GIRET, and L. MALNAR: Geophys. Prosp. **13**, 225—239 (1965).
[6] L. W. PARSONS, and Z. M. WIATR: J. Sci. Instr. **39**, 292—300 (1962).

20 γ under certain conditions for the rubidium magnetometer. In the case of the cesium magnetometer the BACK-GOUDSMIT effect exists but the relevant orientation effect is negligible[5].

γ) Another problem with the single cell magnetometer is that the correct sign for the phase shift also depends upon the angle between the optical axis and the direction of the total field. Both these problems may be overcome to a great extent using a *dual-cell magnetometer* having a common rubidium lamp (see Fig. 3). The further disadvantage of the single-cell instrument, which has dead zones i.e. will not operate, when the field is either parallel with or at right angles to the optical axis, may be overcome in a dual-cell system by having the optical axes of each half non-coaxial.

Compagnie Générale de Telegraphie sans Fils and Compagnie Générale de Géophysique have jointly developed a *cesium airborne magnetometer* (Fig. 4) which utilizes the swept-field optical pumping technique[5,7].

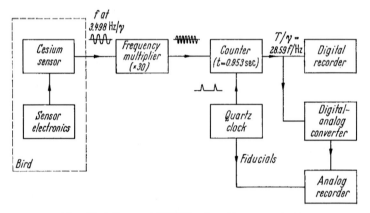

Fig. 4. Block diagram of CSF/CGG cesium airborne magnetometer.

The sensor and associated electronics are contained in a towed bird assembly. In the sensor itself are the light source, the optical system, the cesium cell, and the photo-cell detector. The cesium light source is excited by a 60 MHz source, and because the intensity of the light is sensitive to variations in ambient temperature, the source is thermostated. The pyrex vapour cell contains free atoms of cesium in equilibrium with cesium metal. At this pressure, the mean free path of the Cs atoms is large compared to the dimensions of the vapour cell, so that the atoms suffer numerous collisions with the walls, and therefore tend to revert back to a lower energy state. The walls of the vapour cell are therefore coated with paraffin to provide a buffer and improve the relaxation time. The cell is thermostated because the optimum output is obtained at a temperature near 35° C. This makes possible the operation of the magnetometer between $-20°$ C and $+45°$ C. The resultant cesium frequency, which is approximately 175 kHz for a 50,000 γ field is first multiplied by thirty and then counted over a 0.953 second period by a frequency meter. The magnetic field values are recorded on a digital magnetic tape recorder at ± 0.01 γ sensitivity and are also monitored on an analog recorder with a maximum sensitivity of 0.14 γ/mm. The range of the instrument is 20,000 to 75,000 γ without adjustment. The optical pumping sensor and associated electronics is towed in a bird from a B-17 aircraft, which is equipped with an Airborne Profile Recorder and a statoscope to provide flight elevation and variations of barometric altitude with an accuracy of 1 m.

STROME[8] has described the earlier developments of a *rubidium airborne magnetometer* at the Canadian National Aeronautical Establishment using basic Varian X-4935 dual-cell instruments mounted in a bird and tail boom installations. This high-sensitivity magneto-

---

[7] R. I. GIRET: Geophysics **30**, 883—890 (1965).

[8] W. M. STROME: Magnetometer installation in a North Star aircraft. Nat. Res. Council, Canada, DME/NAE Quart. Bull. 1963 (4), 23—30 (1964).

meter has a digital recording system which utilizes both magnetic tape and printer tape, and also has a seven-channel analog monitor system. More recent modifications have enabled the readings from both magnetometers, Decca navigation co-ordinates, and time from a digital clock to be recorded on a single digital magnetic tape by the use of a shift register.

δ) Varian Associates at Palo Alto (USA) have successfully developed a *self-oscillating airborne rubidium magnetometer*[9,10] which has been designated the Model V-4916.

This transistorized instrument has a sensitivity of 0.05 γ and consists of four units: the sensor assembly, controller, recorder, and data processor. The sensor assembly contains the coaxial dual-cell rubidium-vapour detector and gimbal assembly for orientation of the thermo-

Fig. 5. Graph comparing the relationship between the frequency multiplication factor m, count time t, and sensitivity S (parameter of the different lines) for proton precession and optical absorption magnetometers.

stated magnetometer head, which provides 360° rotation about the vertical axis, and 180° rotation about the horizontal axis. The controller assembly converts the magnetometer frequency to a direct current voltage to drive a Varian chart recorder. This is accomplished using a crystal-controlled oscillator, mixer and discriminator. The controller unit also contains the orientation control system for actuating the gimbal system in the magnetometer head. The rubidium frequency is fed to a phase-lock multiplier which increases the frequency by a factor of eight and this higher frequency is then counted for about 0.5 sec. Navigational data are provided by a BENDIX doppler and ROSEMOUNT pressure altimeter. The magnetometer and navigational data together with time is recorded on a single digital magnetic tape by programming through a shift register.

ε) Fig. 5 shows the relationship between the frequency multiplication factor, count time, and sensitivity for *proton precession and optical absorption magnetometers*. It can be seen that the optical absorption magnetometers have a sensitivity approximately two orders of magnitude greater than that of the proton precession type for the same frequency multiplication factor and count time.

The various applications of optical absorption magnetometers will now be discussed.

---

[9] R. HERBERT, and L. LANGAN: The airborne rubidium magnetometer. Varian Assoc., Geophys. Tech. Mem. **19**, 6 (1965).
[10] H. JENSEN: Geophysics **30**, 875—882 (1965).

B. Ground applications of optical absorption magnetometers.

5. Magnetic observatories and archaeological applications.

α) Most scientifically advanced nations have *magnetic observatories* in which the total field and various components of Earth's magnetic field are continuously monitored, and there are approximately 236 of these permanent magnetic observatories scattered throughout the world[1]. Both the U.S. Coast and Geodetic Survey and British magnetic observatories have used optical absorption magnetometers as station instruments.

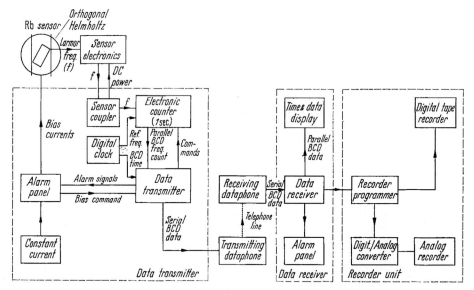

Fig. 6. U.S. Coast and Geodetic Survey automatic standard magnetic observatory (After BLESCH, 1965). [BCD means binary coded decimal.]

Fig. 6 shows the *Automatic Observatory* which has been set up by the U.S. Coast and Geodetic Survey[2]. The sensor employs a standard rubidium oscillator whose output LARMOR frequency is counted for a one-second period.

This observatory instrument is located at Castle Rock in the Santa Cruz mountains of California which is a fairly remote and inaccessible location. The magnetic data in the form of binary coded decimal is transmitted through a conventional 160 km telephone link to the recording equipment which is located on the Berkeley campus of the University of California. There the magnetic data is recorded on digital magnetic tape and is therefore available for direct processing by a high speed digital computer[3].

β) *Archaeological investigations of water-covered areas* mostly involve the search for sunken ships.

Wooden warships from the 17th, 18th, and 19th centuries would be expected to be carrying cannon which would either be brass or cast iron. The ammunition would in either case be cast iron balls, which if present in large numbers would be readily detectable. Iron anchors and fittings would add to the detectability. The ballast carried by such vessels would in general be expected to add to the concomitant anomaly.

---

[1] N. A. OSTENSO: Trans. Am. Geophys. Un. **47**, 303—332 (1966).

[2] J. BLESCH: An automatic standard observatory. Varian Assoc., Geophys. Tech. Mem. **21**, 6 (1965).

[3] L. R. ALLDREDGE, and I. SALDUKAS: The automatic standard magnetic observatory. US Coast and Geodetic Survey, Tech. Bull. **31**, 35 (1966).

The detection of steel ships is a much easier problem and Varian Associates in their commercial literature, have reported that their proton magnetometers played a useful rôle in the search for the U.S. submarine Thresher which went down in 2,560 m of water some 350 km east of Cape Cod in the northeastern United States.

Hall[4] has quoted the following approximate formula for the *detectability range* of a sunken steel ship

$$\frac{R}{m} = \left(10^4 \frac{W}{t} \frac{l/w}{\Delta \mathcal{M}/\gamma}\right)^{\frac{1}{3}}. \tag{5.1}$$

Fig. 7. Cesium gradiometer for ground magnetic surveys; Varian Associates, Palo Alto, California.

Where R is the greatest distance between the ship and magnetometer in meters at which the ship would be detectable, W is the weight of the ship in tons, l/w is the length to width ratio of ship, $\Delta \mathcal{M}$ is the least magnetic anomaly in gammas that is measurable. Thus for $\Delta \mathcal{M} = 5\gamma$ a 100 ton ship which is 5 times as long as it is broad should be detectable at 100 m. This formula may be used to establish the line spacing of the search pattern if the depth of water is also known.

γ) The successful application of *ground magnetic methods to archaeological investigations* depends on the existence of a concomitant magnetic anomaly being associated with a buried structure of archaeological importance such as a wall or foundations. It should be borne in mind that an anomaly is any feature on a magnetic map produced by readings which are either above or below the normal background value for the area.

---

[4] E. T. Hall: Archaeometry **9**, 32—44 (1966).

For instance because air has zero magnetic susceptibility, a tomb buried in a medium having some magnetic properties would be delineated by the mere absence of flux which is also often the case with walls. Other archaeological features which have been detected by magnetic methods include pottery kilns, hearths and baked rocks because they usually possess a significant remanent magnetization due to their heating history; in fact any location where there has been a fire will potentially have a concomitant magnetic anomaly. Humus-rich soil

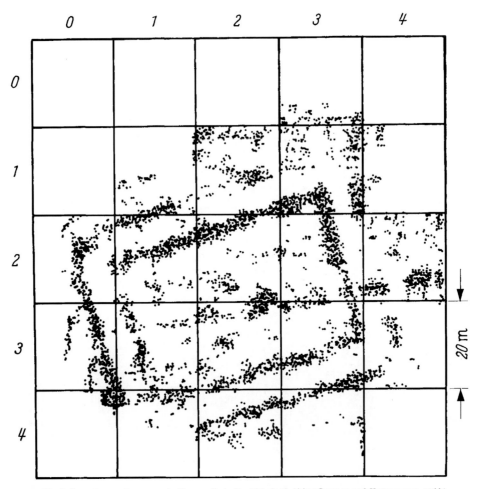

Fig. 8. Ground magnetic survey of a Roman site in the Rhineland. (After SCOLLAR and KRÜCKEBERG, 1966.)

also often possesses significant magnetic properties because of its magnetic susceptibility so that ditches that have been filled-in often produce slightly positive linear anomalies.

Varian Associates at Palo Alto (USA) have been involved in both waterborne and ground archaeological investigations using their high resolution optical absorption magnetometers. Varian's cesium ground magnetometer, the V-4920, is shown in Fig. 7. It has a range of 20,000 to 80,000 γ and has a 0.1 γ sensitivity.

The instrument, which has a digital readout, may be operated either as a magnetometer or as a dual-sensor gradiometer. This magnetometer has been used, for instance, in archaeological investigations of the ruins of the ancient city of Sybaris, a 2,500-year old settlement in southern Italy, by the University Museum of the University of Pennsylvania. Parts of the survey were carried out at two heights so that a vertical gradient map could be compiled.

The increase in resolution of the gradient map is readily apparent and it contains more definitive detail of the buried buildings which comprised the city than does the regular total intensity map[5]. Similar portable optical absorption magnetometers to the above have been developed by Varian Associates for the detection of skiers buried by snow avalanches[6].

Fig. 8 shows one type of presentation of magnetic data obtained by the National Museum of the Rhine from a Roman site in the Rhineland area of Germany[7].

For this investigation measurements were obtained on a one metre grid system so that 10,000 observations were taken for this survey. The results have been digitally filtered using a computer and the positive values only have been automatically plotted using a maximum of 8 dots per data point to represent the amplitude of the magnetic reading at the survey station.

6. Mineral prospecting.

α) There is a definite possibility that in some *electrical ground prospecting* methods for mineral deposits, high resolution magnetometers may be employed in the near future as the detector elements. For instance, in electromagnetic prospecting instruments coils of wire are commonly used to detect the magnetic effects of induced eddy currents flowing in subsurface electrically-conductive orebodies. The main advantages to using a magnetometer detection system is that they are more compact and lightweight, and that they measure the change in field directly rather than the rate of change of field. There is also a possibility that in the induced polarization method of geophysical prospecting, which is currently enjoying great popularity in the United States and Canada, high resolution magnetometers may replace the ground electrodes which are normally used in the receiver system.

β) Magnetometers have been used in *magnetic ground prospecting* since the 17th century.

The early instruments were essentially dip circles, i.e. magnetized needles suspended at their centre of gravity. In order to achieve greater sensitivity, instruments were constructed in which the centre of gravity and the axis of suspension differed. These instruments were, of course, the familiar dip needles which were used extensively in the latter part of the 19th and early part of the 20th century for finding deposits of iron ore.

A more sophisticated version of the dip needle, the magnetic balance, was introduced by ADOLF SCHMIDT in 1915 and was the instrument commonly used in ground magnetic prospecting surveys until its displacement a few years ago by electronic magnetometers such as the fluxgate[1,2] and proton free-precession varieties. These types of magnetometers (see Table 1), which have a sensitivity of around 5 γ, appear to be eminently satisfactory for most types of magnetic prospecting surveys. The increased sensitivity of the optical absorption magnetometers does not appear to have much practical value in ground prospecting surveys except perhaps in a gradiometer configuration.

An example of the application of vertical gradiometer surveys in mineral prospecting is shown in the next two figures. Fig. 9 shows the total intensity map obtained as a result of a winter survey over the ice at Savant Lake in northern Ontario[3]. The map has been drawn with a contour interval of 100 γ from readings obtained along the survey lines shown, which were spaced 30 m apart, using a Barringer Research Ltd. GM-102 proton-precession magnetometer. Several magnetite lenses occur in the survey area and these have been outlined on the

---

[5] S. BREINER: Science **150**, 185—193 (1965).

[6] L. LANGAN: Avalanche victim detection — the magnetic method, 1964—1965. Varian Assoc., Geophys. Tech. Mem. **25**, 4 (1966).

[7] I. SCOLLAR, and F. KRÜCKEBERG: Archaeometry **9**, 61—71 (1966).

[1] P. HOOD: The ground fluxgate magnetometer — a new versatile prospecting tool. Can. Min. J. **85**, (6) 59—64 (1964).

[2] See contribution by H. SCHMIDT and V. AUSTER, p. 346, in this volume.

[3] P. J. HOOD, and D. J. McCLURE: Geophysics **30**, 403—410 (1965).

Table 1. *Characteristics of commercially-available electronic ground prospecting magnetometers.*

Type of magnetometer	Principle of operation	Manufacturer	Model	Field measured	Readout device	Sensitivity	Instrument mount
Fluxgate	Saturable core	Sharpe Instruments (Canada)	MF-1 MF-1-100 PMF-3	Vertical component (All instruments except PMF-3 use a vertical freely-suspended fluxgate element)	Meter scales	20 γ/div. 2 γ/div. 100 γ/div.	Handheld with carrying strap around operator's neck. All are transistorized.
		McPhar Geophysics (Canada)	M 700			20 γ/div.	
		Optillinen Tehdas Oy (Finland)	Jalander Type 46—65 Type 46—66			50 γ/div. 25 γ/div.	
		Levanto Oy (Finland)	Finnmaster			20 γ/div.	
		IZMIRAN (USSR)	M-17			10 γ/div.	
Proton Precession	Free Precession of hydrogen Proton	Barringer Research (Canada)	GM 102A SM-103		Counter —3 sec —2 sec	10 γ 1 γ	
		Varian Associates (USA)	M 49A	Total field	Reed freq. meter —6 sec	10 γ	
		Littlemore Scientific Engineering (UK)	Elsec Type 592		Meter —1 sec	1 γ	Handheld, transistorized
		IZMIRAN (USSR)	PM-5		Meter	1 γ	
	Overhauser-forced precession of hydrogen proton	Sud-Aviation (France)	MP-102	Total field	Meter	0.1 γ	
Optical Absorption	Optical pumping	Varian Associates (USA)	Model 4920-Cesium	Total field and vertical gradient	Continuous and digital	0.1 γ	

map by the dotted areas[4]. It can be seen that magnetic highs occur over the lens as one might expect. Fig. 10 is the vertical-gradient map obtained from pairs of readings taken at each survey peg with a vertical separation of 2.5 m. Examination of the resultant map shows that there is a striking similarity between the outline of the lenses and the zero-gradient contour enclosing the negative-gradient values.

Fig. 9. Total intensity map, Float Bay anomaly, Savant Lake, N. Ontario. (After Hood and McClure, 1965.) Compare Fig. 10.

γ) A high-sensitivity gradiometer/magnetometer *waterborne system* has been developed by Varian Associates[5].

The two magnetometer sensors used in the system are housed in separate hulls of an unmanned, nonmagnetic 3.6 m long catamaran which is towed 60 to 120 m behind the survey vessel. The 136 kg twin-hull platform (catamaran) for the sensors is shown in Fig. 11. The difference between the outputs of the two sensors, the gradiometer output, and the output from one of the magnetometer sensors is recorded simultaneously on strip-chart recorders. The catamaran-gradiometer system and 11 m motor launch for towing the catamaran were carried aboard the U.S. Bureau of Mines mining research ship R/V Virginia City, a 62 m

[4] Barringer Research Ltd., Toronto (private communication.
[5] Anonymous: A new exciting tool for mineral exploration. Ocean Industry, pp, 17—19 (1967a).

Fig. 10. Vertical gradient map, Float Bay anomaly, Savant Lake, N. Ontario. (After Hood and McClure, 1965.) Compare Fig. 9.

Fig. 11. Catamaran for marine magnetic gradiometer system [Varian Associates, Palo Alto, California.]

converted ocean-going tug, on an expedition to Norton Sound, Alaska. The primary purpose of the geophysical activities of the expedition was to investigate methods for locating and characterizing off shore heavy-metal placers, which would probably contain gold in this area [6].

7. Determination of magnetic parameters of rock formations.

α) GOLDSTEIN and WARD [1] at the University of California have demonstrated that *in situ determinations* are possible. The relative proportions of induced and remanent magnetism in a given magnetic rock formation can be established using high resolution magnetometers. It is necessary to use two magnetometers and place one close to the peak of the total field anomaly (positions A in Fig. 12)

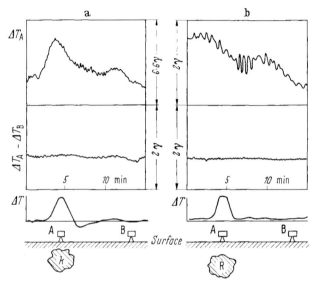

Fig. 12 a and b. In situ determination of induced and remanent magnetism. (After GOLDSTEIN and WARD, 1966.) See text for explanation.

produced by the magnetic body under investigation, and the other adjacent to the anomaly at a point where the field is representative of the background value for the area (positions B in Fig. 12). Fig. 12 shows both an induced and a remanent magnetism case. At the top of the figure are the two magnetic records showing the variation with time of the total field at the respective points A. Below these are the records showing the variation with time of the difference between the outputs of the magnetometers at positions A and B. For the remanent magnetism case, the difference trace does not change very much with time. However, for the induced magnetism case there is a significant variation of the differential trace with time due to the fact that the susceptibility of the magnetic body increases the amplitude of the diurnal magnetic field at position A. Thus magnetic susceptibility has an amplifying effect on the diurnal magnetic field in its immediate vicinity, and this is expressed mathematically in the equation

$$\Delta T_A = \Delta T_B (1 + k / (\text{geom.})), \tag{7.1}$$

---

[6] Anonymous: Undersea ores charted with magnetic measuring system. Mineral. Mag. **57**, 16 (1967b).
[1] N. E. GOLDSTEIN, and S. H. WARD: Geophysics **31**, 779—796 (1966).

where f(geom.) means a function of the geometry of the causative magnetic body. In the computation, this geometrical factor must be computed to obtain a value of the susceptibility k. An estimate of the remanent magnetism R may then be made from an analysis of the non-time varying total-field anomaly produced by the causative magnetic body.

β) Fig. 13 shows the remanent magnetometer constructed at the Massachusetts Institute of Technology[2,3], which uses an optical absorption magnetometer for the *determination of the remanent magnetism of rock samples*. The rock

Fig. 13. Rubidium remanent magnetometer constructed at MIT. (After LANGAN, 1966.) See text for explanation.

sample is spun a short distance, r, from the magnetic sensor with its axis of rotation aligned in the direction of the total field vector of Earth's magnetic field. The maximum value of the field, ΔF, due to the sample occurs when the component of the remanent magnetism vector at right angles to the axis of rotation points in the direction of the sensor.

A second sensor is used to cancel time variations of the geomagnetic field, and a reference signal is also obtained from a suitable device mounted on the shaft of the sample spinner so that a sinusoidal signal is fed to the lock-in amplifier. The phase of the reference signal may be varied in order to determine the direction of the remanent magnetism component in the rock sample. The intensity of the remanent magnetism is, of course, obtained from the amplitude of the peak-to-peak excursion, $2\Delta F_r$, in the sinusoidal magnetometer record. Sensitivities for the remanent magnetization determinations of $0.1\,\gamma = 10^{-6}\,\Gamma$ and $\pm\,5°$ for the angular measurements have been obtained. Susceptibility can be obtained to a few parts in $10^{-5}\,\Gamma\,(=1\gamma)$ in an unshielded environment; and to a few parts in $10^{-6}\,\Gamma$ in a magnetically shielded room.

C. Airborne and satellite applications of optical absorption magnetometers.

8. Extra-terrestial- and aeromagnetic surveying.

α) *Satellite-borne rubidium magnetometers* appear to have been first used by NASA in the Explorer 10 satellite which was launched in March 1961[1]. Currently

---

[2] P. T. DI TORITTO: Magnetic measurements of rocks with a rubidium vapour magnetometer. Mass. Inst. Tech., B.S. thesis 1966.
[3] L. LANGAN: Geophys. Prospecting **14**, 487—503 (1966).
[1] J. P. HEPPNER, N. F. NESS, C. S. SCEARCE, and T. L. SKILLMAN: J. Geophys. Res. **68**, 1—46 (1963).

NASA are using high resolution magnetometers in their OGO (standing for Orbiting Geophysical Observatories) series of satellites. Fig. 14 shows the OGO-A satellite which was equipped with a 3-axis, dual range, fluxgate magnetometer for vector field measurements over the ranges 0 to 30 γ, and 0 to 500 γ, and also a 4-cell rubidium-vapour magnetometer for scalar field measurements over the range 3 to 14,000 γ. The OGO-A satellite was launched on September 5th, 1964[2].

β) A dozen years ago there was essentially only one magnetometer in standard use for *aeromagnetic prospecting* surveys, namely the fluxgate type[3]. In 1955

Fig. 14. OGO-A satellite launched September 4th, 1964. (After HEPPNER et al, 1967.)

Varian Associates at Palo Alto (USA) introduced the first airborne proton free-precession nuclear magnetometer. At the present time there are four basically different types of magnetometers being used in aeromagnetic surveys, which are

        fluxgate,
        electron beam,
        proton precession, including the OVERHAUSER type[4],
        optical absorption magnetometers[5-7].

All of these measure the total intensity of the Earth's magnetic field (see Table 2). The magnetometers are by no means equivalent, because the optical-absorption variety, namely the rubidium, cesium and metastable helium types,

---

[2] J. P. HEPPNER, M. SUGUIRA, T. L. SKILLMAN, B. G. LEDLEY, and M. CAMPBELL: J. Geophys. Res. **72**, 5417—5471 (1967).
[3] See contributions by P.H. SERSON and K. WHITHAM, p. 384, and by J. BALSLEY, p. 395, in this volume.
[4] J. THOMAS: Geophys. Prospecting **13**, 22—36 (1965).
[5] R. I. GIRET: Geophysics **30**, 883—890 (1965).
[6] R. GIRET, and L. MALNAR: Geophys. Prosp. **13**, 225—239 (1965).
[7] H. JENSEN: Geophysics **30**, 875—882 (1965).

Table 2. *Characteristics of airborne magnetometers used in mineral exploration.*

Magnetometer	Principle of operation	Type	Manufacturer & Model designation	Field measured	Readout (A-Analog, D-Digital)	Sensitivity	Mount (B-Bird, I-Inboard)
Fluxgate (vector variometer)	Saturable core	Peak voltage	Gulf Oil Co. Mark III (USA)	T	All continuous A	$<1\,\gamma$	I
		Second harmonic	McPhar 555 AB (Can)	V	A	$50\,\gamma$	I
			AN/ASQ-3, 8, 10 (USA)	T	A	$<1\,\gamma$	I
			AM-13 (USSR)	T	A	$2\,\gamma$	B
Electron beam or aspect tube (vector variometer)	Deflection of moving electron		Elliott Bros (UK) EMD 13	T	A Continuous	$10\,\gamma$	I
Proton precession (Scalar absolute)	Free precession of hydrogen proton	Reciprocal	Elliott Bros (UK) EMD 14 & EMD 21	T	D & A Every 0.5 sec	$1\ldots3\,\gamma$	B or I
			Varian Assoc. (USA) V-4914	T	D & A Every 0.6 sec	$<1\,\gamma$	B or I
		Direct reading	Barringer (Can) AM-101	T	D, A, & Visual Every 1 sec	1 or $5\,\gamma$	B or I
			GSC (Can) Telmag	T	D & A Every 2 sec	$0.1\,\gamma$	B or I
			Bureau of Mineral Resources (Australia)	T	A Every 0.5 sec	$1\,\gamma$	B or I
			Prackla (F.R. Germany) PM-22 & PM-24	T	D & A Every 1 sec	$2\,\gamma$	B
			Varian Assoc. (USA) V-4937A	T	D & A Every 1 sec or 0.5 sec	1 or $2\,\gamma$	B
	Forced precession of hydrogen proton	Overhauser	МАРМ (АЯаАМ)6 (USSR) Sud-Aviation (France) MP 121 & MP 122	T	A Every 1 sec D & A Every 1 sec	$1\,\gamma$ $0.1\,\gamma$	B B
Optical absorption (Scalar absolute)	Optical pumping	Cesium vapour	CSF/CGG (France)	T	D & A Continuous	$0.01\,\gamma$	B
		Rubidium vapour	Varian Assoc. (USA) V-4916	T	D & A Continuous	$0.05\,\gamma$	B
		Metastable helium	Texas instr. (USA)	T	D & A Continuous	$0.05\,\gamma$	B

have better than an order of magnitude greater sensitivity than the fluxgate. Thus there are many more magnetometers used in aeromagnetic surveying at the present time than there were twelve years ago. It can be seen from Table 2 that only the OVERHAUSER and optical absorption magnetometers rate as high resolution instruments using the definition given in the Introduction to this paper.

Some geophysicists will query the need for magnetometers of increased sensitivity and maintain that the fluxgate magnetometer, which is usually considered to be a half-gamma instrument, is good enough for aeromagnetic surveying purposes. As far as the production of standard one-mile aeromagnetic maps is concerned, this is perfectly true. But there appears to be a number of interesting possibilities for the new generation of magnetometers. Of interest in petroleum prospecting is the evidence that magnetic effects seem to be present in at least

Fig. 15. Block diagram of the basic aeromagnetic survey system.

some sedimentary rock formations, and these give rise to fractional gamma anomalies in the recorded magnetic profiles. Another interesting application of high-resolution magnetometers is their use in gradiometer systems.

However, the use of high resolution airborne magnetometers necessitates a much better figure of merit for the magnetic compensation of the survey aircraft, and the navigational requirements are much more stringent. JENSEN[7] has considered the flight-path restrictions necessary in the conduct of high-sensitivity aeromagnetic surveys. He concludes that all high-resolution magnetic surveying should be under the control of DOPPLER navigation, precision pressure altimeters, and modern autopilots.

γ) The basic *aeromagnetic survey system* (Fig. 15) has not really changed since the first fluxgate survey was flown using a modified ASQ-1 in the northern United States in 1944[8]. The heart of any airborne magnetometer is the sensing element (or sensor) which detects the small variations in the earth's magnetic field as the aircraft flies along. The various magnetometer sensors which are presently being used for aeromagnetic surveying purposes have been described in the preceding sections. Nowadays the magnetometer head is usually mounted in an inboard installation in the tail of the aircraft (see Fig. 16). Inboard installations have been made possible by the improvement in the magnetic compensation of aircraft, and either an active or passive compensation system may be used.

The output of the magnetometer sensor is electronically processed in some way to give a signal, usually a voltage, which is directly proportional to the spatial total field variations ΔT measured by the sensor. These are recorded in analog and/or digital form on a suitable recorder. The ΔT variations have to be related to geographic co-ordinates so that a *positioning system* is required. This is com-

---

[8] H. JENSEN: The airborne magnetometer. Sci. Am. **204**, 151—162 (1961).

monly a tracking camera and/or DOPPLER for surveys over land. It is standard practice to either maintain a constant terrain clearance or a constant elevation above mean sea level so an altimeter is an essential part of the aeromagnetic surveying system and its output is often recorded simultaneously with ΔT. In order to be able to relate the ΔT variations to the output of the positioning system in the aeromagnetic map compilation process, a control unit is used to provide fiducial marks to the recorder and/or the positioning system. The control unit may also program the magnetometer to acquire ΔT readings. There are various possibilities all of which have been indicated by the dotted lines on Fig. 15. In order to relate the airborne record to the magnetic diurnal variation recorded on the ground, time must also appear in some fashion in the aeromagnetic survey data.

Fig. 16. Douglas A-26 aircraft equipped for high-sensitivity aeromagnetic surveys; Aero Service Corp., Philadelphia.

Many types of aircraft have been used in aeromagnetic surveys since 1945. However since about 1960 there has been a trend to the use of light twin-engined aeroplanes such as the Aero Commander, Beechcraft Queenair, Cessna 310, and Piper Aztec in aeromagnetic surveys because of their cheapness to operate. Fig. 16 shows the medium-size twin-engined Douglas A26 aircraft which has been equipped by Aero Service of Philadelphia for high sensitivity aeromagnetic surveys.

δ) For inboard installations there are three different sources of *magnetic interference due to the aircraft* itself.

The first is the permanent magnetism of the various components made of steel, such as the engines, whose direction remains fixed with respect to the aircraft.

The second source is the induced field due to the magnetic susceptibility of these same components and the earth's field. Its polarity and magnitude depends upon the orientation of the aircraft with respect to the Earth's magnetic field.

The third source of interference is that due to the magnetic effect of eddy currents generated in the skin and other conducting parts of the aircraft by their motion in the Earth's magnetic field.

In passive compensation systems, the permanent magnetism is eliminated by the use of a set of three-orthogonal compensating coils mounted near the magnetometer head on the roll axis of the aircraft[9] through which the appropriate

---

[9] Bureau of Naval Personnel: Aviation electronics technician 3 & 2. Navpers 10317-A, Washington, Sup. of Documents, 464, 1959.

direct currents may be passed. The induced components are eliminated by the use of Permalloy strips, and the eddy currents are compensated by the use of coils of wire mounted in close proximity to the sensitive element.

It is necessary to carry out a set of pitches and rolls of the aircraft in low gradient areas in order to separate out the effects of the various components[10]. The excellence of compensation of a given aircraft is measured by its 'figure of merit'. This index is obtained by summing, without regard to sign, the amplitudes of the 12 magnetic anomalies recorded when the aircraft carries out 20° rolls, 10° pitches, and 10° yaws peak-to-peak on north, east, south, and west headings.

With the advent of high sensitivity magnetometers, there was a partial reversion to the use of towed bird systems in aeromagnetic surveying. The reason for this apparent step backwards was that the magnetic compensation of aircraft was not, in general, good enough to cause manoeuvre noises to be unimportant. However active 9- and 16-term magnetic compensation systems have been developed by Canadian Aviation Electronics (CAE) Industries Ltd. at Montreal for use in inboard installations. They drastically cut down the time required for aircraft compensation, and improve the 'figure of merit' of aircraft considerably.

CAE's 9-term compensator consists of three saturable-core magnetometer probes, an amplifier-power supply, an electronic unit containing the main control circuitry, and the compensation coils which are mounted near the magnetometer head. The system can be used in any aircraft with any magnetometer at any magnetic latitude. The compensator provides active and semi-automatic compensation for the permanent, induced and eddy-current aircraft interfering fields.

In order to compensate for these fields at the magnetometer detector, *synthetic fields* are generated which simulate and oppose those due to the aircraft. A permanent field component is simulated by sending a constant current through the appropriate compensation coil. An induced field component is simulated by passing a current proportional to the output of a compensation vector magnetometer into the appropriate compensation coil to produce a field proportional to the earth's field component along the compensation magnetometer axis. The eddy-current field component is synthesized by sending a current which is proportional to the differentiated output signal from a compensation magnetometer into the appropriate compensation coil.

9. Aeromagnetic survey results.

α) The basic objective of any aeromagnetic survey is *mapping of the geomagnetic field* in a given horizontal plane above the earths' surface. In prospecting surveys the only anomalies of interest are those due to the induced and remanent magnetization of the crustal rocks, whereas in geomagnetic surveys, such as the U.S. Naval Oceanographic Office Project Magnet, measurement of the spatial variation of the earth's main magnetic field is of primary interest. For this reason prospecting surveys are flown at low elevations, usually 300 m or less, and geomagnetic surveys are flown at higher elevations usually in excess of 3,000 m. In both cases the short-term time variations of the earth's field produce unwanted interference which have to be removed. These are often the limiting factor affecting the final accuracy of a given aeromagnetic survey.

Government agencies, usually the Geological Surveys or the equivalent organization, in a number of countries carry out routine aeromagnetic surveys on a continuing basis in an attempt to obtain as much aeromagnetic survey coverage of their respective countries as possible. The resultant total intensity maps are usually published so that they are available to the general public for a nominal charge, and are consequently in great demand by those

---

[10] P. LELIAK: Identification and evaluation of magnetic field sources of magnetic airborne detector equipped aircraft. IRE Trans. Aerospace Navigational Electron. **8**, 95—105 (1961).

engaged in mineral exploration. Fig. 17 shows the standard sensitivity one-mile aeromagnetic coverage of Canada by the Federal and Provincial Governments to the end of 1968. The coverage is mostly over the Canadian Shield which is, of course, one of the richest mineral-producing areas in the world. Coverage in the United States is less systematic, and many aeromagnetic surveys appear to have been carried out in an attempt to solve a specific geological problem. The cost per square mile depending on the line-spacing used varies from about \$6 to \$30 per square mile (\$3 to \$12 per km^2).

Aeromagnetic surveys are usually carried out in order to throw light on the underlying geology of the surveyed areas. In areas where igneous and metamorphic rocks outcrop, such as the Canadian Shield, the main use is to help in a

Fig. 17. Aeromagnetic and shipborne magnetometer coverage of Canada to December 31, 1968.

subsequent geological mapping program. Any mineral deposits such as magnetite (an oxide of iron), ilmenite (an iron-titanium oxide), chromite (an iron-chromium oxide), and pyrrhotite (a sulphide of iron) found because of their anomalous magnetic properties are an added bonus.

β) Aside from the development of more sensitive magnetometers, there is a pronounced trend towards the *digital recording* of aeromagnetic survey data in order that the compilation of the final maps might be automated as much as possible. One of the main difficulties to complete automation is the fact that the cheapest methods of recovering the track of the survey aircraft is to use 35 mm vertical photography with or without the aid of a DOPPLER navigation system, so that this will probably remain the basic track location method for some years to come. By recording in a digital format it is possible to produce at very little additional cost maps other than the total intensity variety, such as first and second derivatives, digitally filtered, and no doubt a number of other more 'exotic' possibilities.

Digital filters appear to be useful in separating anomalies due to near-surface effects from those due to more deeply-buried causative bodies. ANDERS et al.[1] have presented the necessary background theory for a large class of digital filters. The basic formula for digital filtering is

$$\bar{f}_m = \sum_{n=-N}^{n=N} h_n g_{m+n}, \qquad (9.1)$$

where g_m are the input data values,

$\bar{f}_m$ are the smoothed output values for each g_m,

h_n are the weights of which there are $(2N+1)$.

One of the better zero-phase smoothing filters[1] is the MARTIN-GRAHAM filter. In this filter, the weights (or filter operator) are obtained from the expression

$$h_n = \frac{f_s^2 \left(\sin \frac{2\pi n f_c}{f_s} + \sin \frac{2\pi n f_t}{f_s} \right)}{2\pi n \left(f_s^2 - 4n^2 (f_t - f_c)^2 \right)} \qquad (9.2)$$

where f_c and f_t are the cut-off and termination frequencies of the filter, and f_s is the sampling frequency which is the reciprocal of the sampling interval. A bandpass filter may be obtained by combining two of these smoothing filters.

γ) In areas underlain by sedimentary rocks, the main application of aeromagnetic surveys is as a *reconnaissance method in petroleum prospecting*. Because sedimentary rocks usually have negligible magnetic properties compared to the igneous and metamorphic rocks forming the crystalline basement beneath the sedimentary rocks, the resultant aeromagnetic data will reflect the differing magnetic properties of the various geological formations comprising the crystalline basement. Because the inverse square law applies to magnetic effects, the greater the distance from the magnetometer aircraft to the crystalline basement rocks, the less will be the amplitude of the magnetic anomalies recorded. This means that in areas of sedimentary rock cover, the thicker the sediments the less 'sharp' will the anomalies be.

This is well illustrated in Fig. 18, which is an aeromagnetic profile flown at right angles to the coast of Nova Scotia across its continental shelf just east of Halifax[2]. Included on the figure is the total intensity profile with the regional gradient removed, a MARTIN-GRAHAM high-frequency bandpass digitally filtered trace, a low-frequency bandpass digitally filtered second-derivative trace[1], and the underlying bathymetric contour and interpreted geological cross-section. It can be seen that the amplitude of the anomalies on the total intensity profile gradually decrease with the increasing depth to the crystalline basement.

The high-frequency bandpass filter is designed to enhance the magnetic effects of the near-surface sediments, whereas the low-frequency second-derivative signature discriminates against shallow causative bodies, and augments the anomalies due to more deeply-buried sources such as the crystalline basement.

Thus the edge of the Meguma Group of metamorphic rocks may be inferred from the second-derivative trace, and this interpretation is supported by the fact that the characteristic pattern of linear total intensity contours on the published Geological Survey of Canada aeromagnetic maps which parallel the coastline terminates at the same location. The left-hand part of the high-frequency filtered trace has been omitted because of the large amplitude spikes which are evident where the crystalline basement is shallow. Some idea of the background noise level of the system can be estimated from the amplitude of this trace over the deep water on the right-hand side of the profile. Note that there is an increase in the amplitude of the high-frequency anomalies close to the edge of the continental shelf.

---

[1] E. B. ANDERS, J. J. JOHNSON, A. D. LASAIN, P. W. SPIKES, and J. T. TAYLOR: Digital filters. NASA Contractor Report **136**, 132 (1964).

[2] P. HOOD, and M. E. BOWER: Aeromagnetic survey of the Scotian Shelf southeast of Halifax, Nova Scotia. Geological Assoc. Canada Ann. Meeting, Halifax, N.S., 1966.

δ) Fig. 19 shows aeromagnetic profile data across the Gloucester Fault which is in southern Ontario, a few miles south of Ottawa[3]. At the top of the figure is the total intensity data and it can be seen that there is no particular indication of the fault except perhaps for the presence of a small anomaly. Below the total intensity profile is the same profile which has been filtered to pass anomalies having a wavelength between 0.3 and 0.89 km long using a MARTIN-GRAHAM filter[1]. There is a noticeable reduction in the amplitude of the filtered total intensity trace on the downthrown side of the fault.

Fig. 18. Magnetic profile data and geological cross section, Scotian shelf, Eastern Canada. (After HOOD and BOWER, 1966.)

If one compares the two profiles closely, remembering that there is a 125 to 1 difference in the vertical scales, it is possible to see the high resolution anomalies which appear on the filtered profile in the original untouched total intensity profile. This shows that the high resolution anomalies have not been generated by the filter itself and must be due to changes in the magnetic properties of the underlying igneous basement rocks which are of Precambrian age.

Fig. 20 is a geological map which shows the digitally-filtered total intensity profiles across the Gloucester Fault. The sedimentary formations shown are either limestone (4, 6, and 7), shale (5, 10 and 11) or sandstone (2 and 3) and these formations appear to have negligible magnetic contrast between them. The Gloucester Fault is delineated quite well by the change in amplitude of the profiles, and there is some indication from the profiles that the fault immediately to the north of the Gloucester Fault continues in a westerly direction.

ε) During a combined Sparkarray/Varian V-4937 proton magnetometer survey carried out by EDGERTON, GERMESHAUSEN, and GRIER International Inc. using the motor vessel

---

[3] P. J. HOOD, P. SAWATZKY, and M. E. BOWER: Aeromagnetic investigations of Baffin Bay, the North Atlantic Ocean, and the Ottawa area. Geol. Surv. Canada, Paper 68-1A, 79—84 (1968).

Sect. 9. Aeromagnetic survey results. 449

Fig. 19. Aeromagnetic profile data across the Gloucester Fault, SE Ontario. (After Hood, Sawatzky, and Bower, 1968.)

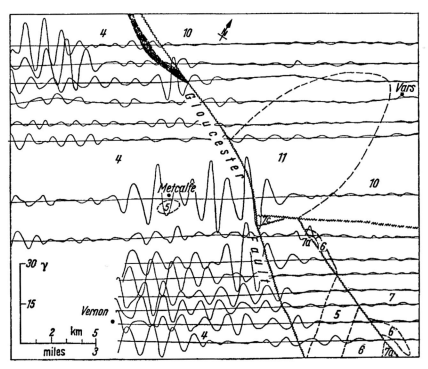

Fig. 20. Digitally-filtered aeromagnetic profiles across the Gloucester Fault, SE Ontario. (After Hood, Sawatzky, and Bower, 1968.)

"Mrs. Maggie", a 40 km *profile* was obtained *across* the Way Salt Dome which is located about 216 km southeast of Galveston, Texas in the Gulf of Mexico. Fig. 21 shows the total field profile with the regional and diurnal variation removed and the underlying geological structure and bottom topography as deduced from the 3,000-Joule seismic profiling system.

PRINDLE[4] has concluded that the negative and positive fluctuations in the magnetic profile at Stations 90 and 93 could have their origin in either (1) faulting, or (2) an irregular contact caused by dome-shaped and truncated shale/sandstone formations overlying the salt dome, or (3) possibly both. These subtle differences are difficult to distinguish because of the small magnetic susceptibility contrast which is to be expected between the salt and cap rock, which would have a negative susceptibility[5] around $-1 \cdot 10^{-6}$ and the surrounding sediments which would probably have a susceptibility between 10 and $50 \cdot 10^{-6}$. A seismic traverse

Fig. 21. Total intensity magnetic and gravity profiles over the Way salt dome, Gulf of Mexico. (After PRINDLE, 1967.) An unconformity appears near Station 92.

obtained by EDGERTON, GERMESHAUSEN and GRIER International in 1966 using a 95,000 Joule sparker system, indicated that the top of the salt dome is approximately 500 m below sea level, which is below the depth scale shown on Fig. 21. The sharpness of the 15 γ negative anomaly indicates that the causative body is within the sedimentary section because the depth to the crystalline basement is in excess of 10,000 m in this area. The assymetry of the anomaly is probably a reflection of the assymetry of the salt mass itself. NETTLETON[6] has reported a ten-milligal negative gravity anomaly from the same feature.

10. Aeromagnetic gradiometer surveys.

α) The recent development of high resolution magnetometers, such as the optical absorption variety, has also made feasible the *measurement of the first-vertical derivative* of the total field in aeromagnetic surveys[1-3]. This is accomplished by using two sensitive magnetometer heads separated by a constant vertical distance and recording the difference in outputs (Fig. 22). The effect of diurnal is thus eliminated in the resultant differential output, and this is an especially desirable feature in such areas as northern Canada where the diurnal variation is usually much greater than is found in more southerly magnetic latitudes.

Fig. 23 is a block diagram of the helicopter aeromagnetic gradiometer system built by Varian Associates at Palo Alto (USA) for Pure Oil Co. (now Union Oil Co.)[3]. The dual-cell

---

[4] O. PRINDLE: Geophys. Prospecting 15, 551—563 (1967).
[5] Numerical indications are valid for (non-rationalized) electromagnetic cgs-units. The numerical values must be multiplied by 4π in order obtain the value in SI-units.
[6] L. L. NETTLETON: Geophysics 22, 630—642 (1957).
[1] P. HOOD: Geophysics 30, 891—902 (1965).
[2] L. LANGAN: Geophys. Prospecting 14, 487—503 (1966).
[3] H. A. SLACK, V. M. LYNCH, and L. LANGAN: Geophysics 32, 877—892 (1967).

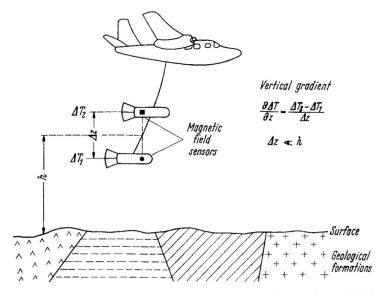

Fig. 22. Definition of the first vertical derivative of the total intensity of the earth's magnetic field.

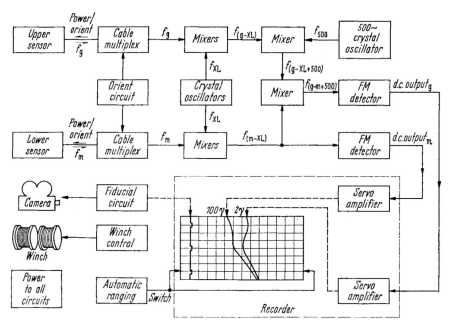

Fig. 23. Block diagram of Varian/Union Oil Company aeromagnetic gradiometer system. (After SLACK et al., 1967.) [See text for explanation.]

rubidium sensors and associated electronics are flown in two birds which are vertically separated by 30 m. The birds are raised and lowered using the winch which has a hydraulic brake to allow the cable to reel off slowly should a winch failure occur. In an emergency, the cable may be cut either electrically or manually using a guillotine system. The sensor electronics are powered by 28 V direct current flowing down the tow cable together with the sensor orient signal, and the rubidium frequencies, f_g and f_m, are returned along the same conductors. These frequencies are beaten with that from a crystal oscillator f_{XL} to produce the difference frequencies f_{g-XL} and f_{m-XL}. The latter (lower sensor) frequency is converted to an analog voltage (d.c. output$_m$) suitable for presentation on the recorder. The signal passing through the upper sensor channel is increased by a bias frequency of 500 Hz to avoid polarity changes and then mixed with the lower sensor frequency. The resultant difference frequency, $f_{g-m+500}$, is then converted to analog form and this gradiometer signal is also displayed on the strip chart recorder using either a 2 or 10 γ full-scale sensitivity. A camera mounted on the side of the helicopter is actuated at eight-second intervals and a fiducial mark is made on the strip chart with each frame for data reduction purposes.

β) The *data reduction* necessary to produce a vertical gradient map is much simpler than with the total-field case because no datum levelling is necessary. As the aircraft track will be available from the main compilation it is only necessary to plot the resultant vertical gradient values on the track map and contour. Thus two maps will be obtained for little more than the cost for one but with a greatly increased gain in geophysical information concerning the geometry of the causative bodies. With further improvements in the sensitivity of optical absorption magnetometers it may prove feasible to measure the second vertical derivative directly in aeromagnetic surveys.

11. Vertical gradient theory.

Many parts of Precambrian Shield areas consist of large rock units which are separated from one another by steeply dipping contacts which extend to great depths. However this model is not true of sedimentary areas, except for the underlying crystalline basement, for oil-bearing reef structures such as occur in Western Canada, and for salt domes such as occur on the Gulf Coast, and where large vertical displacements due to faulting have occurred.

α) It may be shown that the *first vertical derivative* of the total field, $\partial \Delta T/\partial z$ at height h over a dipping contact [1] is given by

$$\frac{1}{J\,bc\,\sin\theta}\frac{\partial \Delta T}{\partial z} = -\frac{2(h\cos\alpha + x\sin\alpha)}{h^2 + x^2}. \tag{11.1}$$

J is the intensity of magnetization due to remanent and induced magnetism; the argument

$$\alpha = (\lambda + \psi - \theta), \tag{11.2}$$

where

$$\tan\lambda = \frac{\tan I}{\cos A}, \tag{11.3}$$

$$\tan\psi = \frac{\tan i}{\cos a}, \tag{11.4}$$

and θ is the dip of the contact. The angles I and A are the dip and declination of the Earth's magnetic field T referred to the axes shown in Fig. 27 and the angles i and a are the inclination and declination of the resultant magnetization $\boldsymbol{J}$ referred to the same axes. Also

$$b = (\sin^2 I + \cos^2 I \cos^2 A)^{\frac{1}{2}}, \tag{11.5}$$

and

$$c = (\sin^2 i + \cos^2 i \cos^2 a)^{\frac{1}{2}}. \tag{11.6}$$

Hence

$$\frac{h}{J\,bc\,\sin\theta}\frac{\partial \Delta T}{\partial z} = -\frac{2(\cos\alpha + X\sin\alpha)}{1 + X^2}, \tag{11.7}$$

where $X \equiv x/h$ is the normalized abscissa value.

This formula has been obtained from the expression for the vertical gradient over a wide dipping dyke, see Eq. (11.25), with an expression consisting of two terms, and each of the terms is negligible at the other contact. It is, therefore, a simple matter to obtain the expression

---
[1] P. Hood: Geophysics **30**, 891—902 (1965).

Vertical gradient theory.

for a dipping contact by changing the origin from the centre of the dyke to its edge after removing the appropriate term.

Fig. 24 shows a normalized family of curves for the vertical gradient over a dipping contact for values of the argument α at 15° intervals. It is easy to show from Eq. (11.1) that

$$\left(\frac{\partial \Delta T}{\partial z}\right)_{X,\alpha} = -\left(\frac{\partial \Delta T}{\partial z}\right)_{-X, 180-\alpha}$$
$$= -\left(\frac{\partial \Delta T}{\partial z}\right)_{X, 180+\alpha} \qquad (11.8)$$
$$= \left(\frac{\partial \Delta T}{\partial z}\right)_{-X, 360-\alpha}.$$

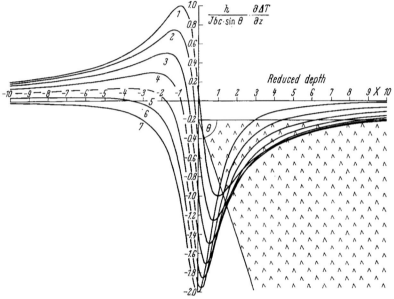

Fig. 24. Vertical gradient curves over the dipping geological contact. (After HOOD, 1965.) [See text and Table 3 for explanation.]

Table 3. *Values of the argument α for the vertical gradient curves in Fig. 24. Induced only; $\theta = 90°$, $A = 0°$, $I = (\alpha + \theta)/2$.*

Curve	$\frac{\partial \Delta T}{\partial z}$	$-\frac{\partial \Delta T}{\partial z}$	$\frac{\partial \Delta T}{\partial z}$	$\frac{\partial \Delta T}{\partial z}$	$\frac{\partial \Delta T}{\partial z}$	$\frac{\partial \Delta T}{\partial z}$	$-\frac{\partial \Delta T}{\partial z}$	$\frac{\partial \Delta T}{\partial z}$
	$+x$	$-x$	$+x$	$-x$	$+x$	$-x$	$+x$	$-x$
	α	α	α	α	I	I	I	I
1	90°	90°	270°	270°	90°	$-90°$	0°	0°
2	75°	105°	255°	285°	$82\frac{1}{2}°$	$-82\frac{1}{2}°$	$-7\frac{1}{2}°$	$7\frac{1}{2}°$
3	60°	120°	240°	300°	75°	$-75°$	$-15°$	15°
4	45°	135°	325°	315°	$67\frac{1}{2}°$	$-67\frac{1}{2}°$	$-22\frac{1}{2}°$	$22\frac{1}{2}°$
5	30°	150°	210°	330°	60°	$-60°$	$-30°$	30°
6	15°	165°	195°	345°	$52\frac{1}{2}°$	$-52\frac{1}{2}°$	$-37\frac{1}{2}°$	$37\frac{1}{2}°$
7	0°	180°	180°	0°	45°	$-45°$	$-45°$	45°

Table 3 shows the equivalent values of α for which the curves are applicable. The negative signs for $(\partial \Delta T/\partial z)$ and X change the signs of the values on the axes. The right-hand side of the table is for the special case where the remanent magnetization is aligned along the Earth's magnetic field or is negligible with respect to the induced magnetization, i.e., $i = I$ and $a = A$, and for an east-west striking vertical contact, i.e., $A = a = 0$ and $\theta = 90°$.

It is readily apparent from Fig. 24 that at high magnetic latitudes ($\alpha = 90°$) and near the magnetic equator ($\alpha = -90°$), steeply dipping contacts are outlined by the zero-gradient contour where the remanent magnetism does not cause the resultant magnetization J to be at a large angle to Earth's magnetic field. It should be noted from Eqs. (11.3, 4) that $\lambda > I$ and $\psi > i$, so that the condition that the contact strike east-west does not apply.

β) It is easy to show that the positive maximum and the negative *maximum values of the gradient* occur at the points.

$$\frac{h}{J b c \sin \theta} \left(\frac{\partial \Delta T}{\partial z}\right)_{+\max} = 2 \sin^2 \frac{\alpha}{2} \tag{11.9}$$

when

$$X_{+\max} = -\cot \frac{\alpha}{2}, \tag{11.10}$$

and

$$\frac{h}{J b c \sin \theta} \left(\frac{\partial \Delta T}{\partial z}\right)_{-\max} = -2 \cos^2 \frac{\alpha}{2} \tag{11.11}$$

when

$$X_{-\max} = \tan \frac{\alpha}{2}. \tag{11.12}$$

The line $(\partial \Delta T/\partial z) = 0$ will be known from a calibration of the aeromagnetic gradiometer, so that using Eqs. (11.9) and (11.11) the argument α may be obtained from

$$\left(\frac{\partial \Delta T}{\partial z}\right)_{+\max} \div \left(\frac{\partial \Delta T}{\partial z}\right)_{-\max} = -\tan^2 \frac{\alpha}{2}. \tag{11.13}$$

It also follows from Eqs. (11.10) and (11.12) that

$$X_{-\max} - X_{+\max} = \tan \frac{\alpha}{2} + \cot \frac{\alpha}{2} = \frac{2}{\sin \alpha}.$$

This means in the field case that the *depth to the top of the contact* is given by

$$h = \frac{(x_{-\max} - x_{+\max}) \sin \alpha}{2} \tag{11.14a}$$

or

$$h = (x_{-\max} - x_{+\max}) \frac{\tan \alpha/2}{1 + \tan^2 \alpha/2}. \tag{11.14b}$$

The *position of the top* of the vertical face of the contact may then be obtained from Eq. (11.12)

$$x_{-\max} = h \tan \frac{\alpha}{2}. \tag{11.15}$$

Actually the line joining the maximum and minimum values of $\partial \Delta T/\partial z$ crosses the curve at the ordinate. This property of the dipping contact curves is true for all values of α and may be proven mathematically.

γ) Alternatively, since the $x_{+\max}$ and $x_{-\max}$ distances are often poorly defined on the magnetic profile, the *cross-over point* on the X axis may be used. Thus at $\partial \Delta T/\partial z = 0$, and using Eq. (11.1), the distance x_0 between the cross-over point and the origin is given by

$$x_0 = h \cot \alpha,$$

i.e.,

$$x_0 = h \frac{\tan^2 \alpha/2 - 1}{2 \tan \alpha/2}. \tag{11.16}$$

For the dipping contact we have that

$$\frac{1}{Jbc\sin\theta}\frac{\partial \Delta T}{\partial x} = 2\frac{(h\sin\alpha - x\cos\alpha)}{h^2 + x^2}. \tag{11.17}$$

Combining Eqs. (11.1) and (11.17),

$$x\frac{\partial \Delta T}{\partial x} + h\frac{\partial \Delta T}{\partial z} = -2Jbc\sin\theta\cos\alpha, \tag{11.18}$$

hence

$$\frac{\partial \Delta T}{\partial z} = -\frac{1}{h}\left(x\frac{\partial \Delta T}{\partial x}\right) - \frac{2Jbc\sin\theta\cos\alpha}{h}. \tag{11.19}$$

Plotting $(\partial \Delta T/\partial z)$ as the ordinate, and $(x\,\partial \Delta T/\partial x)$ as the abscissa using Cartesian coordinates will give a straight line whose slope is $-1/h$, and whose intercept on the ordinate will be $-(2Jbc/h)\sin\theta\cos\alpha$. For the case of curve 1 ($\alpha = 90°$) in Fig. 24, the straight line will pass through the origin since $\cos\alpha = 0$. This is, therefore, an alternative method of obtaining the value of h, although the position of the origin must be determined.

The position of the origin may be obtained by the following consideration. At the peak of the ΔT curve, $\dfrac{\partial \Delta T}{\partial x} = 0$, so that from Eq. (11.18)

$$\frac{\partial \Delta T}{\partial z} = \frac{2Jbc\sin\theta\cos\alpha}{h}. \tag{11.20}$$

Also from consideration of Eq. (11.18) this is also the value of the vertical gradient at $x=0$ i.e. the origin. It is therefore only necessary to ascertain this value on the vertical gradient curve to establish the position of the origin, which for the dipping contact case always lies between the points where the vertical $(\partial \Delta T/\partial z)$ and horizontal $(\partial \Delta T/\partial x)$ gradients are zero.

Substituting Eqs. (11.9), (11.10) into Eq. (11.1) we obtain

$$\frac{h}{Jbc\sin\theta}\left[\left(\frac{\partial \Delta T}{\partial z}\right)_{+\max} - \left(\frac{\partial \Delta T}{\partial z}\right)_{-\max}\right] = 2, \tag{11.21}$$

which relationship is readily apparent from Fig. 24.

δ) This equation may be used to decide on a *desirable sensitivity* for an aeromagnetic gradiometer. Let us consider the case of steeply dipping contacts at high magnetic latitudes where the resultant magnetization of the rock units has a high inclination, then the factor $bc\sin\theta$ will approach the value of unity.

Then

$$\left[\left(\frac{\partial \Delta T}{\partial z}\right)_{+\max} - \left(\frac{\partial \Delta T}{\partial z}\right)_{-\max}\right] = 2\frac{J}{h}, \tag{11.21a}$$

and Fig. 25 has been drawn to illustrate the variation of the maximum values of the vertical gradient with the resultant magnetization J and the height h between the aircraft and the top of the contact. It may also be used for the horizontal-gradient case. Now the resultant magnetization $\boldsymbol{J}$ is the vector sum of the induced magnetization $k\boldsymbol{T}$, where k is the susceptibility contrast across the contact, and the remanent magnetization $\boldsymbol{R}$, i.e.,

$$\boldsymbol{J} = k\boldsymbol{T} + \boldsymbol{R} \tag{11.22a}$$

and

$$\boldsymbol{J} = k_e\,\boldsymbol{T}, \tag{11.22b}$$

where k_e is the effective susceptibility of the rock units due to both remanent and induced magnetization.

Thus, for a susceptibility contrast of $k_e = 4 \cdot 10^{-4}$, for a vertical separation of 500 ft (152 m) between the aeroplane and the top of the vertical contact and using Fig. 25, we have $T = 2 \cdot 10^{-4} \Gamma$

$$\left(\frac{\partial \Delta T}{\partial z}\right)_{+\max} - \left(\frac{\partial \Delta T}{\partial z}\right)_{-\max} = 0.08 \ \gamma/\text{ft} = 0.27 \ \gamma/\text{m}$$

If the vertical separation of the magnetometer heads forming the gradiometer were 50 ft. (15 m), the maximum difference in the readings would be 4 γ. There is a distinct advantage to "biassing" one of the heads so that no polarity changes of the vertical gradient occur; this arrangement, which was previously described would increase the measured difference but the maximum variation for the above case would still be 4 γ.

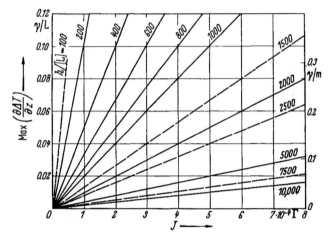

Fig. 25. Graph showing the maximum vertical gradient over the dipping geological contact plotted against the resultant magnetization, J, for various depths, h, to the top of the contact. (After Hood, 1965.) The lower abscissa scale gives the "effective susceptibility", k_e, defined by $k_e = J/T$ for $T = 50,000$ γ, the average value on Earth. [L = unit of length in ft.]

ε) EULER's *Differential Equation* is a useful tool in gradiometer measurements. It applies when the mathematical expression for the anomaly produced by a given geometrical model is a homogenous function, such as that due to a point pole, point dipole, or thin dyke. Thus

$$x \frac{\partial \Delta T}{\partial x} + y \frac{\partial \Delta T}{\partial y} + h \frac{\partial \Delta T}{\partial z} = -n \Delta T \qquad (11.23)$$

and

$n = 1$ for the thin dyke,
$\quad = 2$ for the point pole,
$\quad = 3$ for the point dipole.

At

$$\Delta T_{\max}, \quad \frac{\partial \Delta T}{\partial x} = 0 = \frac{\partial \Delta T}{\partial y},$$

hence it follows that

$$\left(\frac{\partial \Delta T}{\partial z}\right)_{\max} = -\frac{n \Delta T_{\max}}{h}. \qquad (11.24)$$

The reader should note that n is the fall-off exponent of the total field anomaly which is of the general form $\Delta T = K/r^n$ where K is a constant involving a magnetization contrast, and r is distance from the causative body. Fig. 26 shows how the maximum gradient varies with $\Delta T_{\max}$ and h for the point pole, point dipole,

Sect. 11. Vertical gradient theory.

and dipping dyke cases. It can also be used to obtain h from a measurement of ΔT_{max} and $(\partial \Delta T/\partial z)_{max}$, because

$$h = \frac{-n \Delta T_{max}}{\left(\frac{\partial \Delta T}{\partial z}\right)_{max}}. \tag{11.24a}$$

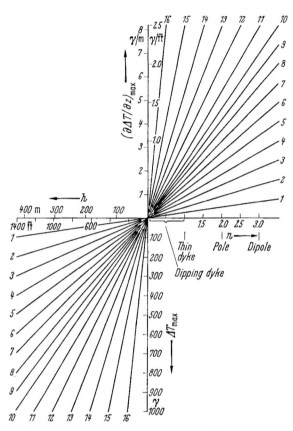

Fig. 26. Graph showing relationship between EULER's n, vertical gradient, $\partial \Delta T/\partial z_{max}$, maximum total intensity, ΔT_{max}, and depth to causative body, h, for various geometrical models. (After HOOD, 1965.)

Thus, at 500 ft (152 m) separation between the aeroplane and the causative body producing a maximum total-intensity anomaly of 500 γ, the maximum gradient will be 2 γ/ft (6.6 γ/m) for the point pole and 3 γ/ft (10 γ/m) for the point dipole (curve 12 in Fig. 26). Hence, the maximum difference in readings at two heads separated by 50 ft (15 m) would be 100 and 150 γ, respectively, for the foregoing cases. Curves for the vertical gradient over the point pole and the finite dipole have been published[2]. Actually, EULER's equation does not apply to the finite-dipole case, and moreover consideration of Eq. (11.18) shows that the left-hand side of Eq. (11.23) goes to zero at high-magnetic latitudes for the dipping-contact case.

ζ) The expression for the *dipping dyke* case, which terminology is here used to denote geometrical shape only and does not necessarily have a geological

---

[2] P. J. HOOD, and D. J. McCLURE: Geophysics 30, 403—410 (1965).

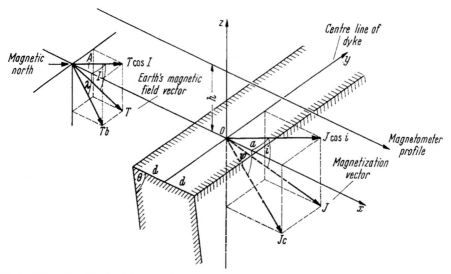

Fig. 27. Oblique view of dipping dyke model showing nomenclature used in the equation for the total intensity anomaly.

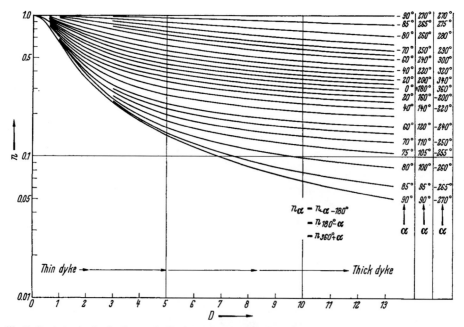

Fig. 28. Graph showing family of curves for EULER's n plotted against the half-width, D, in depth units of a dipping dyke for various values of the parameter α. (After HOOD and SKIBO, 1968.)

genesis connotation, is

$$\frac{h}{J}\frac{\partial \Delta T}{\partial z} = 2b\ c\ \sin\theta \left[\frac{\cos\alpha + (X-D)\sin\alpha}{1+(X-D)^2} - \frac{\cos\alpha + (X+D)\sin\alpha}{1+(X+D)^2} \right] \quad (11.25)$$

using the notation of Fig. 27, where $D = d/h$, d being the half-width of the top of the dyke and $X = x/h$. It can be seen that this expression is not homogenous, so that EULER's differential equation is not strictly valid although a value of n can, of course, be obtained. The values of n calculated for the dipping dyke vary between 0 and 1, see Fig. 28[3], depending on the values of α and D. For small values of D, which is the thin dyke case, $n \approx 1$.

SLACK et al.[4] consider that anomalies with $n = 2$ or less will be mostly due to intrabasement intrusive-type features whereas those with $n > 2$ are due to suprabasement features and arise principally when the basement is uplifted. Thus the latter offer the distinct possibility of oil accumulation in the overlying sediments.

12. Gradiometer field results. Fig. 29 shows the total intensity ΔT and vertical gradient $\partial \Delta T/\partial z$ profiles over iron formation in the Labrador Trough obtained using a rubidium gradiometer flown in the North Star aircraft of the Canadian National Aeronautical Establishment[1].

Positioning of the profile AB was by celestial and DOPPLER navigation, so it is to be expected that some error exists between the plotted and actual aircraft track. Thus the magnetic and geological data cannot be compared in detail. However, several features are

Fig. 29. Aeromagnetic total intensity and gradiometer profiles across the Labrador Trough, E. Canada. (After HOOD and BOWER, 1967.)

apparent in the profiles. The iron formation immediately to the south of Kaniapiskau river gives the highest total-intensity anomaly (1,110 γ) and this corresponds to the highest gradient anomaly (0.138 γ/ft i.e. 0.454 γ/m). The iron formation to the south also gives distinct anomalies. The resolution of separate anomalies is also very much better for the vertical gradient profile. Where inflections occur on the sides of the total-intensity anomalies, these are often resolved into two separate anomalies on the vertical gradient profile. Perhaps the

---
[3] P. J. HOOD, and D. N. SKIBO: Vertical gradient studies; dipping dyke case and EULER's differential equation. Geol. Surv. Canada, Paper 68-1B, 14—18 (1968).

[4] H. A. SLACK, V. M. LYNCH, and L. LANGAN: Geophysics 32, 877—892 (1967).

[1] P. HOOD, and M. E. BOWER: Non-military use of high resolution magnetometers. DRB Symposium on Magnetic Anomaly Detection, Ottawa, 1967.

most surprising fact concerning this profile is that the terrain clearance of the aircraft was 8,000 ft (2,440 m) as measured on a radar altimeter. Actually a component of the horizontal gradient was measured along with the vertical gradient because of the relative horizontal displacements between the magnetometer heads.

Fig. 30 shows the total magnetic field and vertical gradient maps for the Coyanosa Field, Pecos County, Texas (USA)[2]. It can be seen that the total intensity map indicates little or no

Fig. 30 a and b. a) Total magnetic intensity map; b) Vertical magnetic gradient map, Coyanosa Field, Pecos County, Texas. (After SLACK et al., 1967.)

concomitant anomaly associated with the oilfield whose areal extent is indicated by the stipling. However for the vertical gradient map the perimeter of the field corresponds in a general way with the zero gradient contour, except for the eastern portion, and the calculated value of EULER's n was 2.1, indicating that the causative body was of the point pole type[*].

[2] H. A. SLACK, V. M. LYNCH, and L. LANGAN: Geophysics **32**, 877—892 (1967).

[*] The author of this article wishes to thank E. A. GODBY, R. C. BAKER, and W. M. STROME of the National Aeronautical Establishment, Ottawa, and Margaret E. BOWER and D. N. SKIBO of the Geological Survey of Canada for their essential contributions to the acquisition of some of the results presented in this article. The author also wishes to thank the authors of the articles in which Figs. 8, 13, 14, 21, 23 and 30 appeared for permission to use them in this article.

Phénomènes T.B.F. d'origine magnétosphérique.

Par

R. Gendrin*.

Avec 36 Figures.

Les phénomènes dont nous donnons ici une description succinte, constituent une illustration saisissante des théories relatives à la propagation des ondes dans un magnétoplasma [1]–[4]. Cette illustration est d'autant plus frappante qu'il s'agit de phénomènes naturels dont les caractéristiques (fréquence, polarisation, direction du vecteur d'onde) couvrent plus ou moins l'ensemble des cas particuliers envisagés par la théorie. Inversement, l'application de celle-ci aux phénomènes observés, permet de calculer certaines grandeurs définissant le milieu qui entoure notre planète, jusqu'à des distances de l'ordre de huit rayons terrestres.

1. Les phénomènes T.B.F. naturels dans le plasma magnétosphérique.

α) La magnétosphère est constituée principalement d'un *champ magnétique* qui est, en première approximation, celui d'un dipôle, et d'un *plasma* presque *complètement ionisé*, constitué d'électrons et de protons en premier lieu.

La ligne de force du champ magnétique est définie par le paramètre L (ou paramètre de Mac Ilwain) qui, dans l'approximation dipolaire, est la mesure, en rayons terrestres, de la distance au centre de la terre de son point le plus éloigné. Si ϕ est la latitude d'un point quelconque de la ligne de force et R sa distance au centre de la terre (en rayons terrestres), l'intensité du champ magnétique est donnée par**

$$B = B_0 \cdot \frac{(1 + 3\sin^2\phi)^{\frac{1}{2}}}{R^3} \qquad (1.1)$$

avec

$$B_0 = 3{,}1 \cdot 10^{-5}\ \text{T}. \qquad (1.2)$$

Nous désignons par N la densité numérique des électrons ou ions, q et m étant leur charge et masse. *Le plasma*[1] est défini par les gyrofréquences, électronique ou ionique,

$$f_B = \frac{|q_e|}{2\pi m_e} \frac{B_\circ}{c_0\sqrt{\varepsilon_0\mu_0}}. \qquad (1.3)$$

Soit

$$f_B/\text{Hz} = 2{,}8 \cdot 10^{10}\ B/\text{T}, \qquad (1.3\text{a})$$

$$F_B = f_B/m^{*\prime}\ {*{*}*} \qquad (1.3\text{b})$$

et les fréquences de plasma, électronique ou ionique†,

$$f_N^2 \equiv \frac{u\ q^2\ N}{(2\pi)^2\ \varepsilon_0\ m_e} \quad \text{et} \quad F_N^2 \equiv \frac{u\ q^2\ N}{(2\pi)^2\ \varepsilon_0\ m_P}.$$

---

* Manuscrit reçu en juin 1967.
** L'unité dans le Système International est: $1\ \text{T} \equiv 1\ \text{Tesla} = 10^4\ \text{Gs}$.
*** Nous définissons $m^* = m_P/m_e \sim 1840$ comme rapport des masses du proton et de l'électron.
† $u = 1$ dans un système d'unités rationalisées, mais $= 4\pi$ dans un système non-rationalisé. Voir les remarques générales p. 1.

[1] K. Rawer et K. Suchy: Cette encyclopédie, tome 49/2, p. 2. — V. L. Ginzburg et A. A. Ruhadze: Contribution dans le tome 49/4 de cette encyclopédie.

Introduisant la fréquence effective de plasma

$$f_p^2 = f_N^2 + F_N^2, \tag{1.4}$$

on trouve

$$(f_p/\text{Hz})^2 \approx 81 \; N/\text{m}^{-3}. \tag{1.4a}$$

Dans la magnétosphère, la répartition de la densité électronique en différents points est très voisine du «modèle magnétique» dans lequel la densité électronique est proportionnelle au champ magnétique, en sorte que l'on a

$$f_p^2 \approx f_a \cdot f_B \tag{1.5}$$

ou f_a est de l'ordre du MHz. Nous définissons alors le rapport entre la fréquence de plasma et la gyrofréquence électronique par

$$\alpha \equiv f_N/f_B.$$

En conséquence, *le plasma est dense* [1] puisque la quantité

$$A = (1 + m^*)\alpha^2 \gg 1. \tag{1.6}$$

Il est également confiné, car la température T (comprise entre 1000 et 10000 °K) est telle que l'énergie cinétique thermique du plasma (décrite par la vitesse thermique moyenne des particules V)

$$\mathcal{E}_c = \tfrac{1}{2} N \, m_e \, V_e^2 + \tfrac{1}{2} N \, m_p \, V_p^2 \tag{1.7a}$$

est faible devant son énergie magnétique

$$\mathcal{E}_m = \frac{B^2}{2 \, u \, \mu_0}, \tag{1.7b}$$

$$\mathcal{E}_c/\mathcal{E}_m \ll 1. \tag{1.7c}$$

Enfin, pour des altitudes supérieures à 150 km, les *fréquences de collision sont négligeables* par rapport aux fréquences qui nous intéressent.

On voit donc que l'on a affaire à un plasma très particulier dont le Tableau 1 résume les caractéristiques principales en fonction de la ligne de force étudiée [2].

Tableau 1. *Ordres de grandeur des divers paramètres magnétosphériques dans le plan équatorial, pour diverses altitudes*

	À l'altitude de la couche F_2	$L=2$	$L=4$	$L=8$*
N/m^{-3}	10^{12}	$1{,}4 \cdot 10^9$	$1{,}7 \cdot 10^8$	$2{,}1 \cdot 10^7$
B_δ/T	$3{,}1 \cdot 10^{-5}$	$3{,}9 \cdot 10^{-6}$	$4{,}9 \cdot 10^{-7}$	$6 \cdot 10^{-8}$
f_N/MHz	9	0,33	0,12	0,04
f_B/kHz	880	109	13,8	1,7
F_B/Hz	455	60	7,5	0,9
$\alpha = f_N/f_B$	10,2	3,0	8,7	23,6
V_A/ms^{-1}	$6{,}9 \cdot 10^5$	$2{,}3 \cdot 10^6$	$8{,}0 \cdot 10^5$	$3 \cdot 10^5$
$\gamma = \mathcal{E}_c/\mathcal{E}_m$**	$2{,}2 \cdot 10^{-4}$	$1{,}8 \cdot 10^{-5}$	$1{,}5 \cdot 10^{-4}$	$1{,}2 \cdot 10^{-3}$
ν/s^{-1}***	~ 50	—	—	—

* Les chiffres donnés pour cette valeur de L ne tiennent pas compte de l'existence de la faille d'ionisation (ou plasmapause) au voisinage de $L = 6$ (voir Sect. 3).
** Calculée dans l'hypothèse où le seul constituant ionique est l'hydrogène.
*** Pour une température de 2000° K.

---

[2] R. Gendrin: Rev. CETHEDEC (Paris) **3**, 1 (1966).

β) *Il y a en fait deux catégories de phénomènes T.B.F. naturels.* Les phénomènes que l'on se propose de décrire comportent les *sifflements radioélectriques* et les *émissions radioélectriques* de très basse fréquence (0,1 ... 30 kHz).

Les phénomènes de la *première catégorie* sont la conséquence de la propagation dans la magnétosphère d'une partie de l'énergie radioélectrique rayonnée par les éclairs atmosphériques. L'instant origine et la nature spectrale de l'éclair initiateur étant le plus souvent connus, il est possible de calculer, à partir des enregistrements obtenus, le temps de propagation à chaque fréquence et *d'en déduire des caractéristiques relatives au milieu traversé par l'onde*. On distingue les sifflements observés au sol de ceux détectés en satellite. Les premiers font presque exclusivement intervenir la propagation quasi longitudinale (parallèle au champ magnétique); la dispersion observée (différence entre les temps de propagation de chaque fréquence) est due principalement à la contribution électronique. Les seconds sont très influencés par la composition ionique du milieu traversé; ils permettent de mettre en évidence un certain nombre d'effet dûs à la propagation transverse.

Les phénomènes de la *seconde catégorie* ont une origine purement magnétosphérique et sont vraisemblablement dûs à des interactions entre le plasma ambiant et les faisceaux des particules énergétiques qui sont présents dans la magnétosphère. Bien que le choix n'ait pas été définitivement fait entre les différents mécanismes qui peuvent être envisagés comme responsables de ces émissions, l'étude de ces phénomènes doit permettre d'acquérir un certain nombre de *donnés relatives aux flux des particules énergétiques qui traversent la magnétosphère*.

γ) Notre *présentation* ne suivra pas un exposé chronologique des différents types de sifflements observés [5]. Notre point de départ sera l'équation de propagation des ondes dans un magnétoplasma, assortie des conditions suivantes:
— le plasma est dense, la relation (1.6) est vérifiée.
— il est confiné suivant l'inéquation (1.7); les effets de température seront généralement négligés.
— il est sans collision.
— les fréquences étudiées sont inférieures à la gyrofréquence électronique, elle-même inférieure à la fréquence de plasma ($\alpha > 1$).

A partir de l'importance relative des différents termes de cette équation, nous déduirons une classification des divers types de sifflements dont quelques caractéristiques seront rassemblés dans un tableau (Sect. 2). La description des sifflements observés au sol ou de ceux qui ont été détectés en satellite sera faite dans les deux sections suivantes. Nous montrerons comment il est possible d'en déduire des profils de densité électronique (Sect. 3) ou d'évaluer la composition ionique de la basse magnétosphère (Sect. 4). Les phénomènes d'anisotropie et de guidage seront évoqués à la Sect. 5. Nous donnerons ensuite un aperçu sur les émissions T.B.F. naturelles (Sect. 6). La Sect. 7 sera consacrée à des remarques sur les méthodes expérimentales d'analyse, particulières à ce type de phénomènes. Les problèmes relatifs à la polarisation des ondes seront abordés dans la Sect. 8 tandis que la théorie détaillée des sifflements ioniques sera donnée en Annexe (Sect. 9).

2. Equations générales et classification.

La théorie complète de la dispersion des ondes dans un magnétoplasma à plusieurs composants ioniques se trouve dans l'article de RAWER et SUCHY [3]. Il nous suffit de rappeler ici quelques unes des définitions et formules qui nous seront utiles par la suite et qui peuvent différer légèrement de celles utilisées dans la référence [3]. En effet, nous avons préféré utiliser quelques unes des notations de SMITH et BRICE [6] dont l'article a servi de base à un grand nombre de travaux sur la propagation des ondes T.B.F. dans la magnétosphère et la basse ionosphère. Afin de faciliter la comparaison avec les autres contributions nous écrirons la plupart des équations dans les deux systèmes de notation.

Remarquons tout de suite que, pour les fréquences inférieures à la gyrofréquence électronique, le plasma devient de nouveau *transparent*: l'ionosphère ne joue plus le rôle d'écran vis-à-vis des ondes radioélectriques qui proviennent de l'exosphère[1]. En revanche, des effets

---
[1] K. RAWER et K. SUCHY: Cette encyclopédie, tome 49/2, p. 84.

nouveaux de *résonance* apparaissent, qui sont d'une grande utilité pour la détermination de la densité et de la composition du plasma magnétosphérique.

Nous ne tiendrons pas compte de l'effet éventuel des *ions négatifs* sur la propagation[2].

α) *Notations*. On pose*

$$X_h = \frac{u q_h^2 N_h}{\varepsilon_0 m_h} \cdot \frac{1}{\omega^2}; \quad Y_h = \frac{q_h}{m_h} \frac{B_0}{c_0 \sqrt{\varepsilon_0 \mu_0}} \cdot \frac{1}{\omega}. \tag{2.1}$$

N_h, q_h, m_h étant la densité numérique, la charge (avec son signe) et la masse de la h-ième espèce de particules.

Si l'on néglige la contribution des ions négatifs et si i est un indice s'appliquant seulement aux ions, on introduit les masses relatives m_i^* des ions par rapport à celle de l'électron, ainsi que la *composition* fractionnelle A_i de chaque espèce *d'ion*.

$$N_i = N_e A_i \quad \text{avec} \quad \sum_i A_i = 1. \tag{2.2}$$

On a donc:

$$X_i = \frac{A_i X_e}{m_i^*} \quad Y_i = -\frac{Y_e}{m_i^*} \tag{2.3}$$

avec $X_e = \left(\frac{\omega_N}{\omega}\right)^2$; $Y_e = -\frac{\omega_B}{\omega}$.

β) *Modes principaux*. Il existe trois modes limites de propagation, dans lesquels n'intervient aucune fluctuation de densité électronique:

$$\nabla \cdot \boldsymbol{E} \equiv \text{div } \boldsymbol{E} = 0 \quad \text{ou} \quad \boldsymbol{k} \cdot \boldsymbol{E} = 0, \tag{2.4}$$

$\boldsymbol{k}$ étant le vecteur d'onde.

Il s'agit de deux modes polarisés circulairement en sens inverses, dont les vecteurs électrique et magnétique sont perpendiculaires à $\boldsymbol{B}_0$, et d'un mode polarisé linéairement (parallèlement à $\boldsymbol{B}_0$). Ces trois modes sont désignés par le suffixe s ($= +1, -1$, ou 0) et ont des indices de phase définis par les équations

$$\mu_s^2 = 1 - \sum_h \frac{X_h}{s Y_h + 1}. \tag{2.5}$$

Leur vitesse de phase est

$$W_s = c_0/\mu_s. \tag{2.6}$$

Si l'on abandonne la condition (2.4), c.à.d. en admettant des fluctuations de densité électronique, on trouve l'équation de dispersion générale, ou équation des tangentes[3]:

$$(W^2 - W_e^2)(W^2 - W_0^2) \sin^2 \Theta + (W^2 - W_{+1}^2)(W^2 - W_{-1}^2) \cos^2 \Theta = 0 \tag{2.7}$$

où l'on a introduit

$$W_e^2 = \tfrac{1}{2}(W_{+1}^2 + W_{-1}^2). \tag{2.8}$$

Sous cette forme, on voit que:

— pour une propagation strictement parallèle au champ magnétique («longitudinale» $\Theta = 0$), on a la propagation des deux *modes* circulaires *droit* (**R** wave) et *gauche* (**L** wave)

— pour une propagation perpendiculaire au champ magnétique («transversale» $\Theta = \pi/2$), on retrouve bien le mode ordinaire ($s = 0$); mais, *aux fréquences qui nous*

---

* $u = 1$ en unités rationalisées, mais $= 4\pi$ en unités non-rationalisées. (Voir les remarques générales, p. 1.) Remarquer aussi que Y est une quantité négative pour les électrons, alors que f_B est une quantité positive, de sorte que $x = f/f_B = -1/Y_e$.

[2] J. SMITH: J. Geophys. Res. **70**, 53 (1965).
[3] K. RAWER et K. SUCHY: Cette encyclopédie, tome 49/2, p. 33.

Sect. 2. Equations générales et classification.

intéressent ($f < f_B$), et compte tenu du fait que dans la magnétosphère $\alpha = X_e/Y_e^2$ est toujours plus grand que 1, l'indice de phase de ce mode est imaginaire. Il ne reste donc que le mode dont la vitesse de phase W_e est défine par l'équation (2.8) et que nous appellerons *mode extraordinaire ou transverse**.

γ) *Courbes de dispersion.* La Fig. 1 représente la variation du carré de la vitesse de phase en fonction de la fréquence réduite $x = f/f_B \equiv 1/|Y_e|$, dans le cas d'un mélange de deux ions (66% d'hydrogène et 33% d'hélium)[4]. On a choisi la valeur 2,5 pour le rapport α.

Fig. 1. Variation de la vitesse de phase réduite $(1/\mu^2)$ en fonction de la fréquence réduite $(x=1/|Y_e|=f/f_B)$ pour un plasma à deux constituants ioniques, à savoir H+ (66%) et He+ (33%); $\alpha=X^{\frac{1}{2}}/|Y_e|=f_N/f_B$ précise la densité électronique réduite. On a représenté les deux modes qui se propagent parallèlement au champ magnétique, de polarisation circulaire droite (R) et gauche (L) et le mode extraordinaire (e) qui se propage transversalement. En pointillé, l'approximation d'ECKERSLEY-STOREY pour le mode droit. Les zônes hachurées représentent les quatre régions du plan liées à quatre types de phénomènes exosphériques particuliers. (Adapté de SMITH et BRICE [6].)

On remarque l'existence de *fréquences de résonance* ($W = 0$) et de *fréquences de coupure* ($W = \infty$). Les résonances ont lieu, en propagation longitudinale, aux gyrofréquences des électrons (pour le mode droit) et des ions (pour le mode gauche). En propagation transversale, il s'agit des résonances hybrides dont la fréquence est fonction de la fréquence de plasma, de la gyrofréquence et des concentrations des différents ions. On remarque de plus l'existence d'une *fréquence de croisement* f_c, à laquelle les vitesses des 3 modes (**R**, **L**, et e) sont égales; des transitions entre les modes peuvent se produire au voisinage de cette fréquence.

---

* Nous utilisons les termes «longitudinal» et «transversal» par rapport an champ magnétique, pas par rapport à la normale de l'onde. Cet emploi se distingue de la nomenclature «optique» préférée par K. RAWER et K. SUCHY dans leur contribution au tome 49/2 de cette encyclopédie. L'onde dite «transverse» est d'un type quasi-acoustique.

[4] Voir référence [6] pour le cas d'un mélange de trois ions.

Tableau 2. *Classification des différents types de sifflements T.B.F.*

	Dénomination des sifflements	Nature	Lieu d'observation	Figure	Forme spectrale	Explication	Utilisation	Réf.
a	A un bond (One-Hop Whistler)	électronique (electronic)	au sol (ground)			Le sifflement est reçu dans l'hémisphère opposé de celui où a eu lieu l'éclair. L'énergie a suivi approximativement une ligne de force.		(4), [7]
b	A deux bonds (Two-Hop Whistler)	électronique (electronic)	au sol (ground)			Le sifflement, après réflection dans l'hémisphère opposé, est reçu dans le même hémisphère que celui où a eu lieu l'éclair.	Détermination de la densité électronique en des régions éloignées	
c	A trajets multiples (Multi-Path Whistler)	électronique (electronic)	au sol (ground)			L'énergie engendrée par un même éclair se propage suivant des trajets légèrement différents.		
d	A "nez" (Nose Whistler)	électronique (electronic)	au sol (ground)			Le sifflement emprunte une ligne de force de haute latitude. L'effet de la gyrofréquence électronique minimale sur la dispersion se fait sentir.	Détermination précise de la ligne de force suivie par le sifflement et donc meilleur calcul des densités électroniques	(5), (6)
e	"Coudé" (Knee Whistler)	électronique (electronic)	au sol (ground)			Le sifflement emprunte deux types de trajets dont l'un, à l'intérieur de la plasmasphère correspond à des régions de haute densité électronique, tandis que l'autre correspond à des régions de faible densité.	Etude de la plasmapause	(7)
f	A trajet partiel (Fractional Hop Whistler)	électronique (electronic)	en satellite (satellite)			Le sifflement ne parcourt qu'une portion de trajet magnétosphérique.		(8)
g	Subprotonosphérique	électronique (electronic)	en satellite (satellite)			Le sifflement se réfléchit d'une part sur les irrégularités de la couche E, d'autre part au niveau de la protonosphère où la variation rapide de la composition provoque un gradient important d'indice.		(9)

Sect. 2. Equations générales et classification. 467

	Dénomination des sifflements	Nature	Lieu d'observation	Figure	Forme spectrale	Explication	Utilisation	Réf.
h	Protonique (Proton Whistler)	ionique (ionic)	en satellite (satellite)			Le spectre présente une trace horizontale pour une fréquence égale à la gyrofréquence des protons au niveau du satellite.	Détermination de la composition et de la température ionique au niveau du satellite	(10)
i	Hélionique (Helium Whistler)	ionique (ionic)	en satellite (satellite)			Même phénomène pour la gyrofréquence de l'hélium.		(11)
j	Transverse (transverse)	transverse (transverse)	en satellite (satellite)			Le sifflement se réfléchit à une basse altitude et subit une propagation transverse sur une certaine distance.		(12), (12a)
k	Grimpant (Riser)	transverse (transverse)	en satellite (satellite)			Pour certaines valeurs initiales de la direction de la normale à l'onde, on peut avoir une réflection à haute altitude (2...3000 km). La dispersion est alors fortement affectée par le trajet en propagation transverse.		(13)
l	Dos-d'âne (Check Whistler)	transverse (transverse)	en satellite (satellite)			Le même satellite peut recevoir les deux signaux celui qui monte et celui qui descend.		(13)
m	Crochet (Hook Whistler)	transverse (transverse)	en satellite (satellite)			Pour chaque couple de position d'une source et d'un satellite, il y a deux trajets possibles de l'onde, dont l'un correspond au trajet usuel du sifflement. L'autre fait intervenir une propagation très fortement transverse.		(14)
n	Piégé (Trapped Whistler or L.H.R. Whistler)	transverse (transverse)	en satellite (satellite)			Un piégeage des ondes entre deux altitudes peut se produire pour des fréquences voisines de la fréquence hybride basse. Un phénomène de résonance apparaît souvent à cette fréquence.	Détermination de la masse efficace des constituants ioniques	(15), (16)

Références citées dans le Tableau 2.
- (4) T.L. ECKERSLEY: Phil. Mag. **49**, 1250 (1925).
- [7] L.R.O. STOREY: Phil. Trans. Roy. Soc. A **246**, 113 (1953).
- (5) G.R.A. ELLIS: J. Atmosph. Terr. Phys. **8**, 338 (1956).
- (6) R.A. HELLIWELL, J.H. CRARY, J.H. POPE, and R.L. SMITH: J. Geophys. Res. **61**, 139 (1956).
- (7) D.L. CARPENTER: J. Geophys. Res. **68**, 1675, 3727 (1963).
- (8) J.C. CAIN, I.R. SHAPIRO, J.D. STOLARIK, and J.P. HEPPNER: J. Geophys. Res. **66**, 2677 (1961).
- (9) D.L. CARPENTER, N. DUNKEL, and J.F. WALKUP: J. Geophys. Res. **69**, 5009 (1964).
- (10) R.L. SMITH, N.M. BRICE, J. KATSUFRAKIS, D.A. GURNETT, S.D. SHAWHAN, J.S. BELROSE, and R.E. BARRINGTON: Nature **204**, 274 (1964).
- (11) R.E. BARRINGTON: Nature **210**, 80 (1966).
- (12) D.L. CARPENTER, and N. DUNKEL: J. Geophys. Res. **70**, 3781 (1965).
- (12a) I. KIMURA, R.L. SMITH, and N.M. BRICE: J. Geophys. Res. **70**, 5961 (1965).
- (13) G.W. PFEIFFER, D.A. GURNETT, S.D. SHAWHAN, and R. SHAW: Nature **210**, 827 (1966).
- (14) S.D. SHAWHAN: Ph. D. Thesis, University of Iowa, aug. 1966.
- (15) N.M. BRICE, R.L. SMITH, J.S. BELROSE, and R.E. BARRINGTON: Nature **203**, 927 (1964).
- (16) R.L. SMITH, I. KIMURA, J. VIGNERON, and J. KATSUFRAKIS: J. Geophys. Res. **71**, 1925 (1966).

On a représenté en zones hachurées 4 régions du plan f, W qui sont chacune liées à un type de phénomène magnétosphérique déterminé:

Région 1. Il s'agit des ultra-basses fréquences dont nous ne parlerons pas ici[5]. La grandeur qui joue un rôle fondamental dans cette région du plan est la vitesse de Alfven V_A[6]:

$$V_A = \left(\frac{B^2}{u\,\mu_0\,\varrho} \right)^{\frac{1}{2}}. \tag{2.9}$$

Région 2. A cette région sont liés les sifflements ioniques reçus à bord des satellites. La théorie complète de leur propagation dans un milieu *inhomogène* montre que non seulement la gyrofréquence des ions intervient mais également la fréquence de croisement, f_c.

Région 3. C'est le domaine des phénomènes liés à la propagation transverse et en particulier du phénomène de résonance hybride basse détecté en satellite.

Région 4. C'est celle qui correspond aux sifflements usuels ou «électroniques» qui, historiquement, ont été étudiés les premiers.

δ) *Classification.* Il est maintenant possible d'établir une classification des différents types de sifflements observés (Tableau 2). Ces types se distinguent par l'aspect de leur spectre de fréquence en fonction du temps, ou «sonagramme». Les sonagrammes constituent la représentation la plus utilisée et la plus pratique de phénomènes dont la fréquence varie en fonction du temps (voir Sect. 7). L'échelle horizontale des temps est d'environ 1 à 2 sec, l'échelle verticale des fréquences est d'environ 6 kHz.

Nous avons également représenté sur le Tableau 2 le trajet de l'énergie suivi par l'onde entre le moment de son émission par l'éclair originel et celui de sa réception. On remarque que, dans la première partie du tableau, les trajectoires font principalement intervenir la propagation quasi longitudinale, tandis que dans la seconde, les propagations transverses (principalement au voisinage des points de réflexion) font leur apparition.

Une colonne du tableau est réservée à quelques indications sur les mesures qui ont pu être déduites de l'observation de ces sifflements, tandis que référence est donnée des premiers travaux concernant les différents types de phénomènes.

---
[5] Voir l'exposé de E. SELZER dans le tome 49/4 de cette encyclopédie.
[6] Voir V.L. GINZBURG et A.A. RUHADZE: Contribution dans le tome 49/4 de cette encyclopédie. u = 1 en unités rationalisées, mais = 4π en unités non-rationalisées. (Voir les remarques générales p. 1.)

3. Sifflements observés au sol.

Nous avons dit (Sect. 1β) que les sifflements usuels observés au sol étaient le résultat de la propagation entre deux hémisphères de l'énergie de basse-fréquence rayonnée par un éclair atmosphérique. Les raisons pour lesquelles cette énergie suit, avec une bonne approximation, le trajet d'une ligne de force seront expliquées en Sect. 5. Cette hypothèse étant admise, la forme spectrale d'un certain nombre de sifflements peut être interprétée, et il est possible d'en déduire des chiffres concernant la densité électronique des régions éloignées de la magnétosphère.

α) *Equation d'indice.* Il est utile de mettre les équations (2.5) et (2.7) sous une forme plus appropriée aux conditions spécifiques dans lesquelles nous nous plaçons. Nous nous intéressons uniquement au mode droit ($s = +1$) et utilisons les quantités réduites

Fréquence réduite:
$$x = f/f_B \equiv 1/|Y_e|, \qquad (3.1\text{a})$$

fréquence de plasma réduite:
$$\alpha = f_N/f_B \equiv X_e^{\frac{1}{2}}/|Y_e|. \qquad (3.1\text{b})$$

Il est alors facile de voir que l'équation (2.5) s'écrit pour un plasma comportant des électrons et un seul constituant ionique, H^+:

$$\mu_{+1}^2 = 1 + \frac{\alpha^2 \left(1 + \frac{1}{m^*}\right)}{(1-x)\left(x + \frac{1}{m^*}\right)} \equiv 1 + X_e \frac{(m^* + 1)}{(|Y_e| - 1)(|Y_e| + m^*)}. \qquad (3.2)$$

Si l'on néglige $1/m^* = 1/1840$ devant l'unité et si l'on ne s'intéresse qu'aux fréquences grandes devant la gyrofréquence de protons ($f \gg F_B$, ou $x \equiv 1/|Y_e| \gg 1/m^*$), on a alors l'équation

$$\mu^2 = 1 + \frac{\alpha^2}{x(1-x)} \equiv 1 + X_e/(|Y_e| - 1) \qquad (3.3)$$

Comme, dans la magnétosphère, $\alpha^2 = f_N^2/f_B^2 \gg 1$ et que l'on ne s'intéresse qu'à des fréquences inférieures à la gyrofréquence électronique ($f < f_B$, ou $x \equiv 1/|Y_e| < 1$), on peut écrire

$$\mu \approx \frac{\alpha}{x^{\frac{1}{2}}(1-x)^{\frac{1}{2}}} \equiv \frac{X_e^{\frac{1}{2}}}{(|Y_e| - 1)^{\frac{1}{2}}} \qquad (3.4)$$

qui est l'équation fondamentale de dispersion des ondes T.B.F. se propageant longitudinalement dans la magnétosphère.

β) *Loi d'*Eckersley. Cette loi approximative, d'un intérêt fondamental, est la première qui fut obtenue[1]. Elle est valable lorsqu'on ne s'intéresse qu'à des *fréquences faibles devant la gyrofréquence électronique.*

Alors $x \equiv 1/|Y_e| \ll 1$ et l'équation (3.4) peut s'écrire

$$\mu = \frac{\alpha}{x^{\frac{1}{2}}} \equiv (X_e/|Y_e|)^{\frac{1}{2}}. \qquad (3.5)$$

L'indice de groupe, μ_g, défini par

$$\mu' = \frac{d(\mu x)}{dx} \qquad (3.6)$$

---
[1] T. L. Eckersley: Nature **135**, 104 (1935).

vaut alors

$$\mu' = \frac{\alpha}{2x^{\frac{1}{2}}} \equiv \frac{1}{2}(X_e/|Y_e|)^{\frac{1}{2}}. \tag{3.7}$$

Le temps de parcours[2] d'une fréquence $f \equiv xf_B$, entre le point d'émission A et le point de réception B, est alors défini par l'intégrale

$$\tau = \frac{1}{c_0} \int_A^B \mu' \, dl \tag{3.8}$$

dans laquelle dl représente l'élément de longueur de la ligne de force. Remplaçant x et α par leur valeur, il vient,

$$\tau = Df^{-\frac{1}{2}} \tag{3.9}$$

avec

$$D = \frac{1}{2c_0} \int_A^B \frac{f_N}{\sqrt{f_B}} \, dl. \tag{3.10}$$

L'expression (3.9) est connue sous le nom de *loi* d'ECKERSLEY. Elle exprime la proportionnalité du temps de parcours d'une fréquence déterminée à l'inverse de la racine carrée de cette fréquence. La constante D, ou constante de dispersion, est fonction des valeurs des paramètres magnétosphériques le long de la ligne de force suivie par le sifflement [7].

La loi (3.9) explique bien l'allure des sonagrammes: les fréquences élevées apparaissent d'abord, puis les fréquences basses. Comme les fréquences mises en jeu correspondent aux fréquences audibles, on peut brancher un haut parleur à la sortie du détecteur. On entend alors un son dont la hauteur décroit très rapidement, d'où le nom de sifflement donné au phénomène.

γ) *Application à la détermination du contenu électronique de la magnétosphère.* Lorsque le temps d'arrivée de chaque fréquence est porté en fonction de $1/\sqrt{f}$, on obtient une droite dont la pente permet de calculer D. Quelquefois on reçoit simultanément l'impulsion originale de l'éclair, qui parvient de A en B, à la vitesse de la lumière, par un trajet sub-ionosphérique. Le temps t est alors déterminé sans ambiguité. On vérifie bien (Fig. 2) que les points expérimentaux s'alignent sur une droite. Il arrive également qu'une partie de l'énergie du sifflement se réfléchisse en B, effectue un nouvel aller-et-retour dans la magnétosphère avant d'être reçu au sol. On a alors un *sifflement à échos multiples* (deuxième trace de la Fig. 2) dont la dispersion est triple de la précédente).

La formule (3.10) montre que la contribution la plus importante dans l'intégrale, vient des régions pour lesquelles f_B est faible: autrement dit, D donne principalement une indication sur la valeur de f_N au sommet de la ligne de force, $f_{N\Lambda}$. Toutefois, c'est en fait une densité intégrée que l'on mesure, et l'on est obligé d'admettre un modèle de variation de f_N avec l'altitude.

D permet alors d'atteindre $f_{N\Lambda}$. Tous les résultats s'accordent pour montrer que la valeur trouvée pour $f_{N\Lambda}$ dépend en fait assez peu du modèle choisi*. Toutefois il semble que le *modèle magnétique*, dans lequel la densité électronique est proportionnelle en chaque point à l'intensité du champ, représente assez bien

---

* En pratique, il faut tenir compte également de la contribution due à l'ionosphère, importante parce que f_N y est élevé. Cette contribution peut atteindre 5 à 10 s$^{\frac{1}{2}}$ pour une dispersion totale de 50 à 60 s$^{\frac{1}{2}}$ à nos latitudes.

[2] Voir K. RAWER et K. SUCHY: Cette encyclopédie, tome 49/2, p. 498 et suivantes.

Sect. 3. Sifflements observés au sol. 471

Fig. 2. Différents aspects d'un sifflement électronique. *Les deux figures du bas* représentent des enregistrements de l'amplitude en fonction du temps (l'échelle de temps étant fortement dilatée dans la partie inférieure). *Au-dessus*, représentation du spectre fréquence-temps (sonagramme) du phénomène. Le noircissement correspond à un signal intense à l'instant et pour la fréquence considérés. L'instant 0 correspond à l'éclair atmosphérique initial. *Au sommet*, représentation graphique de $1/\sqrt{f}$ en fonction du temps, obtenue à partir du sonagramme. On remarque que la pente de la deuxième trace est 3 fois plus faible que la pente de la première: la «dispersion» $D = t/\sqrt{f}$ est triple. (D'après HELLIWELL [5].)

les conditions moyennes de la magnétosphère. Un tel modèle conduit à la loi approchée (1.5). Différentes variantes ont été proposées [5].

Quoiqu'il en soit, l'étude des sifflements radioélectriques a fourni les ordres de grandeur des variations de la densité électronique avec l'altitude, bien avant que d'autres expériences (en particulier à bord de satellites) aient permis d'atteindre cette quantité. La Fig. 3 donne un exemple des résultats obtenus. La mesure entre deux points conjugués du temps de parcours d'une onde émise par un émetteur à fréquence fixe permet également d'obtenir des renseignements sur le contenu électronique de la magnétosphère[3].

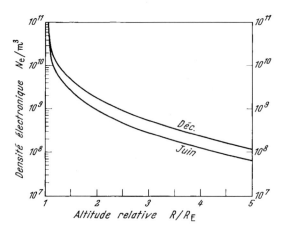

Fig. 3. Courbe de répartition de la densité électronique en fonction de la distance géocentrique réduite, telle qu'on la déduit de la dispersion des sifflements. (D'après Helliwell [5].)

δ) *Le phénomène du «nose-whistler»*. Lorsque les observations se sont étendues à de plus hautes latitudes, un phénomène nouveau est apparu. Pour les fréquences élevées, le temps de parcours augmente avec la fréquence. Le sonagramme a alors l'aspect d'un nez, d'où le nom de «nose-whistler» (Fig. 4). Ce phénomène apparait pour les fréquences f supérieures à f_n, fréquence du nez; les temps de parcours sont tous supérieurs à un temps minimal, τ_n.

L'interprétation en est simple. Lorsque le sifflement traverse des régions dans lesquelles la condition $x = 1/|Y_e| \ll 1$ n'est plus vérifiée, il faut utiliser l'équation complète (3.4). L'indice de phase croit alors avec la fréquence dès que x s'approche de l'unité. Il en est de même de l'indice de groupe

$$\mu' = \frac{1}{2} \alpha\, x^{-\frac{1}{2}} (1-x)^{-\frac{3}{2}} \equiv \frac{1}{2} X_e^{\frac{1}{2}} \frac{|Y_e|}{(|Y_e|-1)^{\frac{3}{2}}} \qquad (3.11\,a)$$

dont on vérifie qu'il passe par un minimum

$$\mu'_{\min} = \frac{8\alpha}{\sqrt{27}} \approx 1{,}54\,\alpha \qquad (3.12)$$

pour une fréquence

$$f_m = \tfrac{1}{4} f_B. \qquad (3.13)$$

---

[3] J. Delloue, M. Garnier et J. M. Comarmond: Compt. rend. **260**, 6169 (1965). L'expérience a été effectuée à l'aide de la station de Sainte-Assise (F.U.B.) sur fréquence ultra-basse de 16,8 kHz.

Le sonagramme théorique d'un tel sifflement correspondant à une propagation en milieu homogène est représenté sur la Fig. 5. La trace verticale correspond à une propagation différente[4], dont il sera question à la Sect. 5.

Mais f_B n'est pas constant le long d'une ligne de force. C'est donc l'expression (3.8) qu'il faut calculer, avec la valeur (3.11 a) pour μ'. Le calcul, effectué avec différents modèles de répartition de la densité électronique, montre que τ est minimum pour une valeur de la fréquence, approximativement indépendante du modèle choisi, et égale à

$$f_n = \sim 0{,}4\, f_{B\min}. \qquad (3.14)$$

$f_{B\min}$ étant la valeur de f_B au sommet de la ligne de force. En mesurant f_n, on peut donc avoir une indication précise de la ligne de force suivie par le sifflement. τ_n permet alors de déterminer la densité au sommet de cette ligne.

Les sifflements reçus aux basses latitudes, ne représentent en général pas le phénomène du nez. Mais une extrapolation de la courbe de dispersion est possible[5]. Cette méthode a permis l'étude détaillée du contenu électronique de la magnétosphère et l'établissement de la loi la plus probable de répartition en fonction de l'altitude[6-8].

ε) *Effet des ions.* Les termes de l'équation d'indice dûs aux ions, négligés dans l'établissement de la relation (3.3) deviennent importants lorsque $f \approx F_B$: leur influence est déjà notable pour $f \approx 2F_B$.

Lorsqu'on s'intéresse aux fréquences comprises entre 0,5 et 1 kHz, f n'est égal à $2F_B$ que pour des faibles altitudes (inférieures à 700 km). S'il s'agit d'un sifflement de haute latitude, la portion du trajet pendant laquelle l'effet des ions

---

[4] R. GENDRIN: Planet. Space Sci. **5**, 274 (1961).
[5] R. L. SMITH, and D. L. CARPENTER: J. Geophys. Res. **66**, 2582 (1961).
[6] R. L. SMITH: J. Geophys. Res. **66**, 3709 (1961).
[7] J. H. POPE: J. Geophys. Res. **66**, 67, 2580 (1961).
[8] Y. CORCUFF: Thèse, Fac. des Sciences de Poitiers, 1965.

Fig. 4. Sonagramme d'un «nose-whistler», à trajets multiples. Le sifflement, pour lequel l'éclair originel s'est produit au temps 0, a suivi un très grand nombre de parcours différents. Ceux de latitudes plus élevées correspondent à un temps de propagation plus long et à une fréquence de nez plus faible. (Les traces courbes visibles à l'extrême gauche sont dues à un éclair antérieur.) (D'après CARPENTER et SMITH [8].)

est sensible est négligeable. Si au contraire on étudie des sifflements de basse latitude, dont le trajet culmine à des faibles altitudes (~ 1500 km) l'effet des ions sera sensible sur une longue portion du parcours et la dispersion apparente sera modifiée.

L'équation (3.11a) définissant l'indice de groupe est à remplacer par l'équation

$$\mu' = \frac{\alpha}{2x^{\frac{1}{2}}} \cdot \frac{1}{(1-x)^{\frac{3}{2}}} \cdot \frac{1 + 2/(m^* x)}{[1 + 1/(m^* x)]^{\frac{3}{2}}} \equiv \frac{1}{2} X_e \frac{|Y_e|}{(|Y_e| - 1)^{\frac{3}{2}}} \cdot \sqrt{m^*} \frac{(m^* + 2|Y_e|)}{(m^* + |Y_e|)^{\frac{3}{2}}} \qquad (3.11\,\text{b})$$

que l'on déduit de (3.2) et (3.6) en supposant $\alpha^2 \equiv X_e/Y_e^2 \gg 1$ et $x \equiv 1/|Y_e| < 1$.

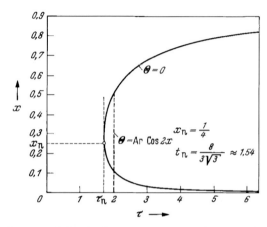

Fig. 5. Aspect théorique d'un «nose-whistler». La figure représente le temps de parcours de chaque fréquence dans un milieu homogène pour une propagation longitudinale ($\Theta = 0$). En ordonnées $x = 1/|Y_e|$, le rapport de la fréquence considérée à la gyrofréquence électronique; en abscisse, le temps mis à parcourir une longueur égale à c_0/α. Lorsque chaque fréquence est émise suivant une direction qui fait l'angle $\Theta = \arccos 2x$ avec le champ magnétique directeur (lire ceci au lieu de ArCos) on obtient la droite brisée verticale: toutes les fréquences arrivent simultanément. (D'après Gendrin [4].)

Sous cette forme, on voit que l'indice de groupe comprend trois termes:
— le terme d'Eckersley,
— le terme de Storey qui interprète les «nose-whistlers»,
— le terme dû à l'effet des ions.

Il est toujours possible de définir une dispersion D par l'équation (3.9). La dispersion n'est plus indépendante de la fréquence; elle est la somme de trois quantités.

— une quantité D_0, qui ne dépend pas de la fréquence et correspond au premier terme de l'équation (3.11b);

— une quantité ΔD_h, qui représente la correction haute fréquence (deuxième terme de l'équation.)

— une quantité ΔD_b, qui n'est importante qu'aux basses fréquences (troisième terme de l'équation).

La Fig. 6 représente les quantités D_0, ΔD_h, ΔD_b et

$$D = D_0 + \Delta D_h + \Delta D_b$$

pour un modèle de répartition électronique et ionique donné. Les points expérimentaux correspondent à un sifflement enregistré à la station de Wakkanai (35,3°

latitude géomagnétique). On remarque que la «dispersion» croit effectivement lorsque la fréquence décroit[9], réssultat confirmé par une autre méthode d'analyse[10]*.

ζ) *Le phénomène du «knee-whistler»*. Lorsqu'un sifflement de haute latitude, qui présente le phénomène du «nez», comporte plusieurs traces (multi-path whistlers), celles-ci s'emboitent en général les unes dans les autres (Fig. 4). Ceci est dû au fait que la partie du sifflement qui a suivi un parcours plus éloigné (donc de temps de propagation plus élevé), a traversé une région dans laquelle

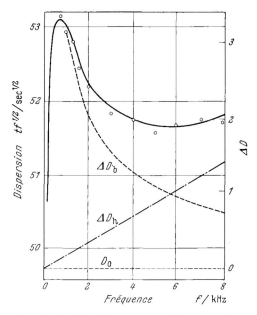

Fig. 6. Variation de la dispersion en fonction de la fréquence pour un sifflement reçu à basse latitude. D_0 est la dispersion correspondant à l'approximation d'ECKERSLEY-STOREY; ΔD_h la correction «haute fréquence» qui permet d'interpréter le phénomène du «nose-whistler». ΔD_b, la correction basse fréquence due à l'effet des ions. Les points sont des points expérimentaux obtenus à partir d'un enregistrement de sifflement effectué au Japon. (D'après OUTSU et IWAÏ[7].)

$f_{B\min}$ est plus faible; la fréquence du nez est donc plus petite. *Si τ_n croît, f_n décroit.* Ceci dans l'hypothèse où la densité électronique décroit de façon régulière avec l'altitude [8].

Mais si cette densité présente brusquement une diminution importante à partir d'une certaine valeur de L, les trajets de L élevé (c.à.d. de f_n faible) vont correspondre à des temps de parcours faibles (car f_N intervient au numérateur de l'équation (3.10) ou de toute équation équivalente). Autrement dit, les différents sifflements présenteront d'abord une croissance normale de τ_n lorsque f_n diminue, puis d'autres traces apparaissent avec des τ_n beaucoup plus faibles.

C'est ce que CARPENTER[11], a réellement observé. La courbe de répartition de la densité électronique que l'on peut déduire d'un tel phénomène (Fig. 7) présente alors l'aspect d'un genou, d'où le nom donné au sifflement.

La Fig. 8 reproduit les valeurs de densité que l'on peut déduire de tels phénomènes, en fonction de la distance radiale au centre de la terre. La courbe en trait

---

* La figure 34 constitue également une illustration de ce phénomène.
[9] J. OUTSU, and A. IWAI: Proc. Res. Inst. Atmosph., Nagoya **6**, 44 (1959).
[10] R. E. BARRINGTON, and T. NISHISAKI: Canad. J. Phys. **38**, 1942 (1960).
[11] D. L. CARPENTER: J. Geophys. Res. **68**, 1675 et 3727 (1963).

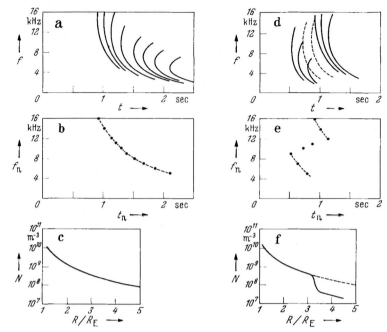

Fig. 7a—f. Phénomène du «knee-whistler». *A gauche*: sifflement normal (a): les différentes traces s'emboitent les unes dans les autres (voir Fig. 4). (b): la variation de la fréquence de nez f_n, en fonction du temps de parcours t_n, décroit régulièrement. (c): on en déduit une répartition de la densité électronique en fonction de l'altitude qui décroit de façon continue. *A droite*: «knee-whistler». (d): les différentes traces ne s'emboitent plus les unes dans les autres; celles qui correspondent à des parcours de latitude élevée ont des temps de propagation très courts. (e): la variation de f_n en fonction de t_n présente une discontinuité. (f): il en est de même de la répartition de la densité électronique en fonction de l'altitude.
(D'après Carpenter[11].)

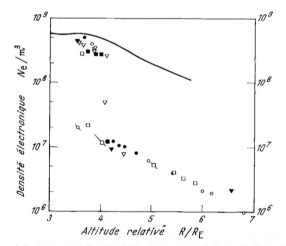

Fig. 8. Mise en évidence de la plasmapause. En trait plein, valeur moyenne de la densité électronique, déduite de l'étude des sifflements, en période magnétiquement calme. Les points expérimentaux correspondent aux résultats qui ont été déduits de l'étude des «knee-whistlers» au cours de périodes magnétiquement agitées ($2 < K_p < 4$). Jusqu'à 4 rayons terrestres, la densité est la même que précédemment; au-delà, elle est fortement réduite (d'un facteur 50). ■ □ 31.7.1963, 1650 et 1710 h TL; ▼ ▽ 2. 8. 1963, 1250 h TL; ● ○ 3. 8. 1963, 1450 h TL. (D'après Angerami et Carpenter[13].)

Fig. 9a—d. Sifflement subprotonosphérique. En (a) et (b), sonagrammes d'enregistrements effectués à bord d'une fusée «Aerobee». En (d), sonagramme d'enregistrement effectué simultanément à bord du satellite «Alouette 1». En (c), représentation schématique du trajet des ondes. (D'après Carpenter et al.[15].)

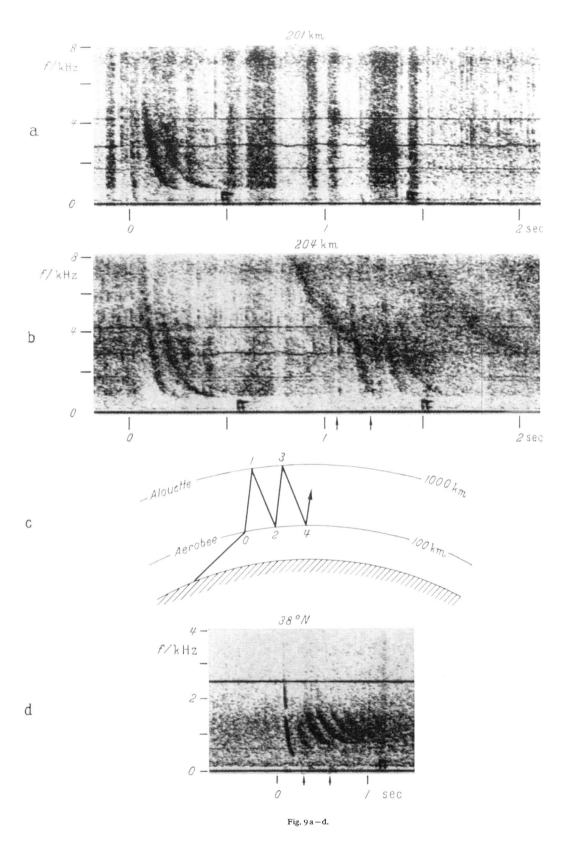

Fig. 9a–d.

plein donne cette même variation au cours d'une période magnétiquement très calme.

Il a été en effet possible de montrer que cette frontière appelée *plasmapause* se déplaçait vers la terre, en cas de situation magnétique agitée[12,13]. Cette faille dans l'ionisation a même été observée à des valeurs aussi faibles que $L = 2,5$ (correspondant approximativement à la latitude de Poitiers), après des orages magnétiques de K_p élevé[14].

Fig. 10. Propagation sub-ionosphérique d'un sifflement. Les deux enregistrements ont été obtenus à la station de Poitiers (50° géomagnétique). Celui du bas est un sifflement normal dont la dispersion est voisine de 65 s$^{\frac{1}{2}}$. Celui du haut, de dispersion beaucoup plus petite, a suivi un trajet magnétosphérique de très faible latitude et s'est propagé sous l'ionosphère (approximativement à la vitesse de la lumière) jusqu'à la station d'observation. (Aimablement communiqué par Mlle Corcuff.)

L'étude de la plasmapause est d'une grande importance du point de vue théorique car son origine est certainement liée aux mouvements «fluides» entraînés sur les bords internes de la magnétosphère par la pénétration des particules de basse énergie qui composent le vent solaire.

η) *Autres sifflements.* Il s'agit ici de sifflements électroniques dont les caractéristiques ont été modifiées par des conditions de réflexion ou de propagation particulières.

La Fig. 9 représente un sifflement «sub-protonosphérique»[15] enregistré simultanément à bord d'une fusée (figures du haut) et d'un satellite (figure du bas). Le trajet de l'onde comporte

---

[12] D. L. Carpenter: J. Geophys. Res. **71**, 693 (1966).
[13] J. J. Angerami, and D. L. Carpenter: J. Geophys. Res. **71**, 711 (1966).
[14] Y. Corcuff et M. Delaroche: Compt. rend. **258**, 650 (1964).
[15] D. L. Carpenter, N. Dunkel et J. F. Walkup: J. Geophys. Res. **69**, 5009 (1964).

plusieurs réflexions soit sur les irrégularités de la couche E, soit au niveau de départ de la protonosphère. Un calcul de tracé des rayons a montré que de telles réflexions étaient effectivement prévisibles aux basses fréquences[16].

Un deuxième phénomène observé est le sifflement «sub-ionosphérique». Il s'agit d'un sifflement engendré à très basse latitude qui, après un court trajet dans la magnétosphère, s'est propagé sur une grande distance, *sous* l'ionosphère, avant d'atteindre la station d'observation. Ce dernier trajet n'étant partiellement pas dispersif (mais présentant par contre une fréquence nette de coupure), la dispersion observée est très faible[17]. Le sonagramme de la partie supérieure de la Fig. 10 représente un tel sifflement. A titre de comparaison, on donne, dans la partie inférieure, le sonagramme d'un sifflement ordinaire.

4. Sifflements observés en satellites.

Ces sifflements, qui ont été découverts récemment, complètent la gamme des phénomènes naturels de fréquence inférieure à la gyrofréquence électronique. Ils font apparaitre l'importance de la propagation au *voisinage des résonances*, comme moyen d'étude de la composition ionique du milieu. Ces études ont été faites principalement à l'aide des satellites Injun 3 et Alouette 1, et la majeure partie des résultats obtenus a déjà été présentée dans quelques articles de synthèse [9], [10].

α) *Les sifflements ioniques.* Ce phénomène se traduit, sur les sonagrammes par une double trace dont les prolongements se coupent à des fréquences très basses[1]. La première, presque verticale s'apparente aux sifflements électroniques à trajets partiels[2]. La deuxième, dont la dispersion est plus importante à mesure que la fréquence augmente, présente une asymptote horizontale pour une fréquence égale à la gyrofréquence des protons au voisinage du satellite (Fig. 11 et 12).

Il était normal de considérer qu'il pouvait s'agir de la propagation dans le mode L d'un éclair atmosphérique. Ce mode présente en effet un indice croissant avec la fréquence, qui tend vers l'infini pour une fréquence égale à la gyrofréquence des ions positifs, justifiant ainsi qualitativement l'allure des sonagrammes. Un premier calcul était fait qui confirmait approximativement cette hypothèse[3].

Mais la théorie correcte de ces sifflements (voir Annexe, Sect. 9) est celle de GURNETT et al. [11], qui mirent en évidence l'existence d'une *fréquence de croisement* et d'une *fréquence de coupure* (Fig. 12).

— En dessous de la fréquence de croisement au niveau du satellite, f_{cS}, l'onde se propage suivant le mode **R** des sifflements électroniques.

— Au-dessus de la fréquence de croisement et jusqu'à une fréquence égale à la gyrofréquence des protons au niveau du satellite F_{BS} elle se propage d'abord dans le mode **R**, puis dans le mode **L**.

— Entre la gyrofréquence des protons et la fréquence de coupure, l'onde ne peut se propager sauf s'il existe un couplage[4] entre les modes **R** et **L**.

Les informations que l'on peut tirer de l'analyse d'un sifflement protonique se divisent en deux catégories: celles que l'on tire de l'étude de la fréquence de croisement, et celles qui sont données par l'étude du comportement asymptotique de la dispersion au voisinage de la gyrofréquence.

β) *Etude de la fréquence de croisement des sifflements protoniques.* Celle-ci a lieu (voir Sect. 2)[5] lorsque

$$W_e = W_{+1}. \tag{4.1}$$

---

[16] I. KIMURA: Radio Sci. **1**, 269 (1966).
[17] P. CORCUFF, Y. CORCUFF et R. RIVAULT: Compt. rend. **261**, 1372 (1965).
[1] R. L. SMITH, N. M. BRICE, J. KATSUFRAKIS, D. A. GURNETT, S. D. SHAWHAN, J. S. BELROSE et R. E. BARRINGTON: Nature **204**, 274 (1964).
[2] J. C. CAIN, I. R. SHAPIRO, J. D. STOLARIK et J. P. HEPPNER: J. Geophys. Res. **66**, 2677 (1961).
[3] R. GENDRIN et J. VIGNERON: Compt. rend. **260**, 3129 (1965).
[4] Voir K. RAWER et K. SUCHY: Cette encyclopédie, tome 49/2, p. 126.
[5] Voir également K. RAWER et K. SUCHY: Cette encyclopédie, tome 49/2, p. 55.

Fig. 11. Sifflements protoniques enregistrés à l'aide du satellite «Injun III» à différentes altitudes. *De haut en bas*: 2600, 1300, 1000, 600 km. Au fur et à mesure que l'altitude décroit, le rapport entre la fréquence de croisement (point d'intersection des deux traces) et la gyrofréquence des protons (aysmptote horizontale) croit; on en déduit que la proportion d'He^+ devient importante (85% à 600 km). Les mesures ont été faites entre 12 et 14 h TL. Les deux traces horizontales sont dûes à des interférences produites par d'autres équipements scientifiques à bord du satellite. (Aimablement communiqué par M. M. Gurnett et Shawhan.)

Alors $W_{-1} = W_e$ et les trois modes ont, pour cette fréquence, la même vitesse de phase.

Compte tenu des définitions (2.5) et (2.8) la fréquence de croisement vérifie:

$$\sum_h \frac{X_h Y_h}{1 - Y_h^2} = 0. \tag{4.2}$$

Fig. 12 a et b. Fréquences caractéristiques d'un sifflement protonique. Les notations de la figure diffèrent de celles du texte: il faut remplacer ω_{12} par f_c et Ω_1 par F_B. La trace en pointillé correspond à une partie du signal qui est en général absente. L'origine de cette bande de coupure du sifflement électronique est expliquée dans l'Annexe (Sect. 9). (D'après Gurnett et al. [11].)

Supposons un mélange à deux constituants seulement. Sachant à priori que $f_c < F_B$, l'hypothèse $\alpha > 1$ entraîne $Y_e^2 \gg 1$. Dans ces conditions, il est facile de vérifier que (4.2) définit f_c par

$$f_c^2 = f_B^2 \left(\frac{A_1}{m_2^{*2}} + \frac{A_2}{m_1^{*2}} \right), \tag{4.3}$$

A_1 et A_2 étant les concentrations relatives des deux ions $(A_1 + A_2 = 1)$; m_1^* et m_2^* étant le rapport de leur masse à celle de l'électron. Dans le cas où l'on a hydrogène et hélium $(m_2^* = 4 m_1^* = 4 m^*)$, on trouve comme rapport Λ de la fréquence de croisement à la gyrofréquence des protons*.

$$\Lambda = \frac{f_c}{F_B} = \sqrt{1 - \frac{15}{16} A_{H^+}}.\qquad(4.4)$$

Les grandeurs f_c et F_B étant directement mesurables sur le sonagramme, il est possible de calculer la condentration relative en ion H^+:

$$A_{H^+} = N_{H^+}/N_e.$$

Des résultats partiels sont rassemblés sur la Fig. 13. On voit que la concentration relative croit d'autant plus vite avec l'altitude que la température est plus basse, ainsi qu'il est normal[6].

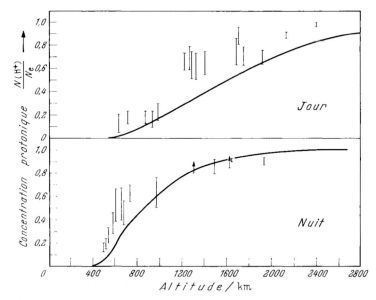

Fig. 13. Détermination de la concentration relative des protons en fonction de l'altitude à partir de l'analyse des sifflements protoniques. Les tirets correspondent aux résultats expérimentaux; les courbes, à un modèle théorique qui tient compte de la différence de température (donc d'échelle de hauteur) entre le jour et la nuit. (D'après Shawhan et Gurnett[6]).

γ) *Détermination de la densité absolue N_{H^+} à partir des sifflements protoniques.* Au voisinage d'une gyrorésonance, la trace d'un sifflement ionique a un comportement asymptotique qui dépend de la concentration absolue de l'ion considéré. En effet, dans le cas d'un sifflement protonique, pour les fréquences f voisines de la gyrofréquence F_{BS} au niveau du satellite, la contribution à l'indice de phase du mode gauche ($s = -1$) vient principalement du terme relatif aux protons dans l'expression (2.5):

$$\mu^2 \sim 1 - \frac{X_{H^+}}{-Y_{H^+} + 1}\qquad(4.5)$$

soit, puisque $1 - Y_{H^+}$ est très petit:

$$\mu \sim \frac{F_N}{f^{\frac{1}{2}} (F_B - f)^{\frac{1}{2}}}\qquad(4.6)$$

---

* La formule (4.4) est valable en présence d'ions O^+ pour peu que ceux-ci n'aient pas une concentration trop importante [11].

[6] S. D. Shawhan, and D. A. Gurnett: J. Geophys. Res. **71**, 47 (1966).

F_N étant la fréquence de plasma *ionique*. On en déduit:

$$\mu' \sim \frac{F_N}{2} \cdot \frac{F_B}{f^{\frac{1}{2}}(F_B-f)^{\frac{3}{2}}}. \tag{4.7}$$

Le temps de propagation est alors défini par une intégrale du type (3.8) dans laquelle μ' a la valeur (4.7). Cette intégrale est calculée le long d'une ligne de force, entre le niveau du sol et la position du satellite.

Mais, le long du trajet, F_B varie. On peut remplacer l'intégrale sur l par une intégrale sur F_B en écrivant:

$$dl = \frac{dF_B}{(\partial F_B/\partial l)}. \tag{4.8}$$

Le temps de parcours τ d'une fréquence f voisine de F_{BS} est alors donné par la formule:

$$\tau \sim \frac{1}{2c_0} \int_{F_{BO}}^{F_{BS}} \frac{F_N}{f^{\frac{1}{2}}(\partial F_B/\partial l)} \cdot \frac{F_B}{(F_B-f)^{\frac{3}{2}}} \cdot dF_B. \tag{4.9}$$

La principale contribution à cette intégrale viendra des régions pour lesquelles F_B est très voisin de la borne supérieure de l'intégrale: F_{BS}. Dans cette zone d'altitude réduite, F_N et $(\partial F_B/\partial l)$ peuvent être considérés comme constants. Il est alors facile d'évaluer (4.9) et l'on obtient[7,8]

$$\tau = \frac{F_{NS} \cdot F_{BS}^{\frac{1}{2}}}{c_0|(\partial F_B/\partial l)|_S} \cdot \frac{1}{(F_{BS}-f)^{\frac{1}{2}}} + A + B. \tag{4.10}$$

A est un terme constant, B un terme qui tend vers zéro lorsque $f \to F_{BS}$. Autrement dit, l'expression aymptotique de la courbe de dispersion au voisinage de la gyrofréquence des protons est de la forme

$$\Delta\tau = \frac{k}{(F_{BS}-f)^{\frac{1}{2}}} \tag{4.11}$$

k étant proportional à la racine carrée de la densité des protons au voisinage du satellite.

A partir d'un enregistrement expérimental, on essaye de déterminer quelles sont les valeurs de F_{BS} et de k qui donnent la meilleure coïncidence avec la loi théorique (4.11)*. On en déduit $N(H^+)$ par la relation

$$N_{H^+} = 4{,}7 \cdot 10^3 \cdot (F_{BS})^{\frac{3}{2}} \cdot \Phi(\phi) \cdot k^2 \tag{4.12}$$

$\Phi(\phi)$ étant une fonction voisine de 1 qui dépend très peu de la latitude[9].

L'étude de nombreux sifflements protoniques a permis à GURNETT et SHAWHAN de déterminer la variation de N_{H^+} avec l'altitude et la latitude. Si l'on utilise de plus la fréquence de croisement (Sect. 4.2), on peut également déterminer la densité électronique N_e. Les résultats sont en bon accord avec ceux que l'on déduit des sondages en contre-haut, obtenus à l'aide des satellites Alouette.

Les méthodes qui permettent, à partir d'un enregistrement, de calculer les grandeurs F_{BS} et k seront exposées en Sect. 7. Ces méthodes sont également applicables aux «hélium whistlers» récemment découverts[10,11] et doivent per-

---

* Voir Fig. 35.
[7] D. A. GURNETT: Ph. D. Thesis, State University of Iowa, 1965.
[8] D. A. GURNETT, and S. D. SHAWHAN: J. Geophys. Res. **71**, 741 (1966).
[9] C. BEGHIN, R. GENDRIN, C. SIREDEY et J. VIGNERON: Ann. Géophys. **22**, 110 (1966).
[10] R. E. BARRINGTON: Nature **210**, 80 (1966).
[11] S. D. SHAWHAN: Ph. D. Thesis, State University of Iowa, 1966.

mettre de déterminer la concentration des ions hélium. La démonstration des formules données dans les Sect. 4β et 4γ est faite dans l'Annexe, Sect. 9.

δ) *Le phénomène de résonance hybride basse.* Nous allons aborder maintenant des phénomènes qui mettent en jeu la propagation transverse des ondes T.B.F. dans la magnétosphère. Nous commencerons par le phénomène de résonance hybride basse[12], bien qu'il s'agisse plutôt d'une *émission* que d'un sifflement.

Des sonagrammes de ces émissions ont été publiés dans un certain nombre d'articles descriptifs[13-15]. La Fig. 14 en donne un exemple. On remarque une bande de bruit au voisinage de 10 kHz dont la fréquence varie en fonction du temps. Des observations répétées ont montré que cette fréquence variait, pour tous les passages du satellite, de façon consistante avec la latitude géomagnétique; il s'agit donc bien d'une variation spatiale. La possibilité de connaître simultanément, sur le satellite Alouette, la densité électronique au niveau du satellite, a permis de s'assurer que la fréquence du phénomène correspondait à la fréquence de résonance hybride basse.

Celle-ci s'obtient en écrivant que $W_e = 0$, ce qui entraîne par l'équation (2.8):

$$W_{+1}^2 = -W_{-1}^2 \qquad (4.13)$$

soit, en tenant compte de (2.5) et (2.6):

$$\sum_h \frac{X_h}{1-Y_h^2} = 1. \qquad (4.14)$$

Comme la fréquence de résonance f_r vérifie

$$F_B \ll f_r \ll |f_B| \qquad (4.15)$$

on a

$$Y_e^2 \gg 1 \quad \text{et} \quad Y_i^2 \ll 1. \qquad (4.16)$$

Dans ces conditions, il est facile de vérifier que

$$\frac{1}{f_r^2} \sum_i \frac{A_i}{m_i^*} = \frac{1}{f_N^2} + \frac{1}{f_B^2}, \qquad (4.17)$$

voir par exemple l'Annexe B de la référence [9].

Si l'expérience permet de mesurer f_r et f_N (f_B étant connu par la trajectoire du satellite), l'équation (4.17) permet de déterminer une masse efficace m_{eff}^* (en unité de masse de l'électron):

$$\frac{1}{m_{\text{eff}}^*} = \sum_i \frac{A_i}{m_i^*}. \qquad (4.18)$$

Lorsque le milieu est composé uniquement d'hydrogène $m_{\text{eff}}^* = m^*$, lorsqu'il est composé uniquement d'oxygène $m_{\text{eff}}^* = 16\,m^*$. On peut étudier la variation de m_{eff}^* avec l'heure du jour et la latitude. On trouve qu'à l'altitude de 1000 km la proportion d'hydrogène est souvent très inférieure à l'unité [11], contrairement à ce que l'on pensait il y a quelques années.

ε) *Les sifflements transverses.* Détectés également à bord du satellite Alouette 1, ces sifflements présentent une dispersion qui est une fonction croissante de la

---

[12] «Basse», par opposition à la résonance hybride «haute» dont la fréquence est comprise entre la gyrofréquence et la fréquence de plasma *électroniques*, voir K. RAWER et K. SUCHY: Cette encyclopédie, tome 49/2, Sect. 8, p. 67.
[13] N. M. BRICE, R. L. SMITH, J. S. BELROSE et R. E. BARRINGTON: Nature **203**, 927 (1964).
[14] R. E. BARRINGTON, J. S. BELROSE, and D. A. KEELEY: J. Geophys. Res. **68**, 6539 (1963).
[15] N. M. BRICE, and R. L. SMITH: J. Geophys. Res. **70**, 71 (1965).

Fig. 14. Emissions à la fréquence de résonance hybride basse (environ 8 kHz) enregistrées à bord du satellite «Alouette 1». (D'après Brice et al.[13].)

fréquence. Mais il est possible de montrer que le temps de parcours observé est la somme de deux termes dont le premier vérifie la loi d'Eckersley:

$$\tau = D_0 f^{-\frac{1}{2}} + \theta(\phi). \tag{4.19}$$

La dispersion D_0 est à peu près constante et égale à 17 s$^{\frac{1}{2}}$ pour tous les sifflements observés, et correspond à la dispersion que l'on peut attendre pour une latitude ϕ d'environ 30°. θ est une fonction croissante de la latitude d'observation, nulle pour $\phi = 30°$, égale à 0,22 s pour $\phi = 44°$ [16].

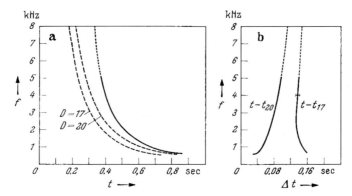

Fig. 15a et b. Sifflement à propagation partiellement transverse reçu à bord du satellite «Alouette 1». A gauche, en trait plein, le sifflement observé; en tirets, deux sifflements théoriques de dispersion 17 et 20 s$^{\frac{1}{2}}$. A droite, différences entre le temps de parcours du sifflement observé et des sifflements théoriques. Cette différence ne dépend pas de la fréquence, comme il est normal pour une propagation transverse. (D'après Carpenter et Dunkel[16].)

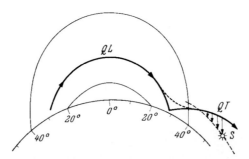

Fig. 16. Trajet supposé des sifflements du type montré sur la Fig. 15. En tirets, dans l'hypothèse d'une courbure progressive du rayon; en trait plein, dans l'hypothèse d'une réflexion sur une couche E sporadique. La propagation est quasi-transverse entre les latitudes de 30° et 50°. (D'après Kimura et al.[17].)

La Fig. 15 donne un exemple de l'écart existant par rapport à la loi d'Eckersley pour deux valeurs de la constante de dispersion. On voit que le choix $D_0 = 17$ s$^{\frac{1}{2}}$ entraîne la constance de θ dans tout le spectre. Il est naturel d'admettre que tous ces sifflements se sont propagés dans la magnétosphère suivant le même trajet, par suite de l'existence d'une fibre d'ionisation. A la base de l'ionosphère, après avoir subi une réflexion, le sifflement suit un mode de propagation transverse, pour lequel la vitesse de propagation est indépendante de la fréquence (Fig. 16)*.

Il est en effet facile de montrer que la vitesse de phase (donc de groupe) est constante dans le mode transverse sur une grande plage de fréquence (voir Fig. 1). Dans le cas d'un plasma à

---

* Mais la longeur du parcours transverse est d'autant plus grande que le satellite est éloigné du point de réflexion, ce qui explique que θ soit une fonction croissante de ϕ.

[16] D. L. Carpenter et N. Dunkel: J. Geophys. Res. **70**, 3781 (1965).
[17] I. Kimura, R. L. Smith et N. M. Brice: J. Geophys. Res. **70**, 5961 (1965).

Sect. 4. Sifflements observés en satellites. 487

Fig. 17a et b. Phénomène de piégeage des ondes de très basse fréquence dans la haute ionosphère. (a): Fréquence de résonance hybride basse en fonction de l'altitude pour un modèle ionosphérique donné. (b): valeur de l'indice de propagation strictement transverse en fonction de l'altitude pour une fréquence de 5,25 kHz dans les mêmes conditions. Aux altitude-proches de 600 et 800 km, μ_t devient infini, et l'on peut avoir un piégeage des ondes dans la region II. (D'après SMITH et al.[18].)

Fig. 18a—d. «Risers» et «checks» détectés à bord du satellite «Injun III». (a) Riser enregistré le 24. 8. 63 à 11 h TL et à une altitude de 1740 km. (b) Riser enregistré le 30. 4. 63 à 21 h TL et à une altitude de 2660 km. (c) Courbe théorique obteunue dans l'hypothèse où la source est décalée d'environ 17° en latitude (voir Tableau 2). (d) Check enregistré le 17. 4. 63 à 23 h TL et à une altitude de 2720 km. (D'après SHAWHAN [19].)

[18] R. L. SMITH, I. KIMURA, J. VIGNERON et J. KATSUFRAKIS: J. Geophys. Res. **71**, 1925 (1966).

[19] S. D. SHAWHAN: Ph. D. Thesis, State University of Iowa, 1966.

constituant unique, on peut même établir rigoureusement (voir référence [9], Annexe B) la formule de dispersion:

$$\mu_\perp^2 = \frac{m^* \alpha^2}{1 - f^2/f_r^2} \equiv m^* \frac{f_N^2}{f_B^2 - f^2(f_B/f_r)^2} \qquad (4.20)$$

avec*

$$\frac{1}{m^* f_r^2} = \frac{1}{f_N^2} + \frac{1}{f_B^2} \qquad (4.21)$$

Fig. 19a—c. «Hook-Whistlers» détectés à bord du satellite «Injun III». L'enregistrement (a) a été effectué à une altitude beaucoup plus élevée (2700 km) que les deux autres (∼1200 km). La latitude géomagnétique à laquelle il a été observé est également beaucoup plus faible (7° contre 29° et 21°). (D'après Shawhan [19].)

qui montre que, sauf au voisinage de f_r, la vitesse de phase W_e, donc de groupe V_g

$$V_{g\perp} = V_A (1 - f^2/f_r^2)^{\frac{3}{2}} \qquad (4.22)$$

ne dépend pratiquement pas de la fréquence**.

L'équation (4.20) a une autre conséquence: la fréquence f_r variant avec l'altitude et présentant deux maximums séparés par une vallée (partie gauche de la Fig. 17), l'indice

---

* L'équation (4.21) n'est autre que l'équation (4.17) pour un milieu à constituant unique: l'hydrogène.

** Toutefois, vu les ordres de grandeurs de f_r à 1000 km, (voir Fig. 17) il est difficile d'admettre que $f = 5$ kHz soit une fréquence faible devant f_r.

transverse est imaginaire entre certaines altitudes, dans un domaine relativement restreint de fréquences. On montre qu'un piégeage de l'onde est possible entre les deux niveaux pour lesquels $f \approx f_r$[18].

ξ) *Autres sifflements faisant intervenir la propagation tranverse.* Si, pour une certaine raison, une onde électromagnétique T.B.F. est amenée à se propager transversalement à des altitudes élevées (de l'ordre de 1000 à 2000 km), son point représentatif se trouve dans la région 3 du diagramme de la Fig. 1. La vitesse de phase (et aussi la vitesse de groupe) est une fonction décroissante de la fréquence. Le temps de propagation sera donc plus élevé pour les fréquences hautes.

Des sifflements répondant à ces caractéristiques ont été découverts récemment[19,20]. Il s'agit des «risers» et des «checks» (Fig. 18). Dans le premier cas, le sifflement parvient au

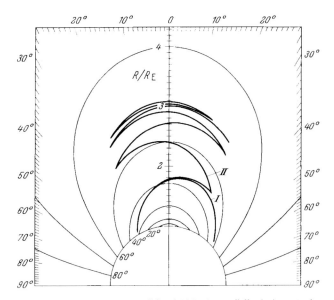

Fig. 20. Trajet théorique d'un rayon pour certaines conditions initiales, lorsque l'effet des ions est pris en compte. Ce trajet, calculé par intégration numérique des équations de HASELGROVE, correspond à une onde de fréquence égale à 1 kHz dont la normale est initialement dirigée verticalement. La courbe I correspond au cas où l'on ne fait pas intervenir les ions dans le calcul du tracé des rayons. La courbe II tient compte de la présence des trois ions O^+, He^+ et H^+. La démonstration de l'existence des points de rebroussement a été faite par CERISIER (réf. [17], Sect. 5). (D'après KIMURA[21].)

satellite *après* avoir été réfléchi et propagé transversalement à une altitude supérieure à celle du satellite. Dans le second cas, en plus de la trace précédente on reçoit le sifflement également à la montée, alors que la vitesse de groupe est une fonction croissante de la fréquence *.

La méthode de tracé des rayons montre également que deux trajets au moins sont probables entre l'éclair originel et le satellite [19]. Un de ces trajets est effectué en propagation presque transverse (et ne peut par conséquent être reçu au sol). L'autre correspond au trajet quasi longitudinal habituel. On peut donc recevoir des sifflements possédant deux traces (Fig. 19). Il peut même arriver [21] que pour certaines directions de la normale à l'onde, à l'entrée dans la magnétosphère, le sifflement effectue un grand nombre d'aller-et-retour avant d'être absorbé dans la magnétosphère (Fig. 20). De tels sifflements périodiques auraient été récemment observés [18].

5. Phénomènes d'anisotropie et de guidage. Bien qu'elle n'ait pas permis d'obtenir jusqu'à présent dans le domaine T.B.F. des résultats concluants, l'aniso-

---

* Il faut noter qu'en propagation oblique, la courbe donnant la vitesse de phase diffère notablement de la courbe «e» de la Fig. 1 dès que $\Theta \neq 90°$; l'explication simple que nous avons donnée n'est donc pas rigoureuse.

[20] G. W. PFEIFFER, D. A. GURNETT, S. D. SHAWHAN et R. SHAW: Nature **210**, 827 (1966).
[21] I. KIMURA: Radio Sci. **1**, 269 (1966).

tropie présentée par par le milieu pour la propagation de ces ondes est un phénomène trop important pour que nous n'en n'exposions pas les caractéristiques particulières.

α) *Données générales.* Le comportement général d'un milieu anisotrope est décrit dans la référence [3]. Nous rappelons ici, sans démonstration, la définition des grandeurs qui nous seront utiles.

Les angles sont mesurés à partir du champ magnétique directeur. Si Θ est l'angle de la normale à l'onde par rapport au champ, l'énergie se propage dans une direction d'angle polaire ϑ définie par[1]:

$$\vartheta = \Theta - \gamma \tag{5.1}$$

$$\tan \gamma = \frac{1}{W} \cdot \frac{\partial W}{\partial \Theta} \tag{5.2}$$

avec une vitesse

$$V = \frac{V_g}{\cos \gamma}. \tag{5.3}$$

V_g étant la vitesse de groupe après projection dans la direction de la normale, Θ, définie par

$$V_g = c_0/\mu'.$$

Les courbes (normalisées) de coordonnées polaires:

$$\mu/\alpha \, (\equiv \mu |Y_e|/X_e^{\frac{1}{2}}), \;\; \Theta \;\; \text{et} \;\; V\alpha/c_0 \, (\equiv V X_e^{\frac{1}{2}}/|Y_e| \, c_0), \;\; \vartheta$$

sont appelées respectivement *courbes d'indice* et *courbes des rayons.* Ces courbes forment une famille à un paramètre: la fréquence; nous les désignerons par C_x et Γ_x.

β) *Application aux conditions spécifiques de la magnétosphère dans le cas des très basses fréquences.* Nous nous limitons dans tout ce qui suit à un milieu contenant une seule espèce d'ion. Nous allons chercher l'équivalent de la quartique de Booker[2] dans le cas où $\alpha^2 \equiv f_N^2/f_B^2 \gg 1$ et où $x \equiv f/f_B < 1$. Dans ces conditions, les différentes vitesses sont définies par[*]

$$W_{+1}^2 = \frac{c_0^2}{\alpha^2}(1-x)\left(\frac{1}{m^*}+x\right) \equiv \frac{c_0^2}{X_e}(|Y_e|-1)\frac{|Y_e|+m^*}{m^*}, \tag{5.4}$$

$$W_{-1}^2 = \frac{c_0^2}{\alpha^2}(1+x)\left(\frac{1}{m^*}-x\right) \equiv \frac{c_0^2}{X_e}(|Y_e|+1)\frac{|Y_e|-m^*}{m^*}, \tag{5.5}$$

$$W_e^2 = \frac{c_0^2}{\alpha^2}\left(\frac{1}{m^*}-x^2\right) \equiv \frac{c_0^2}{X_e}\frac{(Y_e^2-m^*)}{m^*}, \tag{5.6}$$

$$W_0^2 = -\frac{c_0^2}{\alpha^2} \cdot x^2 \equiv -\frac{c_0^2}{X_e}. \tag{5.7}$$

En reportant ces expressions dans l'équation (2.7), on obtient l'expression générale de la vitesse de phase.

$$\frac{W^2 \alpha^2}{c_0^2} \equiv \frac{W^2 X_e}{c_0^2 Y_e^2} = \frac{1}{m^*}\left(1-\frac{\sin^2\Theta}{2}\right) - x^2 \pm \sqrt{\frac{\sin^4\Theta}{4m^{*2}} + x^2 \cos^2\Theta\left(1-\frac{1}{m^*}\right)^2}. \tag{5.8}$$

---
[*] L'équation (5.4) est l'équivalente de (3.2), et (5.6) de (4.20), pour les conditions indiquées.
[1] Voir K. Rawer et K. Suchy: Cette encyclopédie, tome 49/2, p. 104.
[2] Voir K. Rawer et K. Suchy: Cette encyclopédie, tome 49/2, p. 128.

Les approximations de l'équation (5.8), dont il existe d'autres formes équivalentes [*13*], sont simples dans deux cas:
— celui des fréquences nulles, $(x \equiv 1/|Y_e| \ll 1/m^*)$, pour lequel on retrouve les deux modes dont l'un est isotrope:

mode d'ALFVEN
$$W_+ = c_0/\alpha \sqrt{m^*} = V_A \tag{5.8a}$$

et l'autre anisotrope:

mode modifié d'ALFVEN
$$W_- = c_0 \cos \Theta/\alpha \sqrt{m^*} = V_A \cos \Theta \tag{5.8b}$$

— celui des fréquences supérieures à la fréquence de résonance hybride basse $(x^2 \equiv 1/Y_e^2 \gg 1/m^*)$, pour lequel on obtient l'équation simple

$$W_+ \alpha/c_0 = x^{\frac{1}{2}}(\cos \Theta - x)^{\frac{1}{2}} \tag{5.9}$$

soit
$$\frac{W_+}{c_0} \frac{X_e^{\frac{1}{2}}}{|Y_e|} = \frac{(|Y_e| \cos \Theta - 1)^{\frac{1}{2}}}{|Y_e|}$$

d'où
$$\mu = \frac{\alpha}{x^{\frac{1}{2}}(\cos \Theta - x)^{\frac{1}{2}}} \equiv \frac{X_e^{\frac{1}{2}}}{(|Y_e| \cos \Theta - 1)^{\frac{1}{2}}} \tag{5.10}$$

qui redonne l'Equ. (3.4) lorsque $\Theta = 0$. (La solution W_- est imaginaire: lorsque $f_r < f < f_B$, seul le mode droit (R wave) défini par $W = W_+$ peut se propager.)

En utilisant les équations (5.1) à (5.3) et (5.10), on peut alors définir les courbes Γ_x par[3]

$$\frac{V\alpha}{c_0} = \frac{x^{\frac{1}{2}}(\cos \Theta - x)^{\frac{1}{2}} [\sin^2 \Theta + 4(\cos \Theta - x)^2]^{\frac{1}{2}}}{\cos \Theta} \tag{5.11}$$

soit
$$\frac{V}{c_0} X_e^{\frac{1}{2}} = \frac{(|Y_e| \cos \Theta - 1)^{\frac{1}{2}} [Y_e^2 \sin^2 \Theta + 4(|Y_e| \cos \Theta - 1)^2]^{\frac{1}{2}}}{|Y_e| \cos \Theta} \tag{5.11a}$$

$$\tan \vartheta = \frac{\sin \Theta (\cos \Theta - 2x)}{1 + \cos \Theta (\cos \Theta - 2x)} \equiv \frac{\sin \Theta (|Y_e| \cos \Theta - 2)}{|Y_e| + \cos \Theta (|Y_e| \cos \Theta - 2)}. \tag{5.12}$$

γ) *Courbes d'indice et courbes de rayon*. A l'aide des équations (5.10) à (5.12), il est possible de tracer les faisceaux de courbe C_x et Γ_x (Fig. 21 et 22)[4].

Sur la première de ces figures apparaît l'asymptote «gyrorésonance» électronique en propagation oblique définie par

$$\cos \Theta = x \equiv 1/|Y_e| \equiv f/f_B. \tag{5.13}$$

On remarque également — et l'on démontre[5] — que pour les fréquences $f < f_B/2$, les courbes d'indice sont tangentes à une droite perpendiculaire au champ magnétique directeur. Cette remarque nous servira ultérieurement.

L'examen de la deuxième figure fait apparaître le phénomène de *guidage* que STOREY [*7*] avait découvert, en utilisant l'approximation

$$\mu = \frac{\alpha}{(x \cos \Theta)^{\frac{1}{2}}} \equiv \frac{X_e^{\frac{1}{2}}}{|Y_e|^{\frac{1}{2}} \cos^{\frac{1}{2}} \Theta} \tag{5.14}$$

qui est l'équivalent de (3.5) pour une propagation non longitudinale:

---
[3] R. GENDRIN: Compt. rend. **251**, 1085 (1960).
[4] Des courbes plus détaillées se trouvent dans les références [*5*] et [*13*], voir aussi K. RAWER et K. SUCHY: Cette encyclopédie, tome 49/2, p. 490, Fig. 312.
[5] R. GENDRIN: Planet. Space Sci. **5**, 274 (1961).

Pour les fréquences très basses, la direction du rayon fait un angle faible avec le champ magnétique directeur, quelle que soit la direction de la normale à l'onde. Cet angle est toujours inférieur à la valeur $\vartheta_{\max}$, qui est une fonction décroissante

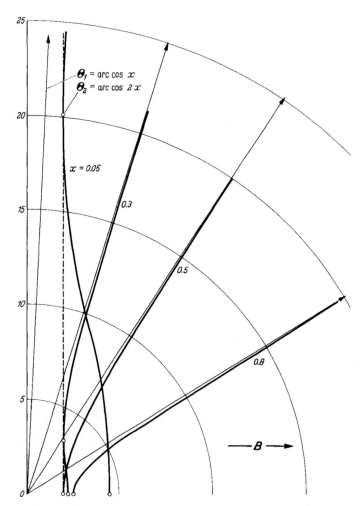

Fig. 21. Courbes d'indice, en fonction de l'angle, pour différentes fréquences. Pour chaque fréquence $f = x f_B$, on a porté, en fonction de l'angle que fait la direction de la normale à l'onde et celle du champ magnétique, le rapport entre l'indice de phase et le paramètre caractéristique du milieu, $\alpha = f_N/f_B = X^{\frac{1}{2}}/|Y_e|$. Toutes les courbes d'indice admettent pour enveloppe une droite perpendiculaire à la direction du champ magnétique. Le point de contact de chaque courbe avec cette enveloppe a lieu pour un angle $\Theta_2 = \arccos 2x$. (D'après GENDRIN [5].)

de la fréquence, et vaut,

$$\vartheta_{\max} = \arctan \frac{[(1-x^2)^{\frac{1}{2}} - \sqrt{3}\,\overline{x}]^{\frac{3}{2}}}{2^{\frac{3}{2}}(1-x^2)^{\frac{3}{2}}} \equiv \left(\frac{1}{2}\,|Y_e|\right)^{\frac{3}{2}} \frac{[(Y_e^2-1)^{\frac{1}{2}} - \sqrt{3\,|Y_e|}]^{\frac{3}{2}}}{(Y_e^2-1)^{\frac{3}{2}}}. \quad (5.15)$$

Pour $x = 0$

$$\vartheta_{\max} = \arctan \frac{1}{\sqrt{8}} = 19°\,29', \quad (5.16)$$

C'est la valeur donnée par STOREY [7].

Lorsque la fréquence augmente, l'angle $\vartheta_{\max}$ n'est plus celui qui donne l'inclinaison maximale du rayon sur le champ magnétique[6]. C'est l'angle

$$\vartheta_1 = \pi/2 - \operatorname{arc\,cos} x \tag{5.17}$$

En effet, pour $\Theta = \operatorname{arc\,cos} x$, la courbe d'indice présente une direction asymptotique. La direction de propagation de l'énergie (qui est toujours parallèle à la normale à la courbe d'indice) devient alors perpendiculaire à la normale d'onde. L'angle ϑ_1 croît lorsque x croît, alors que $\vartheta_{\max}$ défini par Eq. (5.15) décroît lorsque x croît.

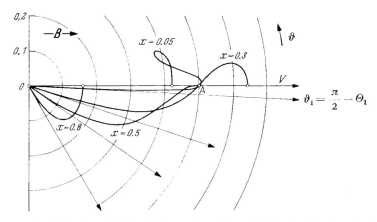

Fig. 22. Vitesse de propagation de l'énergie pour différentes fréquences. Ces courbes sont les lieux des points de coordonnées polaires: $V\alpha/c_0, \vartheta$ (voir texte pour la définition de ces grandeurs). On remarque que toutes les courbes pour lesquelles $f < \tfrac{1}{2} f_B$, passent par le point A tel que $\vartheta = 0$, $V = c_0/2\alpha$. (D'après GENDRIN[5].)

Le guidage est donc le plus efficace lorsque ces deux angles sont égaux. Ceci a lieu pour

$$x \equiv 1/|Y_e| = 1/2\sqrt{7} \approx 0{,}189 \tag{5.18}$$

et l'on a alors:

$$\vartheta_1 = \vartheta_{\max} \approx 10°\,54' \tag{5.19}$$

δ) *Le mode non dispersif.* On remarque, sur la Fig. 22, que toutes les courbes Γ_x, correspondant à $x < 0{,}5$, soit $|Y_e| > 2$, passent toutes par le point A de coordonnées polaires $(\tfrac{1}{2}, 0)$. Il existe en effet, pour chaque fréquence, deux valeurs de Θ pour lesquelles $\vartheta = 0$. L'équation (5.12) montre que cela se produit lorsque

$$\Theta = 0 \quad \text{propagation strictement longitudinale}$$

ou lorsque

$$\Theta = \Theta_2 = \operatorname{arc\,cos} 2x. \tag{5.20}$$

C'est pour cette valeur de Θ que les courbes d'indice sont tangentes à leur enveloppe. Lorsque $\Theta = \Theta_2$, l'équation (5.11) montre alors que

$$V = V_0 = c_0/2\alpha \equiv \frac{c_0}{2} \frac{|Y_e|}{X_e^{\frac{1}{2}}} \equiv \frac{c_0}{2} \frac{f_B}{f_N}. \tag{5.21}$$

Cette vitesse est indépendante de la fréquence, ce qui explique l'origine du segment de droite vertical représenté sur la Fig. 5. La vitesse V_0 joue par ailleurs

---

[6] R. L. SMITH: J. Res. Nat. Bur. Standards **64** D, 505 (1960). Voir aussi K. RAWER et K. SUCHY: Cette encyclopédie, tome 49/2, p. 108, Fig. 47.

un rôle particulier dans les mécanismes d'émission d'ondes T.B.F. (voir Sect. 6). Les raisons géométriques de ce phénomène ont été expliquées[7,8].

ε) *Guidage par les gradients d'ionisation.* Le mécanisme que nous venons d'étudier est néanmoins insuffisant pour expliquer les nombres très grands d'échos (jusqu'à 40) qu'il est quelquefois possible d'observer sur les enregistrements pour un seul éclair initiateur. Le seul guidage par anisotropie n'est pas assez focalisant pour entraîner un tel phénomène. C'est pourquoi deux autres mécanismes ont été invoqués, que nous étudierons brièvement. Supposons d'abord que la densité électronique présente une symétrie de révolution autour d'une ligne de force particulière, mais possède un faible gradient perpendiculairement à cette ligne de force. On a ainsi réalisé une *fibre d'ionisation*, l'ionisation au centre peut être plus grande (fibre dense) ou plus faible (fibre creuse) que sur les bords. On peut montrer que de telles fibres jouent le rôle de guide pour les ondes T.B.F.[9,10]. Nous donnons la démonstration dans deux cas simples et renvoyons à un ouvrage plus détaillé pour les cas complexes [5].

Admettons que le rayon se propage dans un plan de symétrie et supposons une fibre dense. A la surface de séparation de deux couches de densité N_0 et N_1, la normale à l'onde suit la loi de Descartes, qui, compte tenu de la définition de Θ, et de l'orientation de la surface de séparation, s'exprime

$$\mu \cos \Theta = \text{const} \tag{5.22}$$

N_1 étant $< N_0$ on a

$$\mu_1(\Theta=0) < \mu_0(\Theta=0).$$

La condition (5.22) montre alors que le rayon, perpendiculaire à la surface des indices, se rapproche de la normale si l'on est dans le cas où $f \ll |f_B|/2$ (Fig. 23a). On obtient ainsi le trajet indiqué à droite de la figure.

Supposons au contraire une fibre creuse et intéressons-nous à des fréquences supérieures à $|f_B|/2$. La Fig. 21 montre que les surfaces des indices ont une concavité qui n'est plus dirigée vers l'origine. Si $N_1 > N_0$,

$$\mu_1(\Theta=0) > \mu_0(\Theta=0).$$

Mais l'application des lois de Descartes montre que dans ce cas aussi le rayon a tendance à se rapprocher du champ magnétique lorsqu'on s'éloigne du centre de la fibre, parce que rayon et normale à l'onde sont de part et d'autre du champ directeur (Fig. 23b).

L'excès ou le défaut de densité électronique nécessaire pour que de tels guidages existent dépend de la fréquence et de l'angle Θ_0 de départ. Pour le cas les plus courants, il est de l'ordre de 2%. De tels gradients d'ionisation sont fort probables. Ils ont leur prolongement dans les irrégularités des couches E et F, alignées le long du champ magnétique[11]. C'est vraisemblablement parmi ces irrégularités que l'énergie de l'éclair trouve son chemin vers la magnétosphère. Le satellite FR 1, conçu pour étudier des propagations semblables à partir d'émetteurs puissants, apportera certainement des renseignements relatifs au guidage par de telles fibres[12].

ξ) *Guidage par les discontinuités d'ionisation et la courbure des lignes de force.* Lorsque le milieu est homogène, du point de vue de la densité électronique, le seul fait que le champ magnétique change de direction le long d'une ligne de force, entraîne un changement de la direction de la normale à l'onde, et donc une courbure de rayon. La trajectoire de celui-ci peut-être calculée. On peut montrer

---

[7] R. Gendrin: Planet. Space Sci. **5**, 274 (1961).
[8] R. Gendrin, in: Cosmic rays, solar particles and space research (B. Peters, ed.), p. 125—134. New York: Academic Press 1963.
[9] R. L. Smith, R. A. Helliwell, and I. Yabroff: J. Geophys. Res. **65**, 815 (1960).
[10] R. L. Smith: J. Geophys. Res. **66**, 3699 (1961).
[11] Voir K. Rawer et K. Suchy: Cette encyclopédie, tome 49/2, p. 353—355.
[12] Voir Sect. 8.

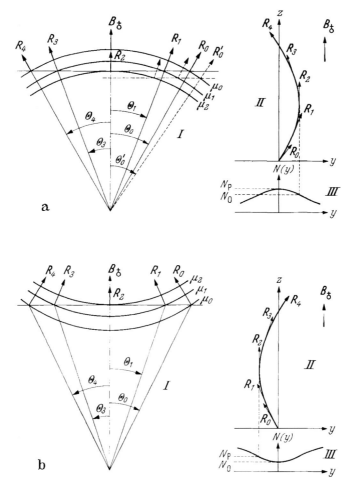

Fig. 23a et b. Guidage dans une fibre dense pour les fréquences extrêmement basses (a), ou dans une fibre creuse, pour les fréquences supérieures à la demi-gyrofréquence électronique (b). $\Theta_0, \Theta_1 \ldots$: directions successives des normales au plan d'onde au cours de la propagation du rayon; $\mu_0, \mu_1, \ldots$: courbes d'indice correspondant aux différents points sur le trajet du rayon; $\boldsymbol{R_0}, \boldsymbol{R_1}, \ldots$: directions de propagation de l'énergie. (a) Par suite de l'allure de la répartition de densité électronique en fonction de la distance au centre de la fibre (III) et lorsque la fréquence considérée est faible, les courbes μ_0, μ_1, μ_2 ont l'aspect indiqué en I (voir Fig. 21). L'application des lois de Descartes à chaque interface ($\|\boldsymbol{B}_{\dot{\circ}}$) montre que $\mu \cos \Theta$ doit rester constant. $\boldsymbol{R_0}, \boldsymbol{R_1}$ et $\boldsymbol{R_2}$ étant perpendiculaires aux surfaces d'indice, le rayon est rabattu vers la direction du champ magnétique et l'énergie suit le trajet de la figure II. Pour un angle initial Θ'_0 plus grand que Θ_0, ce rabattement est insuffisant et l'onde n'est pas piégée dans la fibre. (b) Même explication que précédemment mais, pour ces valeurs de la fréquence ($f > f_B/2$), la concavité de la courbe d'indice, pour les valeurs faibles de Θ, est dirigée dans l'autre sens (voir Fig. 21). Le rayon et la normale à l'onde sont donc de part et d'autre de $\boldsymbol{B}_{\dot{\circ}}$. Pour que le rayon soit rabattu vers $\boldsymbol{B}_{\dot{\circ}}$ (II), il faut que la densité croisse lorsque l'on s'éloigne du centre de la fibre (III). Il y a aussi une valeur limite de Θ_0, non représentée sur la figure: c'est celle au-delà de laquelle la propagation n'est plus possible, même en milieu homogène: $\cos \Theta'_0 = f/f_B$. (D'après Helliwell [5].)

qu'il est impossible d'obtenir un guidage parfait, au sens où la trajectoire du rayon coïnciderait avec une ligne de force.

D'un autre côté, lorsque l'on suppose une discontinuité d'ionisation, répartie le long d'une coquille magnétique, on peut montrer que l'on a un guidage parfait des ondes, même en l'absence d'anisotropie du milieu. Pour cela, on remplace le milieu réel par un milieu stratifié horizontalement, en introduisant un indice modifié, fonction de la courbure de la ligne de force*. On applique alors la théorie des modes et l'on calcule le saut minimal de densité (pour une fréquence donnée) qui permet ce guidage parfait[13,14].

Le tableau 3 indique, dans le cas d'un feuillet, les différentes possibilités de guidage. Le feuillet peut être dense ($\Delta N > 0$ au centre), ou creux ($\Delta N < 0$ au centre). Dans ces calculs[15], il a été tenu compte de l'anisotropie, simplement en introduisant des valeurs différentes pour l'indice de phase suivant que $\Theta = 0$ ou $\Theta = \Theta_2$ [Equ. (5.20)]. En effet, c'est uniquement lorsque la normale à l'onde est voisine d'une de ces deux directions au bord des discontinuités que le guidage est possible.

Tableau 3. *Conditions de guidage par les feuillets magnétosphériques pour une latitude géomagnétique voisine de 50°*

Nature du guidage	$f < f_B/2$	$f_B/2 < f < f_B$	$f > f_N$ (guidage H.F pour $f = 10$ MHz)
	$\Delta N > 0$		
Guidage au dessous du feuillet (par courbure)			Possible, mais conditions très sévères $\Delta N/N \sim 15\%$
Guidage à l'intérieur du feuillet (ondes de surface)	Possible Angles faibles ($\Theta \sim 0$) conditions moins critiques ($\Delta N/N \sim 5\%$)		
Guidage au dessus du feuillet (réflexions successives)	Possible Angles voisins de Θ_2 vitesse indépendante de la fréquence très faible supplément d'ionisation ($\Delta N/N \sim 0,1\%$)	Possible Angles faibles $\Delta N/N \sim 3\%$	
	$\Delta N < 0$		
Guidage au dessous du feuillet	Possible $\Delta N/N \sim 1\%$		
Guidage à l'intérieur du feuillet	Possible $\Delta N/N \sim 0.5\%$	Possible	Possible $\Delta N/N \sim 15\%$

Pas de guidage au-dessus du feuillet

Des valeurs de $\Delta N/N$ beaucoup plus faibles que celles calculées dans les références[9,15] ont été obtenues, lorsqu'on tient compte de façon correcte de l'anisotropie importante du milieu[16].

Signalons que dans un travail récent[17], on a pu, en partant des équations de Haselgrove[18], établir l'expression de $d\Theta/ds$ (Θ étant l'angle de la normale à l'onde et du champ magnétique, ds étant l'élément de parcours le long du trajet de l'énergie).

La variation de Θ le long du trajet du rayon est due principalement à trois termes: le changement de direction du champ magnétique, la variation de son intensité, et celle de la densité électronique. L'étude de l'importance relative de ces trois termes permet de montrer de façon simple pourquoi l'on peut, au cours de certains trajets, atteindre une propagation transversale et rebrousser chemin

---

* Comme pour l'étude du guidage troposphérique.
[13] H. G. Booker: J. Geophys. Res. **67**, 4135 (1962).
[14] H. G. Booker, and M. S. V. Gopal Rao: J. Geophys. Res. **68**, 387 (1963).
[15] J. Voge: Ann. Télécom. **16**, 288 (1961); **17**, 34 (1962).
[16] A. D. M. Walker: J. Atm. Terr. Phys. **28**, 807 (1966).
[17] J. Cerisier: Ann. Géophys. **23**, 249 (1967).
[18] J. Haselgrove: Report of Conference on the Physics of the Ionosphere, London Phys. Soc. 1954, p. 355.

(voir Fig. 20). Certaines conditions sont également favorables pour que l'on puisse atteindre les résonances obliques définies par $\cos \Theta = f/|f_B|$.

6. Données sommaires sur les émissions T.B.F. Dans la même bande de fréquence que les sifflements, une autre catégorie de phénomènes radioélectriques naturels sont observés, que l'on désigne sous le nom d'émissions de très basse fréquence. Pas plus que pour les phénomènes U.B.F.*[1], une théorie définitive de l'origine de ces phénomènes n'a pu être établie.

Ces phénomènes ont fait l'objet d'un grand nombre de travaux tant théoriques qu'expérimentaux, qu'il n'est pas question de résumer ici (voir [5], [9], [15], [16] et [17]). Nous voulons simplement décrire quelques-unes des caractéristiques essentielles de ces phénomènes et exposer les fondements des mécanismes qui ont été proposés pour en interpréter l'origine.

α) *Structure spectrale des émissions T.B.F.* Les émissions T.B.F. détectées au sol se divisent en quatre grandes classes suivant l'allure de leur spectre de fréquence en fonction du temps:

— le souffle,
— le choeur de l'aube,
— les émissions discrètes,
— les émissions périodiques.

Le Tableau 4, emprunté à Helliwell [5], donne les principales formes spectrales de ces différents phénomènes. L'échelle horizontale des temps couvre environ 10 à 20 sec, l'échelle verticale est celle des fréquences comprises approximativement entre 0 et 10 kHz. Les Figs. 24, 25, et 26, donnent quelques exemples de ces émissions.

Lorsqu'on disposera de plus d'enregistrements effectués à bord de satellites, il est vraisemblable que d'autres types d'émission apparaitront, qui ne peuvent parvenir jusqu'au sol. Il s'agit en particulier des émissions au voisinage des résonances, dont nous avons un exemple, voir Sect. 4.4.

Les caractéristiques spectrales (fréquence en fonction du temps) des émissions reçues au sol sont la conséquence de deux phénomènes qu'il est impossible d'isoler

— l'émission: chaque fréquence peut être émise en des points différents d'une même ligne de force, à des temps différents
— la propagation: une fois engendrée, l'onde doit se propager entre son point d'émission et la station d'observation. Cette propagation s'effectue suivant une loi du type Equ. (3.8).

Par suite de la forte dispersion du milieu, on conçoit que le spectre (f, t) reçu, soit très différent du spectre (f, t) initial et que l'interprétation en soit rendue d'autant plus difficile.

β) *Relations avec d'autres phénomènes géophysiques.* De nombreuses études statistiques ont été effectuées dans le but de déterminer les conditions d'apparition des différents phénomènes en fonction de l'heure, de la latitude, de l'indice d'activité magnétique ... Nous n'en voulons citer que quelques unes, dont les résultats montrent la complexité des phénomènes mis en jeu [9].

Le phénomène de souffle (principalement le souffle «auroral» de fréquence plus élevée et de nature plus impulsive) est souvent lié aux précipitations de particules. Celles-ci entraînent souvent une ionisation supplémentaire dans les basses couches de l'ionosphère, où les fréquences de collision sont élevées, ionisation qui entraîne un supplément d'absorption mesurée au «riomètre»[2].

Au moyen des mécanismes que nous décrivons ultérieurement, les mêmes flux de particules engendrent des émissions T.B.F. Mais les ondes ainsi engendrées sont absorbées dans

---

* Dûes à la magnéto-dynamique des gaz, appelées souvent émissions «hydromagnétiques» tout simplement.
[1] Voir E. Selzer: Contribution dans le tome 49/4 de cette encyclopédie.
[2] Voir K. Rawer et K. Suchy: Cette encyclopédie, tome 49/2, Sect. 46.

Tableau 4. *Les différents types d'émissions T.B.F. détectées au sol. (d'après* HELLIWELL [5]).

Type (Name)	Spectre typique (Model spectral form)
I. Souffle (Hiss)	
II. Emissions discrètes (Discrete emissions)	
A. Ton croissant (Rising tone)	
B. Ton décroissant (Falling tone)	
C. Crochet (Hook)	
D. Combinaisons (Combinations)	
III. Emissions périodiques (Periodic emissions)	
A. Dispersives (Dispersive)	
B. Non-dispersives (Non-dispersive)	
C. A plusieures composantes (Multiphase)	
D. A fréquence variable (Drifting)	
IV. Chœur de l'aube (Chorus)	
V. Emissions quasi-périodiques (Quasi-periodic emissions)	
VI. Emissions déclenchées (Triggered emission)	

Sect. 6. Données sommaires sur les émissions T.B.F. 499

Fig. 24. Emissions T.B.F.: Emissons discrètes enregistrées en différentes stations. (D'après Helliwell [5].)

les couches ionosphériques nouvellement créées. Il est donc habituel qu'au cours d'un phénomène entraînant *un peu* d'absorption, on observe des émissions T.B.F. Mais si ce phénomène entraîne *beaucoup* d'absorption, l'émission n'est plus détectable au sol[3,4]. (Ceci explique aussi le fait que — dans un autre domaine — la valeur de K_p pour laquelle on obtient l'activité maximum des sifflements décroît nettement avec un acroissement de la "latitude invariante"[5].)

Un certain nombre de travaux ont été faits dans le but de savoir si les conditions dans lesquelles les émissions T.B.F. pouvaient être engendrées étaient les mêmes que celles au

---

[3] L. Harang, and R. Larsen: J. Atm. Terr. Phys. **27**, 481 (1965).
[4] W. L. Ecklund, J. K. Hargreaves, and J. H. Pope: J. Geophys. Res. **70**, 4285 (1965).
[5] G. McAllcock: J. Inst. Telecom. Engrs. **12**, 158 (1966).

32*

Fig. 25a. Emissions T.B.F.: exemples de souffle mélangé à du chœur de l'aube. (D'après Helliwell [5].)

cours desquelles les oscillations hydromagnétiques étaient produites[6]. Il semble que le souffle soit souvent associé aux pulsations irrégulières, mais que le chœur de l'aube ne se produise jamais en présence d'agitation magnétique rapide[7].

Sur la Fig. 27 est représenté un ensemble de phénomènes se produisant simultanément. De haut en bas sont portés les phénomènes suivants: agitation magnétique lente (baie), amplitude de l'agitation magnétique rapide, intensité de l'aurore, absorption riométrique, valeur de $foF2$ et indications des couches Es, intensité du bruit radioélectrique V.L.F.[8].

---

[6] H. M. Morozumi: Rept. Iono. and Space Res. in Japan **19**, 286 (1965).
[7] R. Gendrin et J. Vigneron: Compt. rend. **263**, 1147 (1966).
[8] M. Dassac, R. Gendrin, N. Kleimenova, C. Ponsot, V. A. Troitskaya et J. Vigneron Ann. Géophys. **22**, 519 (1966).

Sect. 6. Données sommaires sur les émissions T.B.F. 501

Fig. 25b. Emissions T.B.F.: deux bandes de fréquence légèrement croissante enregistrées à Kuerguelen (57° = S géomagn.) le 1er Avril 1964. On remarque, moins intenses, des traces de sifflements à échos multiples (vers 19, 39, 20 ;19.40.50 et 19.44.30). (D'après DASSAC et al.[7].)

Fig. 26a. Différents types d'émissions périodiques intenses. (D'après Helliwell [5].)

On remarque sur ce dernier enregistrement la présence d'impulsions très intenses et courtes au moment du déclenchement du phénomène. De 2200 T.U. à 0100 T.U. l'absorption riométrique est très forte et l'on note la disparition, sur l'enregistrement, de toute émission T.B.F.

Enfin, il faut indiquer que certaines émissions T.B.F. semblent être déclenchées par le passage d'ondes radioélectriques dans la magnétosphère. Des expériences effectuées avec des stations émettant en «Morse» ont même montré que ce passage devait avoir une certaine durée pour déclencher l'émission T.B.F.[9]. La conséquence des ces observations du point de vue théorique est exposée par ailleurs [5], [14].

[9] R. A. Helliwell: J. Geophys. Res. **68**, 5387 (1963); R. A. Helliwell, J. Katsufrakis, M. Trimpi, and N. Brice: J. Geophys. Res. **69**, 2391 (1964).

Sect. 6. Données sommaires sur les émissions T.B.F. 503

Fig. 26b. Emissions T.B.F.: exemples d'émissions périodiques. On remarque la déformation progressive due à la dispersion (principalement pour les émissions débutant aux environs de 10 h 11 min 50 sec et de 10 h 12 min 5 sec). (Aimablement communiqué par M. VIGNERON.)

Fig. 27. Apparition d'émissions T.B.F. à Kerguelen au cours d'un évènement géophysique important. De haut en bas: variation de la composante H du champ magnétique (enregistrement La Cour). Amplitude du bruit magnétique aux environs de 1 Hz (pulsations irrégulières de périodes décroissantes). Intensité lumineuse de l'aurore (bande de N_2^+ à 3914 Å). Absorption ionosphérique mesurée à l'aide d'un riomètre ($f=20,5$ MHz), Fréquence critique de la couche F (en trait plein pour le jour indiqué; en tirets pour la valeur médiane), nature des couches E sporadiques et valeur de f_{min}. Bruit T.B.F. intégré dans la bande de 1,5 à 3,0 kHz. (D'après Dassac et al. [*].)

γ) *Relations entre deux points géomagnétiquement conjugués.* Les caractéristiques de conjugaison d'un phénomène magnétosphérique constituent un élément important pour son interprétation. Depuis quelques années un certain nombre d'études ont été consacrées à cet aspect du problème. On constate toujours[10] que les émissions arrivent l'une après l'autre aux deux extrémités d'une même ligne de force (Fig. 28), Le temps de parcours est identique à celui d'un sifflement de même fréquence[11], à condition de ne pas considérer que la période de répétition est le temps qui sépare deux émissions successives. Dans les *émissions multipériodiques* ce temps est en effet égal à la moitié, ou même au tiers de la période de répétition[12].

Si le phénomène de souffle présente une similitude de forme frappante entre deux points conjugués (Fig. 29), il n'en est pas du même pour le choeur de l'aube qui n'est détecté simultanément aux deux stations que dans de très rares occasions liées à des perturbations ionosphériques mondiales[13].

L'ensemble des observations que nous venons de décrire montre la richesse et la complexité du phénomène. Nous allons maintenant aborder les différentes théories qui ont été proposées pour en expliquer l'origine.

δ) *Théorie des interactions. Mécanismes fondamentaux.* Depuis GALLET[14], qui le premier a introduit la notion d'interaction entre un faisceau de particules et le plasma ambiant pour interpréter certaines émissions discrètes, la théorie des mécanismes d'interaction n'a cessé de se développer.

1° Le premier mécanisme invoqué a été celui des *tubes à ondes progressives*. On admet qu'une interaction prend naissance lorsque la vitesse longitudinale de la particule, u, est égale à la vitesse de phase W de l'onde. L'équation

$$u = W \tag{6.1}$$

jointe à une équation du type (3.4) permet alors de déterminer pour chaque vitesse longitudinale de la particule, une fréquence d'interaction. Si l'on introduit la quantité sans dimension

$$p = \alpha\, u/c_0 \equiv \frac{u}{c_0} \frac{X_e^{\frac{1}{2}}}{|Y_e|} \equiv \frac{u}{c_0} \frac{f_N}{f_B} \tag{6.2}$$

l'équation donnant les fréquences émises est du second degré

$$x^2 - x + p^2 = 0. \tag{6.3}$$

soit

$$1 + Y_e + \left(\frac{u}{c_0}\right)^2 X_e = 0 \tag{6.3a}$$

ou

$$f^2 - f\, f_B + \left(\frac{u}{c_0}\right)^2 f_N^2 = 0. \tag{6.3b}$$

Le lieu des points représentatifs de (6.3) est un cercle du plan x, p [15]. Pour les trop grandes valeurs de p, il n'y a plus d'émission possible. La valeur limite de u est la valeur V_0 défine par l'équation (5.21). Si l'on s'intéresse à la plus petite des deux racines de l'équation (6.3) elle vérifie $x = p^2$ lorsque $p \ll V_0$, soit

$$f = \frac{f_N^2}{f_B} \cdot \frac{u^2}{c_0^2}. \tag{6.4}$$

---

[10] W. B. CARSON, J. A. KOCH, J. H. POPE, and R. M. GALLET: J. Geophys. Res. **70**, 4293 (1965).
[11] R. A. HELLIWELL, and N. M. BRICE: J. Geophys. Res. **69**, 4704 (1964).
[12] N. M. BRICE: Radio Sci. **69D**, 257 (1965).
[13] R. GENDRIN, N. KLEIMENOVA, C. PONSOT, V. TROITSKAYA et J. VIGNERON: Compt. rend. **262**, 1108 (1966).
[14] R. M. GALLET: Proc. I.R.E. **47**, 211 (1959).

Fig. 28. Emissions à double périodicité enregistrée en deux stations conjuguées. Les émissions AA', BB', etc., sont décalées d'un temps égal au temps de propagation entre les deux hémisphères. L'amplitude totale du phénomène présente une deuxième périodicité d'environ 40 s. (D'après CARSON et al.[19])

Dans la magnétosphère, où la répartition du plasma suit les lignes de force [l'équation (1.5) est approximativement vérifiée], des particules de vitesse constante, émettent donc des fréquences constantes quel que soit le point de la ligne de force où a lieu l'interaction. On aurait là l'explication des émissions discrètes, dont la fréquence est constante pendant un certain temps[14].

2° Au lieu d'admettre le parallélisme de u et W, on peut songer que l'onde est émise obliquement. On a alors un *effet* ČERENKOV qui prend naissance lorsque

$$W = u \cos \Theta. \tag{6.5}$$

Fig. 29. Variation d'amplitude d'un phénomène de souffle enregistré en deux stations conjuguées. Il s'agit d'un bruit intégré dans la bande de fréquence 1,5...3,0 kHz. En ordonnée, la puissance (en échelle arbitraire). (D'après GENDRIN et al.[13])

En utilisant l'équation (5.10) pour W, on obtient alors la relation[15,16]

$$x^2 - x \cos \Theta + p^2 \cos^2 \Theta = 0 \tag{6.6}$$

soit

$$1 + Y_e \cos \Theta + \left(\frac{u}{c_0}\right)^2 X_e \cos^2 \Theta = 0 \tag{6.6a}$$

ou

$$f^2 - f f_B \cos \Theta + \left(\frac{u}{c_0}\right)^2 f_N^2 \cos^2 \Theta = 0 \tag{6.6b}$$

qui montre que pour chaque valeur de p on a émission de deux fréquences dans une direction donnée, ou émission d'une fréquence dans deux directions différentes. Lorsque

$$u = V_0 = \frac{c_0}{2\alpha} \equiv \frac{c_0}{2} \frac{|Y_e|}{X_e^{\frac{1}{2}}} \equiv \frac{c_0}{2} \frac{f_B}{f_N} \tag{6.7}$$

ces deux directions sont confondues. Elles vérifient l'Equ. (5.20) et l'on sait (voir Sect. 5.4) que dans ce cas, l'énergie se propage strictement le long du champ

---

[15] R. GENDRIN: Compt. rend. **251**, 1122 (1960).
[16] J. ETCHETO et R. GENDRIN: Compt. rend. **262**, 531 (1966).

magnétique et avec une vitesse indépendante de la fréquence, qui est justement égale à V_0 [15, 16].

Le premier des deux mécanismes invoqués fait effectivement intervenir un couplage de l'onde engendrée avec les oscillations de plasma du faisceau (lorsque celui-ci a une densité n finie). Ce couplage ne peut être que faible, car, en propagation longitudinale, l'onde T.B.F. n'a pas de composante électrique parallèle au faisceau.

Quant au second mécanisme, il correspond à un rayonnement incohérent de particules isolées, dont la puissance est trop faible si l'on fait des hypothèses vraisemblables quant au flux des particules responsables. Pour les grands angles d'émission, il semble toutefois[17] que sa puissance soit supérieure à celle de l'interaction de gyrorésonance dont nous allons maintenant donner le principe.

ε) *Les interactions de gyrorésonance.* Les faisceaux de particules chargées qui parcourent la magnétosphère, tournent autour du champ magnétique directeur $\boldsymbol{B}_\delta$ avec une fréquence f_B ou F_B et un sens, droit ou gauche, suivant qu'il s'agit d'électrons ou de protons. Les ondes T.B.F. et U.B.F. qui peuvent s'y propager* ont une polarisation circulaire, droite ou gauche, suivant le type d'onde considéré. Lorsque les vitesses u et W sont différentes, l'onde est affectée, dans un système de référence lié aux particules, d'un certain effet Doppler. Si cet effet est tel que, dans ce système, l'onde paraisse avoir une fréquence (et un sens de rotation) égaux à la gyrofréquence (et au sens de rotation) des particules, on peut avoir un échange d'énergie. En effet l'angle formé par le champ électrique de l'onde et la vitesse de la particule, au cours du mouvement de celle-ci, reste constant [17].

Un raisonnement simple montre que ces conditions sont remplies lorsque l'équation suivante est satisfaite pour la pulsation ω de l'onde:

$$\omega\left(1 - \frac{u}{W}\right) = \omega_H \tag{6.8}$$

à condition d'introduire un axe de fréquence $f = \omega/2\pi$ positif et négatif en adoptant les conventions suivantes[18]

$$\left.\begin{array}{l} \omega > 0 \text{ pour le mode R} \\ \omega < 0 \text{ pour le mode L} \\ \omega_H = \omega_B \text{ s'il s'agit d'un faisceau d'électrons} \\ \omega_H = -\Omega_B \text{ s'il s'agit d'un faisceau de protons} \end{array}\right\} \tag{6.9}$$

L'équation (6.8) est équivalente aux deux expressions**

$$\omega = \omega_H + k u, \tag{6.10}$$

$$W = \frac{u}{1 - \dfrac{\omega_H}{\omega}} . \tag{6.11}$$

En combinant l'une des équations (6.8), (6.10) ou (6.11) avec une équation de dispersion, on trouve les fréquences émises en fonction des vitesses des particules. Certaines émissions discrètes ont été expliquées par une interaction de ce type entre un faisceau d'électrons et le mode droit (interaction e, R)[19, 20].

---

* A partir de maintenant nous ne considérons plus que des ondes dont le vecteur $\boldsymbol{k}$ est parallèle à $\boldsymbol{B}_\delta$.
** Sous la forme (6.10), on reconnaît l'équation des ondes cyclotrons d'un faisceau; Equ. (6.11) exprime l'égalité des vitesses de phase de ces ondes et des ondes propagées dans le plasma.

[17] P. C. W. Fung: Planet. Space Sci. **14**, 335 (1966).
[18] R. Gendrin: Ann. Géophys. **21**, 414 (1965).
[19] R. L. Dowden: J. Geophys. Res. **67**, 1745 (1962).
[20] S. F. Hansen: J. Geophys. Res. **68**, 5925 (1963).

Lorsque les deux premières des conventions Equ. (6.9) sont adoptées, il est facile de voir que les deux équations de dispersion (2.5) pour $s = +1$ et $s = -1$, sont identiques. Il est commode, pour la suite de l'exposé de transcrire l'équation unique qui les remplace, dans le système des variables ω et $\mathbf{k}$, et d'y adjoindre l'équation (6.10). En d'autres termes, les fréquences d'interaction sont définies par les solutions du système*:

$$F(\omega, \mathbf{k}) \equiv c_0^2 k^2 - \omega^2 + \frac{\omega_N^2 \omega}{\omega - \omega_B} + \frac{\Omega_N^2 \omega}{\omega + \Omega_B} = 0, \qquad (6.12)$$

ou
$$\omega = \omega_B + ku \qquad \text{faisceau d'électrons} \qquad (6.13)$$

$$\omega = -\Omega_B + ku \qquad \text{faisceau de protons}. \qquad (6.14)$$

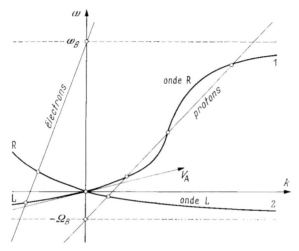

Fig. 30. Représentation graphique dans le plan ω, k des interactions de gyrorésonance. Les courbes continues 1 et 2 représentent dans ce système d'axes l'équation de dispersion pour la propagation longitudinale, dans les deux sens, des ondes T.B.F. et U.B.F. dans un magnétoplasma à un seul constituant ionique. Par convention, les fréquences positives (ou négatives) sont caractéristiques des ondes polarisées à droite (ou à gauche). Les droites représentent l'équation de dispersion des ondes cyclotrons des faisceaux (électrons ou protons); leur pente est proportionnelle à la vitesse longitudinale du faisceau. Une interaction de gyrorésonance est possible pour chacun des points d'intersection. (La partie des fréquences négatives du graphique a été agrandie pour donner plus de clarté au dessin). (D'après GENDRIN [30].)

Une représentation graphique de ce système est donnée sur la Fig. 30; la partie négative des fréquences y a été fortement agrandie par souci de clarté.

On peut y voir immédiatement comment varient les fréquences émises avec la vitesse longitudinale du faisceau. Ces variations, qui sont représentées sur la Fig. 31 pour une interaction avec le mode des sifflements, ont fait l'objet d'un grand nombre de travaux (voir par exemple [21-24]). Une application numérique détaillée suivant la nature des particules, leur énergie et la région d'interaction dans la magnétosphère, a été donnée [25].

ζ) *Instabilités de plasma au voisinage des gyrorésonances.* Jusqu'à présent, nous n'avons considéré que l'aspect phénoménologique de l'interaction sans nous

---
* Nous rappelons l'hypothèse introduite plus haut suivant laquelle $\mathbf{k}$ est parallèle au vecteur du champ magnétique, $\mathbf{B}_0$.

[21] M. A. GINZBURG: Phys. Rev. Letters **7**, 399 (1961).
[22] J. NEUFELD, and H. WRIGHT: Phys. Rev. **129**, 1489 (1963).
[23] R. STEFANT et G. VASSEUR: Compt. rend. **260**, 1465 (1965).
[24] R. STEFANT: Ann. Géophys. **21**, 402 (1965).
[25] R. GENDRIN et B. DE LA PORTE DES VAUX: Note Technique No. 51, Groupe de Recherches Ionosphériques, Paris, 1965.

510   R. GENDRIN: Phénomènes T.B.F. d'origine magnétosphérique.   Sect. 6.

préoccuper des échanges possibles d'énergie. En fait, ceux-ci sont d'autant plus grands que la densité n du faisceau est plus grande (on a toujours dependant $n \ll N$). Mais il faut considérer l'équation de dispersion totale du système plasma-faisceau. On s'aperçoit qu'il existe alors des valeurs de $\boldsymbol{k}$ auxquelles correspondent des

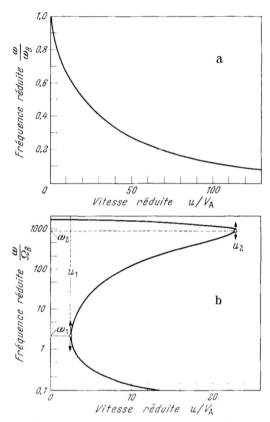

Fig. 31 a et b. Fréquences émises dans une interaction de gyrorésonance donnant naissance à une onde polarisée à droite (mode des sifflements). En abscisse, la vitesse longitudinale du faisceau; en ordonnées, la fréquence émise. (a) cas d'une interaction avec un faisceau d'électrons: la fréquence émise est une fonction continûment décroissante de la vitesse longitudinale du faisceau (d'après GALLET [15]). (b) Cas d'une interaction avec un faisceau de protons: pour certaines valeurs de la vitesse longitudinale du faisceau, on peut avoir trois fréquences émises. Les valeurs limites ont pour expressions

$v_1/V_A \cong 3\sqrt{3}/2, \quad v_2/V_A \cong \sqrt{m^*}/2, \quad \omega_1 \cong 2\Omega_B, \quad \omega_2 = \tfrac{1}{2}\omega_B.$

D'après GINZBURG [21].)

valeurs complexes de ω

$$\omega = \tilde{\omega} + i\delta. \tag{6.15}$$

Il peut alors y avoir instabilité. L'expression littérale de δ n'est aisée à obtenir que dans quelques cas simples [22, 26, 27] mais des cas plus complexes ont été étudiés [28, 29]. A titre d'exemple, nous donnons l'équation de dispersion totale* dans le cas où le faisceau, ici des protons, comporte uniquement une composante de

---

* On suppose $\boldsymbol{k} \| \boldsymbol{B}_0$, voir plus haut.
[26] T. F. BELL, and O. BUNEMANN: Phys. Rev. **133**, 1300 (1964).
[27] J. A. JACOBS, and T. WATANABE: J. Atm. Terr. Phys. **28**, 235 (1966).
[28] J. M. CORNWALL: J. Geophys. Res. **70**, 61 (1965); **71**, 2185 (1966).
[29] A. HRUSKA: J. Geophys. Res. **71**, 1377 (1966).

vitesse longitudinale u

$$D(\omega, k) = F(\omega, k) + \frac{\Omega_n^2 (\omega - ku)}{\omega + \Omega_B - ku} = 0. \tag{6.16}$$

Ω_n^2 est la pulsation de plasma du faisceau*

$$\Omega_n^2 = \frac{u\, q^2\, n}{\varepsilon_0\, m_P}.$$

Les deux valeurs de k pour lesquelles δ est maximal correspondent aux valeurs de $\tilde{\omega}$ définies par les équations (6.12) et (6.14). Les coefficients d'amplification correspondants valent alors

$$\delta^2 = -\frac{\Omega_n^2\, \Omega_B}{\delta F/\delta \omega}.$$

Les valeurs de δ ont été calculées dans le cas des interactions (e, R) et (P, R). Avec des valeurs raisonnables des différents paramètres, l'amplification calculée reste faible (~ 1 dB/1000 km)[30].

L'action du champ électromagnétique de l'onde sur les particules se traduit en retour par une diffusion de celles-ci dans l'espace des vitesses. Cette diffusion entraine la disparition des particules et donc de l'onde engendrée, à moins que d'autres particules ne soient constamment injectées. On montre qu'un tel processus conduit à un état d'équilibre, dit équilibre marginal, qui définit à la fois le flux des particules piégées et l'intensité moyenne du bruit T.B.F. dans la magnétosphère. Cette théorie très prometteuse n'en est encore qu'à ses débuts[28,31]. Notons que ce processus suppose non pas une interaction avec un faisceau monocinétique, mais avec l'ensemble des particules piégées de haute énergie. C'est l'anisotropie de la fonction de distribution angulaire de ces particules, causée principalement par l'existence d'un cône de pertes, qui entraîne l'amplification des ondes — et donc la diffusion de ces mêmes particules — en différentes régions de la magnétosphère.

Signalons enfin que des interactions de gyrorésonance peuvent même avoir lieu avec les particules d'énergie thermique du plasma lui-même. On a alors un phénomène analogue à l'amortissement LANDAU pour les fréquences voisines de la fréquence de plasma. Ce phénomène a été observé sur les sifflements électroniques[32] et sur les sifflements protoniques, pour lesquels on a une diminution brusque de l'amplitude du signal pour des fréquences inférieures de 2 à 3 Hz à la gyrofréquence protonique à proximité du satellite, F_{BS}. Cette coupure permet de déterminer avec une certaine précision la température des ions au niveau du satellite[33].

7. Méthodes d'analyse. Qu'il s'agisse de sifflements ou d'émissions T.B.F., que l'observation ait eu lieu au sol ou en satellite, on a pu remarquer que le spectre de fréquence en fonction du temps constitue la donnée indispensable, servant de base à l'étude des phénomènes que nous avons évoqués. L'amélioration des moyens *d'analyse d'un signal dont la fréquence varie en fonction du temps* conditionne donc le progrès des recherches dans ce domaine. Nous allons passer en revue et discuter quelques-unes des méthodes utilisées.

α) *Méthode originale de Storey.* Nous décrivons cette méthode car elle présente un intérêt plus qu'historique. Le sifflement passe à travers une batterie de filtres (une douzaine), de fréquence centrale f_n. La sortie de ces filtres sert à exciter des petites lampes au néon équidistantes et disposées suivant une direction verticale. Si n est le numéro du filtre, l'ordonnée de la lampe correspondante mesurée vers le bas est $z_n = na$, a étant la distance entre deux lampes successives. Une caméra,

---

* On ne confondra pas la vitesse u et le facteur u charactérisant le système d'unités utilisé (u = 1 dans un système rationalisé, $= 4\pi$ dans un système non-rationalisé).

[30] R. GENDRIN: J. Geophys. Res. **70**, 5369 (1965).
[31] C. F. KENNEL, and H. E. PETSCHEK: J. Geophys. Res. **71**, 1377 (1966).
[32] F. L. SCARF: Phys. Fluids, **5**, 6 (1962).
[33] D. A. GURNETT, and N. M. BRICE: J. Geophys. Res. **71**, 3639 (1966). Voir aussi Sect. 9.

Fig. 32. Méthode originale de Storey pour la mesure de la dispersion des sifflements. La luminosité du film est représentative de l'amplitude du signal à un instant donné dans la bande de fréquence considérée. La trace verticale représente l'éclair originel. L'espacement des filtres a été choisi de telle sorte qu'un sifflement qui obéit à la loi d'Eckersley donne une trace linéaire qui permet de mesurer directement la dispersion (figure du bas). En haut, deux autres exemples de sifflements (court à gauche, long à droite). (D'après Storey [7].)

dont le film défile horizontalement photographie les lampes qui s'allument successivement. L'originalité de la méthode consiste à prendre des fréquences centrales f_n inversement proportionnelles au carré de l'ordonnée des lampes correspondantes

$$f_n = \frac{f_1}{z_n^2} = \frac{f_1}{(na)^2}. \quad (7.1)$$

Alors, si le sifflement suit la loi (3.9), le temps t_n auquel s'allume la lampe de rang n est:

$$t_n = \frac{D}{f_n^{\frac{1}{2}}} = n\, Da/f_1^{\frac{1}{2}}. \quad (7.2)$$

On obtiendra une trace lumineuse rectiligne sur le film, dont la pente permet de déterminer la constante D (Fig. 32).

β) *Analyseurs analogiques.* En fait, la loi (3.9) n'étant pas toujours vérifiée, même pour les seuls sifflements électroniques, il est préférable d'utiliser des appareils où les axes f, t sont linéaires. Ces appareils travaillent de plus à largeur de bande constante B_a. Le temps de réponse du filtre à une fréquence pure, ou temps de résolution minimal est donc

$$t_1 = \frac{1}{\pi B_a}. \quad (7.3)$$

Deux types d'appareils de ce genre existent, dont l'un travaille en temps différé (type 1) et l'autre en temps réel (type 2)*.

Sur le premier appareil (Fig. 33 a) une certaine durée T du signal (de l'ordre de 2 sec) est enregistrée et gardée en mémoire. Le signal est ensuite relu un grand nombre de fois (N) et passe dans un filtre unique dont la fréquence varie. Il est évidemment inutile que le signal soit analysé deux fois dans la même bande de fréquence. C'est pourquoi au bout du temps T la fréquence centrale du filtre s'est déplacée approximativement d'une valeur égale à la largeur de bande d'ana-

Fig. 33a et b. Balayage comparé des analyseurs du type 1 et du type 2 (voir texte).

lyse B_a. Si F_a est la gamme totale de fréquence que l'on veut analyser (non limitée en théorie) le nombre de tours nécessaires sera $N = F_a/B_a$ et le temps d'analyse sera

$$t_1 = NT = F_a T/B_a. \quad (7.4)$$

Le second type d'analyseur, qui peut opérer en temps réel, effectue l'analyse d'une bande de fréquence à priori limitée, F_a (Fig. 33b). Le signal est envoyé sur une batterie de N' filtres en parallèle placés sur un tambour (les filtres se recouvrent légèrement en sorte que $N' > F_a/B_a$). On lit à l'aide d'un rotor, suc-

---

* L'appareil du type 1 le plus communément utilisé est le «Sona-graph»; citons, pour le type 2, le «Ray-span» et le «Spectran».

cessivement le signal à la sortie présent sur chacun des filtres. Les N' filtres sont lus en un temps t_2, limité inférieurement uniquement par des impératifs techniques. Il est inutile que le temps de réponse de chaque filtre soit supérieur à t_2 ($t_2 = 1/\pi B_a$).

Le signal de sortie d'un tel système sert généralement à moduler un Wehnelt d'oscilloscope dont le balayage horizontal est supprimé, et dont le balayage vertical, synchronisé sur la lecture des filtres, représente l'axe des fréquences. Le défilement d'un film dans une caméra qui photographie l'écran de l'oscilloscope joue le rôle d'axe des temps. Le procédé est d'une grande souplesse car on peut comprimer ou étaler le phénomène suivant l'échelle de temps à laquelle on s'intéresse, mais, comme pour les appareils de type 1, la dynamique en amplitude est faible.

γ) *Remarques sur l'analyse spectrale de phénomènes dont la fréquence varie en fonction du temps.* Les questions théoriques relatives à ces problèmes ont été traitées dans un certain nombre de publications[1,2], et nous ne traiterons ici que des conséquences pratiques dont l'influence sur l'analyse des signaux T.B.F. est importante.

Tout signal dont la fréquence f varie en fonction du temps t possède une largeur de bande intrinsèque b et une durée intrinsèque τ tels que:

$$b = \left(\frac{1}{\pi}\frac{df}{dt}\right)^{\frac{1}{2}}, \qquad (7.5\text{a})$$

$$\tau = \left(\pi\frac{df}{dt}\right)^{-\frac{1}{2}}. \qquad (7.5\text{b})$$

L'expression de ces deux grandeurs peut être obtenue par le raisonnement intuitif suivant. Si b est cette largeur intrinsèque, il lui est associé une durée $\tau = 1/\pi b$. Or, entre les instants t et $t+\tau$, on observe par hypothèse, des fréquences comprises entre f et $f + \frac{df}{dt}\tau$. En écrivant que la largeur de bande ainsi trouvée est la largeur de bande intrinsèque du signal, on obtient les relations ci-dessus. Storey [7] a montré que l'intervalle de temps τ était celui pendant lequel les fréquences comprises entre f et $f+b$ restaient en phase, donnant ainsi une définition plus précise de la durée et de la largeur de bande intrinsèque.

Lorsqu'un tel signal passe dans un filtre de largeur de bande B_a on recueille à la sortie un signal dont l'enveloppe présente une certaine durée à mi-hauteur t_3 et dont le maximum apparaît un temps t_4 après celui pour lequel le signal avait exactement la même fréquence que la fréquence centrale du filtre utilisé. Barber et Ursell ont montré que le temps t_3 est minimal lorsque l'on choisit

$$B_a \approx \tfrac{1}{2} b \qquad (7.6)$$

il vaut alors $t_3 \approx 2{,}3\,\tau$ et le décalage est $t_4 \approx \tau$.

Pour des valeurs de B_a différentes de la valeur optimale, le temps t_3 devient plus grand et le signal s'épaissit, sur les représentations de f en fonction de t. On voit donc:

1° — qu'il est nécessaire de choisir la largeur de bande d'analyse en fonction du signal à étudier. Le même analyseur spectral n'est pas automatiquement adapté à l'étude de tous les signaux naturels lorsque les pentes df/dt diffèrent trop l'une de l'autre.

2° — qu'il existe toujours une imprécision dans la mesure du temps d'arrivée d'une fréquence déterminée. Cette imprécision est dûe à la durée fine de la réponse du filtre qui dépasse toujours la durée intrinsèque τ du signal. On peut l'estimer au mieux à $t_3/10$, soit environ $\tau/4$. S'y ajoute le retard systématique, t_4, dont il faut tenir compte.

---

[1] N. F. Barber, and F. Ursell: Phil. Mag. **39**, 355 (1948).
[2] L. R. O. Storey, and J. K. Grierson: Electron. Engrs. **30**, 586; **30**, 648 (1959).

A titre d'application, considérons l'analyse d'un sifflement de dispersion $tf^{\frac{1}{2}} = D \approx 60 \text{ s}^{\frac{1}{2}}$.

$$\frac{df}{dt} = 2f^{\frac{3}{2}}/D \tag{7.7}$$

et l'on peut calculer les valeurs de B_a optimales pour différentes fréquences (Tableau 5). On voit que la largeur optimale des filtres dépend fortement de la bande de fréquence choisie. On remarque de même que le retard à la réponse lorsque $B_a = B_{opt}$ peut atteindre des valeurs importantes.

Tableau 5. *Grandeurs caractéristiques d'analyse pour un sifflement de dispersion* $D \approx 60 \text{ s}^{\frac{1}{2}}$

f/kHz	b/Hz	B_a/Hz	τ/msec	t_3/msec
0,5	10	5	30	70
10	80	40	4	9

Ce retard est supprimé lors de l'emploi de filtres symétriques en temps que l'on peut réaliser de façon analogique[2] ou numérique[3]. L'avantage de la dernière méthode est que l'on peut, pour chaque portion du signal, «adapter» le filtre de façon que sa largeur de bande soit toujours égale à B_{opt}.

δ) *Méthode d'analyse numérique.* Supposons que l'on dispose d'une sorte de ligne à retard dont la loi de dispersion soit $t = -D_0 f^{-\frac{1}{2}}$. Un sifflement de dispersion D constante (c'est-à-dire suivant rigoureusement la loi d'ECKERSLEY), passant à travers cette ligne, donnera à la sortie un signal reconstituant l'atmosphérique initial. Si D n'est pas constant, ou si D_0 n'a pas été choisi égal à D le signal de sortie ne sera pas une impulsion, mais présentera de toute façon un spectre $f(t)$ de pente df/dt très grande. On aura donc une grande précision en temps, $\tau \approx (\pi \, df/dt)^{-\frac{1}{2}}$.

On réalise en pratique l'équivalent de la ligne à retard par la convolution des spectres de Fourier du signal expérimental et d'un signal théorique qui suit la loi d'ECKERSLEY avec une constante D_0 que l'on détermine par l'examen du sonagramme[4].

Cette méthode appliquée à sifférents difflements électroniques a permis d'en déterminer la structure hyperfine. Sur la Fig. 34 sont représentées les courbes de dispersion en fonction de la fréquence, pour trois sifflement reproduits sur la partie droite de la figure. On voit qu'apparaissent une multitude de trajets correspondant chacun à une dispersion légèrement différente. On note de plus l'augmentation de la dispersion pour les basses fréquences, introduite par l'effet des ions (voir Sect. 3.5) dont on peut ainsi déterminer la concentration[5].

Une méthode similaire peut être appliquée aux sifflements protoniques pour lesquels df/dt est voisin de zéro: le problème est alors de connaitre avec précision la fréquence f_0 du signal à un temps t_0 déterminé. L'adaptation de la méthode précédente à ce problème est fondée sur la symétrie qui existe dans le rôle joué par les deux variables fréquence et temps[6].

On peut ainsi déterminer avec une précision de l'ordre de 0,2 Hz la fréquence vraie du signal à chaque instant. En possession de la courbe $f(t)$, on peut tracer t en fonction de $(F_{BS} - f)^{\frac{1}{2}}$ en donnant à F_{BS}, la gyrofréquence à proximité du satellite, différentes valeurs. Pour l'une de ces valeurs, la courbe obtenue est une droite, dont la pente permet de déterminer k (voir Sect. 4.3). La Fig. 35 montre que l'on peut déterminer F_{BS} à 0,2 Hz près[7], et donc k à 5%.

---

[3] C. SIREDEY: Note Technique No. 61, Groupe de Recherches Ionosphériques, 1966.

[4] C. BEGHIN et C. SIREDEY: Ann. Géophys. **20**, 301 (1964).

[5] C. BEGHIN, in: Electron density profiles in ionosphere and exosphere (J. FRIHAGEN, ed.), p. 587—601. Amsterdam: North-Holland Publ. Co. 1966.

[6] C. BEGHIN, R. GENDRIN, C. SIREDEY et J. VIGNERON: Ann. Géophys. **22**, 110 (1966).

[7] La méthode analogique de battement entre le signal et un oscillateur très stable dont on déplace la fréquence, ne permet de déterminer F_{BS} qu'à 2 Hz près. Voir D. A. GURNETT, and N. M. BRICE: J. Geophys. Res. **71**, 3639 (1966).

Fig. 34a—c. Résultat d'une analyse numérique fine des sifflements électroniques. A droite, sont représentés les sonagrammes des différents sifflements étudiés. A gauche, le résultat de l'analyse numérique effectuée après digitalisation du signal. Ce résultat se traduit par un ensemble de traces dans un plan dont les coordonnées sont fréquence et dispersion. Chacune de ces traces a une forme identique à celle de la Fig. 6. Lorsque les traces sont très nettement distinctes, il s'agit d'un sifflement à traject multiple (figure du milieu). La structure hyperfine est dûe, soit à des irrégularités rencontrées à la traversée de l'ionosphère, soit à la multiplicité des modes à l'intérieur d'une même fibre. (D'après BEGHIN[5].)

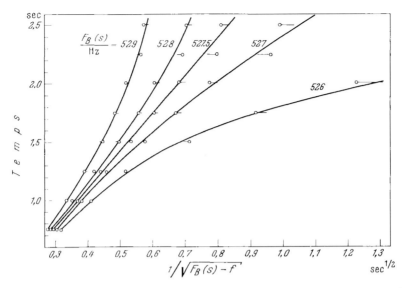

Fig. 35. Résultat d'une analyse numérique fine effectuée sur un sifflement protonique. La théorie montre que la forme asymptotique de l'équation de dispersion d'un sifflement protonique est donnée par la formule Equ. (4.11). Une analyse numérique après digitalisation permet de déterminer avec une précision meilleure que 0,2 Hz la fréquence vraie du signal à chaque instant. Mais la valeur de la gyrofréquence des protons à l'altitude du satellite F_{BS}, n'est pas connue avec cette précision. [On a écrit $F_B(s)$ sur le dessin au lieu de F_{BS}.] Aussi fait-on plusieurs hypothèses concernant la valeur de F_{BS}. Pour l'une d'entre elles, on obtient une droite dont la pente permet de calculer la densité absolue des protons au niveau du satellite. (D'après BEGHIN et al.[6].)

Ces méthodes numériques nécessitent un certain temps de calcul sur machine, mais les résultats qu'elles permettent d'obtenir sont d'une très grande précision.

8. Polarisation. La figure de polarisation présente un intérêt différent suivant qu'on l'obtient à l'intérieur même du plasma ou qu'on la détermine au sol. Dans le premier cas, elle permet de connaître la direction de la normale à l'onde et de calculer l'indice de réfraction; dans le second cas, elle fournit des renseignements précieux sur les perturbations apportées à cette figure par la traversée de l'ionosphère, et donc sur l'ionosphère elle-même.

α) *Etudes expérimentales.* Jusqu'à ces dernières années, l'étude de la polarisation, dans le domaine des ondes radio-électriques de très basse fréquence (T.B.F.), était limitée aux émissions naturelles, que ce soit à l'aide de sifflements[1] ou à l'aide d'émissions T.B.F.[2]. Ces études n'ont jamais conduit à des résultats très nets, le problème se compliquant du fait de la localisation mal connue du phénomène, de son spectre complexe en fréquence. Depuis quelques années, l'utilisation d'émetteurs à fréquence fixe (notamment celui de Sainte-Assise, France, à 16,8 kHz) permet d'étudier avec précision ces phénomènes de polarisation, tant au sol (propagation entre deux hémisphères, guide d'onde terre-ionosphère) qu'en satellite. Nous ajoutons à ce paragraphe[3] quelques précision relatives à la polarisation des ondes T.B.F. telle qu'elle a pu être étudiée grâce au satellite «FR 1».

Ce satellite, lancé le 6 décembre 1965, en coopération avec la NASA, possède une orbite presque circulaire d'altitude voisine de 750 km et inclinée de 75° sur le plan équatorial. Il comporte, pour l'étude des ondes T.B.F., 3 antennes magnétiques orthogonales et 2 antennes

---

[1] J. DELLOUE: Thése de Doctorat ès Sciences, Paris, 1960.
[2] L. HARANG and K. N. HAUGE: J. Atm. Terr. Phys. **27**, 499 (1965).
[3] Au moment de la correction des épreuves.

électriques, accordées sur la fréquence de l'émetteur de Sainte-Assise. Lors de son passage dans l'hémisphère Nord au voisinage de cette station, ou dans l'hémisphère Sud au voisinage de son point conjugé, il reçoit le champ électromagnétique émis et en mesure l'intensité et la fréquence sur chacune des composantes. Les travaux préliminaires de dépouillement ont été orientés de la façon suivante.

A l'aide des 3 composantes magnétiques, il est possible de reconstituer la figure de polarisation de la composante magnétique $\boldsymbol{b}$ de l'onde, et donc d'en déduire la direction $\boldsymbol{n}^0$ de la normale au plan d'onde puisque l'ellipse de polarisation de $\boldsymbol{b}$ est dans le plan d'onde.

Un recoupement est obtenu par la mesure de l'effet Doppler Δf produit par le mouvement du satellite: le fait d'utiliser un émetteur à fréquence fixe et bien déterminée permet en effet, non seulement de réduire le rapport signal à bruit à l'analyse, mais aussi d'obtenir une indication supplémentaire sur la normale à l'onde. On sait en effet que $\Delta f = - \boldsymbol{v} \cdot \boldsymbol{n}/\lambda$, λ étant la longueur d'onde, $\boldsymbol{v}$ la vitesse de l'engin et $\boldsymbol{n} = \mu + i\chi$ le vecteur d'indice[4]. Lorsque, sur un intervalle de temps donné, les déterminations de $\boldsymbol{n}$ par les deux méthodes sont cohérentes, il est possible, non seulement d'être certain de la direction du vecteur $\boldsymbol{n}$, mais aussi de calculer λ, c'est-à-dire l'indice de réfraction μ (χ étant négligeable) et donc, à l'aide de formules du type de l'Eq. (5.10), de déduire la densité électronique au niveau du satellite (puisqu'on connaît Θ, f, f_B et μ)[5].

Comme il n'y avait pas, à bord du satellite, 3 antennes électriques[6], certaines hypothèses ont été faites sur la constitution du champ électrique de l'onde. Ces hypothèses — qui sont en fait des conséquences de la théorie magnéto-ionique — sont les suivantes:

— l'ellipse de polarisation du champ électrique est approximativement perpendiculaire à $\boldsymbol{B}_\xi$.

— à chaque instant $\boldsymbol{E}$ est perpendiculaire à $\boldsymbol{b}$.

— le rapport du grand axe au petit axe est le même pour l'ellipse «électrique» que pour l'ellipse «magnétique».

L'indice de réfraction μ se déduit alors aisément du rapport entre les composantes magnétiques et électriques reconstituées. Le bon accord obtenu avec l'indice calculé par l'autre méthode prouve la validité des hypothèses admises. Il prouve également que l'estimation de la hauteur effective d'antenne était correcte[7].

Les conclusions principales de ces études sont les suivantes:

— Dans de nombreux cas, la figure de polarisation observée au niveau du satellite est en bon accord avec celle prévue par la théorie magnéto-ionique.

— L'indice μ que l'on peut en déduire permet de calculer une densité électronique en bon accord avec celle mesurée par une sonde d'ionisation placée sur l'engin.

— Les normales aux ondes ne sont généralement pas verticales, comme elles devraient l'être si le milieu était stratifié horizontalement. L'existence de gradients horizontaux de densité électronique est cause d'une inclinaison des normales vers l'équateur qui peut atteindre 50°.

— Les fluctuations rapides de la figure de polarisation et de la direction de la normale à l'onde sont corrélées soit avec l'existence d'une couche F diffuse, soit avec celle d'irrégularités ionosphériques dont l'étude est actuellement en cours[8].

β) *Formules*. Le problème de la polarisation des ondes électromagnétiques se propageant dans un plasma a été traité dans de nombreux ouvrages [*1* à *4*, *19*]. Il est néanmoins utile de donner les formules correspondant au cas où $\alpha \gg 1$ et $x < 1$ ($X_e \gg Y_e^2$ et $Y_e > 1$). Pour cela, nous partons des formules de SMITH et BRICE [*6*] qui ont l'avantage de donner directement et de façon symétrique l'expression des trois composantes du champ électrique (définies à une constante

---

[4] K. RAWER et K. SUCHY: Cette encyclopédie, tome 49/2, p. 25 et 444.

[5] M. P. AUBRY: Ann. Géophys. **23**, 467 (1967); — J. Atm. Terr. Phys. **30**, 1161 (1968).

[6] Pour des questions de stabilité de rotation, seules deux antennes électriques orthogonales entre elles et à l'axe du spin, ont été installées.

[7] R. J. L. GRARD: Conference on M. F. L.F. and V.L.F. Radio Propagation 8th—10th Novemb. 1967), p. 25, IEE Conference publication 36, 1968.

[8] M. P. AUBRY: Ann. Géophys. **24**, 39 (1968).

près), à partir des vitesses des modes principaux.

$$\left.\begin{aligned} E_x &\propto \frac{\sin\Theta}{2}\left[\frac{W_{+1}^2}{W_{+1}^2-W^2}+\frac{W_{-1}^2}{W_{-1}^2-W^2}\right], \\ iE_y &\propto \frac{\sin\Theta}{2}\left[\frac{W_{+1}^2}{W_{+1}^2-W^2}-\frac{W_{-1}^2}{W_{-1}^2-W^2}\right], \\ E_z &\propto \cos\Theta\left[\frac{W_0^2}{W_0^2-W^2}\right]. \end{aligned}\right\} \quad (8.1)$$

Dans ces formules, l'axe z^0 est pris le long du champ magnétique $\boldsymbol{B}_0$, et le plan x^0, z^0 contient la normale à l'onde, qui fait avec $\boldsymbol{B}_0$ l'angle Θ. Le trièdre x^0, y^0, z^0 est un trièdre direct.

On remarque de suite que le champ électrique est contenu dans un plan perpendiculaire à x^0, z^0 car E_x et E_z sont en phase.

On utilise souvent des «coordonnées polarisées», liées à la normale d'onde[9]. Définissons un trièdre direct dont les axes 1, 2, 3 sont tels que 3 soit aligné le long de $\boldsymbol{k}$, 2 soit opposé à y^0 et 1 soit, dans le plan x^0, z^0, perpendiculaire à $\boldsymbol{k}$[10]. On a les correspondances suivantes:

$$\left.\begin{aligned} E_1 &= E_z \sin\Theta - E_x \cos\Theta, \\ E_2 &= E_y, \\ E_3 &= E_z \cos\Theta + E_x \sin\Theta. \end{aligned}\right\} \quad (8.2)$$

L'avantage de ce nouveau système de coordonnées réside dans le fait qu'il permet de définir une polarisation $\varrho = E_2/E_1$ dépendant de façon simple de l'indice μ et telle que le produit $\varrho_+ \cdot \varrho_-$ des polarisations des deux ondes pouvant se propager dans une même direction Θ soit égal à 1 [19].

Pour calculer ϱ, on utilise les Equs. (8.1) et (8.2) et l'on remplace W^2 par son expression tirée de l'Equ. (2.7). Une multiplication par la quantité conjuguée permet d'éliminer les radicaux du dénominateur.

Tous calculs faits, l'on trouve:

$$\frac{1}{\varrho_\pm} = \frac{i}{\cos\Theta} \cdot \frac{(W_e^2 - W_0^2)\sin^2\Theta \mp \sqrt{(W_e^2-W_0^2)^2\sin^4\Theta + (W_{+1}^2 - W_{-1}^2)^2\cos^2\Theta}}{(W_{-1}^2 - W_{+1}^2)}. \quad (8.3)$$

L'indice $+(-)$ correspond à l'utilisation du signe supérieur (inférieur) dans les Equs. (8.3) et (8.4) donnant la vitesse de phase

$$W_\pm^2 = W_e^2 + (W_0^2 - W_e^2)\frac{\sin^2\Theta}{2} \pm \frac{1}{2}\sqrt{(W_0^2-W_e^2)^2\sin^4\Theta + (W_{+1}^2-W_{-1}^2)^2\cos^2\Theta}. \quad (8.4)$$

Application au cas $\alpha^2 \gg 1$ *et* $x < 1$.

W^2 est donné par la formule Equ. (5.8) que l'on récrit pour bien voir les signes qui se correspondent

$$W_\pm^2 = \frac{c_0^2}{\alpha^2}\left[\frac{1}{m^*}\left(1 - \frac{\sin^2\Theta}{2}\right) - x^2 \pm \sqrt{\frac{\sin^4\Theta}{4m^{*2}} + x^2\cos^2\Theta\left(1 - \frac{1}{m^*}\right)^2}\right] \quad (8.5)$$

et ϱ par la formule

$$\frac{1}{\varrho_\pm} = \frac{i}{x\cos\Theta\left(1 - \frac{1}{m^*}\right)}\left[-\frac{\sin^2\Theta}{2m^*} \pm \sqrt{\frac{\sin^4\Theta}{4m^{*2}} + x^2\cos^2\Theta\left(1 - \frac{1}{m^*}\right)^2}\right]. \quad (8.6)$$

Les approximations sont simples dans deux cas:
— celui de fréquences supérieures à la fréquence de résonance hybride basse $(x^2 \gg 1/m^*)$:

$$\varrho_+ = \frac{1}{\varrho_-} = -i. \quad (8.7)$$

---
[9] K. RAWER et K. SUCHY: Cette encyclopédie, tome 49/2, p. 36 et suivantes.
[10] Les axes 1, 2, 3 correspondent aux axes $\boldsymbol{m}, -\boldsymbol{l}, \boldsymbol{n}$, de la référence[9]. Les notations adoptées ici sont celles des références [18] et [19].

Seul le mode correspondant au signe $+$ peut se propager [voir Eq. (5.9)]. La projection du vecteur électrique dans le plan d'onde est polarisée circulairement à droite, et tourne dans le même sens que les électrons autour, du champ magnétique.

— celui des fréquences très basses ($x \ll 1/m^*$), dans lequel il faut distinguer l'approximation Q.T. de l'approximation Q.L.[11] [V_A est la vitesse d'ALFVEN].

Approximation Q.T.:

$$W_+^2 = V_A^2 = \frac{c_0^2}{\alpha^2 m^*}, \qquad \varrho_+ = -\frac{i \sin^2 \Theta}{m^* x \cos \Theta \left(1 - \frac{1}{m^*}\right)}, \qquad (8.8)$$

$$W_-^2 = V_A^2 \cos^2 \Theta = \frac{c_0^2 \cos^2 \Theta}{\alpha^2 m^*}, \qquad \varrho_- = \frac{i m^* x \cos \Theta \left(1 - \frac{1}{m^*}\right)}{\sin^2 \Theta}. \qquad (8.9)$$

La condition de validité de l'approximation Q.T. montre alors que ϱ_+ est égal au produit de $-i$ par un nombre très grand; on a donc pour ce mode une polarisation elliptique droite de très grande excentricité, dont le grand axe est confondu avec l'axe 2 $(-y^0)$. Le mode anisotrope est lui polarisé à gauche, presque linéairement le long de l'axe 1.

Approximation Q.L.:
Les deux vitesses de phases W_+ et W_- sont égales à V_A. La polarisation est définie par

$$\varrho_+ = \frac{1}{\varrho_-} = -i. \qquad (8.10)$$

On a deux modes polarisés circulairement.

γ) *Direction du plan de polarisation.* Le plan de polarisation est défini par le rapport E_1/E_3 ou E_x/E_z. C'est sous cette dernière forme que les calculs sont les plus simples.

On trouve

$$\xi \equiv \frac{E_x}{E_z} = \frac{1}{W_0^2 \sin \Theta \cos \Theta} \left\{ -W_e^2 + \frac{1}{2}[(W_0^2 + W_e^2) \sin^2 \Theta] \right. \\ \left. \mp \sqrt{(W_0^2 - W_e^2)^2 \sin^4 \Theta + (W_{+1}^2 - W_{-1}^2)^2 \cos^2 \Theta} \right\}. \qquad (8.11)$$

Lorsque $\alpha^2 \gg 1$ et $x < 1$, on a

$$\xi = -\frac{1}{x^2 \sin \Theta \cos \Theta} \left\{ x^2 - \frac{1}{m^*} - \left[\left(x^2 - \frac{1}{2m^*}\right) \sin^2 \Theta\right] \right. \\ \left. \pm \sqrt{\frac{\sin^4 \Theta}{4 m^{*2}} + x^2 \cos^2 \Theta \left(1 - \frac{1}{m^*}\right)^2} \right\}. \qquad (8.12)$$

Dans le cas des basses fréquences, et pour la propagation Q.T., on trouve

$$\xi_- = \frac{1}{m \, x^2 \tan \Theta}, \qquad (8.13)$$

$$\xi_+ = \frac{1}{m \, x^2 \sin \Theta \cos \Theta}. \qquad (8.14)$$

Le plan de polarisation est presque toujours perpendiculaire à $\boldsymbol{B_0}$.

---
[11] Dans le domaine des fréquences ultra basses, l'approximation quasi-transversale Q.T. ($(\sin^2\Theta/2m^*)^2 \gg x^2\cos^2\Theta$) est presque toujours valable dès que Θ dépasse une certaine valeur Θ_0 faible. Dans le domaine T.B.F., c'est l'approximation quasi-longitudinale, Q.L. qui est presque toujours valable, tant que Θ reste inférieur à une valeur Θ_1 grande. K. RAWER et K. SUCHY: Cette encyclopédie, tome 49/2, p. 56 et suivantes, utilisent les désignations LP ("linearly polarized") et CP ("circularly polarized") an lien de QT et QL, respectivement.

Dans le cas où $x^2 \gg 1/m^*$, pour lequel seul le mode droit se propage, avec la vitesse de phase définie par l'Equ. (5.9), on trouve[12]

$$\xi_+ = \frac{1 - x \cos \Theta}{x \sin \Theta}, \tag{8.15}$$

d'ou

$$\frac{E_3}{E_1} = \frac{\sin \Theta}{x - \cos \Theta}. \tag{8.16}$$

Lorsque $x \ll 1$, on voit que le plan de polarisation est perpendiculaire à $\boldsymbol{B}_0$ [18], ce qui justifie les hypothèses du paragraphe 8α. Toutefois, le domaine de fréquence dans lequel on a à la fois $x^2 \gg 1/m^*$ et $x < 1$ est très restreint. En pratique, pour le mode des sifflements, le plan de polarisation se rapproche de $\boldsymbol{B}_0$ lorsque Θ augmente. *A la limite, sur le cone de resonance* ($\cos \Theta = x$), *il contient la normale à l'onde* (lorsque $x = \cos \Theta$, $\xi_+ = \tan \Theta$ d'après l'Eq. (8.15)].

Ceci n'est valable — comme les autres formules — que dans le cas d'un plasma froid et sans collision. Lorsqu'il y a des collisions, il suffit de remplacer la masse m_h des particules par une masse effective $m_h (1 - iZ_h)$ avec $Z_h = \nu_h/\omega$, c'est-à-dire, remplacer X_h par $X_h \cdot (1 - iZ_h)^{-1}$ et Y_h par $Y_h (1 - iZ_h)^{-1}$. Cette correction est inexistante dans la haute ionosphère et la magnétosphère; elle joue un rôle important dans la basse ionosphère, surtout dans les domaines E.B.F. et U.B.F., où cette correction peut complètement changer l'indice de réfraction et la figure de polarisation. [$\omega = 2\pi f$ est la pulsation de l'onde.]

Les corrections de température, elles, sont *toujours* importantes dès que l'on s'approche des résonances, car il faut alors tenir compte de la vitesse thermique qui devient plus grande que la vitesse de phase[13] [20].

9. Annexe: Théorie détaillée des sifflements ioniques.

α) *Forme des sonagrammes.* Dans leur article de base, GURNETT et al. [11] démontrent l'existence des différentes fréquences dont il a été question à la Sect. 4.1. Pour cela, ils partent d'un modèle de répartition des différents ions en fonction de l'altitude (Fig. 36a). A partir de ce modèle, il est possible de calculer pour chaque altitude les différentes fréquences caractéristiques (Fig. 36b). Les courbes $L = \infty$ représentent par exemple les gyrofréquences des ions hydrogène et helium. Les courbes $L = 0$ représentent les fréquences de coupure pour chacun des modes gauches correspondants. Les courbes $D = 0$ correspondent aux fréquences de croisement ($W_{+1} = W_{-1}$) à chaque altitude.

Considérons une fréquence de 400 Hz et cherchons comment elle peut se propager du sol vers les altitudes élevées. Pour cela, on calcule la vitesse de phase pout cette fréquence en fonction de l'altitude dans les deux modes, droit et gauche (Fig. 36c). Partant du sol, on voit que le signal peut se propager dans le mode droit jusqu'à l'altitude pour laquelle $D = 0$. Dans l'hypothèse d'une très faible inclinaison sur le champ magnétique ($\Theta \neq 0$), on poursuit de façon continue sur la branche L (polarisée à gauche). La vitesse de phase décroît de plus en plus et l'onde est absorbée à l'altitude pour laquelle la gyrofréquence protonique est égale à 400 Hz.

Il est alors possible de comprendre l'allure générale d'un sifflement reçu à bord d'un satellite. Fixons nous l'altitude de celui ci, 1000 km par exemple, et limitons-nous aux fréquences supérieures à 200 Hz pour ne pas compliquer inutilement l'interprétation. Les fréquences comprises entre 200 et 350 Hz environ parviennent au satellite après s'être propagées dans le mode droit (mode des sifflements usuels). Au-dessus de 350 Hz, les ondes vont passer par une altitude (inférieure à 1000 km) pour laquelle le croisement avec le mode gauche est possible. Entre cette altitude et l'altitude du satellite, la vitesse de phase est beaucoup plus faible que si l'onde avait continué son parcours dans le mode droit. Ceci sera d'autant plus vrai que la fréquence considérée est plus élevée; jusqu'à ce qu'on atteigne la gyrofréquence des protons à l'altitude du satellite, F_{BS}, (550 Hz environ dans notre exemple). On a l'explication de la forme horizontale de la trace du sonagramme.

---

[12] Cette expression peut se trouver beaucoup plus rapidement en reportant directement les valeurs données par les Equ. (5.4) à (5.7) et (5.9) dans les Equ. (9.1). Elle diffère de la formule $E_3/E_1 = -\tan \Theta$ donnée par STOREY [18] valable uniquement lorsque $x \ll 1$.

[13] V. L. GINZBURG et A. A. RUHADZE: Contribution dans le tome 49/4 de cette encyclopédie.

Au-delà de 550 Hz et jusqu'à 750 Hz environ*, aucune onde ne peut parvenir au satellite. En effet, choisissons une fréquence de 600 Hz par exemple. Elle se propage dans le mode droit jusqu'à une altitude de 550 km environ. Là le croisement avec le mode gauche s'effectue, l'onde se propage dans le mode gauche jusqu'à 650 km environ, altitude pour laquelle $F_B = 600$ Hz. Pour que des ondes puissent de nouveau parvenir au satellite, il faut que leur fréquence soit supérieure à la gyrofréquence des protons au niveau du sol. Alors elles parviennent au satellite dans le mode usuel des sifflements électroniques.

Fig. 36a—c. Interprétation des sifflements ioniques. (a) Courbe de densité (nombre de particules par m³) des différents ions et des électrons en fonction de l'altitude. (b) Fréquences caractéristiques en fonction de l'altitude; $L=\infty$ correspond à la gyrofréquence des ions; $L=0$ à la fréquence de coupure des modes *gauche* (L) et *transverse* (e) (voir Sect. 2 et Fig. 1); $D=0$ correspond à la fréquence de croisement $W_{+1}=W_{-1}=W_e$). (c) Vitesse de phase d'une onde de fréquence égale à 400 Hz issue du sol et se propageant vers l'exosphère. (D'après Gurnett et al. [11].)

β) L'explication que nous avons donnée ci-dessus n'est valable que dans ses grandes lignes. Deux phénomènes viennent perturber la *propagation au voisinage du croisement entre les deux modes*:

1° *en l'absence de collisions*, un phénomène de couplage entre les deux modes R et L peut apparaître et un peu d'énergie continuer à se propager malgré tout dans le mode droit. *La trace du sifflement électronique est alors ininterrompue* sur le sonagramme. Ce phénomène n'a pas été étudié quantitativement.

---

* Cette fréquence est la fréquence de coupure, visible sur la Fig. 12.

2° *en l'absence de couplage*, les collisions modifient sensiblement le comportement des ondes au voisinage de la fréquence de croisement. Un angle critique de propagation s'introduit*. Dans certains cas, *la trace protonique du sifflement n'existe pas.*[1]

γ) *Développement asymptotique*. Pour démontrer l'équation (4.11) nous partons de l'équation (4.9):

$$\tau = \frac{1}{2c_0} \int_{F_{BO}}^{F_{BS}} \frac{F_N}{f^{\frac{1}{2}}(\partial F_B/\partial l)} \cdot \frac{F_B \, dF_B}{(F_B - f)^{\frac{3}{2}}}. \qquad [4.9]$$

On s'intéresse aux fréquences voisines de F_{BS}; fixons 20 Hz comme ordre de grandeur de la gamme de fréquence à laquelle nous nous intéressons. $dF_B \approx$ 20 Hz, cela correspond à une variation d'altitude de l'ordre de 100 km. Autrement dit nous intéressons au parcours du sifflement dans les 100 derniers kilomètres. A cette échelle

1. $\partial F_B/\partial l$ est une constante
2. F_N est sensiblement constant parce que l'échelle de hauteur à ces altitudes est supérieure à 1000 km. Il reste donc à évaluer

$$\tau \approx \frac{F_N}{2c_0 f^{\frac{1}{2}}(\partial F_B/\partial l)} \cdot \mathscr{I} \qquad (9.1)$$

avec

$$\mathscr{I} = \int_{F_{BO}}^{F_{BS}} \frac{F_B \, dF_B}{(F_B - f)^{\frac{3}{2}}}. \qquad (9.2)$$

On pose

$$F_B = F_{BS} + w, \qquad (9.3)$$

$$f = F_{BS} - y \qquad (9.4)$$

w est la variable d'intégration, et y sera l'infiniment petit principal qui servira pour l'expression asymptotique, *après* intégration.

$$\mathscr{I} = \int_{F_{BO}-F_{BS}}^{0} \frac{(f+y+w) \, dw}{(w+y)^{\frac{3}{2}}} = \int \frac{f \, dw}{(w+y)^{\frac{3}{2}}} + \int \frac{dw}{(w+y)^{\frac{1}{2}}}, \qquad (9.5)$$

$$\mathscr{I} = -2[F_{BS} - y] \cdot \left[\frac{1}{y^{\frac{1}{2}}} - \frac{1}{[y + F_{BO} - F_{BS}]^{\frac{1}{2}}} \right] + 2[y^{\frac{1}{2}} - \{y + F_{BO} - F_{BS}\}^{\frac{1}{2}}], \qquad (9.6)$$

$$\mathscr{I} = -\frac{2F_{BS}}{y^{\frac{1}{2}}} + \text{termes constants} + 4y^{\frac{1}{2}} + O(y^{\frac{3}{2}}) \qquad (9.7)$$

soit, en remarquant que $\partial F_B/\partial l$ est négatif,

$$\tau = \frac{F_N F_{BS}^{\frac{1}{2}}}{c_0 |(dF_B/\partial l)_S|} \cdot \frac{1}{[F_{BS} - f]^{\frac{1}{2}}} + A + B[F_{BS} - f]^{\frac{1}{2}} + \cdots. \qquad (9.8)$$

Le premier terme de cette expression permet de déterminer F_N si l'on connaît F_{BS}, $(\partial F_B/\partial l)_S$ et si l'on a la relation expérimentale centre t et $[F_{BS} - f]^{\frac{1}{2}}$:

$$\Delta t = -\frac{k}{[F_{BS} - f]^{\frac{1}{2}}}. \qquad (9.9)$$

---

* Cet angle est défini comme celui pour lequel, à la fréquence de croisement, les parties réelles et imaginaires des indices des deux modes sont égales. Pour les angles de propagation inférieurs à cet angle critique, l'onde continue à se propager dans le mode R; pour les angles supérieurs, on a le passage en mode L. L'absorption, dans ce dernier mode due aux collisions est très importante. Voir K. RAWER et K. SUCHY: Cette encyclopédie, tome 49/2, p. 70.

[1] D. A. GURNETT: Ph. D. Thesis, State University of Iowa, 1965.

Mais dans un champ dipolaire, F_B et $\partial F_B/\partial l$ sont liées entre elles et il est plus pratique de ne garder qu'une seule de ces deux variables. On a l'expression du champ magnétique (champ d'un dipôle, voir 'Introductory Remarks', p. 3):

$$B = \mu_0 \frac{u}{4\pi} \frac{M}{r^3} (1 + 3 \sin^2 \phi)^{\frac{1}{2}} \qquad (9.10)$$

et

$$\frac{dB}{dl} = - \frac{3 B^{\frac{4}{3}}}{\left(\mu_0 \dfrac{u}{4\pi} M\right)^{\frac{1}{3}}} \frac{(3 + 5 \sin^2 \phi) \sin \phi}{(1 + 3 \sin^2 \phi)^{\frac{5}{3}}} . \qquad (9.11)$$

Si l'on pose

$$F_B = \lambda B \quad \text{avec} \quad \lambda = 1,52 \cdot 10^7 \text{ Hz} \cdot \text{T}^{-1}, \qquad (9.12)$$

$$F_N^2 = \nu N \quad \text{avec} \quad \nu = 4,38 \cdot 10^{-2} \text{ Hz}^2 \cdot \text{m}^3 \qquad (9.13)$$

on obtient facilement l'expression de

$$k^2 = \frac{F_N^2 \, F_{BS}}{c_0^2 (\partial F_B/\partial l)_S^2}$$

et donc l'expression de N en fonction de λ, ν, M, F_{BS}, ϕ et k. On trouve[2]

$$N = \frac{9 c_0^2}{\nu \left(\lambda \mu_0 \dfrac{u}{4\pi} M\right)} F_B^{\frac{5}{3}} \; \Phi \, \phi) \cdot k^2 \qquad (9.14)$$

avec

$$\Phi(\phi) = \frac{(3 + 5 \sin^2 \phi)^2 \sin^2 \phi}{(1 + 3 \sin^2 \phi)^{\frac{10}{3}}} . \qquad (9.15)$$

$\Phi(\phi)$ est une fonction qui varie peu dès que ϕ est supérieur à $40°$ et dont la valeur reste voisine de 0,63 et le premier terme $9 c_0^2 / \nu \left(\lambda \mu_0 \dfrac{u}{4\pi} M\right)^{\frac{2}{3}}$ est un coefficient numérique dont la valeur (dans le système MKSA) est égale à $7,44 \cdot 10^3 \text{ m}^{-3} \cdot \text{sec}^{2/3}$.

La détermination expérimentale de k et de F_{BS} permet donc de déduire immédiatement la densité des protons au voisinage du satellite.

δ) *Effets de température.* Ainsi qu'il a été dit à la fin de la Sect. 6.6, l'absorption cyclotron se manifeste vers la fin de la trace du sonagramme.

GURNETT et BRICE[3] ont utilisé la théorie des plasmas chauds pour calculer la partie imaginaire du nombre d'onde, k_i, au voisinage de la résonance. Ils calculent ainsi

$$k_i = \frac{\sqrt{\pi}}{2} \cdot \frac{\Delta \omega}{V_i} \cdot \exp\left[- \frac{V_A^2}{V_i^2} \left(\frac{\Delta \omega}{\omega}\right)^2 \right] \qquad (9.16)$$

dans laquelle V_A et V_i sont la vitesse de ALFVEN et la vitesse thermique des ions $\Delta \omega$ est l'écart par rapport á la résonance:

$$\frac{\Delta \omega}{2\pi} = \Delta f = F_{BS} - f .$$

Dans un travail récent[4], il a été montré que l'absorption au voisinage de la résonance était en fait un phénomène beaucoup plus complexe.

1° L'approximation qui permet d'aboutir à l'équation (9.16) n'est valable qu'assez loin de la résonance ($\Delta \omega / k V_i \gg 1$). Près de la résonance, il n'y a pas de formule simple qui permette de déterminer l'absorption. Les intégrations ne peuvent s'effectuer que numériquement. On met d'ailleurs en évidence le fait que la partie réelle du nombre d'onde reste finie, alors qu'elle est infinie dans le cas des plasmas froids[5].

[2] C. BEGHIN, R. GENDRIN, C. SIREDEY et J. VIGNERON: Ann. Géophys. **22**, 110 (1966).
[3] D. A. GURNETT et N. BRCE: J. Geophys. Res. **71**, 3639 (1966).
[4] D. QUEMADA, P. VELUT et J. VIGNERON: Note Technique No. 35, Groupe de Recherches Ionosphériques, Paris 1968.
[5] Voir par example D. QUEMADA: Ondes dans les Plasmas, p. 290. Paris: Hermann 1968.

2° Du fait de l'absorption différentielle subie par les différentes fréquences, un paquet d'ondes émis du sol se déformera au cours de son trajet; la vitesse de propagation du maximum de l'amplitude n'est plus la vitesse de groupe. On retrouve là la distinction bien connue entre vitesse de signal et vitesse de groupe au voisinage des résonances dans un milieu absorbant et dispersif. Dans le cas des sifflements protoniques, le calcul montre que la vitesse de propagation du signal passe par un minimum pour une fréquence légèrement inférieure à F_{BS}. *La loi (9.9) n'est donc pas valable très près de la résonance.*

Suivant les conditions de température, de concentration électronique et ionique, la coupure brusque observée sur l'amplitude du sifflement protonique peut se produire avant ou parès le passage par ce maximum de temps de propagation. L'application numérique de cette théorie à la détermination précise des températures ioniques en est encore à ses débuts, mais les premiers résultats en semblent très prometteurs.*

Bibliographie.

[1] Denisse, J.F., et J.L. Delcroix: La propagation des ondes dans les plasmas. Paris: Dunod 1961.
[2] Stix, T.H.: The theory of plasma waves. New York: McGraw Hill Book Co. 1962.
[3] Rawer, K., and K. Suchy: Radio Observations of the Ionosphere. Handbuch der Physik, Vol. 49/2, p. 1—534. Berlin-Heidelberg-New York: Springer 1967.
[4] Ginzburg, V.L., and A.A. Ruchadze: Waves in plasmas. Handbuch der Physik, Vol. 49/4 (à l'impression). Berlin-Heidelberg-New York: Springer.
[5] Helliwell, R.A.: Whistlers and related phenomena. New York: Plenum Press. 1965.
[6] Smith, R.L., and N.M. Brice: Propagation in multicomponent plasmas. J. Geophys. Res. **69** (23), 5029—5040 (1964).
[7] Storey, L.R.O.: An investigation of whistling atmospheries. Phil. Trans. Roy. Soc. A **246**, 113—141 (1953).
[8] Carpenter, D.L., and R.L. Smith: Whistler measurements of electron density in the magnetosphere. Rev. Geophys. **2**, 415 (1964).
[9] Gendrin, R.: Progrès récents dans l'étude des ondes T.B.F. et E.B.F. Space Sci.
[10] Rivault, R.: Sifflements radioélectriques. (Exposé de synthèse.) Progress in Radio Science. Proc. XVth Gen. Assembly of URSI, pp. 1042—1066. Brussels: International Scientific Radio Union 1967,
Assemblée Cénérale de l'U.R.S.I., München, Sept. 1966.
[11] Gurnett, D.A., S.D. Shawhan, N.M. Brice, and R.L. Smith: Ion cyclotron whistlers. J. Geophys. Res. **70**, 1665 (1965).
[12] Barrington, R.E., J.S. Belrose, and G.L. Nelms: Ion composition and temperature at 1000 km as deduced from simultaneous observations of a V.L.F. plasma resonance and topside sounding data from the Alouette 1 satellite. J. Geophys. Res. **70**, 1647 (1965).
[13] Booker, H.G., and R.B. Dyce: Dispersion of waves in a cold magneto-plasma from hydromagnetic to whistler frequencies. Radio Sci. **69** D, 463 (1965).
[14] Helliwell, R.A.: V.L.F. Noise of magnetospheric origin. (Exposé de synthèse.) Progress in Radio Science. Proc. XVth Gen. Assembly of URSI, pp. 1073—1096. Brussels: International Scientific Radio Union 1967.
[15] Gallet, R.M.: Whistler mode and theory of V.L.F. emissions, in: Natural Electromagnetic phenomena below 30 kc/s, p. 167—204. New York: Plenum Press 1964.
[16] Gendrin, R.: E.B.F. et micropulsations. Ann. Géophys. **23**, 145, 299 (1967).
[17] Brice, N.M.: Fundamentals of very low frequency emissions generation mechanisms. J. Geophys. Res. **69**, 4515 (1964).
[18] Storey, L.R.O.: A method for measuring local electron density from an artificial satellite. J. Res. N.B.S. **63** D, 325 (1959).
[19] Budden, K.G.: Radio Waves in the ionosphere. Cambridge: Cambridge University Press 1961.
[20] Quemada, D.: Ondes dans les plasma. Paris: Editions Hermann 1968.

* Je tiens à remercier mon collègue Jacques Vigneron, qui m'a considérablement aidé dans la recherche bibliographique; c'est à lui que je dois aussi la conception des Tableaux 2 et 3.
Mes remerciements s'adressent également aux différents chercheurs qui ont bien voulu m'adresser des épreuves de leurs enregistrements expérimentaux, ou m'autoriser à publier des figures extraites de leurs articles.

Sachverzeichnis.

(Deutsch-Englisch.)

Bei gleicher Schreibweise in beiden Sprachen sind die Stichwörter nur einmal aufgeführt.

Abklingen der Störung, *decay of the disturbance* 88.
Ablenkung, *deflection* 303.
Abnahme der kosmischen Strahlung, *cosmic ray decrease* 133.
absoluter magnetischer Theodolit, *absolute magnetic theodolite* 309.
Abtasten der Resonanzlinie, *scanning of resonance line* 372.
adiabatische Kompression, *adiabatic compression* 156.
äquatorialer Elektrojet, *equatorial electrojet* 17, 113.
— Ringstrom, *ring current* 77, 100.
Äquinoktial-Hypothese, *equinoctial-hypothesis* 257.
äquivalente planetarische Amplitude, *equivalent planetary amplitude* 226.
äquivalenter Flächenstrom, *equivalent overhead electric current* 14, 47.
— Ionosphärenstrom, *ionospheric current* 19.
äquivalentes Stromsystem, *equivalent overhead current-system* 18.
äußeres äquivalentes Stromsystem, *equivalent overhead electric current-system* 76, 114.
Anfangsphase des magnetischen Sturmes, *initial phase of the magnetic storm* 160, 131.
Anomalien der magnetischen Feldstärke, *anomalies of magnetic field intensity* 414.
Anstiegszeit von SC, *rise-time of SC's* 46, 55.
Aufblähung des magnetischen Feldes, *inflation of the geomagnetic field* 182.
Aufspaltungsfaktor, *splitting factor* 370.
Aufzeichnung der Pulsation, *recording of pulsations* 345.
— schneller Variationen, *of rapid variations* 294.
Aug- und Ohr-Methode, *eye and ear method* 338.
Ausdehnung des Aufnahmebereiches, *extension of recording range* 290.
Ausrichtung von Variometer-Magneten, *orientation of variometer magnets* 288, 289.
Austausch magnetischer Feldlinien, *interchange of magnetic lines of force* 146.
automatisches Observatorium, *automatic observatory* 325.
Axial-Hypothese (CORTIE), *axial-hypothesis (Cortie)* 257.

Balance-Magnet, *balancing magnet* 285, 316.
BARTELSsche Kennziffer K, *Bartels' index* 208, 212.
Basiswert, *base line value* 279, 298.
Besetzungszahlverteilung der ZEEMAN-Terme, *number distribution of Zeeman-terms* 369.
Bestimmung der Horizontalintensität durch die GAUSS-LAMONT-Methode, *determination of the horizontal force by the method of Gauss-Lamont* 301.
— der Inklination, *determination of the inclination* 300.
— der magnetischen Deklination, *determination of the magnetic declination* 298.
— der Skalenwerte, *determination of scale values* 288.
— der Vertikalintensität, *determination of the vertical force* 315.
Bewegte Spulen, *moving coils* 339.
Blackout 82.
BMZ 315.
BOBROVs tragbarer Magnetograph, *Bobrov's portable magnetograph* 291.

c.g.s-System 1, 2.
charakteristische Maßzahl, *characteristic value* 207.
Charakterzahl, *character-figure* 27, 28.
Charakterzahlen Ci, *character figures* 207.
COVINGTON-Index 251.

Dämpfung, *damping* 295.
Datenfluß geomagnetischer Observatorien, *data flux of geomagnetic observatories* 382.
D_{CF} 77.
Digitalisierung von Kurvenschrieben, *digitalisation of registrations curves* 382.
— von Wertfolgen, *of data sequences* 389.
Dipol, *dipole* 3.
direktanzeigendes Magnetometer, *magnetometer direct-reading* 425.
direkte Bestimmung geomagnetischer Elemente, *direct determination of geomagnetic elements* 297.

directe Bestimmung von $\mathcal{H}$, *direct determination of* $\mathcal{H}$ 301, 305, 307, 308, 309, 314.
— — von J, *of* J 300.
— — der Skalenwerte, *of scale values* 289.
— — von Z, *of* Z 308, 309, 315.
D_m 60, 62, 63.
Doppelkernsonde, *double core probe* 348.
D_P 77, 97.
D_P-Teilsturm, D_P *substorm* 100.
D_R 59, 77.
dreikomponenten Flugmagnetometer, *magnetometer three-component airborne* 384, 441.
dreistündliche, äquivalente, planetarische Amplitude, *three-hourly equivalent planetary amplitude* 63.
D_S (SC), D_{St} (SC) 48, 74, 99.
D_{St}- und S_D-Stromsysteme, *electric currentsystems for* D_{St}- *and* S_D-*fields* 68.
D-Variometer, *variometer for* D 277, 280.
D_W 77.
DYE-Magnetometer, *Dye magnetometer* 314.
dynamischer Skalenwert, *dynamic scale value* 294.
Dynamo-Theorie, *dynamo theorie* 24.

Eigenfrequenz der Variometers, *variometer eigenfrequency* 333.
Eingefangene Partikel, *trapped particles* 158, 165.
Einheit, *unit* 1.
elastische Nachwirkung, *elastic after-effect* 312.
elektrische Leitfähigkeit, *electric conductivity* 6.
Elektrojet, 245.
Empfindlichkeit von Aerogradientenmagnetometern, *sensitivity of aeromagnetic gradiometers* 455.
— der Sonden, *probe sensitivity* 352.
empfohlene Skalenwerte, *recommended scale values* 290.
energiereiche Elektronen, *energetic electrons* 154.
Entmagnetisierungsfaktor, *demagnetizing factor* 320.
Entmagnetisierungskurve, *demagnetisation curves* 319.
Erdinduktor, *earth-inductor* 300.
erdmagnetische Bayströmung, *geomagnetic bay* 104.
— Pulsation, *pulsations* 115ff., 119, 129.
erdmagnetischer Schweif, *geomagnetic tail* 140, 143, 146.
Erd-Sturm, *earth storm* 12.
Erholungsphase, *non-cyclic change* 63.

Feldkomponenten, *field components* 277.
Feldkompression, *field compression* 179.
Feldstärkedifferenzmesser, *fieldstrength difference measuring device* 351.
Feldstärkemesser, *fieldstrength meter* 350.
Feldwaage nach SCHMIDT, *Schmidt vertical magnetometer* 315.

FERMI-Beschleunigung, *Fermi acceleration* 155.
Feuchtigkeitseinfluß bei Quarzaufhängungen, *humidity effect* on *quartz suspensions* 313.
Flugmagnetometer, *magnetometer airborne* 408.
Fluxmeter 297.
FÖRSTER-Sonde, *fluxgate magnetometer (Förster-Probe)* 348, 395, 399.
freie Präzession, *free precession* 363.
frequenzabgebender Magnetometer, *frequency delivering magnetometer* 324.
Frequenzkennlinienverfahren, *frequency characteristic method* 332.
Frequenzmessung, *frequency measurement* 365.

Gamma (γ) 277.
gasmagnetische Wellen, *gasmagnetic waves* 129, 141.
GAUSS-LAMONT-Methode, *Gauss-Lamont method* 301.
geomagnetische Aktivität, *geomagnetic activity* 27, 137.
— Koordinaten, *coordinates* 39.
— Observatorien, *magnetic observatories* 276.
— Restaktivität, *residual geomagnetic agitation* 38.
— Stürme, *geomagnetic storms* 131.
geomagnetisches Feld, *geomagnetic field* 277.
Gradientenmagnetometer, *gradiometer magnetic* 405.
gyromagnetisches Verhältnis, *gyromagnetic ratio* 360.

HALL-Leitfähigkeit, *hall conductivity* 7, 25.
— Strom, *hall current* 7.
Hauptphase, *main phase* 58.
— des magnetischen Sturmes, *main phase of the magnetic storm* 131, 161, 162, 202.
Hauptphasen-Aufblähung, *main phase inflation* 164, 168.
Hauptphasen-Relaxation, *main phase relaxation* 165.
HELMHOLTZ-GAUGAIN-Spule, *Helmholtz-Gaugain coil* 289.
$\mathcal{H}_{gm}$ 58.
hochempfindliches Gradiometer, *high-sensitivity gradiometer* 437.
höhenintegrierte elektrische Leitfähigkeit, *height-integrated electric conductivity* 8.
Horizontalkomponente ($\mathcal{H}$), *horizontal force* ($\mathcal{H}$) 277.
$\mathcal{H}$-Variometer, *variometer for* $\mathcal{H}$ 282.
Hyperfeinstruktur, *hyperfinestructure* 37(
Hystereseschleifen hochpermeabler Legierungen, *hysteresis curve of high permeability alloy* 347.

Induktionsfaktor, *induction factor* 302, 304, 312.

Sachverzeichnis.

Induktionsspule mit Galvanometer, *induction-coil with galvanometer* 342.
— mit Verstärker, *with amplifier* 342.
Inklination (J), *inclination* (J) 277.
integrale Regelung, *integrated control* 328.
integrierte elektrische Leitfähigkeit der Ionosphäre, *integrated electric conductivity of the ionosphere* 7.
internationale ruhige Tage, *international quiet days* 56.
internationales Maßsystem, *"système international"* 1.

Jahresgang der Aktivität, *annual variation of activity* 255.

kernlose Spule, *coreless coil* 334.
Kernresonanz im Erdmagnetfeld, *nuclear resonance in earth's magnetic field* 361.
Kernspule, *core coil* 344.
klassische geomagnetische Meßtechnik *classical geomagnetic measuring technique* 323.
Koerzitivkraft, *coercive force* 318.
Kollimationswinkel, *collimation angle* 299, 300.
Komponentenmessung mit Zusatzfeldern, *measurement of components with additional fields* 376.
Kompression des geomagnetischen Feldes, *compression of the geomagnetic field* 169.
Kontrollmagnet, *control magnet* 281.
Konvektion der Magnetosphäre, *convection of the magnetosphere* 146, 166.
Korpuskular-Hypothese, *corpuscular hypothesis* 200.
korrigierte geomagnetische Koordinaten, *corrected geomagnetic coordinates* 41.
kreiselgesteuerte Sondentripel, *gyro-controlled tripel probe* 357.
kreiselstabilisierte Plattform, *platform gyro-stabilized* 386.
Kreuzstrahlenanordnung, *crossing beam array* 374.
K-Variationen, K-*variations* 210.

LA COUR-Magnetograph, *La Cour magnetograph* 294.
Ladungsaustausch, *charge exchange* 165.
Länge des erdmagnetischen Schweifs, *length of the geomagnetic tail* 151.
LARMOR-Frequenz, *Larmor frequency* 360.
L-B-Koordinaten, L-B *coordinaten* 41.
lichtelektrisch kompensierende Magnetometer, *magnetometers with photoeffect compensation* 327.
lokale Aktivitätsmaße, *local activity indices* 17.
lokalzeitabhängige Störung, *disturbance local-time variation* 56.
longitudinale Störungsdissymmetrie, *disturbance longitudinal inequality* 56.

magnetische Materialien, *magnetic materials* 284, 318.

magnetische Parameter von Gesteinsformationen, *magnetic parameters of rock formation* 439.
— Sturmvariation, *storm variation* 56.
magnetischer Meridian, *magnetic meridian* 277.
— Schalenparameter, *shell parameter* 42.
— Sturm, *storm* 33, 42.
— Theodolit, *theodolite magnetometer* 304.
magnetisches Moment, *magnetic moment* 279, 303, 321.
magneto-gasdynamische Gleichung, *magneto gas-dynamic equation* 183 ff.
Magnetograph 277, 289, 291.
Magnetopause 139, 173.
Magnetosphäre, *magnetosphere* 8, 10 ff., 54, 138.
magnetosphärische Übergangsregion, *magnetosheath* 138.
Magnetwechselwirkung, *interaction of magnets* 303.
MARTIN-GRAHAN-Filter, *Martin-Grahan filter* 447, 448.
Maßeinheiten, *units of measurement* 277.
Maßzahl, *value* 206.
MAXWELL-Gleichungen, *Maxwell's equations* 2.
Meßbereich, *detectability range* 433.
mittlerer täglicher Sturmzeitgang, *daily mean storm-time effect* 60.
mondentägliche Variationen, *lunar-daily variation* 20.
M-Region 32, 254, 266.

negativer plötzlicher Sturmbeginn, *negativ sudden commencement* 160.
neutrale Schicht, *neutral sheet* 11, 154.
neutraler Punkt, *neutral point* 150.
NICHOLS-Diagramm, *Nichols diagram* 332.
Nicht-K-Variationen, *Not-K-variations* 210.
numerischer magnetischer Charakter, *digital magnetic character* 235.

optischer Skalenwert, *optical scale value* 279.
optisches Absorptionsmagnetometer, *magnetometer optical absorption* 428.
— Pumpen, *optical pumping* 368.
Orientierung der Variometermagnete, *orientation of variometer magnets* 289.
Orientierungssonden, *orientation probe* 355.
Oszillationen, *oscillations* 301.
OVERHAUSER-Effekt, *Overhauser effect* 379.
OVERHAUSER-Magnetometer, *Overhauser magnetometer* 427.

paramagnetische Elektronenresonanz, *paramagnetic electron resonance* 379.
pc, pi 116.
PEDERSEN-Leitfähigkeit, *Pedersen conductivity* 7, 25.
PEDERSEN-Strom, *Pedersen current* 7.
Perioden-Amplituden Diagramm für geomagnetische Variationen, *period-amplitude diagram of geomagnetic variations* 341.

Handbuch der Physik, Bd. XLIX/3.

Permanent-Magnet, *permanent magnets* 284, 318
Phasendifferenz, *phase difference* 295.
photoelektrisches Registriergerät, *photoelectric recording device* 295.
photographische Aufzeichnung, *photographic recording* 277, 289, 291.
planetarische Kennziffer K_p, *planetary index K_p* 209, 222ff.
planetarischer K_p-Wert, *planetary K_p-value* 209, 222, 253
plötzlicher Impuls (si), *sudden impulse* (si) 42, 46, 53, 54, 212.
— Sturmbeginn (ssc) *storm commencement* 212.
polarer Elektrojet, *auroral electrojet* 97.
— Elementarsturm, *polar elementary storms* 99.
— Teilsturm, *substorm* 134.
polares Elektrojet-Aktivitäts-Maß, *auroral Elektrojet Activity Index* 245.
— Störungsfeld, *polar disturbance field* (D_p) 77.
Polarisationsstrom, *polarisation current* 365.
Polarisationszustand, *polarisation state* 368.
Polarlicht, *Aurora* 124, 152, 158.
— -Gürtel, *auroral oval belt* 79.
— -Radarecho, *auroral radar echo* 123.
Problem des Maßsystems, *rationalization problem* 2.
proportionale Regelung, *proportional control* 328.
Protonenmagnetometer, *protonmagnetometer* 364.
Protonen-Präzessionsmagnetometer, *proton precession magnetometer* 423.
Pulsationsregistriergerät nach GRENET, *pulsation recorder, type Grenet* 296.

QHM-Magnetometer 309.
Q-Index 101, 104, 243.
Quantisierungsfehler, *quantisation error* 381.
Quick-Run-Magnetogramme, *quick-run magnetograms* 44.

Regelschwingung, *oscillation of the servo-system* 332.
Regelverstärkung, *servo-system amplifier* 331.
Registriergerät, *recorder* 277.
Registrierung fluktuierender Ortsgradierung, *registration of fluctuating local gradients* 335.
— kurzperiodischer erdmagnetischer Pulsationen, *recording of rapid geomagnetic pulsation* 296.
— schneller geomagnetischer Variationen, *recording of rapid geomagnetic variations* 294.
— der Zeitableitung, *of time derivatives* 297.
— der zeitlichen Ableitung erdmagnetischer Größen, *of the time derivatives of magnetic elements* 297.

regulärer Magnet, *regular magnet* 303.
Remanenz, *remanent magnetic induction* 318.
Remanenzmagnetometer, *remanent magnetometer* 440.
Resonanzbeschleunigung, *resonance acceleration* 156.
Resttorsion, *residual torsion* 299.
R-Index 244.
Ringstrom, *ring current* 59.
Ringstromwerte nach KERTZ, *ring current values after* KERTZ 240.
— nach SUGIURA, *after Sugiura* 242.
— nach VESTINE, *after Vestine* 240.

Satelliten-Rubidium-Magnetometer, *satellite-borne rubidium magnetometer* 440.
Saturationskernsonde, *saturation core probe* 348.
schematischer Magnet, *schematic magnet* 303.
Schnellaufregistriergerät nach LA COUR, *quick-run recorder, type La Cour* 394.
schnellschreibender Magnetograph, *qucikrun magnetograph* 291.
Schrumpfung von Photopapier, *shrinkage of photographic paper* 289.
SCHUSTER-SMITH-Spulenmagnetometer, *Schuster-Smith coil magnetometer* 307.
Schweif der Magnetosphäre, *magnetospheric tail* 12, 32.
Schwingungen (GAUSS-LAMONT), *oscillations (Gauss-Lamont)* 301.
Schwingzeit Meßanlage, *period length measuring device* 337.
S_D 56, 64, 65, 70.
Sektorenstruktur des interplanetaren Magnetfeldes, *sector structure of the interplanetary magnetic field* 136, 266, 269.
selbstorientierendes Nachführsystem, *self orientating follower system* 323.
selbstschwingende Einzelanordnung, *self oscillating single suspension* 374.
„SHORAN" 406.
Signal-Rausch-Verhältnis, *signal to noise ratio* 362.
Sinus-Galvanometer, *sine-galvanometer* 305.
Skalenwert, *scale value* 290.
Skalenwertbestimmung, *scale value determination* 288.
solare Teilchenstrahlung, *solar corpuscular radiation* 169, 204.
Solar-Flare-(Eruptions-) Effekt, *solar-flareeffect* 211, 215.
solarer Plasmastrom, *solar plasma stream* 54.
— Wind, *wind* 11, 33, 54, 135ff., 138, 140, 204.
Sonden-Vektor-Magnetometer, *vector magnetometer probe* 356.
Sonnenaktivität, *solar activity* 135.
Sonnen-Flare-Effekt, *solar-flare effect (s.f.e.)* 20, 26, 211.

Sonnenfleckenrelativzahl, *relativ sun spot number* 251.
sonnentägliche Variationen an ruhigen Tagen, *solar-daily variations on quiet days* 16.
Spektrum der Sondenausgangsspannung, *spectrum of probe output voltage* 349.
Spinpräzessionsmagnetometer, *spin-precession magnetometer* 289.
Spiralstruktur geomagnetischer Störung, *spiral pattern of geomagnetic disturbance* 80.
S_q-Strom, S_q *current systems* 17.
S_q-Wert, S_q-*value* 211.
S_q^p-Verstärkung, S_q^p *enhancement* 86.
ssc*, ssc*, SC, 42, 45, 46, 54.
Störfeld, *disturbance field* 26.
Störvektor, *disturbance force vector* 62.
Stoßfront, *shock* 140.
Stoßwelle (stehende), *bow shock* 138, 178
Strömungswiderstand an der Magnetopause, *magnetopause drag* 149.
Sturmaktivität, *stormines* 243.
Sturm-Magnetograph, *storm magnetograph* 290.
Sturm-Zeit, *storm time* 56.
Sturmzeitvariation, *storm time variation* 55, 56, 60, 62.
Sudden commencement magnetischer Stürme, *sudden commencement of magnetic storms* 42, 127, 131, 160.

tägliche internationale Charakterzahl, *daily international charakter figure* 237.
— planetarische Charakterzahl, *planetary character figure* 227.
Tagesgang der Aktivität, *diurnal variation of activity* 270.
— der Störung, *disturbance-daily variation* 55, 56, 78.
Teilchenbeschleunigung, *particle acceleration* 155.
Teilchenstrom von der Sonne, *corpuscular flux from the sun* 77.
Temperatureffect (GAUSS-LAMONT), *temperature effect (Gauss-Lamont)* 302, 304.
Temperaturgang von Variometern, *temperature effect on variometers* 279.
Torsionskonstante, *torsion constant* 279, 281.
tragbarer Magnetograph, *portable magnetograph* 290.
Transit-Magnetometer, *transit magnetometer* 300.
transportabler Magnetograph (Typ Askania), *portable magnetograph, type Askania* 290.
transportabler Magnetograph (Typ BOBROV), *type Bobrov* 291.
transversale Relaxationszeit, *transverse relaxation time* 361.
TSUBOKAWA magnetometer 308.

umgekehrter SC, SC* (D), *inverted SC, SC** (D) 43.
Unifilarvariometer (Theorie), *unifilar variometer (theory)* 279.
Universal-Torsionsmagnetometer, *universal-torsion-magnetometer* 313.
unsymmetrische Aufblähung, *nonsymmetric inflation* 193.

Variometer, *variometers* 277.
—, D-, *variometer for* D 277, 280.
—, H-, *for* H 282.
— mit Direktsichtanzeige, *with direct view registration* 336.
—, X-, *for* X 285.
—, Y-, *for* Y 285.
—, Z-, *for* Z 285.
Variometertemperaturkoeffizient, *temperature coefficient of variometer* 279, 283.
Veränderlichkeit von S_p, *variability of* S_p 18.
Vergleich von Instrumten, *intercomparison of instruments* 321.
Verschiebung der Polarlichtzone, *shifting of the auroral zone* 89.
Verteilung, *distribution* 303.
Verteilungskoeffizienten, *distribution coefficients* 303, 304.
Vertikalintensität (Z), *vertical force* (Z) 277.
vorausgehender gegenphasiger Impuls, *preliminary reverse impulse* 43, 47

wiederkehrender Sturm, *recurring storm* 133.

X- oder Y-Variometer, *variometer for* X *or* Y 284.
X-Variometer 285.

Y-Variometer 285.

ZEEMAN-Effekt, *Zeeman effect* 358.
Zeitmarken, *time marks* 289.
zusätzliche sonnentägliche Variationen an ruhigen Tagen, *additional solar-daily variations on quiet days* 18.
Z-Variometer, *variometer for* Z 285.
zweite adiabatische Invariante, *second adiabatic invariant* 42.

Subject Index.

(English-German.)

Where English and German spelling of a word is identical the German version is omitted.

Absolute magnetic theodolite, *absoluter magnetischer Theodolit* 309.
Additional solar-daily variations on quiet days, *zusätzliche sonnentägliche Variationen an ruhigen Tagen* 18.
Adiabatic compression, *adiabatische Kompression* 156.
Annual variation of activity, *Jahresgang der Aktivität* 255.
Anomalies of magnetic field intensity, *Anomalien der magnetischen Feldstärke* 414.
Aurora, *Polarlicht* 124, 152, 158.
Auroral electrojet, *polarer Elektrojet* 97.
— Activity Index, *polares Aktivitäts-Maß* 245.
— oval belt, *Polarlichtoval* 79.
— radar echo, *Polarlicht-Radarecho* 123.
Automatic observatory, *automatisches Observatorium* 325.
Axial-hypothesis (CORTIE), *Axial-Hypothese (Cortie)* 257.

Balancing magnet, *Balance-Magnet* 285, 316.
BARTELS' index, *Bartelssche Kennziffer K* 208, 212.
Base line value, *Basiswert* 279, 298.
Blackout, 82.
BMZ 315.
BOBROV's portable magnetograph, *Bobrov's tragbarer Magnetograph* 291.
Bow shock, *Stoßwelle (stehende)* 138, 178.

c.g.s.-system 1, 2.
Character-figure, *Charakterzahl* 27, 28.
Character figures, *Charakterzahlen Ci* 207.
Characteristic value, *charakteristische Maßzahlen* 207.
Charge exchange, *Ladungsaustausch* 165.
Classical geomagnetic measuring technique, *klassische geomagnetische Meßtechnik* 323.
Coercive force, *Koerzitivkraft* 318.
Collimation angle, *Kollimationswinkel* 299, 300.
Compression of the geomagnetic field, *Kompression des geomagnetischen Feldes* 169.
Control magnet, *Kontrollmagnet* 281.

Convection of the magnetosphere, *Konvektion der Magnetosphäre* 146, 166.
Core coil, *Kernspule* 344.
Coreless coil, *kernlose Spule* 344.
Corpuscular flux from the sun, *Teilchenstrom von der Sonne* 77.
— hypothesis, *Korpuskular-Hypothese* 200.
Corrected geomagnetic coordinates, *korrigierte geomagnetische Koordinaten* 41.
Cosmic ray decrease, *Abnahme der kosmischen Strahlung* 133.
COVINGTON-Index 251.
Crossing beam array, *Kreuzstrahlenanordnung* 374.

Daily international character figure, *tägliche internationale Charakterzahl* 237.
— mean storm-time effect, *mittlerer täglicher Sturmzeitgang* 60.
— planetary character figure, *planetarische Charakterzahl* 227.
Damping, *Dämpfung* 295.
Data flux of geomagnetic observatories, *Datenfluß geomagnetischer Observatorien* 382.
D_{CF}, 77.
Decay of the disturbance, *Abklingen der Störung* 88.
Deflection, *Ablenkung* 303.
Demagnetisation curves, *Entmagnetisierungskurve* 319.
Demagnetizing factor, *Entmagnetisierungsfaktor* 320.
Detectability range, *Meßbereich* 433.
Determination of the horizontal force, *Bestimmung der Horizontalintensität* 315.
— — — by the method of GAUSS-LAMONT, *Bestimmung der Horizontalintensität durch die Gauss-Lamont-Methode* 301.
— of the inclination, *der Inklination* 300.
— of the magnetic declination, *der magnetischen Declination* 298.
— of scale values, *der Skalenwerte* 288.
— of the vertical force, *Vertikalintensität* 315.
Digital magnetic character, *numerischer magnetischer Charakter* 235.
Digitalisation of data sequences, *Digitalisierung von Wertfolgen* 389.
— of registrations curves, *von Kurvenschrieben* 382.

Subject Index.

Dipole, *Dipol* 3.
Direct determination of geomagnetic elements, *direkte Bestimmung geomagnetischer Elemente* 297.
— — of $\mathcal{H}$, *von* $\mathcal{H}$ 301, 305, 307, 308, 309, 314.
— — of J, *von* J 300.
— — of scale values, *der Skalenwerte* 289.
— — of Z, *von* Z 308, 309, 315.
Distribution, *Verteilung* 303.
— coefficients, *Verteilungskoeffizienten* 303, 304.
Disturbance-daily variation, *Tagesgang der Störung* 55, 56, 78.
Disturbance field, *Störfeld* 26.
— force vector, *Störvektor* 62.
— local-time variation, *lokalzeitabhängige Störung* 56.
— longitudinal inequality, *longitudinale Störungsdisymmetrie* 56.
Diurnal variation of activity, *Tagesgang der Aktivität* 270.
D_m 60, 62, 63.
Double core probe, *Doppelkernsonde* 348.
D_p 77, 97.
D_p substorm, D_p-*Teilsturm* 100.
D_R 59, 77.
$D_S(SC)$, $D_{st}(SC)$ 48, 74, 99.
D_W 77.
DYE magnetometer, *Dye-Magnetometer* 314.
Dynamic scale value, *dynamischer Skalenwert* 294.
Dynamo theorie, *Dynamo Theorie* 24.

Earth-inductor, *Erdinduktor* 300.
Earth storm, *Erd-Sturm* 12.
Elastic after-effect, *elastische Nachwirkung* 312.
Electric conductivity, *elektrische Leitfähigkeit* 6.
— current-systems for D_{st}- and S_D-fields, D_{st}- *und* S_D-*Stromsysteme* 68.
Electrojet 245.
Energetic electrons, *energiereiche Elektronen* 154.
Equatorial electrojet, *Äquatorialer Elektrojet* 17, 113.
— ring current, *Ringstrom* 77, 100.
Equinoctial-hypothesis, *Äquinoktial-Hypothese* 257.
Equivalent ionospheric current, *äquivalenter Ionosphärenstrom* 19.
— overhead current-system, *äquivolentes Stromsystem* 18.
— — electric current, *äquivalenter Flächenstrom* 14, 47.
— — — current-system, *äußeres äquivalentes Stromsystem* 76, 114.
— planetary amplitude, *äquivalente planetarische Amplitude* 226.
Extension of recording range, *Ausdehnung des Aufnahmebereiches* 290.
Eye and ear method, *Aug- und Ohr-Methode* 338.

FERMI acceleration, *Fermi-Beschleunigung* 155.
Field components, *Feldkomponenten* 277.
— compression, *Feldkompression* 179.
Fieldstrength difference measuring device, *Feldstärkedifferenzmesser* 351.
— meter, *Feldstärkemesser* 350.
Fluxgate magnetometer (FÖRSTER-Probe), *Förster-Sonde* 348, 395, 399.
Fluxmeter 297.
Free precession, *freie Präzession* 363.
Frequency characteristic method, *Frequenzkennlinienverfahren* 332.
— delivering magnetometer, *frequenzabgebende Magnetometer* 324.
— measurement, *Frequenzmessung* 365.

Gamma (γ) 277.
Gasmagnetic waves, *gasmagnetische Wellen* 129, 141.
GAUSS-LAMONT method, *Gauss-Lamont-Methode* 301.
Geomagnetic activity, *Geomagnetische Aktivität* 27, 137.
— bay, *Bayströmung* 104.
— coordinates, *Koordinaten* 39.
— field, *Feld* 277.
— pulsations, *Pulsation* 115 ff., 119, 129.
— storms, *Stürme* 131.
— tail, *Erdmagnetischer Schweif* 140, 143, 146.
Gradiometer magnetic, *Gradientenmagnetometer* 405.
Gyro-controlled tripel probe, *kreiselgesteuerte Sondentripel* 357.
Gyromagnetic ratio, *gyromagnetisches Verhältnis* 360.

HALL conductivität, *Hall Leitfähigkeit* 7.
— current, *Strom* 7.
Height-integrated electric conductivity, *höhenintegrierte elektrische Leitfähigkeit* 8.
HELMHOLTZ-GAUGAIN coil, *Helmholtz-Gaugain-Spule* 289.
$\mathcal{H}_{gm}$, 58.
High-sensitivity gradiometer, *hochempfindliches Gradiometer* 437.
Horizontal Force ($\mathcal{H}$), *Horizontalkomponente* ($\mathcal{H}$) 277.
Humidity effect on quartz suspensions, *Feuchtigkeitseinfluß bei Quarzaufhängungen* 313.
Hyperfinestructure, *Hyperfeinstruktur* 370.
Hysteresis curve of high permeability alloy, *Hystereseschleifen hochpermeabler Legierungen* 347.

Inclination (J), *Inklination* (J) 277.
Induction-coil with amplifier, *Induktionsspule mit Verstärker* 342.
— with galvanometer, *Induktionsspule mit Galvanometer* 342.
Induction factor, *Induktionsfaktor* 302, 304, 312.

Inflation of the geomagnetic field, *Aufblähung des magnetischen Feldes* 182.
Initial phase of the magnetic storm, *Anfangssphase des magnetischen Sturmes* 131, 160.
Integrated control, *integrale Regelung* 328.
— electric conductivity of the ionosphere, *integrierte elektrische Leitfähigkeit der Ionosphäre* 7.
Interaction of magnets, *Magnetwechselwirkung* 303.
Interchange of magnetic lines of force, *Austausch magnetischer Feldlinien* 146.
Intercomparison of instruments, *Vergleich von Instrumenten* 321.
International quiet days, *internationale ruhige Tage* 56.
Inverted SC, SC* (D), *umgekehrter SC, SC* (D)* 43.

K-Variations, *K-Variationen* 210.

La Cour magnetograph, *La Cour Magnetograph* 294.
Larmor frequency, *Larmor Frequenz* 360.
L-B coordinates, *L-B Koordinaten* 41.
Length of the geomagnetic tail, *Länge des erdmagnetischen Schweifs* 151.
Local activity indices, *lokale Aktivitätsmaße* 217.
Lunar-daily variation, *mondentägliche Variationen* 20.

Magnetic materials, *magnetische Materialien* 284, 318.
— meridian, *magnetischer Meridian* 277.
— moment, *magnetisches Moment* 279, 303, 321.
— observatories, *Observatorien* 276.
— parameters of rock formation, *Parameter von Gesteinsformationen* 439.
— shell parameter, *Schalenparameter* 42.
— storm, *Sturm* 33, 42.
— — variation, *Sturmvariation* 56.
Magneto gas-dynamic equation, *Magnetogasdynamische Gleichung* 183 ff.
Magnetograph, *Magnetograph* 277, 289, 291.
Magnetometer airborne, *Flugmagnetometer* 408.
— direct-reading, *direktanzeigendes Magnetometer* 425.
— optical absorption, *optisches Absorptionsmagnetometer* 428.
— Overhauser, *Overhauser Magnetometer* 427.
— proton precession, *Protonen-Präzessionsmagnetometer* 423.
— rubidium, *Rubidium Magnetometer* 400.
— spin-precession, *Spinpräzessionsmagnetometer* 427.
— theodolite, *magnetischer Theodolit* 304.
— three-component airborne, *dreikomponenten Flugmagnetometer* 384, 441.
— transit, *Transit-Magnetometer* 300.

Magnetometers with photoeffect compensation, *lichtelektrisch kompensierende Magnetometer* 327.
Magnetopause, *Magnetopause* 139, 173.
— drag, *Strömungswiderstand an der Magnetopause* 149.
Magnetosheath, *magnetosphärische Übergangsregion* 138.
Magnetosphere, *Magnetosphäre* 8, 10ff., 54, 138.
Magnetospheric tail, *Schweif der Magnetosphäre* 12, 32.
Main phase, *Hauptphase* 58.
— — inflation, *Hauptphasen-Aufblähung* 164, 168.
— — of the magnetic storm, *des magnetischen Sturmes* 131, 161, 162, 202.
— — relaxation, *Hauptphasen-Relaxation* 165.
Martin-Grahan filter, *Martin-Grahan-Filter* 447, 448.
Maxwell's equations, *Maxwell-Gleichungen* 2.
Measurement of components with additional fields, *Komponentenmessung mit Zusatzfeldern* 376.
Moving coils, *bewegte Spulen* 339.
M-Region, *M-Region* 32, 254, 266.

Negative sudden commencement, *negativer plötzlicher Sturmbeginn* 160.
Neutral point, *neutraler Punkt* 150.
— sheet, *neutrale Schicht* 11, 154.
Nichols Diagram, *Nichols-Diagramm* 332.
Non-cyclic change, *Erholungsphase* 63.
Nonsymmetric inflation, *unsymmetrische Aufblähung* 193.
No-*K*-variations, *Nicht-K-Variationen* 210.
Nuclear resonance in Earth's magnetic field, *Kernresonanz im Erdmagnetfeld* 361.
Number distribution of Zeeman-terms, *Besetzungszahlverteilung der Zeeman-Terme* 369.

Optical pumping, *optisches Pumpen* 368.
— scale value, *optischer Skalenwert* 279.
Orientation probe, *Orientierungssonden* 355.
— of variometer magnets, *Ausrichtung von Variometer Magneten* 288, 289.
— — —, *Orientierung der Variometermagnete* 289.
Oscillation of the servo-system, *Regelschwingung* 332.
Oscillations, *Oszillationen* 301.
— (Gauss-Lamont), *Schwingungen (Gauss-Lamont)* 301.
Overhauser effect, *Overhauser-Effekt* 379.
— magnetometer, *-magnetometer* 427.

Paramagnetic electron resonance, *Paramagnetische Elektronenresonanz* 379.
Particle acceleration, *Teilchenbeschleunigung* 155.

pc, pi, 116.
PEDERSEN conductivity, *Pedersen-Leitfähigkeit* 7, 25.
— current, *Pedersen-Strom* 7.
Period-amplitude diagram of geomagnetic variations, *Perioden-Amplituden Diagramm für geomagnetische Variationen* 341.
Period length measuring device, *Schwingzeit Meßanlagen* 337.
Permanent magnets, *Permanent-Magnet* 284, 318.
Phase difference, *Phasendifferenz* 295.
Photo-electric recording device, *photoelektrisches Registriergerät* 295.
Photographic recording, *photographische Aufzeichnung* 277, 289, 291.
Planetary index Kp, *planetarische Kennziffer Kp* 209, 222 ff.
— Kp-value, Kp-*Wert* 209, 222, 253.
Platform gyro-stabilized, *kreiselstabilisierte Plattform* 386.
Polar disturbance field (D_p), *polares Störungsfeld* 77.
— elementary storms, *Elementarsturm* 99.
— substorm, *Teilsturm* 134.
Polarisation current, *Polarisationsstrom* 365.
— state, *Polarisationszustand* 368.
Portable magnetograph, *tragbarer Magnetograph* 290.
— —, type Askania, *transportabler Magnetograph (Typ Askania)* 290.
— —, type BOBROV, *(Typ Bobrov)* 291.
Preliminary reverse impulse, *vorausgehender gegenphasiger Impuls* 43, 47.
Probe sensitivity, *Empfindlichkeit der Sonden* 352.
Proportional control, *proportionale Regelung* 328.
Proton magnetometer, *Protonenmagnetometer* 364.
Pulsation recorder, type GRENET, *Pulsationsregistriergerät nach Grenet* 296.

QHM-magnetometer, *QHM-Magnetometer* 309.
Q-Index, Q-*Index* 101, 104, 243.
Quantisationerror, *Quantisierungsfehler* 381.
Quick-run magnetograms, *Quick-Run-Magnetogramme* 44.
— magnetograph, *schnellschreibender Magnetograph* 291.
— recorder, type LA COUR, *Schnellaufregistriergerät nach La Cour* 394.

Rationalization problem, *Problem des Maßsystems* 2.
Recommended scale values, *empfohlene Skalenwerte* 290.
Recorder, *Registriergerät* 277.
Recording of pulsations, *Aufzeichnung der Pulsation* 345.

Recording of rapid geomagnetic pulsation, *Aufzeichnung kurzperiodischer erdmagnetischer Pulsationen* 296.
— — variations, *schneller geomagnetischer Variationen* 294.
— — variations, *Aufzeichnung schneller Variationen* 294.
— of time derivatives, *Registrierung der Zeitableitung* 297.
— — of magnetic elements, *der zeitlichen Ableitung erdmagnetischer Größen* 297.
Recurring storm, *wiederkehrender Sturm* 133.
Registration of fluctuating local gradients, *Registrierung fluktuierender Ortsgradierung* 335.
Regular magnet, *regulärer Magnet* 303.
Relativ sun spot number, *Sonnenfleckenrelativzahl* 251.
Remanent magnetic induction, *Remanenz* 318.
— magnetometer, *Remanenzmagnetometer* 440.
Residual geomagnetic agitation, *geomagnetische Restaktivität* 38.
— torsion, *Resttorsion* 299.
Resonance acceleration, *Resonanzbeschleunigung* 156.
R-Index 244.
Ring current, *Ringstrom* 59.
— — values after KERTZ, *Ringstromwerte nach Kertz* 240.
— — — after SUGIURA, *nach Sugiura* 242.
— — — after VESTINE, *nach Vestine* 240.
Rise-time of SC's, *Anstiegszeit von SC* 46, 55.

Satellite-borne rubidium magnetometer, *Satelliten-Rubidium-Magnetometer* 440.
Saturation core probe, *Saturationskernsonde* 348.
Scale value, *Skalenwert* 290.
— — determination, *Skalenwertbestimmung* 288.
Scanning of resonance line, *Abtasten der Resonanzlinie* 372.
Schematic magnet, *schematischer Magnet* 303.
SCHMIDT vertical magnetometer, *Feldwaage nach Schmidt* 315.
SCHUSTER-SMITH coil magnetometer, *Schuster-Smith Spulenmagnetometer* 307.
S_D 56, 64, 65, 70.
Second adiabatic invariant, *zweite adiabatische Invariante* 42.
Sector structure of the interplanetary magnetic field, *Sektorenstruktur des interplanetaren Magnetfeldes* 136, 266, 269.
Self orientating follower system, *selbstorientierendes Nachführsystem* 323.
— oscillating single suspension, *selbstschwingende Einzelanordnung* 374.

Sensitivity of aeromagnetic gradiometers, *Empfindlichkeit von Aerogradientenmagnetometern* 455.
Servo-system amplifier, *Regelverstärkung* 331.
Shifting of the auroral zone, *Verschiebung der Polarlichtzone* 89.
Shock, *Stoßfront* 140.
"SHORAN" 406.
Shrinkage of photographic paper, *Schrumpfung von Photopapier* 289.
Sine-galvanometer, *Sinus-Galvanometer* 305.
Signal to noise ratio, *Signal-Rausch-Verhältnis* 362.
Solar activity, *Sonnenaktivität* 135.
— corpuscular radiation, *solare Teilchenstrahlung* 169, 204.
Solar-daily variations on quiet days, *sonnentägliche Variationen an ruhigen Tagen* 16.
Solar-flare effect (s.f.e.), *Sonnen-Flare-Effekt* 20, 26, 211.
—, *Solar-Flare-(Eruptions-)Effekt* 211, 215.
Solar plasma stream, *solarer Plasmastrom* 54.
— wind, *Wind* 11, 33, 54, 135 ff., 138, 140, 204.
Spectrum of probe output voltage, *Spektrum der Sondenausgangsspannung* 349.
Spiral pattern of geomagnetic disturbance, *Spiralstruktur geomagnetischer Störungen* 80.
Splitting factor, *Aufspaltungsfaktor* 370.
S_q^p enhancement, S_q^p *Verstärkung* 86.
S_q current systems, S_q *Strom* 17.
S_q-Value, S_q-*Wert* 211.
ssc, ssc*, SC, SC* 42, 45, 46, 54.
Stormines, *Sturmaktivität* 243.
Storm magnetograph, *Sturm-Magnetograph* 290.
— time, -*Zeit* 56.
— — variation, *Sturmzeitvariation* 55, 56, 60, 62.
Sudden commencement of magnetic storms, *Sudden commencement magnetischer Stürme* 42, 127, 131, 160.
— impulse (si), *Impuls* (si) 42, 46, 53, 54, 212.
— storm commencement, *Sturmbeginn* (ssc) 212.
"Système International", *internationales Maßsystem* 1.

Temperature coefficient of variometer, *Variometertemperaturkoeffizient* 279, 283.

Temperature effect (Gauss-Lamont), *Temperatureffekt (Gauss-Lamont)* 302, 304.
— — on variometers, *Temperaturgang von Variometern* 279.
Theodolite magnetometer, *magnetischer Theodolit* 304.
Three-hourly equivalent planetary amplitude, *dreistündliche, äquivalente, planetarische Amplitude* 63.
Time marks, *Zeitmarken* 289.
Torsion constant, *Torsionskonstante* 279, 281.
Transit Magnetometer, *Transit Magnetometer* 300.
Transverse relaxation time, *transversale Relaxationszeit* 361.
Trapped particles, *eingefangene Partikel* 158, 165.
Tsubokawa magnetometer 308.

Unifilar variometer (theory), *Unifilarvariometer (Theorie)* 279.
Unit, *Einheit* 1.
Units of measurement, *Maßeinheiten* 277.
Universal-torsion-magnetometer, *Universal-Torsionsmagnetometer* 313.

Value, *Maßzahl* 206.
Variability of S_q, *Veränderlichkeit von* S_q 18.
Variometer with direct view registration, *Variometer mit Direktsichtanzeige* 336.
— eigenfrequency, *Eigenfrequenz des Variometers* 333.
— for *D*, *D-Variometer* 277, 280.
— for *H*, *H-Variometer* 282.
— for *X*, *X-Variometer* 285.
— for *Y*, *Y-Variometer* 285.
— for *X* or *Y*, *X*- oder *Y-Variometer* 284.
— for *Z*, *Z-Variometer* 285.
Variometers, *Variometer* 277.
Vector magnetometer probe, *Sonden-Vektor-Magnetometer* 356.
Vertical force (Z), *Vertikalintensität* (Z) 277.

X-Variometer 285.

Y-Variometer 285.

Zeeman effect 358.
Z-Variometer 285.

Index

pour la contribution écrite en français:

Roger Gendrin: Phénomènes T.B.F. d'origine magnétosphérique.

Constante de dispersion 470.
Couplage de modes 523.
Courbe de dispersion 465.
— des rayons 490.
— d'indice 490.

Densité électronique 472.

Éclaire atmosphérique 463.
Émission radioélectrique 463.
— T.B.F. 497.
Equation d'indice 469.

Fibre d'ionisation 486, 494.
Fréquence de plasma 461.

Gradient d'ionisation 494.
Guidage des sifflements 491.
Gyrofréquence 461.

«Hélium whistler» 483.

Instabilité de plasma 509.
Interaction de gyrorésonance 508.

«Knee-whistler» 475.

Loi d'Eckersley 469.

Mode d'Alfven 491.
— non dispersif 493.
— ordinaire, extraordinaire, circulaire 465.

«Nose-whistler» 472.

Paramètre de MacIlwain 461.
Plasmapause 476, 478.

Quartique de Booker 490.

Résonance hybride basse 484, 487.

Sifflement protonique 479.
— radioelectrique 463.
— «sub-ionosphérique» 479.
— «sub-protonsphérique» 478.
«Sonagramme» 468.
«Système International» 1.

Theodolite magnetometer 304.

Printed by Books on Demand, Germany